POLICY OPTIONS FOR STABILIZING GLOBAL CLIMATE

POLICY OPTIONS FOR STABILIZING GLOBAL CLIMATE

Editors

Daniel A. Lashof and Dennis A. Tirpak
US Environmental Protection Agency
Office of Policy, Planning, and Evaluation
Washington, DC

HEMISPHERE PUBLISHING CORPORATION
A member of the Taylor & Francis Group

New York Washington Philadelphia London

DISCLAIMER: This material was circulated for review and comment and does not necessarily reflect the official position of the U.S. Environmental Protection Agency. Mention of trade names does not constitute an endorsement.

POLICY OPTIONS FOR STABILIZING GLOBAL CLIMATE

1 2 3 4 5 6 7 8 9 0   E B E B   9 8 7 6 5 4 3 2 1 0

Cover design by Renee Winfield
A CIP catalog record for this book is available from the British Library.

Library of Congress Cataloging-in-Publication Data

Policy options for stabilizing global climate / editors, Daniel A. Lashof and Dennis A. Tirpak.
p. cm.
Includes bibliographical references.

1. Climatic changes. 2. Environmental policy. 3. Environmental protection. 4. Greenhouse effect, Atmospheric.
I. Lashof, Daniel A. II. Tirpak, Dennis A.
QC981.8.C5P654 1990
363.73'9255—dc20 90-30125
ISBN 1-56032-072-9 CIP

CONTENTS

CHAPTER III CLIMATIC CHANGE

CHAPTER IV HUMAN ACTIVITIES AFFECTING TRACE GASES AND CLIMATE

CHAPTER V THINKING ABOUT THE FUTURE

CHAPTER VI SENSITIVITY ANALYSES

CHAPTER VII TECHNICAL OPTIONS FOR REDUCING GREENHOUSE GAS EMISSIONS

CHAPTER VIII POLICY OPTIONS

CHAPTER IX INTERNATIONAL COOPERATION TO REDUCE GREENHOUSE GAS EMISSIONS

ACKNOWLEDGEMENTS

This report would not have been possible without the tireless efforts of the primary chapter authors:

Executive Summary	Daniel Lashof Dennis Tirpak
Chapter I. Introduction	Joel Scheraga Irving Mintzer
Chapter II. Greenhouse Gas Trends	Inez Fung Michael Prather
Chapter III. Climatic Change	Daniel Lashof Alan Robock
Chapter IV. Human Activities Affecting Trace Gases and Climate	Barbara Braatz Craig Ebert
Chapter V. Thinking About the Future	Daniel Lashof Leon Schipper
Chapter VI. Sensitivity Analyses	Craig Ebert
Chapter VII. Technical Control Options	Paul Schwengels (Energy Services) Michael Adler (Renewable Energy) Dillip Ahuja (Biomass) Kenneth Andrasko (Forestry) Lauretta Burke (Agriculture) Craig Ebert (Energy Supply) Joel Scheraga (Energy Supply) John Wells (Halocarbons)
Chapter VIII. Policy Options	Alan Miller
Chapter IX. International Cooperation to Reduce Greenhouse Gas Emissions	Alan Miller Jayant Sathaye
Appendix A. Model Descriptions	William Pepper
Appendix B. Scenario Definitions	Craig Ebert
Appendix C. Results of Sensitivity Analyses	Craig Ebert

Model integration was coordinated by William Pepper and Craig Ebert, with assistance from Rossana Florez. Models and/or analysis were prepared by Irving Mintzer; Jayant Sathaye, Andrea Ketoff, Leon Schipper, and Sharad Lele; Klaus Frohberg and Phil Vande Kamp; Richard Houghton; Berrien Moore, Chris Ringo, and William Emmanuel; Michael Prather; Ivar Isaksen, Terje Berntsen, and Sverre Solberg; and Anne Thompson.

Document integration was coordinated by Craig Ebert and Barbara Braatz. Editorial assistance was provided by Susan MacMillan. Technical, graphics, and typing assistance was provided by Courtney Dinsmore, Katey Donaldson, Donald Devost, Michael Green, Karen Zambri, Judy Koput, Donna Whitlock, Margo Brown, and Cheryl LaBrecque.

Literally hundreds of other people have contributed to this report, including the organizers and attendants at four workshops sponsored by EPA to gather information and ideas, and dozens of formal and informal reviewers. We would like to thank this legion for their interest in this project, and apologize for not doing so individually. Particularly important comments were provided by, among others, Thomas Bath, Deborah Bleviss, Gary Breitenbeck, William Chandler, Robert Friedman, Howard Geller, James Hansen, Tony Janetos, Stan Johnson, Julian Jones, Michael Kavanaugh, Andrew Lacis, Michael MacCracken, Elaine Matthews, William Nordhaus, Steven Piccot, Marc Ross, Stephen Schneider, Paul Steele, Pieter Tans, Thomas Wigley, Edward Williams, and Robert Williams.

This work was conducted within EPA's Office of Policy Analysis, directed by Richard Morgenstern, within the Office of Policy Planning and Evaluation, administered by Linda Fisher. Technical support was provided by the Office of Research and Development, administered by Eric Brethauer.

CHAPTER I

INTRODUCTION

INTRODUCTION

The Earth's Climate and Global Change

The **greenhouse effect** is a natural phenomenon that plays a central role in determining the Earth's climate. Sunlight passes through the atmosphere and warms the Earth's surface. The Earth then radiates infrared energy, some of which escapes back into space. But certain gases (known as greenhouse gases) that occur naturally in the atmosphere absorb most of the infrared radiation and emit some of this energy back toward the Earth, warming the surface. This effect is, to a great extent, responsible for making the Earth conducive to life. In its absence, the Earth would be approximately 30°C colder.

Concerns about the greenhouse effect arise out of apprehension that anthropogenic (man-made) emissions of greenhouse gases will further warm the Earth. Greenhouse gases -- primarily carbon dioxide (CO_2), methane (CH_4), nitrous oxide (N_2O), chlorofluorocarbons (CFCs), and tropospheric ozone (O_3) -- are produced as by-products of human activities. When these gases are emitted into the atmosphere and their concentrations increase, the greenhouse effect is compounded. The result is an increase in mean global temperatures.

There is scientific consensus that increases in greenhouse gas emissions will result in climate change (Bolin et al., 1986; NAS, 1979, 1983, 1987; WMO, 1985). The Council on Environmental Quality concluded in 1981 that the potential long-term risks of societal disruption caused by increased atmospheric concentrations of CO_2 (aside from the other greenhouse gases) are significant. However, considerable uncertainty exists with regard to the ultimate magnitude of the warming, its timing, and the regional patterns of change. In addition, there is great uncertainty about changes in climate variability and regional impacts.

CONGRESSIONAL REQUEST FOR REPORTS

EPA has studied the effects of global warming for several years. The goal of its efforts has been to use the best available information and models to assess the effects of climatic change and to evaluate policy strategies for both limiting and adapting to such change.

In 1986, Congress asked EPA to develop two reports on global warming. In one of these studies Congress directed EPA to include:

> *"An examination of policy options that if implemented would stabilize current levels of atmospheric greenhouse gas concentrations. This study should address the need for and implications of significant changes in energy policy, including energy efficiency and development of alternatives to fossil fuels; reductions in the use of CFCs; ways to reduce other greenhouse gases such as methane and nitrous oxide; as well as the potential for and effects of reducing deforestation and increasing reforestation efforts."*

These issues are the focus of this report.

This report differs from most previous studies of the climate change issue in that it is primarily a policy assessment. Although some aspects of the relevant scientific issues are reviewed, this document is not intended as a comprehensive scientific assessment. A recent review of the state of the science is contained in the U.S. Department of Energy's State of the Art series (MacCracken and Luther, 1985a, 1985b; Strain and Cure, 1985; Trabalka, 1985).

Congress also asked EPA to prepare a companion report on the health and environmental effects of climate change in the U.S., which would examine the impact of climate change on agriculture, forests, and water resources, as well as on other ecosystems and society. In response to the latter request, EPA produced its report entitled, *The Potential Effects of Global Climate Change on the United States* (Smith and Tirpak, 1989). That report provides insights into the ranges of possible future effects that may occur under alternative climate change scenarios, and establishes qualitative sensitivities of different environmental systems and processes to changes in climate. The report also examines potential changes in hydrology, agriculture, forestry, and infrastructure in the Southeast, Great Lakes, California, and Great Plains regions of the United States.

Goals of this Study

Congress presented EPA with a very challenging task. From a policy perspective, it is not enough to know how emissions would have to change from current levels in order to stabilize the atmosphere. Instead, policy options must be evaluated in the context of expected economic and technological development and the uncertainties that prevent us from knowing precisely how a given level of emissions will impact the rate and magnitude of climate change. It is also necessary for the scope of this study to be global and the time horizon to be more than a century, because of the long lags built into both the economic and climatic systems (we chose 1985-2100 as the time frame for the analysis). Predictions with such a scope cannot be attempted, but scenarios can be developed to explore policy options.

Based on these considerations EPA established four major goals:

- To assemble data on global trends in emissions and concentrations of all major greenhouse gases and activities that affect these gases.

- To develop an integrated analytical framework to study how different assumptions about the global economy and the climate system could influence future greenhouse gas concentrations and global temperatures.

- To identify promising technologies and practices that could limit greenhouse gas emissions.

- To identify policy options that could influence future greenhouse gas concentrations and global warming.

To achieve these goals EPA conducted an extensive literature review and data gathering process. The Agency held several informal panel meetings, and enlisted the help of leading experts in the governmental, non-governmental, and academic research communities. EPA also conducted five workshops, which were attended by over three hundred people, to gather information and ideas regarding factors affecting atmospheric composition and options related to greenhouse gas emissions from agriculture and land-use change, electric utilities, end-uses of energy, and developing countries. Experts in NASA, the Department of Energy, and the Department of Agriculture were actively engaged.

Report Format

The structure of this report is designed to answer the following questions in turn: What is the greenhouse effect? What evidence is there that the greenhouse effect is increasing? How will the Earth's climate respond to changes in greenhouse gas concentrations? What activities are responsible for the greenhouse gas emissions? How might emissions and climate change in the future? What

TERMINOLOGY OF CLIMATE CHANGE

An attempt has been made throughout this report to avoid technical jargon, yet some specialized terminology is inevitable. The specialized terms used in this Report are defined below.

Climate System

The interactive components of our planet which determine the climate. This includes the atmosphere, oceans, land surface, sea ice, snow, glaciers, and the biosphere. Climate change can be measured in terms of any part of the system, but it is most convenient to use surface air temperature as a measure of climate, since it is the parameter for which we have the best record and it is most directly relevant to the component of the biosphere that we know best -- humans.

Radiative Forcing (also called "external forcing," "forcing," or "perturbation")

A change imposed on the climate system (as opposed to generated by the internal dynamics of the climate system) that modifies the radiative balance of the climate system. Examples include: changes in the output of the sun or the orbit of the Earth about the sun, increased concentrations of particles in the atmosphere due to volcanoes or human activity, and increased concentrations of greenhouse gases in the atmosphere due to human activity. Radiative forcing is often specified as the net change in energy flux at the tropopause (W/m^2) or the equilibrium change in surface temperature in the absence of feedbacks (°C).

Climate Feedbacks

Processes that alter the response of the climate system to radiative forcings. We distinguish between physical climate feedbacks and biogeochemical climate feedbacks. Physical climate feedbacks are processes of the atmosphere, ocean, and land surface, such as increases in water vapor, changes in cloudiness, and decreases in land- and sea-ice accompanying global warming. Biogeochemical feedbacks involve changes in global biology and chemistry, such as the effect of changes in ocean circulation on carbon dioxide concentrations and changes in albedo from shifts in ecosystems. The impact of climate feedbacks is generally measured in terms of their effect on climate sensitivity. Positive feedbacks increase climate sensitivity, while negative feedbacks reduce it.

Climate Sensitivity (or equilibrium sensitivity)

The ultimate change in climate that can be expected from a given radiative forcing. Climate sensitivity is generally measured as the change in global average surface air temperature when equilibrium between incoming and outgoing radiation is reestablished following a change in radiative forcing. A common benchmark, which we use in this report, is the equilibrium temperature increase associated with a doubling of the concentration of carbon dioxide from preindustrial levels. The National Academy of Sciences has estimated that this sensitivity is in the range of 1.5-4.5°C, with a recent analysis suggesting 1.5-5.5°C; a reasonable central uncertainty range is 2-4°C.

TERMINOLOGY OF CLIMATE CHANGE
(continued)

Transient Response

The time-dependent response of climate to radiative forcing. Climate responds gradually to changes in radiative forcing, primarily because of the heat capacity of the oceans. The transient mode is characterized by an imbalance between incoming and outgoing radiation. Given the changing concentrations of greenhouse gases the Earth's climate will be in a transient mode for the foreseeable future. Most GCMs (see below), however, have so far examined equilibrium conditions because transient effects are much more difficult to analyze.

Albedo

The fraction of incoming solar radiation that is reflected back into space.

Flux

Flow per unit time per unit area. The flow can be of energy (e.g., watts per square meter [W/m^2]) or mass (e.g., grams per square meter per day [$g\ m^{-2}d^{-1}$]).

General Circulation Model (GCM)

A computer model of the Earth's climate based on equations that describe, among other things, the conservation of energy, momentum, and mass, and which explicitly calculates the distribution of wind, temperature, precipitation and other climatic variables. Such models are applied to the atmosphere, to the oceans, or to both coupled together.

Solar Luminosity, Solar Constant

Solar luminosity is the total amount of energy emitted by the sun. The so-called "solar constant" is the average amount of energy received at the top of the Earth's atmosphere at the mean Earth-sun distance; this amount varies with changes in solar luminosity.

Troposphere, Tropopause, Stratosphere

The troposphere is the lower atmosphere, from the ground to an altitude of about 8 kilometers (km) at the poles, 12 km in midlatitudes and 18 km in the tropics. The tropopause marks the top of the troposphere; temperature decreases with altitude below the tropopause and increases with altitude above the tropopause to the top of the stratosphere. The stratosphere extends from the tropopause to about 50 km. The troposphere and stratosphere together contain more than 99.9% of the mass of the atmosphere.

technologies are available for limiting greenhouse gas emissions? And what domestic and international policy options, if implemented, would help to stabilize global climate?

This chapter provides a general introduction to the climate change issue and reviews selected previous studies. Chapter II discusses the greenhouse gases, their sources and sinks, chemical properties, current atmospheric concentrations and distributions, and related uncertainties. Chapter III relates the greenhouse gases to the process of climatic change. Once this link is made, Chapter IV examines those human activities that affect trace-gas emissions and ultimately influence climate change. Chapter V discusses the scenarios developed for this report to assist us in thinking about possible future emissions and climate change. Chapter VI then presents sensitivity analyses of the modeling results. Chapter VII gives a detailed description of existing and emerging technologies that should be considered in the formulation of a comprehensive strategy for mitigating global warming. Chapter VIII outlines domestic policy options, and the concluding chapter (Chapter IX) discusses international mechanisms for responding to climate change.

THE GREENHOUSE GASES

Congress presented EPA with an extremely challenging task. Once emitted, greenhouse gases remain in the atmosphere for decades to centuries. As a result, if emissions remained constant at 1985 levels, the greenhouse effect would continue to intensify for more than a century. Carbon dioxide concentrations would reach 440-500 parts per million (ppm) by 2100, compared with about 350 ppm today, and about 290 ppm 100 years ago. CFC concentrations would increase by more than a factor of three from current levels, while nitrous oxide concentrations would increase by about 20%, and methane concentrations might remain roughly constant. Indeed, in many cases drastic cuts in emissions would be required to stabilize atmospheric composition.

Carbon dioxide

Carbon dioxide (CO_2) is the most abundant and single most important greenhouse gas in the atmosphere. Its concentration has increased by about 25% since the industrial revolution. Detailed measurements since 1958 show an increase from 315 to 350 parts per million (ppm) by volume (Figure 1-1). These data clearly demonstrate that human activities are now of such a magnitude as to produce global consequences. Current emissions are estimated at 5.5 billion tons of carbon (Pg C) from fossil-fuel combustion and 0.4-2.6 Pg C from deforestation.[1] Most of this CO_2 remains in the atmosphere or is absorbed by the ocean. Even though only about half of current emissions remain in the atmosphere, currently available models of CO_2 uptake by the ocean suggest that substantially more than a 50% cut in emissions is required to stabilize concentrations at current levels (Figure 1-2).

Methane

The concentration of methane (CH_4) has more than doubled during the last three centuries. Methane, which is currently increasing at a rate of 1% per year, is responsible for about 20% of current increases in the greenhouse effect. Of the major greenhouse gases, only CH_4 concentrations can be stabilized with modest cuts in anthropogenic emissions: a 10-20% cut would suffice to stabilize concentrations at current levels due to methane's relatively short atmospheric lifetime (assuming that the lifetime remains constant, which may require that hydrocarbon and carbon monoxide emissions be stabilized).

---

[1] One billion tons of carbon = 10^{15} grams of carbon = 1 petagram of carbon (Pg C).

FIGURE 1-1

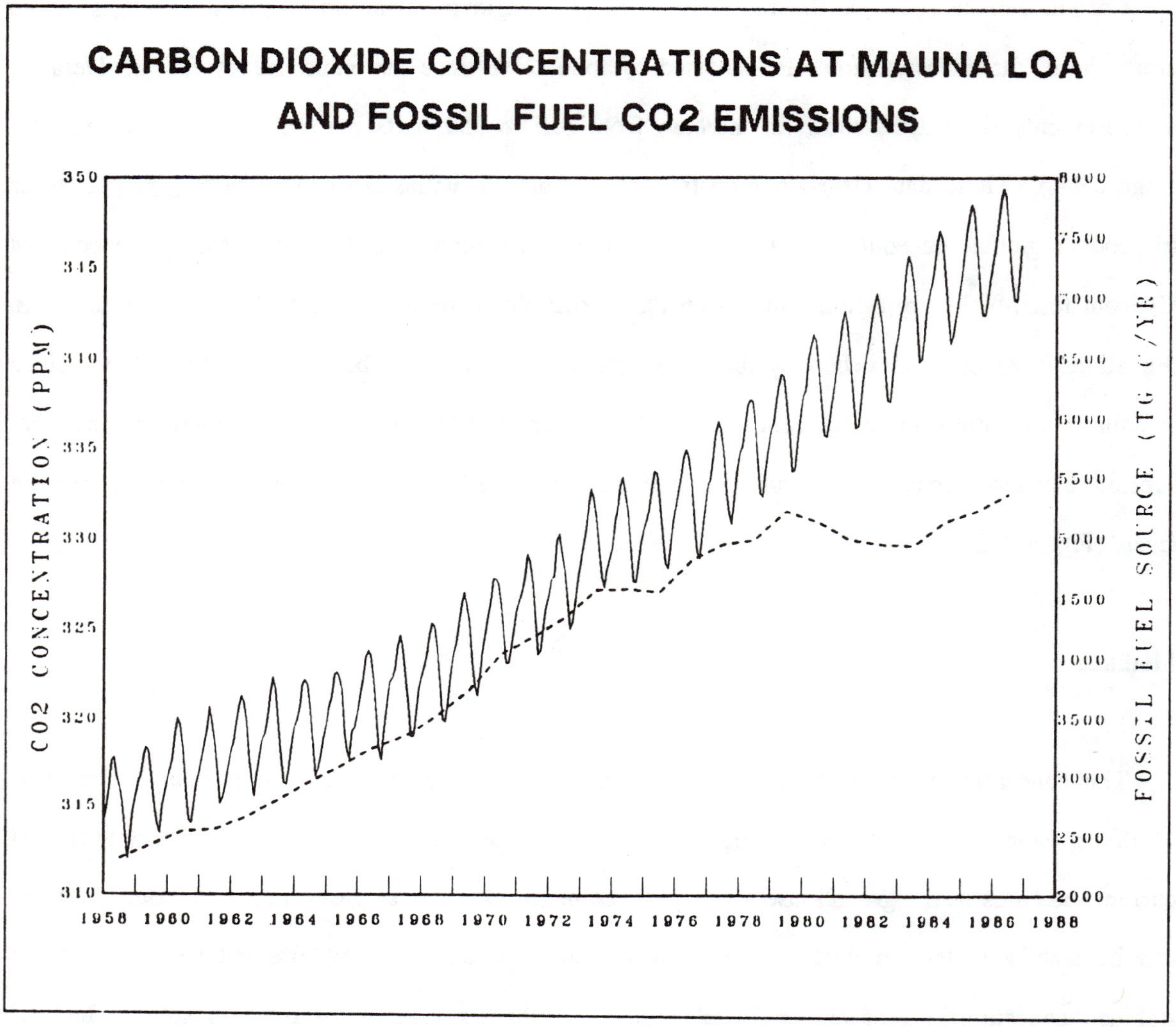

Figure 1-1. The solid line depicts monthly concentrations of atmospheric CO_2 at Mauna Loa Observatory, Hawaii. The yearly oscillation is explained mainly by the annual cycle of photosynthesis and respiration of plants in the northern hemisphere. The steadily increasing concentration of atmospheric CO_2 at Mauna Loa since the 1950s is caused primarily by the CO_2 inputs from fossil fuel combustion (dashed line). Note that CO_2 concentrations have continued to increase since 1979, despite relatively constant emissions; this is because emissions have remained substantially larger than net removal, which is primarily by ocean uptake. (Sources: Keeling, 1983, pers. communication; Komhyr et al., 1985; NOAA, 1987; Conway et al., 1988; Rotty, 1987, pers. communication.)

FIGURE 1-2

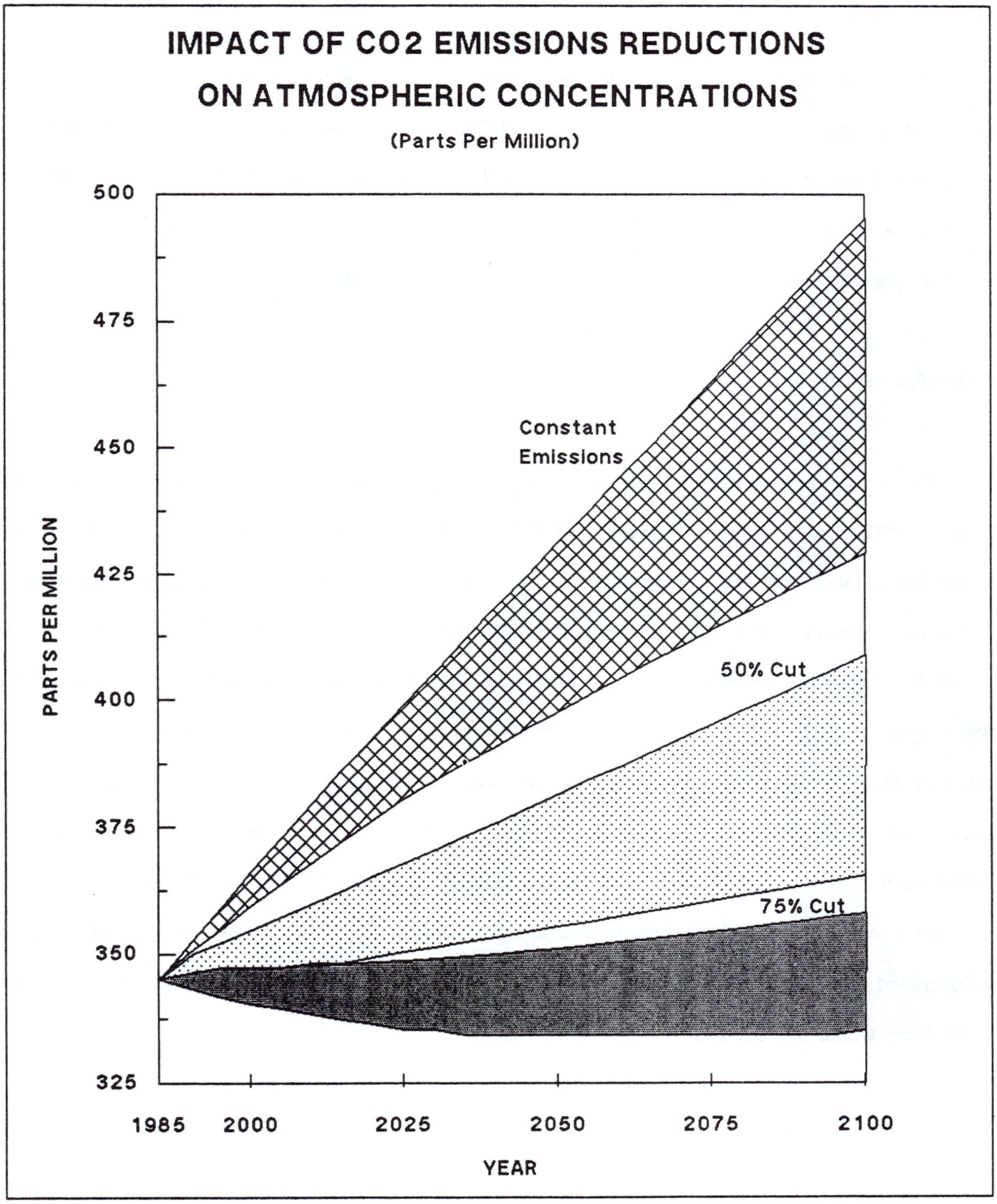

Figure 1-2. The response of atmospheric CO_2 concentrations to arbitrary emissions scenarios based on two one-dimensional models of ocean CO_2 uptake. See Chapter VI for a description and models. (Sources: Hansen et al., 1984; Lashof, 1988; Siegenthaler, 1983).

Nitrous oxide

The concentration of nitrous oxide (N_2O) has increased by 5-10% since preindustrial times. Nitrous oxide is currently increasing at a rate of 0.25% per year, which represents an imbalance of about 30% between sources and sinks. Assuming that the observed increase in N_2O concentrations is due to anthropogenic sources and that natural emissions have not changed, then an 80-85% cut in anthropogenic emissions would be required to stabilize N_2O at current levels.

Chlorofluorocarbons

Chlorofluorocarbons (CFCs) were introduced into the atmosphere for the first time during this century. The most common species are CFC-12 (CF_2Cl_2) and CFC-11 ($CFCl_3$), which had atmospheric concentrations in 1986 of 392 and 226 parts per trillion (ppt) by volume, respectively. While these concentrations are tiny compared with that of CO_2, each additional CFC molecule has as much as 20,000 times more impact on climate, and CFCs are increasing very rapidly--more than 4% per year since 1978. A focus of attention because of their potential to deplete stratospheric ozone, the increasing concentrations of CFCs also account for about 15% of current increases in the greenhouse effect. For CFC-11 and CFC-12, cuts of 75% and 85%, respectively, of current global emissions would be required to stabilize concentrations. However, in order to stabilize stratospheric chlorine levels -- of particular concern for stratospheric ozone depletion -- a 100% phaseout of fully-halogenated compounds (those that do not contain hydrogen) and a freeze on the use of methyl chloroform would be required.

Other gases influencing composition

Emissions of carbon monoxide (CO), nitrogen oxides (NO_x), and other species, in addition to the greenhouse gases just described, are also changing the chemistry of the atmosphere. This change in atmospheric chemistry alters the distribution of ozone and the oxidizing power of the atmosphere, changing the atmospheric lifetimes of the greenhouse gases. If the concentrations of the long-lived gases were stabilized, it might only be necessary to freeze emissions of the short-lived gases at current levels to stabilize atmospheric composition.

PREVIOUS STUDIES

Evidence that the composition of the atmosphere is changing has led to a series of studies analyzing the potential magnitude of future greenhouse gas emissions. A few of these studies have carried the analysis further, making projections of the timing and severity of future global warming. The first generation of these studies focussed principally on energy use and CO_2 emissions. (See, for example, Arrhennius, 1896; NAS, 1979; Clark et al., 1982; IIASA, 1983; Nordhaus and Yohe, 1983; Rose et al., 1983; Seidel and Keyes, 1983; Edmonds and Reilly, 1983b, 1984; Legasov, et al. 1984; Goldemberg et al., 1985, 1987; and Keepin et al., 1986). Subsequent studies have recognized that other radiatively-active trace gases significantly amplify the effects of CO_2. (See, for example, Lacis et al., 1981; Ramanathan et al., 1985; Dickinson and Cicerone, 1986; WMO, 1985; and Mintzer, 1987). In the following sections, some of the most important of these earlier analyses are reviewed in order to provide a basis for comparison with this study.

Estimates of the Climatic Effects of Greenhouse Gas Buildup

The first serious analysis of the effect of increasing CO_2 concentrations on global warming was conducted by the Swedish chemist Svante Arrhennius (1896). Arrhennius, concerned about the rapidly increasing rate of fossil-fuel use in Europe, recognized that the resulting increase in the atmospheric concentration of CO_2 would alter the thermal balance of the atmosphere. Using a simplified, one-dimensional model, Arrhennius estimated that if the atmospheric concentration of CO_2 doubled, the surface of the planet would warm by about 5°C. (The expected equilibrium climate change associated with a doubling of CO_2 has become a benchmark. That is, many studies examine the consequences of greenhouse gas increases with a total warming effect equivalent to that from a doubling of the concentration of CO_2.)

In 1979, a study by the U.S. National Academy of Sciences (NAS) evaluated the impact on global climate of doubling the concentration of CO_2 relative to the preindustrial atmosphere (NAS, 1979). The NAS study concluded that the planet's surface would be 1.5-4.5°C warmer under such conditions. Subsequent re-evaluations by NAS (1983, 1987) as well as the "State-of-the-Art" report issued by the Department of Energy (MacCracken and Luther, 1985a) have reaffirmed this estimate.

Recent work by Dickinson (1986) suggests that the effects of a greenhouse gas buildup radiatively equivalent to doubling the preindustrial concentration of CO_2 might warm the planet to a greater extent than had previously been expected. Focusing on the uncertainties in current understanding of atmospheric feedback processes, Dickinson estimated that the warming effect of such a buildup was likely to be between 1.5° and 5.5°C. Dickinson's "best guess" was that the actual equilibrium warming would be between 2.5° and 4.5°C.

Studies of Future CO_2 Emissions

For the next eighty years after Arrhenius issued his warning, little additional scientific attention was directed toward understanding the factors that contribute to future greenhouse gas emissions. By the mid-1970s, measurements of atmospheric CO_2 concentrations at Mauna Loa begun by Keeling during the International Geophysical Year (1957-1958) provided indisputable evidence of a long-term increasing trend (see Figure 1-2), while the oil embargo of 1973 and the nuclear power debate focussed attention on future energy supplies. Increasing interest was placed on the problems of projecting future global energy use and on estimating the resulting CO_2 emissions.

A major international study of future energy use was conducted by the International Institute for Applied Systems Analysis (IIASA, 1981 and 1983). Employing an international group of almost 200 scientists, the IIASA team developed a set of computer models to estimate regional economic growth, energy demand, energy supply, and future CO_2 emissions. Although the models were never completely integrated, the first phase of the IIASA study produced two complete scenarios of global energy use. The IIASA low scenario generated CO_2 emissions of about 10 petagrams of carbon per year (Pg C/yr) in 2030. The IIASA high scenario projected emissions of about 17 Pg C/yr in 2030. In the second phase of the IIASA study a third scenario was outlined, emphasizing increased use of natural gas. In this third scenario, CO_2 emissions in 2030 were only about 9.4 Pg C/yr.

In 1983 Edmonds and Reilly, two U.S. economists, developed a detailed partial equilibrium model to investigate the effects of alternative energy policies and their implications for future CO_2 emissions (Edmonds and Reilly, 1983a). This model disaggregates the world into nine geopolitical regions. It offers a highly detailed picture of the supply side of the world's commercial energy business but only limited detail on the demand side. It considers nine primary and four secondary forms of commercial energy (including biomass grown on plantations) but ignores non-commercial uses of biomass for fuel. Using explicit assumptions about regional population changes and economic growth and combining them with assumptions about technological change and the costs of extracting various grades of fuel resources in each region, the model calculates supply and demand schedules for each type of fuel.

For their first major report, Edmonds and Reilly (1983b) developed a Base Case energy future for the period 1975 to 2050. In this scenario, CO_2 emissions in 2050 were approximately 26.3 Pg C/yr. The authors generated several other scenarios in this study that reflected the effect of various taxes imposed on fuel supply and use. These taxes reduced CO_2 emissions by varying amounts, with emissions in some scenarios falling as low as 15.7 Pg C/yr in 2050. In 1984 Edmonds and Reilly produced a new set of scenarios for the U.S. Department of Energy by varying other key parameters in the model (Edmonds and Reilly, 1984). In these new scenarios, CO_2 emissions in 2050 vary from about 7 to 47 Pg C/yr, with a new "Base Case" value of about 15 Pg C/yr. The principal force contributing to the difference between the results of the two studies conducted by Edmonds and Reilly is the higher coal price applied in the second study.

A number of other studies have used the Edmonds-Reilly (ER) model to project future energy use and CO_2 emissions. The most important of these were studies conducted by the U.S. Environmental Protection Agency (Seidel and Keyes, 1983) and Rose et al. (1983). The EPA study used the ER model to generate 13 scenarios for the period 1975-2100, which were used as a basis for investigating whether actions taken now to reduce fossil-fuel consumption could significantly delay a future global warming. Six baseline and seven policy-driven scenarios were investigated in this study. The scenarios generated in the EPA study projected CO_2 emissions in 2050 at levels of 10-18 Pg C/yr. The authors concluded from these scenarios that the timing of a 2°C warming is not very sensitive to the effects of the energy policies they tested.

Rose and his colleagues at the Massachusetts Institute of Technology (MIT) also used the ER model to study the effect of various energy policy options on the timing and extent of future CO_2 emissions (Rose et al., 1983). Eleven scenarios were investigated, covering the period from 1975 to 2050 and incorporating a much wider range of assumptions and policies than those tested in the EPA study. Rose et al. studied the effects of increased energy efficiency, increased fossil-fuel prices, higher nuclear energy supply costs and a moratorium on building nuclear plants, lower photovoltaic costs, higher oil prices, and a cutoff of oil imports from the Middle East. The MIT study went beyond the ER model results to provide detailed estimates of the materials required for construction and operation of energy facilities in each scenario. In the MIT scenarios, emissions of CO_2 in 2050 ranged from less than 3 to about 15 Pg C/yr. The most important new conclusion of Rose et al. was that a feasible "option space exists in which the CO_2/climate problem is much ameliorated" through energy policy choices and improvements in technology.

In 1983 the National Academy of Sciences completed a Congressionally-mandated study to evaluate, among other things, the effects of fossil-fuel development activities authorized by the Energy Security Act of 1980 (NAS, 1983). One of the chapters in this study, authored by energy economists Nordhaus and Yohe, used a compact model of global economic growth and energy use to analyze CO_2 emissions between 1975 and 2100 (Nordhaus and Yohe, 1983). Unlike the partial equilibrium approach employed in the ER model, the Nordhaus and Yohe (NY) model used a generalized Cobb-Douglas production function to estimate future energy demand. In this approach global GNP is estimated as a function of assumptions about average rates of change in labor productivity, population, and energy consumption. Demand for energy is separated into two categories, fossil and non-fossil. Projections of CO_2 emissions (based on the weighted average release rate from fossil fuels) were used as inputs to a simple airborne fraction model of the carbon cycle.

The NY analysis used an approach called "probabilistic scenario analysis," to evaluate the effects on CO_2 emissions of alternative assumptions used in the model. The results of 1000 cases were examined. The CO_2 emissions trajectories in these cases were presented as percentiles in the overall distribution among the 1000 scenarios. Using this approach to uncertainty analysis, Nordhaus and Yohe concluded that the 50th percentile for carbon emissions in 2050 was approximately 15 Pg C/yr. The 95th percentile case suggested that emissions in 2050 would likely be less than 26 Pg C/yr, while the 5th percentile case indicated that emissions would likely be greater than 5 Pg C/yr.

The probabilistic approach was subsequently applied to the more detailed ER model using Monte Carlo analysis (Edmonds et al., 1986; Reilly et al., 1987). The results of this analysis suggest a larger total range of uncertainty and a substantially lower median emissions estimate compared with the Nordhaus and Yohe (1983) results. When the likely correlations between model parameters are taken into account Edmonds et al. obtain emissions of 7.7 Pg C/yr in 2050 for the 50th percentile case with 5th and 95th percentile bounds of 2.3 and 58.1 Pg C/yr, respectively. Note that the median result is about half of the Base Case scenario obtained in earlier analysis by Edmonds and Reilly (1984).

In 1984 Legasov et al. published one of a continuing series of Soviet analyses of future global energy use and its environmental implications. Legasov et al. analyzed two scenarios in which energy demand reaches 6 and 20 kilowatts per capita by the end of the next century. Annual per capita energy consumption is treated as a logistic function, approaching these levels asymptotically in 2100. Assuming a global population of 10 billion persons, the minimal variant implies a global energy demand of 60 terawatts (TW), about six times the current level by 2100.[2] CO_2 emissions in this scenario follow a bell-shaped trajectory, peaking at about 13.3 Pg C/yr in 2050.

---

[2] 1 terawatt = 10^{12} watts = 31.5×10^{18} joules per year = 31.5 exajoules (EJ) per year = 29.9 Quadrillion British Thermal Units (Quads) per year.

Goldemberg and his colleagues have used a completely different approach to projecting future energy use and its consequences for CO_2 emissions (Goldemberg et al., 1985, 1987, 1988). The Goldemberg et al. analysis is based on an end-use oriented approach to evaluating the demand for energy services, rather than the availability of energy supply. Based on detailed studies of energy demand in four countries (U.S., Sweden, India, and Brazil), Goldemberg and his colleagues developed a scenario of future energy requirements in both industrialized and developing countries. Although the study does not represent a forecast of future energy demand, it provides an "existence proof," demonstrating the feasibility of a world economy that continues to grow while consuming much less energy than it would if historical trends continue.

Emphasizing the potential to improve the efficiency of energy supply and use, per capita energy demand in the industrialized countries is cut by 50% in the Goldemberg et al. scenarios. During the same 40-year period, per capita demand for energy in the developing countries grows by about 10%, with commercial fuels displacing traditional biomass fuels at a rapid and increasing rate. Global energy demand remains essentially constant in the base case with CO_2 emissions in 2020 of 5.9 Pg C/yr, only about 5% higher than today's level.

A limitation of the Goldemberg et al. studies is that the impact of market imperfections and the rate of capital stock turnover are not fully addressed. Nonetheless, these studies, along with the Rose et al. analysis, demonstrate that economic growth can be decoupled from increases in CO_2 emissions. Experience over the last 15 years in the U.S., Western Europe, and Japan suggests that this conclusion is correct.

A study by Keepin et al. (1986) reviewed and re-evaluated the range of previous energy and CO_2 projections, including those summarized here. It concluded that the feasible range for future energy in 2050 was somewhere between about 10 and 35 TW, with CO_2 emissions between 2 and 20 Pg C/yr.

Studies of the Combined Effects of Greenhouse Gas Buildup

In the last few years a number of analysts have investigated the *combined* effects on global surface temperature of a buildup of CO_2 and other trace gases. Preliminary analysis of the impact of concentration increases during the 1970s was presented by Lacis et al. (1981) and estimates of future impacts were included in Seidel and Keyes (1983). A seminal article by Ramanathan et al. (1985) focused attention on the subject. This study used a one-dimensional radiative-convective model to estimate the impact of a continuation of current trends in the buildup of more than two dozen radiatively active trace gases between 1980 and 2030. Ramanathan and his colleagues calculated an expected value for the equilibrium warming of about 1.5°C over this period, with a little less than half of that amount due to the buildup of CO_2 alone. (The Ramanathan et al. analysis included the effects of water-vapor feedback, but not the other known feedback mechanisms; see Chapter III.) The most important conclusion of the analysis by Ramanathan et al. is that, if current trends continue and uncertainties in the future emissions projections are accounted for, the warming effects of the non-CO_2 trace gases will amplify the warming due to the buildup of CO_2 alone by a factor of between 1.5 and 3.

In 1986, Dickinson and Cicerone extended the work of Ramanathan et al. to evaluate a range of trace-gas scenarios covering the period from 1985 to 2050. Using the radiative-convective model developed by Ramanathan et al., and considering a range of emissions growth rates for the most important greenhouse gases, Dickinson and Cicerone (1986) estimated that equilibrium global average surface temperatures would rise at least 1°C and possibly more than 5°C by 2050, when the full range of atmospheric feedback processes was considered.

Each of the analyses described above was based on the assumption that historical trends in the growth of greenhouse gas emissions continue for the next 40-50 years. Mintzer (1987) has developed a model to consider the alternative: that policy and investment choices made in the next several decades will substantially alter the growth rates of future emissions. Mintzer's analysis uses a composite tool called the Model of Warming Commitment to link future rates of economic growth to the increasing atmospheric concentrations of carbon dioxide, nitrous oxide, chlorofluorocarbons, methane, and tropospheric ozone. The results are reported as the date of atmospheric commitment to a warming equivalent to doubling preindustrial CO_2 concentrations and as the magnitude of warming commitment in 2075.

Mintzer's initial analysis considered four policy-driven global scenarios, including a Base Case representing a continuation of current trends. All four scenarios support a global population of about 10 billion people and the same levels of regional economic growth. Most recent analyses, including the ones cited above and Mintzer's Base Case, indicate that a continuation of current trends would lead to a warming commitment equivalent to doubling the preindustrial concentration of CO_2 by about 2030. In Mintzer's Base Case, by 2075, the planet is committed to an eventual warming of about 3-9°C. Alternatively, in the High Emissions case, policies that increase coal use, spur deforestation, extend the use of the most dangerous CFCs and limit improvements in energy efficiency, will accelerate the onset of the "doubled CO_2 equivalent" atmosphere to about 2010 and commit the planet to a warming of about 5-15°C in 2075. By contrast, in Mintzer's Slow Buildup scenario, a warming associated with the doubled CO_2 equivalent atmosphere is postponed beyond the end of the simulation period in 2075. In the Slow Buildup scenario this level of risk reduction is achieved by aggressively pursuing policies to increase energy efficiency, limit tropical deforestation, reduce the use of the most dangerous CFCs, and shift the fuel mix from carbon-intensive fuels like coal to hydrogen-intensive fuels like natural gas, and ultimately, to energy sources that emit no CO_2.

More recently, Rotmans et al. (1988) have used a framework similar to the Model of Warming Commitment to develop scenarios of greenhouse warming based on alternative policy assumptions. Also, Rotmans and Eggink (1988) have analyzed the role of methane in greenhouse warming.

Major Uncertainties

Major uncertainties underlie many aspects of our understanding of the climate change problem. These uncertainties encompass our understanding of the geophysical processes underlying the sensitivity of the climate to perturbations, that is, the processes that control how fast greenhouse gases flow into and out of the atmosphere and biosphere, including feedback processes that may affect future concentrations of greenhouse gases, and socio-economic uncertainties that are inherent in any energy/economic model used to forecast long-term emissions. The physical uncertainties include uptake of heat and CO_2 by the ocean and any other sinks, geophysical and biogeochemical feedback mechanisms, and natural rates of emission of the greenhouse gases. The social and economic uncertainties include population growth, GNP growth, structural changes in economic systems, rates of technological change, future reliance on fossil fuels, and future compliance with the Montreal Protocol. Future rates of greenhouse gas emissions cannot be predicted with certainty. Future emissions rates will be determined by the emerging pattern of human industrial and agricultural activities as well as by the effects of feedback processes in the Earth's biogeophysical system whose details are not well understood at the present time.

All existing climate models encompass large uncertainties that limit the accuracy of the models and the level of geographic detail that can be considered. Even the best General Circulation Models (GCMs) are limited by the assumptions necessarily made about the influence of clouds, vegetation, ice and snow, soil moisture, and terrain, all of which affect the energy balance of the Earth's surface.

Two of the largest uncertainties involve our limited understanding of the role that clouds, and ocean uptake and transport of heat, play in the climate system.

Conclusions From Previous Studies

Despite the significant uncertainties that underlie our understanding of climate change, several important conclusions emerge from the existing literature. First, emissions of a number of other trace gases will amplify the future warming effect of any further buildup in the atmospheric concentration of CO_2. Second, it is too late to prevent all future global warming. Trace gases released over the last century have already committed the planet to an ultimate warming (of up to 2°C) that may be greater than any other in the period of written human history. Finally, policy choices and investment decisions made during the next decade that are designed to increase the efficiency of energy use and shift the fuel mix away from fossil fuels could slow the rate of buildup sufficiently to avoid the most catastrophic potential impacts of rapid climate change. Alternatively, decisions to rapidly expand the use of coal, extend the use of the most dangerous CFCs, and rapidly destroy the remaining tropical forests could "push up the calendar," accelerating the onset of a dangerous global warming.

The rate at which climate may change must be of particular concern to policy makers. The temperature increases resulting from doubling the concentration of CO_2 that are predicted by most GCMs are comparable to the increase that has occurred since the last ice age. The difference is that the period of time within which this increase could happen is much shorter. Atmospheric scientists predict that within approximately 100 years we could experience temperature increases equivalent to those that have occurred over the past 18,000 years (about 5°C; see Chapter III). It is not clear that our ecosystems and economic systems will be able to adjust to such a rapid change in global mean temperatures. Increases in world population, coupled with limited environmental and agricultural resources, increase the vulnerability of social systems to climatic change.

The potential impacts of climatic change are highly uncertain and are beyond the scope of this report. They are addressed in the companion volume, *The Potential Effects of Global Climate Change on the United States* (Smith and Tirpak, 1989). The findings of this study collectively suggest that the climatic changes associated with a global warming of roughly 2-4°C would result in "a world that is different from the world that exists today. Global climate change will have significant implications for natural ecosystems; for when, where, and how we farm; for the availability of water to drink and water to run our factories; for how we live in our cities; for the wetlands that spawn our fish; for the beaches we use for recreation; and for all levels of government and industry." Although sensitivities were identified in this report, detailed regional predictions of climate change cannot be made at this time. Thus potential responses to the greenhouse gas buildup must be viewed in the context of risk management, or buying insurance.

A second major concern is that the greenhouse gases have very long lifetimes once they are introduced into the atmosphere. Although there is a substantial lag between the time when a greenhouse gas is introduced into the atmosphere and when its full impact on climate is realized, once the gases are in the atmosphere they will remain there for a long time. The longer the delay before mitigating action is taken, the larger will be the commitment to further global warming.

Policy makers must determine how best to minimize the costs of global warming to the peoples of the world and the damage to ecosystems. But global warming is a complex problem for which there is no single, simple solution. No single policy initiative will completely mitigate man-made climate change. The sources, sectors, and countries contributing to the emissions of greenhouse gases are numerous (Chapter IV).

Compounding the difficulty of identifying solutions to the greenhouse problem is that the greenhouse gases do not all have the same forcing effect on global temperatures. In fact, CO_2 is the least effective absorbent of infrared radiation of all of the greenhouse gases per additional molecule added to the atmosphere. Because the combined effect of the other greenhouse gases is comparable to the effect of CO_2, mitigatory policies cannot be directed solely at reducing CO_2 emissions. The sources of methane, CFCs, nitrous oxide, and other gases must therefore be carefully considered.

As we explore the options for limiting greenhouse gas emissions in this report, it is important to remember two salient points: (1) Global warming is an international problem whose solution will require extensive cooperation between both industrialized and developing countries; and (2) No single economic sector can be held entirely responsible for the greenhouse effect. In focusing on strategies to stabilize climate in this Report, we recognize that the optimal mix of adaptation and prevention is uncertain. The Earth is already committed to some degree of climatic change, so adaptation to some level of change is essential. On the other hand, the highest rates of potential change may be considered unacceptable, requiring some degree of prevention. Stabilizing strategies would require global cooperation of an unprecedented nature and could be costly for some countries. The activities responsible for greenhouse gas emissions are economically valuable, the distribution of emissions is large, and the responsible countries reflect diverse economies and a variety of interests. Adaptation strategies, on the other hand, can be adopted unilaterally. They may also be less burdensome because the costs will be spread out into the future when countries may be better able to afford them. Imposing climatic change on our grandchildren, however, raises serious concerns regarding intergenerational equity.

CURRENT NATIONAL AND INTERNATIONAL ACTIVITIES

Subsequent to the Congressional request to produce this report and the companion document on potential effects of climate change there have been a wide variety of new domestic and international initiatives related to climatic change.

National Research and Policy Activities

The Global Climate Protection Act of 1987 requires that:

> *The President, through the Environmental Protection Agency, shall be responsible for developing and proposing to Congress a coordinated national policy on global climate change.*

This Act is a very broad mandate that will require close cooperation between EPA and other agencies (including NASA, NOAA, the Corps of Engineers, and the Departments of Energy, Agriculture, and the Interior, the National Climate Program Office, and the Domestic Policy Council).

The Global Climate Protection Act also requires that the Secretary of State and the EPA Administrator jointly submit, by the end of 1989, a report analyzing current international scientific understanding of the greenhouse effect, assessing U.S. efforts to gain international cooperation in limiting global climate change, and describing the U.S. strategy for seeking further international cooperation to limit global climate change. This report, along with those being developed by other Federal agencies, will provide a foundation upon which a national policy can be formulated.

International Activities

The greenhouse gas problem is an international issue. In order to respond effectively to this problem, the nations of the world must act in concert. Several international organizations have recognized the need for multilateral cooperation and have become involved with the global climate change issue. The United Nations Environment Programme (UNEP) is responsible for conducting climate impact assessments. The World Meteorological Organization (WMO) is supporting research on and monitoring of atmospheric and physical sciences. The International Council of Scientific Unions (ICSU) is developing an international geosphere-biosphere program.

The U.S. Government is supporting the Intergovernmental Panel on Climate Change (IPCC) established under the auspices of UNEP and WMO. The IPCC, which held its first meeting in November 1988, will help ensure an orderly international effort in responding to the threat of global climate change. At its first meeting the IPCC established three working groups: the first, to assess the state of scientific knowledge on the issue, will be chaired by the United Kingdom; the second, to assess the potential social and economic effects from a warming, will be chaired by the Soviet Union; and the third, to examine possible response strategies, including options for limiting emissions and adapting to change, will be chaired by the United States.

In addition, several countries have held or plan to hold international conferences on global climate change. These include Canada, The Federal Republic of Germany, Italy, Japan, India, Egypt, and the Netherlands. The Netherlands and the Federal Republic of Germany (through the Enquete Commission) are undertaking analyses of policy options.

The efforts of all of these organizations may be hindered if some countries perceive themselves as winners instead of losers, as a result of climate change. For example, both Canada and the Soviet

Union, which have vast land areas that are currently largely unusable because of their severe climates, could benefit from increased agricultural productivity as those areas become warmer. But the notion that a global warming might be beneficial to some may prove fallacious. Two particular problems might limit anticipated benefits. First, if the shift in climatological zones happens too quickly, ecosystems may not be able to keep up and may be severely disrupted. Although the Canadian climate may become more conducive to certain types of forests, if the forests can't migrate fast enough and therefore die back as the climatological zones shift northward, benefits to the Canadians will be reduced. Second, although shifts towards more favorable climatic conditions may be a necessary condition for increased agricultural productivity, a warmer climate in itself may not be sufficient. For example, the northern areas of Canada might not have the proper soil composition for high agricultural yields. The conclusion must therefore be drawn that it is difficult to predict what the net costs and benefits of climate change will be for any one country.

The global warming issue is an international concern. In order to develop a responsible program, the U.S. government must consider the feasibility of achieving both domestic and international acceptance and implementation of policy initiatives. Otherwise, the effectiveness of programs instituted by any one country could be compromised by the lack of participation by other countries. International collaboration must be pursued.

REFERENCES

Arrhennius, S. 1896. On the influence of carbonic acid in the air upon the temperature on the ground. *Phil. Mag.* 41:237.

Bolin, B., J. Jager, and B.R. Doos. 1986. The greenhouse effect, climatic change, and ecosystems: A synthesis of present knowledge. In Bolin, B., B.R. Doos, J. Jager, and R.A. Warrick, eds. *The Greenhouse Effect, Climatic Change, and Ecosystems.* Scope 29. John Wiley & Sons, Chichester 1-34.

Clark, W.C., K.H., Cook, G., Marland, A.M. Weinberg, R.M. Rotty, P.R., Bell, L.J. Allison, and C.L. Cooper. 1982. The carbon dioxide question: A perspective in 1982. In Clark, W.C. ed. *Carbon Dioxide Review 1982.* Oxford University Press, New York. 3-43.

Conway, T.J., P. Tans, L.S. Waterman, K.W. Thoning, K.A. Masarie, and R.H. Gammon. 1988. Atmospheric carbon dioxide measurements in the remote global troposphere. 1981 - 1984. *Tellus* 40:81-115.

Council on Environmental Quality. 1981. *Global Energy Futures and the CO_2 Problem.* The Council on Environmental Quality, Washington, D.C.

Dickinson, R. 1986. The Climate System and modelling of future climate. In Bolin, B., B. Doos, J. Jager, R. Warrick, eds. *The Greenhouse Effect, Climatic Change, and Ecosystems.* Scope 29. John Wiley & Sons, Chichester. 207-270.

Dickinson, R.E., and R.J. Cicerone. 1986. Future global warming from atmospheric trace gases. *Nature* 319:109-115.

Edmonds, J.A., and J.M. Reilly. 1983a. A long-term global energy-economic model of carbon dioxide release from fossil fuel use *Energy Economics* 5:74-88.

Edmonds, J.A., and J.M. Reilly. 1983b. Global Energy and CO_2 to the Year 2050. *The Energy Journal* 4:21-47.

Edmonds, J.A., and J.M. Reilly. 1984. *The IEA/ORAU Long-Term Global Energy CO_2 Model.* Carbon Dioxide Information Center, Oak Ridge National Laboratory, Oak Ridge, Tennessee.

Edmonds, J., J. Reilly, R. Gardner, and A. Brenhert. 1986. Uncertainty in future global energy use and fossil fuel CO_2 emissions 1975-2075. U.S. Department of Energy, Washington, D.C. DOE/NBB-0081 (TR036).

Goldemberg, J., T.B. Johansson, A.K.N. Reddy, and R.H. Williams, 1985. An end-use oriented global energy strategy. *The Annual Review of Energy* 10:613-688.

Goldemberg, J., T.B. Johansson, A.K.N. Reddy, and R.H. Williams. 1987. *Energy for a Sustainable World.* World Resources Institute, Washington, D.C. 119 pp.

Goldemberg, J., T.B. Johansson, A.K.N. Reddy, and R.H. Williams. 1988. *Energy for a Sustainable World.* Wiley Eastern Limited, New Delhi, India. 517 pp.

Hansen, J., A. Lacis, D. Rind, G. Russell, P. Stone, I. Fung, R. Ruedy, and J. Lerner. 1984. Analysis of feedback mechanisms. In J. Hansen, and T. Takahashi, eds. *Climate Processes and Climate Sensitivity*. Geophysical Monograph 29, Maurice Ewing Volume 5. American Geophysical Union, Washington, D.C. 130-163.

IIASA (International Institute for Applied Systems Analysis). 1981. *Energy in a Finite World.* Hafele, W., ed. Ballinger, Cambridge.

IIASA (International Institute for Applied Systems Analysis). 1983. *'83 Scenario for Energy Development: Summary.* Rogner, H.H., ed. IIASA, Laxenburg, Austria.

Keeling, C.D. 1983. The global carbon cycle: What we know and could know from atmospheric, biospheric, and oceanic observations. In *Proceedings of the CO_2 Research Conference: Carbon Dioxide, Science and Consensus*. DOE CONF-820970, pp. II.3-II.62.

Keepin, W., I. Mintzer, and L. Kristoferson. 1986. Emission of CO_2 into the atmosphere. In Bolin, B., B.R. Doos, J. Jager, and R.A. Warrick, eds. *The Greenhouse Effect, Climatic Change, and Ecosystems.* Scope 29. John Wiley and Sons, Chichester. 35-91.

Komhyr, W.D., R.H. Gammon, T.B. Harris, L.S. Waterman, T.J. Conway, W.R. Taylor, and K.W. Thoning. 1985. Global atmospheric CO_2 distribution and variations from 1968-1982 NOAA/GMCC CO_2 flask sample data. *Journal of Geophysical Research* 90:5567-5596.

Lacis, A., J. Hansen, P. Lee, T. Mitchell, and S. Lebedeff. 1981. Greenhouse effect of trace gases, 1970-1980. *Geophysical Research Letters* 8:1035-1038.

Lashof, D. 1988. *The Dynamic Greenhouse: Feedback Processes That May Influence Future Concentrations of Atmospheric Trace Gases*. U.S. Environmental Protection Agency, Washington, D.C.

Legasov, V.A., I.I. Kuzmin, and A.I. Chernoplyokov. 1984. The influence of energetics on climate. *Fizika Atmospheri i Okeana* 11:1089-1103. USSR Academy of Sciences.

MacCracken, M. C., and F. M. Luther, eds. 1985a. *Projecting the Climatic Effects of Increasing Carbon Dioxide.* U.S. Department of Energy, Washington, D.C.

MacCraken, M.C. and F.M. Luther, eds. 1985b. *Detecting the Effects of Increasing Carbon Dioxide.* U.S. Department of Energy, Washington, D.C.

Mintzer, I.M. 1987. *A Matter of Degrees: The Potential for Controlling the Greenhouse Effect.* World Resources Institute, Washington, D.C.

NAS (National Academy of Sciences). 1979. *Carbon Dioxide and Climate: A Scientific Assessment.* National Academy Press, Washington, D.C.

NAS (National Academy of Sciences). 1983. *Changing Climate.* National Academy Press, Washington, D.C.

NAS (National Academy of Sciences). 1987. *Current Issues in Atmospheric Change.* National Academy Press, Washington, D.C.

NOAA (National Oceanographic and Atmospheric Administration). 1987. *Geophysical Monitoring for Climatic Change No. 15, Summary Report 1986.* Schnell, R.C., ed. U.S. Department of Commerce, NOAA Environmental Research Laboratories, Boulder. 155 pp.

Nordhaus, W.D., and G. Yohe. 1983. Future paths of energy and carbon dioxide emissions. In *Changing Climate*. National Academy Press, Washington, D.C.

Ramanathan, V., R. J. Cicerone, H. B. Singh, and J. T. Kiehl. 1985. Trace gas trends and their potential role in climate change. *Journal of Geophysical Research 90:5557-5566.*

Reilly, J., J. Edmonds, R. Gardner, and A. Brenkeri. 1987. Uncertainty analysis of the IEA/ORAU CO_2 emissions model. *The Energy Journal* 8:1-29.

Rose, D.J., M.M. Miller, and C. Agnew. 1983. *Global Energy Futures and CO_2-Induced Climate Change.* MITEL 83-015. Prepared for National Science Foundation. Massachusetts Institute of Technology, Cambridge.

Rotmans, J., H. de Boois, and R.J. Swart. 1988. An integrated model for the assessment of the greenhouse effect: The Dutch approach. Working paper. National Institute of Public Health and Environmental Protection, Bilthoven, The Netherlands.

Rotmans, J., and E. Eggink. 1988. Methane as a greenhouse gas: A simulation model of the atmospheric chemistry of the CH_4 - CO - OH cycle. Working paper. National Institute of Public Health and Environmental protection, Bilthoven, The Netherlands.

Rotty, R.M. 1987. A look at 1983 CO_2 emissions from fossil fuels (with preliminary data for 1984). *Tellus* 39B:203-208.

Seidel, S., and D. Keyes. 1983. *Can We Delay a Greenhouse Warming?* Office of Policy and Resources Management, U.S. Environmental Protection Agency, Washington, D.C.

Siegenthaler, U. 1983. Uptake of excess CO_2 by an outcrop-diffusion model of the ocean. *Journal of Geophysical Research* 88:3599-3608.

Smith, J. and D. Tirpak, eds. 1989. *The Potential Effects of Global Climate Change on the United States*. U.S. Environmental Protection Agency, Washington, D.C.

Strain, B.R. and J.D. Cure, eds. 1985. *Direct Effects of Increasing Carbon Dioxide on Vegetation*. U.S. Department of Energy, Washington, D.C.

Trabalka, J.R., ed. 1985. *Atmospheric Carbon Dioxide and the Global Carbon Cycle*. U.S. Department of Energy, Washington, D.C.

WMO (World Meteorological Organization). 1985. *Atmospheric Ozone 1985: Assessment of Our Understanding of the Processes Controlling its Present Distribution and Change*. Volume 1. WMO, Geneva. 392+ pp.

CHAPTER II

GREENHOUSE GAS TRENDS

FINDINGS

The composition of the atmosphere is changing as the result of human activities. Increases in the concentration of carbon dioxide, methane, nitrous oxide, and chlorofluorocarbons (CFCs) are well documented. In addition, tropospheric (lower atmospheric) and stratospheric (upper atmospheric) chemistry are being modified due to the addition of these gases as well as emissions of carbon monoxide, nitrogen oxides, and other compounds. Specifically, we find that:

- The concentration of carbon dioxide in the atmosphere has increased by 25% since the industrial revolution. Detailed measurements since 1958 show an increase of about 35 parts per million by volume. Both land clearing and fossil fuel combustion have contributed to this rise, but the fossil fuel source has dominated in recent years. Carbon dioxide is increasing at a rate of about 0.4% per year and is responsible for about half of the current increases in the greenhouse effect. Carbon cycle models indicate that the oceans are responsible for the uptake of most of the fossil fuel CO_2 that does not remain in the atmosphere. The total net uptake of CO_2 by the oceans and the net uptake/release of CO_2 by the terrestrial biosphere cannot be precisely determined at this time.

- The concentration of methane has more than doubled during the last three centuries. There is considerable uncertainty about the total emissions from specific sources of methane, but the observed increase is probably due to increases in a number of sources as well as changes in tropospheric chemistry. Agricultural sources, particularly rice cultivation and animal husbandry, have probably been the most significant contributor to historical increases in concentrations. But there is the potential for rapid growth in emissions from landfills, coal seams, permafrost, natural gas exploration and pipeline leakage, and biomass burning associated with forest clearings in the future. Methane is increasing at a rate of 1% per year and is responsible for about 20% of the current increases in the greenhouse effect.

- The concentration of nitrous oxide has increased by 5-10% since preindustrial times. The cause of this increase is highly uncertain, but it appears that the use of nitrogenous fertilizer, land clearing, biomass burning, and fossil fuel combustion have all contributed. Nitrous oxide is over 200 times more powerful, on a per molecule basis, than carbon dioxide as a greenhouse gas, and can also contribute to stratospheric ozone depletion. Nitrous oxide is currently increasing at a rate of about 0.25% per year, which represents an imbalance between sources and sinks of about 30%. Nitrous oxide is responsible for about 6% of the current increases in the greenhouse effect.

- CFCs were introduced into the atmosphere for the first time during this century; the most common species are CFC-12 and CFC-11 which had atmospheric concentrations in 1985 of 380 and 220 parts per trillion by volume, respectively. While these concentrations are tiny compared with that of carbon dioxide, these compounds are about 30,000 times more powerful, on a per molecule basis, than carbon dioxide as a greenhouse gas and are increasing very rapidly -- 5% per year from 1978 to 1983. Of major concern because of their potential to deplete stratospheric ozone, the CFCs also represent about 15% of the current increases in the greenhouse effect.

- The chemistry of the atmosphere is changing due to emissions of carbon monoxide, nitrogen oxides, and volatile organic compounds, among other species, in addition to the changes in the greenhouse gases just described. This alters the amount and distribution of ozone and the oxidizing power of the atmosphere, which changes the lifetimes of methane and other greenhouse gases. Changes in global ozone are quite uncertain, and may have contributed to an increase or decrease in the warming commitment during the last decade.

INTRODUCTION

The composition of the Earth's atmosphere is changing. Detailed background atmospheric concentration measurements combined with analyses of ancient air trapped in Antarctic and Greenland ice now give a compelling picture, not only of recent trends, but also of major changes that have occurred since preindustrial times. Mounting evidence that the atmosphere is changing has increased the urgency of understanding the processes that control atmospheric composition and the significance of the changes that are taking place. In this chapter we examine what is known and not known about the gases expected to be most important in altering climate during the coming decades. For each gas, we present data regarding its concentration history and geographic distribution, its sources and sinks, and its chemical and radiative interactions in the atmosphere. This information is summarized at the end of the Chapter in Table 2-2.

The concentrations of a number of greenhouse gases have already increased substantially over preindustrial levels. The estimated relative radiative forcing from the major gases (excluding water vapor and clouds) is illustrated in Figure 2-1 for the period 1880-1980 and for the expected concentration changes during the 1980s. Carbon dioxide accounted for about two-thirds of the total forcing over the last century, but its relative importance has declined to about half the total in recent years due to more rapid growth in other gases during the last few decades (see Chapter IV). Particularly important has been the recent growth in chlorofluorocarbon (CFC) concentrations. Methane (CH_4) has remained the second most important greenhouse gas, responsible for 15-20% of the forcing. With the recent signing of the Montreal Protocol on Substances that Deplete the Ozone Layer, growth in CFC concentrations is likely to be substantially restrained compared with what has been assumed until recently (e.g., Ramanathan et al., 1985; see Chapters IV and V). The relative

FIGURE 2-1

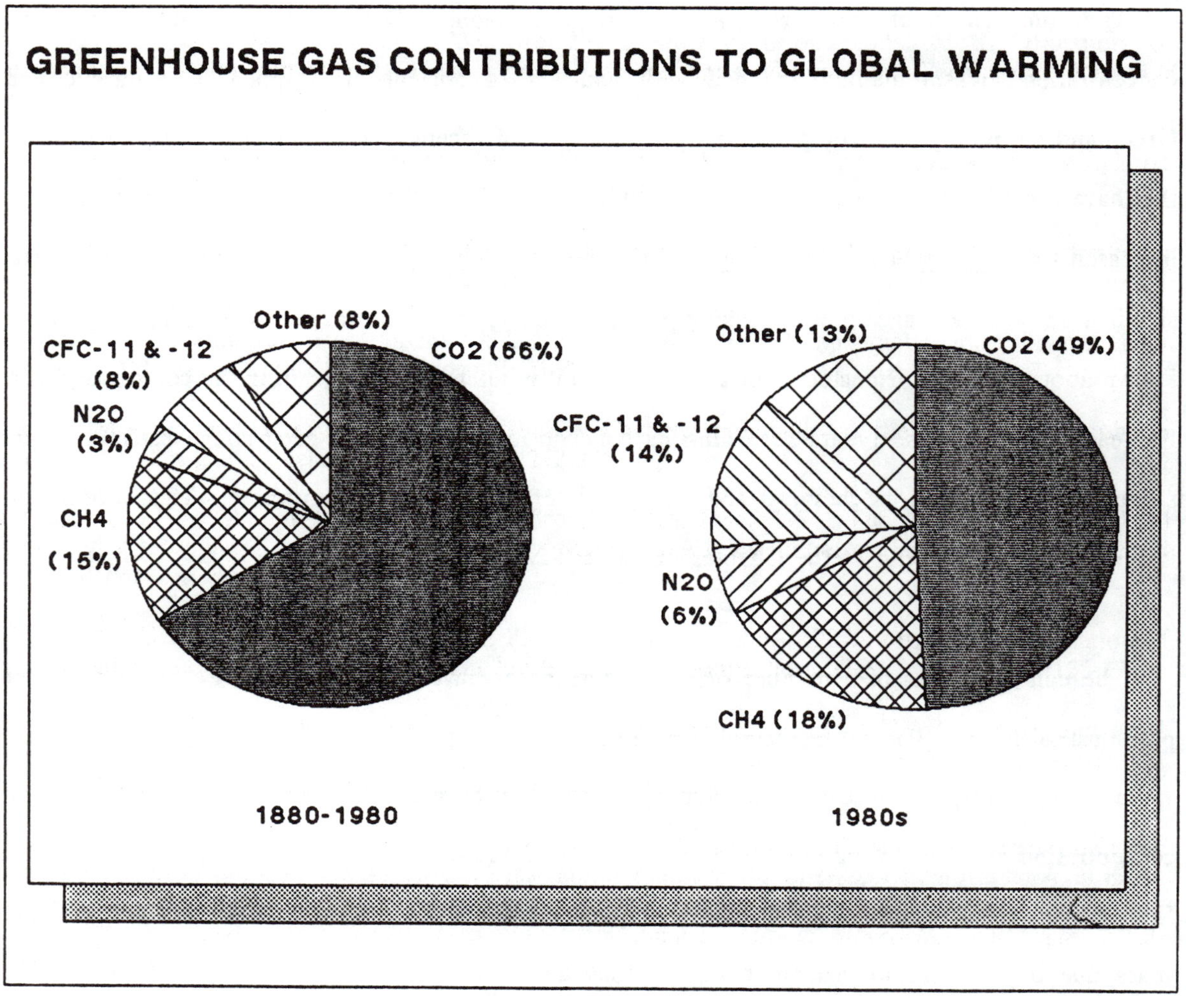

Figure 2-1. Based on estimates of the increase in concentration of each gas during the specified period. The "Other" category includes other halons, tropospheric ozone, and stratospheric water vapor. The contribution to warming of the "Other" category is highly uncertain. (Sources: 1880-1980: Ramanathan et al., 1985; 1980s: Hansen et al., 1988.)

importance of CO_2 is therefore likely to increase again in the future unless these emissions are also restricted (Chapter V).

The radiative impact of greenhouse gases is characterized here in terms of the effect of concentration changes on surface temperatures in the absence of climate feedbacks. Climate feedbacks are defined and discussed in Chapter III, where the climatic effects of changes in greenhouse gases are put into the broader context of other factors that influence climate. The human activities that are apparently responsible for the concentration trends documented in this chapter are described in Chapter IV.

CARBON DIOXIDE

Concentration History and Geographic Distribution

Carbon dioxide (CO_2) is the most abundant and single most important greenhouse gas (other than water vapor) in the atmosphere. Its role in the radiative balance, and its potential for altering the climate of the Earth have been recognized for over a hundred years. Chemical measurements of atmospheric CO_2 were made in the 19th Century at a few locations (Fraser et al., 1986a; From and Keeling, 1986). However, the modern high-precision record of CO_2 in the atmosphere did not begin until 1958, the International Geophysical Year (IGY), when C.D. Keeling of Scripps Institution of Oceanography pioneered measurements of CO_2 using an infrared gas analyzer at Mauna Loa Observatory (MLO) in Hawaii and at the South Pole. Since 1974, background measurements of atmospheric CO_2 have been made continuously at four stations (Pt. Barrow, Alaska; Mauna Loa, Hawaii; American Samoa; and the South Pole) as part of the Geophysical Monitoring for Climatic Change (GMCC) program of the National Oceanic and Atmospheric Administration (NOAA) of the

U.S. Department of Commerce. In addition to the continuous monitoring stations, NOAA/GMCC also operates a cooperative sampling network. Flask samples of air are collected weekly from these sites and are shipped to the GMCC facility in Boulder, Colorado, for analysis. The sampling network began before 1970 at a few initial sites, expanded to a network of 15 stations in 1979, and, as of 1986, consisted of ~26 stations (Komhyr et al., 1985; Conway et al., 1988). In addition to the U.S. programs, surface measurements of atmospheric CO_2 around the globe are made by many countries including Australia, Canada, France, Italy, Japan, New Zealand, West Germany, and Switzerland.

<u>Mauna Loa</u>

The MLO CO_2 record is shown in Figure 1-2 in Chapter I. CO_2 steadily increased from 315 parts per million by volume (ppm) in 1958 to 346 ppm in 1986. This corresponds to an increase at the rate of 0.4% per year, or a mean increase of 1.5 ± 0.2 ppm per year. From 1958 to 1986, CO_2 at Mauna Loa increased by 31 ppm; over the same period, fossil fuel combustion (shown also in Figure 1-2) was a source of 117 petagrams (Pg)[1] of carbon (C) as CO_2 to the atmosphere, which is equivalent to 56 ppm of CO_2. The apparent fraction of the fossil fuel sources of CO_2 that remained in the atmosphere during this period is thus 55%. As other net sources of CO_2, particularly deforestation (see below), may have been important during this period, the actual fraction of anthropogenic carbon emissions remaining in the atmosphere is uncertain. Superimposed on the increasing secular trend of atmospheric CO_2 are regular seasonal oscillations: the concentration peaks in May/June, decreases steadily through the summer, and reaches a minimum in September/October. The seasonal peak-to-trough amplitude is ~5.8 ppm. The seasonal cycle of CO_2 at Mauna Loa and at other northern hemispheric locations is caused primarily by the natural dynamics of the terrestrial biosphere: There is net removal of CO_2 from the atmosphere via

---

[1] peta = 10^{15}, giga = 10^9, 1 ton = 10^6 grams. Thus, 1 petagram (Pg) = 1 gigaton (Gt).

photosynthesis during the growing season, and net return of CO_2 to the atmosphere via respiration and decomposition processes during the rest of the year.

Despite its regular appearance, there are interannual variations in the CO_2 concentration measured at MLO. Annual mean concentration changes do not remain uniform throughout the duration of the record, but have large fluctuations around the mean (Keeling, 1983). These excursions of atmospheric CO_2 from the mean generally occur during El Niño-Southern Oscillation events, where the large-scale perturbations of atmospheric temperature, precipitation, and other circulation statistics also alter the biological, chemical, and physical aspects of carbon cycling between the atmosphere, land, and ocean reservoirs. These El Niño excursions highlight the possibility of climatic feedbacks in the carbon cycle. They do not mask the increasing secular trend, which mainly tracks the trend in fossil fuel combustion.

The seasonal amplitude also does not remain constant and has a ±10% variation about the mean. Recent analysis reveals a statistically significant positive trend in the seasonal amplitude since 1976 (Bacastow et al., 1985; Enting, 1987). The causes of this amplitude trend have not been unambiguously identified; hypotheses involve shifts in the seasonality of photosynthesis and respiration, faster cycling of carbon as a result of climatic warming, and the direct effects of CO_2 on plants (also referred to as the CO_2 fertilization effect).

Ice-core Data

Bubbles in natural ice contain samples of ancient air. Analysis by gas chromatography and laser infrared spectroscopy of gases occluded in gas bubbles in polar ice has provided a unique reconstruction of atmospheric CO_2 history prior to the modern high-precision instrumental record

(Oeschger and Stauffer, 1986). Deep ice cores have been drilled from many locations in both Greenland and Antarctica.

From the ice-core data, it is deduced that in pre-industrial times, ~1800, the CO_2 concentration was 285 ± 10 ppm and has increased at an accelerating rate since the industrial era (Neftel et al., 1985; Raynaud and Barnola, 1985; Pearman et al., 1986) (Figure 2-2). The ice-core data reveal the possible existence of natural fluctuations of the order of ± 10 ppm occurring at decadal time scales during the last few thousand years (Delmas et al., 1980; Neftel et al., 1982; Stauffer et al., 1985; Raynaud and Barnola, 1985; Oeschger and Stauffer, 1986).

Recent analysis of the 2083-meter-deep ice core from Vostok, East Antarctica, provides for the first time information on CO_2 variations in the last 160,000 years (Barnola et al., 1987; Figure 3-3 in Chapter III). Large CO_2 changes were associated with the transitions between glacial and interglacial conditions. CO_2 concentrations were low (~200 ppm) during the two glaciations and high (~285 ppm) during the two major warm periods. The Vostok ice-core data also emphasize that current levels of atmospheric CO_2 are higher than they have ever been in the past 160,000 years. The CO_2 increase since 1958 is larger than the natural CO_2 fluctuations seen in the Greenland and Antarctic ice-core record.

GMCC Network

The CO_2 concentrations from the ~26 globally distributed sites in the NOAA/GMCC cooperative flask sampling network have been reviewed in Komhyr et al. (1985) and Conway et al. (1988). The distribution for 1981-1985 is shown in Figure 2-3. There are large-scale, coherent, temporal and

FIGURE 2-2

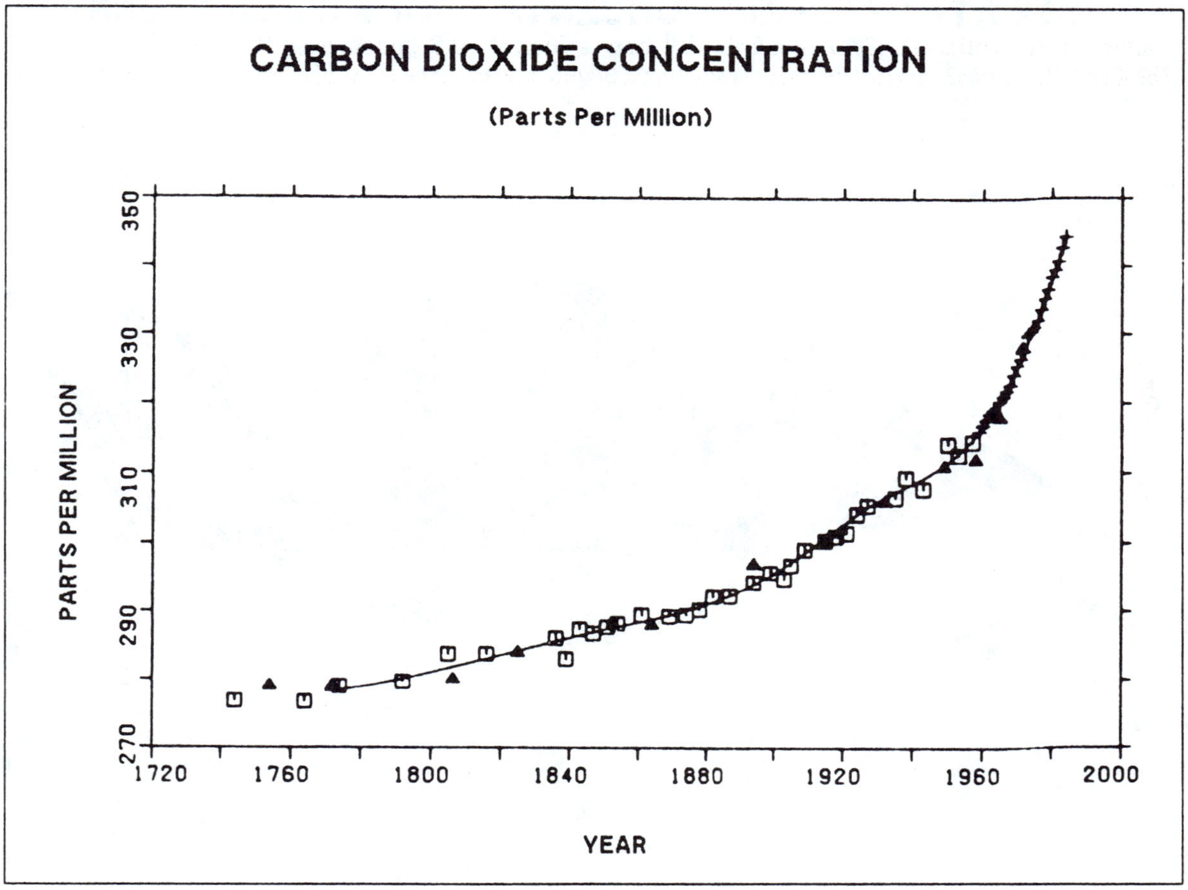

Figure 2-2. The history of atmospheric CO_2 presented here is based on ice core measurements (open spaces, closed triangles) and atmospheric measurements (crosses). The data show that CO_2 began to increase in the 1800s with the conversion of forests to agricultural land. The rapid rise since the 1950s, due primarily to fossil fuel combustion, is at a rate unprecedented in the ice core record, (Sources: Neftel et al., 1985; Friedli et al., 1986; Keeling, pers. communication; as cited in: Siegenthaler and Oeschger, 1987).

FIGURE 2-3

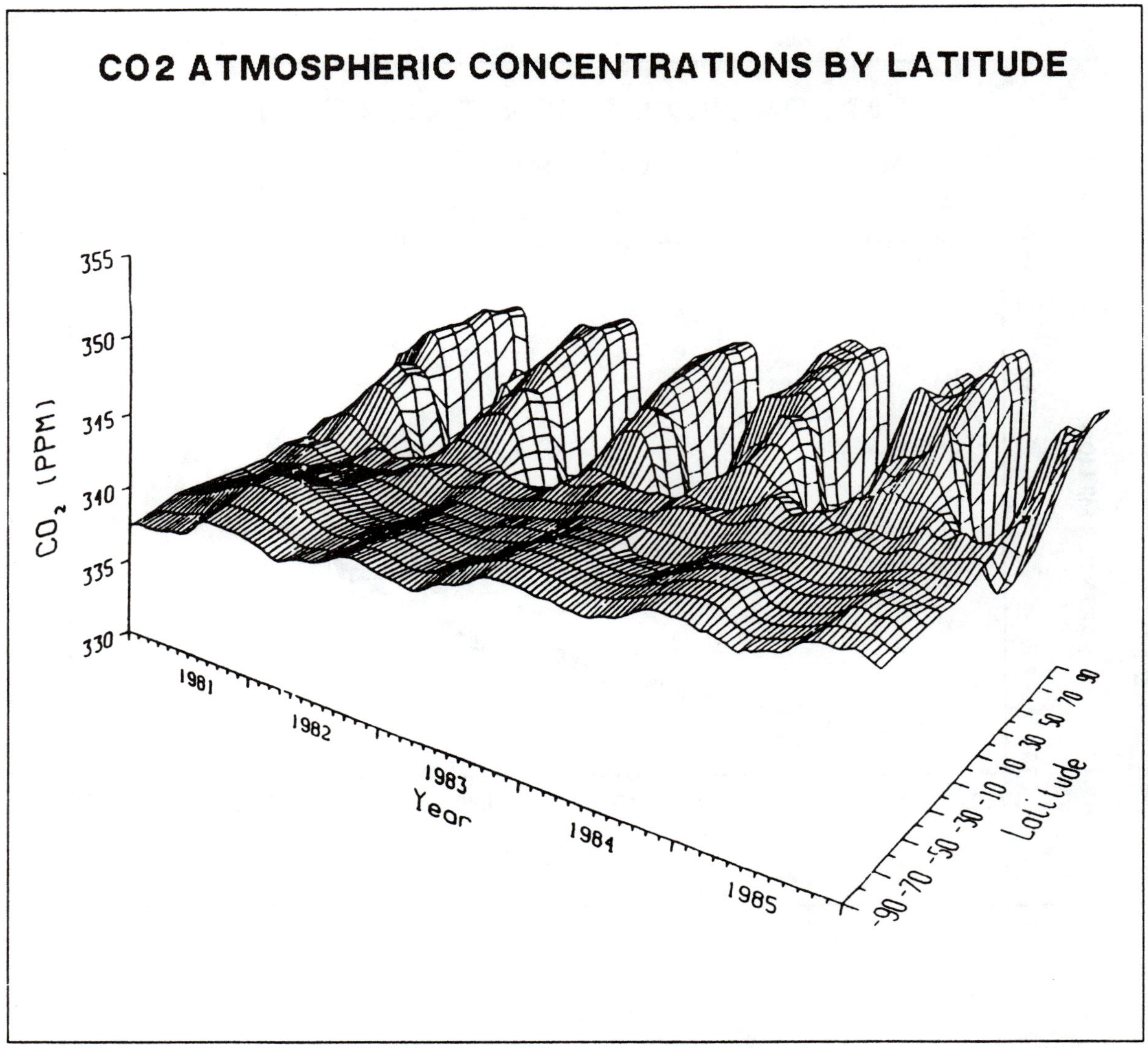

Figure 2-3. The distribution of CO_2 by latitude from 1981-1985 shows that CO_2 is increasing globally. Superimposed on the increasing trend are coherent seasonal oscillations reflective of seasonal dynamics of terrestrial vegetation. The seasonal cycle is strongest at high Northern latitudes, and is weak and of opposite phase in the Southern Hemisphere, reflecting the distribution of terrestrial vegetation. The data are from the NOAA/GMCC flask sampling network. (Sources: Komhyr et al., 1985; NOAA, 1987; Conway et al., 1988.).

spatial variations of CO_2 in the atmosphere. Concentrations of CO_2 at all the stations are increasing at the rate of ~1.5 ppm per year (ppm/yr), similar to the rate of increase at Mauna Loa.

Annually averaged CO_2 concentrations are higher in the Northern Hemisphere than in the Southern Hemisphere. The interhemispheric difference was ~1 ppm in the 1960s and is ~3.2 ppm now, reflecting the Northern Hemisphere mid-latitude source (about 90%) of fossil fuel CO_2. This gradient has remained approximately constant in the past decade. Also evident in the north-south distribution of atmospheric CO_2 is the relative maximum of ~1 ppm in the equatorial regions, caused mainly by the outgassing of CO_2 from the super-saturated surface waters of the equatorial oceans. Although tropical deforestation may also contribute to the equatorial maximum in atmospheric CO_2, models of the global carbon cycle suggest that the observations are inconsistent with a net deforestation source greater than approximately 1.5 Pg C/yr (Pearman et al., 1983; Keeling and Heimann, 1986; Tans et al., 1989).

There is a coherent seasonal cycle at all the observing stations: the Northern Hemisphere cycles resemble that at Mauna Loa. The seasonal amplitude is largest, ~16 ppm, at Pt. Barrow, Alaska, and decreases toward the equator to ~6 ppm at Mauna Loa (Figure 2-3). The CO_2 concentration is flat through the year in the equatorial region and is of opposite seasonality in the Southern Hemisphere. The seasonal cycle in the Northern Hemisphere is caused primarily by seasonal exchanges with the terrestrial biosphere (Fung et al., 1987; Pearman and Hyson, 1986), while in the Southern Hemisphere, oceanic and terrestrial exchanges are equally important in determining the seasonal oscillations in the atmosphere (Pearman and Hyson, 1986). The CO_2 seasonal cycle shows a consistent amplitude increase with time for some sites (Cleveland et al., 1983; Thompson et al., 1986).

The geographical variations of CO_2 growth rates at the GMCC sites show more clearly the El Niño perturbations, as noted already in the Mauna Loa data. For example, the El Niño-caused cessation of upwelling that resulted in the devastation of the fishing industry and marine wildlife in the eastern equatorial Pacific is also evidenced by reduced outgassing of CO_2 to the atmosphere (Feely et al., 1987) and a concomitant decrease in the global CO_2 growth rate (Conway et al., 1988). These variations in the growth rate contain information about the response of the carbon system to climatic perturbations, some of which are under investigation currently.

Sources and Sinks

The atmosphere exchanges CO_2 with the terrestrial biosphere and with the oceans. Averaged over decades, sources must approximately equal sinks if the system is to remain in quasi-steady state; however, the individual flux in each direction may be large (50-100 Pg C/yr). The fluxes of carbon to the atmosphere associated with anthropogenic activities are roughly ten times smaller than the natural fluxes of carbon. However, the anthropogenic fluxes are unidirectional and are thus net sources of carbon to the atmosphere (Figure 2-4).

<u>Fossil Carbon Dioxide</u>

The combustion of fossil fuels, in liquid, solid, or gas forms, is the major anthropogenic source of CO_2 to the atmosphere. A recent documentation and summary of the fossil fuel source of CO_2 is given by Rotty (1987a, 1987b). In 1985, about 5.2 Pg C were released in the form of CO_2 as a result of fossil fuel combustion. Of this, the USA, USSR, and China contributed 23%, 19%, and 10%, respectively (Rotty, pers. communication). The emissions for 1987 were 5.5 Pg C. The history and mix of activities and fuels giving rise to these emissions are discussed in detail in Chapter IV.

FIGURE 2-4

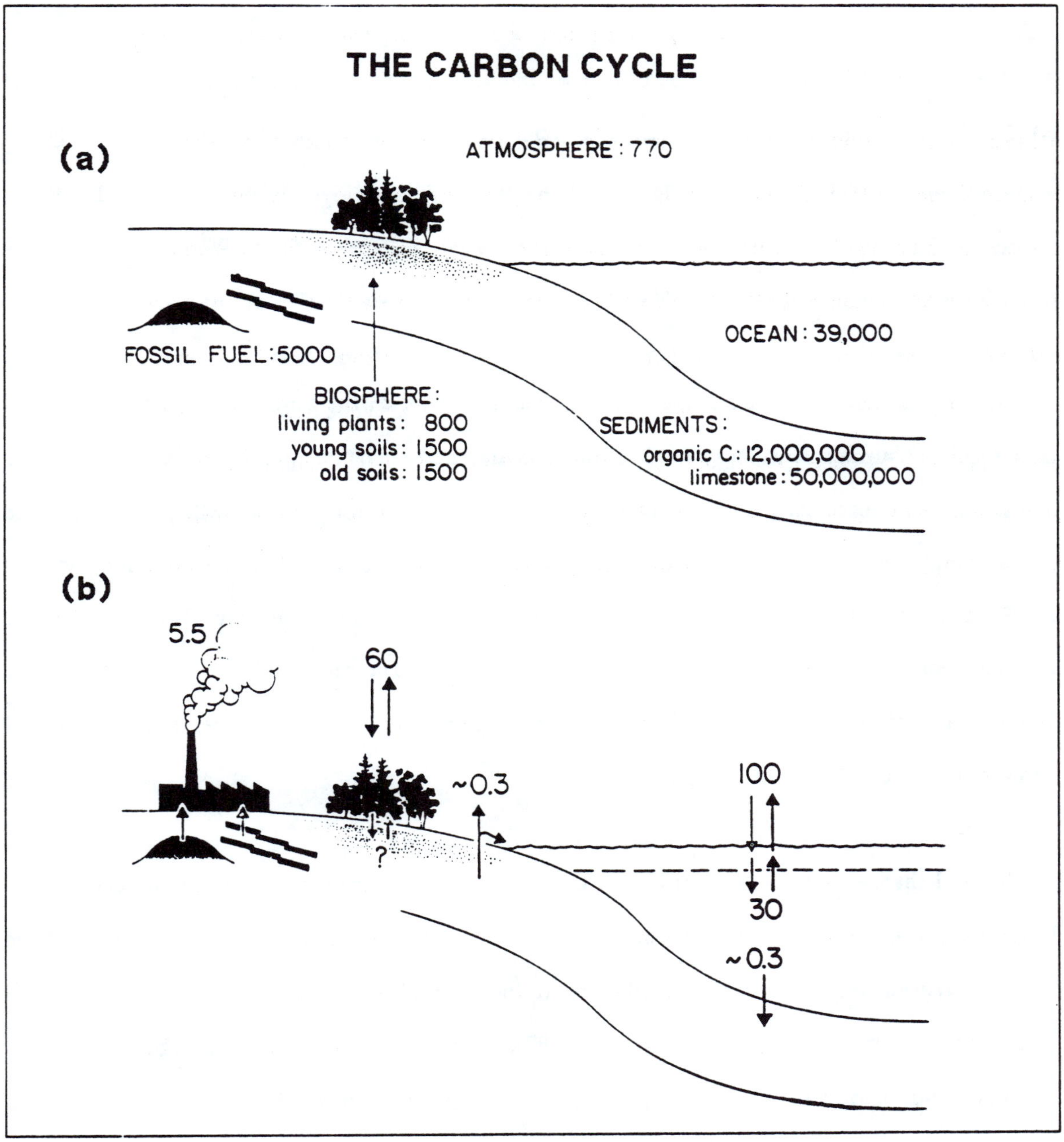

Figure 2-4. (a) Major reservoirs of the global carbon cycle. Reservoirs (or stocks) are in units of 10^{15} grams of carbon (Pg C).

(b) Fluxes of carbon, in 10^{15} grams of carbon per year (Pg C/yr).

Source: Adapted from Keeling, 1983.

Biospheric Cycle

The terrestrial biosphere absorbs CO_2 from the atmosphere via photosynthesis on the order of 60 Pg C/yr. Approximately the same amount is returned to the atmosphere annually via heterotrophic respiration and decomposition processes. While the net exchange of the unperturbed biosphere is close to zero over a period of one year, the seasonal asynchronicity of the exchange gives rise to the regular oscillations seen in the atmospheric CO_2 records.

In general, land-use modification is a net source of CO_2 to the atmosphere. CO_2 is released as a result of burning and decay of dead plant matter and oxidation of soil organic matter. The amount of this release exceeds the amount of CO_2 absorbed as a result of regrowth of live vegetation and accumulation of soil organic matter. Recently, Houghton et al. (1987) and Detwiler and Hall (1988) estimated a net source of 0.4-2.6 Pg C/yr to the atmosphere from land-use changes. Deforestation in the tropics accounted for nearly all the flux: The release of carbon from temperate and boreal regions was only 0.1 Pg C/yr. The regional and temporal patterns and causes of deforestation are taken up in Chapter IV.

Natural changes in terrestrial biospheric dynamics may result from climate warming and/or from increased CO_2 concentrations in the atmosphere. The possibility of such natural changes is suggested by the increasing amplitude of CO_2 oscillations in the atmosphere (Bacastow et al., 1985; Cleveland et al., 1983; Thompson et al. 1986; Enting, 1987). The amplitude change may signal a tendency towards a biospheric sink of CO_2, as photosynthesis responds to increasing temperatures and CO_2 concentrations (Pearman and Hyson, 1981; D'Arrigo et al., 1987; Kohlmaier et al., 1987). The amplitude change can also mean increased sources via respiration and decay, which are strongly temperature-dependent processes (Houghton, 1987). Because growth and decay cycles are intimately linked, it is difficult to tell whether atmosphere-biosphere interactions will act as a positive or a negative feedback without further theoretical and field studies (see the discussion of biogeochemical feedbacks in Chapter III).

Ocean Uptake

The exchange of CO_2 across the air-sea interface depends on the degree of CO_2 supersaturation in the surface waters of the oceans and the rate at which CO_2 is transferred across the interface itself. Because of the very nature of shipboard measurements, data on oceanic CO_2 partial pressure (pCO_2) are sparse, both spatially and temporally. Most of the data have come from oceanographic research programs, mainly Scripps Institution of Oceanography in the 1960s (Keeling, 1968), Geochemical Sections (GEOSECS) in the 1970's (Takahashi et al., 1980, 1981), and Transient Tracers in the Oceans (TTO) in the early 1980s (Brewer et al., 1986) and more recently from NOAA survey cruises and from ships of opportunity.

Depending on the regional interplay between temperature, carbon supply from upwelling, and carbon consumption by biological activities, the seasonal cycle of CO_2 in surface water may peak at different times of the year in different oceanic regions (Peng et al., 1987; Takahashi et al., 1986, 1988). This makes it extremely difficult to interpret the sparse oceanic carbon data in the context of the global carbon cycle. The interpretation is aided by data from carbon-14 and other transient tracers in the ocean.

Based on the available data and an understanding of carbon dynamics in the ocean, it is estimated that, on an annual basis, about 90 Pg C/yr is exchanged between the atmosphere and the ocean. This exchange results in a net outgassing of approximately 1 Pg C/yr from the equatorial oceans and a net absorption of about the same amount by the middle to high latitude oceans.

Superimposed on this exchange of 90 Pg C/yr in either direction is the penetration of fossil fuel CO_2 into the oceans, estimated to be ~3 Pg C/yr currently. This oceanic uptake of fossil fuel CO_2 is corroborated by the observations of anthropogenic tracers penetrating gradually into the oceanic thermocline. These tracers include tritium and carbon-14, by-products of nuclear testing in the 1960s, and the chlorofluorocarbons, a recent man-made compound. The magnitude of the oceanic uptake of fossil fuel CO_2 is estimated using numerical models calibrated by tracers. Because of the variability of the oceanic carbon system and the precision of ocean carbon measurements, the oceanic signature of fossil fuel CO_2 has not been demonstrated unambiguously. Takahashi et al. have demonstrated that in the Atlantic, the oceanic pCO_2 increased by 8 ± 8 microatmospheres (μatm) from 1958 to the mid-1970s, an increase that is consistent with estimates from numerical models.

Chemical and Radiative Properties/Interactions

Carbon dioxide is chemically inert in the atmosphere, but it has a very important impact on the Earth's radiation budget and hence on climate and the chemistry of the atmosphere. After water vapor, CO_2 is the most abundant and most significant infrared (IR) absorbing gas in the atmosphere. As discussed in Chapter III, the Earth's climate is determined by the point at which incoming solar (short-wave) radiation is balanced by IR (long-wave) emissions to space from the warm surface and atmosphere. Increasing the concentration of CO_2 and other greenhouse gases in the atmosphere elevates the average surface temperature required to achieve this balance. Doubling the atmospheric CO_2 concentration from 315 to 630 ppm would produce a radiative forcing (the equilibrium surface temperature increase in the absence of climate feedbacks) of 1.2-1.3°C. At current concentrations

THE RADIATIVE EFFECTS OF GREENHOUSE GASES

The radiative effects of greenhouse gases have received a great deal of attention over the last decade. Recent reviews are given by Dickinson and Cicerone (1986) and Ramanathan et al. (1987). In the absence of an atmosphere the Earth would radiate energy to space as a black body with a temperature of about 250° K (-23°C). Figure 2-5 shows the actual emissions, indicating the absorption bands of the major greenhouse gases. Not shown is water vapor, which has continuous absorption throughout this spectral range and dominates all other gases at wavelengths < 8 micrometers (μm) and > 18 μm (Dickinson and Cicerone, 1986). The 15 μm band of CO_2 dominates absorption in the spectral range from 12 to 18 μm, and its absorption in the other parts of the spectrum amounts to 15% or less of its impact in this region.

The shaded region in Figure 2-5, between about 7 and 13 μm, is called the atmospheric window because it is relatively transparent to outgoing radiation: 70-90% of the radiation emitted by the surface and clouds in these wavelengths escapes to space (Ramanathan et al., 1987). Many trace gases happen to have absorption bands in this window region and are therefore very effective greenhouse absorbers. For example, CFCs are as much as 20,000 times more effective than CO_2 per incremental increase in concentration (see Table 2-1).

CO_2 already absorbs most of the radiation emitted from the Earth's surface in the wavelengths where it is active. As a result, each additional molecule of CO_2 added to the atmosphere has a smaller effect than the previous one. Hence, radiative forcing scales logarithmically, rather than linearly, with increases in the concentration of atmospheric CO_2. For example, a 50 ppm increase in CO_2 from 350 to 400 ppm yields a radiative forcing of 0.23°C, while the same increment from 550 to 600 ppm yields a radiative forcing of only 0.16°C. Despite the reduced greenhouse effectiveness of each molecule of CO_2 as concentrations increase, CO_2 will remain the dominant greenhouse gas in the future, responsible for 50% or more of the increased greenhouse effect during the next century for plausible scenarios of future trace gas emissions (Hansen et al., 1988; Chapter V).

FIGURE 2-5

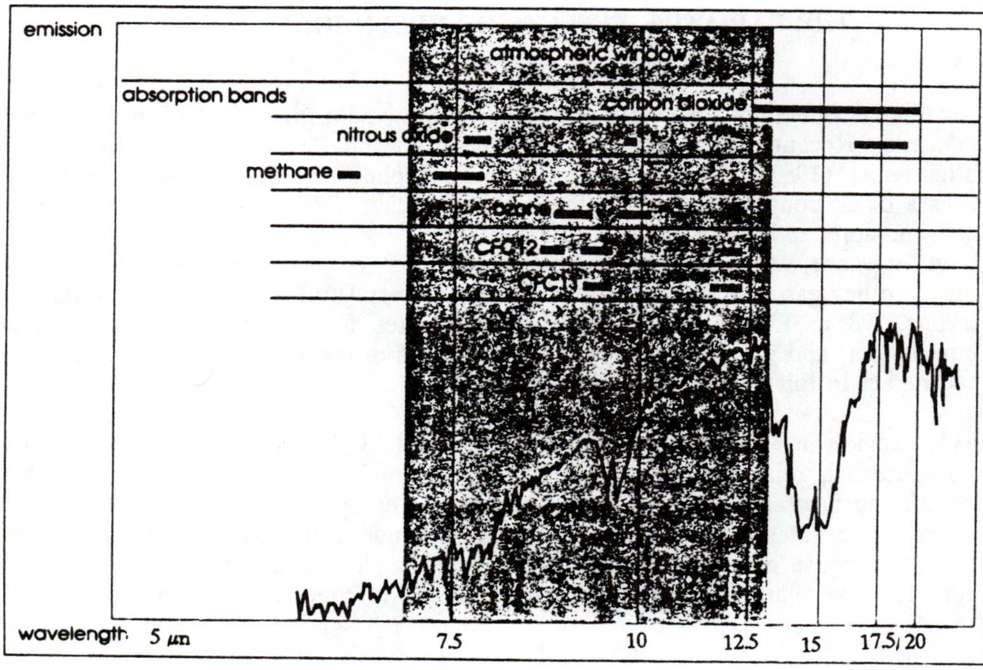

Figure 2-5. Infrared (long-wave) emissions to space from the Earth. Many of the absorption bands of the greenhouse gases fall within the atmospheric window -- a region of the spectrum, between 7 and 13 μm, in which there is little else to prevent radiation from the Earth escaping directly into space. (Source: UNEP, 1987.)

TABLE 2-1

Radiative Forcing for a Uniform Increase in Trace Gases From Current Levels

Compound	Radiative Forcing (No Feedbacks) (°C/ppb)
CO_2	.000004
CH_4	.0001
N_2O	.001
CFC-11	.07
CFC-12	.08
CFC-13	.10
Halon 1301	.10
F-116	.08
CCl_4	.05
$CHCl_3$	.04
F-14	.04
HCFC-22	.03
CH_2Cl_2	.02
CH_3CCl_3	.01
C_2H_2	.01
SO_2	.01

Source: Adapted from Ramanathan et al., 1985.

METHANE

Concentration History and Geographic Distribution

High-precision atmospheric measurements of methane (CH_4) have been made in the past decade at many different locations. The data show clearly that the concentration of methane, 1675 parts per billion by volume (ppb) in 1987, has been increasing at the rate of about 16 ppb per year (Blake and Rowland, 1986, 1988) (Figure 2-6).

Since 1982, air samples from ~25 globally distributed sites of the NOAA/GMCC cooperative network have been analyzed for CH_4 (Steele et al., 1987). In addition to flask sampling, continuous measurements of atmospheric CH_4 are now made at Cape Meares, Oregon (Khalil and Rasmussen, 1983); Pt. Barrow, Alaska; and Mauna Loa, Hawaii (NOAA, 1987).

The data show that CH_4, like CO_2, exhibits very coherent spatial and temporal variations. CH_4 is approximately uniform from mid- to high latitudes in the Southern Hemisphere, and increases northward. The Northern Hemisphere concentration is approximately 140 ppb higher than that in the Southern Hemisphere. The seasonal cycle in the Southern Hemisphere shows a minimum in the summer, consistent with higher summer abundances of the hydroxyl radical (OH) and temperature-dependent destruction rates. In the Northern Hemisphere, the seasonal cycle is more complex, showing the interaction mainly between chemical destruction and emissions from high-latitude peat bogs.

Analysis of air bubbles in ice cores shows that in pre-industrial years, CH_4 was relatively constant at ~700 ppb, from 100,000 years before present (100 kyBP) until the mid-19th century, and exhibited

FIGURE 2-6

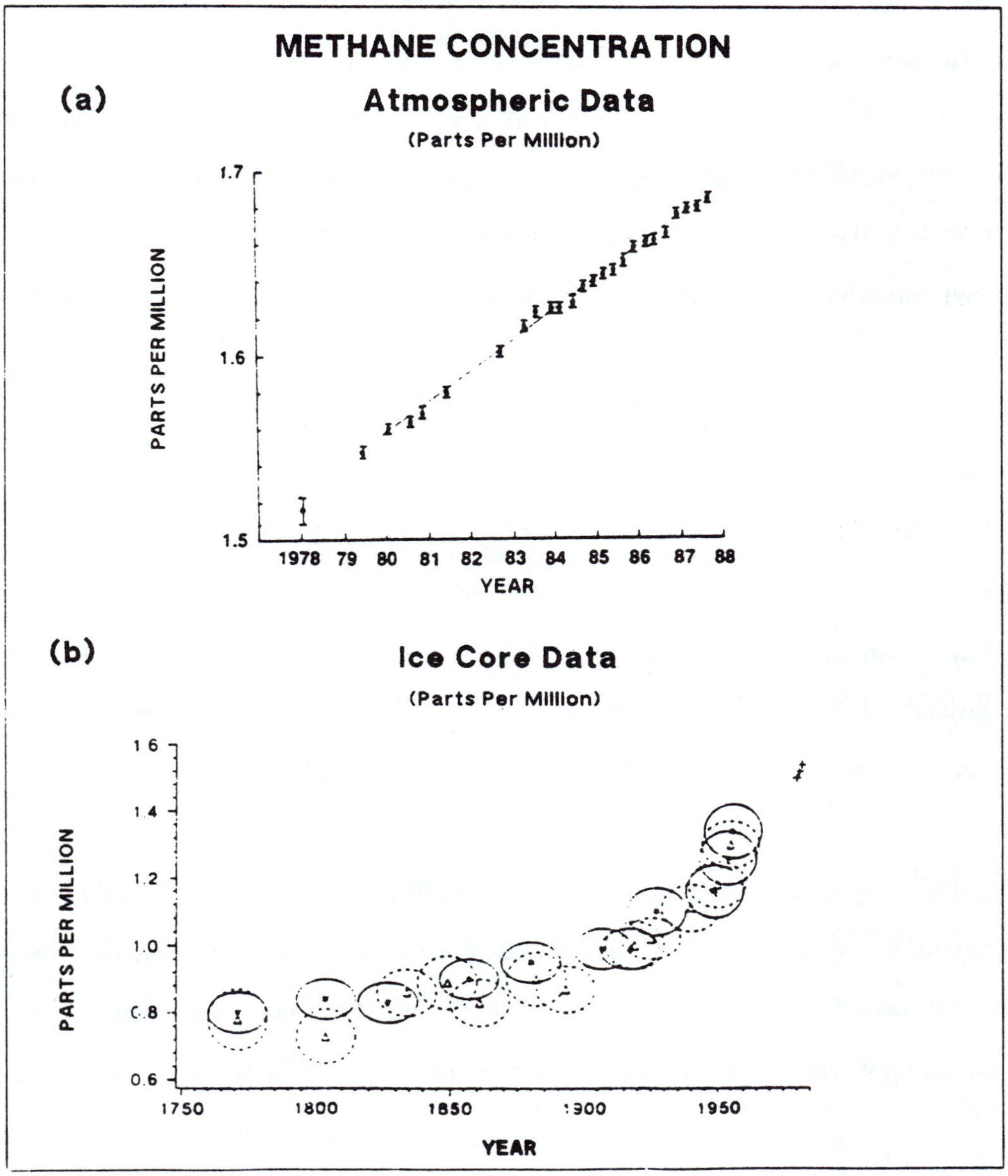

Figure 2-6. Recent measurements of atmospheric CH_4 show that CH_4 has been increasing at the rate of 1%/yr in the last decade (upper panel). Ice core data (lower panel) show that CH_4 was relatively constant in the 1800's, and began to increase rapidly at the beginning of the 20th century. Like CO_2, the recent trend in CH_4 (shown as + + + in the lower panel) is unprecedented in the history of CH_4. The ice core data are from Siple Station, Antarctica. Stars and triangles represent results obtained from melt and dry extraction, respectively. The ellipses indicate the uncertainties in the concentrations as well as in the mean age of the sample. (Sources: Blake and Rowland, 1988; Stauffer et al., 1985.)

a factor of 2.5 increase to its present value in only the last 300 years (Stauffer et al., 1985; Pearman et al., 1986) (Figure 2-6). The 2083-meter ice core recovered by the Soviet Antarctic Expedition at Vostok, Antarctica, shows that the CH_4 concentration was as low as 340 ppb during the penultimate ice age (~155 kyBP) and nearly doubled to 610 ppb in the following interglacial (130 kyBP). The trend in CH_4 closely followed the trend in air temperature deduced from deuterium, confirming the role of CH_4 as an important greenhouse gas (Raynaud et al., 1988). These measurements show that current concentrations of CH_4, like that of CO_2, are higher than they have been in the past 160,000 years.

Sources and Sinks

Methane is produced via anaerobic decomposition in biological systems. It is also a major component of natural gas and of coal gas. While the major sources of CH_4 have been identified, their individual contributions to the global budget are highly uncertain. A recent review of the sources and sinks of CH_4 is given by Cicerone and Oremland (1988) (see Figure 2-7).

The major sink of CH_4 is reaction with OH radicals in the atmosphere.[2] Based on chemical considerations, it is estimated that the global sink of methane is ~510 teragrams (Tg) CH_4/yr.[3] By inference, the annual global source is the sink plus the annual increase, i.e., about 550 Tg CH_4/yr. Cicerone and Oremland (1988) estimate a range of 400 to 640 Tg/yr for the annual global source.

---

[2] A radical is an atom or group of atoms with at least one unpaired electron, making it highly reactive.

[3] 1 Tg = 1 teragram = 10^{12} grams

FIGURE 2-7

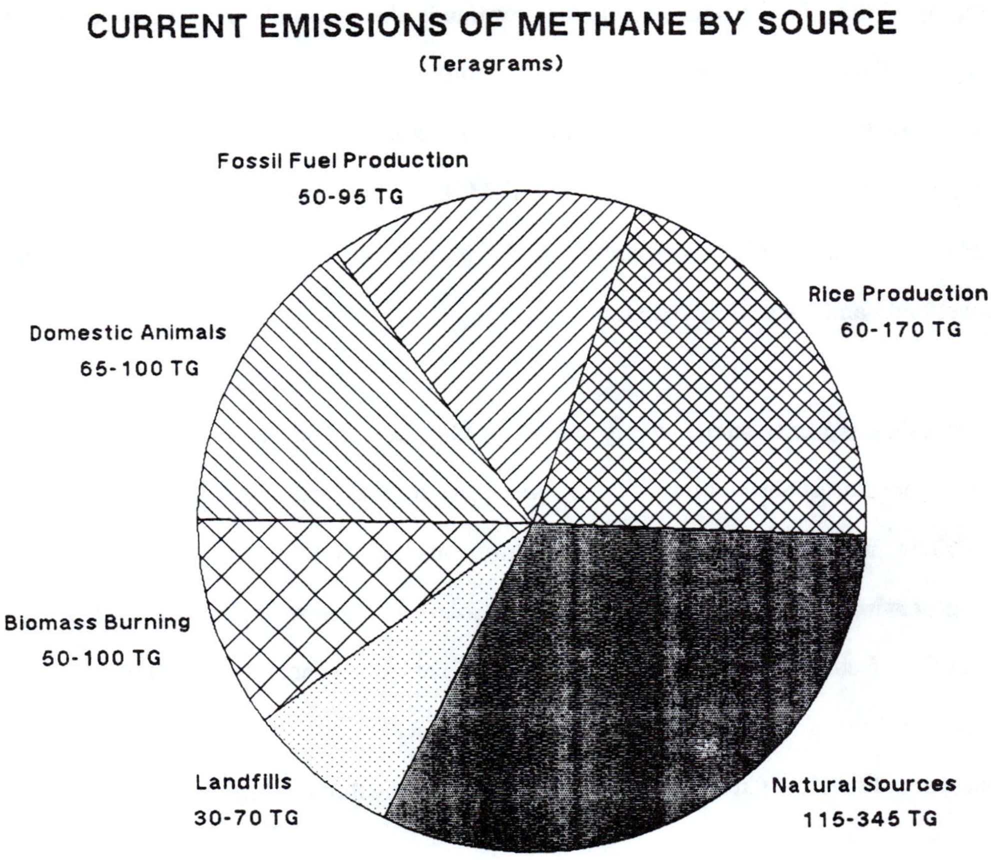

TOP THREE PRODUCERS

Rice Production	Domestic Animals	Fossil Fuel Production
1. India	1. India	1. United States
2. China	2. USSR	2. USSR
3. Bangladesh	3. Brazil	3. China

Figure 2-7. Human activities in the agricultural sector (domestic animals, rice production and biomass burning) and the energy sector (fossil fuel production) are the major sources of atmospheric CH_4. Natural sources, from wetlands, oceans, and lakes, may contribute less than 25% of total emissions. (Sources: Cicerone and Oremland, 1988; Crutzen et al., 1986; Lerner et al., 1988; United Nations, 1987; IRRI, 1986.)

Estimates of methane emissions from natural wetlands have ranged from 11-150 Tg/yr (e.g., Seiler, 1984; Khalil and Rasmussen, 1983). A recent study by Matthews and Fung (1987) estimated that there are 530 million hectares of natural wetlands that account for a global emission of ~110 Tg CH_4/yr. Of this, about 50% of the CH_4 is emitted from productive peat bogs at high latitudes in the Northern Hemisphere, a regional emission that is likely to increase with greenhouse warming. While this study has employed more extensive field data than earlier estimates (e.g., Sebacher et al., 1986; Harriss et al., 1985), uncertainties in the global estimate remain due to the variability of natural wetlands and their CH_4 fluxes.

Rice paddies are environments very similar to natural wetlands in terms of CH_4 production and emission to the atmosphere. In 1984, there were 148 million hectares of rice harvest area globally, with ~50% in India and China. Methane emission studies have been performed in controlled mid-latitude environments (Cicerone et al., 1983; Holzapfel-Pschorn and Seiler, 1986). These studies have identified factors affecting methane fluxes to the atmosphere: inter alia, temperature, soil properties, fertilizer, and irrigation practices. These factors make global extrapolation of methane emissions very difficult. Cicerone and Oremland (1988) estimate a global emission of 60-170 Tg CH_4/yr.

Methane is also produced by enteric fermentation in animals, especially ruminants. The amount of CH_4 produced is dependent on enteric ecology, the composition and quantity of feed, and the energy expenditure of the individual animal. Estimates of emission rates range from 94 kg CH_4/animal/year from West German dairy cattle, to ~35 kg CH_4/animal/year from Indian cattle fed on kitchen refuse, to 5-8 kg CH_4/animal/year from sheep. Using these emission coefficients and population statistics of animals in the world, Crutzen et al. (1986) obtained a global emission of 78 Tg CH_4 for 1983. This emission includes ~5 Tg CH_4 from wild animals and <1 Tg CH_4 from humans. About 75% of the emissions are from cattle and dairy cows. India, the USSR, Brazil, the USA, and China are the five major countries in terms of CH_4 emission from domestic animals (Lerner et al., 1988).

Other natural sources of CH_4 include termites, and exchange with oceans and lakes. The source from termites is highly uncertain and controversial. Estimates of global emissions range from close to zero (Seiler et al., 1984a) to 20 Tg (Fraser et al., 1986b), and as high as 200 Tg CH_4 (Zimmerman et al., 1982, 1984), on the order of half the global emission. The oceanic source is small, estimated to be 5-20 Tg CH_4/yr (Cicerone and Oremland, 1988).

There are several anthropogenic sources of methane. Methane is produced by incomplete combustion during biomass burning. The amount of CH_4 produced depends on the material burned and the degree of combustion. Estimates range from 50-100 Tg CH_4/yr (see Cicerone and Oremland, 1988). While a few studies have attempted to understand and measure CH_4 emission during biomass burning (Crutzen et al., 1979, 1985), extrapolation to a global estimate is difficult because of the lack of global data on area burned, fire frequency, and characteristics of fires. The feasibility of monitoring fires from space (Matson and Holben, 1987; Matson et al., 1987) will improve this estimate significantly.

Methane is also produced in large municipal and industrial landfills, where biodegradable carbon in the refuse is decomposed to form a mixture of CO_2 and CH_4. As in the case of many other CH_4 sources, the fraction of gas produced that escapes to the atmosphere is debated. Recently, Bingemer and Crutzen (1987) estimated that this source produces 45-70 Tg CH_4/yr. These estimates assume that a large fraction of all organic carbon deposited in landfills eventually is subject to methanogenesis and subsequent emission to the atmosphere. Cicerone and Oremland (1988) adopt a range of 30-70 Tg CH_4/yr.

Methane is the major component (~90%) of natural gas, and so the leakage of natural gas from pipelines and the venting of natural gas from oil and gas wells represent sources of CH_4 to the atmosphere. Although natural gas production and consumption statistics are available globally, the nature of this CH_4 source makes it difficult to estimate how much this source contributes to the atmospheric abundance. From U.S. and Canadian natural gas statistics it is estimated that approximately 2-2.5% of the marketable gas is unaccounted for. Assuming that all of the unaccounted for gas is lost to the atmosphere, 25-30 Tg CH_4/yr from line loss is obtained globally (Cicerone and Oremland, 1988). An additional 15 Tg CH_4/yr is estimated from natural gas sources, assuming that ~20% of the gas that is vented and flared at oil and gas wells is not combusted, escaping to the atmosphere as CH_4 (Darmstadter et al., 1987). Together these estimates suggest a source of up to 50 Tg CH_4/yr from natural gas production and consumption. Much of the unaccounted for gas, however, may represent meter discrepancies, and venting of natural gas has been declining in recent years (Darmstadter et al., 1987). Thus a reasonable range for these sources may be 20-50 Tg CH_4/yr.

Methane is also the major component of gas trapped in coal. The percentage of the CH_4 component increases with the age and depth of the coal and is released to the atmosphere during mining and processing/crushing of coal. Globally, the amount of CH_4 in coal is ~0.5% of the mass of coal extracted. This source is estimated to be 15-45 Tg CH_4/yr in 1980 (Darmstadter et al., 1987; Cicerone and Oremland, 1988).

A highly uncertain but potentially large source of CH_4 is methane hydrates in sediments under permafrost and on continental margins (Kvenvolden, 1988). The magnitude of the current CH_4 release from this source is unknown. Climate warming presents the potential for destabilization of the hydrates and subsequent release of CH_4 to the atmosphere (Chapter III).

Chemical and Radiative Properties/Interactions

Methane is the most abundant trace gas in the atmosphere that is active both radiatively and chemically. Although the abundance of methane is 1/200 that of CO_2, CH_4 is a more efficient absorber of thermal radiation than is CO_2: Donner and Ramanathan (1980) and Lacis et al. (1981) estimate that at present levels, an additional molecule of CH_4 will contribute a radiative forcing that is equivalent to that contributed by approximately 25 molecules of CO_2. These radiative transfer calculations suggest that a doubling of atmospheric CH_4 (1.6-3.2 ppm) will contribute a radiative forcing of 0.16°C (Hansen et al., 1988).

The destruction rate of CH_4 is dependent on the amount of OH (and hence water vapor) in the atmosphere as well as on temperature. The lifetime (atmospheric abundance divided by destruction rate) of CH_4 is approximately 10 years, the lifetime being shorter in the tropics. Using estimates of the average concentration of atmospheric hydroxyl radicals derived from measurements of methylchloroform, Prinn et al. (1987) have deduced the average atmospheric lifetime of methane to be 9.6 (+2.2, -1.5) years. The reaction between CH_4 and OH eventually produces CO; CO itself reacts with OH, producing CO_2 (Thompson and Cicerone, 1986). Thus, an increase in the background levels of either CH_4 or CO can reduce OH and the oxidizing power of the entire atmosphere. It is estimated that increases in CO alone from 1960 to 1985 would have lowered OH concentrations in the atmosphere, increased the methane lifetime, and resulted in a 15-20% increase in CH_4 concentrations (Khalil and Rasmussen, 1985; Levine et al., 1985; Thompson and Cicerone, 1986).

Because of the interactions between CO, CH_4, and OH in the atmosphere, it is difficult to predict the effects of climate change on OH destruction of CH_4, as increasing atmospheric water vapor and increased precipitation (and removal of OH reservoirs like nitric acid [HNO_3] and hydrogen peroxide [H_2O_2]) have opposite effects on OH concentrations. Changes in nitrogen oxides (NO_x) and tropospheric ozone (O_3) also strongly affect atmospheric OH (see below).

NITROUS OXIDE

Concentration History and Geographic Distribution

Nitrous oxide (N_2O) is present in minute amounts in the atmosphere but it is nonetheless of great importance. Its concentration is three orders of magnitude less than that of CO_2, but its radiative forcing per molecule is 230 times greater. The first high-precision measurements of atmospheric N_2O in the late 1970s showed an unambiguous increasing trend in its concentration (Weiss, 1981). Continuous measurements at four Atmospheric Lifetime Experiment/Global Atmospheric Gases Experiment (ALE/GAGE) sites have been made since 1979 (Figure 2-8). Flask samples of air from the globally distributed cooperative network of NOAA-GMCC are also being analyzed for N_2O (Thompson et al., 1985; Komhyr et al., 1988).

The ALE/GAGE data show that the mid-1980s concentration of atmospheric N_2O is 310 ppb and that its annual growth rate is ~0.8 ppb per year, or 0.2-0.3% per year. The concentrations at

FIGURE 2-8

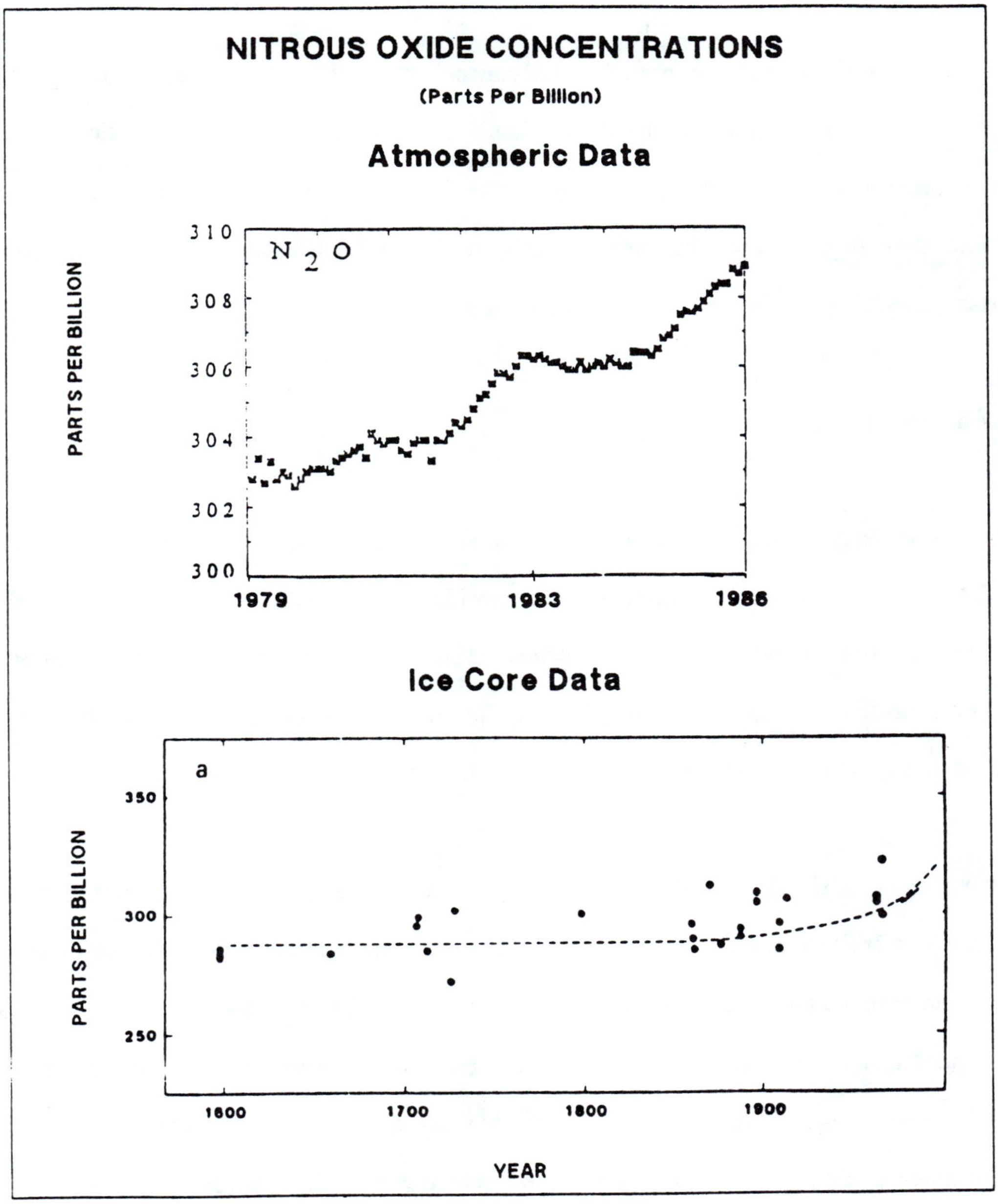

Figure 2-8. Concentration of atmospheric N_2O has been increasing at the rate of 0.25%/yr in the last decade (upper panel). The ice core record shows that N_2O was relatively constant from the 1600's to the beginning of the 20th century, and began increasing rapidly in the last 50 years. (Sources: Khalil and Rasmussen, 1987; Pearman et al., 1986.)

the two Northern Hemisphere sites are 0.8 ppb higher than those at the Southern Hemisphere sites, suggesting the dominance of a northern source.

Ice-core data show that the preindustrial concentration of N_2O was 285 ± 10 ppb averaged between 1600-1800 (Pearman et al., 1986; Khalil and Rasmussen, 1987). Unlike CO_2, whose concentration began to increase significantly in the 1800s, N_2O remained fairly constant until the 1900s, and then began increasing more rapidly in the 1940s (Pearman et al., 1986; Khalil and Rasmussen, 1987). (See the ice-core data in Figure 2-8).

Sources and Sinks

While a lot of progress has been made during the last five years in quantifying the sources and sinks of N_2O in the atmosphere, there remain considerable uncertainties in the global budget and in the contributions of individual source terms. The uncertainties arise not just because of the scarcity of measurements of N_2O fluxes, but also because of the complexity of the biogeochemical interactions and heterogeneous landscape where N_2O is produced.

Nitrous oxide is simultaneously produced and consumed in soils via the metabolic pathways of denitrification, nitrification, nitrate dissimilation, and nitrate assimilation. These processes are affected by various environmental parameters such as temperature, moisture, the presence of plants, and the characteristics and composition of the soils (e.g., Seiler and Conrad, 1987; Sahrawat and Keeney, 1986). The flux of N_2O to the atmosphere also depends on the location of the N_2O-producing and N_2O consuming microorganisms and their relative activity within the soil column (Conrad and Seiler, 1985). Because of the complexity of the N_2O production and destruction processes, and the inherent heterogeneity of soils, it is difficult to estimate the contribution of natural soils to the global N_2O

budget. Slemr et al. (1984) calculated N_2O emissions from natural temperate and subtropical soils to be 4.5 Tg N/yr. Recent measurements (Livingston et al., 1988; Matson and Vitousek, 1987) show that N_2O emission rates from tropical soils are higher than those from temperate soils and that a relationship exists between the N_2O flux and the rate of nutrient cycling in the tropical forest soils. However, no budget of N_2O emissions from tropical soils has been attempted. Seiler and Conrad (1987) give a very tentative estimate of 6 ± 3 Tg N/yr from natural soils globally.

Measurements of supersaturation of N_2O in the oceans indicate that the oceans contribute additional N_2O to the atmosphere, though in smaller quantities (Seiler and Conrad, 1981; Weiss, 1981). Seiler and Conrad (1987) estimated the oceanic contribution to be 2 ± 1 Tg N/yr.

Little is known about N_2O emissions from terrestrial freshwater systems. Extrapolating from measurements in the Netherlands and in Israel of elevated N_2O levels in aquifers contaminated by the disposal of human or animal waste, cultivation, and fertilization, Ronen et al. (1988) estimated a global source 0.8-1.7 Tg N/yr from contaminated aquifers.

Nitrous oxide is also produced during combustion, but the importance of this source is unclear at this time. A recent study of this N_2O source was reported by Hao et al. (1987) who found that the amount of N_2O in flue gases was correlated with the nitrogen content of fuels. N_2O emission rates were highest during coal combustion, lower when oil was used as the fuel, and lowest when the fuel was natural gas. They found that in conventional single-stage boilers, on average, 14% of fuel nitrogen was converted to N_2O during combustion. Conversion of fuel nitrogen to N_2O was much less efficient (2-4%) in a two-stage experimental combuster and in wood fires. They also found consumption of N_2O in fuel-rich flames with low air-fuel ratios, reducing significantly the emission factor. Using statistics on solid- and liquid-fuel production, they estimated an emission of 3.2 Tg

N_2O-N in 1982. Very recent studies, however, suggest that many of the N_2O measurements reported in the literature have been affected by a sampling artifact. A reaction between water, sulfur dioxide (SO_2), and nitrogen oxides (NO_x) generates N_2O in sample cylinders over a period of hours, sometimes increasing N_2O concentrations by more than an order of magnitude unless the samples are carefully dried or N_2O is measured immediately (Muzio and Kramlich, 1988). Reanalysis of measurements made in the U.S., excluding those that were apparently affected by this reaction, found no significant difference between N_2O emissions from gas and coal-fired boilers (Piccot, pers. communication). Recent measurements conducted by EPA with an on-line analyzer confirm this finding: In both utility and small experimental boilers N_2O concentrations in the exhaust gases were always less than 5 ppm and generally less than 2 ppm (Hall, pers. communication). This suggests that the relationship between N_2O and fuel-nitrogen found by Hao et al. (1987) may have actually been due to differences in SO_2 and NO_x emissions. Emissions of N_2O do appear to vary with combustion technology. Preliminary measurements suggest that fluidized-bed combusters and catalyst-equipped automobiles may have substantially elevated N_2O emissions (De Soot, pers. communication). Total N_2O emissions from fossil-fuel combustion cannot be estimated with any confidence at this time, but may be closer to 1 Tg N/yr than to 3 Tg N/yr (Chapter V).

The addition of nitrogenous fertilizers to soils enhances the emission of N_2O and other nitrogen gases to the atmosphere. This emission depends on temperature, soil moisture, rainfall, fertilizer type, fertilizer amount, and the way the fertilizer is applied. It also depends on the properties of the soils and the crops grown. The fraction of fertilizer nitrogen lost to the atmosphere as nitrous oxide ranges from ~0.001-0.05% for nitrate, ~0.01-0.1% for ammonium fertilizers, to ~0.5->5% for anhydrous ammonia. With a global consumption of approximately 70.5 million tons nitrogen as nitrogenous fertilizers in 1984, an N_2O contribution of 0.14-2.4 Tg N/yr is estimated. Although the estimated N_2O emissions associated with the use of nitrogenous fertilizers are small compared with those from natural sources, they are, nonetheless, a source subject to rapid growth.

Land-use modification in the tropics may also contribute N_2O to the atmosphere. N_2O is produced during biomass burning, but because direct estimates of total N_2O emissions are difficult, N_2O emissions are estimated by ratios with emissions of CO_2 or other nitrogen gases. Crutzen (1983) estimated this source to be 1-2 Tg N/yr, although the accuracy of this estimate is highly uncertain.

Recently, Bowden and Bormann (1986) found enhanced N_2O fluxes to the atmosphere from cleared areas in a temperate forest and elevated N_2O concentration in ground water adjacent to the cut watershed. Similarly, tenfold increases in N_2O fluxes are found in pastures and forest clearings in the Amazon (Matson, pers. communication). Robertson and Tiedje (1988) postulate, on the basis of observations in Central America, that the loss of primary tropical rain forest may decrease the emissions of N_2O to the atmosphere. These studies suggest that rapid deforestation in the tropics may significantly alter the N_2O budget, although an estimate of its contribution to the global budget has not been attempted.

Chemical and Radiative Properties/Interactions

N_2O has a low concentration in the atmosphere, and its rate of increase is much smaller than that of the other trace gases. Yet it still plays an important role in the radiative and chemical budgets of the atmosphere. The seemingly small growth rate, ~0.25%/year, is the result of a large imbalance (~30%) between the sources and sinks. The extremely long lifetime of N_2O, ~160 years, means that the system has a very long memory of its emission history.

Nitrous oxide is an effective greenhouse gas. The radiative forcing of one molecule of N_2O is equivalent to that of ~230 molecules of CO_2; a doubling of N_2O and a doubling of CH_4 yield approximately the same radiative forcing, even though the N_2O concentration is less, by a factor of 5, than that of CH_4. A 25% increase in the current burden of N_2O in the atmosphere will yield a radiative forcing of 0.07°C.

Nitrous oxide is not chemically reactive in the troposphere. However, its destruction in the stratosphere, by photolysis and by reaction with atomic oxygen in the excited state [$O(^1D)$], makes N_2O the dominant precursor of odd nitrogen in the stratosphere. Thus, an increase in N_2O would lead to an increase in stratospheric NO_x, which would significantly influence stratospheric ozone chemistry.

CHLOROFLUOROCARBONS (CFCs)

Concentration History and Geographic Distribution

High-precision measurements of CFC-11 ($CC1_3F$) and CFC-12 ($CC1_2F_2$) began in 1971 with the development of electron capture detector/gas chromatograph techniques (Lovelock, 1971). Like CO_2 and CH_4, surface measurements have consisted of high-frequency observations at a few dedicated sites as well as flask samples of air collected from a global network of stations or from irregular global transects.

High-frequency *in situ* measurements of surface concentrations have been or are currently being made at the five coastal/island stations of the Atmospheric Lifetime Experiment/Global Atmospheric Gases Experiment (ALE/GAGE) (Cunnold et al., 1986; Prinn et al., 1983; Rasmussen and Khalil,

1986; Simmonds et al., 1987). In addition, analysis of CFC concentrations in the flask samples of air collected at the NOAA-GMCC globally distributed network of sites have begun at the GMCC facility in Boulder (Thompson et al., 1985; NOAA, 1987).

CFC-12 is the most abundant halocarbon in the atmosphere. Its average tropospheric concentration in 1986 was 392 parts per trillion by volume (ppt), corresponding to a total burden of about 7.6 Tg. Its concentration is increasing at 4%/yr.

With a total burden of about 5.0 Tg, CFC-11 is the second most abundant halocarbon in the atmosphere. Its average concentration in 1986 was 226 ppt, and is also increasing at 4%/yr.

The other important halocarbons include methyl chloroform (CH_3CCl_3), 125 ppt in 1986, increasing at ~6%/yr; carbon tetrachloride (CCl_4), 121 ppt in 1982, increasing at 1.3%/yr; HCFC-22 (formerly denoted CFC-22; $CHClF_2$), ~100 ppt in 1986, increasing at 7%/yr; and Halon-1211 (CF_2C1Br), ~2 ppt in 1986 and increasing at >10%/yr (Prinn, 1988).

Sources and Sinks

CFCs are the product solely of the chemical industry. CFC-11 is used in blowing plastic foams and in aerosol cans. CFC-12 is used primarily in refrigeration and aerosol cans. Comprehensive data on production of CFC-11 and CFC-12 are published by the Fluorocarbon Program Panel (FPP) of the Chemical Manufacturers Association (CMA). The peak year for CFC production was 1974, in which a total of 812.5 gigagrams (Gg) of CFC-11 plus CFC-12 was produced.[4] The annual total CFC production decreased for a number of years following a ban on non-essential aerosol uses in

---

[4] 1 Gg = 10^9 grams = 10^6 Kg

the United States, Canada, and Sweden. Non-aerosol uses have continued to increase, however, and the total has risen rapidly in recent years, reaching 703.2 Gg in 1985. Of this total about 70% was consumed by the U.S., the European Economic Community, and Japan (see Figure 4-9 in Chapter IV).

The CMA data do not cover the USSR. The FPP has estimated Soviet production; however, these estimates are considered unreliable. Data for China and the countries of Eastern Europe are lacking entirely, rendering large uncertainties in the magnitude of world emissions. Cunnold et al. (1986) and Fraser et al. (1983) have found the measured trend of CFC-11 and CFC-12 concentrations is relatively consistent with the CMA estimates of CFC-11 release but not CFC-12 release, which suggests that the USSR and Eastern Europe contribute a substantial amount to CFC-12 emissions.

Methyl chloroform (CH_3CCl_3) is widely used in the manufacturing industry as a solvent for degreasing, CFC-113 is used in the electronics industry, mainly for circuit board cleaning, and HCFC-22 is used mainly in refrigeration. The sources of these gases have been estimated in various studies, but a survey of sources--equivalent to that conducted for CFC-11 and CFC-12 -- has not been done.

Fully halogenated CFCs (those that contain no hydrogen) are destroyed almost solely by photolysis in the stratosphere. The atmospheric lifetimes of CFCs estimated from the ALE/GAGE analyses are 111^{+22}_{-44} years for CFC-12, 73^{+31}_{-17} years for CFC-11, $6.3^{+1.2}_{-0.9}$ years for methyl chloroform, and approximately 40 years for carbon tetrachloride. Compounds containing hydrogen (HCFCs) react with OH in the troposphere and have lifetimes on the order of 20 years or less.

Chemical and Radiative Properties/Interactions

CFCs absorb infrared radiation in the window region of the atmospheric spectrum (Figure 2-5). Although CFCs are present in minute amounts in the atmosphere, together they are one of the dominant greenhouse gases. They have the highest annual fractional increase (~5%/yr) of all the greenhouse gases. Furthermore, the radiative forcing due to each additional molecule of CFC is equivalent to that due to about 30,000 molecules of CO_2, and at present levels increases linearly with the concentration (Ramanathan et al., 1987). A 2 ppb increase in both CFC-11 and CFC-12 will contribute a radiative forcing of 0.3°C, equivalent to that from a 65 ppm increase in CO_2. In the 1980s, CFC-11 and CFC-12 together will contribute about 15% of the increase in global greenhouse forcing.

Chlorinated and brominated compounds contribute chemically active halogens into the atmosphere, where they are broken down by solar ultraviolet radiation or by reaction with OH. In addition to CFC-11 and CFC-12, these species include methyl chloroform (CH_3CCl_3), carbon tetrachloride (CCl_4), and methyl halides (CH_3Cl, CH_3Br, and CH_3I). Also important are two rapidly increasing chlorofluorocarbons, CFC-113 ($C_2Cl_3F_3$) and HCFC-22 ($CHClF_2$). Halocarbons with at least one hydrogen, such as methyl chloroform and HCFC-22 are destroyed primarily by reaction with OH radicals in the troposphere. Long-lived species survive the 1-3 year transport time from the surface up to stratospheric levels to play a critical role in ozone photochemistry. All of these species can contribute to the stratospheric burden of chlorine, but the longer-lived CFCs can accumulate, reaching higher concentrations before a steady state balance is achieved.

The dissociation products of halocarbons are the dominant sources of chlorine (Cl) and fluorine (F) for the stratosphere (WMO, 1985), which are major components in the catalytic cycles that control ozone abundance. Ground-based data and aircraft data of trends for major halocarbon reservoirs (HCl, HF, ClO) are highly uncertain. Within the limits of the uncertainties, the estimated trends in these species were not in disagreement with trends in the source gases themselves.

OZONE

Concentration History and Geographic Distribution

Ozone (O_3) is both produced and destroyed *in situ* in the atmosphere. Unlike the other trace gases, the vertical distribution of O_3 in the atmosphere is of prime importance in determining its radiative and chemical effects (Figure 2-9).

Ozone sondes from a diverse and globally distributed network provide our only record of possible trends in the vertical distribution of tropospheric O_3. A review of ozone sonde and surface data have been given by Logan (1985), Tiao et al. (1986), and more recently by Crutzen (1988). Since the 1970's, surface O_3 concentrations are measured routinely at the four continuous monitoring stations operated by NOAA-GMCC: Pt. Barrow, Alaska; Mauna Loa, Hawaii; American Samoa; and the South Pole (see e.g. NOAA, 1987). NOAA-GMCC also participates in international cooperative ozone sonde profiling activities.

The O_3 data taken near populated and industrial regions in the 1930's to the 1950's generally showed an annually averaged concentration of 10-20 ppb at the surface, with a seasonal cycle that peaked in summer. The data showed a generally increasing trend, especially in the summer, in surface concentrations of O_3 at sites in western Europe, the USA and northern Japan. For example, a factor of two increase, from ~30 ppb in 1933 to ~60 ppb in the 1980s, is found in the summer

FIGURE 2-9

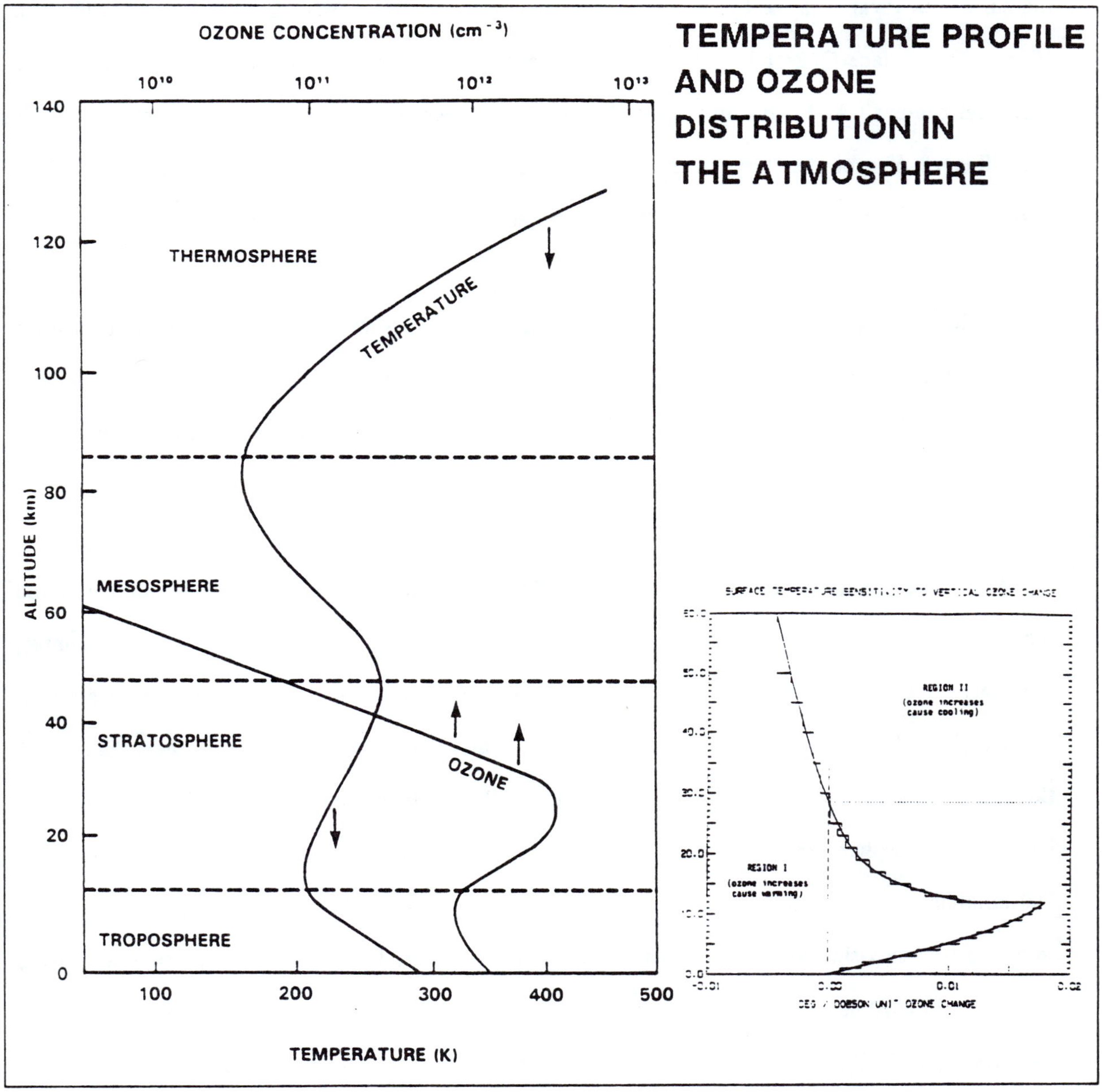

Figure 2-9. On the left, temperature profile and ozone distribution in the atmosphere. On the right, sensitivity of global surface temperature to changes in vertical ozone distribution. Ozone increases in Region I (below ~30 km) and ozone decreases in Region II (above ~30 km) warm the surface temperature. The results are from a 1-D radiative transfer model in which 10 Dobson unit ozone increments are added to each layer. The heavy solid line is a least square fit to step-wise calculations. (Sources: Watson et al., 1986; Lacis et al., no date)

concentrations in south Germany and Switzerland. Similarly, summertime concentrations of O_3 at the surface in rural areas in the Eastern U.S. have increased by 20%-100% since the 1940s (Logan, 1985). The surface O_3 trend is 1%/yr or more at those sites in close proximity to population and industrial centers. At Pt. Barrow and at Mauna Loa, geographically removed from but still under the influence of urban centers, surface O_3 was about 25 ppb in 1986 with summer values of 35-40 ppb. A small positive trend ($0.7 \pm 0.5\%$/yr) is detected at these two sites from 1973-1986.

Analysis of the ozone sonde data at these populated sites show a small but significant positive trend in mid-tropospheric ozone. In general, the mid-tropospheric trends are smaller than those at the surface of the same O_3 profile, and upper-tropospheric and lower-stratospheric trends are negative, ~-0.5%/yr.

At remote locations, surface O_3 exhibits a behavior very different from that near populated and industrial regions. At remote sites in the Canadian Arctic and in Tasmania, Australia, for example, the seasonal cycle of surface O_3 has a minimum, rather than a maximum, in summer or autumn. Surface O_3 at the South Pole was 20 ppb in 1986, similar to that measured in Western Europe in the 1930s. Also, unlike populated sites in the Northern Hemisphere, O_3 at remote sites in the Northern Hemisphere exhibit no significant trends near the surface, but significant positive trends at 700 millibars (mb) and 500 mb. Mid-tropospheric O_3 at Resolute, Canada (75°N), for example, is found to be increasing at 1%/yr, while there is a negative trend in the lower stratosphere. In the Southern Hemisphere, however, there appear to be no significant trends in surface or mid-tropospheric O_3, although O_3 in the lower stratosphere has clearly decreased and the seasonal cycle at the South Pole has doubled in amplitude.

The recent record of O_3 concentrations in the upper atmosphere has been reviewed by a NASA panel of experts (Watson et al., 1988). They report a general decreasing trend in the total O_3 concentration in the entire atmospheric column from 1969 to 1986, the period of examination. Ground-based Dobson instruments, located mainly between 30 and 64 degrees in the Northern Hemisphere, show a total decrease of 1.7-3.0% in 17 years in the annually averaged O_3 concentration of the entire atmospheric column. The decrease was more rapid in the winter months, from December to March. Satellite data, calibrated by coincident Dobson measurements, show a decrease of about 2-3% from October 1978 to October 1985 in the column O_3 concentrations between 53°S and 53°N. The observations of stratospheric O_3 in the Northern Hemisphere indicate that O_3 abundances have declined over the past 20 years. The rate of decrease in the summer is consistent with the predicted change due to CFCs. However, the measured ozone loss poleward of 40°N in winter is greater than that predicted by theory (Rowland, 1989). This depletion of O_3 in the north is not associated with the unusual chemistry of the Antarctic ozone hole and may be the beginning of a truly global decline.

Sources and Sinks

Ozone is not directly emitted by human activity, but its concentration in the troposphere and stratosphere is strongly governed by anthropogenic emissions of NO_x, hydrocarbons, and CFCs, among others. Because of the short lifetimes of NO_x and many of the other chemical species important in tropospheric O_3 chemistry, O_3 concentrations exhibit large variability horizontally, vertically and temporally. Its annual concentration, seasonal cycle, as well its trend have different behaviors in different parts of the globe so that the observations from a few regions cannot be viewed as globally representative. Global trends in tropospheric O_3 cannot be unambiguously extracted from trends in column O_3 either. Stratospheric O_3 dominates the column abundance and its decreasing trend may obscure positive or negative trends in tropospheric ozone. The difficulty in determining the globally representative trend in tropospheric O_3 translates into uncertainties in the O_3 contributions to the greenhouse warming.

Chemical and Radiative Properties/Interactions

Radiative forcing of O_3 is more complex than that of the other greenhouse gases because (1) O_3 is the major source of atmospheric heating due to ultraviolet and visible absorption bands, in addition to being a greenhouse gas, and (2) O_3 trends are not uniform in the atmosphere -- anthropogenic effects are expected to include upper stratospheric losses, lower tropospheric increases, and latitude dependent changes in the lower stratosphere and upper troposphere.

Radiative transfer calculations reveal that the climate forcing due to an O_3 perturbation changes sign at 25-30 km altitude (see Figure 2-9). Ozone increases below this level lead to surface warming because its greenhouse effect dominates its impact on solar radiation, while O_3 added to the stratosphere above ~30 km increases stratospheric absorption of solar energy at the expense of solar energy that would otherwise have been absorbed at lower altitudes. On a per molecule basis, the O_3 changes with by far the largest net effect on surface temperatures are those near the tropopause where the temperature contrast between absorbed and emitted thermal radiation is greatest. O_3 changes near the surface produce little greenhouse forcing since the outgoing thermal radiation absorbed is of nearly the same temperature as the radiation emitted by surface ozone.

While the radiative effects of O_3 are understood theoretically, quantifying surface temperature changes due to O_3 changes is difficult because of the uneven data and the lack of global coverage in the observations. Available ozone trend data are limited to northern mid-latitudes. Based on the observed decreases in O_3 in the upper troposphere and lower stratosphere, which outweighed the warming caused by decreases in the upper stratosphere and by increases in O_3 in the lower troposphere, Lacis et al. (no date) find that during the 1970s, surface cooling resulted from these O_3 changes. This cooling was equal in magnitude to about half of the warming contributed by CO_2 increases during the same time period. These results differ from previous assessments (e.g., Ramanathan et al., 1985; Wang et al., 1988) that were based on one-dimensional photochemical model results which predict ozone increases in the lower stratosphere and upper troposphere, and thus produce surface warming. Predictions of two-dimensional photochemical models for increases in CFCs suggest that ozone should decrease in the lower stratosphere at middle and high latitudes, but increase in the tropics (Ko et al., 1984; WMO, 1985). This implies a strongly latitude-dependent climate forcing for O_3 distributional changes with surface cooling in the middle and high latitudes and warming in the tropics.

The global nature of O_3 changes in the upper troposphere and lower stratosphere cannot be deduced, at this point, from current observations. This makes highly uncertain the evaluation of O_3 contributions to the global greenhouse warming. Lacis et al. (no date) note that even the best sampled O_3 data from mid-latitudes in the northern hemisphere are of uneven quality, and that the associated trends have large uncertainties and may not be globally representative.

OTHER FACTORS AFFECTING COMPOSITION

In addition to those greenhouse gases cited above that have a direct impact on the radiative balance of the Earth, we must consider those forces that control the chemical balance of the atmosphere, in turn controlling the abundance of greenhouse gases. With the exception of ozone, the greenhouse gases are generally not very reactive in the atmosphere; they have long chemical lifetimes, on the order of 10 to 200 years; and they accumulate in the atmosphere until their rate of chemical destruction balances their emissions. The chemistry of the stratosphere and troposphere provides the oxidizing power to destroy the majority of trace pollutants in the Earth's atmosphere (a major exception is CO_2, see above). We outline below those primary and secondary components of the Earth's atmosphere that affect the chemically reactive gases and note changes that may have occurred in the recent past and those possible in the future.

Global Tropospheric Chemistry

In the troposphere, many species are removed in a chain of reactions beginning with the hydroxyl radical, OH, and ending with the deposition or rainout of a soluble compound, or with the complete oxidation of the original compound (i.e., net: $CH_4 + 2O_2 ==> CO_2 + 2H_2O$). For CH_4, most hydrocarbons, and halocarbons containing a hydrogen atom (e.g., anthropogenic HCFCs such as $CHClF_2$), the chemical lifetime will vary inversely with the suitably averaged global OH concentration.

The OH radicals in the troposphere are short-lived (<1 sec) and are produced by sunlight in the presence of O_3 and H_2O; they are consumed rapidly by reaction with CO, CH_4, and other hydrocarbons. Moderate levels of nitrogen oxides (NO_x: NO and NO_2) can play an important role in recycling the odd-hydrogen (HO_x) from HO_2 to OH, thus building up the concentrations of OH; but high levels of NO_x can reduce both OH and O_3. The short lifetime of OH means that, when we integrate the loss of even a well-mixed gas like CH_4 against consumption by reaction with OH, we are integrating over the myriad of conditions of the troposphere in terms of sunlight, O_3, H_2O, CO, CH_4, NO_x, and others. These tropospheric conditions vary over scales that range from smooth in latitude and height, to irregular in plumes downwind from metropolitan areas. At present we do not have adequate models for OH that can describe these varied conditions and accurately integrate the global loss of a greenhouse gas such as CH_4.

In spite of these problems in modeling the chemically complex, heterogenous conditions of the global troposphere, we do understand tropospheric chemistry sufficiently to make some simple generalizations:

- Most loss of CH_4 occurs in marine environments, particularly in the tropics and subtropics, remote from the influence of urban areas and the continental boundary layer,

- Increasing concentrations of CO and CH_4 will reduce global or hemispheric levels of OH,

- Large scale perturbations to tropospheric O_3 and H_2O (from climate change) may have equal impact on OH concentrations,

- Changes in anthropogenic emissions of NO_x are expected to have only a moderate direct impact on globally integrated OH, but may lead to more significant increases in Northern Hemispheric O_3.

<u>Carbon Monoxide</u>

Carbon monoxide (CO) has a lifetime of about one month in the tropics that becomes indefinitely long in winter at high latitudes. The globally averaged destruction of CO corresponds to an estimated lifetime of 3 months. Carbon monoxide is lost almost exclusively through tropospheric reactions with OH (and in this non-linear system, CO is also a major sink for OH). There are some estimates of plant/soil uptake of CO, but these are not of major significance.

Detailed observations of CO concentrations are available over the past decade (Dianov-Klokov and Yurganov, 1981; Khalil and Rasmussen, 1988) and there are sporadic measurements since 1950 (Rinsland and Levine, 1985). These data indicate that CO concentrations have grown modestly, but consistently (1-2%/yr) in the northern mid-latitudes over the last few decades. There is no convincing evidence for growth in the Southern Hemisphere (Seiler et al., 1984b). This pattern is consistent with a growing anthropogenic source, since the short lifetime precludes significant interhemispheric transport. Since CH_4 concentrations have also increased similarly (about 1%, as noted above), we would expect a similar change of opposite sign in tropospheric OH.

Nitrogen Oxides

One form of odd-nitrogen denoted as NO_x is defined as the sum of two species, NO + NO_2. NO_x is created in lightning, in natural fires, in fossil fuel combustion, and in the stratosphere from N_2O. The NO_x levels over the continental boundary layer and in the aircraft flight lanes of the Northern Hemisphere are likely to have increased over the last several decades. Nevertheless, the levels of NO_x in the clean marine environment are so low that they might be accounted for entirely by natural sources (i.e., lightning, fires, stratospheric HNO_3).

The anticipated changes in NO_x levels over limited regions of the Northern Hemisphere are expected to have only a small direct effect on the globally integrated OH concentration. A more important impact of NO_x emissions is likely for tropospheric O_3, where a substantial fraction of the global tropospheric ozone production is predicted to take place in small regions with elevated levels of NO_x and hydrocarbons (Liu et al., 1987). These issues are unresolved and are currently the focus of photochemical studies with multi-dimensional tracer models.

Stratospheric Ozone and Circulation

Some species such as N_2O and CFCs do not react with OH, and these gases are destroyed only in the stratosphere by short-wavelength ultraviolet light and by reactions with the energetic state of atomic oxygen, $O(^1D)$. For CFCs and N_2O the abundances will be perturbed by changes in the rate of stratosphere-troposphere circulation and changes in the stratospheric O_3 that shields the solar ultraviolet radiation. Major perturbations to stratospheric O_3 and circulation may also alter the concentrations of tropospheric O_3, since the stratosphere represents a major source for this gas.

Predictions have been made over the past decade that stratospheric O_3 will change due to increasing levels of CFCs, and that the circulation of the stratosphere may be altered in response to changes in climate induced by greenhouse gases. Recent detection of the Antarctic ozone hole has dramatized the ability of the atmosphere to change rapidly in response to perturbations. There are currently underway many theoretical studies of the impact of the ozone hole on stratospheric circulation, O_3 fluxes, and the mean chemistry of the stratosphere (e.g., N_2O losses). As discussed above, there are also indications of a declining trend in Northern Hemisphere O_3 not associated with the Antarctic hole. In summary, we expect stratospheric O_3 to change in the next few decades, which will lead to alterations in the lifetimes of the long-lived greenhouse gases and will also perturb tropospheric chemistry through the supply of O_3 and through the increase in solar ultraviolet light available to generate OH.

CONCLUSION

Anthropogenic emissions of both long-lived greenhouse gases and short-lived highly reactive species are altering the composition of the atmosphere. The concentrations of CO_2 and CH_4 have increased dramatically since the preindustrial era, and CFCs have been introduced into the atmosphere for the first time. As a result of the rapid pace of human-induced change, neither atmospheric composition nor climate is currently in equilibrium. Thus, significant global change can be anticipated over the coming decades, no matter what course is taken in the future. The rate and magnitude of change, however, are subject to human control, which serves as the motivation for this report.

TABLE 2-2

Trace Gas Data

CO_2 CARBON DIOXIDE	10^{12} kg C
Atmospheric Burden	720
348.4±0.2 ppm in 1987 Not photochemically active	
Annual Trend	2.3
1.15 ppm/yr (0.3%/yr) since 1984	
Annual Sources	6-7
1. Fossil fuel combustion 4.5%/yr since 1984	5.4
2. Land use Modification	0.4 - 2.6
3. Biosphere -- climate feedback Enhanced aerobic decomposition of detrital material due to more favorable climate	?
Annual Sinks	~ 2.5
1. Ocean Ocean's capacity to absorb CO_2 will be altered by changes in temperature, salinity as wells a biological activity of ocean.	~ 2.5
2. Biosphere Enhanced photosynthetic uptake of CO_2 due to more favorable climate and/or due to CO_2 fertilization	?

TABLE 2-2 (Continued)

CH_4 METHANE	10^9 kg CH_4
Atmospheric Burden	
1675 ppb in 1987 Lifetime: 5 - 10 years	4600 - 4800
Annual Trend	
14 - 16 ppb/yr (0.8-1%/yr)	40 - 46
Annual Sources	500 ± 100
1. Fossil fuel	
Coal mining	15 - 45
Natural gas drilling, venting, and transmission loss	25 - 50
2. Biomass burning	50 - 100
3. Natural wetlands	100 - 200
4. Rice Paddies	60 - 170
5. Animals -- mainly ruminants	65 - 100
6. Termites Population unknown	10 - 100
7. Oceans and freshwater lakes	5 - 45
8. Landfills	30 - 70
9. Methane hydrate destabilization	0 - 100 (future)
Annual Sinks	495 ± 145
1. OH destruction	495 ± 145
2. Dry soils absorption by methane - oxidizing bacteria in dry soils	?

TABLE 2-2 (Continued)

N_2O NITROUS OXIDE	10^9 kg N
Atmospheric Burden	1500
340 ppb in 1986 Lifetime: 100 - 175 years	
Annual Trend	3.5 ± 0.5
0.7 ± 0.1 ppb/yr 0.2%/yr	
Annual Sources	14 ± 3 (inferred)
1. Combustion of coal and oil	1 - 3
2. Land use modification	
Biomass burning	1 - 2
Forest clearings	?
3. Fertilized agricultural lands	0.2 - 2.4
4. Contaminated aquifers	0.8 - 1.7
5. Tropical and subtropical forests and woodlands	6 ± 3
6. Boreal and temperate forests	0.1 - 0.5
7. Grasslands	<0.1
8. Oceans	2 ± 1
Annual Sinks	
Stratospheric photolysis and reaction with $O(^1D)$	10.5 ± 3

TABLE 2-2 (Continued)

NO_x	NITROGEN OXIDES	10^9 kg N

NO_x = NO (nitric oxide) + NO_2 (nitrogen dioxide)

$NO_y = NO_x + HNO_2 + HNO_3 + HO_2NO_2 + NO_3 + 2N_2O_5$ + PAN + Particulate Nitrate

Atmospheric Burden -- large variability; lifetime 1-2 days in summer	?
Marine air 4 ppt (NO)	
Continental air non-urban sites 2-12 ppb U.S. & European cities 70 - 150 ppb	
(100 ppt = 240 x 10^9 kg N)	
Annual Trend	?
Annual Sources -- Spatially and temporally concentrated sources	25 - 99
1. Combustion of coal, oil and gas	14 - 28
2. Biomass burning	4 - 24
3. Lightning	2 - 20
4. Oxidation of ammonia	1 - 10
5. Emission from soils (mostly NO)	4 - 16
6. Input from stratosphere (by reaction of $O(^1D)$ with N_2O)	~ 0.5
Sinks	
1. Wet deposition (precipitation scavenging)	24 - 64
ocean	4 - 12
continents	8 - 30
2. Dry deposition	12 - 22

TABLE 2-2 (Continued)

CO CARBON MONOXIDE	10^9 kg CO
Atmospheric Burden	430
~ 90 ppb (150 - 200 ppb Northern Hemisphere, 75 ppb Southern Hemisphere)	
Lifetime: 0.4 year	
Annual Trend	4
1 - 2%/yr Northern Hemisphere 0 - 1%/yr Southern Hemisphere	
Annual Sources	1500 - 4000
1. Fossil fuel combustion 50% for automobiles residential use of coal industrial activities, e.g., steel production	400 - 1000
2. Oxidation of anthropogenic hydrocarbons 45% emissions from automobiles 40% evaporation of fuels and solvents	0 - 180
3. Biomass burning	
Wood used as fuel	25 - 150
Forest wildfires	10 - 50
Forest clearing	200 - 800
Savanna burning	100 - 400
4. CH_4 oxidation	400 - 1000
5. Oxidation of natural hydrocarbons (isoprenes and terpenes)	280 - 1200
6. Emission by plants	50 - 200
7. Ocean	20 - 80
Annual Sinks	3420
1. Soil uptake	250
2. Photochemistry	3170

TABLE 2-2 (Continued)

Chemical Formula	Name	Concentration (pptv)	Annual Trend	Global Prod. (10^6 kg)	Sources	Sinks	Lifetime (years)
$CFCl_3$	CFC-11	226 (1986)	+4%	350 (1986) (CMA rep. co.'s only)	Rigid and flexible foam; Aerosol propellant	Removal in stratosphere	75^{+32}_{-17} (65)
CF_2Cl_2	CFC-12	392 (1986)	+4%	480 (1986) (CMA reporting co.'s plus USSR estimate)	Refrigerant; Rigid and flexible foam; Aerosol propellant	Removal in stratosphere	111^{+289}_{-46} (120)
$CHClF_2$	HCFC-22	~100 (1986)	~+7% (1986)	206 (1984) 163 (1981) 102 (1977)	Refrigerant; production of Teflon polymers (fluoropolymers)	Removal in stratosphere	20
CCl_2FC-ClF_2 (C_2Cl_3F3)	CFC-113	30-70 (1986) (calibration uncertain)	+11%	138-141 (1984) 91 (1979) 79 (1978) 70 (1977)	Electronics solvent	Removal in stratosphere	90
CH_3CCl_3	Methyl-chloroform	125 (1986)	7%	~580 (1985)	Industrial degreasing of metallic or metaplastic pieces; cold cleaning Solvent of adhesives, varnishes and paints	Removal by OH in troposphere	5.5 - 10

TABLE 2-2 (Continued)

Chemical Formula	Name	Concentration (pptv)	Annual Trend	Global Prod. (10^6 kg)	Sources	Sinks	Lifetime (years)
CCl_4	Carbon Tetra-chloride	75-100 (1986) (calibration uncertain)	+1%	~1000 (1985) atmospheric emissions 80-110	Chemical intermediate in CFC-11, 12 production Declining use as: Solvent in chemical & pharmaceutical processes and as grain fumigant	Stratospheric photolysis, perhaps some hydrolysis in ocean	~50
CH_3Cl	Methyl chloride	~600 (1986)		(2000-5000 total, based on OH); 500 industrial	Burning Vegetation Release from oceans Industrial	Troposphere	~1.5
$CBrClF_2$	Halon-1211 (BCF)	~2 (1986) (calibration uncertain)	>10% (1985)	~5 (est. from obs. of atm. increase)	High-tech fire extinguisher (portable)	Photolysis in stratosphere and upper troposphere	25
$CBrF_3$ (CF_3Br)	Halon-1301	~2 (1986) (calibration uncertain)	>10%	7-8? (1984)	Fire extinguisher when CO_2, H_2O are unaccep-table (Refrigerant)	Photolysis in stratosphere	110
CF_3Cl	CFC-13	~3.4(1980)	~5% (avg. of 5 sites)	----		Removal in stratosphere	400
$CClF_2CClF_2$	CFC-114	-----		13-14 (1985) Constant at 13 from 1979-84	Aerosol propellant Refrigerant; Production of CFC-115	Removal in stratosphere	180

TABLE 2-2 (Continued)

Chemical Formula	Name	Concentration (pptv)	Annual Trend	Global Prod. (10^6 kg)	Sources	Sinks	Lifetime (years)
C_2F_5Cl	CFC-115	~4 (1980)				Removal in stratosphere	380
C_2F_6	CFC-116	~4 (1980)	----			Removal in stratosphere	>500
CH_3Br	Methyl bromide	9 (1984)	small	?	Leaded motor fuel; Fumigation 50% natural (anthropogenic sources prob. declining)		~3 (2.3)
$CHBr_3$	Bromoform	~2 (1984)	?	?	Mainly natural		?
$C_2H_4Br_2$	Ethylene dibromide (EDB)	~1 (1984)	?	?	Evaporation of leaded gasoline; Fumigation; anthropogenic sources probably declining		~1
CH_2Br_2 $CHBrCl$ CH_2BrCl_2		1-4 for group; 3.2 (1984)			May well be oceans		
CH_3I	Methyl iodide	~1 (1981)	----				.02

REFERENCES

Bacastow, R.B., C.D. Keeling, and T.P. Whorf. 1985. Seasonal amplitude increase in atmospheric CO_2 concentration at Mauna Loa, Hawaii, 1959-1982. *Journal of Geophysical Research* 90:10529-10540.

Barnola, J.M., D. Raynaud, Y.S. Korotkevich, and C. Lorius. 1987. Vostok ice core provides 160,000-year record of atmospheric CO_2. *Nature* 329:408-414.

Bingemer, H.G., and P.J. Crutzen. 1987. The production of methane from solid wastes. *Journal of Geophysical Research* 92:2181-2187.

Blake, D.R., and F.S. Rowland. 1986. World-wide increase in tropospheric methane, 1978-1983. *Journal of Atmospheric Chemistry* 4:43-62.

Blake, D.R., and F.S. Rowland. 1988. Continuing worldwide increase in tropospheric methane, 1978-1987. *Science* 239:1129-1131.

Bowden, W.B., and F.H. Bormann. 1986. Transport and loss of nitrous oxide in soil water after forest clear-cutting. *Science* 233:867-869.

Brewer, P.G., A.L. Bradshaw, and R.T. Williams. 1986. Measurements of total carbon dioxide and alkalinity in the North Atlantic Ocean in 1981. In Trabalka, J.R., and D.E. Reichle, eds. *The Changing Carbon Cycle: A Global Analysis*. Springer-Verlag, New York. 348-370.

Cicerone, R.J., and R.S. Oremland. 1988. Biogeochemical aspects of atmospheric methane. *Global Biogeochemical Cycles*, in press.

Cicerone, R.J., J.D. Shetter, and C.C. Delwiche. 1983. Seasonal variation of methane flux from a California rice paddy. *Journal of Geophysical Research* 88:11022-11024.

Cleveland, W.S., A.E. Freeny, and T.E. Graedel. 1983. The seasonal component of atmospheric CO_2: Information from new approaches to the decomposition of seasonal time series. *Journal of Geophysical Research* 88:10934-10946.

Conrad, R., and W. Seiler. 1985. Localization of microbial activities relevant to the emission of nitrous oxide from the soil into the atmosphere. *Soil Biol. Biochem* 17:893-895.

Conway, T.J., P. Tans, L.S. Waterman, K.W. Thoning, K.A. Masarie, and R.H. Gammon. 1988. Atmospheric carbon dioxide measurements in the remote global troposphere, 1981-1984. *Tellus* 40:81-115.

Crutzen, P.J. 1983. Atmospheric interactions - homogeneous gas reactions of C, N, and S containing compound. In Bolin, B., and R.B. Cook, eds. *The Major Biogeochemical Cycles and their Interactions*, SCOPE 21. John Wiley & Sons, Chichester. 67-114.

Crutzen, P.J., 1988. Tropospheric ozone: An overview. In Isaksen, I.S.A., ed. *Tropospheric Ozone*. Reidel, Dordrecht. 3-32.

Crutzen, P.J., I. Aselmann, and W. Seiler. 1986. Methane production by domestic animals, wild ruminants, other herbivorous fauna, and humans. *Tellus* 38B:271-284.

Crutzen, P.J., A.C. Delany, J. Greenberg, P. Haagenson, L. Heidt, T. Lueb, W. Pollock, W. Seiler, A. Wartburg, and P. Zimmerman. 1985. Tropospheric chemical composition measurements in Brazil during the dry season. *Journal of Atmospheric Chemistry* 2:233-256.

Crutzen, P.J., L.E. Heidt, J.P. Krasnec, W.H. Pollock, and W. Seiler. 1979. Biomass burning as a source of atmospheric gases CO, H_2, N_2O, NO, CH_3Cl and COS. *Nature* 282:253-256.

Cunnold, D.M., R.G. Prinn, R.A. Rasmussen, P.G. Simmonds, F.N. Alvea, C.A. Cardelino, A.J. Crawford, P.J. Fraser, and R.D. Rosen. 1986. Atmospheric lifetime and annual release estimates for $CFCl_3$ and CF_2Cl_2 from 5 years of ALE data. *Journal of Geophysical Research* 91:10797-10817.

D'Arrigo, R., G. Jacoby, and I. Fung. 1987. Boreal forests and atmosphere-biosphere exchange of carbon dioxide. *Nature* 329:321-323.

Darmstadter, J., L. W. Ayres, R. U. Ayres, W. C. Clark, P. Crosson, P. J. Crutzen, T. E. Graedel, R. McGill, J. F. Richards, and J. A. Tarr. 1987. *Impacts of World Development on Selected Characteristics of the Atmosphere: An Integrative Approach*. Oak Ridge National Laboratory Report ORNL/Sub/86-22033.

Delmas, R.J., J.M. Ascencio, and M. Legrand. 1980. Polar ice evidence that atmospheric CO_2 20,000 yr-BP was 50% of present. *Nature* 284:155-157.

Detwiler, R.P., and C.A. Hall. 1988. Tropical forests and the global carbon cycle. *Science* 239:4247.

Dianov-Klokov, V.I., and L.N. Yurganov. 1981. A spectroscopic study of the global spacetime distribution of atmospheric CO. *Tellus* 33:262-273.

Dickinson, R.E., and R.J. Cicerone. 1986. Future global warming from atmospheric trace gases. *Nature* 319:109-116.

Donner, L., and V. Ramanathan. 1980. Methane and nitrous oxide: Their effects on the terrestrial climate. *Journal of Atmospheric Science* 37:119-124.

Enting, I.G. 1987. The interannual variation in the seasonal cycle of carbon dioxide concentration at Mauna Loa. *Journal of Geophysical Research* 92:5497-5504.

Feely, R.A., R.H. Gammon, B.A. Taft, P.E. Pullen, L.S. Waterman, T.J. Conway, J.F. Gendron, and D.P. Wisgarver. 1987 Distribution of chemical tracers in the eastern equatorial Pacific during and after the 1982-1983 El Niño/Southern Oscillation event. *Journal of Geophysical Research* 92:6545-6558.

Fraser, P.J., W.P. Elliott, and L.S. Waterman. 1986a. Atmospheric CO_2 record from direct chemical measurements during the 19th Century. In Trabalka, J.R., and D.E. Reichle, eds. *The Changing Carbon Cycle, A Global Analysis*. Springer-Verlag, New York. 66-88.

Fraser, P.J., R.A. Rasmussen, J.W. Creffield, J.R. French, and M.A.K. Khalil. 1986b. Termites and global methane - another assessment. *Journal of Atmospheric Chemistry* 4:295-310.

Fraser, P.J., P. Hyson, I. Enting, and G. Pearman. 1983. Global distribution and Southern Hemispheric trends of atmospheric CCl_3F. *Nature* 302:692-695.

Friedi, H., H. Lotscher, H. Oeschger, U. Siegenthaler, and B. Stauffer. 1986. Ice core record of the $^{13}C/^{12}C$ ratio of atmospheric carbon dioxide in the past two centuries. *Nature* 324:237-237.

From, E., and C.D. Keeling. 1986. Reassessment of late 19th century atmospheric carbon dioxide variations in the air of western Europe and the British Isles based on an unpublished analysis of contemporary air masses by G.S. Callendar. *Tellus* 38B:87-105.

Fung, I.Y., C.J. Tucker, and K.C. Prentice. 1987. Application of advanced very high resolution radiometer vegetation index to study atmosphere-biosphere exchange of CO_2. *Journal of Geophysical Research* 92:2999-3015.

Gammon, R.H., W.D. Komhyr, and J.T. Peterson. 1986. The global atmospheric CO_2 distribution 1968-1983: Interpretation of the results of the NOAA/GMCC measurement program. In Trabalka, J.R., and D.E. Reicle, eds. *The Changing Carbon Cycle: A Global Analysis*. Springer-Verlag, New York. 1-15.

Genthon, C., J.M. Barnola, D. Raynaud, C. Lorius, J. Jouzel, N.I. Barkov, Y.S. Korotkevich, and V.M. Kotlyakov. 1987. Vostok ice core: climatic response to CO_2 and orbital forcing changes over the last climatic cycle. *Nature* 329:414-419.

Hansen, J., I. Fung, A. Lacis, D. Rind, S. Lebedeff, R. Ruedy, and G. Russell. 1988. Global climate changes as forecast by Goddard Institute of Space Studies Three-Dimensional Model. *Journal of Geophysical Research* 93:9341-9364.

Hao, W.W., S.C. Wofsy, H.B. McElroy, J.M. Beer, and M.A. Toqan. 1987. Sources of atmospheric nitrous oxide from combustion. *Journal of Geophysical Research* 92:3098-3104.

Harriss, R.C., E. Gorham, D.I. Sebacher, K.B. Bartlett, and P.A. Flebbe. 1985. Methane flux from northern peatlands. *Nature* 315:652-653.

Holzapfel-Pschorn, A., and W. Seiler. 1986. Methane emission during a cultivation period from an Italian rice paddy. *Journal of Geophysical Research* 91:11803-11814.

Houghton, R.A. 1987. Biotic changes consistent with the increased seasonal amplitude of atmospheric CO_2 concentrations. *Journal of Geophysical Research* 92:4223-4230.

Houghton, R.A., R.D. Boone, J.R. Fruci, J.E. Hobbie, J.M. Melillo, C.A. Palm, B.J. Peterson, G.R. Shaver, G.M. Woodwell, B. Moore, D.L. Skole, and N. Myers. 1987. The flux of carbon from terrestrial ecosystems to the atmosphere in 1980 due to changes in land use: Geographic distribution of the global flux. *Tellus* 39B:122-139.

IRRI (International Rice Research Institute). 1986. *World Rice Statistics 1985*. IRI, Los Banos.

Jouzel, J., C. Lorius, J.R. Petit, C. Genthon, N.I. Barkov, V.M. Kotlyakov, and V.M. Petrov. 1987. Vostok ice core: A continuous isotope temperature record over the last climatic cycle (160,000 years). *Nature* 329:403-408.

Keeling, C.D. 1968. Carbon dioxide in surface ocean waters, 4. Global distribution. *Journal of Geophysical Research* 73:4543-4553.

Keeling, C.D. 1983. The global carbon cycle: What we know and could know from atmospheric, biospheric, and oceanic observations. In Institute for Energy Analysis. *Proceedings: Carbon Dioxide Research Conference: Carbon Dioxide, Science and Consensus*. DOE CONF-820970, U.S. DOE, Washington, D.C. II.3-II.62.

Keeling, C.D., and M. Heimann. 1986. Meridional eddy diffusion model of the transport of atmospheric carbon dioxide, 2, Mean annual carbon cycle. *Journal of Geophysical Research* 91:7782-7796.

Khalil, M.A.K., and R.A. Rasmussen. 1983. Sources, sinks, and seasonal cycles of atmospheric methane. *Journal of Geochemical Research* 88:5131-5144.

Khalil, M.A.K., and R.A. Rasmussen. 1985. Causes of increasing atmospheric methane: depletion of hydroxyl radicals and the rise of emissions. *Atmospheric Environment* 19:397-407.

Khalil, M.A.K., and R.A. Rasmussen. 1987. Nitrous oxide: Trends and global mass balance over the last 3000 years. *Annals of Glaciology* 10:73-79.

Khalil, M.A.K., and R.A. Rasmussen. 1988. Carbon monoxide in the Earth's atmosphere: Indications of a global increase. *Nature* 332:242-245.

Ko, M.K.W., N.D. Sze, M. Livshits, M.B. McElroy, and J.A. Pyle. 1984. The seasonal and latitudinal behavior of trace gases and O_3 as simulated by a two-dimensional model of the atmosphere. *Journal of Atmospheric Science* 41:2381-2408.

Kohlmaier, G.H., H. Bröhl, E.O. Siré, M. Plöchl, and R. Revelle. 1987. Modelling stimulation of plants and ecosystem response to present levels of excess atmospheric CO_2 . *Tellus* 39B:155-170.

Komhyr, W.D., R.H. Gammon, T.B. Harris, L.S. Waterman, T.J. Conway, W.R. Taylor, and K.W. Thoning. 1985. Global atmospheric CO_2 distribution and variations from 1968-1982 NOAA/GMCC CO_2 flask sample data. *Journal of Geophysical Research* 90:5567-5596.

Komhyr, W.D., T.M. Thompson, and E.G. Dutton. 1988. Atmospheric nitrous oxide distribution and variations from 1977-1984 NOAA/GMCC flask sample data. *Journal of Geophysical Research,* in press.

Kvenvolden, K. 1988. Methane hydrates and global climate. *Global Biogeochemical Cycles* 2:221-229.

Lacis, A., J. Hansen, P. Lee, T. Mitchell and S. Lebedeff. 1981. Greenhouse effect of trace gases, 1970-1980. *Geophysical Research Letters* 8:1035-1038.

Lacis, A., D.J. Wuebbles, and J.A. Logan. No date. Radiative forcing of global climate by changes in vertical distribution of atmospheric ozone. *Journal of Geophysical Research,* submitted.

Lerner, J., E. Matthews, and I. Fung. 1988. Methane emission from animals: A global high-resolution database. *Global Biogeochemical Cycles* 2:139-156.

Levine, J.S., C.P. Rinsland, and G.M. Tennille. 1985. The photochemistry of methane and carbon monoxide in the troposphere in 1950 and 1985. *Nature* 318:254-257.

Liu, S.C., M. Trainer, F.C. Fehsenfeld, D.D. Parrish, E.J. Williams, D.W. Fahey, G. Hubler and P.C. Murphy. 1987. Ozone production in the rural troposphere and the implications for regional and global ozone distributions. *Journal of Geophysical Research* 92:4194-4207.

Livingston, G.P., P.M. Vitousek, and P.A. Matson. 1988. Nitrous oxide flux and nitrogen transformations across a landscape gradient in Amazonia. *Journal of Geophysical Research* 93:1593-1599.

Logan, J. 1985. Tropospheric ozone: Seasonal behavior, trends, and anthropogenic influence. *Journal of Geophysical Research* 90:10463-10484.

Lovelock, J. 1971. Atmospheric fluorine compounds as indicators of air movements. *Nature* 230:379-381.

Matson, M. and B. Holben. 1987. Satellite detection of tropical burning in Brazil. *International Journal of Remote Sensing* 8:509-516.

Matson, M., G. Stephens and J. Robinson. 1987. Fire detection using data from the NOAA-N satellites. *International Journal of Remote Sensing* 8:961-970.

Matson, P.A., and P.M. Vitousek. 1987. Cross-system comparisons of soil nitrogen transformations and nitrous oxide flux in tropical forest ecosystems. *Global Biogeochemical Cycles* 1:163-170.

Matthews, E., and I. Fung. 1987. Methane emission from natural wetlands: Global distribution, area, and environmental characteristics of sources. *Global Biogeochemical Cycles* 1:61-86.

Muzio, L.J., and J.C. Kramlich. 1988. An artifact in the measurement of N_2O from combustion sources. *Geophysical Research Letters* 15:1369-1372.

Neftel, A., E. Moor, H. Oeschger, and B. Stauffer. 1985. Evidence from polar ice cores for the increase in atmospheric CO_2 in the past two centuries. *Nature* 315:45-47.

Neftel, A., H. Oesghger, J. Schwander, B. Staufferm, and R. Zumbrubb. 1982. Ice core sample measurements give atmospheric CO_2 content during the past 40,000 years *Nature* 295:220-223.

NOAA (National Oceanographic and Atmospheric Administration). 1987. *Geophysical Monitoring for Climatic Change No. 15, Summary Report 1986.* Schnell, R.C., ed. U.S. Department of Commerce, NOAA Environmental Research Laboratories, Boulder. 155 pp.

Oeschger, H. and B. Stauffer. 1986. Review of the history of atmospheric CO_2 recorded in ice cores. In J.R. Trabalka. and D.E. Reichle, eds. *The Changing Carbon Cycle, A Global Analysis*. Springer-Verlag, New York. pp. 89-108.

Pearman, G.I., D. Etheridge, F. deSilva, and P.J. Fraser. 1986. Evidence of changing concentrations of atmospheric CO_2, N_2O and CH_4 from air bubbles in Antarctic ice. *Nature* 320:248-250.

Pearman, G.I., and P. Hyson. 1981. The annual variation of atmospheric CO_2 concentration observed in the northern hemisphere. *Journal of Geophysical Research* 86:9839-9843.

Pearman, G.I., and P. Hyson. 1986. Global transport and inter-reservoir exchange of carbon dioxide with particular reference to stable isotopic distributions. *Journal of Atmospheric Chemistry* 4:181-124.

Pearman, G.I., P. Hyson, and P.J. Fraser. 1983. The global distribution of atmospheric carbon dioxide, 1, Aspects of observations and modeling. *Journal of Geophysical Research* 88:3581-3590.

Peng, T.-H., T. Takahashi, W.S. Broecker, and J. Olafsson. 1987. Seasonal variability of carbon dioxide, nutrients and oxygen in the northern North Atlantic surface water: Observations and a model. *Tellus* 39B:439-458.

Prinn, R.G. 1988. How have the atmospheric concentrations of halocarbons changed? In Rowland, F.S., and I.S.A. Isaksen, eds. *The Changing Atmosphere*. John Wiley & Sons, Chichester. 33-48.

Prinn, R., D. Cunnold, R. Rasmussen, P. Simmonds, F. Alyea, A. Crawford, P. Fraser, and R. Rosen. 1987. Atmospheric trends in methylchloroform and the global average for the hydroxyl radical. *Science* 238:945-950.

Prinn, R.G., P. Simmonds, R. Rasmussen, R. Rosen, F. Alyea, C. Cardelino, A. Crawford, D. Cunnold, P. Fraser, and J. Lovelock. 1983. The Atmospheric Lifetime Experiment, I: Introduction, instrumentation, and overview. *Journal of Geophysical Research* 88:8353-8367.

Ramanathan, V., L. Callis, R. Cess, J. Hansen, I. Saksen, W. Kuhn, A. Lacis, F. Luther, J. Mahlman, R. Reck, and M. Schlesinger. 1987. Climate-chemical interactions and effects of changing atmospheric trace gases. *Reviews of Geophysics* 25:1441-1482.

Ramanathan, V., R.J. Cicerone, H.B. Singh, and J.T. Kiehl. 1985. Trace gas trends and their potential role in climate change. *Journal of Geophysical Research* 90:5547-5566.

Rasmussen, R.A., and M.A. Khalil. 1986. Atmospheric trace gases: Trends and distributions over the last decade. *Science* 232:1623-1625.

Raynaud, D., and J.M. Barnola. 1985. An Antarctic ice core reveals atmospheric CO_2 variations over the past few centuries. *Nature* 315:309-311.

Raynaud, D., J. Chappellaz, J.M. Barnola, Y.S. Korotkevich, and C. Lorius. 1988. Climatic and CH_4 cycle implications of glacial-interglacial CH_4 change in the Vostok ice core. *Nature* 333:655-657.

Rinsland, C.P., and J.S. Levine. 1985. Free tropospheric carbon monoxide concentrations in 1950 and 1951 deducted from infrared total column amount measurements. *Nature* 318:250-254.

Robertson, G.P., and J.M. Tiedje. 1988. Reforestation alters denitrification in a lowland tropical rainforest. *Nature* 336:756-759.

Ronen, D., M. Magaritz, and E. Almon. 1988. Contaminated aquifers are a forgotten component of the global N_2O budget. *Nature* 335:57-59.

Rotty, R.M. 1987a. A look at 1983 CO_2 emissions from fossil fuels (with preliminary data for 1984). *Tellus* 39B:203-208.

Rotty, R.M. 1987b. Estimates of seasonal variation in fossil fuel CO_2 emissions. *Tellus* 39B:184-202.

Rowland, F.S. 1989. The role of halocarbons in stratospheric ozone depletion. In National Research Council, *Ozone Depletion, Greenhouse Gases and Climate Change.* National Academy Press, Washington, D.C. 33-47.

Sahrawat, K.L., and D.R. Keeney. 1986. Nitrous oxide emissions from soils. *Advances in Soil Science* 4:103-148.

Sebacher, D.I., R.C. Harris, K.B. Bartlett, S.M. Sebacher, and S.S. Grice. 1986. Atmospheric methane sources: Alaskan tundra bogs, an alpine fen, and a sub-arctic boreal marsh. *Tellus* 38B:1-10.

Seiler, W. 1984. Contribution of biological processes to the global budget of CH_4 in the atmosphere. In Klug, M., and C. Reddy, eds. *Current Perspectives in Microbial Ecology*. American Society for Microbiology, Washington, D.C. 468-477.

Seiler, W. and R. Conrad. 1981. Field measurements of natural and fertilizer induced N_2O release rates from soils. *Journal of Air Pollution Control Association* 31:767-772.

Seiler, W., and R. Conrad. 1987. Contribution of tropical ecosystems to the global budgets of trace gases, especially CH_4, H_2, CO, and N_2O. In Dickinson, R.E., ed. *Geophysiology of Amazonia*. John Wiley & Sons, New York. 133-162.

Seiler, W., R. Conrad, and D. Scharffe. 1984a. Field studies of methane emission from termite nests into the atmosphere and measurements of methane uptake by tropical soils. *Journal of Atmospheric Chemistry* 1:171-187.

Seiler, W., H. Giehl, E.E. Branke, and E. Halliday. 1984b. The seasonality of carbon monoxide abundance in the Southern Hemisphere. *Tellus* 36B:250-254.

Siegenthaler, V., and H. Oeschger. 1987. Biospheric CO_2 emissions during the past 200 years reconstructed by deconvolution of ice core data. *Tellus* 39B:140-154.

Simmonds, P., D. Cunnold, F. Alyea, C. Cardelino, A. Crawford, R. Prinn, P. Fraser, R. Rasmussen, and R. Rosen. 1987. Carbon tetrachloride lifetimes and emissions determined from daily global measurements during 1978-1985. *Journal of Atmospheric Chemistry* 7:35-58.

Slemr, F., R. Conrad and W. Seiler. 1984. Nitrous oxide emissions from fertilized and unfertilized soils in a subtropical region (Andalusia, Spain). *Journal of Atmospheric Chemistry* 1:159-169.

Stauffer, B., G. Fischer, A. Neftel, and H. Oeschger. 1985. Increase of atmospheric methane recorded in Antarctic ice cores. *Science* 229:1386-1389.

Steele, L.P., P.J. Fraser, R.A. Rasmussen, M.A.K. Khalil, T.J. Conway, A.J. Crawford, R.H. Gammon, K.A. Masarie, and K.W. Thoning. 1987. The global distribution of methane in the troposphere. *Journal of Atmospheric Chemistry* 5:125-171.

Tans, P.P., T.J. Conway, and T. Nakazawa. 1989. Latitudinal distribution of the sources and sinks of atmospheric carbon dioxide derived from surface observations and an atmospheric transport model. *Journal of Geophysical Research*, in press.

Takahashi, T., W.S. Broecker, and A.E. Bainbridge. 1981. The alkalinity and total carbon dioxide concentration in the world oceans. In Bolin, B., ed. *Carbon Cycle Modeling*. Scope 16. John Wiley & Sons, Chichester. 271-286.

Takahashi, T., W.S. Broecker, S.R. Werner and A.E. Bainbridge. 1980. Carbonate chemistry of the surface waters of the world oceans. In Goldberg, E., Y. Horibe, and K. Saruhashi, eds. *Isotope Marine Chemistry*. Uchida Rokakuho, Tokyo. 291-326.

Takahashi, T., J. Goddard, S. Sutherland, D.Chipman and C. Breeze. 1986. *Seasonal and Geographic Variability of Carbon Dioxide Sink/Source in the Oceanic Areas: Observations in the North and Equatorial Pacific Ocean, 1984-1986 and Global Summary*. Final Technical Report for MM 19X-89675C, Lamont-Doherty Geological Observatory, Palisades, New York, 66 pp.

Takahashi, T., J. Goddard, S. Sutherland, G. Mathieu, and D. Chipman. 1988. *Assessment of carbon dioxide sink/source in the North Pacific basin: Seasonal and geographic variability, 1986-1987*. Progress Report for MM 19X-89675C, Lamont-Doherty Geological Observatory, Palisades, New York, 66 pp.

Thompson, A.M., and R.J. Cicerone. 1986. Possible perturbations to atmospheric CO, CH_4, and OH. *Journal of Geophysical Research* 91:10853-10864.

Thompson, M.L., I.G. Enting, G.I. Pearman, and P. Hyson. 1986. Interannual variation of atmospheric CO_2 concentration. *Journal of Atmospheric Chemistry* 4:125-156.

Thompson, T., W. Komhyr, and E. Dutton. 1985. *Chlorofluorocarbon-11, -12, and Nitrous Oxide Measurements at the NOAA/GMCC Baseline Stations (1973-1979)*. NOAA Technical Report ERL 428-ARL8. NOAA, Boulder.

Tiao, G.C., G.C. Reinsel, J.H. Pedrick, G.M. Allenby, C.L. Mateer, A.J. Miller, and J.J. DeLuisi. 1986. A statistical trend analysis of ozone sonde data. *Journal of Geophysical Research* 91:13121-13136.

UNEP (United Nations Environment Programme). 1987. *The Greenhouse Gases*. UNEP, Nairobi. 40 pp.

United Nations. 1987. *1985 Energy Statistics Yearbook*. United Nations, New York.

Wang, W.C., N.D. Sze, G. Molnar, M. Ko, and S. Goldenberg. 1988. Ozone-climate interactions with increasing atmospheric trace gases. In Isaksen, I.S.A., ed. *Tropospheric Ozone*. D. Reidel, Hingham, Massachusetts. 147-159.

Watson, R.T., M.A. Geller, R.S. Stolarski, and R.F. Hampson. 1986. *Present State of Knowledge of the Upper Atmosphere: An Assessment Report*. NASA Reference Publication 1162. Washington, D.C. 134 pp.

Watson, R.T. and Ozone Trends Panel, M.J. Prather and Ad Hoc Theory Panel, and M.J. Kurylo and NASA Panel for Data Evaluation. 1988. *Present State of Knowledge of the Upper Atmosphere 1988: An Assessment Report*. NASA Reference Publication 1208. Washington, D.C., 200 pp.

Weiss, R.F. 1981. The temporal and spatial distribution of tropospheric nitrous oxide. *Journal of Geophysical Research* 86:7185-7195.

WMO (World Meteorological Organization). 1985. *Atmospheric Ozone 1985: Assessment of our Understanding of the Processes Controlling its Present Distribution and Change.* Volume 1. WMO, Geneva, 392+pp.

Zimmerman, P.R., J.P. Greenberg, and J.P.E.C. Darlington. 1984. Response to: Termites and atmospheric gas production (Technical Comment by N.M. Collins, and T.G. Wood). *Science* 224:84-86.

Zimmerman, P.R., J.P. Greenberg, S.O. Wandiga, and P.J. Crutzen. 1982. Termites: A potentially large source of atmospheric methane, carbon dioxide and molecular hydrogen. *Science* 218:563-565.

CHAPTER III

CLIMATIC CHANGE

FINDINGS

- Climate exhibits natural variability on all time scales, from years to millions of years. This variability is caused by a combination of changes in external factors, such as solar output, and internal dynamics and feedbacks, such as the redistribution of heat between the atmosphere and the oceans.

- The ultimate warming that can be expected for a given increase in greenhouse gas concentrations is uncertain due to our inadequate understanding of the feedback processes of the climate system. For the benchmark case of doubling carbon dioxide concentrations, the National Academy of Sciences has estimated that the equilibrium increase in global average temperature would most likely be in the range of 1.5-4.5°C; recent analyses suggest that the warming could be as much as 5.5°C; a reasonable central uncertainty range is 2-4°C.

- A variety of geophysical and biogenic feedbacks exist that have generally been ignored in quantifying the temperature change that could occur for any given initial increase in greenhouse gases. In particular, the potential of future global warming to increase emissions of carbon from northern latitude reservoirs in the form of both methane and carbon dioxide, to alter uptake of CO_2 by oceans, and a variety of other temperature dependent phenomena indicate that the true sensitivity of the Earth's climate system to increased greenhouse gases could exceed 5.5°C for an initial doubling of CO_2. While there are biogenic and geochemical feedbacks that could decrease greenhouse gas concentrations -- enhanced photosynthesis due to higher CO_2, for example -- it appears that the risk of all biogenic and geochemical feedbacks raising the Earth's true temperature sensitivity to anthropogenic emissions of greenhouse gases is greater than the possibility of such feedbacks decreasing this sensitivity.

- Uncertainties about ocean circulation and heat uptake, and about future internal climate oscillations and volcanic eruptions, make it difficult to predict the time-dependent response of climate to changes in greenhouse gas concentrations. Because the oceans delay the full global warming that would be associated with any increase in greenhouse gases, significant climatic change could continue for decades after the composition of the atmosphere were stabilized. The Earth already is committed to a total warming of about 0.7-1.5°C (assuming that the climate sensitivity to doubling CO_2 is 2-4°C). The Earth has warmed by 0.3-0.7°C during the last century, which is consistent with expectations given the uncertain delay caused by ocean heat uptake.

INTRODUCTION

The increasing concentrations of greenhouse gases documented in the last chapter are expected to alter significantly the Earth's climate in the coming decades. The magnitude and timing of actual climatic change will be determined by future emissions (Chapter V), by changes in other climatic forcings, and by the sensitivity of the climate system to perturbations. Weather and climate (the time-average of weather) are determined by complex interactions between the atmosphere, land surface, snow, sea ice, and oceans, involving radiative and convective exchange of energy within and among these components. As is readily apparent, this system exhibits considerable variability from day to day, month to month, and year to year.

Systematic diurnal (day-night) and seasonal variations are driven by changes in the distribution and amount of solar energy reaching the top of the Earth's atmosphere as the Earth rotates on its axis and orbits around the sun. Changes in the amount of energy emitted by the sun, changes in the atmospheric composition (due to volcanic eruptions and human input of aerosols and greenhouse gases), and changes in the earth's surface (such as deforestation) can also affect the earth's energy balance. Such factors are considered "external forcings" because they do not depend on the state of the climate system itself.

In contrast, much of the day to day and year to year variation results from the internal dynamics of the climate system. For example, the polar front may be unusually far south in North America during a given year, producing colder than normal weather in the northern Great Plains, but there can be warmer than average weather somewhere else, leaving the global average more or less unchanged. Similarly, upwelling of cold water off the Pacific Coast of South America may fail for several years. This irregularly recurring event, referred to as El Niño, leads to various regional weather anomalies, impacts like the collapse of the Peruvian anchovy fishery, and warmer global temperatures. In this case there is a temporary net release of heat from the ocean to the atmosphere, which is usually followed by a reversal, sometimes referred to as La Niña (Kerr, 1988). Such random variations of the atmospheric and oceanic circulation can produce anomalous redistributions of energy in the climate system resulting in random climate variations, with amplitudes and time scales which may be comparable to climate changes expected from past increases in greenhouse gases (Lorenz, 1968; Hasselmann, 1976; Robock, 1978; Hansen et al., 1988).

In order to determine precisely the potential effects of the input of greenhouse gases on future climate, it would be necessary not only to be able to understand all the physics of the climate system and the effects of each potential cause of climate change, but also to be able to predict the future changes of these forcings. If we could do this, we could explain past climate change and separate the effects of greenhouse gases from the other factors that have acted during the past 100 years for which we have instrumental temperature records. We could also use theoretical climate models to calculate the future size and timing of climate changes due to greenhouse gases. Since our measurements of past climate are incomplete, our understanding of the climate system is incomplete, and some (not well known) portion of climate change is random and unpredictable, we can only estimate the impact of greenhouse gas buildup within a broad range of uncertainty.

In this chapter we discuss in brief the magnitude and rate of past changes in climate and examine the various factors influencing climate in order to place the potential warming due to increasing greenhouse gas concentrations in context. Feedback mechanisms that can amplify or lessen imposed climate changes are discussed next. The overall sensitivity of climate to changes in forcing is then considered, followed by a discussion of the time-dependent response of the Earth system. The focus is on global temperature as an indicator for the magnitude of climatic change. Regional climate and the potential impacts of climatic change are not discussed here, but are considered in the companion report *Potential Effects of Global Climate Change on the United States* (Smith and Tirpak, 1989).

CLIMATIC CHANGE IN CONTEXT

The most detailed information on climate is, of course, from the modern instrumental record, but even this data set is quite sparse in the Southern Hemisphere and over the oceans. Wigley et al. (1986) reviewed a number of recent analyses, noting that independent groups (including Hansen et al., 1981 and Vinnikov et al., 1980; more recent publications are Hansen and Lebedeff, 1988 and Vinnikov et al., 1987), necessarily relying on the same basic data sources, but using different data selection and averaging approaches, have obtained very similar results. Given the various uncertainties due to factors such as poor spatial coverage in some regions, changes in the number and location of stations, growth of urban heat islands, and changes in instrumentation, Wigley et al. conclude that the warming since 1900 has been in the range of 0.3-0.7°C. The most complete and up-to-date global surface air temperature record available (Jones et al., 1988) is displayed in Figure 3-1a, which shows a global warming of about 0.3°C from 1900 to 1940, a cooling of about 0.1°C from 1940 to 1975 and a warming of about 0.2°C from 1975 to 1987. The four warmest years in the record occurred during the 1980s: 1980, 1981, 1983, and 1987. The overall warming is similar in the land air temperature record of the Northern and Southern Hemispheres (Figures 3-1b,c), though the long-term trend is steadier in the Southern Hemisphere where the 1940-1975 cooling is less evident. While the gradual warming seen in Figure 3-1 during the past century is not inconsistent with the increasing greenhouse gases during this period (Chapter II), the large interannual variations and the relatively flat curve from 1940 to 1975 show that there are also other important causes of climate change. The differences between the two hemispheres also show that there are regional differences in the climate response to a global forcing (greenhouse gases), that important other forcings (such as large volcanic eruptions) are not global in their effects, or that internal climate variations produce regional differences. Because of past and potential future emissions of greenhouse gases (see below and Chapter V), climate changes during the next century may be larger than the variations shown for the past 100 years.

FIGURE 3-1

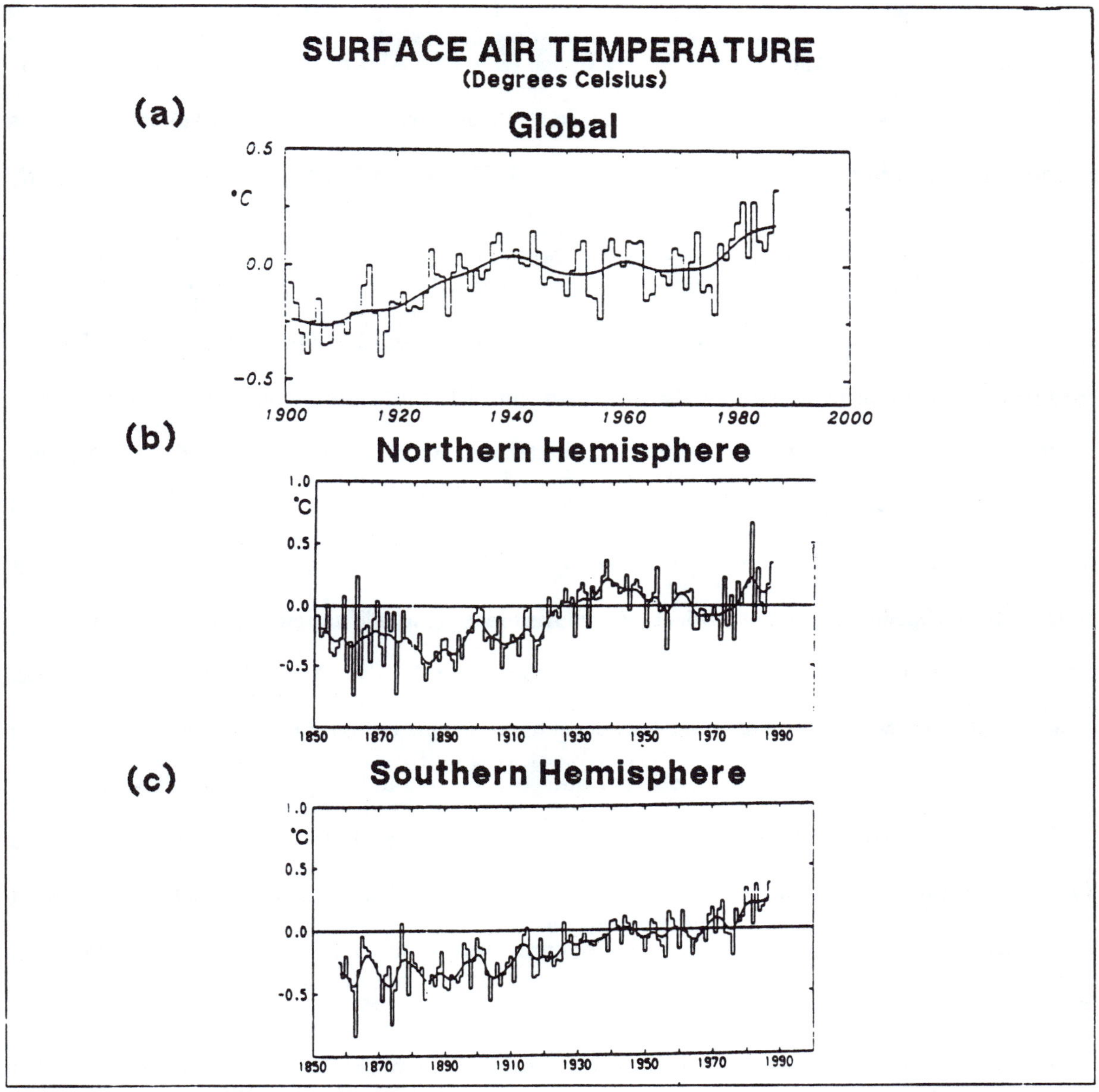

Figure 3-1. (a) Global surface air temperature, 1901-1987. This record incorporates measurements made both over land and from ships. The smooth curve shows 10-year Gaussian filtered values. The gradual warming during this period is not inconsistent with the increasing greenhouse gases during this period, but the large interannual variations and the relatively flat curve from 1940 to 1975 show that there are also other important causes of climate change. (Source: Jones et al., 1988.)

(b,c) Land surface air temperatures, 1851-1987 for the Northern Hemisphere (NH) and 1857-1987 for the Southern Hemisphere (SH). Note the larger interannual variability before 1900, when data coverage was much more sparse. (Source: Jones, 1988.)

Recent climate variations are put in a longer-term perspective in Figures 3-2 through 3-4. The amplitude of climatic change over the last millennium (Figure 3-2) is similar to what has been seen during the last century. The Medieval Warm Epoch (800-1200 AD) may have been restricted to the North Atlantic Basin (Wigley et al., 1986), and in any case appears to have been about as warm as the present. The Little Ice Age (1430-1850; Robock, 1979) appears to have been as cool as the early 20th Century in parts of Europe. The peak of the most recent glaciation is generally given as 18 thousand years before the present (kyBP) (Figures 3-3, 3-4) with globally averaged temperatures about 5°C cooler than today (Hansen et al., 1984) between 15 and 20 kyBP. Even over the 700,000 year period illustrated in Figure 3-4 the maximum global temperature swing appears to have been no greater than about 5°C, with the periods of greatest warmth being the present and the interglacial peaks which occurred approximately every 100,000 years for the past million years (Figure 3-4). The temperature change shown in Figure 3-3 is for Antarctica and is substantially greater than what is believed to represent the globe as a whole. (Such high-latitude amplification of temperature increases can be expected for greenhouse-induced warming in the future). The CO_2 variations are, in general, in step with the temperature variations deduced from deuterium variations in the same ice core (Jouzel et al., 1987), demonstrating the importance of CO_2 in amplifying the relatively weak orbital forcings during past climate variations (Genthon et al., 1987; see Orbital Parameters). While it is difficult to assign a cause for these past changes, it is reasonable to conclude that, given current greenhouse gas concentrations, global temperatures will soon equal or exceed the maximum temperatures of the past million years.

CLIMATE FORCINGS

The patterns of climate variations discussed in the last section are the result of a combination of external forcings, internal feedbacks, and unforced internal fluctuations. The strictly external forcings are changes in solar output and variations in the Earth's orbital parameters, while changes

FIGURE 3-2

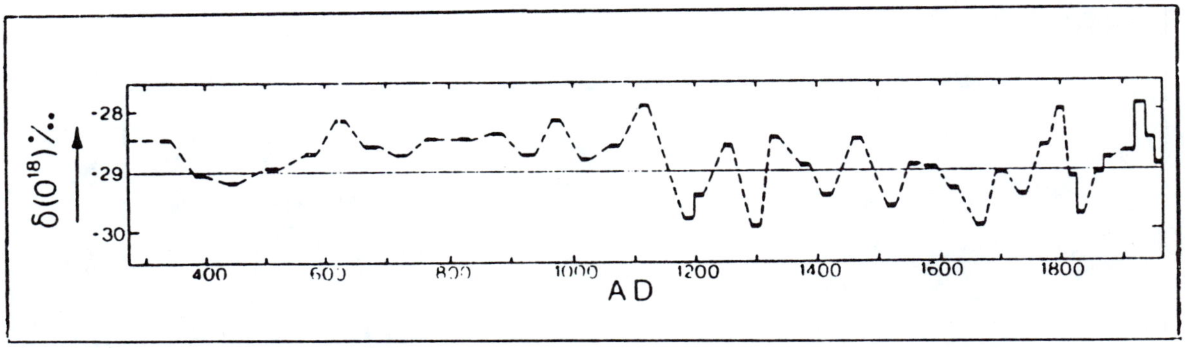

Figure 3-2. Oxygen isotope ($\delta^{18}O$) variations from ice cores in Greenland. This is an index of Northern Hemisphere temperature, with the maximum range equal to about 1°C. (Source: record of Dansgaard as given by Lamb, 1977.)

FIGURE 3-3

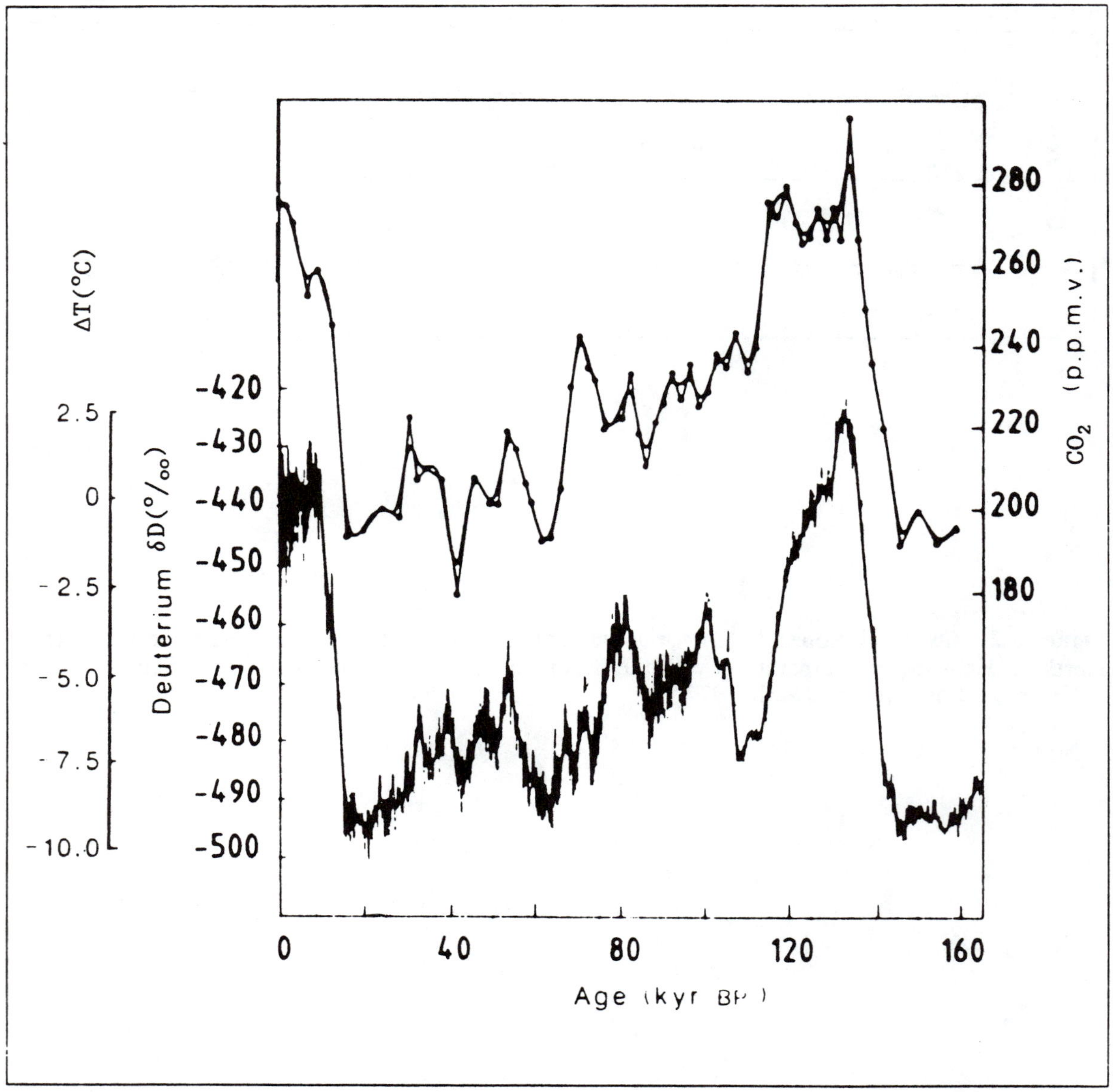

Figure 3-3. Carbon dioxide levels and temperatures over the last 160,000 years from Vostok 5 Ice Core in Antarctica. The temperature scale is for Antarctica; the corresponding amplitude of global temperatures swings is thought to be about 5°C. (Source: Barnola et al., 1987.)

FIGURE 3-4

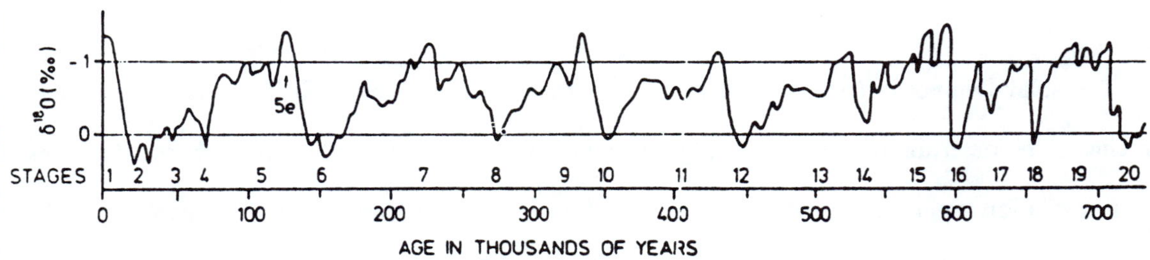

Figure 3-4. Composite $\delta^{18}O$ record of Emiliani (1978) as given by Berger (1982). This comes from deep sea sediment cores and is an index of global temperature, with the temperature range from stage 1 (present) to stage 2 (18,000 years ago) equal to about 5°C.

in aerosols and greenhouse gas concentrations may be viewed as external forcings or internal feedbacks, depending on the time scale and processes considered. The sensitivity of the climate system is determined by the feedbacks that modify the extent to which climate must change to restore the overall energy balance of the Earth as external forcings change.

Solar Luminosity

The solar luminosity (or total energy output from the sun) has an obvious and direct influence on climate by determining the total energy reaching the top of the Earth's atmosphere. Theories of stellar evolution suggest that solar output was 25% lower early in Earth history, but geologic evidence and the fact that life was able to evolve on Earth shows that the Earth was not an inhospitable ice-covered planet. An important part of the explanation for this "faint young sun paradox" now appears to be that the CO_2 content of the atmosphere was many times higher than it is at present. The enhanced greenhouse effect from CO_2 was probably the main factor in counteracting the lower solar luminosity (see below). Geochemical models suggest that over millions of years CO_2 has acted as an internal feedback that has kept the Earth's climate in a habitable range (Walker et. al., 1981; Berner and Lasaga, 1988; Figure 3-2b).

Solar luminosity also varies by small but significant amounts over shorter time periods. Various attempts have been made to explain past climate variations by assuming a link between solar luminosity and observed parameters, such as sunspot activity, solar diameter, and the umbral-penumbral ratio (Wigley et al., 1986). Unfortunately, measurements with sufficient precision to detect solar luminosity changes have only been available since 1979 -- too short a time period to be able to definitively confirm or refute the proposed relationships. These measurements show a decline in solar luminosity between 1980 and 1986; whereas the most recent data show a reversal of this trend (Willson and Hudson 1988; Willson et al., 1986). The luminosity data are positively correlated with sunspot number and suggest an 11-year cycle with an amplitude of 0.04% or 0.1 W/m^2 at the top of the atmosphere (Willson and Hudson, 1988).

Orbital Parameters

Cyclic changes in Earth orbital characteristics are now widely accepted as the dominant trigger behind the glacial/interglacial variations evident in Figure 3-3 and extending back to at least 1.7 million years before present (e.g., Wigley et al., 1986; COHMAP, 1988). While causing only small changes in the total radiation received by the Earth, the orbital changes (known as the Milankovitch cycles) significantly alter the latitudinal and seasonal distribution of insolation. For example, Northern Hemisphere summer insolation was about 8% greater 9 kyBP than it is now, but winter insolation was 8% lower. Changes of this type, in combination with internal feedbacks as discussed below, are presumed to have determined the pattern of glaciations and deglaciations revealed in the geologic record. Attempts have been made to compare model predictions with paleoclimatic data. There has been reasonably good agreement between the two, given specified ice sheet extent and sea surface temperatures (COHMAP, 1988; Hansen et al., 1984). To the extent that the Milankovitch explanation of ice ages is correct, one would expect the Earth to be heading toward a new ice age over the next 5000 years, but the very gradual changes in orbital forcings expected in this period will be overwhelmed if current trends in greenhouse gas concentrations continue (Wigley et al., 1986).

Volcanoes

Large volcanoes can significantly increase the stratospheric aerosol concentration, increasing the planetary albedo and reducing surface temperatures by several tenths of degree for several years (Hansen et al., 1978, 1988; Robock, 1978, 1979, 1981, 1984). Because of the thermal inertia of the climate system, discussed below, volcanoes can even be responsible for climate changes over decades, and in fact the warming shown in Figure 3-1 from 1920 to 1940 can be attributed to a period with very few volcanic eruptions (Robock, 1979). Since large eruptions occur fairly frequently and cannot now be predicted, this component of climate change will have to be considered when searching past climate for a greenhouse signal and when projecting future climate change.

Surface Properties

The Earth's radiative balance can also be changed by variations of surface properties. While interactions with the ocean, which covers 70% of the Earth's surface, are considered internal to the climate system and are discussed below, land surfaces also exert a strong influence on the climate. Human activities, such as deforestation, not only provide a source of CO_2 and CH_4 to the atmosphere, but also change the surface albedo and moisture flux into the atmosphere. Detailed land surface models, incorporating the effects of plants, are now being developed and incorporated into GCM studies of climate change (Dickinson, 1984; Sellers et al., 1986).

The Role of Greenhouse Gases

The greenhouse effect does not increase the total energy received by the Earth, but it does alter the distribution of energy in the climate system by increasing the absorption of infrared (IR) radiation by the atmosphere. If the Earth had no atmosphere, its surface temperature would be strictly determined by the balance between solar radiation absorbed at the surface and emitted IR. The amount of IR emitted by any body is proportional to the fourth power of its absolute temperature, so that an increase in absorbed solar radiation (due to increased solar luminosity or decreased albedo, for example) would be balanced by a small increase in the surface temperature, increasing IR emissions until they are again equal to the absorbed solar radiation. The role of greenhouse gases can be understood by thinking about the atmosphere as a thin layer that absorbs some fraction of the IR emitted by the surface (analogous to the glass in a greenhouse). The energy absorbed by the atmosphere is then remitted in all directions, and the downward half of this energy flux warms the surface (Figure 3-5). Higher concentrations of greenhouse gases increase the IR absorption in the atmosphere, raising surface temperatures.

Changes in the atmosphere's radiative properties can result from external perturbations (such as anthropogenic emissions of CO_2) or from internal adjustments to climatic change. The amount of water vapor, the dominant greenhouse gas, is directly determined by climate and contributes the largest positive feedback to climatic change (Hansen et al., 1984; Dickinson, 1986). Similarly, clouds are an internal part of the climate system that strongly influence the Earth's radiative balance (Ramanathan et al., 1989). Changes in the concentrations of other greenhouse gases may be imposed by human activity or may result from changes in their sources and sinks induced by climatic change. Such feedbacks are discussed below.

Internal Variations

As discussed in the introduction, even with no changes in external forcings, climate still exhibits variations due to internal rearrangements of energy within the atmosphere and between the atmosphere and the ocean. The total amplitude and time scales of these internal stochastic climate variations are not well known; these variations therefore pose an additional difficulty in interpreting the past record and projecting the level of future climate change.

PHYSICAL CLIMATE FEEDBACKS

Any imposed imbalance in the Earth's radiative budget, such as discussed above, will be translated into a changed climate through feedback mechanisms which can act to amplify or decrease

FIGURE 3-5

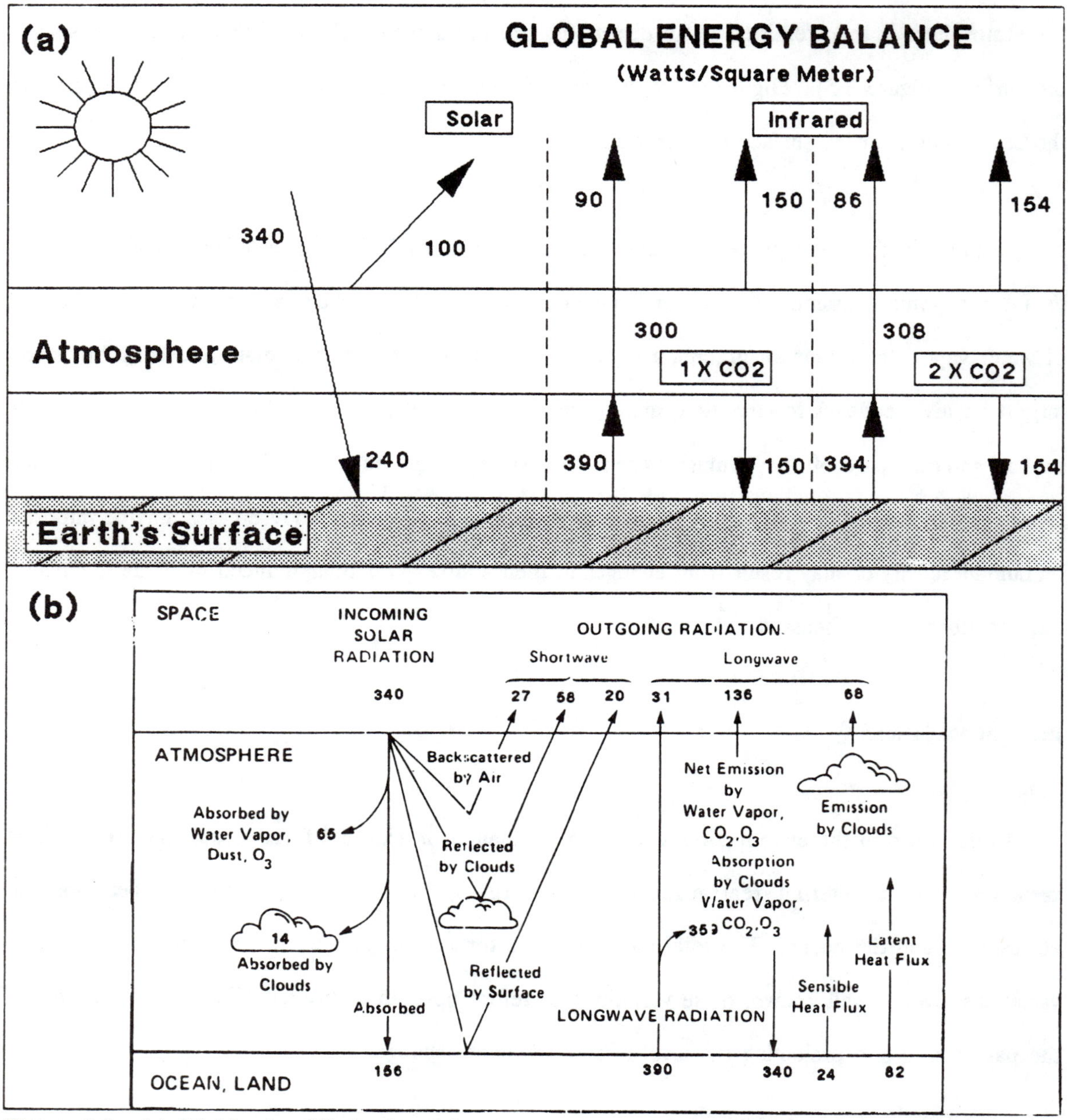

Figure 3-5. (a) Highly simplified schematic of the global energy balance illustrating the mechanism by which increased greenhouse gas concentrations warm the Earth's surface. The atmosphere is treated as a thin layer that does not absorb solar radiation; the role of convective and latent heat transfer is also neglected. Doubling the concentration of CO_2 increases the absorption and emission of infrared radiation by the atmosphere, increasing the total energy absorbed at the surface. In the equilibrium depicted, total emissions to space remain unchanged.

(b) A more realistic schematic of the global energy balance for current conditions. (Source: adapted from MacCracken, 1985.)

the initial imposed forcing. In this section, several of these mechanisms which are internal to the physical climate system are discussed. In the next section, several recently quantified mechanisms involving the planet's biology and chemistry are described.

By no means do we understand or even know about all the mechanisms involved in climate feedbacks. Figure 3-6 shows some of the physical climate feedbacks involved in changing surface temperature. Current state-of-the-art climate models attempt to incorporate most of the physical feedbacks that have been identified, but are forced, for example, to provide a very crude treatment for one of the most important -- changes in clouds -- because of inadequate understanding of cloud physics and because of the small spatial scale on which clouds form compared to the resolution of climate models.

Water Vapor - Greenhouse

When the climate warms, the atmosphere can hold more water vapor. This enhances the warming because it increases the greenhouse effect from water vapor, producing still more evaporation from the warmed surface. This positive feedback acts to approximately double imposed forcings.

Snow and Ice

When climate warms, snow and ice cover are reduced, exposing land or ocean with a lower albedo than the snow or ice. In addition, the albedo of the remaining snow and ice is reduced due to meltwater puddles and debris on the surface. This acts to absorb more energy at the surface, further enhancing the warming. This albedo feedback was originally thought to be the dominant positive feedback effect of snow and ice, but we now understand that the thermal inertia feedback

FIGURE 3-6

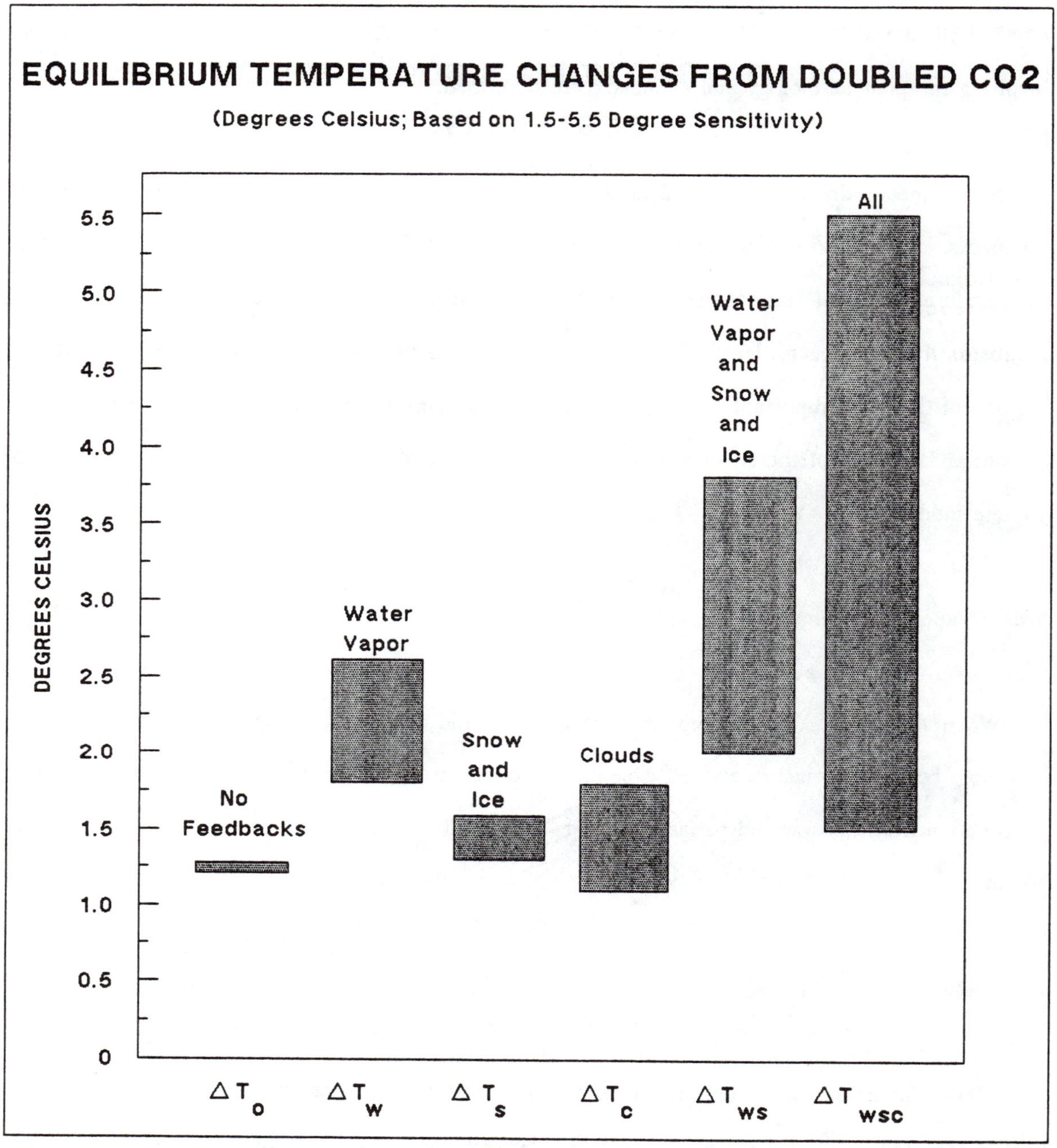

Figure 3-6. Equilibrium temperature changes from doubling CO_2 (ΔT_{2x}) inferred from a review of the strength of individual feedback processes in various climate models. ΔT_0 is the temperature increase expected from doubling CO_2 with no feedbacks. The subscripts w, s, and c, refer to feedbacks due to water vapor and lapse rate, sea ice and surface albedo, and clouds, respectively. Each bar shows the estimated two-standard deviation range of equilibrium global warming with the indicated feedbacks included. (Source: adapted from Dickinson, 1986.)

of sea ice plays a much more important role (Manabe and Stouffer, 1980; Robock, 1983). The albedo feedback requires that the sun be shining, and since the maximum ice and snow extent is in the winter, it plays a small role in influencing the albedo except in the spring, when the snow and ice are present along with high insolation.

The thermal inertia feedback acts to increase the thermal inertia of the oceans when climate warms by melting sea ice, reducing its insulating effect and increasing the transfer of heat from the ocean to the atmosphere at high latitudes. This acts to reduce the seasonal cycle of surface temperature and is the prime reason for the enhancement of imposed climate change in the polar regions in the winter (Robock, 1983). If sea ice retained its current seasonal cycle, there would be no preferential latitude or time of year for climate change.

Clouds

Clouds respond directly and immediately to changes in climate and may represent the most important uncertainty in determining the sensitivity of the climate system to the buildup of greenhouse gases. Fractional cloud cover, cloud altitude and cloud optical depth can all change when climate changes (Schlesinger, 1985). It has not been possible to calculate the net effect of cloud feedbacks because all these properties of clouds can change simultaneously, because clouds affect long-wave radiation, short-wave radiation, and precipitation (which affects soil moisture and hence albedo, thermal inertia, and moisture flux of land), and because the net effect depends on the location of the cloud (in 3 dimensions), the underlying surface albedo, and the time of day and year of the changes. The current net effect of clouds has only recently been measured (Ramanathan et al., 1989). Current climate models include crude calculations of clouds and have difficulty even reproducing the current cloud distribution.

BIOGEOCHEMICAL CLIMATE FEEDBACKS

In addition to the climatic processes discussed above, a number of biogeochemical feedback processes will influence future concentrations of greenhouse gases and climatic change. Increased greenhouse gas concentrations will alter not only the climate, but also biogeochemical processes that affect sources and sinks of radiatively important gases. Climatically important surface properties, such as albedo and evapotranspiration, will also be modified by vegetation changes. The major biogeochemical feedback links, illustrated in Figure 3-7, can be categorized as follows: physical effects of climatic change, changes in marine biology, and changes in terrestrial biology. Potential physical effects of climatic change include release of methane hydrates and changes in ocean chemistry, circulation, and mixing. Changes in marine biology may alter the pumping of carbon dioxide from the ocean surface to deeper waters and the abundance of biogenic cloud condensation nuclei. Potential biological responses on land include changes in surface albedo, increased flux of CO_2 and CH_4 from soil organic matter to the atmosphere due to higher rates of microbial activity, increased sequestering of CO_2 by the biosphere due to CO_2 fertilization, and changes in moisture flux to the atmosphere.

Release of Methane Hydrates

Potentially the most important biogeochemical feedback is the release of CH_4 from near-shore ocean sediments. Methane hydrates are formed when a methane molecule is included within a lattice of water molecules; the ratio can be as small as 1:6, that is, one methane molecule for every six water molecules (Bell, 1982). The hydrate structure is stable under temperature and pressure conditions that are typically found under a water column of a few hundred meters or more in the Arctic and closer to a thousand meters in warmer waters; the region where hydrates are found can start at the sea floor and extend up to a few hundred meters into the sediment, depending on the

FIGURE 3-7

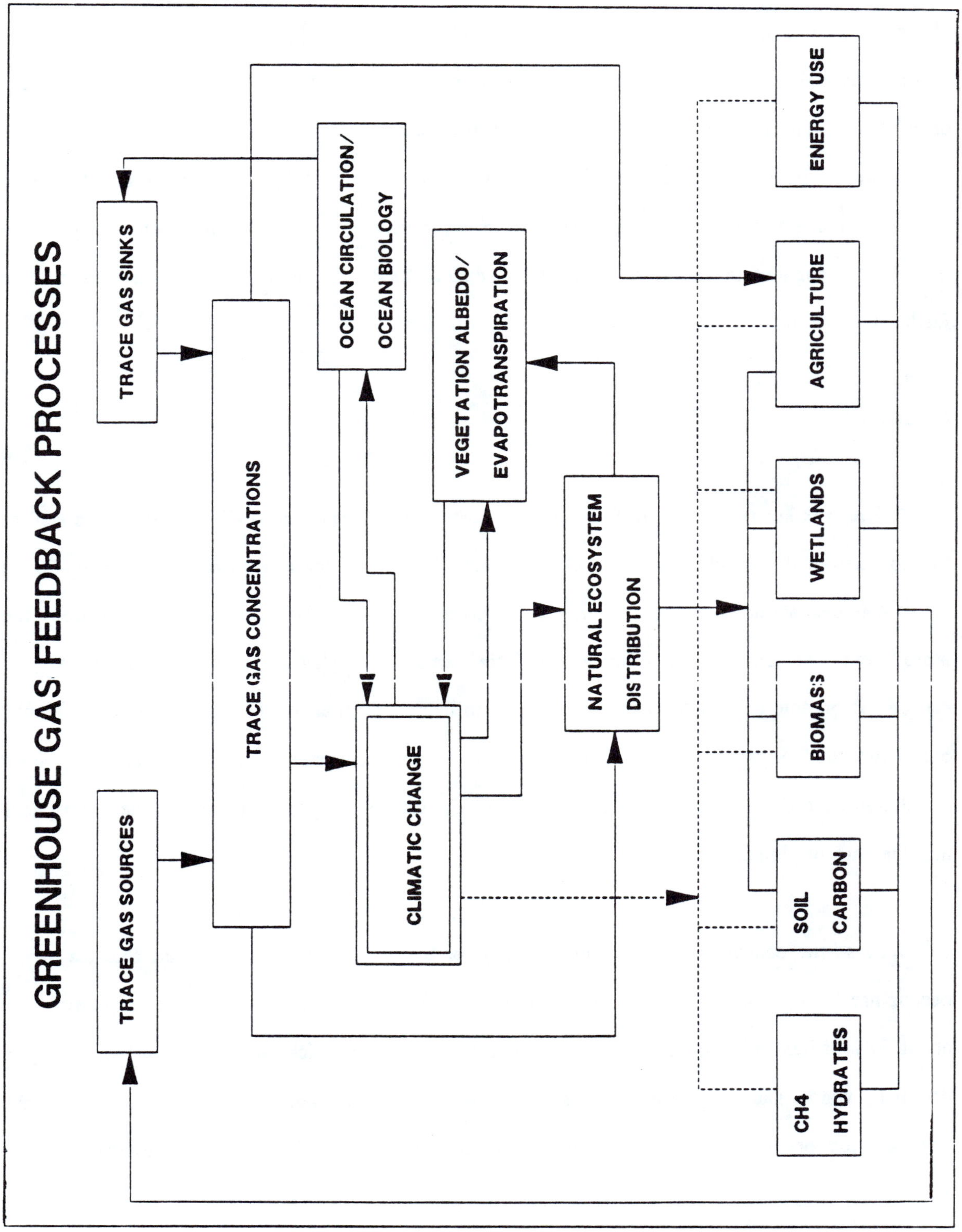
GREENHOUSE GAS FEEDBACK PROCESSES
TRACE GAS SINKS
TRACE GAS SOURCES
TRACE GAS CONCENTRATIONS
OCEAN CIRCULATION/ OCEAN BIOLOGY
VEGETATION ALBEDO/ EVAPOTRANSPIRATION
CLIMATIC CHANGE
NATURAL ECOSYSTEM DISTRIBUTION
ENERGY USE
AGRICULTURE
WETLANDS
BIOMASS
SOIL CARBON
CH4 HYDRATES

geothermal temperature gradient (Kvenvolden and Barnard, 1984). Estimates of the total quantity of CH_4 contained in hydrates range from $2x10^3$ to $5x10^6$ Pg (Kvenvolden, 1988). Given the climatic change associated with a doubling of CO_2, Bell (1982; as corrected by Revelle, 1983) estimated that there could be a release of ~120 Tg CH_4/yr from Arctic Ocean sediments, and Revelle (1983) calculated global emissions of ~640 Tg CH_4/yr from continental slope hydrates. These estimates are, however, highly uncertain both because the total quantity of hydrates potentially subject to destabilization is not known and because bottom water may be insulated from surface temperature increases throughout much of the ocean (Kvenvolden, 1988). Nonetheless, a very strong positive feedback from this source cannot be excluded at this time.

Oceanic Change

The oceans are the dominant factor in the Earth's thermal inertia to climate change as well as the dominant sink for anthropogenic CO_2 emissions. The mixed layer (approximately the top 75 m) alone contains about as much carbon (in the form of H_2CO_3, HCO_3, and $CO_3^=$) as does the atmosphere (see Chapter II). Furthermore, the ocean biota play an important role in carrying carbon (as organic debris) from the mixed layer to deeper portions of the ocean (see, for example, Sarmiento and Toggweiler, 1984). Thus, changes in ocean chemistry, biology, mixing, and large-scale circulation have the potential to substantially alter the rate of CO_2 accumulation in the atmosphere and the rate of global warming.

Because the oceans are such an integral part of the climate system, significant changes in the oceans are likely to accompany a change in climate. For example, the oceans are responsible for about 50% of heat transport from the equator toward the poles (Dickinson, 1986), surface mixing is driven by winds, and deep circulation is driven by thermal and salinity gradients. The feedbacks involving the ocean can be divided into three categories: the direct effect of temperature on carbonate chemistry, reduced mixing due to increased stability of the thermocline, and the possibility of large-scale reorganization of ocean circulation and biological activity[1].

Ocean Chemistry

The most straightforward feedback is on ocean carbonate chemistry. As the ocean warms, the solubility of CO_2 decreases and the carbonate equilibrium shifts toward carbonic acid; these effects combine to increase the partial pressure of CO_2 (pCO_2) in the ocean by 4-5%/°C for a fixed alkalinity and total carbon content. Because the total carbon content would only have to decrease by about one-tenth this amount to restore pCO_2 to its previous level, the impact of this feedback is to increase atmospheric CO_2 by about 1%/°C for a typical scenario (Lashof, 1989; Chapter VI).

Ocean Mixing

As heat penetrates from the mixed layer of the ocean into the thermocline the stratification of the ocean will increase and mixing can be expected to decrease, resulting in slower uptake of both CO_2 and heat. This feedback raises the surface temperature that can be expected in any given year for two reasons: First, the atmospheric CO_2 concentration will be higher because the oceans will take up less CO_2. Second, the realized temperature will be closer to the equilibrium temperature due to reduced heat transport into the deep ocean (see the discussion of the transient response below).

---

[1] The thermocline starts at the base of the mixed layer and extends to a depth of about 1000m. It is characterized by a rapid decrease in temperature with increasing depth, which inhibits mixing in the water column because the colder deeper water is denser than the warmer overlying water.

Ocean Biology and Circulation

A more speculative, but potentially more significant, feedback involves the possibility of large-scale changes in the circulation of the atmosphere-ocean system as suggested by Broecker (1987). This possibility is illustrated by the apparently very rapid changes in the CO_2 content of the atmosphere during glacial-interglacial transitions as revealed by ice-core measurements (e.g., Jouzel et al., 1987; Figure 3-2a). Only shifts in carbon cycling in the ocean are thought to be capable of producing such large, rapid, and sustained changes in atmospheric CO_2. A number of papers have attempted to model the changes in ocean circulation and/or biological productivity required to account for the change in pCO_2, emphasizing the importance of high-latitude processes (Kerr, 1988; Sarmiento and Toggweiler, 1984; Siegenthaler and Wenk, 1984; Knox and McElroy, 1984). Given that continuation of current trends could lead to a climate change during the next century of the same magnitude as that which occurred between glacial and interglacial periods, one must take seriously the possibility of sudden changes in ocean circulation. Should this happen, the oceans could even become a CO_2 source rather than a sink -- significantly accelerating climatic change. Such changes in circulation could also cause abrupt changes in climate, a scenario that conflicts with the general assumption that the warming will be gradual (Broecker, 1987).

A different feedback involving ocean biology has been proposed by Charlson et al. (1987). It is also uncertain, but potentially significant. Dimethyl sulfide (DMS) emitted by marine phytoplankton may act as cloud condensation nuclei in remote marine environments, affecting cloud reflectivity and therefore climate (Charlson et al., 1987; Bates et al., 1987). Climate presumably affects biogenic DMS production but the relationship is complex and poorly understood at this time (Charlson et al., 1987). While this mechanism was originally proposed as a potential negative feedback consistent with the Gaia Hypothesis (Lovelock, 1988; Lovelock and Margulis, 1973), ice-core data indicate that aerosol levels were higher during the last glacial maximum, suggesting that biogenic DMS production may act instead as a positive feedback (Legrand et al., 1988). This is only one possible cloud optical property feedback (discussed above), and the net effect cannot be determined because other cloud properties (amount, elevation) would also change in a complex way.

Changes in Terrestrial Biota

The terrestrial biota interact with climate in a wide variety of important ways (Figure 3-7). The most significant effects on climate may result from large-scale reorganization of terrestrial ecosystems as well as the direct effects of temperature and CO_2 increases on carbon storage.

Vegetation Albedo

Probably the most significant global feedback produced by the terrestrial biota, on a decades-to-centuries time scale is due to changes in surface albedo (reflectivity) as a result of changes in the distribution of terrestrial ecosystems. Changes in moisture flux patterns are probably also important. Cess (1978) argued that vegetation albedo feedback could have played a major role in explaining the glacial-interglacial temperature change. Dickinson and Hanson (1984) reanalyzed this problem and found a much smaller, but still significant effect, i.e., that the planetary albedo was 0.0022 higher at the glacial maximum due to differences in mean annual vegetation albedo. A similar result was obtained by Hansen et al. (1984) using a prescriptive scheme to relate vegetation type to climate in GCM simulations for current and glacial times. (The much larger effect found by Cess was due to differences in the albedo assigned to similar vegetation types for 18 kyBP versus the present [Dickinson and Hanson, 1984].) This feedback may be less important in the future than it was during the last deglaciation because of direct human effects on the surface, such as deforestation, and because the pattern of vegetation change will be different.

Carbon Storage

Other significant feedbacks are related to the role of the terrestrial biosphere as a source and sink for CO_2 and CH_4. The carbon stored in live biomass and soils is roughly twice the amount in atmospheric CO_2, and global net primary production (NPP) by terrestrial plants absorbs about 10% of the carbon held in the atmosphere each year. On average this is nearly balanced by decay of organic matter, about 0.5-1% of which is anaerobic and thus produces CH_4 rather than CO_2. Small shifts in the balance between NPP and respiration, and/or changes in the fraction of NPP routed to CH_4 rather than CO_2, could therefore have a substantial impact on the overall greenhouse forcing, because CH_4 has a much larger greenhouse effect than CO_2 per molecule. Both NPP and respiration rates are largely determined by climate and NPP is directly affected by the CO_2 partial pressure of the atmosphere. Thus the potential for a substantial feedback exists.

Other Terrestrial Biotic Emissions

The biosphere plays an important role in the frequency and quantity of emissions of various other atmospheric trace gases, which are also likely to be influenced by climatic change. For example, as much as half of nitrous oxide (N_2O) emissions are attributed to microbial processes in natural soils (Bolle et al., 1986). Emissions of N_2O tend to be episodic, depending strongly on the pattern of precipitation events in addition to temperature and soil properties (Sahrawat and Keeney, 1986). Thus, climatic change could be accompanied by significant changes in N_2O emissions, although there is not sufficient understanding of the microbiology to predict these changes at present. The biosphere is also a key source of atmospheric non-methane hydrocarbons (NMHCs), which play an important role in global tropospheric chemistry; the oxidation of NMHCs generates a substantial share of global carbon monoxide and therefore influences the concentration of OH and the lifetime of CH_4 (Mooney et al., 1987; Thompson and Cicerone, 1986). As much as 0.5-1% of photosynthate is lost as isoprene and terpene (Mooney et al., 1987). Lamb et al. (1987) found that the volume of biogenic NMHC emissions in the United States is greater than anthropogenic emissions by about a factor of two. The ratio for the globe is probably greater. Emissions, at least for isoprene and α-pinene are exponentially related to temperature (Lamb et al., 1987; Mooney et al., 1987). The first-order impact of climatic change, then, would be to increase NMHC emissions, producing a positive feedback through the CO-OH-CH_4 link. The actual impact when changes in ecosystem distribution are considered is uncertain, however, as different species have very different emissions (Lamb et al., 1987).

Summary

Of the feedbacks that will come into play during the next century, the physical climate feedbacks discussed earlier (water vapor, clouds, ice cover, and ice and snow albedo) will almost certainly have the greatest impact. In comparison, the potential individual impacts of the biogeochemical feedbacks discussed here are rather modest. If the physical climate feedbacks approach the strongly positive end of their ranges, then the overall sensitivity of the climate system would be substantially increased by even small additional feedbacks. However, since both the internal and biogeochemical feedbacks are presently so poorly understood, and since other feedbacks may be discovered, the overall equilibrium response of the climate system (discussed below) can only be specified with a fairly wide range.

The perturbations to global biogeochemical cycles reflected in the feedback processes discussed here are of great importance in their own right in addition to whatever warming they may produce. The vegetation albedo feedback, for example, contributed only 0.3°C out of the 3.6°C global cooling in the ice-age analysis of Hansen et al. (1984), but this represented a massive change in terrestrial ecosystems. A better assessment of both the impact of climatic change on biogeochemical cycles and the associated feedbacks is needed. Several aspects of the impact of climatic change on biogeochemical processes are discussed in the companion report *Potential Effects of Global Climate Change on the United States* (Smith and Tirpak, 1989). A quantitative estimate of the impact of some of the feedbacks discussed here is presented in Chapter VI based on incorporating them in the Atmospheric Stabilization Framework developed for this study.

EQUILIBRIUM CLIMATE SENSITIVITY

When any forcing, such as an increase in the concentration of greenhouse gases, is applied to the climate system, the climate will start to change. Since both the imposed forcings and the climatic response are time-dependent, and since the climate system has inertia due to the response times of the ocean, the exact relationship between the timing of the forcings and the timing of the response is complex. In an attempt to simplify the problem of understanding the sensitivity of the climate system to forcings, it has become a standard experiment to ask the question, "What would be the change in global average surface air temperature if the CO_2 concentration in the atmosphere were doubled from the preindustrial level, all other climate forcings were held constant, and the climate became completely adjusted to the new radiative forcing?" This quantity is called the "equilibrium climate sensitivity to doubled CO_2" and is indicated as ΔT_{2X} (see Box 3-1).

The actual path that the climate system would take to approach the equilibrium climate would be determined by the time scales of the forcings and the various elements of the climate system. This is called the "transient response" and is discussed in the next section. Because the climate system response always lags the forcing, there will always be a built-in unrealized warming that will occur in the future, even if no more forcing occurs. Thus, there is certain to be some future climate response to greenhouse gases that were put into the atmosphere in the past, even if no more are put in starting today. Another way of saying this is that societal responses to the greenhouse problem

BOX 3-1. SIMPLIFIED MODELING FRAMEWORK

The concepts discussed in this chapter can be summarized in a simple zero-dimensional or one-box model of climate as discussed by Dickinson (1986):

(1) $$C(d\Delta T/dt) + \lambda\Delta T = \Delta Q$$

where ΔQ is the climate forcing and could be due to changes in solar output, volcanoes, surface properties, stochastic processes or greenhouse gases (as discussed under Climate Forcings and Feedbacks); ΔT is the change in tropospheric/mixed-layer temperature from the preindustrial equilibrium climate; the factor λ, called the "feedback parameter" by Dickinson, gives the change in upward energy flux resulting from a change in surface temperature, ΔT, and is the net result of all the climate feedbacks (as discussed in the section on Climate Sensitivity); t is time; and C is the effective heat capacity of the Earth, which is determined by the rate of heat uptake by the ocean (C must be a function of time to account for the gradual penetration of heat into an increasing volume of the deep ocean and changing sea ice cover). In equilibrium the first term in (1) is zero, so the equilibrium climate sensitivity is simply given by

(2) $$\Delta T = \Delta Q/\lambda.$$

For a doubling of CO_2, ΔQ is about 4.3 W/m^2, so the range 1.5-5.5°C of ΔT_{2x} discussed above corresponds to a range of 2.9-0.8 W m^{-2} $°C^{-1}$ in λ. This conceptual model, with ΔQ calculated from changes in greenhouse gases and C replaced by a diffusive model of the ocean, is incorporated into the Integrating Framework used in the modeling exercises for this report (see Chapter V and Appendix A).

that are undertaken now will be felt for decades in the future, and lack of action now will similarly bequeath climate change to future generations.

Analysis of past climate change, and model calculations of future climate change can both be used to determine ΔT_{2X}. Unfortunately, our knowledge of both past climate change and the responsible forcings are too poor to reliably determine ΔT_{2X} from past data. Wigley and Raper (1987) estimate that if all of the warming of the past 100 years was due to greenhouse gases, then ΔT_{2X} would be approximately 2°C. If however, one allows for other possible forcings, natural variability, uncertainties in ocean heat uptake and the transient response, and for uncertainties in preindustrial greenhouse gas concentrations (see below; Hansen et al., 1985; Wigley and Schlesinger, 1985; Wigley et al., 1986), then the climate record of the last 100 years is consistent with any ΔT_{2X} between 0 and 6°C (Wigley, personal communication).

Due to the various problems with direct empirical approaches, mathematical models of the climate system are the primary tool for estimating climate sensitivity. While they have inherent errors, they can isolate the greenhouse forcing, and many theoretical calculations can be made to test the importance of various assumptions and various proposed feedback mechanisms. The simplest climate model is the zero-dimensional global average model described in the box below. Models that are one-dimensional in the vertical, often called "radiative-convective" models, and that are one-dimensional in the horizontal, often called "energy-balance" models, are very useful for quickly and inexpensively testing various components of the climate system. In order to calculate the location of future climate change, however, and in order to incorporate all the important physical interactions, especially with atmospheric circulation, fully three-dimensional general circulation models (GCMs) are necessary. These sophisticated models solve simultaneous equations for the conservation of energy, momentum, mass, and the equation of state on grids with horizontal resolution ranging from 3 to 8 degrees of latitude by 3 to 10 degrees of longitude and with varying vertical resolution. The radiation schemes attempt to account for the radiatively significant gases, aerosols and clouds. They generally use different schemes for computing cloud height, cover and optical properties. The models also differ in their treatment of ground hydrology, sea ice, surface albedo, and diurnal and seasonal cycles (Schlesinger and Mitchell, 1988). Perhaps the most important differences lie in the treatment of oceans, ranging from prescribed sea surface temperatures, to "swamp" oceans with mixed layer thermal capacity but no heat transport, to mixed layers with specified heat transport, to full oceanic GCMs.

A series of reviews by the National Academy of Sciences (NAS, 1979, 1983, 1987) as well as the "State-of-the-Art" report of the Department of Energy (MacCracken and Luther, 1985) have concluded that the equilibrium sensitivity of climate to a $2xCO_2$ forcing (ΔT_{2X}) is probably in the range of 1.5 to 4.5°C. An independent review by Dickinson (1986) attempts to quantitatively combine the uncertainties indicated by the range of recent GCM results and concludes that the range should be broadened to 1.5-5.5°C. The GCM result of Wilson and Mitchell (1987) giving ΔT_{2X} = 5.2°C was published after all of the reviews cited here. Dickinson's estimates of the contributions of the individual factors to climate sensitivity are shown in Figure 3-6. The largest positive feedback is from changes in the amount and distribution of water vapor. Substantial positive feedback may also be contributed by changes in sea ice and surface albedo and clouds, although the uncertainty range includes the possibility that clouds contribute significant negative feedback. The differences in the strength of these feedbacks between models is the result of different parameterizations of the relevant processes as well as differences in the control ($1xCO_2$) simulation (Cess and Potter, 1988). Even though the exact value of ΔT_{2X} is not known, we can study the potential impact of climatic warming caused by greenhouse gases by choosing scenarios that span the range of theoretical calculations. Thus, we adopt 2-4°C as a putative one standard deviation (1σ) confidence interval about the center of the range proposed by the National Academy of Sciences, and the range proposed by Dickinson (1.5 to 5.5°C) as 2σ bounds for subsequent modeling (Chapter V). When the biogeochemical feedbacks discussed above are also considered, a ΔT_{2X} as great as 8-10°C cannot be ruled out (Lashof, 1989).

THE RATE OF CLIMATIC CHANGE

The Earth's surface does not immediately come to an equilibrium following an increase in radiative forcing. Excess radiation captured by the Earth heats the land surface, the ocean, and the atmosphere. The effective heat capacity of the oceanic part of the climate system, in particular, is enormous. The result is that the warming realized in any given year may be substantially less than the warming that would occur in equilibrium if greenhouse gas concentrations were fixed at their levels in that year. Hundreds of years would be required for the entire ocean to equilibrate with the atmosphere, but only the surface layer (about 100 m) is well mixed by winds and therefore tightly linked to climate in the short term. The heat capacity of the surface layer is about 1/40th that of the entire ocean and this layer by itself would equilibrate with a response time (the time required to reach 1 - 1/e, or 63%, of the equilibrium response) of 2-15 years, depending on the climate sensitivity and assumed mixed layer depth. The equilibration time is longer if the climate sensitivity is greater because the feedback processes that increase climate sensitivity respond to the realized changes in climate, not to the initial change in forcing (Hansen et al., 1985). When the transfer of heat from the mixed layer into the deep ocean is considered, it is impossible to characterize the oceanic response with a single time constant (Harvey and Schneider, 1985; Wigley and Schlesinger, 1985).

While the main features of ocean circulation and mixing, and therefore the rate of heat and carbon uptake, have been identified, they are not well defined or modeled on a global scale. The theory and modeling of ocean circulation are currently limited by the inadequacy of the database (Woods, 1985). The development of Ocean General Circulation Models (OGCMs) lags significantly behind their atmospheric counterparts, mainly because it is difficult and expensive to obtain the necessary data with sufficient temporal and spatial coverage, because fewer scientists have addressed this problem, and because a large amount of computer power is needed to resolve the necessary time and space scales. Due to these problems it may be a decade or more before OGCMs reach the state of development achieved by current atmospheric GCMs.

Lacking well-tested OGCMs, the main tools used so far to investigate ocean heat uptake have been highly parameterized models, very similar to those used for carbon (see Chapter II). These models are calibrated with data on the penetration of tracers such as tritium and ^{14}C produced by atmospheric nuclear weapons tests during the 1950s and early '60s and/or with the steady-state profiles of various ocean parameters, such as natural ^{14}C and temperature. The simplest models that yield a plausible time-dependence for heat (and carbon) uptake lump the entire ocean into two compartments: A well-mixed surface layer and a deep ocean compartment in which mixing is parameterized as a diffusive process (Box-Diffusion or BD model). This approach was introduced by Oeschger et al. (1975) for modeling carbon uptake, and has been applied to ocean heat absorption by Hansen et al. (1985) and Wigley and Schlesinger (1985), among others. A more elaborate version of this model which includes a representation of upwelling implicitly balanced by high-latitude bottom-water formation (Upwelling-Diffusion or UD model), has been used by Hoffert et al. (1980), Harvey and Schneider (1985), and Wigley and Raper (1987). The addition of an upwelling term allows the observed mean thermal structure of the ocean to be approximated (Hoffert et al., 1980), but given the highly parameterized nature of both of these models, there is no convincing reason to favor one approach over the other for modeling small perturbations to heat flows.

The response time, τ, of Box-Diffusion models is proportional to $\kappa(\Delta T_{2x})^2$, where κ is the diffusion constant used to characterize deep ocean mixing (Hansen et al., 1985; Wigley and Schlesinger, 1985). Data on the penetration of tracers into the ocean suggests that κ = 1-2 cm^2/s (Hansen et al., 1985). Hoffert and Flannery (1985) have argued that mixing rates derived from tracers may be too high for heat because mixing rates are highest along constant density surfaces, which are nearly parallel to ocean isotherms. On the other hand, in a preliminary coupled GCM-OGCM run, Bryan and Manabe (1985) found that heat was taken up more rapidly than with a passive tracer because of reduced upward heat convection. Using a range of 0.5-2 cm^2/s for κ and the 1σ range for ΔT_{2x} given above (2-4°C) in the equation derived by Wigley and Schlesinger (with their recommended values for other parameters) yields τ = 6-95 years.[2] Correspondingly, the warming expected by now, based on past increases in greenhouse gases and assuming no other climate forcings, is roughly 40-80% of the equilibrium warming (Wigley and Schlesinger, 1985). In other words, even if greenhouse gas concentrations could be fixed at today's level, the Earth would still be subject to significant climatic change that has yet to materialize. The large uncertainty surrounding ocean heat uptake, combined with uncertainty about potential climate forcings other than those from greenhouse gases, also implies that it is not possible to obtain a useful constraint on ΔT_{2x} from the observed temperature record as discussed above (see also Hansen et al., 1985; Wigley and Schlesinger, 1985).

Experiments with Upwelling-Diffusion models demonstrate the importance of the bottom water formation process for the rate of ocean heat uptake. The impact of using an Upwelling-Diffusion ocean model rather than a Box-Diffusion ocean model is that the heat that diffuses into the thermocline is pushed back toward the mixed layer, which decreases the effective heat capacity of the ocean and the time constant for tropospheric temperature adjustment, *assuming that the upwelling rate and the temperature at which bottom water is formed do not change*. If the initial temperature of the downwelling water is assumed to warm as much as the mixed layer, however, then a UD model actually takes up more heat in the ocean than a BD model, leading to a larger disequilibrium between a given radiative forcing scenario and the expected realized warming. While there are reasons to think that the temperature of Antarctic Bottom Water will not increase as climate changes, the temperature of North Atlantic Deep Water could increase or decrease (Harvey and Schneider, 1985). Furthermore, there is no reason to assume that the rate of bottom-water formation will remain constant as climate changes. The tropospheric temperature could even overshoot equilibrium

[2] It is important to note that the actual response does not correspond to exponential decay with a single time constant, so that while τ gives the time required for one e-folding and is a useful measure, it would not apply to subsequent e-foldings (the time constant would be substantially longer) and must be interpreted with care.

if the average bottom-water temperature cools as the surface temperature warms or if the upwelling rate increases with warming (Harvey and Schneider, 1985). One must also recognize the potential for sudden reorganizations of the ocean-atmosphere circulation system as suggested by Broecker (1987), which could lead to discontinuous, and perhaps unpredictable, changes in climate that cannot be included in the models used in this report.

Another major limitation of the BD and UD models generally used to analyze the climate transient problem is that they have limited or no spatial resolution (at best hemispheric, land-sea) and thus cannot consider spatial heterogeneity in either the magnitude or rate of climatic change. Work at the Goddard Institute for Space Studies (Hansen et al., 1988) has produced one of the few three-dimensional time-dependent analyses of climatic change that have been published to date. This study employed three simple, but reasonably realistic, scenarios of future greenhouse gas concentrations and volcanic eruptions. The results suggest that the areas where warming is initially most prominent relative to interannual variability are not necessarily those where the equilibrium warming is greatest. For example, low-latitude ocean regions warm quickly because ocean heat uptake is limited by strong stratification in these regions. Warming is also prominent in high-latitude ocean areas where a large equilibrium warming is expected due to increased thermal inertia as sea ice melts. Global average temperatures are used in this report as an indicator of the rate and magnitude of global change but, as these results emphasize, it must be recognized that major variations among regions are a certainty.

CONCLUSION

The changing composition of the atmosphere will in turn drive significant changes in the Earth's climate. These changes may have already begun, but because of the uncertainties in temperature data sets and the complexity of the interaction between climate sensitivity and the transient response, definitive predictions are subject to a good deal of controversy at this time. *Whether next year is warmer or cooler than this year, however, has no direct bearing on how the greenhouse effect should be viewed.* Internal fluctuations or countervailing forcings may temporarily mask the warming due to increasing concentrations of greenhouse gases or make the climate warmer than expected solely from greenhouse warming. Therefore, to derive our estimates of the magnitude and rate of change that can be expected during the next century we must continue to rely on model calculations, which indicate that by early in the next century the Earth could be warmer than at any time during the last million years or more, and that the rate of change could be unprecedented in Earth history.

REFERENCES

Barnola, J.M., D. Raynaud, Y.S. Korotkevich, and C. Lorius. 1987. Vostok ice core provides 160,000-year record of atmospheric CO_2. *Nature* 329:408-414.

Bates, T., R. Charlson, and R. Gammon. 1987. Evidence for the climatic role of marine biogenic sulphur. *Nature* 329:319-321.

Bell, P. 1982. Methane hydrate and the carbon dioxide question. In W. Clark, ed. *Carbon Dioxide Review: 1982.* Oxford University Press, New York. 255-277.

Berger, W.H. 1982. Climate steps in ocean history-lessons from the Pleistocene. In Berger, W.H., and J.C. Crowell, Panel Co-chairmen. *Climate in Earth History.* National Academy Press, Washington, D.C. 43-54.

Berner, R. A., and A. C. Lasaga. 1988. Modeling atmospheric O_2 and CO_2 over Phanerozoic time. Paper presented at the Chapman Conference on Gaia Hypothesis, San Diego. March 7-11, 1988.

Bolle, H., W. Seiler, and B. Bolin. 1986. Other greenhouse gases and aerosols: Assessing their role for atmospheric radiative transfer. In Bolin, B., B. Doos, J. Jager, and R. Warrick, eds., *The Greenhouse Effect, Climatic Change, and Ecosystems.* Scope 29. John Wiley and Sons, Chichester. 157-203.

Broecker, W. 1987. Unpleasant surprises in the greenhouse? *Nature* 328:123-126.

Bryan, K., and S. Manabe. 1985. A coupled ocean-atmosphere and the response to increasing atmospheric CO_2. In J. Nihoul, ed. *Coupled Ocean-Atmosphere Models.* Elsevier, Amsterdam. 1-6.

Cess, R., 1978. Biosphere-albedo feedback and climate modeling. *Journal of the Atmospheric Sciences* 35:1765-1769.

Cess, R., and G. Potter. 1988. A methodology for understanding and intercomparing atmospheric climate feedback processes in General Circulation Models. *Journal of Geophysical Research* 93:8305-8314.

Charlson, R., J. Lovelock, M. Andreae, and S. Warren. 1987. Oceanic phytoplankton, atmospheric sulphur, cloud albedo and climate. *Nature* 326:655-661.

COHMAP Members. 1988. Climatic changes of the last 18,000 years: Observations and model simulations. *Science* 241:1043-1052.

Dickinson, R.E. 1984. Modeling evapotranspiration for three-dimensional global climate models. In Hansen, J.E., and T. Takahashi, eds. *Climate Processes and Climate Sensitivity.* Geophysical Monograph 29, Maurice Ewing Volume 5. American Geophysical Union, Washington, D.C. 58-72.

Dickinson, R.E. 1986. The climate system and modelling of future climate. In B. Bolin, B. Doos, J. Jager, and R. Warrick, eds. *The Greenhouse Effect, Climatic Change, and Ecosystems.* Scope 29. John Wiley & Sons, Chichester. 207-270.

Dickinson, R.E., and B. Hanson. 1984. Vegetation-albedo feedbacks. In Hansen, J., and T. Takahashi, eds. *Climate Processes and Climate Sensitivity*, Geophysical Monograph 29, Maurice Ewing Volume 5. American Geophysical Union, Washington, D.C. 180-186.

Emiliani, C. 1978. The cause of the ice ages. *Earth and Planetary Science Letters* 37:349-352.

Genthon, C., J.M. Barnola, D. Raynaud, C. Lorius, J. Jouzel, N.I. Barkov, V.S. Korotkevich, and V.M. Kotlyakov. 1987. Vostok ice core: climatic response to CO_2 and orbital forcing changes over the last climatic cycle. *Nature* 329:414-419.

Hansen, J., I. Fung, A. Lacis, S. Lebedeff, D. Rind, R. Ruedy, G. Russell, and P. Stone. 1988. Global climate changes as forecast by Goddard Institute for Space Studies Three-Dimensional Model. *Journal of Geophysical Research* 93:9341-9364.

Hansen, J.E., Johnson, D., Lacis, A., Lebedeff, S., Lee, P., Rind, D., and Russell, G. 1981. Climatic impact of increasing atmospheric carbon dioxide. *Science*, **213**, 957-966.

Hansen, J., A. Lacis, D. Rind, G. Russell, P. Stone, I. Fung, R. Ruedy, and J. Lerner. 1984. Climate sensitivity: Analysis of feedback mechanisms. In Hansen, J., and T. Takahashi, eds. *Climate Processes and Climate Sensitivity*. Geophysical Monograph 29, Maurice Ewing Volume 5. American Geophysical Union, Washington, D.C. 130-163.

Hansen, J., and S. Lebedeff. 1988. Global surface air temperatures: Update through 1987. *Geophysical Research Letters* 15:323-326.

Hansen, J., G. Russell, A. Lacis, I. Fung, D. Rind. 1985. Climate response times: Dependence on climate sensitivity and ocean mixing. *Science* 229:857-859.

Hansen, J.E., W. Wong, and A.A. Lacis. 1978. Mount Agung eruption provides test of a global climatic perturbation. *Science* 199:1065-1068.

Harvey, L., and S. Schneider. 1985. Transient climate response to external forcing on 100-104 year time scales Part 1: Experiments with globally averaged, coupled, atmosphere and ocean energy balance models. *Journal of Geophysical Research* 90:2191-2205.

Hasselmann, K. 1976. Stochastic climate models, I. Theory. *Tellus* 18:473-485.

Hoffert, M., A. Callegari, and C. Hsieh. 1980. The role of deep sea heat storage in the secular response to climatic forcing. *Journal of Geophysical Research* 85:6667-6679.

Hoffert, M., and B. Flannery. 1985. Model projections of the time-dependent response to increasing carbon dioxide. In MacCracken, M., and F. Luther, eds. *Projecting the Climatic Effects of Increasing Carbon Dioxide*. U.S. Department of Energy, Washington, D.C. (DOE/ER-0237). 149-190.

Jones, P.D. 1988. Hemispheric surface air temperature variations: recent trends and an update to 1987. *Climate Journal*, in press.

Jones, P.D., T.M.L. Wigley, C.K. Folland, D.E. Parker, J.K. Angell, S. Lebedeff, and J.E. Hansen. 1988. Evidence for global warming in the past decade. *Nature* 332:790.

Jouzel, J., C. Lorius, J.R. Petit, C. Genthon, N.I. Barkov, V.M. Kotlyakov, and V.M. Petrov. 1987. Vostok ice core: A continuous isotope temperature record over the last climatic cycle (160,000 years). *Nature* 329:403-408.

Kerr, R. 1988. La Niña's big chill replaces El Niño. *Science* 241:1037-1038.

Knox, F. and M. B. McElroy. 1984. Changes in atmospheric CO_2: Influence of the marine biota at high latitude. *Journal of Geophysical Research* 89:4629-4637.

Kvenvolden, K. 1988. Methane hydrates and global climate. *Global Biogeochemical* Cycles 2:221-229.

Kvenvolden, K., and L. Barnard. 1984. Hydrates of natural gas in continental margins. *American Association of Petroleum Geologists Memoir* 34:631-640.

Lamb, B., A. Guenther, D. Gay, and H. Westberg. 1987. A national inventory of biogenic hydrocarbon emissions. *Atmospheric Environment* 21:1695-1705.

Lamb, H.H. 1977. *Climatic History and the Future*. Princeton University Press, Princeton. 835 pp.

Lashof, D. 1989. The dynamic greenhouse: Feedback processes that may influence future concentrations of greenhouse gases and climate. *Climatic Change*. In press.

Legrand, M.R., R.J. Delmas, and R.J. Charlson. 1988. Climate forcing implications from Vostok ice core data. *Nature* 334:418-420.

Lorenz, E.N. 1968. Climatic determinism. *Meteorological Monographs* 30:1-30.

Lovelock, J.E. 1988. *The Ages of Gaia: A Biography of our living earth*. W. W. Norton & Co., New York. 252 pp.

Lovelock, J.E., and L. Margulis. 1973. Atmospheric homeostasis by and for the biosphere: The Gaia Hypothesis. *Tellus* 23:2-10.

MacCracken, M.C. 1985. Carbon dioxide and climate change: Background and overview. In MacCracken, M.C., and F.M. Luther, eds. *Projecting the Climatic Effects of Increasing Carbon Dioxide*. U.S. Department of Energy, Washington, D.C. (DOE/ER-0237).

MacCracken, M.C., and F.M. Luther, eds. 1985. *Projecting the Climatic Effects of Increasing Carbon Dioxide*. U.S. Department of Energy, Washington, D.C. (DOE/ER-0237). 381 pp.

Manabe, S., and R. Stouffer. 1980. Sensitivity of a global climate model to an increase of CO_2 concentration in the atmosphere. *Journal of Geophysical Research* 85:5529-5554.

Mooney, H., P. Vitousek, and P. Matson. 1987. Exchange of material between terrestrial ecosystems and the atmosphere. *Science* 238:926-932.

NAS (National Academy of Sciences). 1979. *Carbon Dioxide and Climate: A Scientific Assessment*. National Academy Press, Washington, D.C.

NAS (National Academy of Sciences). 1983. *Changing Climate*. National Academy Press, Washington, D.C. 496 pp.

NAS (National Academy of Sciences). 1987. *Current Issues in Atmospheric Change*. National Academy Press, Washington, D.C.

Oeschger, H., U. Siegenthaler, U. Schotterer, and A. Gugelmann. 1975. A box diffusion model to study the carbon dioxide exchange in nature. *Tellus* 27:168-192.

Ramanathan, V., R.J. Cicerone, H.B. Singh, and J.T. Kiehl. 1985. Trace gas trends and their potential role in climate change. *Journal of Geophysical Research* 90:5547-5566.

Ramanathan, V., L. Callis, R. Cess, J. Hansen, I. Isaksen, W. Kuhn, A. Lacis, F. Luther, J. Mahlman, R. Reck, and M. Schlesinger. 1987. Climate-chemical interactions and effects of changing atmospheric trace gases. *Reviews of Geophysics* 25:1441-1482.

Ramanathan, V., R.D. Cess, E.F. Harrison, P. Minnis, B.R. Barkstrom, E. Ahmad, and D. Hartmann. 1989. Cloud-radiative forcing and climate: Results from the Earth Radiation Budget Experiment. *Science* 243:57-63.

Revelle, R. 1983. Methane hydrates in continental slope sediments and increasing atmospheric carbon dioxide. In National Academy of Sciences, *Changing Climate*. National Academy Press, Washington, D.C. 252-261.

Robock, A. 1978. Internally and externally caused climate change. *Journal of Atmospheric Science* 35:1111-1122.

Robock, A. 1979. The "Little Ice Age": Northern Hemisphere average observations and model calculations. *Science* 206:1402-1404.

Robock, A. 1981. A latitudinally dependent volcanic dust veil index, and its effect on climatic simulations. *Journal of Volcanological and Geothermal Research* 11:67-80.

Robock, A. 1983. Ice and snow feedbacks and the latitudinal and seasonal distribution of climate sensitivity. *Journal of Atmospheric Science* 40:986-997.

Robock, A., 1984. Climate model simulations of the effects of the El Chichon eruption. *Geofisica Internacional* 23:403-414.

Sahrawat, K., and Keeney. 1986. Nitrous oxide emission from soils. *Advances in Soil Science* 4:103-148.

Sarmiento, J., and J. Toggweiler. 1984. A new model for the role of the oceans in determining atmospheric pCO_2. *Nature* 308:621-624.

Schlesinger, M.E. 1985. Analysis of results from energy balance and radiative-convective models. In MacCracken, M.C., and F.M. Luther, eds. *Projecting the Climatic Effects of Increasing Carbon Dioxide*. U.S. Department of Energy, Washington, D.C. (DOE/ER-0237). 280-319.

Schlesinger, M., and J. Mitchell. 1985. Model projections of the equilibrium climatic response to increased Carbon Dioxide. In MacCraken, M.C., and F.M. Luther, eds. *Projecting the Climatic Effects of Increasing Carbon Dioxide*. U.S. Department of Energy, Washington, D.C. (DOE/ER-0237). 81-147.

Sellers, P.J., Y. Mintz, Y.C. Sud, and A. Dalcher. 1986. A simple biosphere model (SiB) for use within general circulation models. *Journal of Atmospheric Science* 43:505-531.

Siegenthaler, U., and T. Wenk. 1984. Rapid atmospheric CO_2 variations and ocean circulation. *Nature* 308:624-626.

Smith, J., and D. Tirpak. 1989. *Potential Effects of Global Climate Change on the United States*. U.S. Environmental Protection Agency, Washington, D.C., in press.

Thompson, A., and R. Cicerone. 1986. Possible perturbations to atmospheric CO, CH_4, and OH. *Journal of Geophysical Research* 91:10853-10864.

Vinnikov, K. Ya., P. Ya. Groisman, K.M. Lugina, and A.A. Golubev. 1987. Variations in Northern Hemisphere mean surface air temperature over 1841-1985. *Meteorology and Hydrology*. No. 1:45-55 [In Russian]

Vinnikov, K. Ya., Gruza, G.V., Zakharov, V.F., Kirillov, A.A., Kovyneva, N.P., and Ran'kova, E. Ya. 1980. Current climatic changes in the Northern Hemisphere. *Meteorologiya i Gidrologiya 1983*, no. 6, 5-17 (in Russian).

Walker, J.C. G., P. B. Hays and J.F. Kasting. 1981. A negative feedback mechanism for the long-term stabilization of Earth's surface temperature. *Journal of Geophysical Research* 86:9776-9782.

Wigley, T.M.L., P. D. Jones, and P. M. Kelly. 1986. Empirical climate studies. In Bolin, B., B. Doos, J. Jager, and R. Warrick, eds. *The Greenhouse Effect, Climatic Change, and Ecosystems*. Scope 29. John Wiley & Sons, Chichester. 271-322.

Wigley, T.M.L., and S.C.B. Raper. 1987. Thermal expansion of sea water associated with global warming. *Nature* 330:127-131.

Wigley, T.M.L., and M.E. Schlesinger. 1985. Analytical solution for the effect of increasing CO_2 on global mean temperature. *Nature* 315:649-652.

Willson, R. C., and H. S. Hudson. 1988. Solar luminosity variations in solar cycle 21. *Nature* 332:810-812.

Willson, R. C., H. S. Hudson, C. Frohlich, and R. W. Brusa. 1986. Long-term downward trend in total solar irradiance. *Science* 234:1114-1117.

Wilson, C.A., and J.F.B. Mitchell. 1987. A $2xCO_2$ climate sensitivity experiment with a global climate model including a simple ocean. *Journal of Geophysical Research* 92:13315-13343.

Woods, J.D. 1985. The world ocean circulation experiment. *Nature* 314:501-511.

CHAPTER IV

HUMAN ACTIVITIES AFFECTING TRACE GASES AND CLIMATE

FINDINGS

Various human activities affect the Earth's climate by altering the level of trace gases in the atmosphere. These activities include energy consumption, particularly fossil-fuel consumption; industrial processes; land use change, particularly deforestation; and agricultural practices such as waste burning, fertilizer usage, rice production, and animal husbandry. Economic development and population growth are key factors affecting the level of each activity.

- Population levels and growth rates have increased tremendously over the last 200 years. Between 1650 and 1980, the global population doubling time shrunk from 200 to 35 years. At the beginning of this century, global population was about 1.6 billion; in 1987, it reached 5 billion. By the early part of the next century total population is likely to reach 8 billion. The rate of population increase is most acute in the developing regions, particularly Africa and Asia where annual rates of growth exceed 2%.

- Fossil fuel combustion emits carbon dioxide and other radiatively important gases and is the primary cause of atmospheric warming. Energy consumption currently accounts for more than five of the six to eight billion tons of carbon dioxide emitted to the atmosphere annually from anthropogenic sources. Between 1950 and 1986, annual global fossil fuel consumption grew 3.6-fold and annual carbon dioxide emissions grew 3.4-fold.

- Emissions of other trace gases due to fossil fuel consumption are more uncertain. Approximately 0 to 2 million tons nitrogen as nitrous oxide, 20 million tons nitrogen as nitrogen oxides, and 180 million tons carbon as carbon monoxide are emitted annually from fossil fuel combustion. Leaking and venting of natural gas contributes approximately 20 to

50 million tons methane annually to the atmosphere, and coal mining contributes approximately 25 to 45 million tons methane.

- Three significant non-energy sources of greenhouse gases are associated with industrial activity: Production and use of chlorofluorocarbons (CFCs), halons, and chlorocarbons; waste disposal in landfills; and cement manufacture. Production of CFC-11 and CFC-12 grew 4.7-fold between 1960 and 1985. Consumption of major CFCs and halons reached nearly one million tons in 1985. An international agreement (the Montreal Protocol), however, came into force on January 1, 1989 to reduce future production of certain CFCs and halons. Anaerobic decay of organic wastes in landfills currently contributes approximately 30-70 million tons of methane to the atmosphere annually. Cement production, which has increased sevenfold since the 1950s, contributed approximately 134 million tons carbon as CO_2 to the atmosphere in 1985.

- Land use change has resulted in substantial emissions of greenhouse gases to the atmosphere. Since 1850, approximately 15% of the world's forests have been converted to agricultural and other land uses. Currently, deforestation contributes between one-tenth and one-third of the total anthropogenic carbon dioxide emissions to the atmosphere, i.e., between 0.4 to 2.6 billion tons of carbon. Between one-quarter and one-half of the world's swamps and marshes also have been destroyed by man. Wetlands currently contribute approximately one-fifth of the total methane emissions to the atmosphere; continued changes to wetlands could significantly alter the global methane budget. Biomass burning, in addition to contributing to the atmospheric concentrations of carbon dioxide, contributes approximately 10 to 20% of total annual methane emissions, 5 to 15% of the nitrous oxide emissions, 10 to 35% of the nitrogen oxide emissions, and 20 to 40% of the carbon monoxide emissions.

- Three agricultural activities directly result in major contributions to atmospheric emissions of greenhouse gases: animal husbandry, rice cultivation, and nitrogenous fertilizer use. Domestic animals, which produce methane as a by-product of enteric fermentation, currently contribute approximately 65 to 85 million tons of methane annually. Over the past several decades, domestic animal populations have grown by up to 2% annually. Methane is also produced by anaerobic decomposition in rice paddies. Currently, about one-fifth of annual methane emissions, or between 60 and 170 million tons, comes from rice cultivation. Rice production has grown rapidly since the mid-1900s due to increases in crop acreage, double cropping, and higher yields. Between 1950 and 1984 rice production increased nearly threefold, and harvested area grew by about 70%. Use of nitrogenous fertilizers results in nitrous oxide emissions, either directly from the soil, or indirectly from groundwater. Global use of organic and inorganic fertilizers has risen markedly, and nitrogen-based fertilizers increased their market share of total inorganic fertilizer consumption from 28% in 1950 to 64% in 1981. Nitrogenous fertilizer use may contribute between 0.14 and 2.4 million tons nitrogen as nitrous oxide per year to the atmosphere.

- In addition to the human activities that directly affect trace gas emissions, future concentrations of greenhouse gases will be influenced by feedback processes resulting from humans living in a world that has undergone climatic change. Two potential feedbacks of increased temperatures, which may counteract each other to some extent, are increased energy demand for air conditioning in the summer, and decreased energy demand for heating in the winter.

INTRODUCTION

As discussed in Chapter III, the Earth's climate has been in a constant state of change throughout geologic time due to natural perturbations in the global geobiosphere. However, various human activities have the potential to cause future global warming over a relatively short amount of time. These activities, which affect the Earth's climate by altering the concentrations of trace gases in the atmosphere, include energy consumption, particularly fossil-fuel consumption; industrial processes (production and use of chlorofluorocarbons, halons, and chlorocarbons, landfilling of wastes, and cement manufacture); changes in land use patterns, particularly deforestation and biomass burning; and agricultural practices (waste burning, fertilizer usage, rice production, and animal husbandry). Population growth is an important underlying factor affecting the level of growth in each activity.

This chapter describes how the human activities listed above contribute to atmospheric change, the current pattern of each activity, and how levels of each activity have changed since the early part of this century. Figure 4-1 illustrates the current contributions to the greenhouse gas buildup by region. Almost 50% of the warming is attributable to activities in the United States, the USSR, and the European Economic Community. As background to the discussion of trace-gas producing activities, we first provide an overview of population trends. This historical perspective is meant to serve as a framework for the discussion of possible future scenarios of trace-gas emissions in Chapter V.

HISTORICAL OVERVIEW OF POPULATION TRENDS

One of the major factors affecting trends in greenhouse gas emissions is the increase in human population. As population levels rise, increasing pressures are placed on the environment as the larger population strives to feed and clothe itself and achieve a higher standard of living. Without

FIGURE 4-1

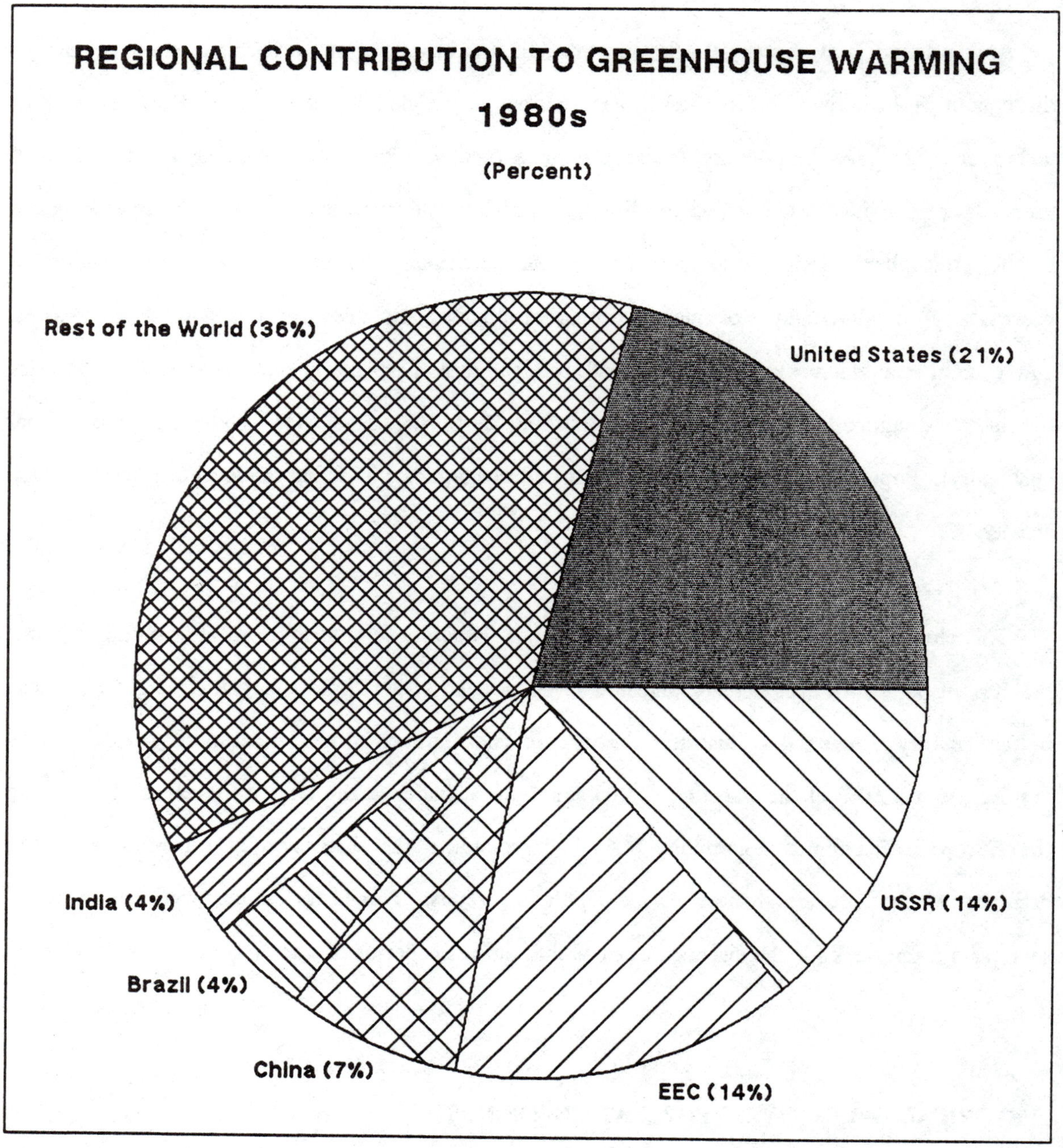

Figure 4-1. Estimated regional contribution to greenhouse warming for the 1980s, based upon regional shares of current levels of human activities that contribute to greenhouse gas emissions. (Sources: U.S. EPA, 1988a; United Nations, 1987; U.S. BOM, 1985; IRRI, 1986; FAO, 1986a, 1987; Bolle et al., 1986; Rotty, 1987; Lerner et al., 1988; Seiler, 1984; WMO, 1985; Hansen et al., 1988; Houghton et al., 1987; Matthews and Fung, 1987.)

changes in the methods used to meet people's needs, higher population levels invariably lead to increased emissions of greenhouse gases.

Global Population Trends

Not only has global population grown rapidly over the past few centuries, but the rate of growth has also increased (see Figure 4-2). World population in the year 1 A.D., approximately 0.25 billion, doubled by 1650 (Wagner, 1971). By 1850 (i.e., 200 years later), global population had roughly doubled again to 1.1 billion. The global population doubling time has continued to decline -- 80 years later, in 1930, world population was 2 billion. By 1975 global population had reached 4 billion, and according to some estimates the population will double once again within 35 years (world population reached 5 billion in 1987). Moreover, despite recent declines in the world's annual population growth rate (IIED and WRI, 1987), world population is expected to continue to grow rapidly. Several studies estimate that world population will exceed 8 billion by 2025 (Zachariah and Vu, 1988; U.S. Bureau of the Census, 1987). Such rapid population growth can be expected to result in increasing pressure on the global environment, particularly as the burgeoning human population strives to improve its living standards through economic growth.

Population Trends by Region

The rapid population growth in recent decades has not occurred uniformly around the world (see Figure 4-2). Between 1950 and 1985, population in developed countries increased by 41%, compared to 117% in developing countries (IIED and WRI, 1987). Recent trends indicate these differences will continue: Annual growth rates in the developed countries are generally less than 1%, while many developing countries continue to experience rates of growth between 2 and 3% (see Table 4-1). These higher growth rates in the 20th century in developing countries have been due primarily to the combined effects of declining death rates and continued high birth rates.

FIGURE 4-2

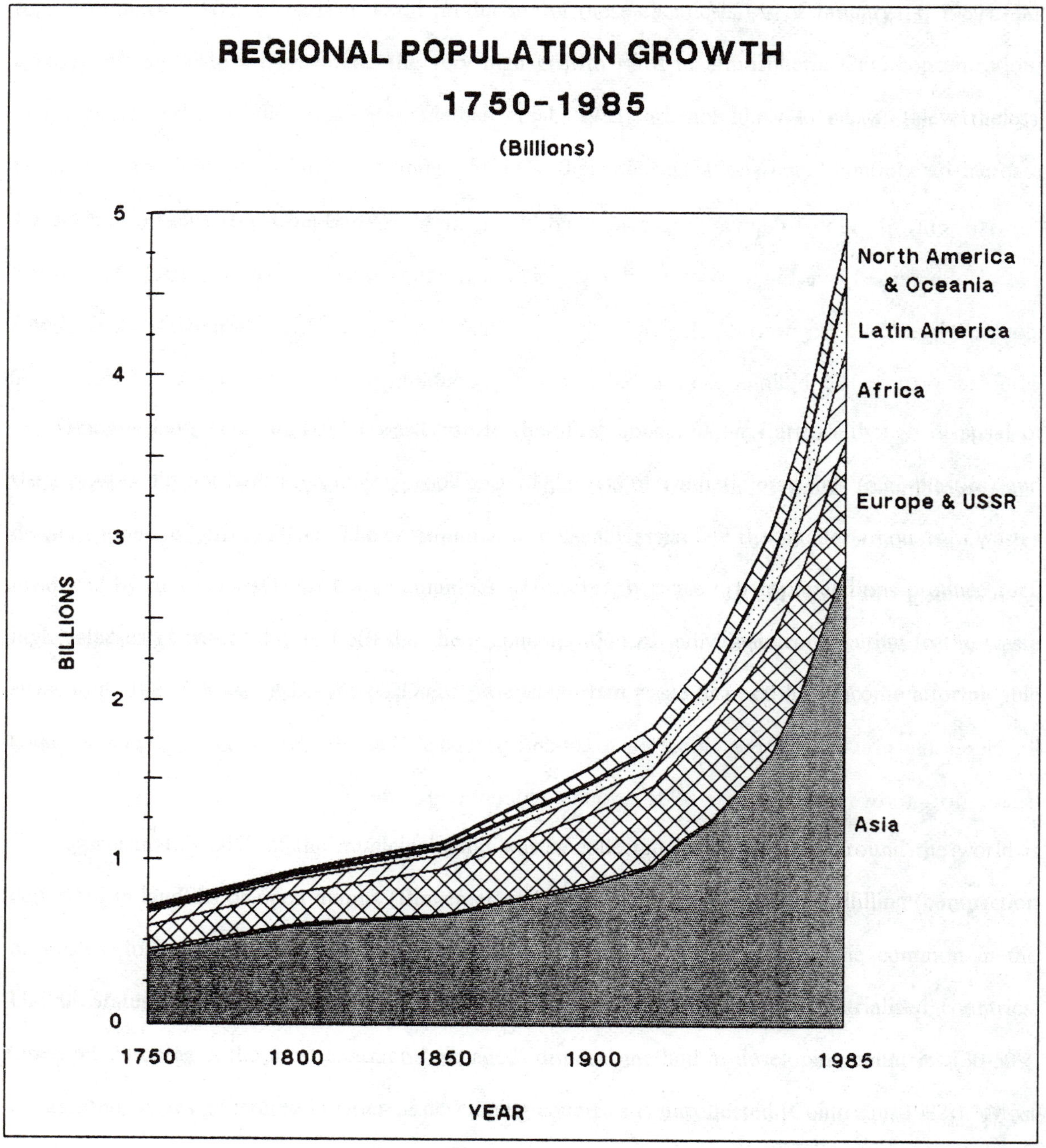

Figure 4-2. Since about 1850, global population has grown at increasingly rapid rates. In 1850, the population doubling time was approximately 200 years; by 1975, the doubling time had declined to approximately 45 years. Most of the growth has occurred in the developing world, particularly Asia. (Sources: Matras, 1973; Hoffman, 1987.)

TABLE 4-1

Regional Demographic Indicators

Region	Total Fertility[a] 1980-85	Infant Mortality[b] 1980-85	Annual Population Growth Rate 1980-85 (percent)
North America	1.83	11	0.90
Europe	1.88	15	0.30
East Asia	2.34	36	1.22
Oceania	2.65	31	1.51
Caribbean	3.34	65	1.53
Southeastern Asia	4.11	73	2.05
Latin America	4.17	61	2.34
Southern Asia	4.72	115	2.14
Western Asia	5.22	81	2.79
Africa	6.34	112	2.92
World Average	3.52	60	1.67

[a] The total fertility rate is the average number of children that a woman bears in a lifetime.

[b] The infant mortality rate is the average number of infant deaths (deaths before the first birthday) per 1,000 live births.

Source: Adapted from IIED and WRI, 1987.

Industrialized Countries

Population growth rates in the industrialized countries are substantially lower compared with growth rates in the developing world. For example, while most developing countries contend with growth rates that will double their populations within 20-40 years, current growth rates in North America and Europe will lead to a doubling within about 100 years and 250 years, respectively (IIED and WRI, 1987). This trend toward lower growth rates is due to many complex economic and social factors, including the changing role of women in the labor force, the higher economic costs of child rearing, and the reduced need for children as a labor pool.

Developing Countries

The highest rates of population growth are in the developing countries: From 1950 to 1985 developing countries increased their share of the world's population from 66.8% to 75.6% (IIED and WRI, 1987). During this time Asia's population grew from 1.3 to 2.7 billion, Africa's from 224 to 555 million, and Latin America from 165 to 405 million. Key trends are summarized below.

Africa. Africa currently has the highest fertility rates and population growth rates in the world. Its growth rate has increased recently: Between 1955 and 1985, Africa's average annual growth rate increased from 2.3% to 2.9%. The total fertility rate (i.e., average number of children that a woman bears in a lifetime) is six or higher in 38 African countries, most of which have experienced declining infant mortality rates (infant deaths per thousand live births) over the past 20 years (IIED and WRI, 1987). For example, in Kenya, where the total fertility rate is 7.8, the infant mortality rate fell from 112 to 91 between 1965 and 1985. Between 1965 and 1985, the crude birth rate (births per thousand population) for Kenya grew by 4.7%, while the crude death rate fell by 37.7%. The average annual growth rate reached 4.1% in the 1980s (World Bank, 1987). The United Nations expects the African population to continue to grow rapidly, with the average annual growth rate increasing to 3% in 1990 (United Nations, 1986).

Asia. From 1850 to 1950, Asia experienced the largest increase in population in the world (Ehrlich and Ehrlich, 1972). Rates of growth have continued at high levels -- annual growth rates since 1960 have exceeded 2%. These rates are likely to remain high in several Asian countries in future years (United Nations, 1986). For example, China currently is the most populous country in the world, with 22% of the world's total (Ignatius, 1988). Although its strong population policy of one child per family helped to halve the 2% annual growth rates of the 1960s, growth rates have recently turned upward, approaching 1.5% annually. This trend of growth could lead to population levels in China in excess of 1.7 billion by 2025.

India's population has also been rapidly expanding. It is the second most populous country in the world (United Nations, 1986), with 765 million people as of 1985. India's rate of growth has been relatively high this century, although it has declined in recent years; in 1960 its annual rate of growth was 2.3%, but has since dropped to 1.7% (IIED and WRI, 1987). Despite this recent decline, its population is expected to grow for many years; e.g., the United Nations estimates that India's population will be over 1.2 billion by 2025 (United Nations, 1986).

Latin America. Latin America currently has one of the highest population growth rates in the world: From 1980 to 1985 the annual rate of growth averaged 2.3% for the region (IIED and WRI, 1987), although these rates of growth varied substantially between countries. Argentina, Chile, and Uruguay have the lowest growth within Latin America, while countries such as Bolivia, Ecuador, El Salvador, Guatemala, Honduras, Nicaragua, Paraguay, and Venezuela have annual population growth rates that exceed 2.5%. Fertility rates have been declining throughout the region due to industrialization, urbanization, rising incomes, and official population policies, although one source estimates that Latin America's share of world population will nonetheless increase from 8.4 to 9.5% between 1985 and 2025 (IIED and WRI, 1987). The two population projection sources used in this report (U.S. Bureau of the Census, 1987, and Zachariah and Vu, 1988; see Chapter V) project that by 2025, Latin America's share of world population will grow to 9.1 and 8.7%, respectively.

ENERGY CONSUMPTION

The major human activity affecting trace-gas emissions is the consumption of energy, particularly energy from carbon-based fossil fuels. As discussed in Chapter II, global carbon dioxide (CO_2) emissions from anthropogenic sources currently range from 6 to 8 petagrams (Pg) of carbon (C) annually, with commercial energy consumption accounting for approximately 65-85% of this total.[1] Non-commercial (biomass) energy consumption accounts for approximately 7%. Energy consumption and production also produce substantial amounts of other greenhouse gases, including carbon monoxide (CO), methane (CH_4), nitrogen oxides (NO_x, or NO and NO_2), and nitrous oxide (N_2O).[2]

This section explores the role of energy consumption in climate change. We first discuss the world's increasing reliance on fossil fuels, the roles that fossil-fuel production (e.g., coal mining and oil drilling) and fossil-fuel combustion play in the emission of trace gases to the atmosphere, and the implications of the continuation of current energy consumption patterns on future global warming.

---

[1] Anthropogenic sources of trace gases are those resulting from human activities, e.g., combustion of fossil fuels. These sources are distinguished from natural sources, since emissions from anthropogenic sources result in unbalanced trace-gas budgets and accumulation of gases in the atmosphere.

[2] Throughout the report these gases are often referred to as greenhouse gases, although strictly speaking, CO and NO_x are not greenhouse gases since they do not directly affect radiative forcing (see Chapter II). These two gases indirectly affect global warming due to their chemical interactions with other gases in the troposphere. As a result, for simplicity, we shall refer to them as greenhouse gases.

History of Fossil-Fuel Use

Prior to the discovery and development of fossil fuels (coal, oil, and natural gas), people relied on readily-available energy resources such as wood and other forms of biomass (i.e., living matter), as well as water and wind, to satisfy their basic energy needs. Since the beginning of the 19th century, fossil fuels have played an increasingly important role in the world economy, particularly for developed countries, by providing the energy required for industrial development, residential and commercial heating, cooling, and lighting, and transportation services. Fossil fuels now provide about 85% of the world's total energy requirements. This dependence on fossil fuels is greatest in industrialized countries, where over 95% of all energy needs are provided by fossil fuels, compared with about 55% in developing countries (Hall et al., 1982).[3]

Global consumption of fossil fuels has increased rapidly over the past century as human populations and their economic activities have grown, resulting in the development of additional fossil-fuel resources. Since 1950, global primary energy consumption has increased nearly fourfold (Figure 4-3), with energy consumption per capita approximately doubling. In 1985, 42% of global energy demand was supplied by liquid fossil fuels (primarily petroleum); solid fuels (coal) supplied 31%, natural gas, 22%, and other fuels combined accounted for 5% of the market share.[4] These relative proportions have changed considerably since 1950, when coal supplied 59% of total commercial energy requirements, liquids, 30%, natural gas, 9%, and other fuels, 2%.

The increase in fossil fuel consumption over the last century has caused a substantial increase in the amount of CO_2 emitted to the atmosphere. Carbon dioxide emissions from fossil fuels grew

---

[3] In some developing countries, the dependence on biomass can approach 95 percent of total energy requirements.

[4] Non-commercial biomass estimates are not included in these figures.

FIGURE 4-3

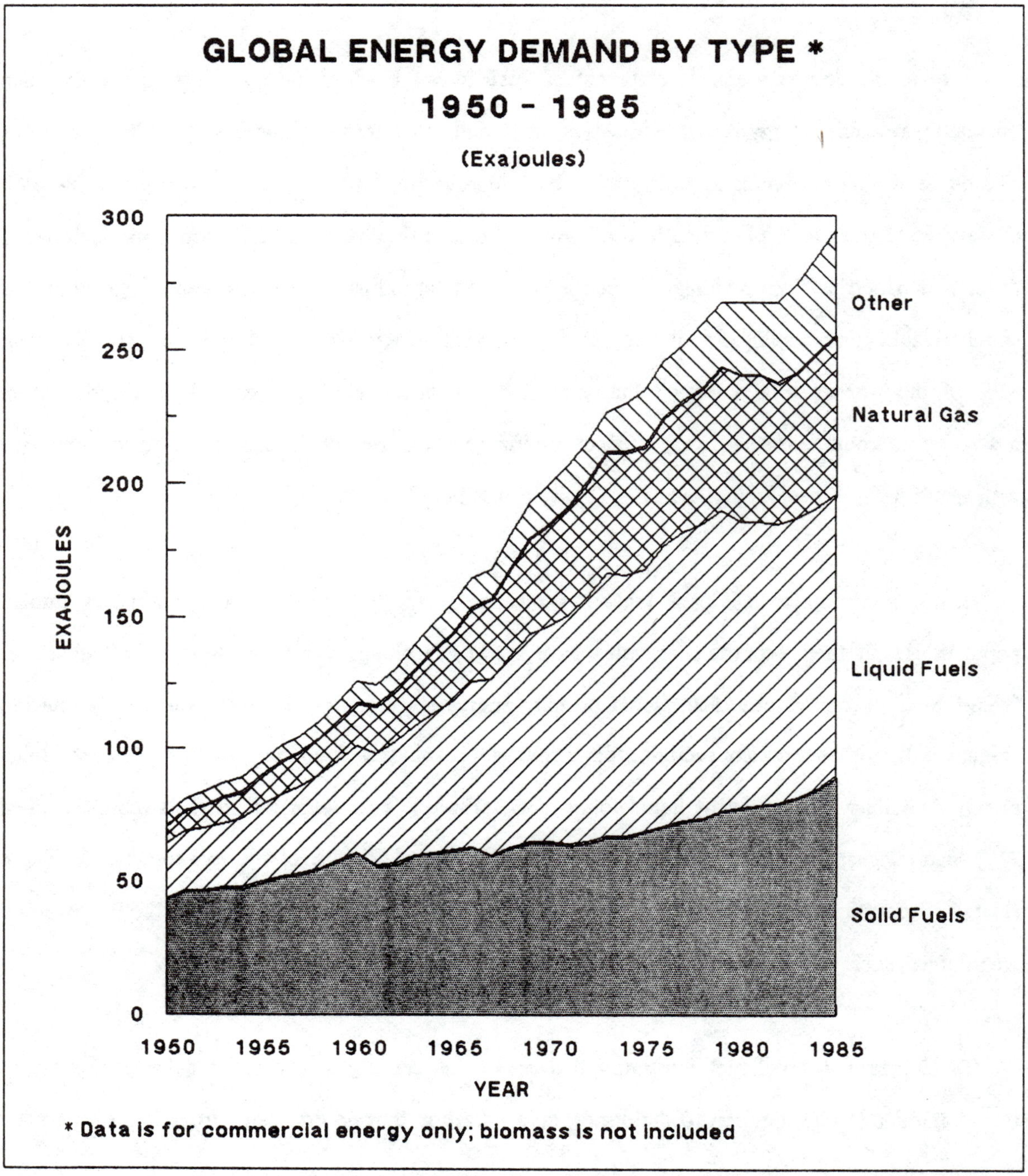

Figure 4-3. Global demand for fossil fuels has more than tripled since 1950. Today, about 85% of the world's energy needs are met by fossil fuels. (Sources: United Nations, 1976, 1982, 1983, 1987.)

from less than 0.1 Pg C annually in the mid-nineteenth century, to about 5.4 Pg C in 1986 (see Figure 4-4).[5] This rate of increase is about 3.6% per year and is the major reason why atmospheric CO_2 concentrations increased from about 290 ppm in 1860 to about 348 ppm as of 1987 (Farman et al., 1985). Currently fossil fuel combustion also contributes approximately 0-2 Tg N as N_2O, 20 Tg N as NO_x and 180 Tg C as CO annually to the atmosphere.

In recent decades there has also been a significant shift in global energy use patterns. In 1950, countries belonging to the Organization for Economic Cooperation and Development (OECD) consumed about three fourths of all commercial energy supplies, the centrally-planned economies of Europe and Asia, 19%, and developing countries, 6% (United Nations, 1976, 1983).[6] By 1985 OECD countries consumed just over one-half of all commercial energy globally, while the European and Asian centrally-planned economies and the developing countries had increased their relative shares to 32% and 15%, respectively (see Figure 4-5). Between 1950 and 1985, commercial energy use per capita in the OECD grew from 93 to 189 gigajoules per capita (GJ/cap) (103%), in centrally-planned economies from 16 to 59 GJ/cap (269%), and in the developing countries from 3 to 18 GJ/cap (500%).[7] The proportion of energy consumed by the OECD is expected to decline further as the developing world continues to experience population growth and economic development and, thus, significantly expands their energy requirements (Chapter V).

---

[5] In 1986 CO_2 emissions from fossil fuels were approximately 5370 million metric tons C, or 5.37 Pg C. 1 billion metric tons = 1 gigaton = 1 Pg = 10^{15} grams.

[6] The OECD countries include the U.S., Canada, Japan, Australia, New Zealand, the United Kingdom, France, Spain, Portugal, the German Federal Republic, Belgium, the Netherlands, Sweden, Norway, Finland, Italy, Ireland, Iceland, Denmark, Austria and Switzerland.

[7] 1 GJ = 1 gigajoule = 10^9 joules. 1055 joules = 1 Btu.

FIGURE 4-4

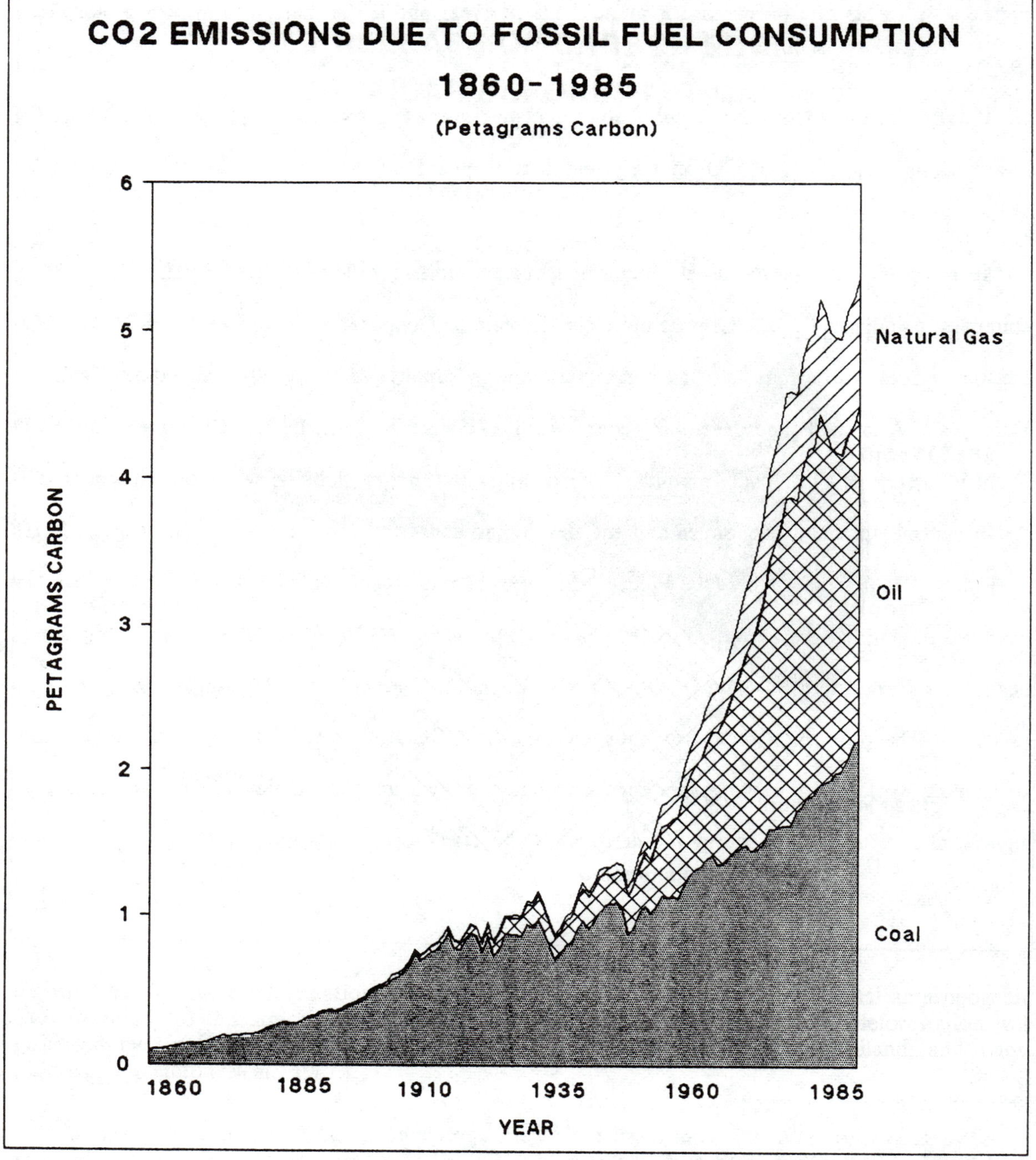

Figure 4-4. Carbon dioxide emissions from fossil fuel combustion have grown from less than 0.1 Pg C in the mid-1850s to approximately 5.4 Pg C in 1986. This is the major reason why the atmospheric concentration of CO_2 increased from approximately 290 ppm in 1860 to approximately 348 ppm in 1987. (Sources: Rotty and Masters, 1985; Rotty, 1987, pers. communication.)

FIGURE 4-5

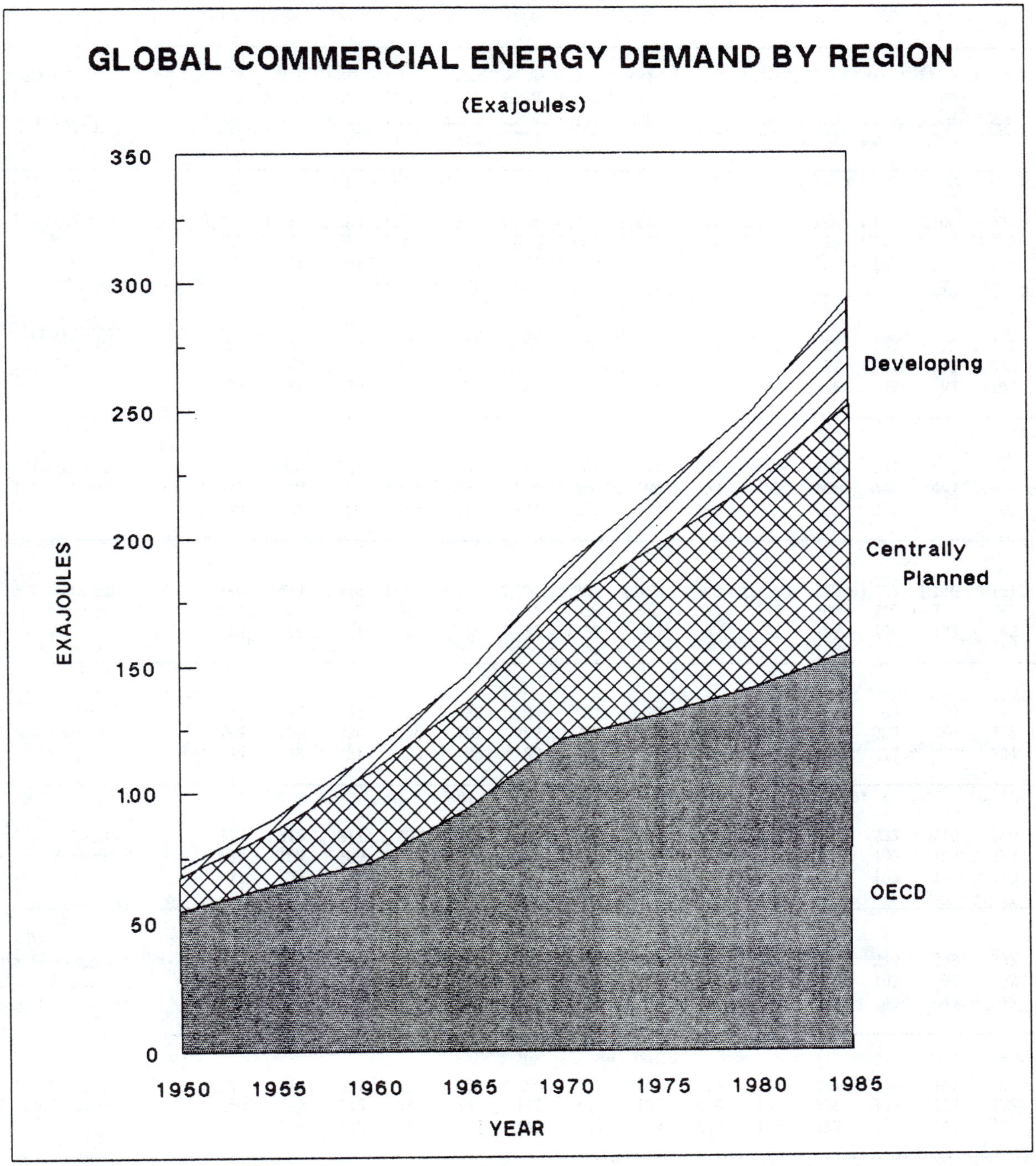

Figure 4-5. Primary energy use by region. Between 1950 and 1985, the share of global energy demand for the OECD declined, while that for the centrally planned and developing economies increased. (Sources: United Nations, 1982, 1987.)

Current Energy Use Patterns and Greenhouse Gas Emissions

The allocation of energy consumption among sectors varies considerably from one region to the next. Figure 4-6 summarizes 1985 end-use energy demand (for both commercial and non-commercial, or biomass, fuels) by sector for the OECD countries, the centrally-planned economies of Asia and Europe (including China and the USSR), and the developing countries. Whereas the OECD split is approximately one-third industrial, one-third transportation, and one-third residential/commercial, centrally-planned economies of Asia and Europe consume more than 50% of their energy in the industrial sector.

These energy consumption patterns partly reflect the basic differences in the structure of economic activity at the current stage of each region's economic development. The centrally-planned economies and the developing countries devote a greater share of their energy requirements to the industrial sector because they are at a stage of economic development where energy-intensive basic industries account for a large share of total output, while infrastructure in the transportation and commercial sectors has not been extensively developed. In the OECD, transportation consumes a larger share of total energy compared with other regions, primarily because of the large number of automobiles in the OECD. For example, in the U.S. there are 550 cars and light trucks/1000 people, compared with 60 cars and light trucks/1000 people in the USSR, and 6 cars and light trucks/1000 people in China. Also, biomass is very important to the residential energy requirements of the developing economies compared with those of industrialized countries; the industrial sector is the major consumer of fossil fuels in most developing countries.

FIGURE 4-6

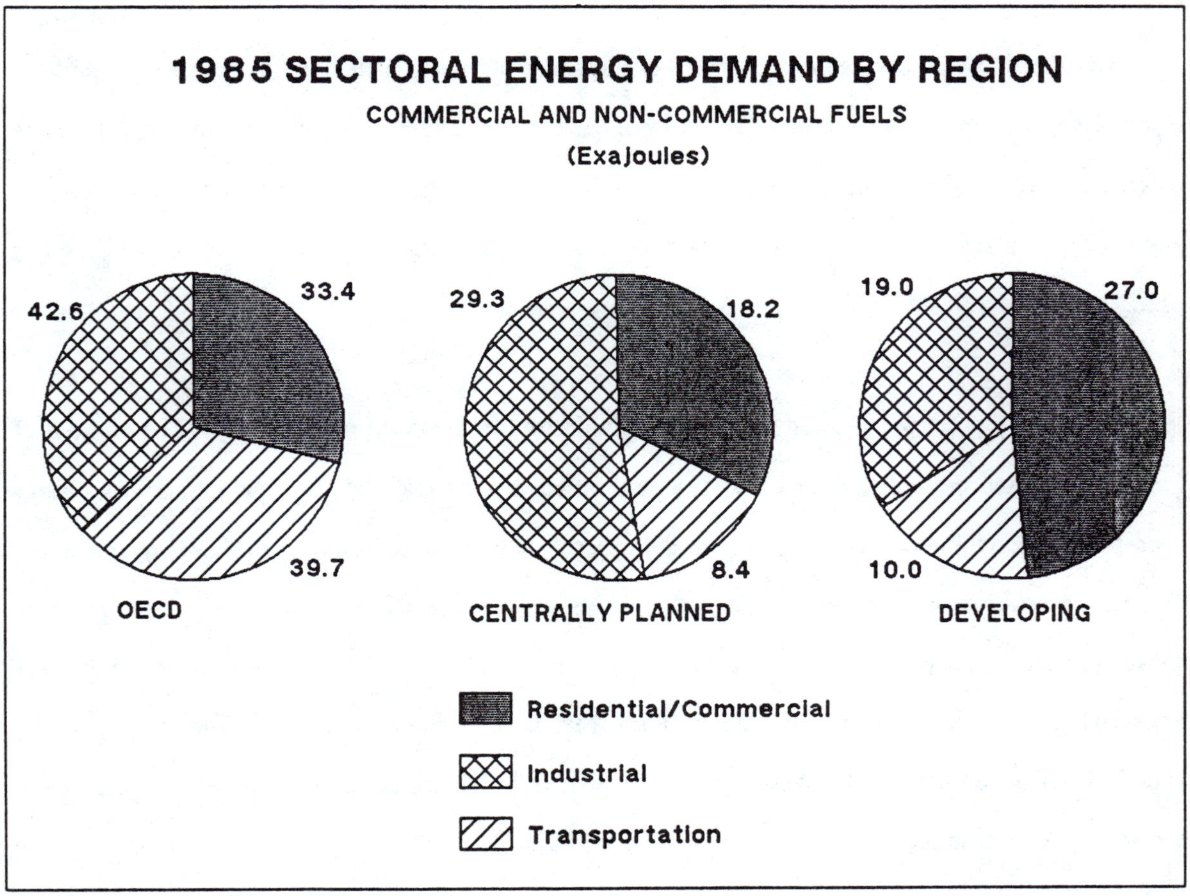

Figure 4-6. End-use energy demand by sector for three global regions. While energy demand in the OECD countries is split almost equally between the three sectors, over 50% of the energy in the centrally planned countries is consumed by the industrial sector, and almost 50% of the energy in the developing countries is consumed by the residential/commercial sector. (Sources: Sathaye et al., 1988; Mintzer, 1988.)

Emissions by Sector

The differences among regions in terms of the share of energy consumed by each sector and the types of applications for which the energy within each sector is used can have a major impact on the amount and types of greenhouse gases emitted. This section discusses how emissions of greenhouse gases vary as a result of differences in type of fossil energy consumed and combustion technology used.

Electric Utility Sector. Energy is increasingly desired in the form of electricity. The amount of greenhouse gases produced from electricity generation is a function of the type of primary energy used to produce the electricity and the production technology. For example, nuclear, hydroelectric, or solar primary energy sources emit little or no greenhouse gases, while fossil fuels generate substantial quantities of CO_2, as well as other gases (see Table 4-2). The amount of greenhouse gas emissions varies according to the type of fossil fuel used because of inherent differences in the chemical structure of the fuels. Additionally, the level of emissions varies as a function of production efficiency. For example -

- Coal-fired power plants produce about two to three times as much CO_2 as natural gas-fired units per unit of electricity generated (330 kg CO_2/GJ for a pulverized coal wall-fired unit compared with 120 kg CO_2/GJ for a combined cycle gas-fired unit). Oil-fired units produce more CO_2 than natural gas units produce, but less CO_2 than coal-fired units produce. Within fuel types the emission levels may vary. For example, when natural gas is used as the fuel, combined cycle or ISTIG units produce about 40% lower CO_2 emissions than simple cycle units (see Chapter VII) because of the greater generating efficiency obtained through the use of these technologies.

TABLE 4-2

Emission Rate Differences by Sector (grams per gigajoule)[a]

Source	Efficiency (%)	CO_2	CO	CH_4	N_2O	NO_x	Total Carbon
Electric Utility (g/GJ delivered electricity)							
Gas Turbine Comb. Cycle	42.0	120,300	70	13	20	400	32,850
Gas Turbine Simp. Cycle	26.4	191,400	110	20	30	640	52,060
Residual Oil Boilers	32.4	230,000	43	2.2	44	590	62,750
Coal - F. Bed Comb. Cycle	35.0	290,000	NA	1.8	40	690	79,090
Coal - PC Wall Fired	31.3	330,000	42	2.0	45	1,400	90,020
Coal - PC Cyclone	31.3	330,000	42	2.0	45	2,600	90,020
Coal - Integrated Gas	27.3	253,600	222	NA	51	760	69,260
Industrial (g/GJ delivered steam for boilers; energy output for others)							
Boilers							
Coal-Fired	80	130,000	110	2.9	18	390	35,510
Residual Oil-Fired	85	88,000	17	3.3	16	180	24,010
Natural Gas-Fired	85	57,000	18	1.5	3.5	71	15,550
Kilns - Coal	65-75	300-350,000	75	1	2	500	81-95,490
Dryer - Natural Gas	30-65	75-170,000	10	1	NA	52	20-46,370
Dryer - Oil	30-65	100-240,000	15	1	NA	160	27-65,460
Dryer - Coal	30-65	155-340,000	170	1	NA	215	42-92,800
Residential/Commercial (g/GJ energy output)							
Wood Stoves	50	[150,000]	17,600	70	NA	190	48,500
Coal Stoves	50	198,000	3,400	NA	NA	170	55,460
Distillate Oil Furnaces	75	111,000	17	7	NA	65	30,290
Gas Heaters	70	101,000	13	1	NA	61	27,550
Wood Boilers	67.5	[138,000]	280	21	6	47	37,770
Gas Boilers	80.9	61,800	10.6	1.4	2.7	53	16,860
Residual Oil Boilers	84.9	86,000	19	1.8	14	183	23,460
Coal Boilers	75.9	135,000	244	13	16	295	36,930
Transportation (g/GJ energy input)							
Rail	NA	69,900	570	13	NA	1,640	19,320
Jet Aircraft	NA	72,800	120	2	NA	290	19,910
Ships	NA	70,000	320	20	NA	830	19,240
Light Duty Gasoline Vehicle	NA	54,900	10,400	36	0.5	400	19,460
Light Duty Diesel Vehicle	NA	73,750	340	2	20	300	20,260
Light Duty Compressed N. Gas Vehicle	NA	50,200	4	120	7	140	13,780

a All emission rates, except for the total carbon estimates in the last column, are based on total molecular weight. Total carbon estimates refer only to the total amount of carbon emitted.

NA = Not Available
[ ] = No Net CO_2 if based on sustainable yield

Source: Radian, 1988; except N_2O data, which is based on unpublished EPA data. N_2O emission coefficients are highly uncertain and currently undergoing further testing and review.

Similarly, coal-fired fluidized bed units produce less NO_x emissions than do coal-fired cyclone units because the higher operating temperatures typical of cyclone-units are more conducive to NO_x formation.

Industrial Sector. The industrial sector includes mining, construction, and manufacturing, which are some of the most energy-intensive economic activities. Energy consumption in this sector can be subdivided into four categories:

- Boilers -- Boilers produce steam for many different purposes, including machine drive, on-site electricity production, high-pressure cleaning, and process requirements. Virtually any fuel can be consumed to produce steam (e.g., fossil fuels, biomass, hazardous wastes, by-product wastes, etc.). In the U.S., boilers consume about 30% of all industrial energy.

- Process Heat -- Many industrial processes that do not use steam require the use of some form of heat during production. Examples of process heat applications include ovens, furnaces, dryers, melters, and kilns. The degree of flexibility in fuel choice a consumer may have depends on the process heat application -- some applications may use technologies or production processes that require a particular fuel.[8] Process heat applications consume about 40% of the energy in the U.S. industrial sector.

---

[8] For example, some food production processes use natural gas because its relatively clean-burning characteristic allows it to be used when product contamination may be an issue. Similarly, melters in the glass industry are often designed to burn natural gas because of the flame characteristics of this fuel. Use of other fuels would tend to produce an inferior product and likely require the redesign of equipment.

- Feedstocks -- Fuels may be used as a raw material for the production process. Examples of such applications include the conversion of metallurgical coal to coke for use in the manufacture of steel, natural gas for fertilizer production, and petroleum for asphalt. It is usually very difficult to switch to alternative fuels with these applications. In the U.S., feedstocks consume about 15% of industrial energy.

- Other -- This category consists primarily of industrial activities requiring electricity, e.g., lighting, motor drive, etc. These applications account for 15% of all energy consumed by U.S. industry.

The amount of greenhouse gas emissions generated from industrial energy consumption is a function of fuel type and the process in which it is consumed (see Table 4-2 for emissions from selected industrial applications).

Residential and Commercial Sectors. In the residential and commercial sectors the main end-use applications for energy are heating, cooling, cooking, and lighting. The form and amount of energy used to meet these needs varies, as summarized in Table 4-3 for the U.S. and South/Southeast Asia. In developing countries, most of the energy in these two sectors is consumed for cooking purposes, with consumers relying on biomass or kerosene for fuel. In industrialized countries, however, space heating and water heating consume the most energy, which is supplied primarily by fossil fuels and, to some extent, electricity; gas and electricity are the primary energy forms for cooking in industrialized countries. Because of the wide variety of end-use applications, types of energy consumed, and conversion efficiencies in the residential and commercial sectors it is difficult to generalize about emission trends in these sectors; for illustrative purposes, emission coefficients for several major applications in industrialized countries are listed in Table 4-2.

TABLE 4-3

End-Use Energy Consumption Patterns for the Residential/Commercial Sectors

(% of Total Energy)

End-Use	Type of Energy: Biomass	Fossil Fuels	Electricity	Total
South/Southeast Asia				
Heating	0	16	NA	NA
Cooling	0	0	NA	NA
Cooking	75	3	NA	NA
Lighting	0	1	NA	NA
TOTAL	75%	20%	5%	100%
United States				
Heating	<1	59	8	67
Cooling	0	0	6	6
Cooking	0	7	3	10
Lighting	0	0	7	7
Other	0	0	10	10
TOTAL	<1%	66%	34%	100%

Sources: Sathaye et al., 1988; Mintzer, 1988; U.S. DOE, 1987; Leon Schipper, pers. communication.

Transportation Sector. As consumers become wealthier, the absolute quantity and share of energy used in the transportation sector increases. For example, as discussed earlier, in many developing countries, such as China, the transportation sector consumes a much smaller portion of the country's energy requirements than the portion consumed by this sector in industrialized countries, such as the United States. Energy requirements in the transportation sector are typically met with fossil fuels, particularly petroleum-based products such as gasoline, diesel, or jet fuel. For example, in 1985 countries belonging to the OECD met 91% of their transportation energy requirements with oil-derived products, 8% with electricity, and the remaining 1% with natural gas and coal (OECD 1987). As countries become wealthier, increased use of petroleum to meet transportation needs can significantly increase greenhouse gas emissions to the atmosphere (see Table 4-2).

The amount and type of greenhouse gases emitted can also be affected by the transportation technology. For example, gasoline vehicles produce about 25% less CO_2 on an energy input basis than do diesel vehicles, while producing substantially more CO. However, the CO is eventually oxidized to CO_2, so the CO_2 emissions attributable to gasoline vehicles are comparable to those of diesel vehicles. Also, the efficiencies of diesel engines are generally greater than those of gasoline engines for a similar vehicle, implying that diesel vehicles would actually have lower effective CO_2 emissions per mile travelled. Similarly, vehicles powered with compressed natural gas would emit CO_2 and CO at lower levels than would either gasoline or diesel vehicles, although CH_4 emissions might be higher.

Fuel Production and Conversion

Significant quantities of greenhouse gases are emitted during the production of energy and its conversion to end-use energy forms. Several major components of these fuel production and conversion processes are discussed below.

Natural Gas Flaring, Venting, and Leaking. During the production of oil and natural gas, some portion of natural gas, which is mostly methane, is typically vented to the atmosphere (as CH_4) or flared (thereby producing CO_2) rather than produced for commercial use. Venting typically occurs during natural gas drilling and well maintenance operations to avoid pressure buildup, to test well drawdown, and during required maintenance at existing production wells. Flaring is most common in conjunction with oil production when no market can be found for the natural gas associated with oil reservoirs. In some circumstances, the gas may be vented rather than flared. On average, the amount of natural gas flared and vented amounts to about 2-3% of global natural gas production, although in some regions virtually all of the natural gas may be vented or flared, while in other regions (like the U. S.) the total amount flared or vented is less than 0.5% of total production (U.S. DOE, 1986).[9] Currently, approximately 50 teragrams (Tg) of CO_2 are released to the atmosphere from flaring of natural gas (Rotty, 1987).[10]

Leaks of natural gas also occur during the refining, transmission, and distribution of the gas. These leaks may occur at the refinery as the gas is cleaned for market, from the pipeline system during transportation to the end-user, or during liquefaction and regasification if liquified natural gas (LNG) is produced. About 20-50 Tg of CH_4 are released to the atmosphere each year from leaking and venting of natural gas (Crutzen, 1987; Cicerone and Oremland, 1989).

Coal Mining. As coal forms, CH_4 produced by the decomposition of organic material, becomes trapped in the coal seam. This CH_4 is released to the atmosphere during coal extraction operations. The amount of CH_4 released by coal mining varies depending on factors such as depth of the coal seam, quality of the coal, and characteristics of the geologic strata surrounding the seam. The

---

[9] U.S. regulations strictly govern the flaring and venting of natural gas. In other parts of the world, however, insufficient data exists to determine whether the natural gas is flared or vented, although safety precautions would strongly encourage flaring rather than venting.

[10] 1 Tg = 1 teragram = million metric tons = 10^{12} grams.

amount of CH_4 emitted as a result of coal mining is highly uncertain, with estimates ranging from 25 to 45 Tg per year (Cicerone and Oremland, 1989). If coal mining operations intensify, the quantity of methane released as an indirect result of mining is expected to increase at a comparable rate.

Synthetic Fuel Production. As conventional petroleum resources are depleted, some of the demand for liquid (oil and natural gas liquids) and gaseous (natural gas) fuels may be met by synthetic fuels. Although there is currently little synthetic fuel produced in the world, processes have been developed to convert relatively abundant solid energy resources such as coal, oil shale, and tar sands to liquid or gaseous products that could be consumed in the same end-use applications as conventional oil and gas.

Significant amounts of energy are typically required to produce synthetic fuels. The conversion process produces greenhouse gas emissions, particularly CO_2, so that the net emissions per unit of energy for synthetic fuels are greater than those for conventional fossil fuels. For example, the CO_2 emissions from production and consumption of synthetic liquid fuels from coal are about 1.8 times the amount from conventional liquid fuels from crude oil (Marland, 1982). Table 4-4 lists emission rates for both conventional fossil fuels and synthetic fuels produced from coal and shale oil.

Future Trends

As shown in Figure 4-4, the quantity of CO_2 emitted to the atmosphere as a result of the combustion of fossil fuels has increased dramatically in the last century. This increase in fossil-fuel-produced CO_2 emissions is the main factor that has led to an increase in atmospheric CO_2 concentrations -- from about 280 ppm in preindustrial periods to about 345 ppm today. As discussed in Chapter II, future CO_2 concentrations will depend on many factors, but most important will be

TABLE 4-4

Carbon Dioxide Emission Rates for Conventional and Synthetic Fuels

Fuel	CO_2 Emission Rate ($g\ C/10^9J$)	Notes
Conventional Fossil Fuels (rates for consumption)		
Natural Gas	13.5-14.2	
Liquid Fuels from Crude Oil	18.2-20.6	Differences are partly attributable to product mix, i.e., gasoline versus fuel oil and gasoline.
Bituminous Coal	23.7-23.9	
Synthetic Fuels (rates for production and consumption)		
Shale Oil	104.3	High temperature, 10 gal/ton shale
	66.4	High temperature, 25 gal/ton shale
	47.6	Modified in situ, 28 gal/ton shale
	28.4	Low temperature retorting
Liquids from Coal	51.3	Gasoline from methanol using Mobil MTG process
	41.8	Sasol-type technology, Eastern coal
	39.9	FHP process
	38.6	Exxon-Donor Solvent, Eastern coal
	37.2	H-coal
	31.9	Generic 75% thermal efficiency
	30.5	SRC-II, liquid and gas products
High-Btu gas from coal	40.7	Lurgi
	40.1	Hygas
	36.2	Generic 66% thermal efficiency
	32.7	Via synthesis gas with by-product credits

Source: Marland, 1982.

the rate of growth in energy demand and the type of energy that is consumed in order to satisfy this demand.

The Fossil-Fuel Supply

Higher levels of energy demand will produce higher levels of greenhouse gas emissions if the demand is satisfied with fossil fuels. As indicated above, fossil fuels currently supply a majority of the world's energy needs, and it seems likely that fossil fuels will continue to play a key role in the world's energy supply picture for decades to come. However, supplies of fossil fuels are not unlimited. Resource and reserve estimates for coal, oil, and gas are listed in Table 4-5. A resource is any presently or potentially extractable mineral supply, a reserve is a presently extractable supply. A resource that is not presently economic to extract, may become economic in the future and then be called a reserve. The estimates of the lifetimes of fossil-fuel reserves are based on current (1985) rates of production. The lifetime estimates of fossil-fuel resources are based on linear and exponential extrapolations of current energy demand (described below). Despite uncertainties about the size of the resource base and the rate at which the resource base may be depleted, it is clear from a technical standpoint that the consumption of fossil fuels could continue for a very longtime. As will be discussed in Chapter V, if the world continues to rely on fossil fuels to meet the majority of its energy needs, the amount of carbon emitted to the atmosphere may be many times greater than current levels.

Future Energy Demand

The future rate of energy demand depends on many variables, including the rate of population growth, the rate of economic growth, energy prices, the types of energy services demanded by consumers, the type and efficiency of technology used, and the type and amount of energy supplies

TABLE 4-5

Estimates of Global Fossil-Fuel Resources[a]

(Exajoules)

	Geological Resources (exajoules)	Reserves (exajoules)	Reserves/ Resources (%)	Reserves Lifetime[b] (Years)	Resource Lifetime (Years)	
					Linear Extrapolation	Exponential Extrapolation
Coal	315,800	20,400	6	229	524	103
Oil	12,800	4,300	34	41	69	39
Gas	10,100	3,700[c]	37	61	86	44

[a] Resources estimates, as of 1985, are from the World Energy Conference (1980), adjusted for global production from 1979-85. Reserve estimates are from DOE/EIA (1986): oil and gas estimates as of January 1, 1986; coal estimates as of 1981.

[b] Based on 1985 rates of production.

[c] Includes estimates for the Middle East and USSR.

Sources: World Energy Conference, 1980; United Nations, 1983, 1987; U.S. DOE, 1986.

available (Chapter V). Two hypothetical cases based on crude extrapolations illustrate potential upper and lower bounds on future energy demand (see Figure 4-7) and the lifetime of fossil fuel resources (Table 4-5). For example, from 1950 to 1973, the average annual growth rate in energy demand was 5.2%. If this rate of growth were exponentially extrapolated to 2050, global energy demand would be about 254 TW (or equivalently about 8,000 EJ), almost 30 times the 1985 level.[11] This amount of energy demand could lead to an increase in annual CO_2 emissions from the current 5.2 Pg C to about 140 Pg C in 2050, assuming that this demand is met by consumption of fossil fuels. Cumulative energy demand for 1985 through 2050 based on this extrapolation represents over five times the amount of fossil fuels in proven reserves and about 45% of the resource estimate. On the other hand, the average annual growth rate in energy demand from 1973 to 1985 was much lower: about 2.2%. If this rate were linearly extrapolated to 2050, global energy demand would be about 23 TW (720 EJ) -- almost 150% greater than the demand in 1985 -- which could increase annual CO_2 emissions from fossil fuels to nearly 13 Pg C. Cumulative energy demand for 1985 through 2050 based on the linear extrapolation represents about 115% of proven fossil fuel reserves, or nearly 10% of estimated resources.

INDUSTRIAL PROCESSES

There are three significant non-energy sources of greenhouse gases associated with industrial activity: the use of chlorofluorocarbons (CFCs), halons, and chlorocarbons (collectively, halocarbons); cement manufacture; and waste disposal in landfills. The use of CFCs, halons, and chlorocarbons, which are man-made chemicals with a variety of applications, results in their release to the atmosphere. Certain uses, such as aerosol propellants and solvents, result in instantaneous release (when the product is used), while others, such as foam-blowing agents and refrigerants, result in a

[11] TW = Terawatt-years per year = 10^{12} watt-years per year; 1 TW = 31.53 EJ; 1 EJ = 1 Exajoule = 10^{18} joules; 1055 joules = 1 Btu.

FIGURE 4-7

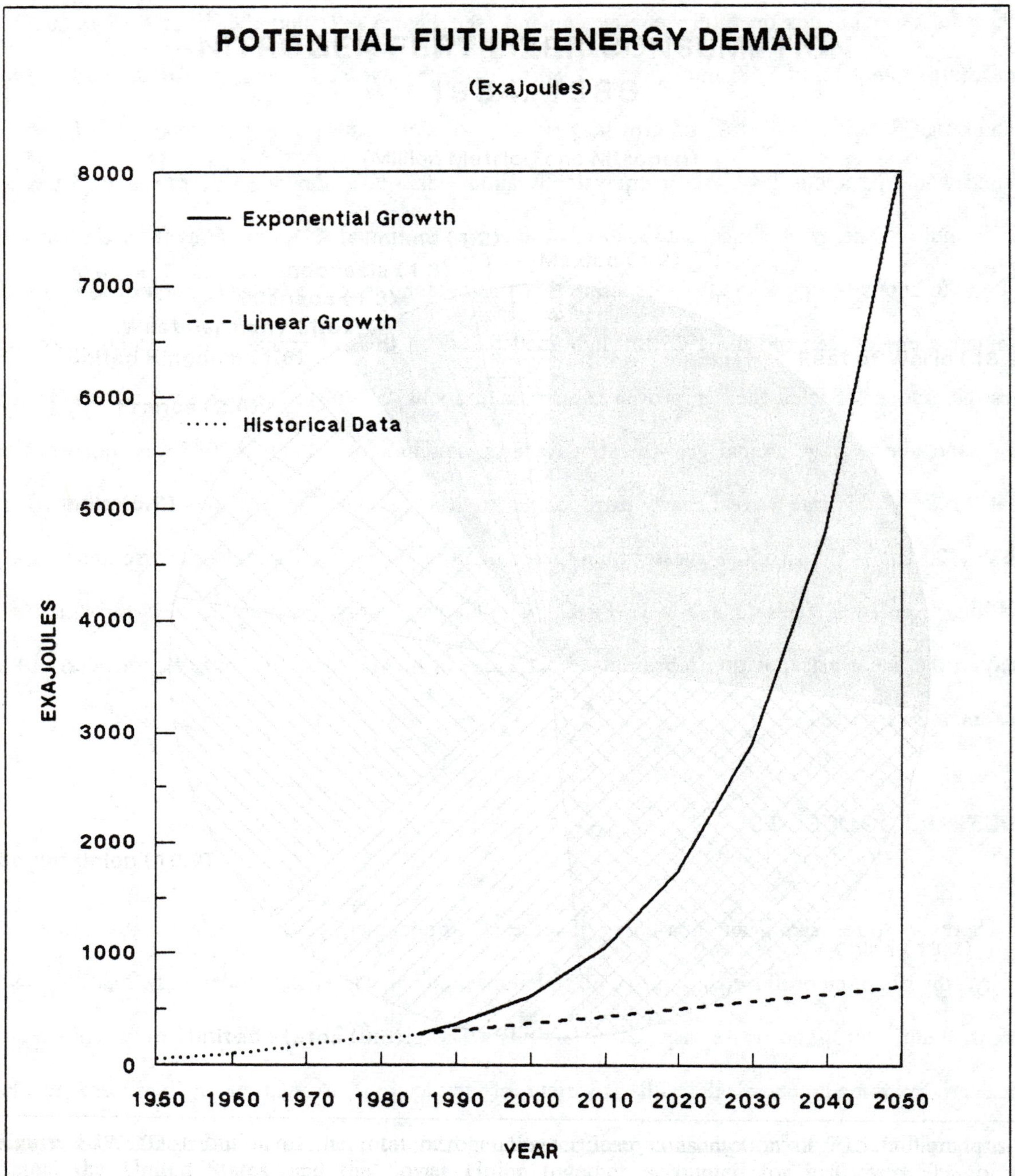

Figure 4-7. Two hypothetical cases of future energy demand. The upper case is based on an exponential extrapolation of the average annual growth rate in energy demand between 1950 and 1973, a period of rapid growth in demand. The lower case is based on a linear extrapolation of the average annual growth rate in energy demand between 1973 and 1985, when the growth rate was much lower. These two cases illustrate potential upper and lower bounds on future energy demand. (Sources for historical data: United Nations, 1976, 1982, 1983, 1987.)

delayed release. Cement manufacture results in CO_2 emissions, and waste disposal in landfills results in CO_2 and CH_4 emissions, although only the CH_4 emissions are significant in terms of the total global source.

Chlorofluorocarbons, Halons, and Chlorocarbons

Historical Development and Uses

Chlorofluorocarbons are man-made chemicals containing chlorine, fluorine, and carbon, hence the name CFCs (HCFCs contain hydrogen as well). Table 4-6 lists the major CFCs with their chemical formulae. CFCs were developed in the late 1920s in the United States as a substitute for the toxic, flammable, refrigerator coolants in use at that time. The chemicals, which are noncorrosive, nontoxic, nonflammable, and highly stable in the lower atmosphere, provided the refrigerator industry with a safe, efficient coolant that soon proved to have numerous other uses as well. Commercial development of CFCs began in 1931. During World War II, CFCs were used as propellants in pesticides against malaria-carrying mosquitos. Since then, CFCs have been used as aerosol propellants in a wide range of substances, from hairsprays to spray paints. In the 1950s, industries began using CFCs as blowing agents for plastic foam and foam insulation products. Chillers, used for cooling large commercial and industrial buildings, as well as cold storage units for produce and other perishable goods, became feasible at this time with the use of CFCs. Mobile air conditioners (in automobiles, trucks, and buses) currently constitute the largest single use of CFCs in the United States. CFCs are also used in gas sterilization of medical equipment and instruments, solvent cleaning of manufactured parts, especially electronic components and metal parts, and miscellaneous other processes and products such as liquid food freezing.

TABLE 4-6

Major Halocarbons: Statistics and Uses

Chemical	1986 Atmospheric Concentration (pptv)	Atmospheric Lifetime (Years)	Current Annual Atmospheric Concentration Growth Rates (%/yr)	Major Uses
Chlorofluorocarbons				
CFC-11 ($CFCl_3$)	226	75 (+32, -17)	4	Aerosols, Foams
CFC-12 (CF_2Cl_2)	392	111 (289, -46)	4	Aerosols, Refrigeration
HCFC-22 ($CHClF_2$)	~100	20	7	Refrigeration
CFC-113 ($C_2Cl_3F_3$)	30-70	90	11	Solvents
Halons (Bromofluorocarbons)				
Halon-1211 ($CBrClF_2$)	~2	25	>10	Fire extinguisher
Halon-1301 ($CBrF_3$)	~2	110	>10	Fire extinguisher
Chlorocarbons				
Carbon tetrachloride (CCl_4)	75-100	~50	1	Production of CFC-11 and CFC-12
Methylchloroform (CH_3CCl_3)	125	5.5-10	7	Solvents

Sources: U.S. EPA, 1988a; Hammitt et al., 1987; Wuebbles, 1983; WMO, 1985.

NA = No data available.

Halons, or bromofluorocarbons, are man-made chemicals containing carbon, fluorine, and chlorine and/or bromine (see Table 4-6 for the chemical formulae of the major halons in use today). These chemicals were developed in the 1970s, and are used primarily as fire extinguishants. Halon-1211 is used almost exclusively for portable (i.e., wheeled or handheld) fire extinguishers, particularly for situations where human exposure to the chemical is possible, such as in airplanes. Halon-1301 is used exclusively for total flooding fire extinguishing systems such as those used to protect computer centers, document rooms, libraries, and military installations. A summary of the 1985 end-use applications for the major CFC and halon compounds is shown in Table 4-6.

Chlorocarbons, man-made chemicals containing chlorine and carbon (see Table 4-6), are used primarily as solvents and chemical intermediates. The primary chlorocarbons are carbon tetrachloride and methylchloroform. In the United States, carbon tetrachloride was once used extensively as a solvent and grain fumigant, but because of its toxicity, only small amounts of it are used in such applications today. Its primary use in the United States is in the manufacture of CFC-11 and CFC-12, a process which consumes or destroys almost all of the carbon tetrachloride, resulting in very small emissions. However, carbon tetrachloride is believed to be used as a solvent in developing countries, resulting in considerable emissions. Methylchloroform is used worldwide as a cleaning solvent in two applications: 1) vapor degreasing (the solvent is heated and the item to be cleaned is suspended in the vapor); and 2) cold cleaning (the part to be cleaned is submerged in a tank of solvent). Small amounts are also used in adhesives, aerosols, and coatings.

Production of CFCs, halons, and chlorocarbons has grown steadily as new uses have developed. Production of the two largest CFC compounds, CFC-11 and CFC-12, increased rapidly in the 1960s and early 1970s (see Figure 4-8). Production peaked in 1974 at 812.5 gigagrams (Gg) and then declined due to a ban on most aerosol use in the United States, Canada, and Sweden in the late

FIGURE 4-8

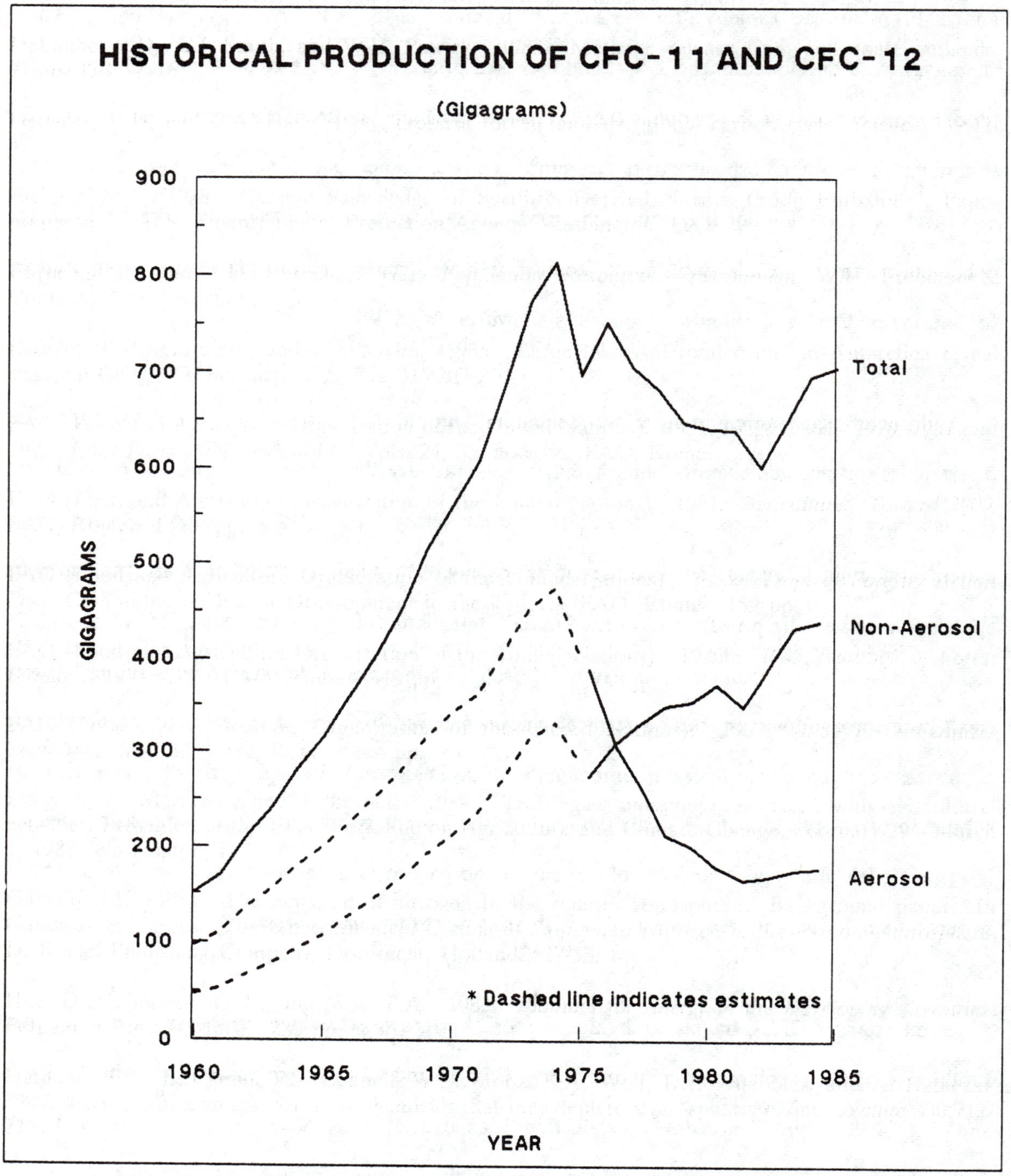

Figure 4-8. While non-aerosol production of CFC-11 and CFC-12 has grown fairly steadily since 1960, aerosol production declined in the 1970s and then leveled off in the 1980s due to a ban on most aerosol use of CFCs in the United States, Canada, and Sweden. (Source: U.S. EPA, 1987.)

1970s.[12] However, non-aerosol use has continued to grow, with 1985 production of 703.2 Gg. Globally, major CFC and halon consumption reached nearly one Tg in 1985 (see Table 4-7). Global production of carbon tetrachloride and methylchloroform in 1985 was estimated at nearly 1,029 Gg and 545 Gg, respectively (Hammitt et al., 1987).

Most CFC and halon consumption occurs in the United States and other industrialized nations. Of the 703.2 Gg of CFC-11 and CFC-12 produced in 1985, about 70% was consumed by the U.S., the EEC, and Japan (see Figure 4-9). Although CFC use is concentrated in the industrialized world, consumption has also increased recently in developing countries.

The Montreal Protocol

Concern over the effect on the Earth's atmosphere of CFCs and related anthropogenically-produced compounds containing chlorine, bromine, and nitrogen began in the 1970s. Because of their stability (i.e., their long lifetimes, see Table 4-6), CFCs are transported to the stratosphere where they contribute to the destruction of ozone. Since the early 1970s, improved understanding of this process, accumulation of data indicating growing atmospheric concentrations of CFCs, and observed depletion of stratospheric ozone, particularly in the Antarctic, have fueled international concern over this issue.

International negotiations to protect the stratosphere began in 1981 under the auspices of the United Nations Environment Programme (UNEP). These negotiations culminated in September 1987 in Montreal, Canada, where a Diplomatic Conference was held, resulting in an international agreement ("The Montreal Protocol on Substances That Deplete the Ozone Layer," or the Montreal Protocol) to begin reducing the use of CFCs and halons (chlorocarbons were not included). The

---

[12] 1 Gg = 10^9 grams = 1 million kilograms.

TABLE 4-7

Estimated 1985 World Use of Potential Ozone-Depleting Substances

(gigagrams)

Chemical	World	United States	Reporting Countries	Other Communist Countries
CFC-11	341.5	75.0	225.0	41.5
CFC-12	443.7	135.0	230.0	78.7
CFC-113	163.2	73.2	85.0	5.0
Halon 1301	10.8	5.4	5.4	0.0
Halon 1211	10.8	2.7	8.1	0.0
Carbon tetrachloride	1029.0	280.0	590.0	159.0
Methylchloroform	544.6	270.0	186.7	87.0

Source: Hammit et al., 1986.

FIGURE 4-9

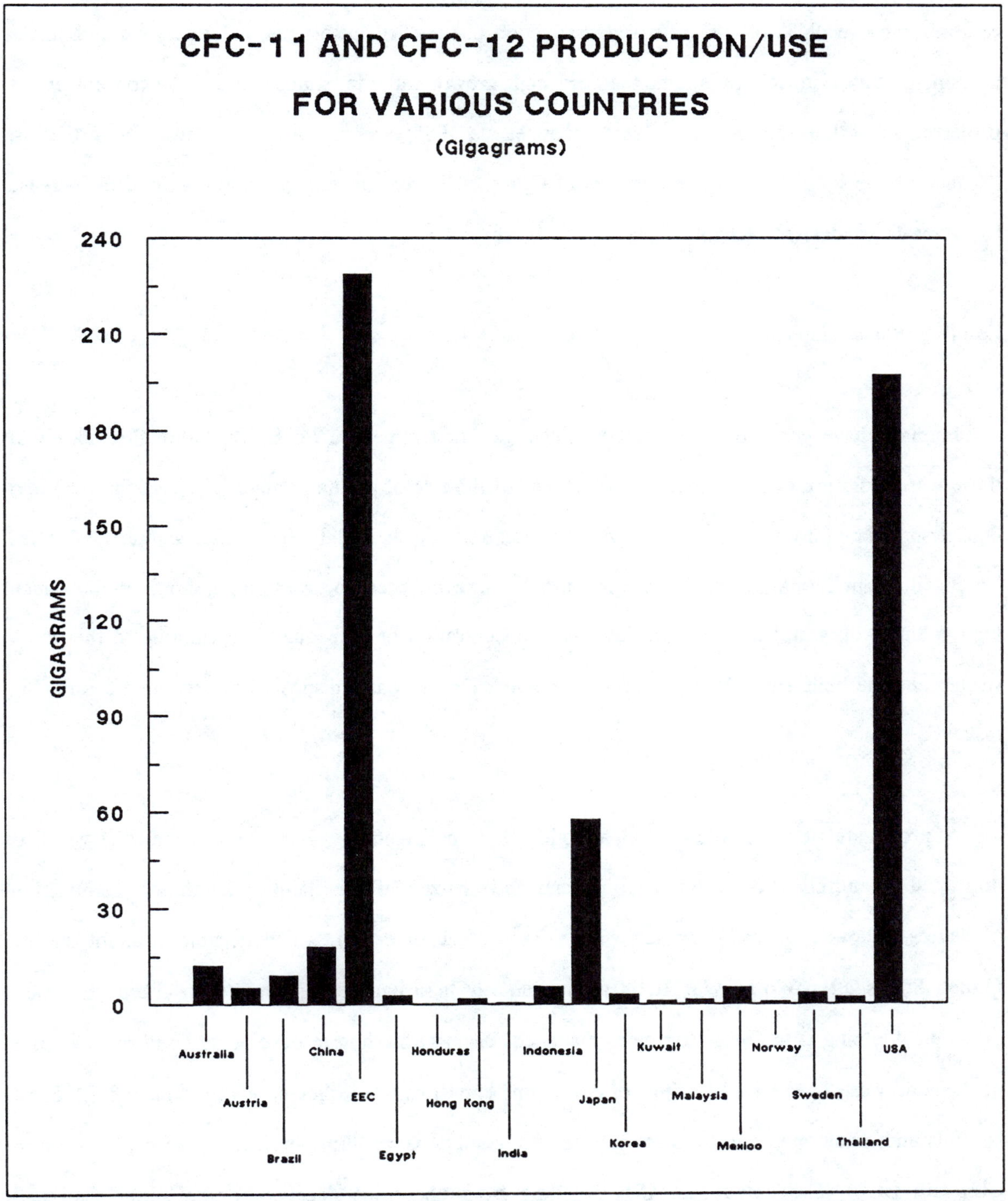

Figure 4-9. The EEC, the United States, and Japan accounted for almost 70% of the 1985 global production of CFC-11 and CFC-12. (Source: U.S. EPA, 1988a.)

Montreal Protocol came into force on January 1, 1989 and has been ratified by 31 countries, representing over 90% of current world production of these chemicals (as of January 11, 1989). As a result of this historic agreement, the very high growth rates in atmospheric CFC concentrations projected in earlier studies (e.g., Ramanathan et al., 1985) are not likely to occur. Nevertheless, because of the long atmospheric lifetimes of CFCs, their concentrations could continue to increase for several decades (see Chapter V).

Landfill Waste Disposal

Humans have generated solid wastes since they first appeared on Earth, although disposal of these wastes did not become a major problem until the rise of synthetic materials (e.g., plastics) and densely-populated urban areas. The environment can usually assimilate the smaller amounts of wastes produced by rural, sparsely-settled communities. However, because urban populations produce such high volumes of waste, due to both the sheer concentration of individuals contributing to the waste stream and the high use of heavily-packaged products, urban waste disposal has become a formidable task.

Approximately 80% of the municipal solid wastes collected in urban areas around the world is deposited in landfills or open dumps (Bingemer and Crutzen, 1987). Sanitary landfilling (compaction of wastes, followed by daily capping with a layer of clean earth), which became common in the United States after World War II, is used primarily in urban centers in industrialized countries. Open pit dumping is the most common "managed" disposal method in developing countries (30-50% of the solid wastes generated in cities of developing countries is uncollected [Cointreau, 1982]). Most landfills and many open dumps develop anaerobic conditions, resulting in decay of organic carbon to CH_4 and CO_2. The amount of CH_4 resulting from anaerobic decay of organic municipal and industrial wastes in landfills is currently about 30-70 Tg per year (Bingemer and Crutzen, 1987), approximately 10% of the total annual CH_4 source.[13]

The primary variable affecting gas generation in landfills is the composition of the refuse. Wastes high in organic material (e.g., food wastes, agricultural wastes, paper products) decompose readily, while inorganics are relatively unaffected by the decomposition process. While agriculture is the largest single source of solid wastes in the U.S. (Berry and Horton, 1974), most of these wastes are not landfilled. Increasing urbanization and demand for "convenience" items, which encourages marketing of single-serving and heavily-packaged products, has resulted in increasingly greater proportions of plastics, glass, metals, and paper products, in the waste stream. Other factors influencing gas generation include inclusion of sewage sludge (which enhances gas generation), oxygen concentration, moisture content, pH, and available nutrients.

Disposal of municipal solid waste in industrial nations increased by 5% per year during the 1960s, and by 2% per year in the 1970s (CEQ, 1982). Currently, per capita waste production in industrialized countries is considerably larger than in developing countries (see Table 4-8), and the largest contribution of landfill CH_4 comes from the industrialized world (Bingemer and Crutzen, 1987). Although current rates of waste disposal in landfills have begun to level off in many industrialized countries, associated CH_4 emissions are probably still growing because the total quantity of waste in place is still increasing. In the developing world, with its high population growth rates and increasing urbanization, municipal solid waste disposal is projected to double by the year 2000 (Kresse and Ringeltaube, 1982), so CH_4 production from waste dumps and/or sanitary landfills can be expected to increase rapidly in developing countries.

---

[13] This estimate does not include methane from anaerobic decomposition of agricultural wastes, which could be a significant quantity. The total amount of carbon in agricultural wastes in the United States alone is already 2.5 times larger than the 113 million metric tons of waste carbon that are generated and dumped in landfills worldwide (Bingemer and Crutzen, 1987).

TABLE 4-8

Refuse Generation Rates in Selected Cities

City	Per Capita Waste Generation Rate (kg Per Day)
Industrial Cities	
New York, United States	1.80
Singapore	0.87
Hong Kong	0.85
Hamburg, West Germany	0.85
Rome, Italy	0.69
Developing Cities	
Jakarta, Indonesia	0.60
Lahore, Pakistan	0.60
Tunis, Tunisia	0.56
Bandung, Indonesia	0.55
Medellin, Colombia	0.54
Surabaya, Indonesia	0.52
Calcutta, India	0.51
Cairo, Egypt	0.50
Karachi, Pakistan	0.50
Manila, Philippines	0.50
Kanpur, India	0.50
Kano, Nigeria	0.46

Source: Cointreau, 1982.

Cement Manufacture

Cement manufacture produces CO_2, as well as numerous other exhaust gases. As demand for cement has grown over the last century, CO_2 emissions associated with this industry have increased from 18 to 134 Tg C between 1950 and 1985 (see Figure 4-10). In recent years CO_2 emissions from cement production have grown at a faster rate than those from fossil fuel combustion: In the early 1950s CO_2 emitted as a result of cement manufacture was approximately 1% of the amount emitted from the consumption of fossil fuels; by the early 1980s this fraction had increased to 2.5% (Rotty, 1987).

The CO_2 emissions resulting from cement manufacture occur during the production of clinker, a material produced midway through the process. After the raw materials (cement rock, limestone, clay, and shale) are quarried and crushed, they are ground and blended to a mixture that is approximately 80% limestone by weight. The mixture is then fed into a kiln for firing, where it is exposed to progressively higher temperatures that cause heating, then drying, calcining, and sintering. Finally, the feed is heated to the point of fusion (approximately 1595°C), and clinker (round, marble-sized particles) is produced. It is during the calcination process, which occurs at approximately 900 to 1000°C, that the limestone ($CaCO_3$) is converted to lime (CaO) and CO_2, and the CO_2 is released. For every million tons of cement produced, approximately 0.137 Tg C as CO_2 is emitted from calcining (Rotty, 1987).[14] An additional 0.165 Tg C is emitted per million ton of cement produced from fossil fuel used for kiln firing and electricity. This CO_2 is accounted for as part of industrial energy use emissions.

World cement production has increased at an average annual rate of approximately 6% since the 1950s, from 133 million tons in 1950 to 972 million tons in 1985 (U.S. BOM, 1949-1986). Cement

[14] 1 ton = 1 metric ton = 1000 kg.

FIGURE 4-10

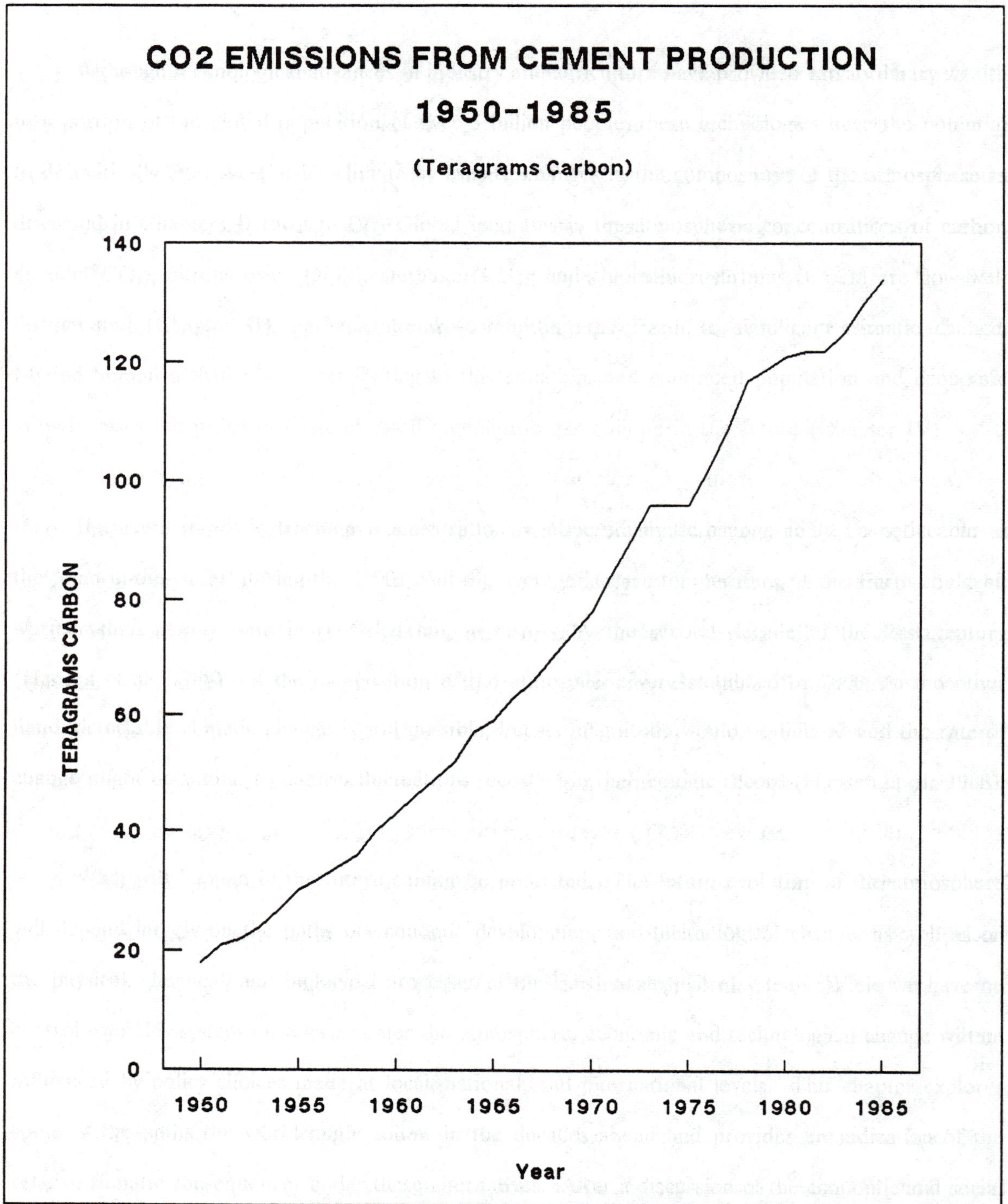

Figure 4-10. Carbon dioxide emissions from cement production grew from 18 to 134 Tg between 1950 and 1985, an average annual rate of growth of about 6%. (Sources: Rotty, 1987; U.S. BOM, 1986.)

production growth rates in individual countries have varied during this period (see Figure 4-11) due to economic fluctuations in cement's primary market, i.e., the construction industry, and competitive shifts internationally among the primary cement-producing countries. For example, in 1951 the United States produced approximately 28% of the global total, while by 1985 its share had shrunk to 7%.[15] During the same time, the production shares for the USSR grew from 8% to 13%, for China, from less than 1% to 15%, and for Japan, from 4% to 8%. Although many national markets, except the United States', experienced low levels of demand during the 1980s, global cement production is expected to continue to grow faster than the GNP for some time.

LAND USE CHANGE

Over the past few centuries, man has significantly changed the surface of the Earth. Forests have been cleared, wetlands have been drained, and agricultural lands have been expanded. All of these activities have resulted in considerable changes in trace gas emissions to the atmosphere. Deforestation results in a net release of carbon from both the biota and the soils (unless the land is reforested) as these organic carbon pools burn or are decomposed. Biomass burning, due to shifting agriculture, burning of savanna, use of industrial wood and fuelwood, and burning of agricultural wastes, is a source of CO_2, as well as CH_4, N_2O, and NO_x.[16] Destruction of wetlands, from either filling or dredging, can alter the atmospheric CH_4 budget, since anaerobic decomposition in wetlands produces CH_4.

[15] The U.S. is currently a net importer of cement; the volume of its imports has grown, representing only a few percent of consumption in the early 1980s but as much as 18 percent in 1986 (International Trade Administration, 1987).

[16] Shifting agriculture is the practice of clearing and planting a new area, farming it until productivity declines, and then moving on to a new plot to start the cycle over again. If the land is allowed to reforest, there are no net CO_2 emissions.

FIGURE 4-11

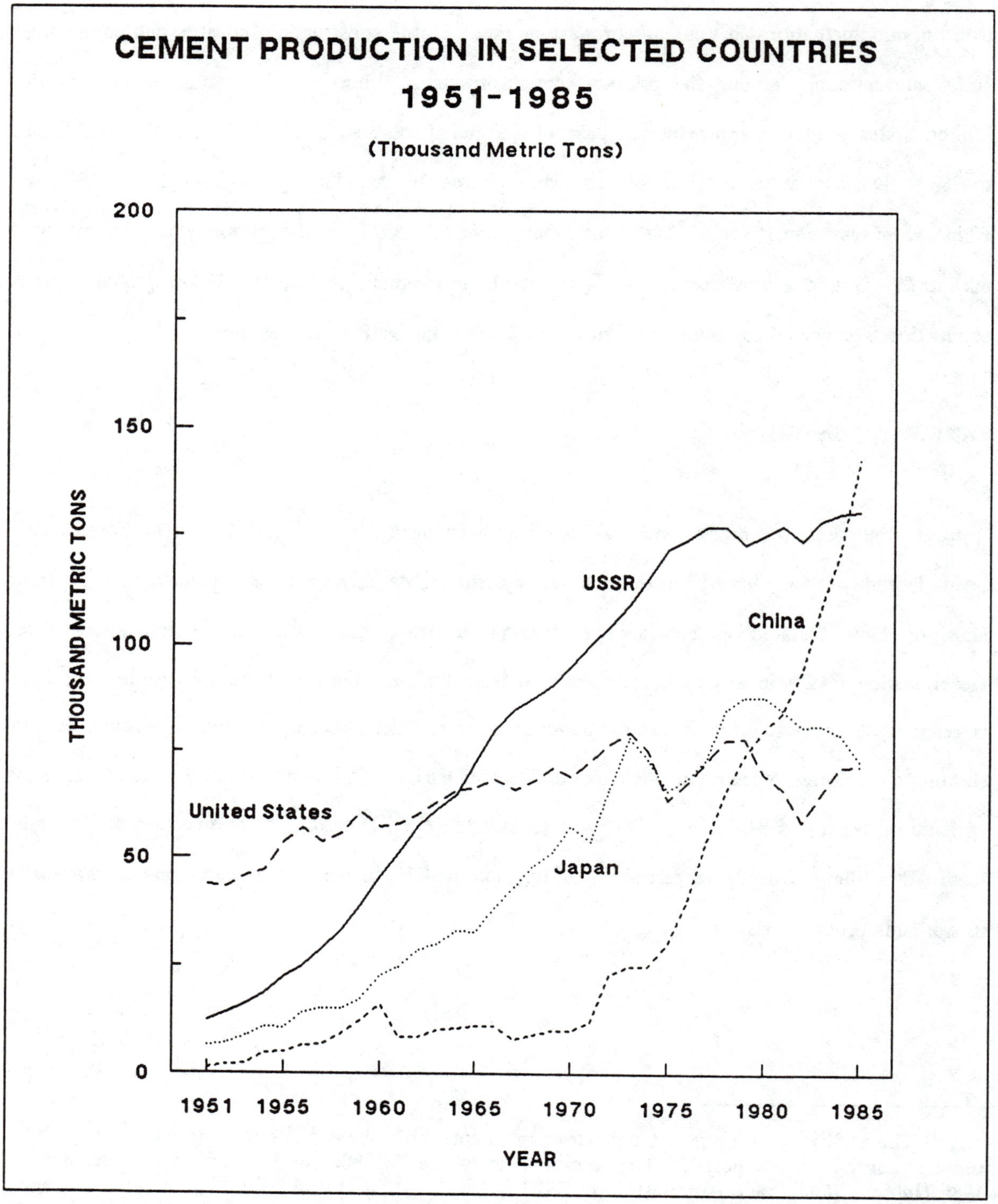

Figure 4-11. World cement production grew at an average annual rate of about 6% between 1950 and 1985. Growth has been particularly rapid in China, the U.S.S.R., and Japan. (Source: U.S. BOM, selected years.)

Deforestation

Estimates of net emissions of CO_2 to the atmosphere due to changes in land use (deforestation, reforestation, logging, and changes in agricultural area) in 1980 range from 0.4 to 2.6 Pg C (Houghton et al., 1987; Detwiler and Hall, 1988), which accounts for approximately 10-30% of annual anthropogenic CO_2 emissions to the atmosphere. Deforestation in the tropics accounted for almost all of the flux; the carbon budget of temperate and boreal regions of the world has been approximately in balance in recent years. Of the net release of carbon from tropical deforestation, 55% was produced by only six countries in 1980: Brazil, Indonesia, Columbia, the Ivory Coast, Thailand, and Laos (see Figure 4-12).

The world's forest and woodland areas have been reduced 15% since 1850, primarily to accommodate the expansion of cultivated lands (IIED and WRI, 1987). The largest changes in forest area during this period have occurred in Africa, Asia, and Latin America. Europe is the only region that has experienced a net increase in forest area over this time interval. Forest area began to increase in Europe in the 1950s and in North America in the 1960s (see Table 4-9). However, recent data from the Food and Agriculture Organization of the United Nations (FAO) and the U.S. Forest Service indicates that net deforestation may be occurring in the United States -- although there are discrepancies between the two data sets. The FAO data indicates that between 1980 and 1985 the area of U.S. forest and woodlands decreased by approximately 3.8 million hectares (million ha) per year, or 1.4% per year (FAO, 1986b).[17] The U.S. Forest Service (Alig, 1988) estimates that between 1977 and 1987 the area of U.S. forests decreased by approximately 0.41 million ha per year, or 0.14% per year.

[17] 1 ha = 1 hectare = 2.471 acres.

FIGURE 4-12

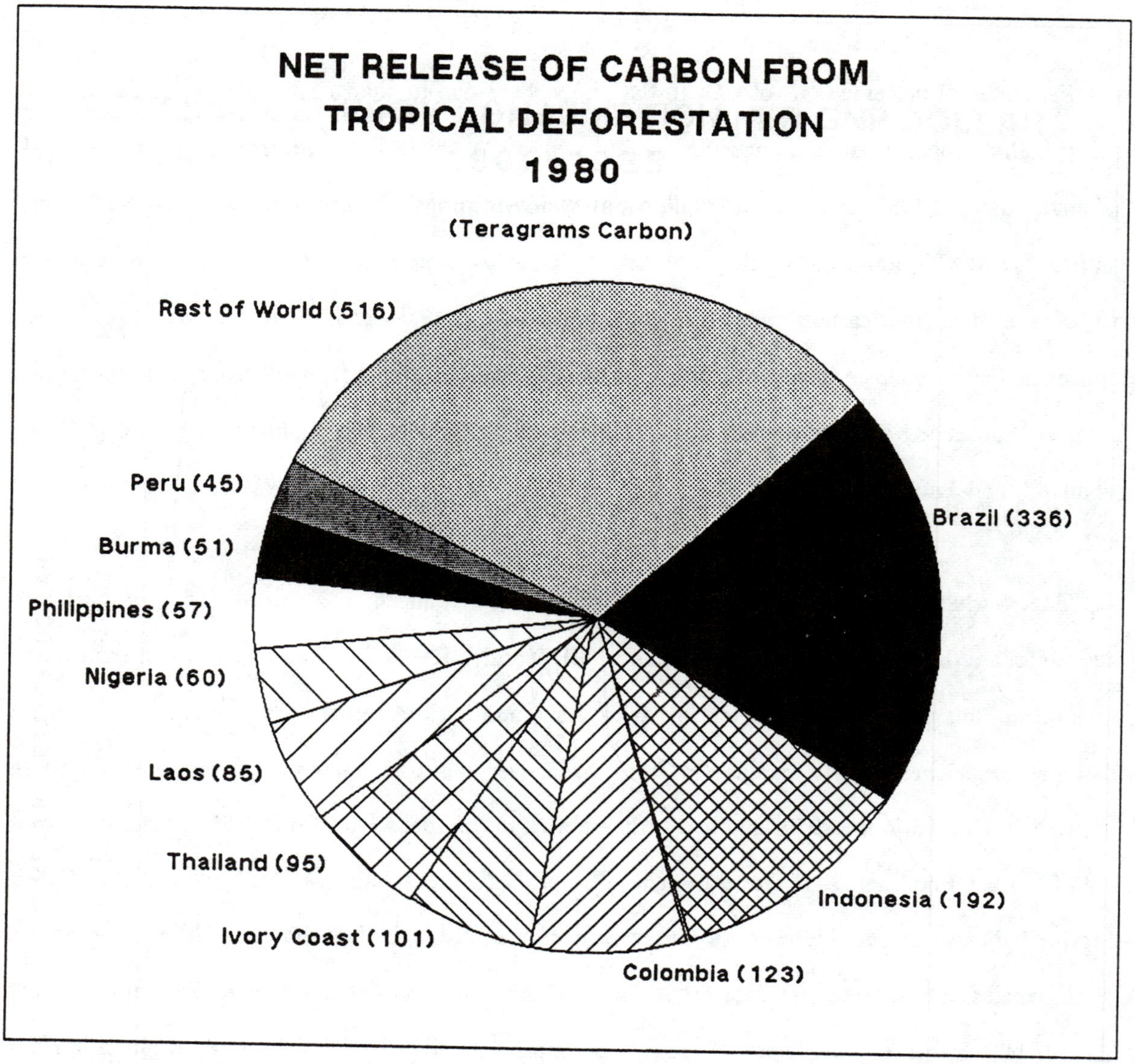

Figure 4-12. Tropical deforestation accounts for approximately 10-30% of the annual anthropogenic CO_2 emissions to the atmosphere. Over half of the 1980 CO_2 emissions from deforestation was produced by six countries: Brazil, Indonesia, Columbia, the Ivory Coast, Thailand, and Laos. (Source: Houghton et al., 1987.)

TABLE 4-9

Land Use: 1850-1980

	Area (million hectares)														Percentage Change 1850 to 1980
	1850	1860	1870	1880	1890	1900	1910	1920	1930	1940	1950	1960	1970	1980	
TEN REGIONS															
Forests and Woodlands	5,919	5,898	5,869	5,833	5,793	5,749	5,696	5,634	5,553	5,455	5,345	5,219	5,103	5,007	-15
Grassland and Pasture	6,350	6,340	6,329	6,315	6,301	6,284	6,269	6,260	6,255	6,266	6,293	6,310	6,308	6,299	-1
Croplands	538	569	608	659	712	773	842	913	999	1,085	1,169	1,278	1,396	1,501	179
Tropical Africa															
Forests and Woodlands	1,336	1,333	1,329	1,323	1,315	1,396	1,293	1,275	1,251	1,222	1,188	1,146	1,106	1,074	-20
Grassland and Pasture	1,061	1,062	1,064	1,067	1,070	1,075	1,081	1,091	1,101	1,114	1,130	1,147	1,157	1,158	9
Croplands	57	58	61	64	68	73	80	88	101	118	136	161	190	222	288
North Africa and Middle East															
Forests and Woodlands	34	34	33	32	31	30	28	27	24	21	18	17	15	14	-60
Grassland and Pasture	1,119	1,119	1,118	1,117	1,116	1,115	1,113	1,112	1,108	1,103	1,097	1,085	1,073	1,060	-5
Croplands	27	28	30	32	35	37	40	43	49	57	66	79	93	107	294
North America															
Forests and Woodlands	971	968	965	962	959	954	949	944	941	940	939	939	941	942	-3
Grassland and Pasture	571	559	547	535	522	504	486	468	454	450	446	446	447	447	-22
Croplands	50	65	80	95	110	133	156	179	196	201	206	205	204	203	309
Latin America															
Forests and Woodlands	1,420	1,417	1,414	1,408	1,401	1,394	1,383	1,369	1,348	1,316	1,273	1,225	1,186	1,151	-19
Grassland and Pasture	621	623	625	627	630	634	638	646	655	673	700	730	751	767	23
Croplands	18	19	21	24	28	33	39	45	57	72	87	104	123	142	677
China															
Forests and Woodlands	96	93	91	89	86	84	82	79	76	73	69	64	59	58	-39
Grassland and Pasture	799	799	798	796	797	797	797	796	796	794	793	789	784	778	-3
Croplands	75	78	81	84	86	89	91	95	96	103	108	117	127	134	79
South Asia															
Forests and Woodlands	317	315	311	307	303	299	294	289	279	265	251	235	210	180	-43
Grassland and Pasture	189	189	189	189	189	189	190	190	190	190	190	190	189	187	-1
Croplands	71	73	77	81	85	89	93	98	108	122	136	153	178	210	196
Southeast Asia															
Forests and Woodlands	252	252	251	251	250	249	248	247	246	244	242	240	238	235	-7
Grassland and Pasture	123	123	122	121	119	118	116	114	111	108	105	102	97	92	-25
Croplands	7	7	8	10	12	15	18	21	25	30	35	40	47	55	670
Europe															
Forests and Woodlands	169	158	157	157	156	156	155	155	155	154	154	156	161	167	4
Grassland and Pasture	150	147	145	144	143	142	141	139	138	137	136	136	137	138	-8
Croplands	132	136	140	142	143	145	146	147	149	150	152	151	145	137	4
USSR															
Forests and Woodlands	1,067	1,060	1,052	1,040	1,027	1,014	1,001	987	973	961	952	945	940	941	-12
Grassland and Pasture	1,078	1,081	1,083	1,081	1,079	1,078	1,076	1,074	1,072	1,070	1,070	1,069	1,065	1,065	-1
Croplands	94	98	103	118	132	147	162	178	194	208	216	225	233	233	147
Pacific Developed Countries															
Forests and Woodlands	267	267	266	265	264	263	262	261	260	259	258	252	247	246	-8
Grassland and Pasture	638	638	638	637	635	634	632	630	629	627	625	617	609	608	-5
Croplands	6	6	7	9	12	14	17	19	22	24	28	42	56	58	841

Source: IIED and WRI, 1987.

Currently, it is estimated that approximately 11.3 million ha of tropical forests are lost each year, while only 1.1 million ha are reforested per year (FAO, 1985). Most of the tropical deforestation is due to transfer of forest land to agricultural use, through shifting agriculture and conversion to pasture. FAO has estimated a demand for an additional 113-150 million ha of cultivated land for the 20-year period between 1980 and 2000 to meet food production needs (FAO, 1981). Most of this land will have to come from areas that were once forested, however there is a large potential to use land currently under shifting cultivation by adapting low-input agricultural techniques (Chapter VII). Fuelwood use also contributes to deforestation, particularly in Africa where fuelwood is a major source of residential energy. Sixty-three percent of the total energy consumption of developing African countries, 17% in the Asian countries, and 16% in the Latin American countries, is provided by fuelwood. In the Sudan, Senegal, and Niger, fuelwood provides 94, 95, and 99%, respectively, of household energy consumption (Anderson and Fishwick, 1984). Rapidly-increasing populations, particularly in developing nations, will result in increasing demands on forest lands to meet growing agricultural and energy needs.

Biomass Burning

Biomass burning, in addition to contributing to the atmospheric CO_2 budget, contributes approximately 10-20% of total annual CH_4 emissions, 5-15% of the N_2O emissions, 10-35% of the NO_x emissions, and 20-40% of the CO emissions (Crutzen et al., 1979; WMO, 1985; Logan, 1983; Stevens and Engelkemeir, 1988; and Andreae et al., 1988). These estimates are for instantaneous emissions from combustion. Recent research has shown that biomass burning also results in longer-term (at least up to 6 months after the burn) emissions of NO and N_2O due to enhancement of biogenic soil emissions (Anderson et al., 1988). Estimates of emissions of trace gases due to biomass burning are very uncertain for two reasons: 1) data on amounts and types of biomass burned are scarce, and 2) emissions per unit of biomass burned are highly variable.

Activities associated with biomass burning include agriculture, colonization, wildfires and prescribed fires, and burning of industrial wood and fuelwood. Currently, agricultural burning, due to shifting agriculture and burning of agricultural wastes is estimated to account for over 50% of the biomass burned annually (Table 4-10). Biomass burning is a particularly important source of trace-gas emissions in the tropics, where forest exploitation is unsurpassed. Continued rapid population growth and exploitation of forests may substantially increase emissions from biomass burning in the future.

Wetland Loss

Annual global emissions of CH_4 from freshwater wetlands are estimated to be 110 Tg, approximately 25% of the total annual source of 400 to 600 Tg (Matthews and Fung, 1987). Of the approximately 530 million ha producing this CH_4, 39% is forested bog, 17% is nonforested bog, 21% is forested swamp, 19% nonforested swamp, and 4% alluvial formations.[18] The bulk of the bog acreage is located between 40°N and 70°N, while swamps predominate between 10°N and 30°S. Alluvial formations are concentrated between 10°N and 40°S (see Figure 4-13). Coastal saltwater and brackish water environments produce minor amounts of CH_4 in comparison, probably due to the inhibitory effects of dissolved sulfate (SO_4) in the interstitial water of salt-marsh sediments (DeLaune et al., 1983; Bartlett et al., 1985).

[18] Bogs are peat- or organic-rich systems, usually associated with waterlogging and seasonal freeze-thaw cycles; swamps are low-organic formations occurring most commonly in the tropics, and alluvial formations are low-organic riverine formations.

TABLE 4-10

Summary Data on Area and Biomass Burned

Activity	Burned and/or Cleared Area (million ha)	Burned Biomass (100 Tg dry matter)
Burning due to shifting agriculture	21-62 (41)	9-25 (17)
Deforestation due to population increase and colonization	8.8-15.1 (12.0)	5.5-8.8 (7.2)
Burning of savanna and brushland	(600)	4.8-19 (11.9)
Wildfires in temperate and boreal forests	4.0-6.5 (5.4)	1.9-3.2 (2.6)
Prescribed fires in temperate forests	2.0-3.0 (2.5)	0.1-0.2 (0.2)
Burning of industrial wood and fuelwood		10-11 (10.5)
Burning of agricultural wastes		17-21 (19)
TOTAL	630-690 (660)	48-88 (68)

Data in parentheses represent average values.

Source: Crutzen et al., 1979.

FIGURE 4-13

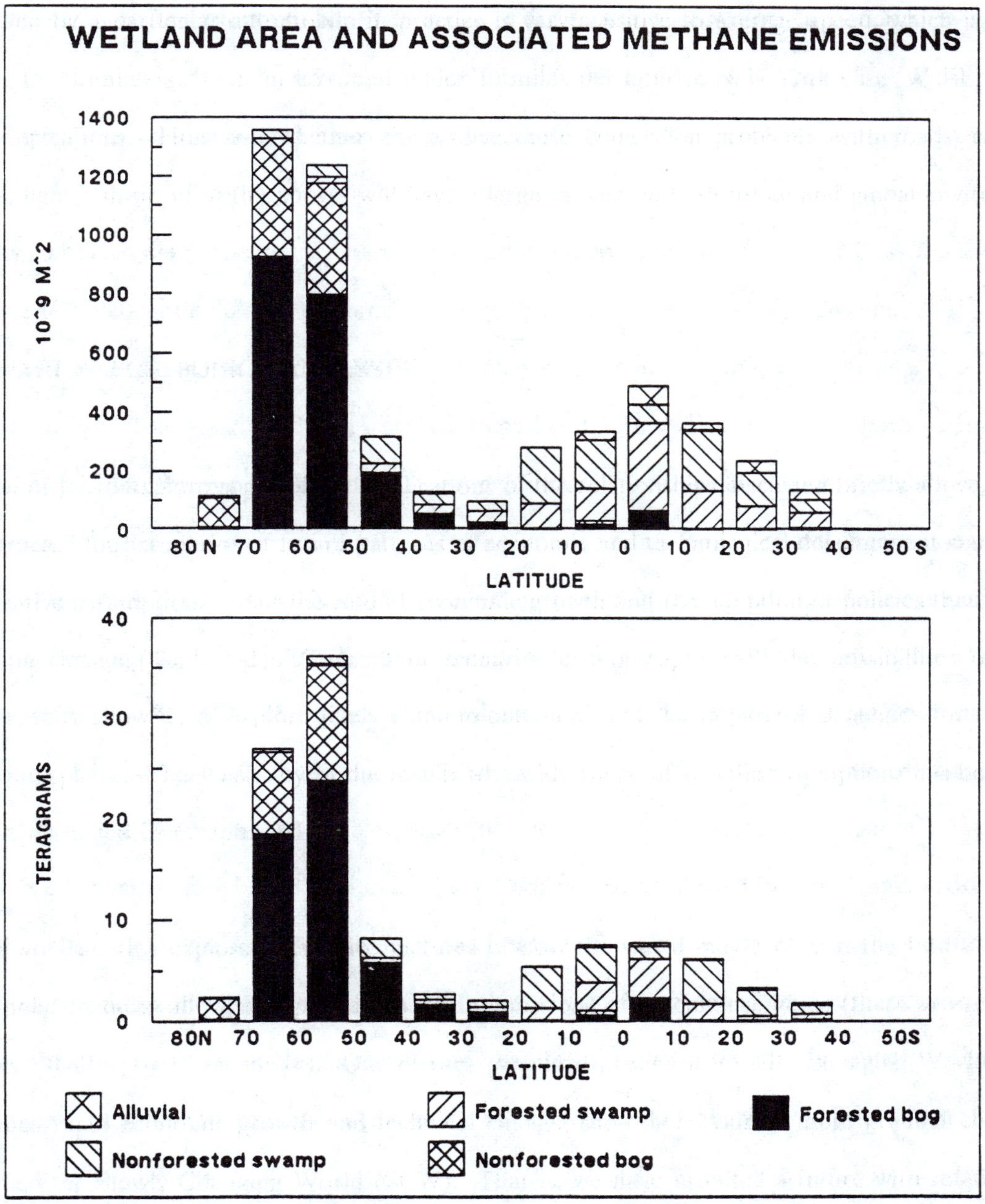

Figure 4-13. Estimated latitudinal distribution of wetland area (top) and associated methane emissions (bottom). Forested and non-forested bogs located between 40° and 70°N account for approximately 50% of the current CH_4 emissions from wetlands. (Source: Matthews and Fung, 1987.)

The latitudinal distribution of wetland CH_4 emissions is estimated to be very similar to the latitudinal distribution of freshwater wetland area. About 50% of the emissions originate between 50°N and 70°N, and about 25% between 20°N and 30°S. The source of the high-latitude emissions is organic-rich bogs, while most of the low-latitude emissions come from swamps (see Figure 4-13).

Between 25 and 50% of the world's original swamps and marshes have been eliminated by human activities (IIED and WRI, 1987). For centuries people have drained and filled marshes and swamps to create dry land for agricultural and urban development. Wetland areas have been converted to open water by dredging and installation of flood-control levees, and have been used as disposal sites for dredge materials and solid wastes. Peat mining and pollution from agricultural and industrial runoff have also contributed to the destruction of wetlands. By 1970, more than half of the original wetland acreage in the United States had been destroyed (IIED and WRI, 1987). Between the mid-1950s and mid-1970s, there was a net loss of wetlands in the United States of approximately 4.6 million ha, 97% of which occurred in inland freshwater areas (U.S. OTA, 1984). Agricultural conversions were responsible for 80% of this freshwater wetland loss.[19] Wetland loss has also been extensive in Europe and the Asia-Pacific region. For example, approximately 40% of the coastal wetlands of Brittany, France, have been lost in the last 20 years, and 8100 ha of wetlands on the east coast of England have been converted to agricultural use since the 1950s. Large-scale wetland losses have not been as prevalent in the developing world, but rising populations will result in increasing demands for agricultural expansion. There is already pressure to develope two large wetland systems in Africa, the Okavango Swamps of Botswana and the Sudd Swamps of southern Sudan, for agricultural use (IIED and WRI, 1987).

[19] For example, drainage of prairie potholes in Iowa to provide new farmland has resulted in the reduction of Iowa's original wetlands by over 98%, from 930,000 ha when settlement began, to 10,715 ha today.

AGRICULTURAL ACTIVITIES

Three agricultural activities contribute directly to atmospheric emissions of greenhouse gases: enteric fermentation in domestic animals, rice cultivation, and use of nitrogenous fertilizer. Global demand for food and agricultural products has more than doubled since 1950, fueled by rising populations and incomes. Agricultural advancements during the post-war years, such as the "Green Revolution," brought improvements in soil management and disease control, new high-yielding varieties of crops, increased application of commercial fertilizers, and increased use of machinery. Between 1950 and 1986, world grain production increased from 624 to 1,661 million tons and average yield more than doubled, from 1.1 to 2.3 tons per ha (Wolf, 1987). Over this same time interval, growth of various domestic animal populations ranged from 20 to 150% (Crutzen et al., 1986) and fertilizer consumption grew approximately 750% (Herdt and Stangel, 1984). According to projections by the Food and Agriculture Organization of the United Nations, by the year 2000, a world population of about 6 billion will require an agricultural output approximately 50 to 60% greater than that required in 1980 (FAO, 1981).

Enteric Fermentation In Domestic Animals

Methane is produced as a by-product of enteric fermentation in herbivores, a digestive process by which carbohydrates are broken down by microorganisms into simple molecules for absorption into the bloodstream. Both ruminant animals (e.g., cattle, dairy cows, sheep, buffalo, and goats) and some non-ruminant animals (e.g., pigs and horses) produce CH_4. The highest CH_4 losses are reported for ruminants (approximately 4-9% of total energy intake), which are able to digest cellulose due to the presence of specific microorganisms in their digestive tracts. The amount of CH_4 that is released from both ruminant and non-ruminant animals depends on the type, age, and weight of the animal, the quality and quantity of feed, and the energy expenditure of the animal.

Of the annual global source of 400-600 Tg CH_4, domestic animals contribute approximately 65-85 Tg (Crutzen et al., 1986; Lerner et al., 1988). Domestic animals that produce the bulk of the CH_4 are (in decreasing order of amount produced) cattle, dairy cows, buffalo, goats, sheep, camels, pigs, and horses. Currently, approximately 57% comes from cattle, and 19% from dairy cows. Domestic animals in six countries, India, the USSR, Brazil, the U.S., China, and Argentina, produce over 50% of the methane by enteric fermentation (Lerner et al., 1988).

The domestic animal population has increased considerably during the last century. Between the early 1940s and 1960s, increases in global bovine and sheep populations averaged 2% per year. Since the 1960s, the rates of increase have slowed somewhat, to 1.2% and 0.6% per year, respectively (see Figure 4-14). The annual increases in global populations of pigs, buffalo, goats, and camels since the 1960s have been comparable: 1.4%, 1%, 1.2%, and 0.5%, respectively. The horse population declined about 0.25% per year. For comparison, the average annual increase in global human population since the 1960s has been about 1.8%.

Rice Cultivation

Anaerobic decomposition in flooded rice fields produces methane, which escapes to the atmosphere by ebullition (bubbling) up through the water column, diffusion across the water/air interface, and transport through the rice plants. Research suggests that the amount of CH_4 released to the atmosphere is a function of rice species, number and duration of harvests, temperature,

FIGURE 4-14

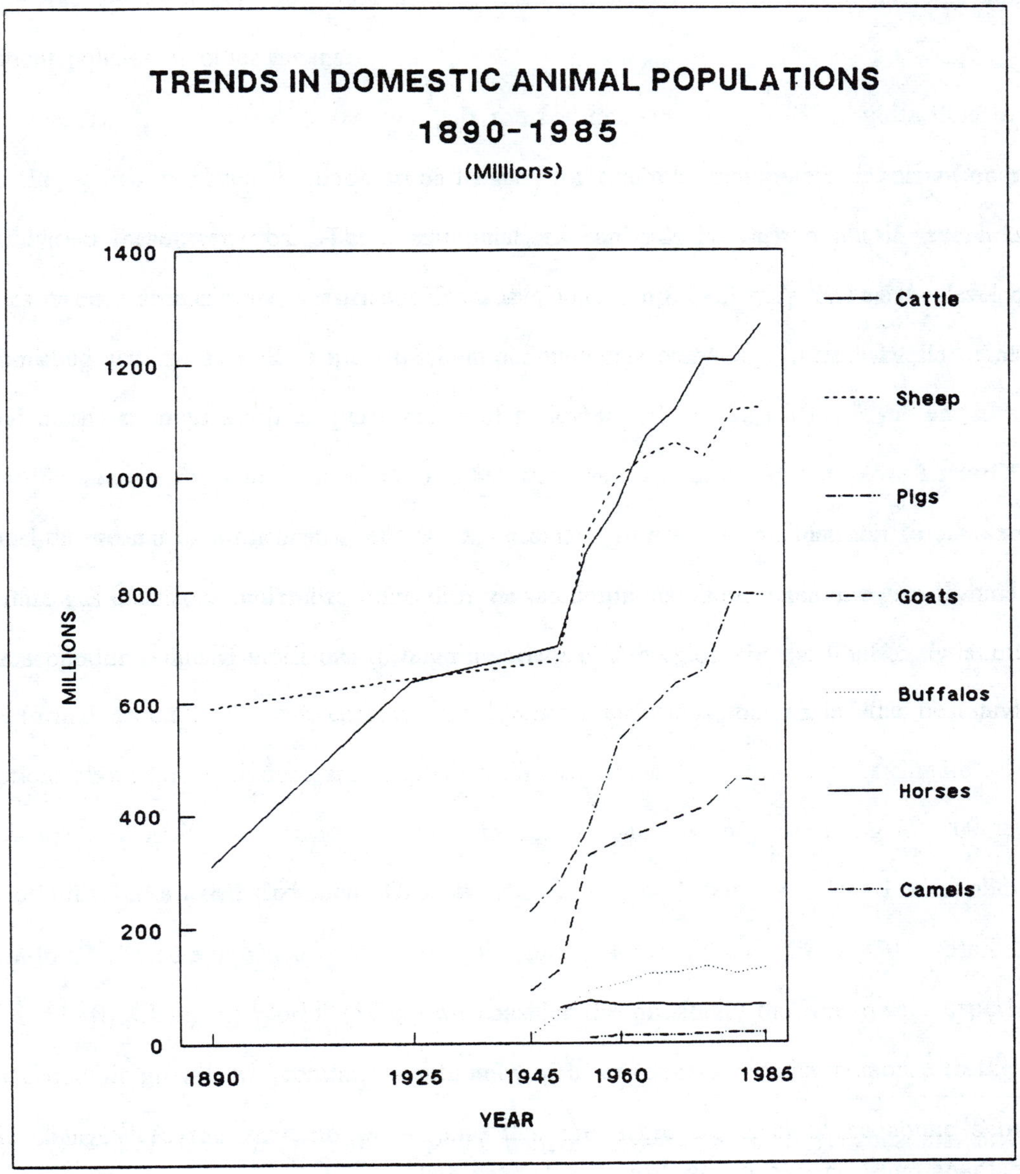

Figure 4-14. Global domestic animal populations have grown by about 0.5 to 2.0% per year during the last century. Currently, domestic animals account for about 15% of the annual anthropogenic CH_4 emissions. Note: The cattle population figures include dairy cows. (Sources: Crutzen et al., 1986; FAO, 1971, 1982, 1986a.)

irrigation practices, and fertilizer use (Holzapfel-Pschorn and Seiler, 1986; Seiler et al., 1984; Cicerone et al., 1983).

Rice cultivation has grown tremendously since the mid-1900s, due both to increases in crop acreage and yields.[20] Between 1950 and 1984, rough rice production grew from 163 to 470 million tons, nearly a 200% increase.[21] During the same time, harvested rice paddy area increased approximately 40%, from 103 to 148 million ha, and average global yields doubled, from 1.6 to 3.2 tons per ha (IRRI, 1985).[22] Average yields higher than 5 tons per ha have already been obtained in parts of the developed world (FAO, 1986a). The increase in rice production has been due both to the "Green Revolution" of the 1960s, which resulted in the development and dissemination of high-yield varieties of rice and an increase in fertilizer use, and to a significant expansion of land area under cultivation. Methane emissions are probably primarily a function of area under cultivation, rather than yield, although yield could influence emissions, particularly if more organic matter is incorporated into the paddy soil.

Over 90% of global rice acreage and production occurs in Asia. Five Asian countries, China, India, Indonesia, Bangladesh, and Thailand, account for 75% of global production and 73% of the harvested area (IRRI, 1986; see Figures 4-15 and 4-16). Rice fields contribute 60-170 Tg of methane

---

[20] Rice statistics are for rice grown in flooded fields, i.e., they do not include upland rice, since methane emissions result only from flooded rice fields.

[21] Rough rice, also called paddy rice, is rice with the hull, or husk, attached. The hull contributes about 20% of the weight of rough rice. The kernel remaining after the hull is removed is brown rice. Milling of brown rice, which removes the bran, followed by polishing, results in white rice.

[22] Harvested area is the area under cultivation multiplied by the number of crops per year. For example, 1 ha that is triple-cropped is counted as 3 ha of harvested area.

FIGURE 4-15

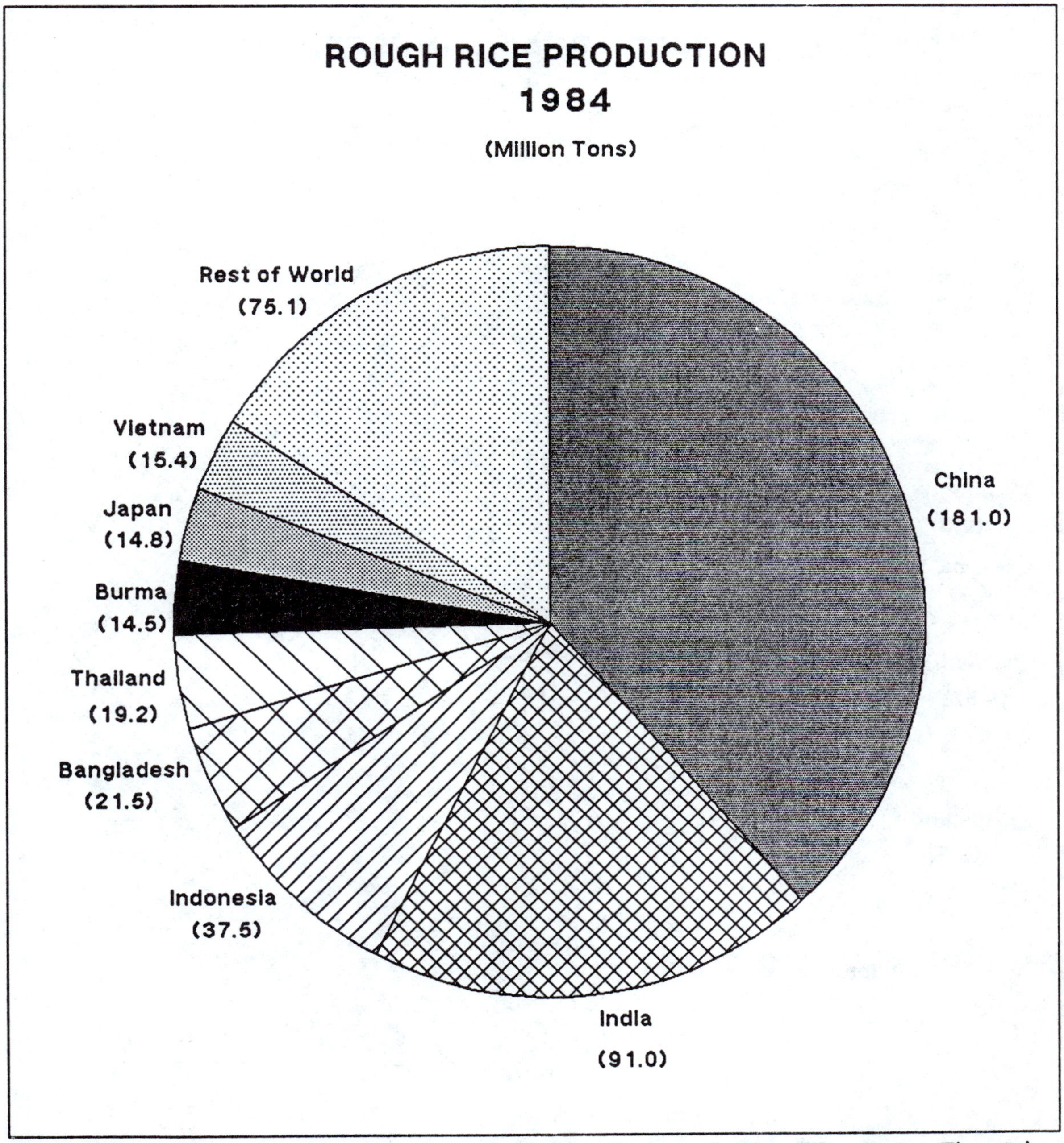

Figure 4-15. Distribution of the total rough rice production of 470 million tons. Five Asian countries, China, India, Indonesia, Bangladesh, and Thailand, accounted for approximately 75% of the 1984 global rice production. (Source: IRRI, 1986.)

FIGURE 4-16

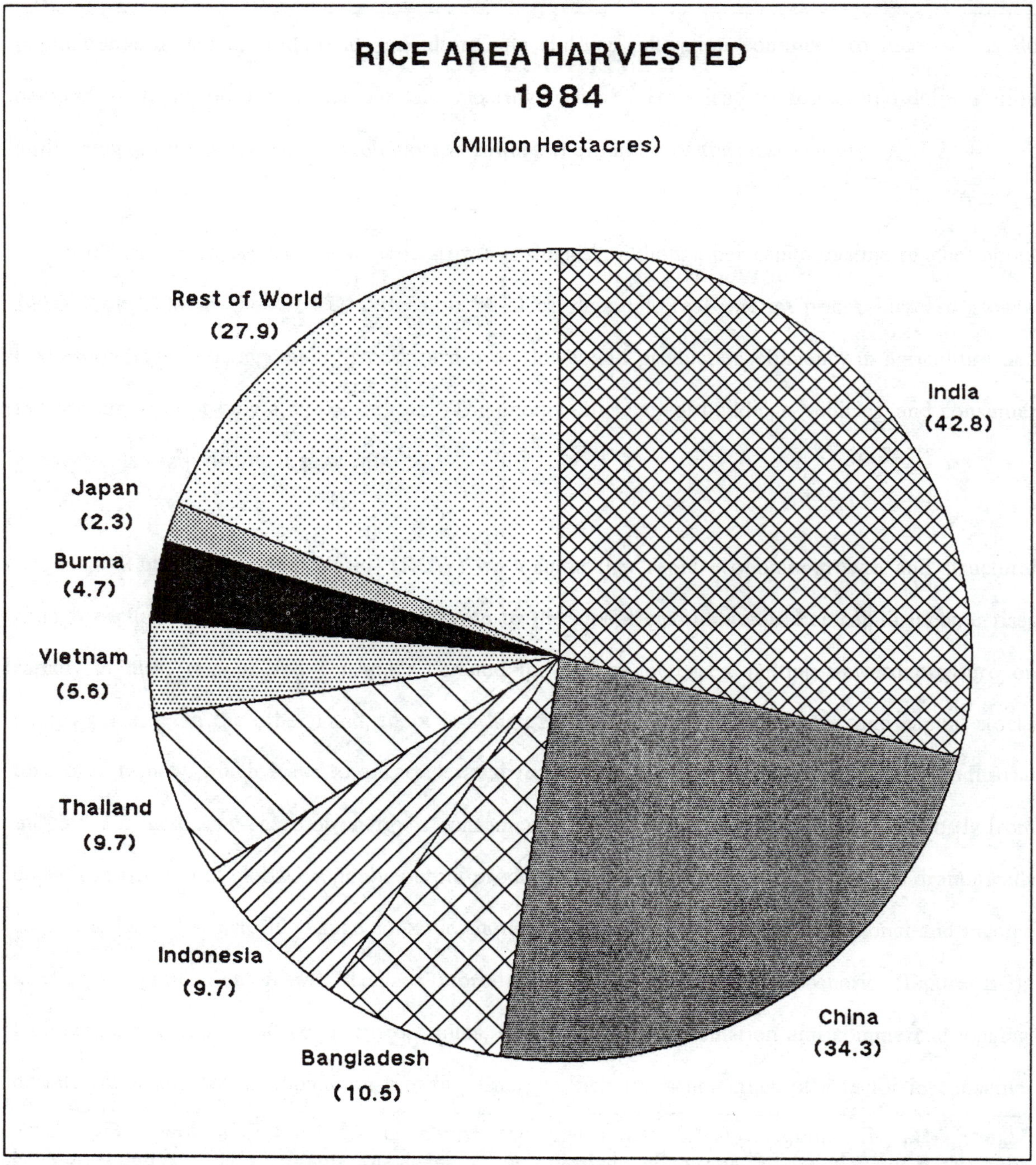

Figure 4-16. Distribution of the total harvested rice paddy area of 148 million ha. Five Asian countries, India, China, Bangladesh, Indonesia and Thailand accounted for 73% of the 1984 rice acreage harvested. (Source: IRRI, 1986.)

per year to the atmosphere, or approximately 20% of the global flux (Cicerone and Oremland, 1989). This estimate is highly uncertain because there have been no comprehensive rice-paddy flux measurements in the major rice-producing countries in Asia.

Use of Nitrogenous Fertilizer

Nitrous oxide is released through microbial processes in soils, both through denitrification and nitrification. Nitrogenous fertilizer application enhances N_2O flux rates, since some of the applied fixed N is converted to N_2O and released to the atmosphere. The amount of N_2O released depends on rainfall, temperature, the type of fertilizer applied, mode of application, and soil conditions.

Nitrogen is currently the most abundant commercial fertilizer nutrient consumed worldwide. Its dominance in the fertilizer markets has increased steadily over the last few decades, from 28% of total nutrients (nitrogen, phosphorus, and potassium) in 1950 to 64% in 1981 (Herdt and Stangel, 1984). Approximately 70.5 million tons N was consumed worldwide in 1984/1985 in the form of nitrogenous fertilizers (FAO, 1987). A preliminary estimate suggests that this produced N_2O emissions of 0.14-2.4 Tg N of the global source of approximately 8-22 Tg N per year (Fung et al., 1988) although this estimate is highly uncertain. Experiments to determine the fraction of fertilizer nitrogen lost to the atmosphere as nitrous oxide have shown a wide range of results (see Table 4-11, and Chapter II). Anhydrous ammonia, which requires sophisticated equipment for application (it is injected under pressure into the soil), is used exclusively in the United States. It comprises about 38% of the U.S. nitrogenous fertilizer consumption. Urea, which is usually broadcast as pellets by hand, comprises about 69% and 58% of nitrogenous fertilizer consumption in Asia and South America, respectively.

TABLE 4-11

Nitrous Oxide Emissions by Fertilizer Type

Fertilizer Type	Percent of Nitrogenous Fertilizer Evolved as N_2O
Anhydrous Ammonia	0.5 to 6.84
Ammonium Nitrate	0.04 to 1.71
Ammonium Type	0.025 to 0.1
Urea	0.067 to 0.5
Nitrate	0.001 to 0.50

Source: Eichner, 1988; Galbally, 1985.

Asia, Western Europe, Eastern Europe, and North America consume the major share of the world's nitrogenous fertilizers (collectively, about 85%). China, the Soviet Union, and the United States together account for approximately one-half of the world's fertilizer consumption. The twelve largest nitrogen fertilizer consumers, all of which consume more than one million tons N annually, are (in decreasing order): China, the United States, the Soviet Union, India, France, the United Kingdom, West Germany, Canada, Indonesia, Poland, Mexico, and Italy (see Figure 4-17). Together, these twelve countries account for approximately 74% of the annual nitrogenous fertilizer consumption.

Although developed nations will probably increase their consumption of commercial fertilizer over the next few decades, most of the increased demand will occur in developing nations. The World Bank estimates that over 90 million tons N will be consumed in 1997/98, a 30% increase over consumption in 1986/87. Almost 50% of the growth between 1986/87 and 1997/98 is expected to occur in the developing nations (World Bank, 1988).

IMPACT OF CLIMATIC CHANGE ON ANTHROPOGENIC EMISSIONS

Climate change will affect human activity in a myriad of ways, and thus influence anthropogenic emissions of greenhouse gases (see Chapter III for a discussion of the biogeochemical feedbacks of climate change). The impact of climatic change on land-use patterns and agricultural practices could be particularly significant in influencing the trace gas emissions from these sources. The magnitude (or even the direction) of such changes have not been examined to date. More information is available regarding the impact of climatic change on electric utilities (Linder et al., 1987). A brief

FIGURE 4-17

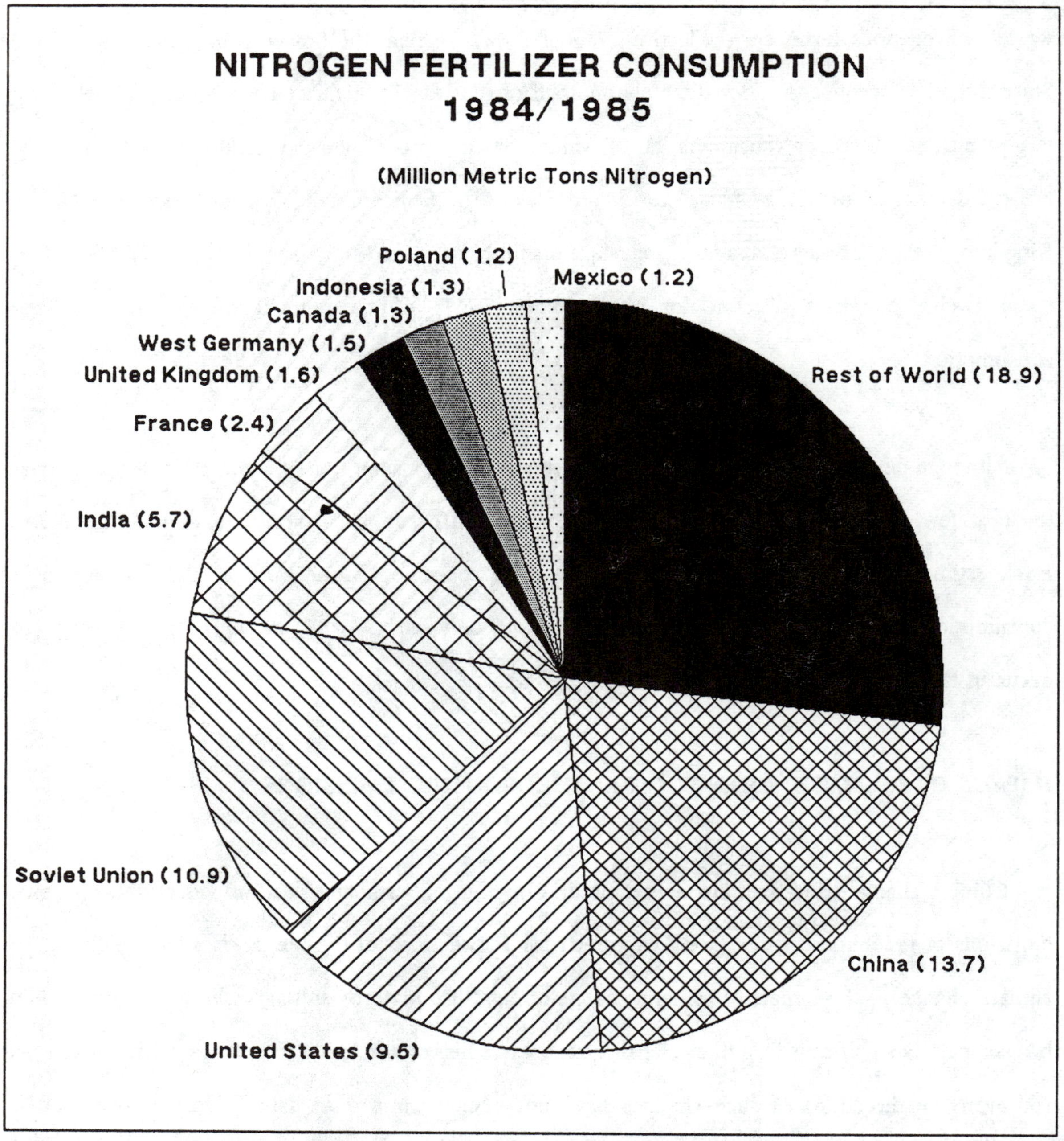

Figure 4-17. Distribution of the total nitrogenous fertilizer consumption of 70.5 million tons N. China, the United States, and the Soviet Union together accounted for just over 50% of the 1984/1985 global fertilizer consumption. Currently, 5-35% of the total anthropogenic N_2O emissions are attributed to nitrogenous fertilizer consumption. (Source: FAO, 1987.)

discussion of this subject is presented here as an illustration of some ways in which climatic change can, in turn, influence trace gas emissions.

Linder and Inglis (1988) estimate that annual electricity consumption increases by 0.5 to 2.7%/°C for utilities in the United States, depending on the local climate and the fraction of buildings with electrical heating and air-conditioning equipment. If climate change leads to increases in ownership levels, then substantially greater sensitivities are possible (Linder et al., 1987). Currently, 37% of total CO_2 emissions from fossil fuels are produced by electric utilities and this share is expected to increase in the future (see Chapter V). Applying the U.S. average sensitivity of 1.0%/°C obtained by Linder and Inglis (1988) to the rest of the world implies a feedback on CO_2 emissions of 0.4%/°C. This feedback would be offset to an extent that has not been estimated by lower fuel use for heating, but as the penetration of air conditioning rises in developing countries this feedback could increase.

Climate change may affect the electricity industry from the supply side as well. When steam is produced to generate electricity in a power plant, either water (usually from a nearby reservoir or river) or air is used as a coolant to condense the steam back into water and start the process over again. Higher atmospheric temperatures will result in warming of these coolants, and reduction in the efficiency of the power plants. This effect is not likely to be as significant as others, however, since seasonal temperature changes are already much greater than the warming predicted for the next century (Linder et al., 1987).

More immediate and acute effects of climate change on electric utilities are likely to occur due to reduced availability of water. The drought of the summer of 1988 resulted in such low river levels in the U.S. Midwest, that some electric plants were forced to reduce generation due to lack of cooling water. More frequent and severe droughts would also result in reduced hydropower for generation of electricity. (This change would also affect barge shipping, since many rivers would become unnavigable, and result in increased trace gas emissions from truck and rail transport.)

Sea-level rise and lowered stream flows resulting from climate change would also have adverse effects on electric utilities. Salinities in rivers and estuaries would increase, and stream chemistry could change, so that the water may become too corrosive to be used as a coolant. A few power plants in the United States use saltwater for cooling purposes, so the technology exists to adapt to more saline coolants, although the conversion process is costly.

These feedback mechanisms are likely to have a smaller influence on future warming than the biogeochemical feedbacks discussed in Chapter III. The impact of climatic change on anthropogenic trace gas emissions may nevertheless prove to be important and should be investigated further.

REFERENCES

Alig, R.J. 1988. Projecting land cover and use changes. (review draft). In Bones, J.T., ed. *An Analysis of the Land Situation in the United States: 1989-2040*. General Technical Report WO-OO. USDA Forest Service, Washington, D.C. in press.

Anderson, D., and R. Fishwick. 1984. *Fuelwood Consumption and Deforestation in African Countries*. World Bank Staff Working Papers, Number 704. The World Bank, Washington, D.C. 52 pp.

Anderson, I.C., J.S. Levine, M.A. Poth, and P.J. Riggan. 1988. Enhanced biogenic emissions of nitric oxide and nitrous oxide following surface biomass burning. *Journal of Geophysical Research* 93:3393-3898.

Andreae, M.O., E.V. Browell, M. Garstang, G.L. Gregory, R.C. Harriss, G.F. Hill, D.J. Jacob, M.C. Pereira, G.W. Sachse, A.W. Setzer, P.L. Silva Dias, R.W. Talbot, A.L. Torres, and S.C. Wofsy. 1988. Biomass-burning emissions and associated haze layers over Amazonia. *Journal of Geophysical Research* 93:1509-1527.

Bartlett, K.B., R.C. Harriss, and D.I. Sebacher. 1985. Methane flux from coastal salt marshes. *Journal of Geophysical Research* 90:5710-5720.

Berry, B.J.L., and F.E. Horton. 1974. *Urban Environmental Management, Planning for Pollution Control*. Prentice-Hall, Inc., Englewood Cliffs. 425 pp.

Bingemer, H.G., and P.J. Crutzen. 1987. The production of methane from solid wastes. *Journal of Geophysical Research* 92:2182-2187.

Bolle, H.J., W. Seiler, and B. Bolin. 1986. Other greenhouse gases and aerosols. In Bolin, B., B.R. Döös, J. Jager, and R.A. Warrick, eds. *The Greenhouse Effect, Climate Change, and Ecosystems*. Scope 29. John Wiley and Sons, Chichester. 157-203.

CEQ (Council on Environmental Quality). 1982. *Environmental Quality 1981, 12th Annual Report*. Washington, D.C.

Cicerone, R.J., and R.S. Oremland. 1989. Biogeochemical aspects of atmospheric methane. *Global Biogeochemical Cycles*. in press.

Cicerone, R.J., J.D. Shetter, and C.C. Delwiche. 1983. Seasonal variation of methane flux from a California rice paddy. *Journal of Geographical Research* 88:11022-11024.

Cointreau, S.J. 1982. *Environmental Management of Urban Solid Wastes in Developing Countries*. The World Bank, Washington, D.C. 214 pp.

Crutzen, P.J. 1987. Role of the tropics in atmospheric chemistry. In Dickinson, R., ed. *Geophysiology of Amazonia*. John Wiley & Sons, New York. 107-133.

Crutzen, P.J., I. Aselmann, and W. Seiler. 1986. Methane production by domestic animals, wild ruminants, other herbivorous fauna, and humans. *Tellus* 38B:271-284.

Crutzen, P.J., L.E. Heidt, J.P. Krasnec, W.H. Pollock, and W. Seiler. 1979. Biomass burning as a source of atmospheric gases CO, H_2, N_2O, NO, CH_3Cl, and COS. *Nature* 282:253-256.

DeLaune, R.D., C.J. Smith, and W.H. Patrick. 1983. Methane release from gulf coast wetlands. *Tellus* 35B: 8-15.

Detwiler, R.P., and C.A. Hall. 1988. Tropical forests and the global carbon cycle. *Science* 239:42-47.

Eichner, M.J. 1988. Current Knowledge of Fertilizer-Derived Nitrous Oxide Emissions. Paper prepared for U.S. Environmental Protection Agency, Washington, D.C.

Ehrlich, P.R., and A.H. Ehrlich. 1972. *Population, Resources, Environment.* W.H. Freeman & Company, San Francisco.

Farman, J., B. Gardiner, and J. Shanklin. 1985. Large losses of total ozone in Antarctica reveal seasonal ClO_x/NO_x interaction. *Nature* 315:207-210.

FAO (Food and Agriculture Organization of the United Nations). 1971, 1982, 1986a. *1970, 1981 and 1985 FAO Production Yearbooks*. Vols. 24, 35, and 39. FAO, Rome.

FAO (Food and Agriculture Organization of the United Nations). 1981. *Agriculture: Toward 2000.* FAO, Rome. 134+ pp.

FAO (Food and Agriculture Organization of the United Nations). 1985. *Tropical Forestry Action Plan*. Committee on Forest Development in the Tropics, FAO, Rome. 159 pp.

FAO (Food and Agriculture Organization of the United Nations). 1986b. *1985 Yearbook of Forest Products*, 1974-1985. FAO, Rome. 348 pp.

FAO (Food and Agriculture Organization of the United Nations). 1987. *1986 FAO Fertilizer Yearbook*. Vol. 36, FAO, Rome. 124 pp.

Fung, I., E. Matthews, and J. Lerner. 1988. Trace gas emissions associated with agricultural activities. Presented at the EPA Workshop on Agriculture and Climate Change, February 29 - March 1, 1988, Washington, D.C.

Galbally, I.E. 1985. The emission of nitrogen to the remote atmosphere: Background paper. In Galloway et al., eds. *The Biogeochemical Cycling of Sulfur and Nitrogen in the Remote Atmosphere.* D. Reidel Publishing Company, Dordrecht, Holland. 27-53.

Hall, D.O., Barnard G.W., and Moss P.A. 1982. *Biomass for Energy in the Developing Countries*. Pergamon Press, Oxford. 220 pp.

Hammitt, J.K., F. Camm, P.S. Connell, W.E. Mooz, K.A. Wolf, D.J. Wuebbles, and A. Bamezai. 1987. Future emission scenarios for chemicals that may deplete stratospheric ozone. *Nature* 330:712-716.

Hammitt, J.K., K.A. Wolf, F. Camm, W.E. Mooz, T.H. Quin, and A. Bamezai. 1986. *Product Uses and Market Trends for Potential Ozone-Depleting Substances, 1985-2000*. Prepared for the U.S. EPA. Rand, Santa Monica. 107 pp.

Hansen, J., I. Fung, A. Lacis, D. Rind, S. Lebedeff, R. Ruedy, and G. Russell. 1988. Global climate changes as forecast by Goddard Institute for Space Studies three-dimensional model. *Journal of Geophysical Research* 93:9341-9364.

Herdt, R.W., and P.J. Stangel. 1984. Population, rice production, and fertilizer outlook. In *Organic Matter and Rice*. International Rice Research Institute, Los Banos. 1-34.

Hoffman, M., ed. 1987. *The World Almanac and Book of Facts*. Ballantine Books, New York. 927 pp.

Holzapfel-Pschorn, A., and W. Seiler. 1986. Methane emission during a cultivation period from an Italian rice paddy. *Journal of Geophysical Research* 91:11803-11814.

Houghton, R.A., R.D. Boone, J.R. Fruci, J.E. Hobbie, J.M. Melillo, C.A. Palm, B.J. Peterson, G.R. Shaver, G.M. Woodwell, B. Moore, D.L. Skole, and N. Myers. 1987. The flux of carbon from terrestrial ecosystems to the atmosphere in 1980 due to changes in land use: Geographic distribution of the global flux. *Tellus* 39B:122-139.

Ignatius, A. 1988. Beijing boom: China's birthrate rises again despite a policy of one-child families. *Wall Street Journal*. July 14, p. 1.

IIED and WRI (International Institute for Environment and Development and World Resources Institute). 1987. *World Resources 1987*. Basic Books, Inc., New York. 369 pp.

IRRI (International Rice Research Institute). 1986. *World Rice Statistics 1985*. IRRI, Los Banos. 271 pp.

International Trade Administration. 1987. *A Competition Assessment of the U.S. Cement Industry.* U.S. Department of Commerce, Washington, D.C.

Kresse, K., and J. Ringeltaube. 1982. How resources recovery and appropriate technologies can cut costs at waste management in developing countries. In Thome-Kozmiensky, K.J., ed. *Recycyling in Developing Countries*. E. Freitag, Berlin.

Lerner, J., E. Matthews, and I. Fung. 1988. Methane emission from animals: A global high-resolution database. *Global Biogeochemical Cycles* 2:139-156.

Linder, K., and M. Inglis. 1988. *The Potential Impacts of Climate Change on Electric Utilities: Regional and National Estimates*. Report of ICF Inc. to the U.S. EPA. in press.

Linder, K., M. Gibbs, and M. Inglis. 1987. *Potential Impacts of Climate Change on Electric Utilities*. Report of the New York State Energy Research and Development Authority 88-2. Albany, New York (December).

Logan, J.A. 1983. Nitrogen oxides in the troposphere: Global and regional budgets. *Journal of Geophysical Research* 88:10785-10807.

Marland, G. 1982. The impact of synthetic fuels on carbon dioxide emissions. In Clark, W.C., ed. *Carbon Dioxide Review 1982*. Oxford University Press, New York. 406-410.

Matras, John. 1973. *Populations and Societies*. Prentice-Hall, Inc., Englewood Cliffs.

Matthews, E. and I. Fung. 1987. Methane emissions from natural wetlands: Global distribution area, and environmental characteristics of sources. *Global Biogeochemical Cycles* 1:61-86.

Mintzer, I.M. 1988. *Projecting Future Energy Demand in Industrialized Countries: An End-Use Oriented Approach*. Prepared for U.S. EPA, Washington, D.C. 40+ pp.

OECD (Organization for Economic Cooperation and Development). 1985. *The State of the Environment 1985*. OECD, Paris. 271+ pp.

OECD (Organization for Economic Cooperation and Development). 1987. *Energy Statistics 1970-1985, Volume I*. International Energy Agency, OECD, Paris. 513+ pp.

Radian Corporation. 1988. *Emissions and Cost Estimates for Globally Significant Anthropogenic Combustion Sources of* NO_x, N_2O, CH_4, *CO, and* CO_2. Prepared for U.S. EPA, Research Triangle Park. 136+ pp.

Ramanathan, V. 1987. Observed increases in greenhouse gases and predicted climatic changes. Testimony to the United States Senate, Committee on Energy and Natural Resources. Washington, D.C. November 9, 1987.

Ramanathan, V., R.J. Cicerone, H.B. Singh, and J.T. Kiehl. 1985. Trace gas trends and their potential role in climate change. *Journal of Geophysical Research* 90:5547-5566.

Revelle, R. 1983. Methane hydrates in continental slope sediments and increasing atmospheric carbon dioxide. In National Research Council. *Changing Climate*. National Academy Press, Washington, D.C. 252-261.

Rotty, R.M. 1987. A look at 1983 CO_2 emissions from fossil fuels (with preliminary data for 1984). *Tellus* 39B:203-208.

Rotty, R.M., and C.D. Masters. 1985. Carbon dioxide from fossil fuel combustion: Trends, resources, and technological implications. In Trabalka, J.R., ed. *Atmospheric Carbon Dioxide and the Global Carbon Cycle*. U.S. DOE, Washington, D.C. 63-80.

Sathaye, J.A., A.N. Ketoff, L.J. Schipper, and S.M. Lele. 1988. *An End-Use Approach to Development of Long-Term Energy Demand Scenarios for Developing Countries*. Prepared for U.S. EPA, Washington, D.C. 35 pp.

Seiler, W. 1984. Contribution of biological processes to the global budget of CH_4 in the atmosphere. In Klug, M., and C. Reddy, eds. *Current Perspectives in Microbial Ecology*. American Society for Microbiology, Washington, D.C. 468-477.

Seiler, W., A. Holzapfel-Pschorn, R. Conrad, and D. Scharffe. 1984. Methane emission from rice paddies. *Journal of Atmospheric Chemistry* 1:241-268.

Stevens, C.M., and A. Engelkemeir. 1988. Stable carbon isotopic composition of methane from some natural and anthropogenic sources. *Journal of Geophysical Research* 93:725-733.

United Nations. 1976. *World Energy Supplies: 1950-1974.* United Nations, New York.

United Nations. 1982. *1980 Yearbook of World Energy Statistics.* United Nations, New York.

United Nations. 1983. *1981 Yearbook of World Energy Statistics.* United Nations, New York.

United Nations. 1986. *World Population Prospects.* Population Studies, No. 98. Department of International Economic and Social Affairs, United Nations, New York.

United Nations. 1987. *1985 Energy Statistics Yearbook.* United Nations, New York.

United Nations. 1988. *1986 Energy Statistics Yearbook.* United Nations, New York.

U.S. BOM (U.S. Bureau of Mines). Each year (1949-1986). *Minerals Yearbook, Volume I, Metals and Minerals.* U.S. Government Printing Office, Washington, D.C.

U.S. Bureau of the Census. 1987. *World Population Profile: 1987.* Department of Commerce, U.S. Bureau of the Census, Washington, D.C. 74 pp.

U.S. DOE (U.S. Department of Energy). 1986. *International Energy Annual: 1985.* Energy Information Administration, U.S. DOE, Washington, D.C.

U.S. DOE (U.S. Department of Energy). 1987. *Annual Energy Review 1986.* Energy Information Administration, U.S. DOE, Washington, D.C. 293 pp.

U.S. EPA (U.S. Environmental Protection Agency). 1987. *Assessing the Risks of Trace Gases That Can Modify the Stratosphere.* Office of Air and Radiation, EPA, Washington, D.C.

U.S. EPA (U.S. Environmental Protection Agency). 1988a. *Regulatory Impact Analysis: Protection of Stratospheric Ozone.* Office of Air and Radiation, EPA, Washington, D.C.

U.S. EPA (U.S. Environmental Protection Agency). 1988b. *How Industry is Reducing Dependence on Ozone-Depleting Chemicals.* Office of Air and Radiation, EPA, Washington, D.C. 24 pp.

U.S. OTA (U.S. Office of Technology Assessment). 1984. *Wetlands, Their Use and Regulation.* U.S. Congress, Office of Technology Assessment, Washington, D.C. OTA-0-206. 208 pp.

Wagner, Richard H. 1971. *Environment and Man.* W.W. Norton & Company, Inc., New York. 491 pp.

Wolf, E. 1987. Raising agricultural productivity. In Starke, L., ed. *State of the World 1987.* W.W. Norton & Company, New York. 139-156.

World Bank. 1987. *World Development Report 1987.* Oxford University Press, New York. 285 pp.

World Bank. 1988. *World Bank, FAO, UNIDO Fertilizer Working Group Nitrogen Supply, Demand and Balances for 1986/87 to 1992/93.* Washington, D.C.

World Energy Conference. 1980. *Survey of Energy Resources 1980.* Prepared for the 11th World Energy Conference, Munich, 8-12 September 1980, World Energy Conference, London. 352+ pp.

WMO (World Meteorological Organization). 1985. *Atmospheric Ozone 1985: Assessment of Our Understanding of the Processes Controlling its Present Distribution and Change.* Volume 1. WMO, Geneva, 392+ pp.

Wuebbles, Donald J. 1983. Chlorocarbon emissions scenarios: Potential impact on stratospheric ozone. *Journal of Geophysical Research* 88:1433-1443.

Zachariah, K.C., and M.T. Vu. 1988. *World Population Projections 1987-88 Edition*. Johns Hopkins University Press, Baltimore. 439 pp.

CHAPTER V

THINKING ABOUT THE FUTURE

FINDINGS

- Decisions made in the next few decades, about how electricity is produced, homes are constructed, and cities are laid out, for example, will have an impact on the climate in 2100 and beyond. While it is not possible to predict the level of greenhouse gas emissions over this time period, it is possible to construct scenarios of economic and technological development, and a reasonable range for resulting greenhouse gas emissions, atmospheric concentrations, and global temperature changes. Global temperature change estimates provide an indicator for the rate and magnitude of climatic change.

- Carbon dioxide emissions are likely to grow by a factor of 2 to 5 during the next century if stabilizing policies are not adopted, primarily due to expansion of global coal consumption. Options are available, however, that could stabilize or reduce carbon dioxide emissions. Despite the Montreal Protocol to control CFCs, global emissions of these compounds could remain constant or even increase significantly unless the agreement is strengthened. Methane emissions could increase by 60-100% during the next century unless measures to control these emissions are taken.

- Although per capita emissions of greenhouse gases are currently very low in developing countries, their share of global emissions will rise significantly in the future.

- The relative contribution of carbon dioxide to greenhouse warming is likely to increase significantly in the future. Carbon dioxide accounts for more than 70% of the increased commitment to global warming between 2000 and 2100 in all of the scenarios analyzed in this report. This represents a significantly higher estimate of the role of CO_2 compared to its roughly 50% contribution to global warming in the last few decades, but is similar to the estimated contribution of CO_2 to increases in the greenhouse effect over the last century.

- If there is no policy response to the risk of climatic change carbon dioxide concentrations are likely to reach twice preindustrial levels sometime in the latter half of the 21st Century, but total greenhouse gas concentrations equivalent to this level may occur by 2030 or even earlier, and are likely to occur before 2050.

- Even with modest economic growth and optimistic assumptions regarding technical progress, the world could be committed to an equilibrium warming of 1-2°C by 2000, 2-4°C by 2050 and 3-6°C by 2100 (assuming the climate sensitivity to doubling CO_2 is 2.0-4.0°C). Realized warming would be about 2°C by 2050 and 3-4°C by 2100.

- With rapid, but not unprecedented rates of economic growth, the world could be committed to an equilibrium warming of 1-2°C by 2000, 3-5°C by 2050 and 5-10°C by 2100 (assuming that the climate sensitivity to doubling CO_2 is 2.0-4.0°C). Realized warming would be 2-3°C by 2050 and 4-6°C by 2100. Estimated warming commitments greater than 5°C may not be fully realized because the strength of some positive feedback mechanisms may decline as the Earth warms.

- The adoption of policies to limit emissions on a global basis, such as simultaneous pursuit of energy efficiency, non-fossil energy sources, reforestation, the elimination of CFCs, and other measures, could reduce the rate of warming during the 21st century by 60% or more. Even under these assumptions, the Earth could ultimately warm by 1-3°C or more relative to preindustrial times. Extremely aggressive policies to reduce emissions would be necessary to ensure that total warming is less than 2°C.

INTRODUCTION

Although technological advances in industry and agriculture have provided extraordinary wealth to a portion of the global population of over 5 billion people, these technologies have the potential to dramatically alter the Earth's climate by causing changes in the composition of the atmosphere as discussed in Chapters II through IV. Global increases in the atmospheric concentrations of carbon dioxide (CO_2), nitrous oxide (N_2O), methane (CH_4), and chlorofluorocarbons (CFCs) are now well documented (Chapter II), perhaps already committing the Earth to significant climatic change. Myriad human activities are contributing to this situation, and continued population and economic growth raises the prospect of accelerated greenhouse gas buildup in the future (Chapter IV).

If current trends in trace-gas concentrations continue, climatic change could be noticeable to the "man-in-the-street" during the 1990s, and the average surface temperature of the Earth could be warmer than at any time in recorded human history by the second decade of the 21st century (Hansen et al., 1988). If the composition of the atmosphere were stabilized by 2000, on the other hand, detectable climatic change is still possible, but its magnitude would be limited and the rate of change might be similar to natural fluctuations recorded in the geologic record (Hansen et al., 1988).

What will happen in the future cannot be predicted. The future evolution of the atmosphere will depend largely on the paths of economic development and technological change, as well as on the physical, chemical, and biological processes of the Earth-atmosphere system. While we have no control over this system once gases enter the atmosphere, economic and technological change will be influenced by policy choices made at local, national, and international levels. This chapter explores some of the paths the world might follow in the decades ahead and provides an indication of the relative climatic consequences under these alternatives. After a discussion of the economic and social factors that determine emissions, four scenarios of economic and technological development are presented. These scenarios cannot capture all the possibilities, of course; rather, they have been developed in order to explore the probable climatic effects under significantly different, but plausible, economic and technological conditions. The climatic implications of these scenarios are analyzed using an integrated framework described briefly in this chapter and in greater detail in Appendix A. The chapter concludes with the results of this analysis and a comparison of these results with other studies.

APPROACH TO ANALYZING FUTURE EMISSIONS

The scope of this analysis must be global, and because of the long lags built into both the economic and climatic systems, this study must consider a time horizon of more than a century--we chose 2100 as the ending year for the analysis. While this is an eternity for most economists and planners, it is but a moment for geologists. And indeed, decisions made in the next few decades, about how electricity is produced, homes are constructed, and cities are laid out, for example, will have an impact on the climate in 2100 and beyond. Decisions about what kinds of automobiles and other industrial products to produce and how to produce them will also have a profound impact. These choices, which will affect the amount and type of fuel we use to travel, to heat and light our homes and offices, and to run our factories, will influence the magnitude of greenhouse gas emissions for many years.

The vast difference between the energy demand projections of the early 1970s and what has actually occurred illustrates the danger inherent in simple trend extrapolations and, indeed, even in making predictions based on results obtained from more complex models. Our approach then is not to attempt to predict the future, but to construct what we believe are logically coherent scenarios of possible paths of economic and technological development. An analytical framework is used to keep track of the assumptions, data, and relationships needed to define the scenarios. Our intent is to define the probable climatic effects under the various economic/social/technological alternatives and, in so doing, increase the likelihood that these consequences will be taken into account when policy decisions are made. If we believe that under a wide variety of assumptions about long-term economic growth and technological change the world will face severe climate problems in the absence of political or economic forces arising from concerns over the greenhouse problem, then it will be necessary to seriously examine the options available for reducing greenhouse gas emissions.

The difficulty we face is that projections of greenhouse gas emissions are very uncertain, because of uncertainties in world economic growth, future fuel prices (which demonstrably affect both the intensities of their use and the substitution amongst alternative energy sources), future rates of land clearing, and rates of technological change, among other factors. For example, both the vagaries of the world oil market in the medium term, as well as true uncertainties regarding the long-term relationship between the cost of producing fossil fuels and the cost of using those fuels in ways that are relatively benign to the local environment, mean that at best we can only guess at future fossil fuel use.

Another avenue of analysis, however, yields information that can guide policy makers faced with these uncertainties. If we can construct scenarios of future energy demand, land-clearing rates, CFC production, etc., that are driven by reasonable assumptions about population, economic growth, technologies, and energy prices, then we can develop a plausible range of future greenhouse gas emissions. To accomplish this task we must consider the structural factors that determine the quantities and patterns of emissions of radiatively-important gases (i.e., those gases whose presence in the atmosphere contribute to a greenhouse warming).

It is conceptually useful to distinguish between production activities and consumption activities. Production emissions arise largely from the processing of bulk materials--steel from ore, plastics from petroleum, cement and glass from limestone and silicate rock--which requires large amounts of energy per unit of industrial value added (i.e., the difference in value between an industry's products and its inputs) and may also be associated with direct emissions of greenhouse gases. For example, during cement making, $CaCO_3$ is reformed to CaO + CO_2, which is released to the atmosphere, and during the making of plastic foams, CFCs are released. Much lower emissions per unit of value added are generally associated with fabrication and finishing. Food production leads to emissions of methane and nitrous oxide as discussed in Chapter IV, as well as to emissions of CO_2 and other gases as a result of the energy used on and, even more, off the farm. The large amount of energy required to move freight is also attributable to production activities. Consumption leads to greenhouse gas emissions as individuals use energy, primarily in pursuit of comfort (heating and air conditioning) and mobility (automobile and air travel). Other major end-uses for energy include refrigeration, lighting, water heating, and cooking.

Production

As societies develop over time, both the quantity and the structure of activities that influence emissions change radically. For example, energy use per unit of Gross National Product (GNP) has declined steadily and dramatically in industrialized countries, even in periods of declining real energy prices (Figure 5-1). This decline is due to a combination of two factors. First, improvements in production processes, which often save capital and labor as well as energy, reduce the energy intensity per ton of physical output. For example, in steel production modern energy recovery and process technology make it possible to produce a ton of steel using only $13 x 10^9$ joules (13 GJ) of final

FIGURE 5-1

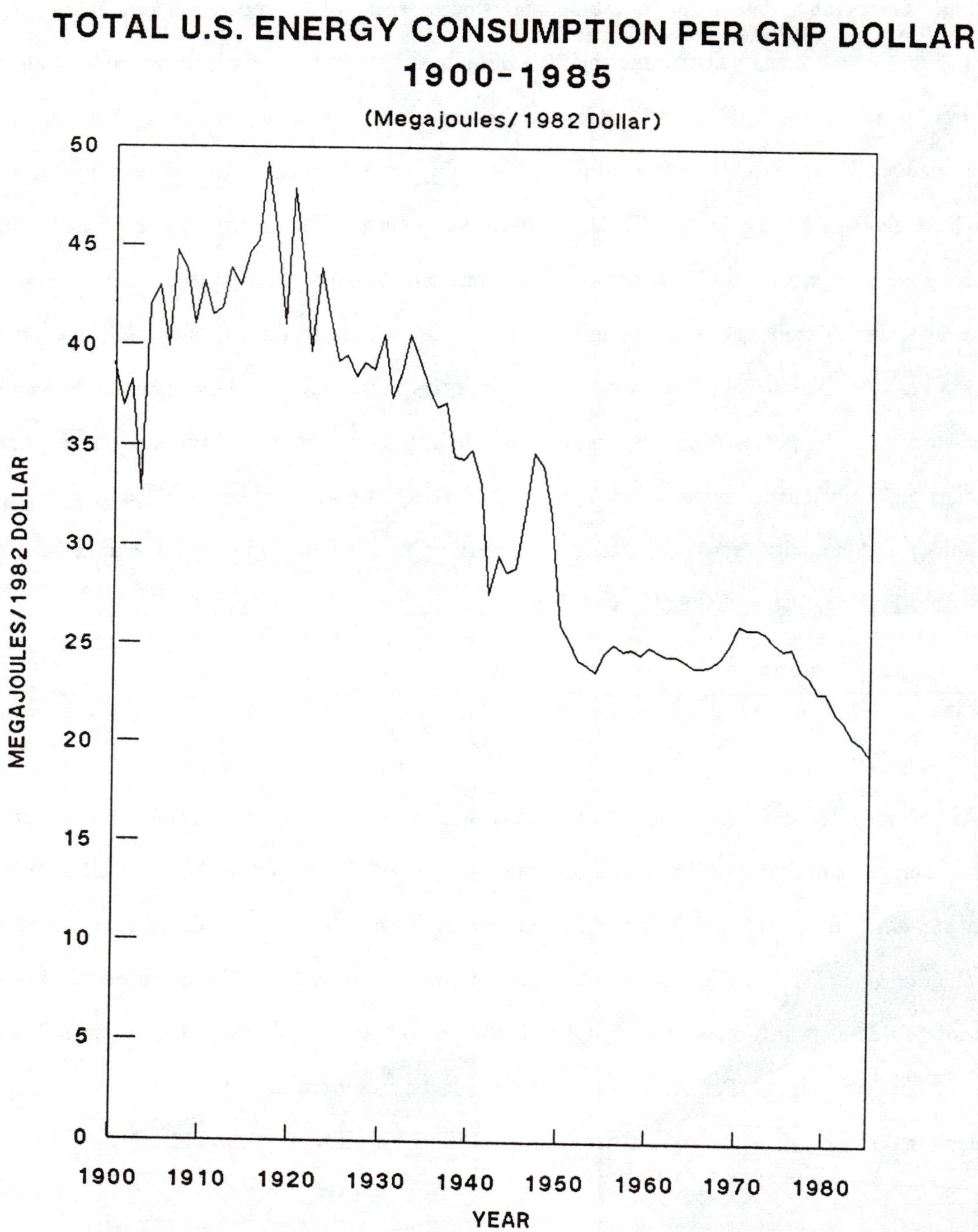

Sources: U.S. DOE, 1987a; U.S. Bureau of the Census, 1975.

energy, less than half of the current U.S. average.[1] New processes under development in Sweden (Elred and Plasma-Smelt) that integrate a number of operations have even lower energy requirements and reduced overall costs (Goldemberg et al., 1988). Second, the bulk of value added by industry tends to shift from basic materials processing to fabrication and finishing as a country's infrastructure matures. Strout (1985) suggests that consumption of steel, cement, and other raw materials begins to decline after income surpasses about $5000 (1985$) per capita (Figure 5-2). These shifts in technology and the mix of products generally increase the share of energy consumed as electricity, but do not significantly increase absolute electricity intensity because efficiency in electric end-uses improves as well (Kahane, 1986). Rapid economic growth over the long term can be expected to accelerate the reduction of industrial energy intensity in wealthier countries by promoting the replacement of old plant and equipment with more efficient technology, as well as by accelerating the shift toward a less energy-intensive product mix.

In industrialized societies services such as public and private administration, health care, and education, are likely to grow faster than GNP, both because much of industry is being redefined as services and because much of our new wealth is being created by the development and transfer of information. Heating, air conditioning, and lighting, which dominate energy and electricity use in buildings today, will become less energy-intensive (even as indoor environmental quality continues to rise) as more efficient technology is adopted. At the same time, information technology is exerting upward pressure on electricity use per square foot in office buildings and schools. The Business Services sector depends more on electricity than does any other sector in the economy, although it still uses less electricity per unit of output than industry. If there is a large increase in electricity use in industrialized countries, it will come from a massive expansion of the service sector.

---

[1] 1 GJ = 0.948 million British Thermal Units (BTU).

FIGURE 5-2

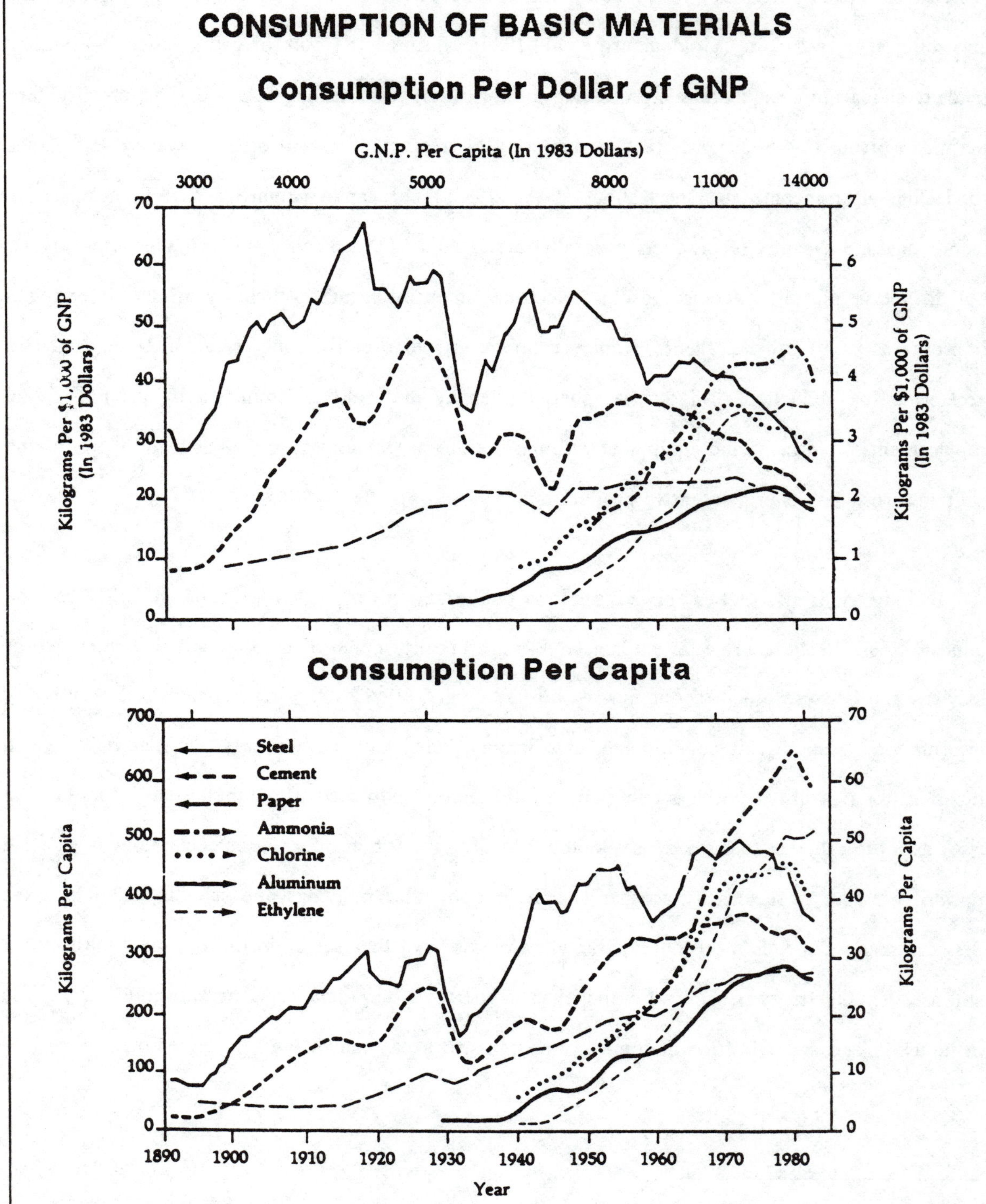

Source: Williams et al., 1987

The greatest potential for large increases in production-related emissions lies in developing countries. As these countries expand their industrial infrastructure the demand for basic materials could skyrocket. But developing countries have the opportunity to take advantage of new processes and materials that sharply reduce the energy required to produce a given level of amenity. As a result, it is unlikely that materials and energy intensity per capita in developing countries would reach the levels of industrialized countries today, even as similar levels of per capita income are achieved. The extent to which developing countries seize these kinds of opportunities will strongly influence future greenhouse gas emissions.

Consumption

The factors influencing emissions arising from consumption are quite different from those that affect production. In developing countries energy use in consumer products can be expected to increase rapidly as the number of households that can afford to acquire fans, televisions, refrigerators, and automobiles grows. Part of the reason that developing country energy demand has historically increased faster than it did in OECD countries is that developing-country households can afford to purchase these products at lower income levels than was the case for industrialized-country households. The declining price-to-income ratios for many of the energy-intensive consumer goods make this possible, with the consequence that developing country energy consumption tends to grow more rapidly than the experience of the industrialized countries might indicate. At the same time, the efficiency of many of these products is increasing, so that per capita energy consumption in developing countries may not reach the levels of industrialized countries today, even if these income levels are surpassed.

As a society becomes wealthier, the penetration of energy-intensive equipment saturates (there are, for example, 600 cars for every 1000 people in the U.S. compared to 6 in Asian countries), and changes in the efficiency of the stock and how the stock is used becomes more important than changes in the levels of ownership alone. For example, as automobile ownership shifts from corporate to private hands the number of vehicles increases dramatically, but the miles driven per vehicle declines. Consumers rarely consider energy use in making major purchases, and many key decisions that determine energy requirements are made by developers rather than the consumers who pay the energy bills (Ruderman et al., 1987). Increased amenity levels can often be achieved while simultaneously reducing energy use and emissions (better insulated houses are more comfortable because they are less drafty and more efficient air conditioners are usually quieter), but more affluent consumers are likely to choose powerful cars and spacious dwellings, paying less attention to the associated operating costs. Further, because a very wide range of efficiency can be achieved with a small impact on total costs over the life-cycle of the product (see Chapter VIII; von Hipple and Levi, 1983; Ruderman et al., 1987), consumers who are concerned about initial cost are unlikely to choose a product whose level of efficiency is optimal from a social perspective.

The level and pattern of mobility may be the most significant uncertainty in future energy use. Will we spend our free time in our air-conditioned homes watching rented movies on the VCR/TV, or are we more likely to drive to the countryside to go for a hike? Not surprisingly, the pattern of automobile use at present (roughly 1/3 of all passenger-kilometers driven in the U.S. are to/from work, 1/3 are for family business, and 1/3 are in pursuit of leisure activities; OTA, 1988) is a function both of distances among where we live, work, and relax, and of how often we choose to move about. Similarly, airline travel, already dominated in the U.S. by personal rather than business travel, is more and more determined by how and where people want to spend their free time. Meanwhile in cities like Hong Kong and Sao Paulo, but also in New York and Los Angeles, congestion is increasingly constraining automobile use. The level of fuel economy and emissions achieved by a particular automobile in practice is very sensitive to average speed, which is down to about 15-20 miles per hour in L.A. and under 10 miles per hour in New York City (Walsh, personal communication). How and whether cities solve these congestion problems--with roads, car pools, buses, light rail, or all of the above--will have a large impact on both urban and global environmental quality.

SCENARIOS FOR POLICY ANALYSIS

In order to explore some of the implications of the relationships discussed briefly above, we have constructed four scenarios of future patterns of economic and technological development starting with alternative assumptions about the rate of economic growth and the adoption of policies that influence climatic change (Table 5-1). These four scenarios cannot capture all the possibilities of course; rather, they allow us to explore likely climatic outcomes and the impact of strategies for stabilizing the atmosphere. The sensitivity of the results to a wide range of specific assumptions has been tested and is discussed in Chapter VI.

Two scenarios explore alternative pictures of how the world may evolve in the future assuming that policy choices allow unimpeded growth in emissions of greenhouse gases (these are referred to as the "No Response" scenarios). One of these scenarios, called a Rapidly Changing World (RCW), assumes rapid economic growth and technical change; the other assumes more gradual change and is called the Slowly Changing World (SCW). That is, we have invented a future with relatively high and robust economic growth, and one representing a more pessimistic view of the evolution of the world's economies. The first world would likely illustrate the upper half of the potential range of future greenhouse gas emissions, because in general higher economic activity means a higher total

TABLE 5-1

Overview of Scenario Assumptions

Slowly Changing World	Rapidly Changing World
Slow GNP Growth	Rapid GNP Growth
Continued Rapid Population Growth	Moderated Poulation Growth
Minimal Energy Price Increases	Modest Energy Price Increases
Slow Technological Change	Rapid Technological Improvements
Carbon-Intensive Fuel Mix	Very Carbon-Intensive Fuel Mix
Increasing Deforestation	Moderate Deforestation
Montreal Protocol/Low Participation	Montreal Protocol/High Participation

Slowly Changing World with Stabilizing Policies	Rapidly Changing World with Stabilizing Policies
Slow GNP Growth	Rapid GNP Growth
Continued Rapid Population Growth	Moderated Population Growth
Minimal Energy Price Increases/Taxes	Modest Energy Price Increases/Taxes
Rapid Efficiency Improvements	Very Rapid Efficiency Improvements
Moderate Solar/Biomass Penetration	Rapid Solar/Biomass Penetration
Rapid Reforestation	Rapid Reforestation
CFC Phase-Out	CFC Phase-Out

energy use and emissions; conversely, the second world could serve as a useful guide to the lower half of the range. In either case, our scenarios are first constructed as if there were no interventions motivated by global climate problems.

In constructing these two worlds/scenarios, we have borne two important ideas in mind. First, evidence is clear that with more rapid economic growth, energy efficiency improves more rapidly than with slower growth (Schurr, 1983). This occurs because innovation proceeds more rapidly and because older, less efficient systems are more rapidly replaced with new technology. History shows, for example, that for almost every country, energy efficiency in industry increases with increasing incomes, as sophistication and scale win over brute force. At the same time, higher incomes allow people to spend more money on two key energy-intensive uses, space conditioning (heating and air conditioning), and automobiles. Thus not all of the technological benefits of rapid economic growth put the brakes on overall energy use. But more rapid economic growth does allow society to put resources aside to improve the efficiency of both space comfort and personal transportation. Similar patterns can be expected in other emissions sectors.

Conversely, slower economic growth retards innovation, in part because both consumers and producers do not see bright economic times that make innovation and expansion into new technologies useful. Comfort and mobility still manage to increase as important drivers of personal energy demand, but at a slower rate. When these two paths are compared, the effect of more rapid efficiency increases in the higher growth world is to narrow the difference in greenhouse gas emissions; that is, the likely difference between emissions in the Rapidly and Slowly Changing Worlds is less than the differences in Gross National Product. This result makes our scenarios somewhat more robust than one might otherwise think.

The second idea concerns energy prices. In a world of high and robust economic growth, which we have **assumed** in the Rapidly Changing scenario, energy demand will likely increase, and in the medium term, so will energy prices. Yet if energy efficiency increases, then energy costs can increase more rapidly than the rate of economic growth and still not consume an increasing share of national wealth and income. In other words, energy prices can rise without putting the brakes on economic growth, as long as the price increases are gradual (CONAES, 1979). But in a world of sluggish economic growth, energy demand rises more slowly, so that energy prices would rise very little. This idea is an additional reason why we believe that energy efficiency increases more rapidly in the high growth scenario (RCW) than in the low growth scenario (SCW).

With these ideas in mind, we can build scenarios of world energy demand by end use and region as well as levels of other activities that emit greenhouse gases. The scenarios are not exact predictions, but serve as guides to the level of emissions associated with each important purpose or end use in the worlds we constructed.

One benefit of using this approach is that we can compare the utilization efficiencies that we assume for the No Response scenarios with those we believe achievable if more than just market forces were acting. Two additional scenarios (referred to as the "Stabilizing Policy" scenarios) start with the same economic and demographic assumptions, but examine the effect that policies could have on global warming. These scenarios are called the Slowly Changing World with Stabilizing Policies (SCWP) and the Rapidly Changing World with Stabilizing Policies (RCWP).

Using our best information about technologies that could become available, or technologies that are already available but not taken up by the market because of market failures or other reasons, we can reconstruct activity patterns that are still consistent with our overriding economic assumptions, but produce much lower levels of greenhouse gas emissions. Key changes are assumed in energy efficiency, the energy supply mix, land-clearing rates, and other factors that might be changed by government policies or other means.

In other words, we keep the basic scenarios but, for example, manipulate important energy use patterns within these scenarios. These manipulations can only be carried out if greenhouse gas emissions in each scenario are constructed from the bottom up, i.e., by specifying the level of each major-emitting activity, as well as the emissions per unit of activity (e.g., total harvested rice paddy area and methane emissions per square meter of paddy).

Thus the scenarios we constructed are a necessary step towards illustrating both ranges of greenhouse gas emissions under two quite different assumptions about economic growth, and where there is scope for reducing emissions through a variety of strategies. In the final analysis, our work can be turned around: we can consider the levels of emissions that under the best and worst assumptions about how emissions are coupled to climatic change leave the world's climate tolerable.

Scenarios with Unimpeded Emissions Growth

In a "Slowly Changing World" (SCW) we consider the possibility that the recent experience of modest economic growth will continue indefinitely, with no concerted policy response to the risk of climatic change. In this scenario we assume that the aggregate level of economic activity (as measured by GNP) increases relatively slowly on a global basis (Table 5-2). Per capita income is stagnant for some time in Africa and the Middle East as rapid population growth continues. Modest increases in per capita income occur elsewhere, and per capita growth rates increase slightly over time in all developing countries as population growth rates slowly decline (Figure 5-3). The share

TABLE 5-2

Economic Growth Assumptions

(percent per year)

			Slowly Changing World		Rapidly Changing World	
	1965-1975	1975-1985	1985-2025	2025-2100	1985-2025	2025-2100
US & OECD	3.9	2.8	1.7	1.0	2.7	1.5
USSR & Eastern Europe	6.2	NA	2.2	1.6	4.3	2.6
Centrally Planned Asia	7.0	7.8	3.2	2.5	5.1	4.0
Other Developing Countries	5.6	3.2	2.7	2.1	4.5	3.3
World	**4.4**	**2.9[a]**	**2.0**	**1.5**	**3.4**	**2.6**

[a] Excludes USSR and Eastern Europe.

Source: IMF, 1988.

FIGURE 5-3

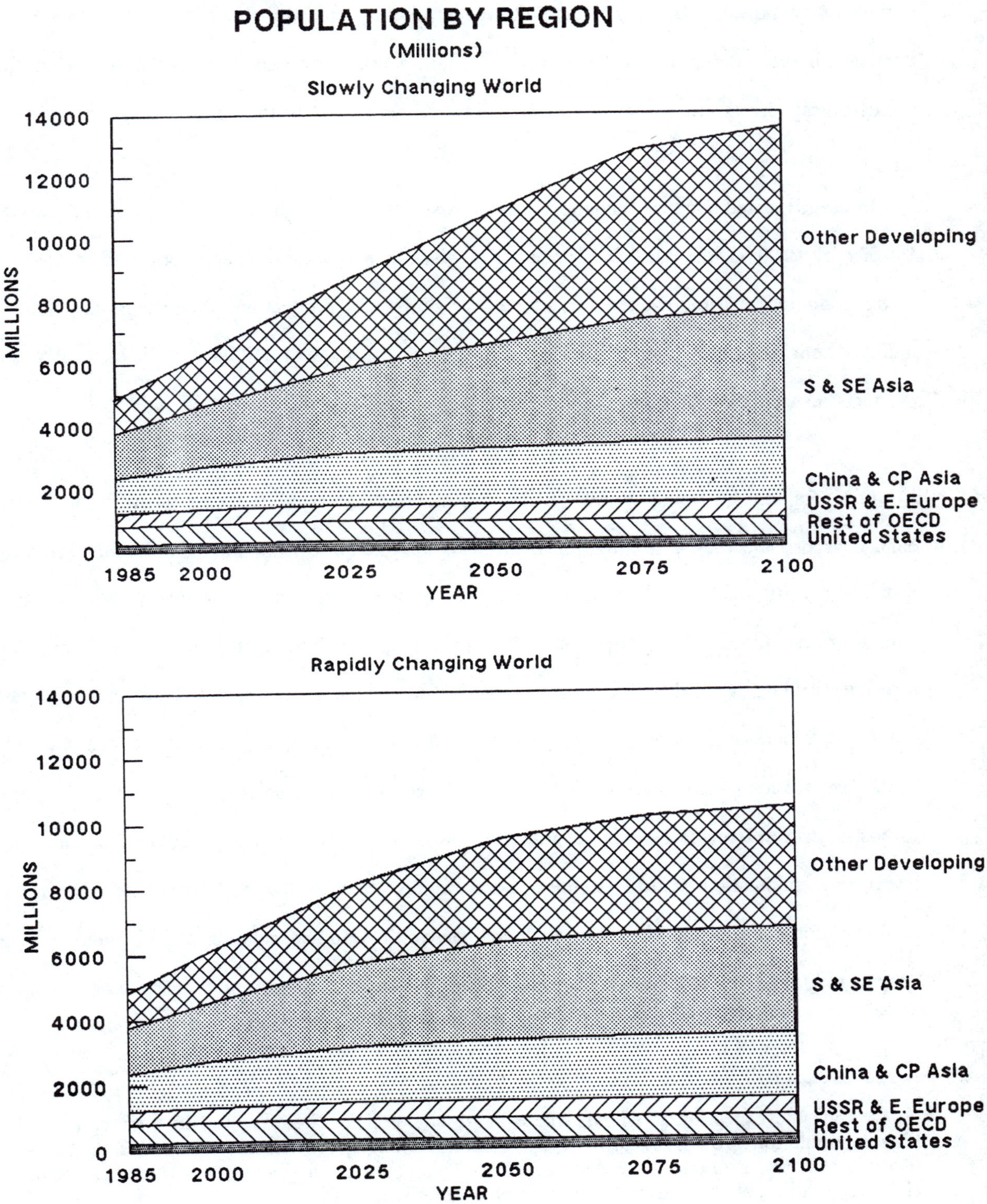

Source: U.S. Bureau of the Census, 1987

of global income going to the developing world does increase with time, but not dramatically. The population engaged in traditional agriculture and shifting cultivation continues to increase, as do demand for fuelwood and speculative land clearing. These factors lead to accelerated deforestation until tropical forests are virtually eliminated toward the middle of the next century.

In industrialized countries economic growth is sluggish, although per capita income reaches about $40,000 by 2100 in the OECD. Because of slack demand, real energy prices increase slowly. Correspondingly, existing capital stocks turn over slowly and production efficiency in agriculture and industry improve at only a moderate rate. The energy efficiency of buildings, vehicles, and consumer products also improve at a slow rate.

In a "Rapidly Changing World" (RCW) we assume that rapid economic growth and structural change occurs and that little attention is given to the global environment. Per capita income rises rapidly in most regions and consumer demand for energy increases, putting upward pressure on energy prices. On the other hand, there is a high rate of innovation in industry, and capital stocks turn over rapidly, which leads to an accelerated reduction in energy required per unit of industrial output. An increasing share of energy is consumed in the form of electricity, produced mostly from coal. The fraction of global economic output produced in the developing world increases dramatically as post-industrial structural change continues in the industrialized world. As educational and income levels rise, population growth declines more rapidly than in the SCW scenario (Figure 5-3).[2] Deforestation continues at about current rates, spurred by land speculation and commercial logging, despite reduced rates of population growth. Energy efficiency is not much of a factor in consumer

---

[2] The sole exception is China, where aggressive policies are assumed in both cases. Slightly higher population growth is shown in the Rapidly Changing World scenario based on the sources of the alternative estimates (see Appendix B). This could be attributed to a relaxation of the one-child-per-family policy in response to greater economic growth.

decisions, as incomes increase faster than real energy prices. Private vehicle ownership increases rapidly in developing countries while air travel increases rapidly in wealthier ones. Nonetheless, significant reductions in energy intensity occur with technological innovation and structural change.

Scenarios with Stabilizing Policies

Two variants of the above scenarios explore the impact of policy choices aimed at reducing the risk of global warming. These scenarios, labelled "Slowly Changing World with Stabilizing Policies" (SCWP) and "Rapidly Changing World with Stabilizing Policies" (RCWP), start with the same economic and demographic assumptions used in the SCW and RCW scenarios, respectively, but assume that government leadership is provided to ensure that limiting greenhouse gas emissions becomes a consideration in investment decisions beginning in the 1990s. We assume that policies to promote energy efficiency in all sectors succeed in substantially reducing energy demand relative to the No Response scenarios and that efforts to expand the use of natural gas increase its share of primary energy supply relative to other fossil fuels in the near term. Research and development into non-fossil energy supply options such as photovoltaics (solar cells) and biomass-derived fuels (fuels made from plant material) assure that these options are available and begin to become competitive after 2000. As a result, non-fossil energy sources meet a substantial fraction of total demand in later periods. The existing protocol to reduce CFC emissions is assumed to be strengthened, leading to a phase-out of fully-halogenated compounds and a freeze on methyl chloroform. A global effort to reverse deforestation transforms the biosphere from a source to a sink for carbon, and technological innovation and controls reduce agricultural, industrial, and transportation emissions.

While these general assumptions apply to both the SCWP and RCWP cases, the degree and speed of improvement are higher in the Rapidly Changing variant because technological innovation and capital stock replacement are greater in this case. The policies considered do not require changes in basic life styles. For example, energy use in buildings is greatly reduced in the Stabilizing Policy scenarios relative to the No Response scenarios, but the floor space available per person and the amenity levels provided are assumed to be the same. The technological strategies and policy options available to achieve the Stabilizing Policy scenarios are discussed in detail in Chapters VII, VIII, and IX.

ANALYTICAL FRAMEWORK

To make it possible to assess the implications of the kinds of scenarios just described, we have developed an integrated analytical framework to organize the data and assumptions required to calculate emissions of radiatively and chemically active gases, concentrations of greenhouse gases, and the rate of climatic change. This framework is described very briefly here, and in more detail in Appendix A.

The analytical framework consists of four emissions modules and two concentration modules as shown in Figure 5-4. The four emissions modules use input data, including scenario specifications for population growth, GNP, energy efficiency, etc., to estimate emissions of greenhouse gases for nine regions of the globe (Figure 5-5). Emissions are calculated every 5 years from 1985 to 2025 and then every 25 years through 2100. Emissions of the greenhouse gases CO_2, CH_4, N_2O, and a number of CFCs are explicitly calculated within the framework. Emissions of CO and NO_x, which are not themselves greenhouse gases, are also explicitly calculated, as these gases can significantly alter the chemistry of the atmosphere and thus affect the concentrations of the greenhouse gases.

FIGURE 5-4

Structure of the Atmospheric Stabilization Framework

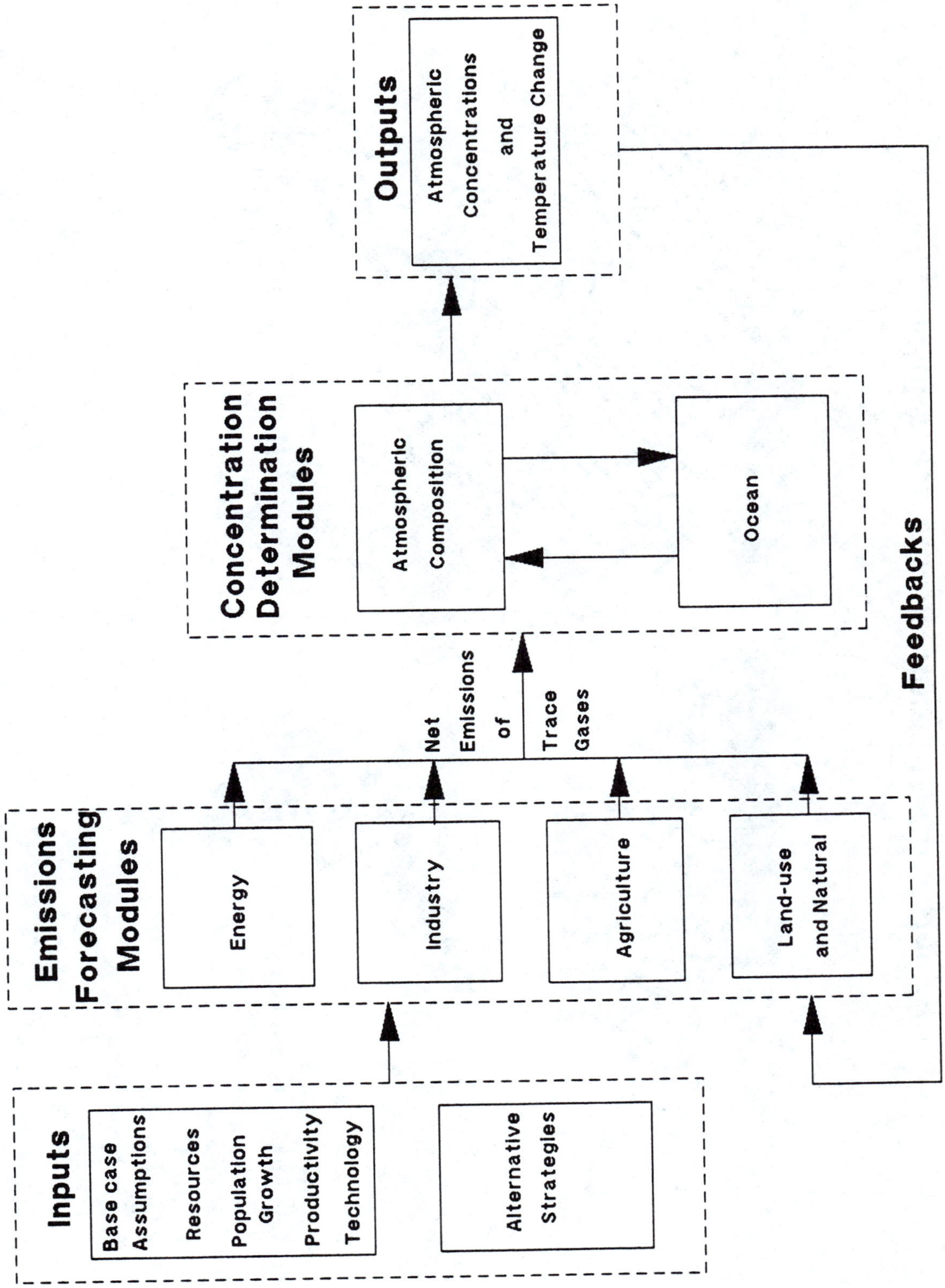

FIGURE 5-5

GEOPOLITICAL REGIONS FOR CLIMATE ANALYSES

Key:

1. USA
2. OECD West
3. OECD Asia
4. Centrally planned Europe
5. Centrally planned Asia
6. Middle East
7. Africa
8. Latin America
9. South and East Asia

Source: Mintzer, 1988.

The concentrations of other greenhouse gases, such as water vapor and ozone, are calculated implicitly or explicitly as a function of the other gases. The atmospheric composition and ocean modules together estimate global concentrations of the greenhouse gases resulting from the projected emissions, and increases in global temperatures resulting from the calculated concentrations. The atmospheric trace gas concentrations and temperatures affect the emissions and concentration modules in the next time period.

Energy Module

The energy module consists of a Global Energy Supply Model (SUPPLY), which is based on the energy-CO_2 model of Edmonds and Reilly (1983a, 1984), and was developed by ICF Inc. for this study; a global energy end-use analysis (DEMAND), conducted by the World Resources Institute and Lawrence Berkeley Laboratories; and combustion emission coefficients developed by Radian (1987).

DEMAND estimates energy consumption based on specific assumptions about the level of energy using activities and technical efficiency by region and sector (industry, transportation, buildings). Although this analysis provided more detail than most previous global studies, this level of aggregation obscures many important variations, particularly for developing countries. For example, per capita incomes vary from $150 for Bangladesh to $7000 for Singapore within the South and East Asia region. The share of energy used by the manufacturing sector, vehicle ownership levels, and types of fuels used (particularly the importance of biofuels), all vary from one economy to another. In conducting the analysis we capture some of this diversity by examining energy use by regions and by income groups within regions. Detailed analysis was performed for 2025 to anchor the demand estimates calculated for other years using SUPPLY.

SUPPLY includes estimates of energy resources and costs by region and can balance supply and demand using a highly-aggregated estimate of demand as a function of price and income. Supply-demand equilibration takes place within SUPPLY, which projects fuel mix and final prices. Trace-gas emissions are calculated by allocating the final fuel consumption among the individual combustion technologies for which emission coefficients are available. Additional emissions associated with fuel production are also estimated.

Industry Module

The industry module consists of a CFC model and a model for other non-combustion trace-gas sources. The CFC model was developed by EPA for use in assessing stratospheric ozone depletion (U.S. EPA, 1987). It projects production and emissions of the following compounds: CFC-11, CFC-12, HCFC-22, CFC-113, CCl_4, CH_3CCl_3, CH_3C1, CH_3Br, CF_4, Halon 1211, and Halon 1301. Other industrial sources of trace gases include landfilling and cement production. Emissions from these activities are estimated as a simple function of population and per capita income.

Agriculture Module

The agricultural module uses the IIASA/IOWA Basic Linked System, or BLS, (Frohberg, 1988) to forecast fertilizer use and agricultural production. These estimates are used with emission coefficients derived from the literature to calculate emissions of N_2O from fertilizer use, CH_4 from rice production, CH_4 from enteric fermentation in domestic animals, and emissions of CH_4, N_2O, NO_x, and CO from burning agricultural wastes.

Land Use and Natural Source Module

This module consists of components dealing with a number of land surface processes and other natural sources of trace gases. The most important of these is carbon dioxide released from land use change, particularly deforestation, which is projected with the Marine Biological Laboratory/Terrestrial Carbon Model, or MBL/TCM, (Houghton et al., 1983) based on assumptions about future rates of land clearing. Other anthropogenic emissions related to land clearing, such as a portion of CO emissions from biomass burning and N_2O emissions from land disturbance, are scaled based on the CO_2 emissions calculated by the MBL/TCM. Natural emissions of CO, CH_4, N_2O, and NO_x from sources such as forest fires, wetlands, soils, oceans, and fresh water are based on values from the literature, and generally are held constant throughout the projection period (biogeochemical feedbacks can be assumed to alter these emissions, see Chapter VI).

Ocean Module

Ocean uptake of heat and CO_2 are modeled using the Box-Diffusion approach introduced by Oeschger et al. (1975) as implemented for the GISS GCM (Hansen et al., 1984). The ocean mixing parameter for heat uptake is chosen to reproduce, as closely as possible, the timescales obtained in the time-dependent calculations with the GISS GCM (Hansen et al., 1988). Alternative values for this parameter can be used to approximate the timescales of other approaches to estimating ocean heat uptake (Chapter VI). Alternative ocean model formulations for CO_2, such as the Advective-Diffusive Model (Bjorkstrom, 1979) and the Outcrop-Diffusion Model (Siegenthaler, 1983), are included in the integrating framework and can be used for alternative estimates of CO_2 uptake. Total carbon uptake is calibrated using estimates of historical emissions of CO_2 from fossil fuels (Rotty, 1987a,b) and deforestation (Houghton, 1988). The atmospheric CO_2 concentration is assumed to be

285 ppm in 1800 and is forced to be equal to the values obtained at Mauna Loa for the period of record (1960-1985). The excess flux required to meet these conditions is calculated and held constant in the future at the average value for 1975-1985. Alternative assumptions are considered in Chapter VI.

Atmospheric Composition and Temperature Module

The atmospheric composition model was developed for this study (Prather, 1988). It estimates changes in the concentration of key atmospheric constituents and the global radiation balance based on the emissions/uptake projected by the other modules. Perturbations to atmospheric chemistry are incorporated based on first-order (and occasionally second-order) relationships derived from more process-based chemical models and observations. The model is essentially zero-dimensional, but it does distinguish between the northern hemisphere, southern hemisphere, troposphere, and stratosphere. Global surface temperature change is calculated based on the radiative forcing of the greenhouse gases derived from Lacis et al. (1981) and Ramanathan et al. (1985) coupled to heat uptake by the ocean model using a specified climate sensitivity parameter. This sensitivity parameter is set to yield a global equilibrium temperature increase of 2 or 4°C when the CO_2 concentration is doubled, reflecting a central estimate of the range of uncertainty; a broader range of possibilities is examined in Chapter VI (see discussion in Chapter III).

Assumptions

Population Growth Rates

The population estimates for the Rapidly Changing World scenario were developed from Zachariah and Vu (1988) of the World Bank; for the Slowly Changing World scenario estimates were taken from U.S. Bureau of the Census (1987) of the U.S. Bureau of the Census. These two sources agree quite closely on the size of the world's population through 2000, then diverge thereafter due to different assumptions on the rate at which the global population will stabilize. Zachariah and Vu (1988) assumes that population growth rates in developing countries will begin to decline markedly after 2000, achieving a net reproduction rate of unity in every country by 2040. (A net reproduction rate of unity indicates that people of child-bearing age have children at a replacement rate; it eventually leads to a stable population level.) U.S. Bureau of the Census (1987) assume that global population stability will occur at a later date, with developing countries experiencing rapid population growth rates until the middle of the next century.

Economic Growth Rates

The primary source for the economic growth rate estimates was the World Bank (1987). In their report, Gross National Product (GNP) forecasts were provided for the 1986-1995 period for several different types of country groups. Most countries could be classified into one of these three general categories--low income, middle income, or industrialized. In addition, the World Bank defined several other more select groups for which separate growth rates were estimated, including oil exporters, exporters of manufactures, highly-indebted countries, and sub-Saharan Africa. The low growth case was used as a starting point for this analysis because these estimates were more consistent with recent historical trends and other forecasts. For the RCW (SCW) scenario these initial values were generally increased (decreased) by one percentage point for developing and East Bloc countries and one-half percentage point for OECD countries to reflect the greater uncertainty regarding future growth in developing and centrally-planned economies. The growth rates were applied for the period 1985-2000, and were generally reduced by one-half percentage point each 25-year period, beginning in 2000, to reflect structural change and the decline in population growth rates over time. Nonetheless, GNP per capita continues to increase throughout the projection period, although the rate of growth is substantially lower in the Slowly Changing World scenario.

Oil Prices

The oil prices used in this analysis were taken from U.S. DOE (1988), which supplied a range of oil price forecasts. The Middle Price forecast from DOE was used for the Rapidly Changing World scenario (by 2000 the world oil price is about $31/barrel in 1987 dollars), while the Low Price forecast was used for the Slowly Changing World scenario (oil prices by 2000 were about $25/barrel in 1987 dollars). Since the DOE price forecasts did not extend beyond 2000, oil prices were derived from the SUPPLY model; in each scenario prices escalated about 0.8% annually from 2000-2100.

Limitations

This analytical framework attempts to incorporate some representation of the major processes that will influence the rate and magnitude of climatic change during the next century within a structure that is reasonably transparent and easy to manipulate. In so doing we recognize a number of major limitations:

- Economic growth rates are difficult to forecast. Our alternative assumptions may not adequately reflect the plausible range of possibilities. In particular, we have assumed that aggregate economic growth rates will generally decline over time from the levels assumed for 1985-2000; this may not be the case.

- Economic linkages are not fully captured. The economic analysis uses a partial-equilibrium framework, making it impossible to ensure that the activity levels assumed in each sector are completely consistent with the aggregate economic assumptions. In addition, capital markets are not explicitly considered. This is particularly significant in examining developing countries as it is unclear if they will be able to obtain the capital investments needed to develop the energy supplies assumed in some of the scenarios.

- Technological changes are difficult to forecast. Substantial improvements in the efficiency of energy using and producing technologies are assumed to occur even in the absence of substantial energy price increases or policy measures. If this assumption proves to be untrue, then greenhouse gas emissions may be substantially underestimated in the No Response scenarios. Similarly, aggressive research and development is assumed to substantially reduce the cost of renewable technologies in the Stabilizing Policy scenarios. The impact of policies may be overestimated if such improvements fail to materialize or if they would have materialized as rapidly even without increased government support.

- Detailed cost analyses have not been conducted. Technological strategies have been screened based on judgments about their potential cost-effectiveness, but no attempt has been made to rank the cost-effectiveness of each strategy or to estimate the government expenditures or total costs associated with the stabilizing strategies.

- The modules of the framework are not fully integrated. Existing models of individual processes that affect greenhouse gas emissions were assembled within the analytical framework and were used with consistent assumptions. However, it was not possible to ensure complete consistency of results. For example, while the biomass energy supplies arrived at in the Energy module do not appear to be inconsistent with the land use patterns calculated in the Agriculture and Land Use and Natural Source modules, there is no explicit coupling among these results.

- The ocean models employed are highly simplified. The ocean plays an important role in taking up both CO_2 and heat. The one-dimensional models used to represent this process may not adequately reflect the underlying physical processes, particularly as climate changes.

- Changes in atmospheric chemistry are calculated in a highly-simplified fashion. Chemical interactions are analyzed based on parameters derived from detailed chemical models. These parameters may not adequately reflect the underlying chemistry, particularly as the atmospheric composition changes significantly from current conditions. Also, it is not possible to explicitly model the heterogeneous conditions that control, for example, tropospheric ozone concentrations. In our analysis we also assume that non-methane hydrocarbon emissions remain constant, which may cause future methane and ozone changes to be underestimated.

SCENARIO RESULTS

We have estimated the implications of the four scenarios described above for emissions of radiatively-important gases arising from energy production and use, industrial processes, changes in land use, and agricultural activities using the integrated analytical framework developed for this study. The resulting changes in atmospheric composition and global climate are also estimated.

Energy Sector

The single most important determinant of greenhouse gas emissions is the level of energy demand and the combination of sources that are used to supply that energy.

End-use Consumption

Government policies that affect demand for energy are likely to be the most important determinant of greenhouse gas emissions in the near term. Figures 5-6 and 5-7 illustrate global end-use energy consumption by region for fuel and electricity, respectively. Total end-use energy consumption increases from 220×10^{18} joules (220 EJ) in 1985 to 320 EJ in 2025 in the SCW versus 420 EJ in the RCW.[3] Greater improvements in energy efficiency in the SCWP and RCWP cases reduce end-use demand in 2025 by 13% and 15%, respectively, relative to the No Response scenarios. Extrapolating these trends to 2100 yields 430 EJ in the SCW and 780 EJ in the RCW scenarios, while in the Stabilizing Policy cases there is 20% and 35% lower demand, respectively.

---

[3] 1 EJ = 0.948 quadrillion BTU (Quad).

FIGURE 5-6

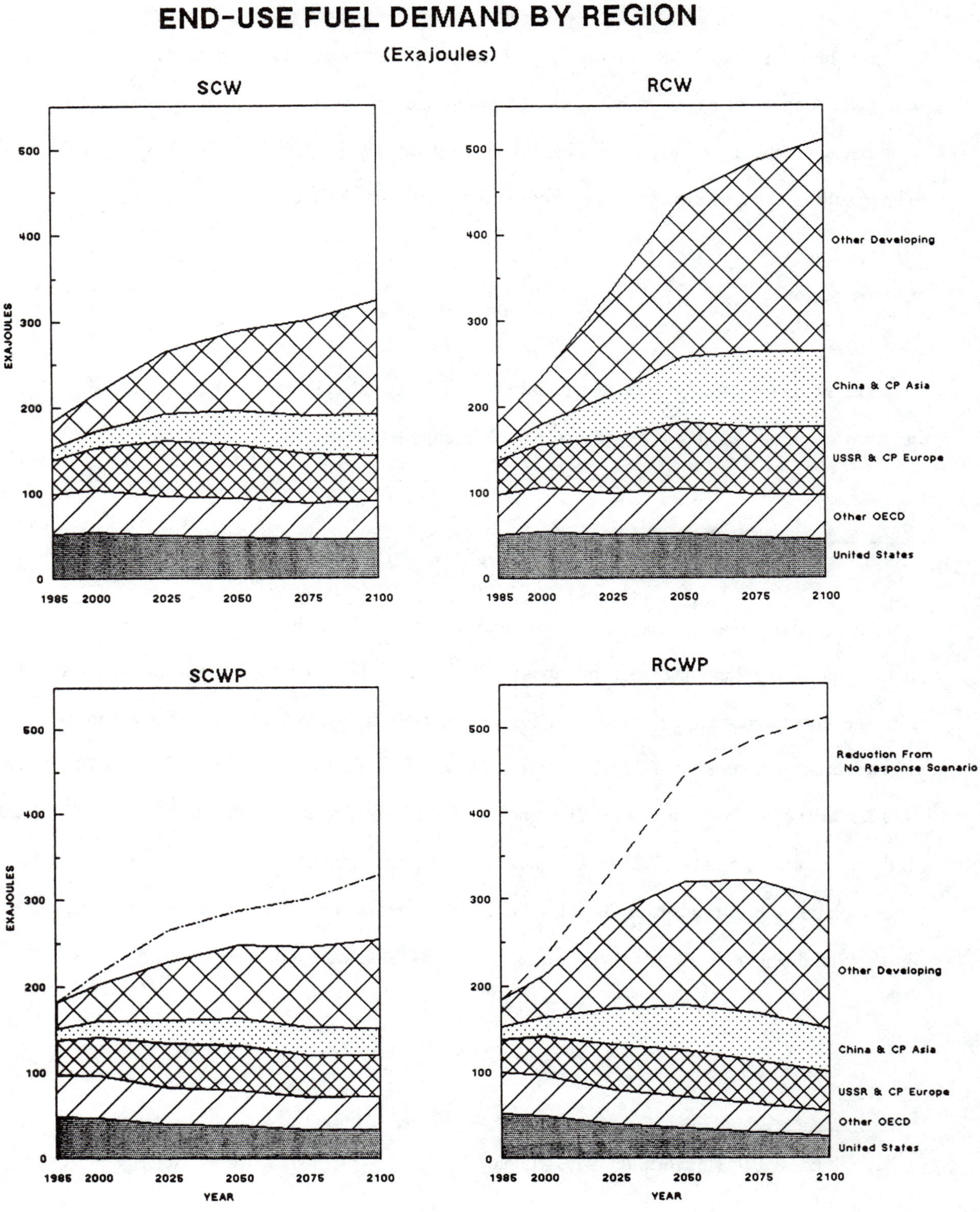
END-USE FUEL DEMAND BY REGION
(Exajoules)
SCW
RCW
SCWP
RCWP
EXAJOULES
YEAR
0
100
200
300
400
500
1985
2000
2025
2050
2075
2100
Other Developing
China & CP Asia
USSR & CP Europe
Other OECD
United States
Reduction From
No Response Scenario

FIGURE 5-7

END-USE ELECTRICITY DEMAND BY REGION

(Exajoules)

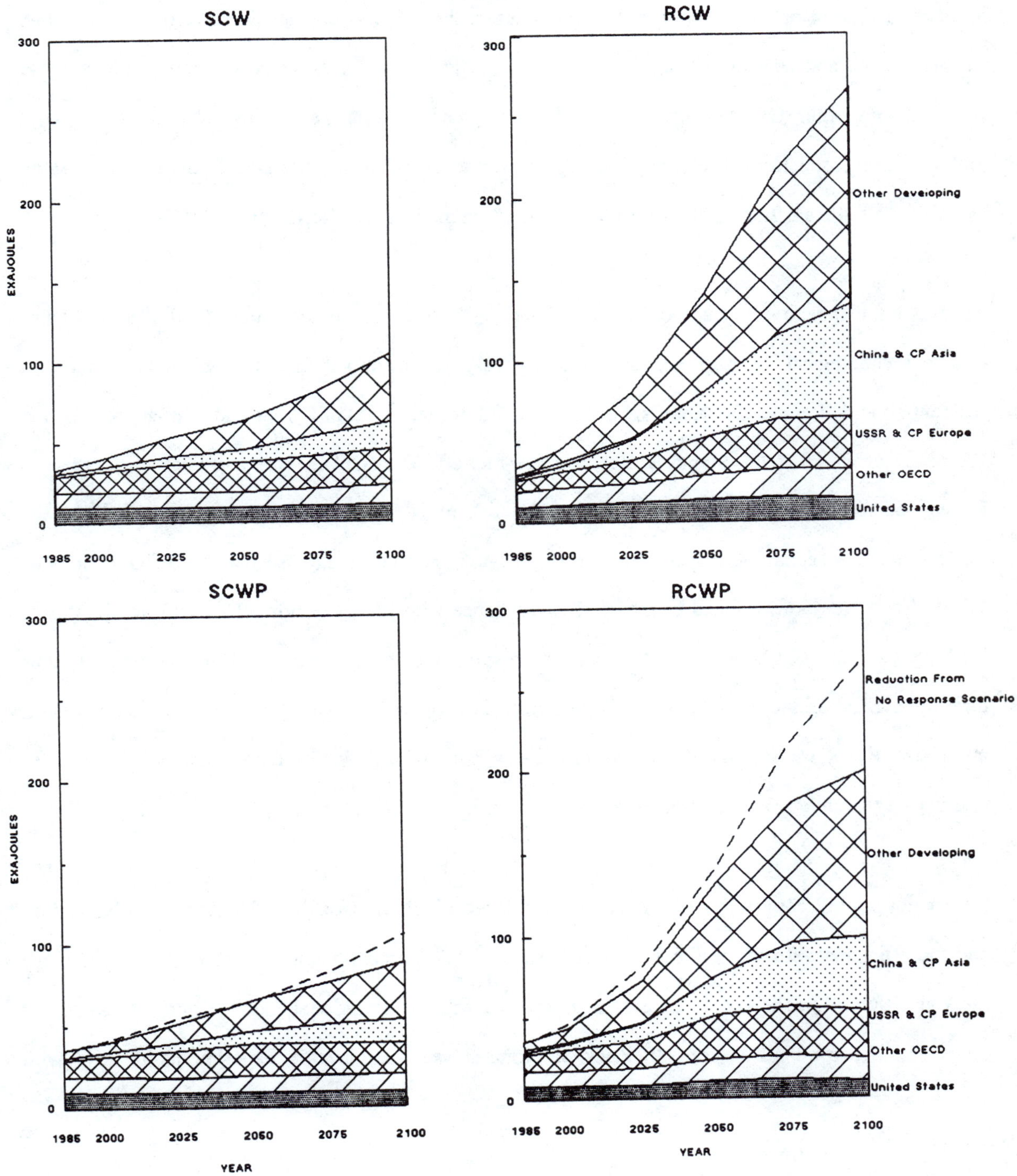

In each scenario the growth in end-use demand is driven almost entirely by countries outside the OECD (USSR, Eastern Europe, China, and other developing countries) as a result of higher rates of economic and population growth in these regions and from more rapid efficiency improvements and the saturation of energy-intensive technologies in the OECD (e.g., steel production, automobile transportation, and central heating). Fuel use, in particular, is not expected to grow significantly in the U.S. and other OECD countries as efficiency gains compensate for increases in floor space, mobility, and production. Electricity use is projected to grow much more rapidly than fuel use in all cases, and significant increases in OECD electricity demand are reflected in the RCW.

It is important to note that both the SCW and RCW scenarios assume substantial efficiency gains due to technological innovation and market forces. For example, fuel use per square meter of residential and commercial floor space is assumed to fall by 45-55% in the United States and Western Europe by 2025. Similarly, fleet average fuel efficiency of U.S. cars and light trucks reaches 7.8 and 6.9 liters per 100 kilometers (liters/100 km), or 30 and 34 miles per gallon (mpg), in the SCW and RCW scenarios, respectively. In the SCW, industrial energy use per unit of GNP falls by 1.5-2%/yr in the industrialized countries, in accordance with recent trends. This rate accelerates to 2-3.5%/yr in the RCW, the highest rate of improvement being for the East Bloc countries as they have the highest initial industrial energy intensities. Less optimistic assumptions about efficiency gains in the No Response scenarios would imply higher rates of associated climatic change and greater relative improvement in the Stabilizing Policy scenarios.

In developing countries the use of biofuels for cooking is strongly influenced by urbanization and the efficiency with which these fuels are used. Urban populations have better access to modern fuels and thus a smaller share of urban households will use traditional fuels. There is substantial scope for improvement of the efficiency of biomass use. Laboratory experiments in Asia with improved cookstoves suggest that it is possible to achieve efficiencies of up to 33% (compared with current averages of 8%). However, experience from the last decade of improved cookstove dissemination projects suggests that efficiencies are unlikely to exceed 20% in the field. We assume the dissemination of efficient cookstoves to almost all users of biomass only in the Stabilizing Policy cases. Thus, the average efficiency of biomass use is assumed to improve to 15-17% in each region in these scenarios. As a result of these efficiency improvements and because an increasingly larger share of the population moves to urban areas, where there is better access to modern fuels, the amount of biofuels consumption declines in the household sector for each scenario.

Important structural shifts underlie the aggregate trends in these scenarios. Electricity's share of end-use consumption more than doubles in the RCW, from 16% in 1985 to 19% in 2025 and 34% in 2100, while it grows less dramatically in the SCW, reaching 24% in 2100. These trends are accentuated in the policy scenarios as there appears to be even greater room for reductions in fuel use than in electricity use, partly because electricity is substituted for fuel in some highly-efficient applications. In particular, electricity accounts for 40% of end-use consumption by 2100 in the RCWP scenario because of dramatic increases in electricity use in developing countries. The distribution of energy use among the industrial, transportation, and residential and commercial sectors also shifts significantly, as shown in Figure 5-8. In the Rapidly Changing World the share of end-use energy going to the residential and commercial sectors declines continuously, while the share going to industry increases until the middle of the 21st Century and then declines. This pattern reflects the increasing importance of developing countries, which generally have low heating demands and a greater percentage of modern energy devoted to the industrial sector. This occurs despite a decline in the share of commercial energy use, particularly electricity, going to the industrial sector **within** developing country regions. As the most intense phase of industrialization is completed the transportation sector begins to take off, its share rising steadily after 2025. In the SCW scenario the share of end-use energy consumed in the industrial sector grows less dramatically and does not peak

FIGURE 5-8

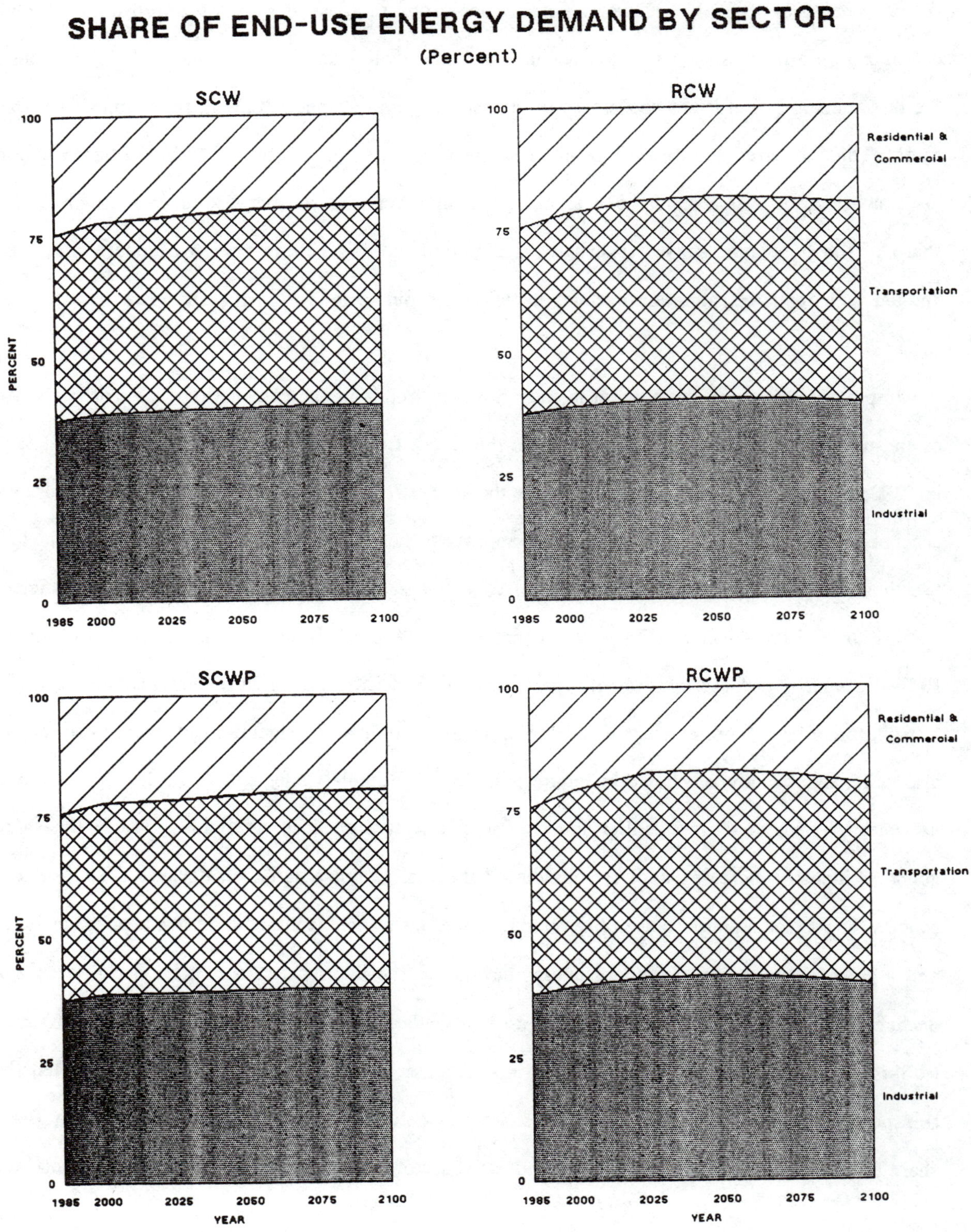
SHARE OF END-USE ENERGY DEMAND BY SECTOR
(Percent)
SCW
RCW
SCWP
RCWP
PERCENT
100
75
50
25
0
1985
2000
2025
2050
2075
2100
YEAR
Residential & Commercial
Transportation
Industrial

until 2075 as industrialization in developing countries is stretched out over a longer time-span and dramatic increases in mobility are delayed. In the Stabilizing Policy scenarios the growth in transportation energy use is suppressed by much higher fuel efficiency and the share of end-use energy going to the residential and commercial sectors increases toward the end of the next century.

Primary Energy Supply

While policies affecting demand will have the largest impact on near-term greenhouse gas emissions, changes in the supply mix will also be very important over the long term. Global primary energy supply is shown by source for the four scenarios in Figures 5-9 and 5-10. Growth in primary energy production is substantially higher than growth in end-use energy consumption because of increased requirements for electricity and synthetic fuel production. This is most dramatic in the RCW, where primary energy production increases from 290 EJ in 1985 to 580 EJ in 2025 and 1410 EJ in 2100; a 100% and 380% increase, respectively, compared with 90% and 260% increases in end-use consumption.

The use of synthetic fuels to supplement conventional oil and gas production becomes particularly important after 2025, influencing both total requirements and the mix of sources (Figure 5-11). In the RCW conventional oil and gas production increases through 2050, then begins to decline due to resource depletion (the share of primary energy supplied by oil and gas declines throughout the projection period). As a result, synthetic fuels are increasingly relied on to supply liquid and gaseous fuel requirements. By 2050 19% of primary energy is used in synthetic fuels production, and this value increases to 40% by 2100. In the SCW heavy dependence on synthetic fuels begins later because conventional oil and gas resources are depleted more gradually. Coal is the dominant feedstock for synfuel production in both of these scenarios.

FIGURE 5-9

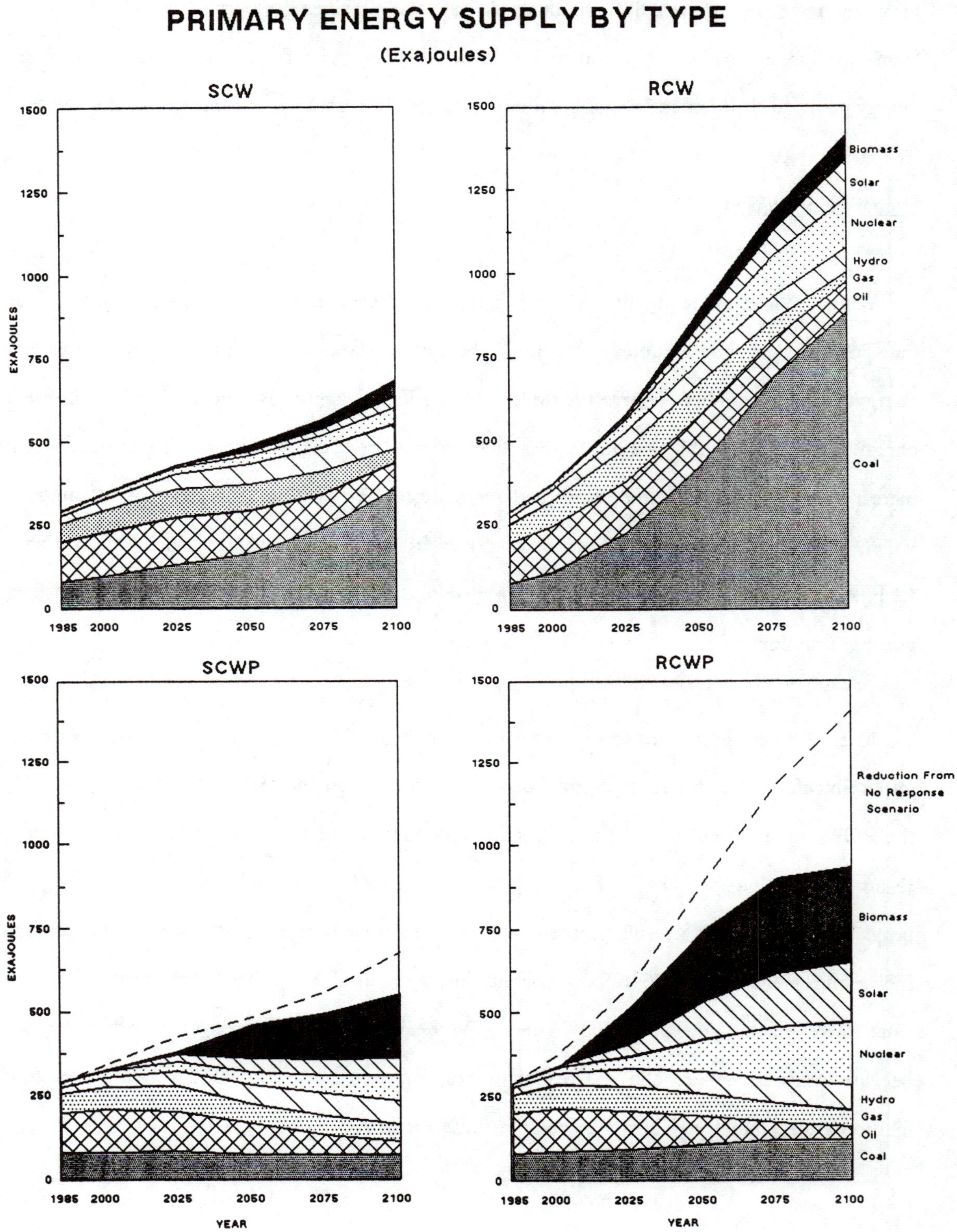
PRIMARY ENERGY SUPPLY BY TYPE
(Exajoules)
SCW
RCW
SCWP
RCWP
EXAJOULES
YEAR
1985
2000
2025
2050
2075
2100
0
250
500
750
1000
1250
1500
Biomass
Solar
Nuclear
Hydro
Gas
Oil
Coal
Reduction From
No Response
Scenario

FIGURE 5-10

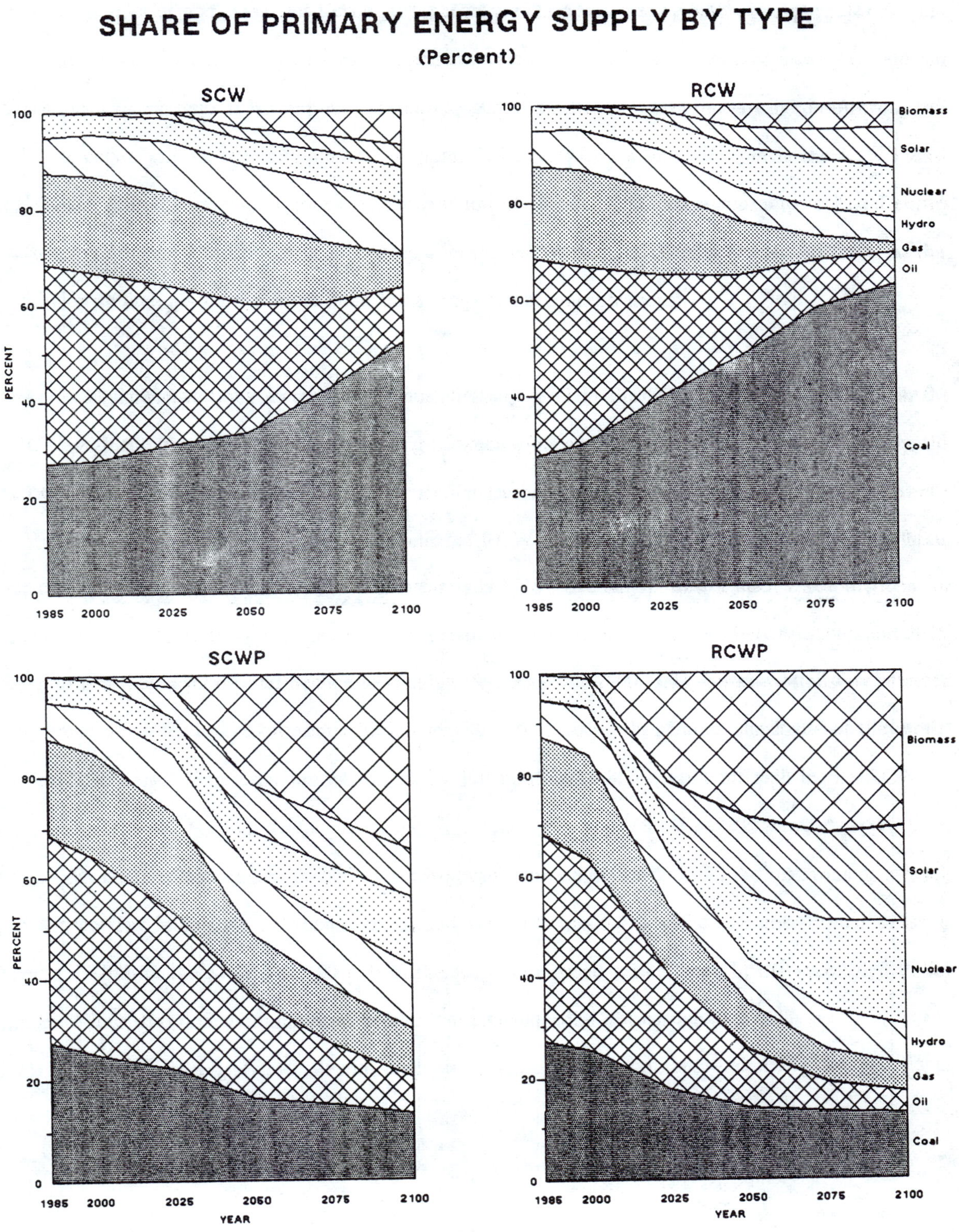
SHARE OF PRIMARY ENERGY SUPPLY BY TYPE
(Percent)
SCW
RCW
SCWP
RCWP
PERCENT
YEAR
0
20
40
60
80
100
1985
2000
2025
2050
2075
2100
Biomass
Solar
Nuclear
Hydro
Gas
Oil
Coal

FIGURE 5-11

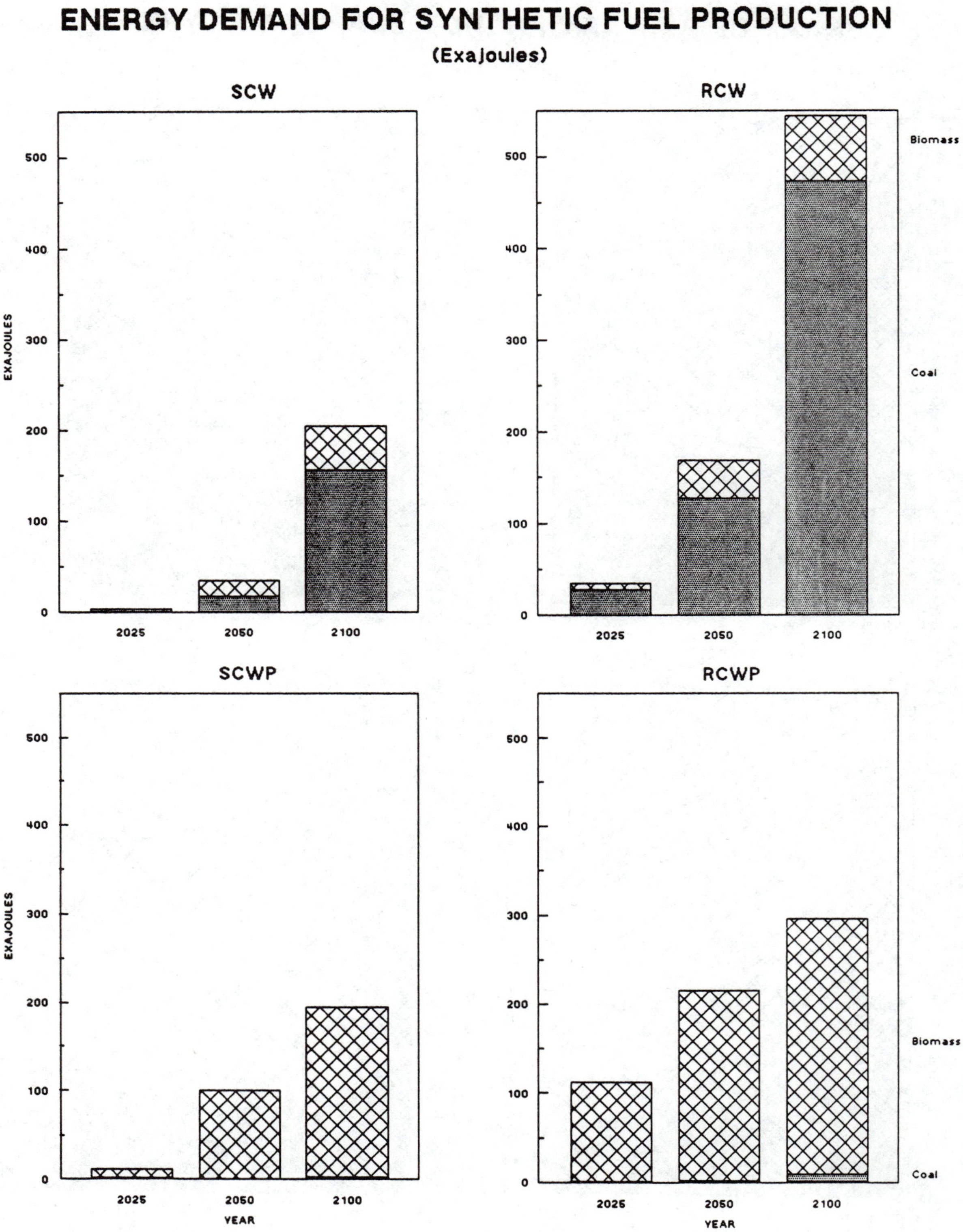
ENERGY DEMAND FOR SYNTHETIC FUEL PRODUCTION
(Exajoules)
SCW
RCW
SCWP
RCWP
EXAJOULES
YEAR
2025
2050
2100
Biomass
Coal
0
100
200
300
400
500

The mix of primary energy resources used to generate electricity is also crucial in determining future greenhouse gas emissions. While non-fossil energy sources (nuclear, solar, and hydro) increase their absolute contribution to primary energy supply in all scenarios, in the absence of policies to limit greenhouse gas emissions, it is likely that future electricity production will be dominated by coal-based technologies over the long term (in the near term, current gas prices make gas-based combustion turbine technology very attractive in many regions). Thus in the RCW, demand for electricity and synfuel production pushes global coal consumption up by more than a factor of 10 between 1985 and 2100. Correspondingly, the share of primary energy supplied by coal increases from 27% in 1985 to 40% in 2025 and 63% in 2100 (Figure 5-10). The same forces are at work in the SCW, but the results are less dramatic: Coal production increases by less than a factor of 5, and its share of primary energy reaches just over 50% by 2100.

In the Stabilizing Policy scenarios natural gas is relied on more heavily in the near term while accelerated research and development and other incentives are assumed to make several non-fossil electricity supply technologies strongly competitive over the long term. In particular, photovoltaics, biomass-based combustion turbines, and advanced nuclear reactors appear to be strong candidates to make a large contribution to future electricity production (these and other options are discussed in some detail in Chapter VII). In the policy scenarios these technologies begin to supply energy after 2000 and become strongly competitive by 2025. By 2050 they supply 60% and 70% of global electricity in the SCWP and RCWP scenarios, respectively.[4] It is also assumed that research priorities and other policies promote the use of biomass-derived fuels rather than coal-based synfuels. In fact, in 2025 and 2050 total synfuel production is higher in the policy scenarios because biomass production and conversion is assumed to become competitive with imported oil and gas in many developing

---

[4] This value includes all of the electricity generated from gas, reflecting the assumption that a little over half of the synthetic gas generated from biomass is actually both produced and consumed in integrated gasifier-combustion turbine units.

regions starting around 2010 (Walter, 1988). The particular mix among the non-fossil supply technologies shown in Figure 5-10 is rather arbitrary, but the type of non-fossil technologies is of little consequence to total greenhouse gas emissions.

Greenhouse Gas Emissions From Energy Production and Use

The heavy reliance on coal in both the SCW and RCW scenarios leads to large increases in both CO_2 and CH_4 emissions (see Figures 5-14 and 5-16 later in the chapter). In the SCW energy-related emissions of CO_2 increase from 5.1 petagrams of carbon (Pg C) in 1985 to 7.2 Pg C in 2025 and 11.1 Pg C in 2100.[5] Emissions reach more than twice this level in the RCW scenario: 10.3 and 24.4 Pg C in 2025 and 2100, respectively. This growth in emissions of 0.5 Pg C per decade in the SCW and 1.3 Pg C per decade in the RCW between 1985 and 2025 compares with average growth of 1.1 Pg C per decade between 1950 and 1980. Emissions of CH_4 from fuel production, predominantly coal mining, grow even more dramatically. The estimated emissions from fuel production in 1985 are 60 teragrams of CH_4 (Tg CH_4) or just over 10% of the total.[6] In the SCW this source increases to 86 Tg CH_4 in 2025 and 160 Tg CH_4 in 2100. The corresponding values for the RCW are 130 Tg CH_4 in 2025 and 360 Tg CH_4 in 2100, about 20% and 30% of the CH_4 total, respectively.

The combination of higher efficiency and greater reliance on non-fossil fuels assumed in the Stabilizing Policy scenarios serves to substantially curtail CO_2 and CH_4 emissions. In both the SCWP and RCWP cases CO_2 emissions from energy use reach only 5.5 Pg C in 2025, after which time they decrease, reaching 3.2 and 4.3 Pg C by 2100 in the two cases, respectively. Similarly, CH_4 emissions from fuel production remain relatively constant in both of these scenarios.

---

[5] 1 petagram = 10^{15} grams.

[6] 1 teragram = 10^{12} grams.

Energy-related emissions, other than of CO_2 and CH_4, are strongly affected by the type of control technology employed in addition to the total amount and type of energy used. Emissions of CO and NO_x associated with energy use can be expected to increase almost as rapidly as primary energy consumption in the absence of new policies. On the other hand, in the Stabilizing Policy scenarios NO_x emissions are roughly constant and CO emissions are cut by more than half. This assumes that the rest of the world gradually adopts control technology similar to that required of new mobile and stationary sources in the United States today, and that industrialized countries adopt standards consistent with the use of Selective Catalytic Reduction technology in utility and industrial applications after 2000, with developing countries following after 2025.

Comparison to Previous Studies

Despite the large range of outcomes illustrated by the four scenarios developed here, none of the global rates of change are unprecedented (Table 5-3). Global reductions in aggregate energy intensity generally fall within the range of 1-2% per year; the lower value is consistent with long-term trends and the higher value is consistent with recent experience. Reductions in the amount of carbon emitted per unit of energy consumed (carbon intensity) varies from 0.0-1.3% per year with significant declines only apparent in the Stabilizing Policy cases. These values are not unprecedented as carbon intensity declined by an average of 1.5% per year between 1925 and 1985 due to increased reliance on oil and gas over coal.

While we know of no previous attempts to develop long-term scenarios for emissions of the full set of gases discussed above based on explicit economic and technological assumptions, there have been a number of previous studies that relate to many of the components examined here. Over the

TABLE 5-3

Key Global Indicators

Parameter	Scenario[a]	Year 1985	Year 2025	Year 2100
GNP/capita (1000 1988 $)	SCW, SCWP	3.0	3.7	7.1
	RCW, RCWP		6.7	35.6
Primary Energy (EJ)[b]	SCW	290	430	680
	RCW		580	1410
	SCWP		380	550
	RCWP		520	940
Fossil Fuel CO_2 (Pg C)[c]	SCW	5.1	7.2	11.1
	RCW		10.3	24.4
	SCWP		5.5	3.2
	RCWP		5.5	4.3

Parameter	Scenario	1985-2025	2025-2100
GNP/capita (%/yr)	SCW, SCWP	0.5	0.9
	RCW, RCWP	2.0	2.3
Energy/GNP (%/yr)	SCW	-1.1	-0.8
	RCW	-1.6	-1.4
	SCWP	-1.3	-1.0
	RCWP	-1.9	-1.8
Fossil Fuel CO_2/Energy (%/yr)	SCW	-0.1	-0.0
	RCW	0.0	-0.0
	SCWP	-0.5	-1.2
	RCWP	-1.3	-1.1

[a] SCW = Slowly Changing World; SCWP = Slowly Changing World with Stabilizing Policies; RCW = Rapidly Changing World; RCWP = Rapidly Changing World with Stabilizing Policies.

[b] EJ = exajoule = 0.948 quadrillion BTUs

[c] Pg C = petagrams of carbon; 1 petagram = 10^{15} grams.

last decade there have been many studies of U.S. energy futures that can be compared to our U.S. results. In addition there have been several recent studies of long-term global energy use and CO_2 emissions (Chapter I). One recent study has developed "conventional wisdom reference scenarios" for CH_4, CO, NO_x, and N_2O emissions related to major energy sources (Darmstadter et al., 1987). This section compares the scenarios presented here to those developed in previous work.

Since the OPEC oil embargo focused the world's attention on energy in 1973 a number of studies have examined the future of energy supply and demand in the United States. Those analyses contain much more detail, particularly in the short term, than is possible in this study, as our focus is necessarily global and long term. Nonetheless, it is useful to compare the results of this study for the U.S. with selected previous work. The National Energy Policy Plan (NEPP) prepared by the Department of Energy (U.S. DOE, 1987b) and *Energy for A Sustainable World* (ESW), an international study supported by the World Resources Institute (Goldemberg et al., 1985, 1987, 1988) are examples of two important recent studies.

The results of these studies for the United States are summarized and compared with our scenarios in Tables 5-4 and 5-5. A key point is that both of the No Response scenarios developed here incorporate much lower growth in energy use and CO_2 emissions than is projected in the NEPP reference and NEPP high-efficiency cases. The largest discrepancies are in demand for electricity and consumption of coal, although all energy sources other than gas and all sectors show higher consumption in the NEPP projections. The NEPP Reference Case projects an increase of almost 40% in U.S. CO_2 emissions between 1985 and 2010, while the High Efficiency case produces about a 20% increase. By contrast, the RCW scenario, which has GNP assumptions similar to those used in NEPP, estimates about a 10% increase in CO_2 emissions, while the SCW scenario predicts essentially flat emissions. Had the NEPP reference case been adopted as one of our No Response

TABLE 5-4

Comparison of No Response Scenarios and NEPP

		End-Use Energy Demand (exajoules)				
			Estimated for 2010			
Sector		1985	SCW[a]	RCW[b]	NEPP-RC[c]	NEPP-HE[d]
Residential/Commercial	Fuel	11	10	11	13	19[e]
	Elec	6	6	7	9	
Transport	Fuel	21	21	19	23	22
	Elec	0	0	0	0	0
Industry	Fuel	18	20	21	26	28[e]
	Elec	3	3	4	7	
Total	Fuel	50	51	51	58	51
	Elec	9	9	10	15	13

	Primary Energy Consumption (exajoules)				
		Estimated for 2010			
Primary Energy	1985	SCW	RCW	NEPP-RC	NEPP-HE
Coal	19	18	22	38	31
Oil	33	32	29	35	33
Gas	19	19	22	19	17
Other[f]	7	8	8	18	16
Total	77	77	82	110	97

	Carbon Dioxide Emissions (petagrams of carbon)				
		Estimated for 2010			
	1985	SCW	RCW	NEPP-RC	NEPP-HE
CO_2	1.3	1.3	1.4	1.8	1.6

[a] Slowly Changing World scenario.
[b] Rapidly Changing World scenario.
[c] National Energy Policy Plan (NEPP) Reference Case (DOE, 1987b)
[d] National Energy Policy Plan (NEPP) High Efficiency Case (DOE, 1987b)
[e] Fuel + Electricity. Separate values not given.
[f] Excludes dispersed wood.

TABLE 5-5

Comparison of Stabilizing Policy Scenarios and ESW

Sector		End-Use Energy Demand (exajoules)				
			Estimated for 2020			
		1985	SCWP[a]	RCWP[b]	ESW-S[c]	ESW-R[d]
Residential/ Commercial	Fuel	11	8	5	5	5
	Elec	6	5	5	4	4
Transportation	Fuel	21	15	12	12	14
	Elec	0	0	0	0	0
Industry	Fuel	18	18	21	14	15
	Elec	3	3	4	5	5
Total	Fuel	50	41	38	31	34
	Elect	9	8	9	9	9

Primary Energy	Primary Energy Consumption (exajoules)				
		Estimated for 2020			
	1985	SCWP	RCWP	ESW-S	ESW-R
Coal	19	11	9	11	13
Oil	33	24	18	13[e]	14[e]
Gas	19	18	22	13[e]	14[e]
Other	7	12	18	14	14
Total	77	65	64	52	56

	Carbon Dioxide Emissions (petagrams of carbon)				
		Estimated for 2020			
	1985	SCWP	RCWP	ESW-S	ESW-R
CO_2	1.3	1.0	0.8	0.7	0.8

[a] Slowly Changing World with Stabilizing Policies.
[b] Rapidly Changing World with Stabilizing Policies.
[c] Energy for a Sustainable World, Goldemberg et al., 1987, 1988. Assumes a 50% increase in per capita GNP from 1980 to 2020. Note that the SCWP case assumes a 50% increase from 1985 to 2020.
[d] Energy for a Sustainable World, Goldemberg et al., 1987, 1988. Assumes a 100% increase in per capita GNP from 1980 to 2020. Note that the RCWP case assumes a 120% increase from 1985 to 2020.
[e] Given as Oil + Gas. A 50% split is assumed following the global supply scenario given by Goldemberg et al., 1987, 1988.

scenarios, the U.S. contribution to global emissions would have been substantially higher than what we have estimated in the SCW and RCW cases, and the difference between the No Response and Stabilizing Policy cases would have been significantly greater.

Comparing low emissions scenarios, U.S. energy use is considerably higher in the Stabilizing Policy cases than in those given in *Energy for a Sustainable World*. The largest differences in consumption are in the industrial sector, with significant differences also in the residential and commercial sectors in the slow-growth cases. We assume that slower turnover of the housing stock leads to higher residential and commercial demand in the slow growth variant, whereas Goldemberg et al. assume that income does not effect demand in this sector. Despite higher energy consumption in our scenario, the two rapid-growth cases have similar CO_2 emissions due to lower consumption of coal and heavier reliance on gas and non-fossil energy sources in the RCWP scenario compared with the ESW cases.

The global energy use and CO_2 emissions calculated for 2050 in the four scenarios developed here are compared to the bounding extrapolations discussed in Chapter IV and the results of selected previous studies in Table 5-6. The total energy use derived in our scenarios falls within the lower end of the range given by trend extrapolation and previous analyses. In those studies that included a "Base Case" that did not assume the implementation of policies to reduce CO_2 emissions, the estimated primary energy demand for the year 2050 ranges from 21 to 52 terawatts (TW)[7]. This level of energy demand is approximately 2.2 to 5.5 times 1985 consumption levels of 9.4 TW. The Rapidly Changing World scenario has total energy demand that is quite similar to the Base Case given by a number of previous studies, including Edmonds and Reilly (1984), Seidel and Keyes (1983), and World Energy Conference (1983). The Slowly Changing World scenario, with almost 50% less

---

[7] 1 terawatt = 10^{12} watts = 31.54 EJ per year.

TABLE 5-6

Summary of Various Primary Energy Forecasts for the Year 2050

Report	"Base Case" TW	Range TW	"Base Case" Pg C/yr	Range Pg C/yr
This Study, No Response cases	-	15.3-28.3	-	7.6-14.9
This Study, Stabilizing Policy cases[a]	-	14.6-23.7	-	4.4-5.1
Trend Extrapolation (Chapter IV)		10-130	-	19-240
Darmstadter et al. (1987)	35.1		22.2	
Edmonds and Reilly (1983)[a]	52.2	--	26.3	15.7-26.3
Edmonds and Reilly (1984)	28.4	17-63	14.5	6.8-47.4
Edmonds and Reilly (1986)[b]	21.3	12.7-38.8	7.7	4.3-18.7
Goldemberg et al. (1985)[a]	11.2 (2020)	11.2-14.8 (2020)	4.6 (2020)	4.6-5.9 (2020)
IIASA (1981)	-	22.4-35.7 (2030)	-	10-17 (2030)
IIASA (1983)	26.3	--	9.4	
Keepin et al. (1986)	-	10-35	-	2-20
Legasov et al. (1984)	42	--	13	
Lovins et al. (1981)[a]	4.6	--	<1 (2030)	
Nordhaus and Yohe (1983)[c]	30.6	18.1-43.1	13.9	7.2-20.6
Reister (1984)	29.9	24.0-41.3	10.7	9.7-27.1
Rose et al. (1983)[a]	-	16.3-32.3	-	2.7-15
Seidel and Keyes (1983)[a]	30.0	--	15	10-18
World Energy Conference (1983)	29.8	18.9-29.8	14.4	10.0-14.4

[a] Policies to limit CO_2 emissions are explicitly considered in some or all of these cases.
[b] Median, 25th and 75th percentile, non-zero correlation between parameters.
[c] Probability weighted mean ± one standard deviation.

Source: Adapted from Keepin et al. (1986). See text, Chapter I, and original source for further discussion and notes.

total energy use in 2050, lies between the median and 25th percentile non-zero correlation scenario of Edmonds et al. (1986) and Reilly et al. (1987). The estimated uncertainty bounds in the systematic uncertainty analysis conducted by Edmonds et al. are not symmetric; their median scenario has significantly lower energy use and CO_2 emissions than both the mean of their results and the result of using the median values for all model parameters. The implication is that very high energy use scenarios may be much less probable than is suggested by simply considering the range given by many studies.

Compared with the energy use estimates there is substantially less, though still considerable, variation in the CO_2 emissions estimates for 2050. None of the studies cited in Table 5-6 approach within a factor of four the result of exponentially extrapolating the pre-1973 rates of energy demand growth, assuming no change in the mix of sources. This reflects the constraint due to the finite size of the fossil fuel resource base (Chapter IV), which implies that very high growth in energy consumption would have to be accompanied by a significant shift away from fossil fuels (but not before atmospheric CO_2 concentrations reached extraordinarily high levels). Considering the full range of values for both energy use and CO_2 emissions represented in Tables 5-4 and 5-5, it does appear that, as intended, the Slowly Changing and Rapidly Changing World scenarios represent very different but not extreme possibilities.

While the general agreement found between this study and previous studies at the aggregate level may be comforting, substantial disagreements are possible when the results are examined more closely. For example, the global increase in energy demand obtained in the Rapidly Changing World scenario is the result of almost level demand in OECD countries coupled with very vigorous demand growth in developing countries. Other scenarios with nearly identical global demand in 2050 may not distinguish among regions (e.g., Nordhaus and Yohe, 1983) or may have a more even pattern of energy demand growth (e.g., Edmonds and Reilly, 1984). Similarly, the GNP growth rate assumed in the RCW scenario is higher than what was assumed by Seidel and Keyes (1983), but because higher rates of technical efficiency improvements were assumed in the RCW case, energy demand and CO_2 emissions are almost identical in 2050.

The results obtained in the policy scenarios developed here are most appropriately compared with the results of Lovins et al. (1981), Rose et al. (1983), and Goldemberg et al. (1985, 1987, 1988). These studies all emphasize the possibility that increased efficiency of energy use could limit energy demand and CO_2 emissions while allowing for sustainable economic growth. They conclude that energy demand in 2050 could be held to between 5 and 16 TW by supplying energy services with advanced cost-effective technology that is either available or nearly commercial today. In these scenarios efficiency improvements combined with shifts in energy supply allow CO_2 emissions to be held at or below today's level, and Lovins et al. (1981) argue that it is technically feasible to reduce fossil fuel CO_2 emissions by about 80% over 50 years. The SCWP and RCWP scenarios have energy consumption of 15 and 24 TW respectively--similar to, but somewhat higher than, what previous studies suggested was feasible. Part of this difference may be explained by the high rate of economic growth assumed in the RCWP case and, particularly in comparison to Lovins et al., our assumption that efficiency measures are not adopted up to their technical potential. The CO_2 emissions in the policy scenarios are 10-20% below current levels, again consistent with some previous analyses. This result is obtained in different ways, however. For example, the lowest CO_2 scenario given by Rose et al. assumes substantially more contribution from non-fossil energy sources than do the policy scenarios developed here, while the Goldemberg et al. high demand scenario has somewhat more oil and gas and less coal than does the RCWP case in 2020.

The estimates of energy-related (fossil fuel and wood use) emissions of CH_4, N_2O, NO_x, and CO developed here are compared in Table 5-7 with results of a study by Darmstadter et al. (1987). While the main purpose of their study was to develop an historical database, reference values for future emissions are presented assuming either constant emission coefficients or coefficients declining by 1% per year. The emissions calculated with constant coefficients by Darmstadter et al. increase much more rapidly than those obtained in any of our scenarios. These differences are not too surprising given our explicit assumptions regarding technological change, including increasing penetration of emission control technologies. The largest discrepancy is for N_2O, reflecting not only our assumptions regarding technical change, but also the much higher initial emission coefficient adopted by Darmstadter et al. based on Hao et al., 1987 (see Chapter II). The initial estimate of CO emissions given by Darmstadter et al. is a factor of two lower than ours, probably due primarily to their extrapolation of the U.S. emission coefficient for gasoline to the rest of the world; we have attempted to account for variations in automobile emission control technology by region, with most regions having higher average CO emissions than the U.S. The closest agreement is for CH_4, probably because these emissions are directly proportional to the total quantity of coal and gas produced, and are not assumed to depend on production technology in our No Response scenarios.

When Darmstadter et al. assume that all emission coefficients decline by 1% per year, they obtain estimates of NO_x emissions that are similar to those occurring in the RCW case and CH_4 emissions estimates closer to those obtained in the RCWP case; their CO emissions estimate falls between these two cases. Overall, the RCWP case has significantly lower emissions than are obtained by Darmstadter et al. even when they decrease their emission coefficients by 1%/y for a full century. This is a result not only of the assumptions regarding emission control technology, but also because our policy scenarios have substantially lower total energy demand and a very different fuel mix.

TABLE 5-7

Comparison of Energy-Related Trace-Gas Emissions Scenarios

Trace Gas	Scenario	Emissions of Trace Gases (teragrams)		
		1985/1980	2025/2030	2075/2080
CH_4 ($Tg\ CH_4$)	RCW	68	141	301
	RCWP		73	74
	* Darmstadter et al.	63	192	432
	** Darmstadter et al.		117	131
N_2O (Tg N)	RCW	1.1	2.1	3.8
	RCWP		1.2	1.4
	* Darmstadter et al.	4.3	16	57
	** Darmstadter et al.		9.5	21
NO_x (Tg N)	RCW	25	43	77
	RCWP		27	21
	* Darmstadter et al.	20	62	184
	** Darmstadter et al.		37	68
CO (Tg C)	RCW	202	318	651
	RCWP		122	82
	* Darmstadter et al.	108	292	614
	** Darmstadter et al.		177	226

* Constant Emission Coefficients.

** Emission Coefficients Decline 1% Per Year.

Industrial Processes

Halocarbon Emissions

The most important industrial source of greenhouse gases not directly associated with energy use is the production and release of CFCs and halons. In both the Slowly Changing World and Rapidly Changing World scenarios, the Montreal Protocol as currently formulated is assumed to come into force and apply throughout the projection period. This agreement (described in Chapter IV, VIII, and IX) calls on developed countries to reduce their emissions of certain CFCs 50% from 1986 levels by 1998, and to freeze the use of halons. Developing countries with low per capita consumption, however, are allowed to increase the use of these compounds for up to 10 years--as a result, emissions of the controlled compounds could actually increase substantially, depending on the number of countries that participate in the Protocol and the rate at which use increases in developing and non-participating nations (Hoffman and Gibbs, 1988). For the Slowly Changing World scenario we adopt the assumptions of the Protocol scenario developed for the Regulatory Impact Assessment of rules to implement the Montreal Protocol in the United States (U.S. EPA, 1988). Namely that, in addition to the U.S., 94% (in terms of current CFC consumption) of developed countries and 65% of developing countries participate in the agreement. In this scenario the global average annual growth rate in demand for products and services that would use CFCs, if they were available, is approximately 4.0% from 1986 to 2000 and 2.5% from 2000 to 2050 (constant production is assumed after 2050). Growth in demand is much higher in certain developing countries, particularly India and China. These growth rates are not applied directly to CFC use in non-participating and developing countries, however, because it is assumed that shifts in technology development away from CFCs in the United States and other participating countries "rechannels" demand in other countries as well.[8] The Stabilizing Policy scenarios assume that the Montreal Protocol is strengthened to produce a complete phase-out of CFCs in participating countries by 2003.[9]

Figure 5-12 shows the estimates for emissions of CFC-11, -12, and -13, and HCFC-22 under the four scenarios. Emissions change more slowly than production because a significant portion of each year's production is "banked" in air conditioners, refrigeration systems, and closed-cell foams. The model keeps track of the size of this bank and estimates the gradual release of these CFCs. In the Slowly Changing World scenario emissions are relatively constant despite the Protocol's requirement of a 50% reduction in participating industrialized countries. After declining to 12% below 1985 levels between 1990 and 2020, emissions of CFC-11 begin to rise again, reaching 1985 levels by the end of the projection period. CFC-113 emissions also fall significantly for a few decades but rise again toward 1985 levels. CFC-12 emissions never decline to 1985 levels: they decline by 11% between 1990 and 2015, reaching a few percent above 1985 values, then they rise slowly, almost reaching the 1990 peak levels towards the end of the 21st Century. Emissions of HCFC-22 grow rapidly as a substitute for the fully-halogenated species that have the highest ozone-depletion potential. Although HCFC-22 has a shorter lifetime and weaker radiative forcing than the fully-halogenated compounds, it could make a significant contribution to global warming during the next century because it is not controlled by the Montreal Protocol.

---

[8] In the Slowly Changing World scenario this rechanneling effect is assumed to decrease growth in CFC demand by 63% in developed countries and by 50% in developing countries. In the Rapidly Changing World scenario the baseline growth rates are increased by a factor of 1.7 to reflect the higher rate of economic growth, but participation is assumed to be 100% in developed countries and 75% in developing countries, and rechanneling reduces the baseline growth rates by 63% in developing countries (rechanneling does not affect developed countries in this scenario as 100% participation in the Protocol is assumed).

[9] Participation is assumed to be 100% in industrialized countries and 85% in developing countries; rechanneling reduces the baseline growth rates of non-participants by 75%.

FIGURE 5-12

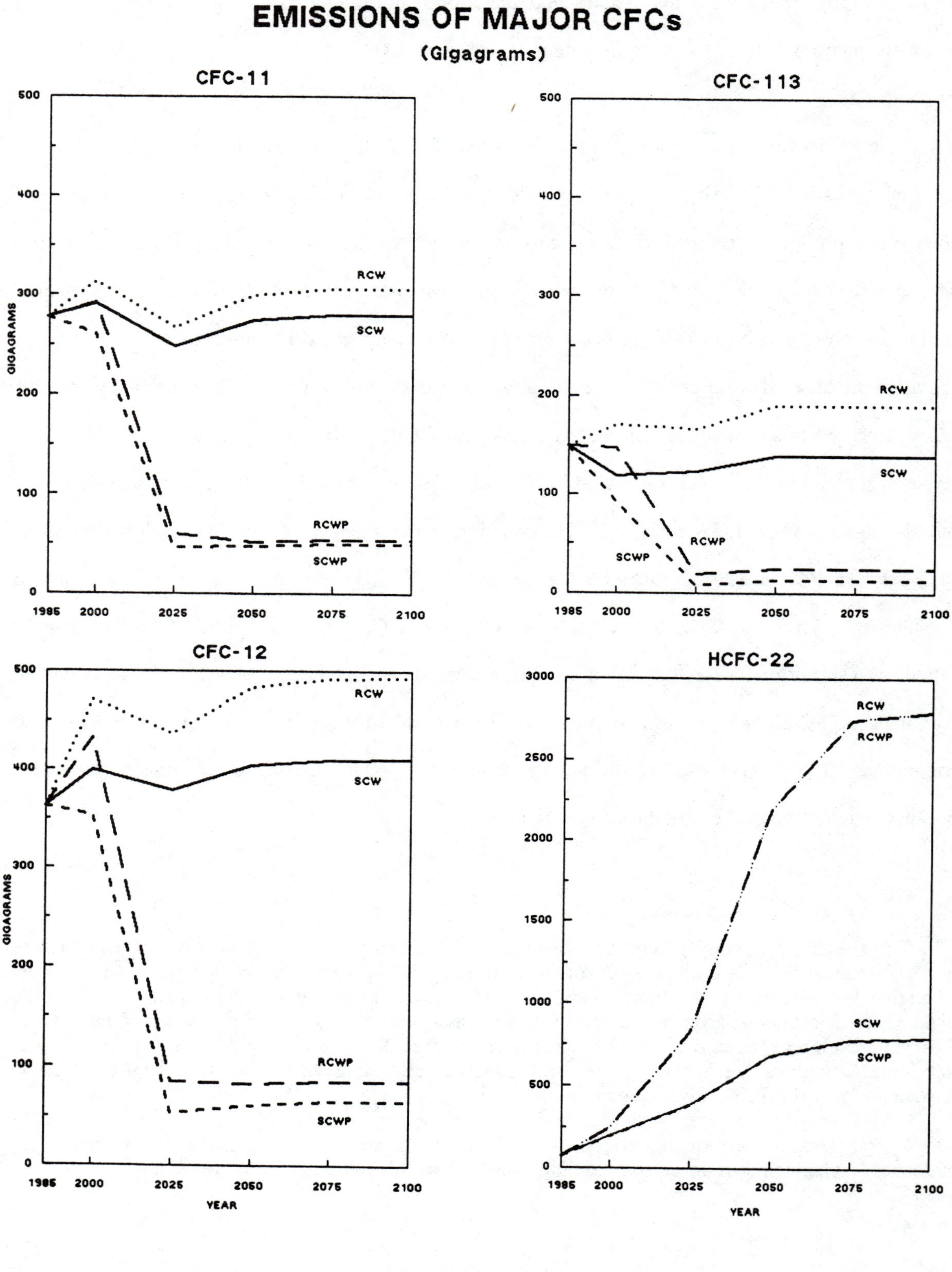
EMISSIONS OF MAJOR CFCs
(Gigagrams)
CFC-11
CFC-113
CFC-12
HCFC-22
GIGAGRAMS
YEAR
RCW
SCW
RCWP
SCWP
500
400
300
200
100
0
3000
2500
2000
1500
1000
500
1985
2000
2025
2050
2075
2100

In the RCW, higher growth rates in developing countries more than compensate for higher participation and rechanneling rates. CFC-11 emissions decline by no more than 6% below 1985 levels, while CFC-12 and -113 each increase by more than 25%. Emissions of HCFC-22 grow dramatically in this scenario. In the SCWP case emissions of the fully-halogenated compounds fall by more than 80% from 1985 levels by 2025, which is sufficient to reverse the trend in concentrations (see Figure 5-18). Emission reductions in the RCWP case are not quite as large, but still lead to declining concentrations after 2025. HCFC-22 emissions are assumed to be the same in the No Response and Stabilizing Policy cases; however, these emissions could rise as a result of a CFC phase-out if chemical substitution is the primary approach to eliminating CFCs, or fall, if product substitution and process redesign are the major approaches (see Chapter VI).

Emissions From Landfills and Cement

Other important activities included in the industrial category are CO_2 emissions from cement making and CH_4 emissions from landfills. The growth of these activities in developing countries is assumed to be related to per capita income in a simple fashion, although growth is curtailed as current per capita levels in industrialized countries are approached. The result is a three- to four-fold increase in CO_2 emissions from cement in the SCW and RCW scenarios, respectively, though emissions remain less than 0.5 Pg C/yr in all cases. Landfill CH_4 emissions increase by more than five-fold in the RCW, reaching 15% of the total by 2100. In the policy scenarios, advanced materials are assumed to reduce the demand for cement (relative to the No Response scenarios), while gas recovery systems and waste reduction policies are assumed to limit emissions from landfills. The result is that emissions from cement making still increase by a factor of two to three, but CH_4 emissions from landfills are held essentially constant.

Changes in Land Use

Deforestation has been a significant source of CO_2 to the atmosphere over the last two centuries, as clearly shown by the measurements of CO_2 concentrations in Greenland and Antarctic ice (Chapter II). If current trends continue, tropical forests could be completely eliminated during the next century, adding significantly to the CO_2 emissions from fossil fuels. On the other hand, efforts are underway to reverse deforestation; if these efforts succeed, reforestation could become a net sink for atmospheric CO_2. The total amount of carbon that can move in either direction between the atmosphere and terrestrial ecosystems is ultimately constrained by the area of forests available for deforestation or by the area of land available to support new forests. The timing and magnitudes of these fluxes of carbon are determined by the timing and extent of changes in land use as influenced by local, national, and international policies.

The causes of deforestation are complex and vary from country to country. This makes it difficult to directly tie assumptions about deforestation rates to the economic and demographic assumptions of the general scenarios. Qualitatively, we assume that in a Slowly Changing World poverty, unsustainable agricultural practices, and rapid population growth lead to continuously increasing pressure on remaining forests. The rate of deforestation is assumed to increase from current levels at the rate of population growth and tropical deforestation increases from 11 million hectares per year (Mha/yr) in 1980 to 34 Mha/yr in 2047, when all the unprotected forests in Asia are exhausted. In a Rapidly Changing World improved agricultural practices and the substitution of modern fuels for traditional uses of wood could ease the pressure on forests. Nonetheless, clearing of forest lands for agriculture, pasture, logging, and speculation could continue apace, even if small areas are set aside as biological preserves. In this scenario tropical deforestation is assumed to increase very gradually, reaching 15 Mha/yr in 2097, before the unprotected forest areas of Latin America are exhausted.

In the Stabilizing Policy scenarios it is assumed that a combination of policies succeed in stopping deforestation by 2025, while more than 1000 Mha is reforested by 2100. Only land that once supported forests and is not intensively cultivated is assumed to be available for reforestation. These lands include 85% of the area currently involved in shifting cultivation (370 Mha) under the assumption that this practice is replaced by sustainable low input agriculture (Sanchez and Benites, 1987). In addition, fallow agricultural land in the temperate zone (250 Mha), planted pasture in Latin America (100 Mha), and degraded land in Africa and Asia (400 Mha) is assumed to be reforested. Of the reforested land, about 380 Mha is assumed to be in plantations (sufficient to produce the biomass energy requirements of the RCWP case with productivity increases expected by the Department of Energy; Walter, 1988), the rest absorbs carbon at a much lower rate but reaches a higher level of average biomass.[10]

The carbon fluxes associated with these deforestation/reforestation scenarios based on Houghton's (1988) low estimates of average biomass are shown in Figure 5-13. In the SCW CO_2 emissions from deforestation increase rapidly from 0.7 Pg C/yr to more than 2 Pg C/yr in 2047 before the Asian forests are exhausted. Latin American and African forests are exhausted by 2075, reducing emissions drastically. Total deforestation emissions are almost the same in the RCW but they are spread out over a longer period. Emissions are close to 1 Pg C/yr from 2000 to 2100. In the Stabilizing Policy scenarios the biosphere becomes a sink for carbon by 2000 and reaches its peak

---

[10] Plantation products decay at various rates at the end of each rotation; no attempt to protect this carbon from oxidation is assumed.

FIGURE 5-13

CO2 EMISSIONS FROM DEFORESTATION

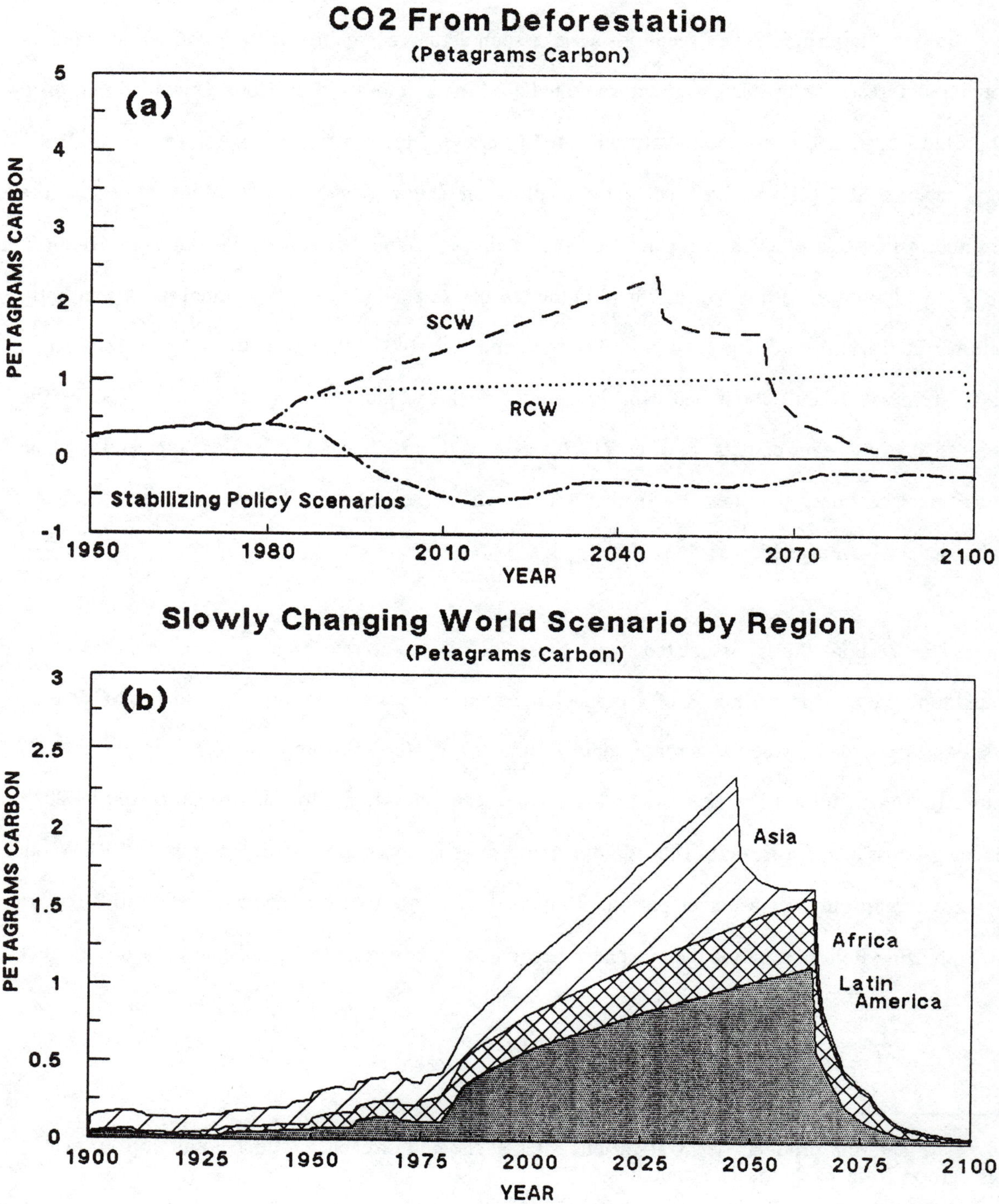

absorption of 0.7 Pg C/yr before 2025. The size of this sink declines gradually after 2025 as forests reach their maximum size and extent.

Agricultural Activities

The demand for agricultural products is a direct function of population, but is not strongly dependent on income levels. Thus, there are only small differences between the scenarios as the much higher incomes largely offset the somewhat lower populations in the RCW compared with the SCW. The land area used for rice production, and thus the methane emissions from this source, increases by only about 50% by 2100 in both the SCW and RCW scenarios (production per hectare increases by 80-100%). Meat production increases more, about 125%, as demand rises with income to some extent. Satisfying the demands of increasing populations with a finite amount of land requires more intensive cultivation, and fertilizer use increases by 160% as a result.

In the Stabilizing Policy scenarios we do not assume changes in the demand for agricultural commodities, but rather changes in technology and production methods that could reduce greenhouse gas emissions per unit of product. Although the impact of specific technologies cannot be estimated at present, a number of techniques have been identified for reducing methane emissions associated with rice and meat production and nitrous oxide emissions related to the use of fertilizer (Chapter VII). For simplicity, we have assumed that CH_4 emissions per unit of rice, meat, and milk production decrease by 0.5% per year (emissions from animals not used in commercial meat or milk production are assumed to be constant). Emissions of N_2O per unit of nitrogenous fertilizer applied are also assumed to decrease by 0.5% per year for each fertilizer type. In addition, fertilizer use is assumed to shift away from those types with the highest emissions after 2000. Based on these assumptions, CH_4 emissions from rice production remain roughly constant until 2075, after which time they fall by about 20% as the global population stabilizes. Methane emissions from enteric fermentation increase by 40-50% by the middle of the 21st Century, before falling to within about 30% of 1985 levels. Similarly, N_2O emissions from fertilizer increase from 0.3 to 0.5 Tg N/yr between 1985 and 2025 and then remain roughly constant.

Total Emissions

Total emissions of the key radiatively-important gases, the aggregate of estimates of the emissions from each activity discussed above and of natural emissions, are shown in Table 5-8. Overall, emissions increase gradually in the Slowly Changing World scenario and more dramatically in the Rapidly Changing World, while in the policy scenarios emissions are reasonably stable or declining.

In the No Response scenarios CO_2 emissions are projected to increase to a much greater extent than emissions of the other gases. This is because all net CO_2 emissions are assumed to be anthropogenic in origin and because CO_2 is a fundamental product of all fossil-fuel combustion. In the SCW increased deforestation contributes significantly to near-term growth in CO_2 emissions, and total emissions are relatively constant between 2025 and 2075 as forests are exhausted (see Figure 5-14). In the RCW CO_2 emissions are dominated by the growth in fossil-fuel combustion and total emissions increase by a factor of three by 2050. In the Stabilizing Policy scenarios increased end-use efficiency and reforestation contribute significantly to producing decreasing emissions in the near-term, while decreased reliance on fossil fuels in conjunction with continued improvements in efficiency allow for further decreases later on.

TABLE 5-8

Trace Gas Emissions

	1985	2000	2025	2050	2075	2100
CO_2 (Pg C)						
SCW	5.9	7.2	9.2	9.8	9.7	11.4
RCW	5.9	7.6	11.5	16.6	22.2	25.5
SCWP	5.9	5.3	5.1	4.2	3.7	3.1
RCWP	5.9	5.5	5.2	5.1	4.8	4.5
N_2O (Tg N)						
SCW	11.3	12.3	13.5	13.7	12.2	12.1
RCW	11.3	12.0	13.0	14.0	15.0	15.0
SCWP	11.3	11.0	10.7	10.6	10.6	10.7
RCWP	11.3	11.0	10.8	10.8	10.8	10.9
CH_4 (Tg CH_4)						
SCW	514.4	569.4	676.4	739.9	773.3	815.9
RCW	510.5	576.7	712.4	879.6	1,025.4	1,089.0
SCWP	514.4	519.0	545.0	534.0	522.2	484.7
RCWP	510.5	527.3	558.7	567.2	542.6	508.0
NO_x (Tg N)						
SCW	53.3	59.3	68.8	70.8	65.2	70.1
RCW	53.2	60.0	72.9	91.7	108.2	118.2
SCWP	53.3	50.9	44.8	40.9	41.0	42.6
RCWP	53.2	51.5	50.7	46.1	44.8	43.8
CO (Tg C)						
SCW	502.3	616.1	842.0	858.7	594.7	603.5
RCW	502.0	571.1	699.1	894.6	1,049.9	1,207.1
SCWP	502.3	382.5	286.1	245.5	250.6	250.9
RCWP	502.0	382.3	290.8	241.7	242.0	244.9
CFC-11 (Gg)						
SCW	278.3	292.0	248.1	275.6	280.4	280.8
RCW	278.3	314.2	267.3	300.3	306.3	306.8
SCWP	278.3	261.8	47.3	47.8	50.3	50.6
RCWP	278.3	293.1	60.2	52.4	54.2	54.3
CFC-12 (Gg)						
SCW	363.8	401.1	379.7	404.8	410.3	410.8
RCW	363.8	471.8	437.5	483.6	492.3	493.1
SCWP	363.8	354.2	54.9	62.3	65.7	66.0
RCWP	363.8	433.3	85.9	83.6	86.4	86.6
CFC-22 (Gg)						
SCW	73.8	192.1	385.0	686.5	785.7	794.8
RCW	73.8	247.4	829.1	2,194.2	2,744.9	2,795.6
SCWP	73.8	192.1	385.0	686.5	785.7	794.8
RCWP	73.8	247.4	829.1	2,194.2	2,744.9	2,795.6
CFC-113 (Gg)						
SCW	150.5	119.3	124.2	140.4	140.4	140.4
RCW	150.5	171.8	167.4	190.3	190.3	190.3
SCWP	150.5	93.2	9.0	13.8	13.8	13.8
RCWP	150.5	148.7	19.7	25.4	25.4	25.4

FIGURE 5-14

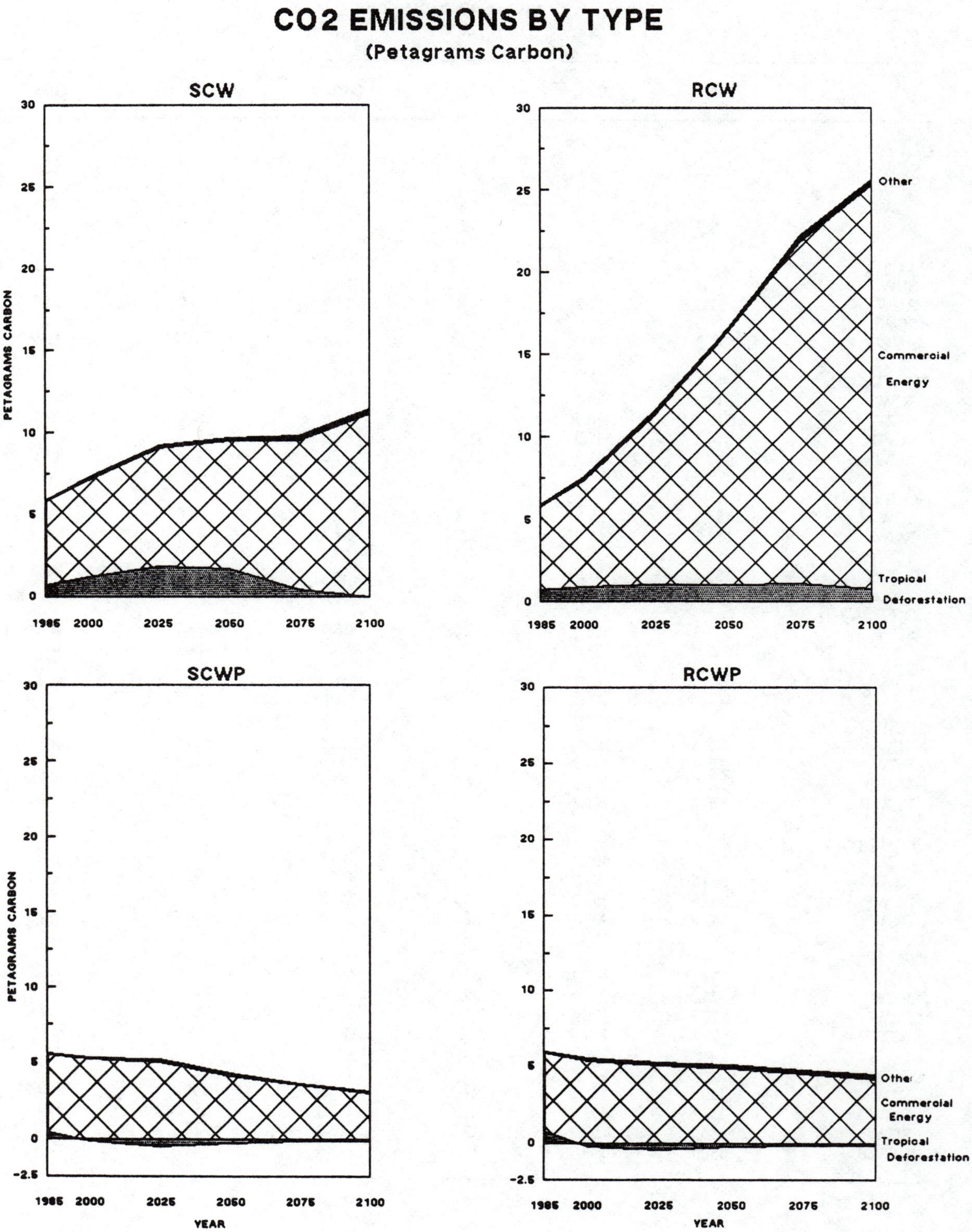
CO2 EMISSIONS BY TYPE
(Petagrams Carbon)
SCW
RCW
SCWP
RCWP
PETAGRAMS CARBON
YEAR
30
25
20
15
10
5
0
-2.5
1985
2000
2025
2050
2075
2100
Other
Commercial
Energy
Tropical
Deforestation

The regional allocation of CO_2 emissions shows a rapid increase in the share attributed to developing countries in all scenarios (Figure 5-15). This share increases from about 34% currently to 57% by 2025 and levels off at about 60% after 2050 in the RCW. The developing countries account for a little over 50% of CO_2 emissions in the SCW after 2025, with the share from developing countries other than China decreasing after 2050 as deforestation emissions decline. China's share of emissions grows most dramatically in the Stabilizing Policy scenarios as deforestation is eliminated in other developing countries and China becomes by far the world's largest coal consumer. About 70% of global CO_2 emissions are from China and other developing countries by 2100 in the RCWP scenario, but only 42% in the SCWP.

The projected increases in CH_4 emissions in the No Response scenarios are a result of growth in a variety of sources (see Figure 5-16). In the SCW almost 60% of the increase between 1985 and 2050 is due to enteric fermentation and rice cultivation, whereas in the RCW these sources account for less than 40% of the increase and the growth in emissions from fuel production accounts for another 40%. In both of these scenarios emissions from landfills increase steadily, becoming quite significant by the end of the period. Reduced growth in each component is responsible for relatively stable CH_4 emissions in the policy scenarios. The total increases gradually until 2025 and declines after 2050, falling below 1985 levels by the end of the period in both the SCWP and RCWP cases.

The regional contributions to CH_4 emissions do not shift as dramatically as for CO_2 (Figure 5-17). The share of CH_4 emissions from industrialized countries increases in the RCW scenario due to rapid growth in coal production, but this share declines somewhat in the other three scenarios.

Total N_2O emissions do not increase dramatically in any of the scenarios, although we note again that current, and therefore future, emissions of N_2O are highly uncertain. These uncertainties,

FIGURE 5-15

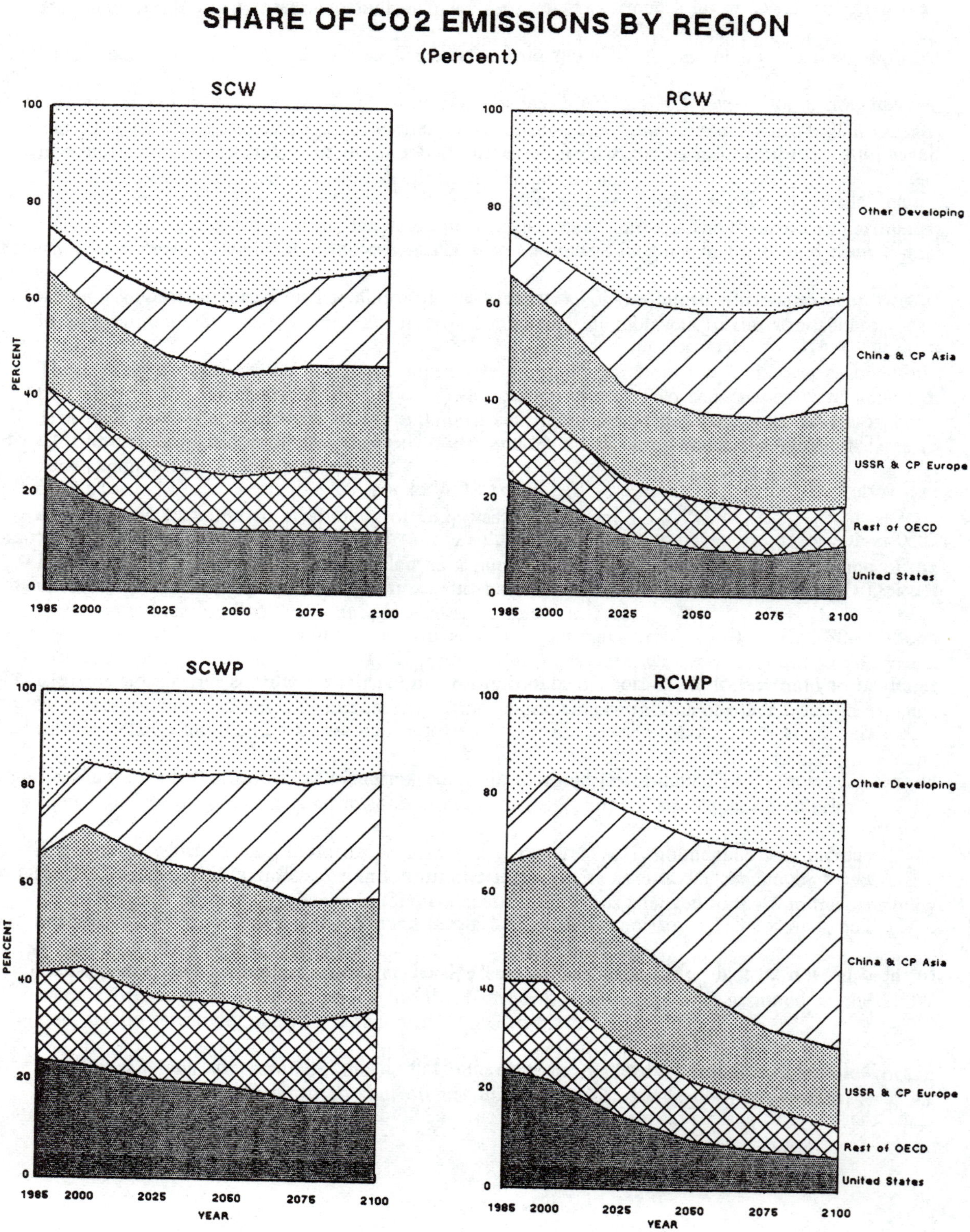
SHARE OF CO2 EMISSIONS BY REGION
(Percent)
SCW
RCW
SCWP
RCWP
PERCENT
YEAR
0
20
40
60
80
100
1985
2000
2025
2050
2075
2100
Other Developing
China & CP Asia
USSR & CP Europe
Rest of OECD
United States

FIGURE 5-16

CH4 EMISSIONS BY TYPE

(Teragrams)

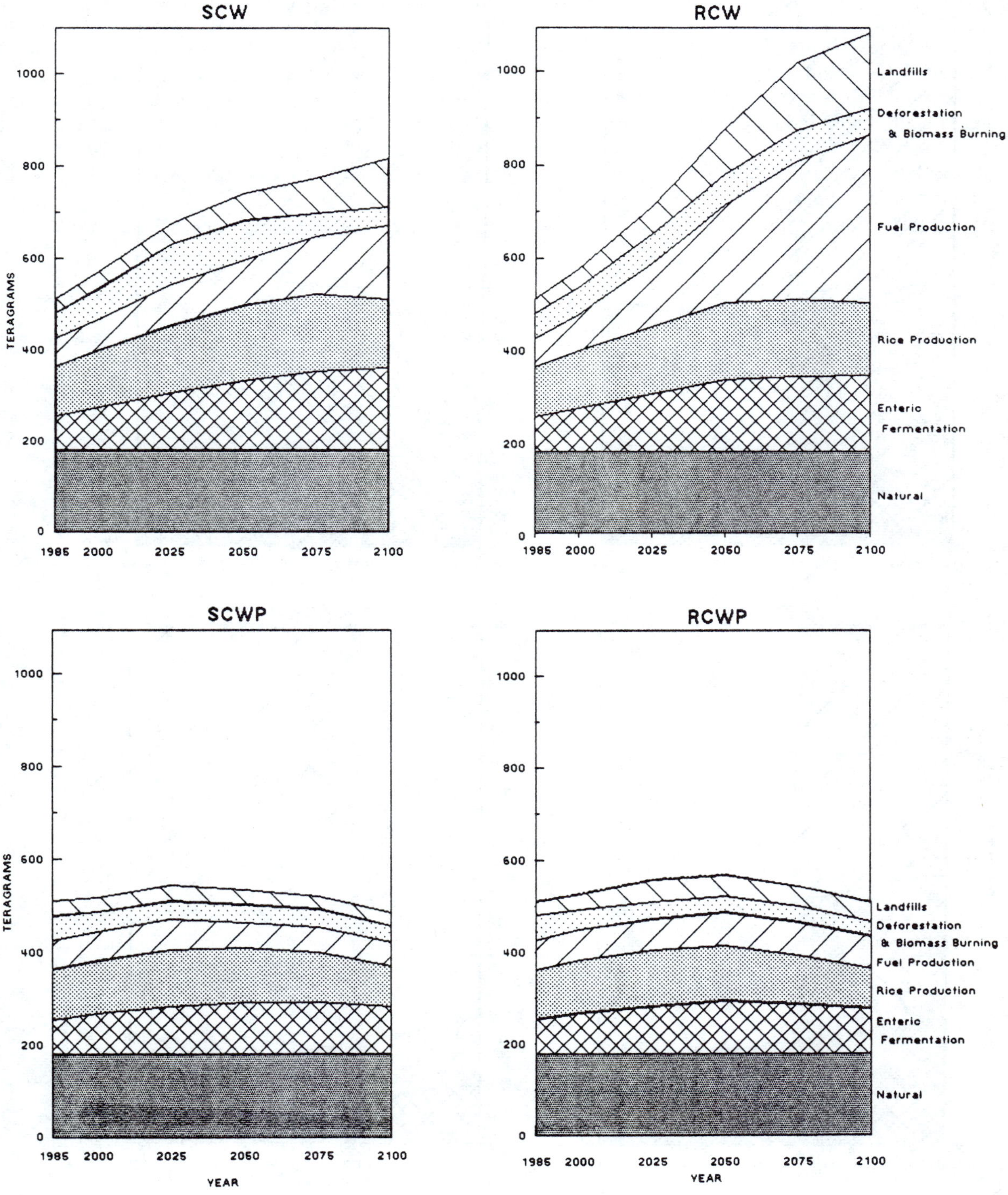

FIGURE 5-17

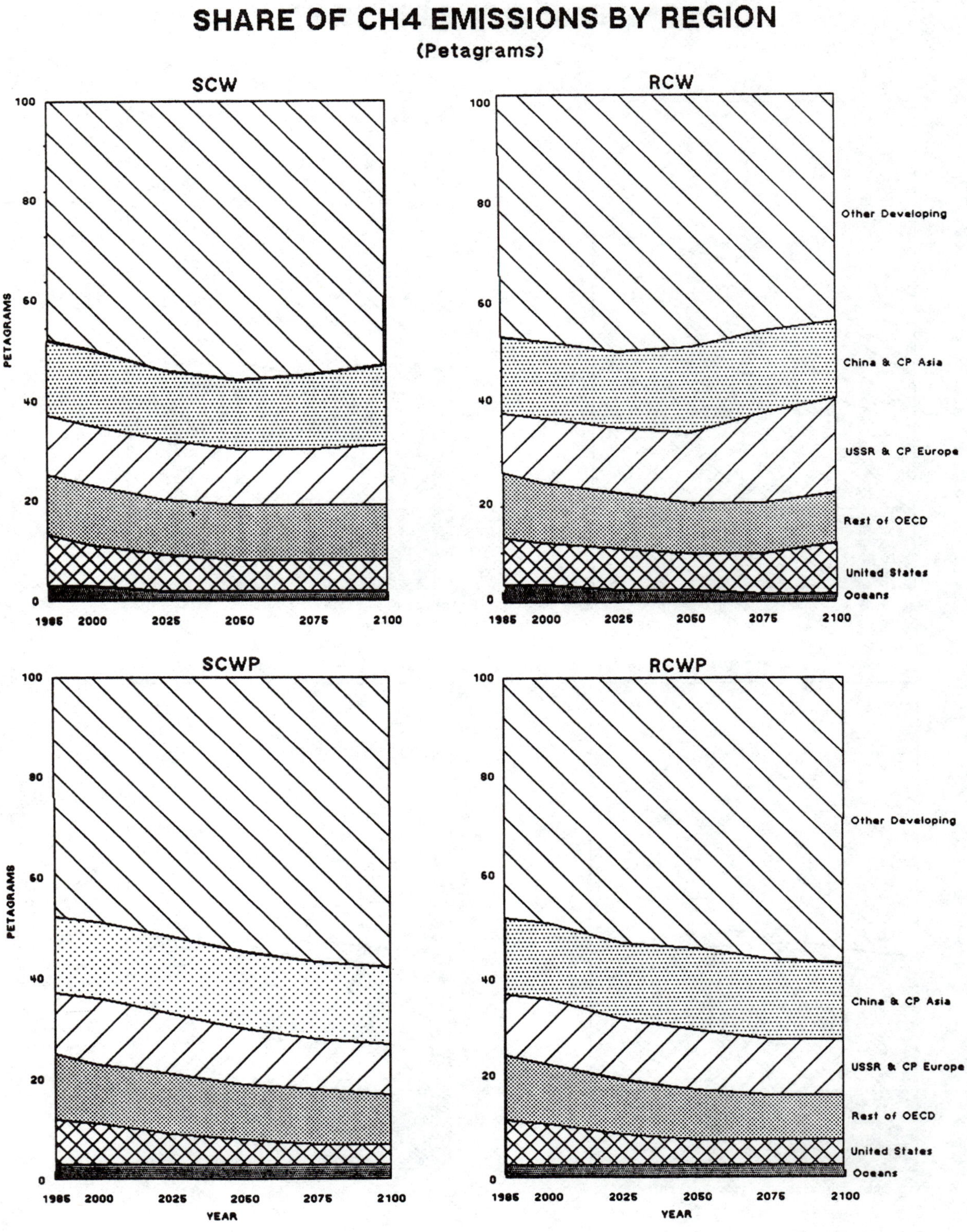
SHARE OF CH4 EMISSIONS BY REGION
(Petagrams)
SCW
RCW
SCWP
RCWP
PETAGRAMS
YEAR
0
20
40
60
80
100
1985 2000 2025 2050 2075 2100
Other Developing
China & CP Asia
USSR & CP Europe
Rest of OECD
United States
Oceans

however, do not appear to have a large impact on the overall rate or magnitude of climatic change in these scenarios (see Chapter VI). In the SCW, emissions related to deforestation and land-clearing, as well as fertilizer-induced emissions, increase significantly through 2025, and total emissions decline after 2050. In the RCW, emissions growth is driven mainly by a four-fold increase in fossil-fuel combustion emissions between 1985 and 2100. In the policy cases, total emissions decline slightly due to the assumed decreases in emissions per unit of fertilizer use and fossil-fuel combustion, and because deforestation is halted.

In the No Response scenarios, emissions of both NO_x and CO increase significantly through 2050. After 2050 declining emissions related to deforestation in the SCW compensate for continued increases in energy-related emissions. The deforestation assumptions have a particularly large impact on CO emissions as deforestation accounts for 40-50% of the total between 2000 and 2050 in this case. In the RCW, deforestation emissions are relatively uniform, and both CO and NO_x emissions continue to increase through 2100. In the Stabilizing Policy cases emission controls produce relatively stable NO_x emissions and declining CO emissions from fossil-fuel combustion sources, while deforestation emissions are eliminated. The result is moderate decline in NO_x emissions and more than a 50% cut in CO emissions by 2050.

Atmospheric Concentrations

Figure 5-18 shows concentrations of greenhouse gases that result from the pattern of emissions discussed above. Because CO_2, N_2O, and CFCs are long-lived in the atmosphere, their concentrations respond gradually to changes in emissions. CH_4 has an intermediate lifetime (about 10 years), which is itself affected by changes in emissions of CO, NO_x, CH_4, and other trace gases, so its atmospheric concentration responds rapidly to changes in emissions.

FIGURE 5-18

ATMOSPHERIC CONCENTRATIONS

(3.0 Degree Celsius Climate Sensitivity)

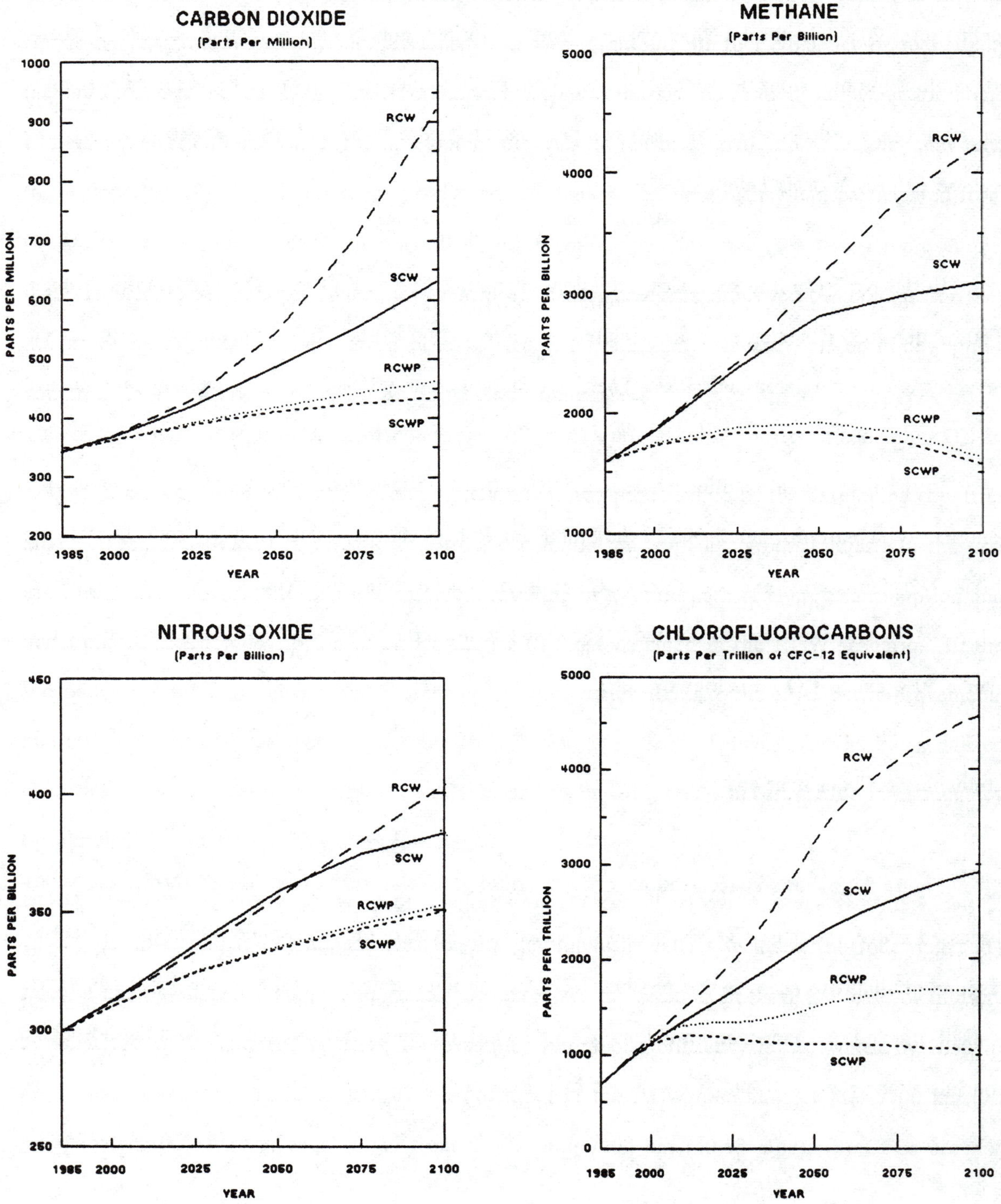

CO_2 concentrations reach twice their preindustrial levels (570 ppm) in about 2080 in the Slowly Changing World scenario; this level is reached by 2055 in the Rapidly Changing World and concentrations more than three times preindustrial values are reached by 2100 (Figure 5-18). Despite declining CO_2 emissions in the policy scenarios, CO_2 concentrations continue to increase throughout the projection period, reaching almost 440 ppm in the SCWP case and 470 ppm in the RCWP case by 2100. It is interesting to note that the fraction of total CO_2 emissions during the 21st Century that remain in the atmosphere in 2100 is 46% in the RCW case and 39% in the RCWP case, so that emission reductions have a more than linear impact on concentrations.

CH_4 concentrations increase by almost a factor of 2 in the SCW and a factor of 2.6 in the RCW, with the most rapid growth occurring between 1985 and 2050 (Figure 5-18). Interestingly, the 2050 concentration obtained in the SCW is similar to the result of linearly extrapolating the currently observed growth rate of 1% per year, whereas the RCW value is close to an exponential extrapolation of current growth; the 2100 values lie substantially below a continuation of such extrapolations for another 50 years. In the policy cases CH_4 concentrations increase by 13-18% between 1985 and 2025, after which they level off and decline to roughly 1985 levels by 2100. CH_4 concentrations are affected by temperature feedbacks on atmospheric chemistry: Increasing the climate sensitivity of the model from 2.0°C to 4.0°C reduces concentrations by 100 ppb in the SCWP and 210 ppb in the RCW.

By contrast with methane, N_2O concentrations increase gradually in all the scenarios as a result of the current imbalance between sources and sinks (Figure 5-18). The concentration increase is about 80 ppb in the SCW and 100 ppb in the RCW compared with 50 ppb in the Stabilizing Policy cases. Thus, the policy assumptions reduce the concentration growth by 40-50%.

CFC concentrations increase dramatically in the No Response scenarios despite the assumption that at least 65% of developing countries and 95% of industrialized countries participate in the Montreal Protocol (Figure 5-18). The total concentration of CFCs weighted by their contribution to the greenhouse effect increases by a factor of 4.2 and 6.5 in the SCW and RCW scenarios, respectively. On the other hand, the phase-out assumed in the policy cases does stabilize CFC concentrations (other than HCFC-22) by 2025, but their total greenhouse forcing still increases by 1.5-2.4 times over current levels.

It is interesting to compare the concentration changes calculated here, on the basis of explicit assumptions linking emissions with activities, to recent studies that have made less formal estimates based primarily on current trends in concentrations and/or emissions (Table 5-9). Our No Response estimates of future concentrations are in good agreement with those of Ramanathan et al. (1985) for 2030 and Dickinson and Cicerone (1986) for 2050. A notable exception is CFCs, for which we expect significantly lower concentrations as a result of the recent Montreal Protocol to control production of these compounds. In addition, our 2030 estimates of N_2O concentrations are somewhat below the lower end of the range given by Ramanathan et al., although they fall within the lower end of Dickinson and Cicerone's range for 2050. The differences between the SCW and RCW scenarios are significantly less than the ranges suggested by these authors for all the compounds listed in Table 5-9--at least partially because the only differences between the SCW and RCW scenarios are assumptions about activity levels and technology, whereas the estimated ranges from the literature also consider uncertainties in current sources, atmospheric chemistry, and ocean carbon uptake (uncertainties in these factors are considered in Chapter VI). Also, the Slowly Changing World and Rapidly Changing World scenarios are not intended to completely bound future possibilities; significant reductions in emissions per unit of GNP are built into the No Response scenarios--if this

TABLE 5-9

Comparison of Estimates of Trace-Gas Concentrations in 2030 and 2050

		Concentrations in 2030						
Trace-Gas	Ramanathan et al. (1985)	GISS A	GISS B	GISS C	SCW	RCW	SCWP	RCWP
CO_2 (ppm)	450	443	427	368	440	450	400	400
CH_4 (ppm)	2.3(1.8-3.3)	3.5	2.5	1.9	2.5	2.6	1.8	1.9
Trop-O_3 (%)	12.5	*	0	0	19	18	-1	-1
N_2O (ppb)	375(350-450)	381	352	314	340	340	330	330
CFC-11 (ppb)	1.1(0.5-2.0)	2.3	0.8	0.2	0.5	0.5	0.3	0.4
CFC-12 (ppb)	1.8(0.9-3.5)	3.9	1.4	0.5	1.0	1.1	0.7	0.8
HCFC-22 (ppb)	0.9(0.4-1.9)				0.4	0.7	0.3	0.6

		Concentrations in 2050						
Trace-Gas	Dickinson & Cicerone (1985)	GISS A	GISS B	GISS C	SCW	RCW	SCWP	RCWP
CO_2 (ppm)	400-600	513	465	368	490	540	410	420
CH_4 (ppm)	2.1-4.0	4.7	2.7	1.9	2.8	3.1	1.8	1.9
Trop-O_3 (%)	15-50	*	0	0	23	26	-2	0
N_2O (ppb)	350-450	480	376	314	360	360	330	340
CFC-11 (ppb)	0.7-3.0	4.2	1.0	0.2	1.2	1.4	0.3	0.3
CFC-12 (ppb)	2.0-4.8	7.3	1.8	0.4	1.2	1.4	0.6	0.7
HCFC-22 (ppb)					0.6	1.7	0.5	1.6

* In this scenario the effect of O_3 and other trace gas changes is approximated by doubling the radiative forcing contributed by CFC-11 and CFC-12.

fails to materialize, and/or if economic growth is more rapid than assumed here, concentrations of a number of greenhouse gases could be significantly higher than is estimated in these scenarios.

Global Temperature Increases

Evaluating the consequences of alternative climate change scenarios is beyond the scope of this report (a variety of potential effects are examined in the companion report *Potential Effects of Global Climate Change on the United States*; Smith and Tirpak, 1988), but an indicator of the relative magnitude of change is needed as a basis for comparing the scenarios considered here. Analysts of trace-gas emissions have often emphasized the date at which carbon dioxide concentrations (or the equivalent combination of trace gases) can be expected to reach twice their pre-industrial level of $2xCO_2$. In the absence of policies to reduce emissions, however, climate change is potentially open-ended. Atmospheric composition and climate would continue to change after the $2xCO_2$ level were reached and the ecological and social consequences may depend as much on what happens after CO_2 doubles (if it does) as on when this benchmark occurs. More relevant to ecological and social systems are the average and maximum rate of climatic change. In order to compare scenarios, we therefore focus on the average rate at which global temperature may increase during the next century as well as the maximum rate of change. We emphasize that these parameters are only indicators of global change; changes at the regional level will vary in both magnitude and timing and changes in precipitation may be as important as changes in temperature. Nonetheless, the global quantities calculated here can be used to compare the scenarios presented here among themselves and with results of more detailed climate models.

The changes in concentrations shown in Figure 5-18 produce the estimated global temperature changes shown in Figure 5-19 for a range of climate sensitivity (2.0-4.0°C equilibrium increase in

FIGURE 5-19

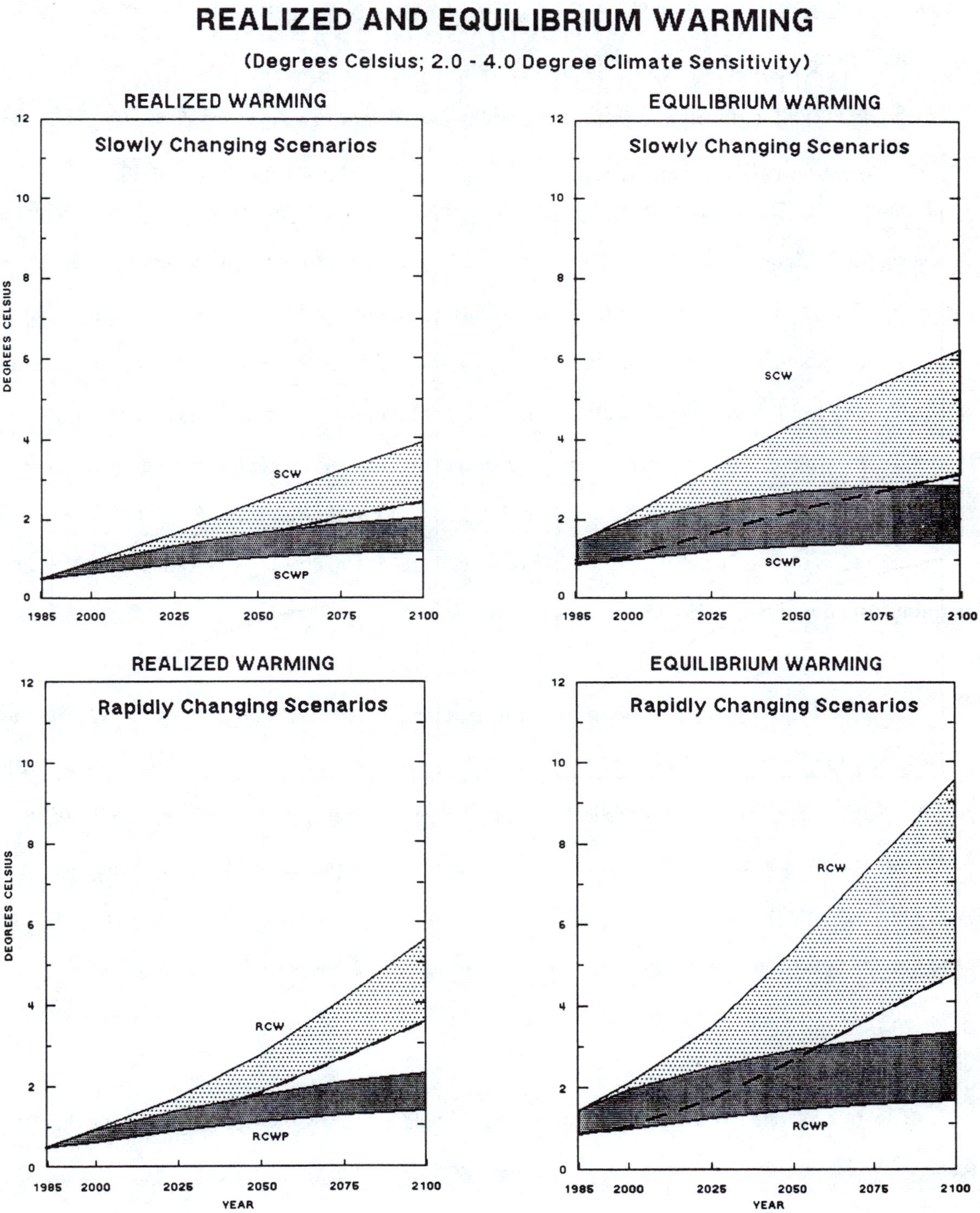
REALIZED AND EQUILIBRIUM WARMING
(Degrees Celsius; 2.0 - 4.0 Degree Climate Sensitivity)
REALIZED WARMING
Slowly Changing Scenarios
SCW
SCWP
EQUILIBRIUM WARMING
Slowly Changing Scenarios
SCW
SCWP
REALIZED WARMING
Rapidly Changing Scenarios
RCW
RCWP
EQUILIBRIUM WARMING
Rapidly Changing Scenarios
RCW
RCWP
DEGREES CELSIUS
YEAR
0
2
4
6
8
10
12
1985
2000
2025
2050
2075
2100

global temperature from doubling the atmospheric concentration of CO_2; see Chapter III). Both the "equilibrium warming commitment" and the "realized warming" are presented as a function of time. The equilibrium warming commitment for any given year is the temperature increase that would occur in equilibrium if the atmospheric composition was fixed in that year. Because the oceans have a large heat capacity the temperature change realized in the atmosphere lags considerably behind the equilibrium level. Realized warming has been estimated with a simple model of ocean heat uptake as discussed in Chapter III. Because the response of the climate system to changes in greenhouse gas concentrations is quite uncertain we also consider a range of "climate sensitivities". Climate sensitivity is defined as the equilibrium warming commitment due to doubling the concentration of carbon dioxide from preindustrial levels. Given a particular emissions scenario and climate sensitivity, the realized warming is much more uncertain than the equilibrium warming commitment because the effective heat storage capacity of the ocean is not known. On the other hand, because the amount of unrealized warming increases with increasing climate sensitivity, for a given scenario realized warming depends less on climate sensitivity than does warming commitment.

Both the SCW and RCW scenarios lead to substantial global warming. In the SCW, estimated realized warming increases 1.0-1.5°C between 2000 and 2050, and 2-3°C between 2000 and 2100 (Figure 5-19). The maximum decadal rate of change associated with this scenario is 0.2-0.3°C sometime in the middle of the next century. The total equilibrium warming commitment is substantially higher, reaching 3-6°C by 2100 relative to preindustrial levels. The equilibrium warming commitment equivalent to doubling the concentration of CO_2 from preindustrial levels is reached by about 2040 in the SCW scenario.

The rate of change during the next century would be more than 50% greater in the RCW scenario, which shows a global temperature increase from 2000 of 1.2-1.9°C by 2050 and 3-5°C by

2100 (Figure 5-19). In this case the maximum rate of change is 0.4-0.6°C per decade, which occurs sometime between 2070 and 2100. The equilibrium warming commitment is 5-10°C by 2100 in this scenario and the 2xCO_2 equivalent level is reached by about 2030.[11]

By contrast, the rate of climatic change in the Stabilizing Policy scenarios would be less than 1.6°C per century. Global temperatures in the SCWP case increase by 0.4-0.8°C from 2000 to 2050 and 0.6-1.1°C from 2000 to 2100; corresponding values are 0.5-0.9°C and 0.8-1.4°C in the RCWP case. The maximum rate of change in these scenarios is less than 0.3°C per decade and occurs before 2010, largely as a result of warming to which the world may already be committed. (Indeed, the 0.3°C per decade figure, obtained assuming a climate sensitivity of 4°C, occurs at the very beginning of the simulation and may be an artifact of how the model is initialized.) In these cases the additional commitment to warming is greater between 2000 and 2050 than it is between 2050 and 2100: 0.3-0.9°C versus 0.1-0.4°C. Total equilibrium warming commitment reaches 1.4-2.8°C in the SCWP and 1.7-3.3°C in the RCWP. While not without some risk, the rate of change represented by the Stabilizing Policy scenarios would give societies and ecosystems much more time to adapt to climatic change than would be the case in the No Response scenarios.

Carbon dioxide accounts for more than 65% of increased commitments to global warming between 2000 and 2100 in all of the scenarios analyzed in this report (Figure 5-20). This represents a significantly higher estimate of the role of CO_2 compared to roughly 50% in the last few decades and in Ramanathan et al.'s scenario for 2030. Much of this difference is due to smaller increases in CFCs in our scenarios due to our assumption that the Montreal Protocol comes into force. In addition, growth in emissions of CH_4 and N_2O is projected to be slower than that of CO_2, particularly

---

[11] Estimates of equilibrium warming commitments greater than 6°C represent extrapolations beyond the range tested in most climate models, and this warming may not be fully realized because the strength of some positive feedback mechanisms may decline as the Earth warms.

FIGURE 5-20

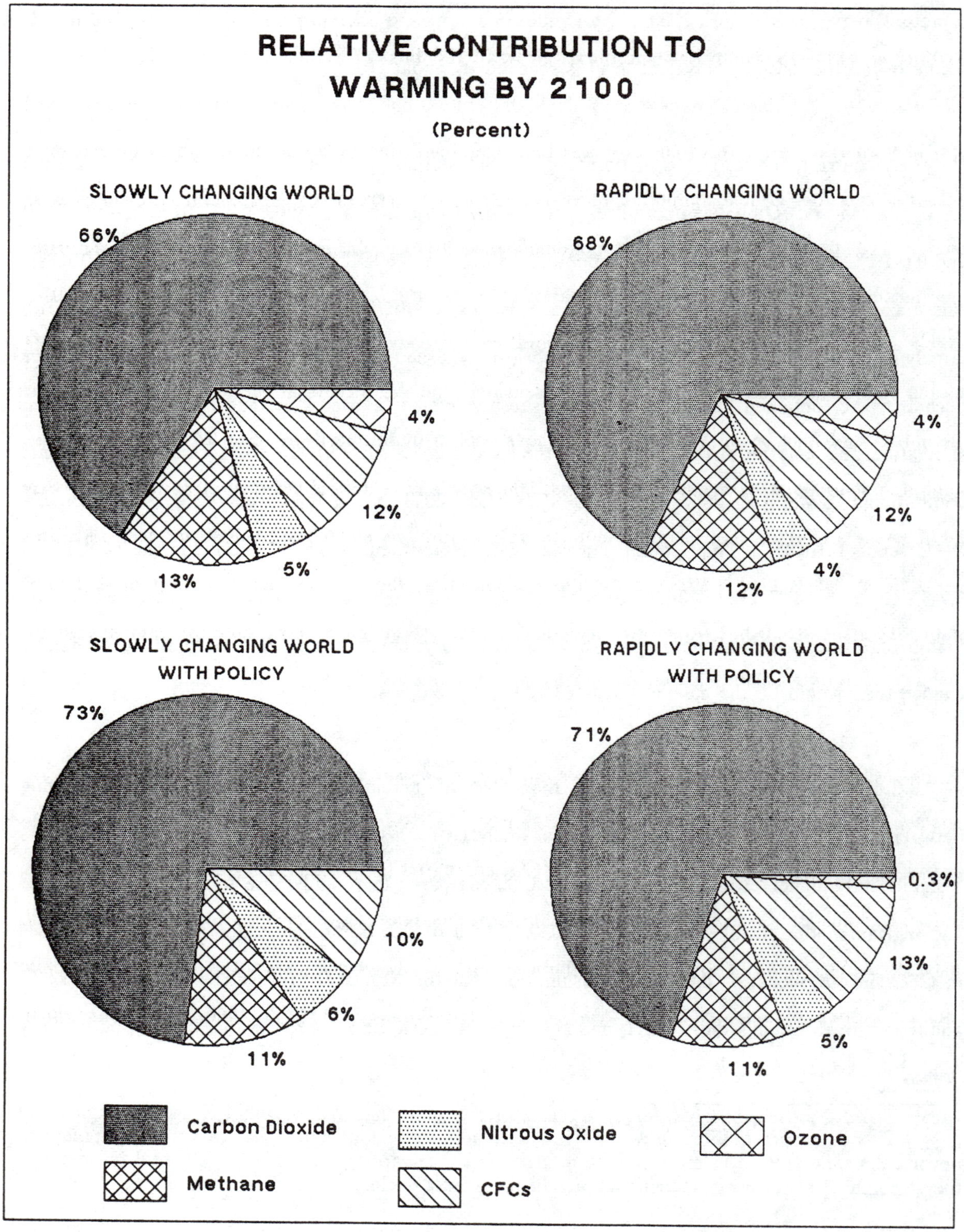
RELATIVE CONTRIBUTION TO
WARMING BY 2100
(Percent)
SLOWLY CHANGING WORLD
66%
4%
12%
5%
13%
RAPIDLY CHANGING WORLD
68%
4%
12%
4%
12%
SLOWLY CHANGING WORLD
WITH POLICY
73%
10%
6%
11%
RAPIDLY CHANGING WORLD
WITH POLICY
71%
0.3%
13%
5%
11%
Carbon Dioxide
Nitrous Oxide
Ozone
Methane
CFCs

after 2030. The role of CO_2 is greatest in the policy scenarios because our assumptions lead to relatively stable concentrations of CH_4 and tropospheric ozone, while CO_2 concentrations continue to increase gradually.

Comparison with General Circulation Model Results

Hansen et al. (1988) analyzed three transient trace-gas scenarios using the GISS GCM. The GISS A scenario, based on exponentially extrapolating current greenhouse gas trends, most closely resembles our RCW with 4.0°C climate sensitivity (the climate sensitivity of the GISS GCM is 4.2°C). Indeed, both the equilibrium and realized global warming in these cases are within 0.1°C in 2025.[12] By 2050 the continuation of exponential growth in trace-gas concentrations in the GISS A scenario leads to an equilibrium warming commitment that is about 40% higher than in the RCW, with a corresponding realized warming of 3.4 versus 2.8°C (all references to realized warming in the GISS scenarios are based on 5-year running means, Figure 3b in Hansen et al., 1988a). By 2060, the end of the GISS simulation, the realized warming in the GISS A scenario is 4.2°C compared with 3.3°C in the RCW. The GISS B scenario, which is based on linearly increasing trace-gas concentrations at current rates, is most similar to the RCWP case (with 4.0°C climate sensitivity). These two cases have very similar equilibrium warming commitment and realized warming in 2030 (the end of the GISS simulation for this scenario). The final scenario examined by GISS (case C) assumes that atmospheric composition is stable after 2000, which leads to realized warming of about 0.9°C by 2040. The policy cases examined here do not achieve this result; realized warming reaches 1.3-1.4°C by 2025 if the climate sensitivity is 4.0°C. Thus, the GISS scenarios bracket the range of the scenarios

---

[12] The path to 2025, however, is not identical. The GISS scenarios are referenced to the atmospheric composition of 1960, whereas our scenarios are referenced to the estimated preindustrial atmospheric composition. Thus, the warming commitment in 2000 is already 2.1°C in the RCW, whereas it is only 1.9°C in the GISS A scenario.

developed here, and may provide some indication of the regional differences in the rates and magnitude of change that might be associated with our cases.

Relative Effectiveness of Selected Strategies

The major assumptions that distinguish the Worlds where governments provide leadership in pursuing stabilizing policies from the Worlds with no such policies have been grouped into eleven categories in order to examine the relative importance of different policy strategies. Each set of options was applied individually to the RCW case; the combination of all the strategies represents the RCWP case. Figure 5-21 presents the results in terms of the effect of each policy strategy in reducing the equilibrium warming commitment in 2050 and 2100. This analysis suggests that accelerated energy efficiency improvements, reforestation, modernization of biomass use, and carbon emissions fees could have the largest near-term impact on the rate of climatic change. In the long run, advances in solar technology and biomass plantations also play an essential role.

CONCLUSIONS

While the future will never be anticipated with certainty, it is useful to explore the consequences of alternative plausible scenarios. The results of this exercise suggest that even with sluggish rates of economic growth and optimistic assumptions regarding technical innovation, the world could experience significant rates of climatic change during the next century. Temperature increases reach 3-4°C by 2100 under our assumptions; and the world would be committed to an additional warming of up to 2°C at this date. With higher rates of economic growth, certainly the goal of most governments, significantly more rapid rates of climatic change are possible. With our assumptions, which involve lower global energy use than considered in many previous studies, a

FIGURE 5-21

STABILIZING POLICY STRATEGIES: DECREASE IN EQUILIBRIUM WARMING COMMITMENT

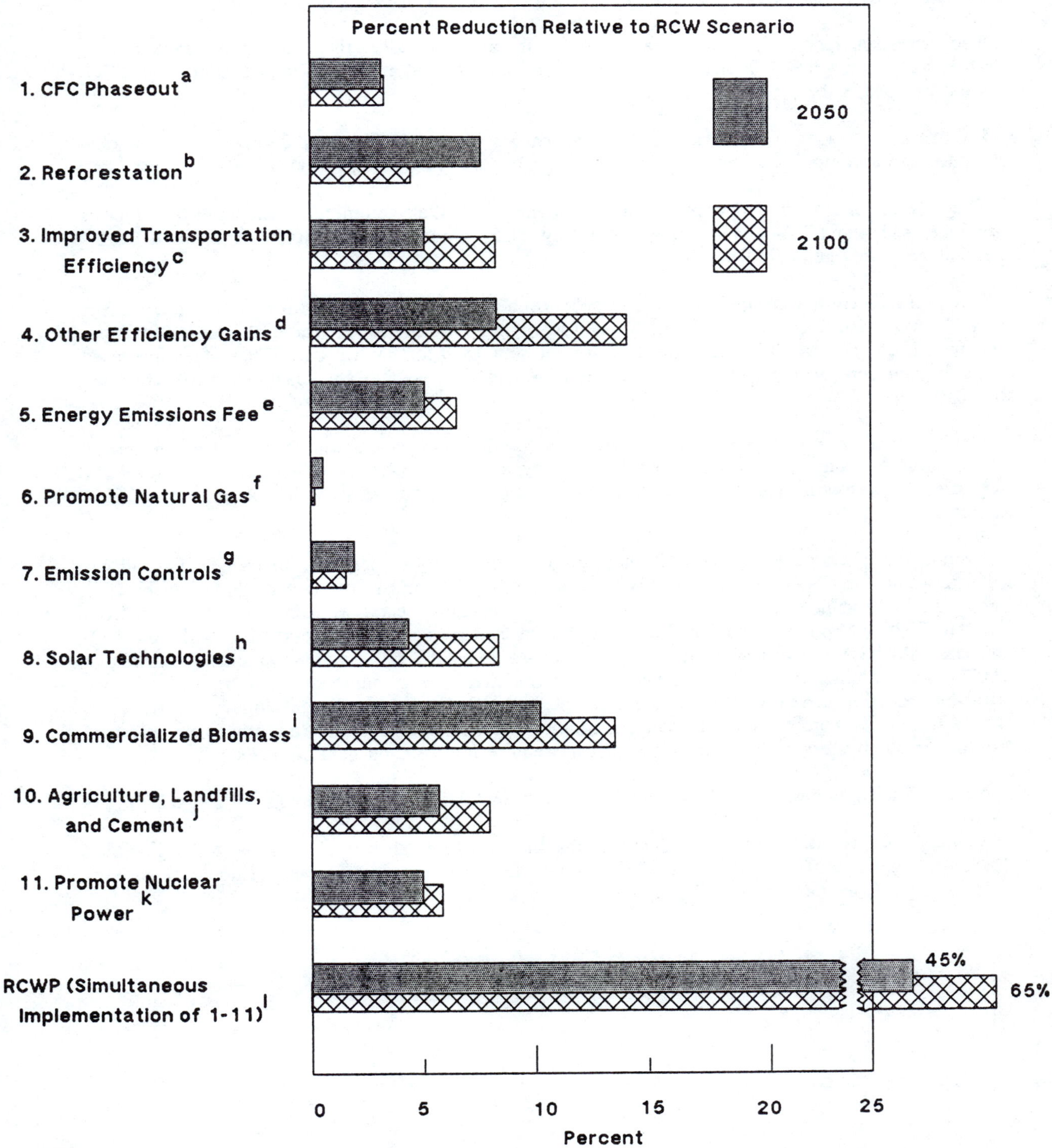

FIGURE 5-21 -- NOTES

Impact Of Stabilizing Policies On Global Warming

[a] A 100% phaseout of CFCs by 2003 and a freeze on methyl chloroform is imposed. There is 100% participation by industrialized countries and 94% participation in developing countries.

[b] The terrestrial biosphere becomes a net sink for carbon by 2000 through a rapid reduction in deforestation and a linear increase in the area of reforested land and biomass plantations. Net CO_2 uptake by 2025 is 0.7 Pg C.

[c] The average efficiency of new cars in the U.S. reaches 40 mpg (5.9 liters/100 km) by 2000. Global fleet-average automobile efficiency reaches 50 mpg by 2025 (4.7 liters/100 km).

[d] The rate of energy efficiency improvements in the residential, commercial, and industrial sectors are increased about 0.1-0.2 percentage points by 2025 compared to the RCW, and about 0.3-0.4 percentage points annually from 2025-2100.

[e] Emission fees are placed on fossil fuels in proportion to carbon content. Maximum production fees (1985$) were $0.50/GJ for coal $0.36/GJ for oil, and $0.23/GJ for natural gas. Maximum consumption fees were 28% for coal, 20% for oil, and 13% for natural gas. These fees increased linearly from zero, with maximum consumption fee changed by 2025 and maximum production fee by 2050.

[f] Assumes that economic incentives accelerate exploration and production of natural gas, reducing the cost of locating and producing natural gas by an annual rate of .5% relative to the RCW scenario. Incentives for gas use for electricity generation increases gas share by 5% in 2025 and 10% thereafter.

[g] Assume more stringent NO_x and CO controls on mobile and stationary sources including all gas vehicles using three-way catalysts in OECD countries by 2000 and in the rest of the world by 2025 (new light duty vehicles in the rest of the world uses oxidation catalysts from 2000 to 2025); from 2000 to 2025 conventional coal boilers used for electricity generation are retrofit with low NO_x burners with 85% retrofit in the developed countries and 40% in developing countries; starting in 2000 all new combustors used for electricity generation and all new industrial boilers require selective catalytic reduction in the developed countries and low NO_x burners in the developing countries and after 2025 all new combustors of these types require selective catalytic reduction; other new industrial non-boiler combustors such as Kilns and Dryers require low NO_x burners after 2000.

[h] Assumes that low cost solar technology is available by 2025 at costs as low as 4.6 cents/kwh.

[i] Assumes the cost of producing and converting biomass to modern fuels reaches $4.00/gigajoule (1985$) for gas and $6.00/gigajoule (1985$) for liquids. The maximum amount of liquid or gaseous fuel available from biomass is 210 exajoules.

FIGURE 5-21 -- NOTES (continued)

Impact Of Stabilizing Policies On Global Warming

[j] Assumes that research and improved agricultural practices result in an annual decline of 0.5% in the emissions from rice production, enteric fermentation, and fertilizer use. CH_4 emissions from landfills assumed to decline at an annual rate of 2% in developed countries due to policies aimed at reducing waste and landfill gas recovery, emissions in developing countries continue to grow until 2025 then remain flat due to incorporation of the source policies. Technological improvements reduce demand for cement by 25%.

[k] Assumes that technological improvements in nuclear design reduce cost by about $4/gigajoule (1985$) by 2050 (about 0.5% per year efficiency improvement).

[l] Impact on warming when all the above measures are implemented simultaneously. The sum of each individual reduction in warming is not precisely equal to the difference between the RCW and RCWP cases because not all the strategies are strictly additive.

warming of 4-6°C could be expected by 2100, with an additional commitment of 1-4°C by that date. On the other hand, by vigorously pursuing a variety of technical and policy options simultaneously, it would be possible to reduce the average rate of warming during the next century by more than 60%. Chapters VII-IX of this report explore these options in more detail.

REFERENCES

Bjorkstrom, A. 1979. A model of CO_2 interaction between atmosphere, oceans, and land biota. In Bolin B., E. Degens, S. Kempe, and P. Ketner, eds. *The Global Carbon Cycle*. Scope 13. John Wiley & Sons, New York.

Bolin, B., B.R. Doos, J. Jager, and R.A. Warrick, eds. 1986. *The Greenhouse Effect, Climatic Change, and Ecosystems.* John Wiley & Sons, Chichester. 541 pp.

CONAES. 1979.

Darmstadter, J., L. Ayres, R. Ayres, W. Clark, P. Crosson, P. Crutzen, T. Graedel, R. McGill, J. Richards, and J. Tarr. 1987. *Impacts of World Development on Selected Characteristics of the Atmosphere: An Integrative Approach*. Oak Ridge National Laboratory, Oak Ridge, Tennessee, vol. 1.

Dickinson, R.E., and R.J. Cicerone. 1986. Future global warming from atmospheric trace gases. *Nature* 319:109-115.

Edmonds, J. and J. Reilly. 1983a. A long-term global energy-economic model of carbon dioxide release from fossil fuel use. *Energy Economics* 5:75-88.

Edmonds, J., and J. Reilly. 1983b. Global energy and CO_2 to the year 2050. *Energy Journal* 4:21-47.

Edmonds, J.A., J. Reilly, J.R. Trabalka, and D.E. Reichle. 1984. *An Analysis of Possible Future Atmospheric Retention of Fossil Fuel CO_2*. Report No. DOE/OR/21400-1, U.S. Department of Energy, Washington, D.C., 109+ pp.

Edmonds, J., J. Reilly, R. Gardner, and A. Brenkert. 1986. *Uncertainty in Future Global Energy Use and Fossil Fuel CO_2 Emission 1975 to 2075*. Report DOE/NBB-0081. U.S. Department of Energy, Washington, D.C. 95 pp.

Frisch, J.R., ed. 1983. *Energy 2000-2020: World prospects and regional stresses*. World Energy Conference, Conservation Commission. Graham and Trotman, London.

Frohberg, Klaus. 1988. *The Basic Linked System: A Tool for National and International Food Policy Analysis.* Unpublished summary paper.

Goldemberg, J., T.B. Johansson, A.K.N. Reddy, and R.H. Williams. 1985. An End-Use Oriented Global Energy Strategy. *The Annual Review of Energy* 10:613-688.

Goldemberg, J., T.B. Johansson, A.K.N. Reddy, and R.H. Williams. 1987. *Energy for a Sustainable World.* World Resources Institute, Washington, DC. 119 pp.

Goldemberg, J., T.B. Johannson, A.K.N. Reddy, and R.H. Williams. 1988. *Energy for a Sustainable World*. Wiley Eastern Limited, New Delhi, India. 517 pp.

Hansen, J., I. Fung, A. Lacis, S. Lebedeff, D. Rind, R. Ruedy, G. Russell, P. Stone. 1988. Global climate changes as forecast by the Goddard Institute for Space Studies Three-Dimensional Model. *Journal of Geophysical Research* 93:9341-9364.

Hansen, J., A. Lacis, D. Rind, G. Russel, P. Stone, I. Fung, R. Ruedy, and J. Lerner. 1984. Climate sensitivity: Analysis of feedback mechanisms. In: Hansen, J. and T.Takahashi, eds. *Climate Processes and Climatę Sensitivity.* Geophysical Monograph 29, Maurice Ewing Volume 5. American Geophysical Union, Washington, DC. 130-163.

Hansen, J., and S. Lebedeff. 1988. Global surface air temperatures: Update through 1987. *Geophysical Research Letters* 15:323-326.

Hao, W.M., S.C. Wofsy, and M.B. McElroy. 1987. Sources of atmospheric nitrous oxide from combustion. *Journal of Geophysical Research* 92:3098-3104.

Hoffman, J.S., and M.J. Gibbs. 1988. *Future Concentrations of Stratospheric Chlorine and Bromine.* U.S. EPA, Office of Air and Radiation, Washington, D.C.

Houghton, R. 1988. Reply to S. Idso, Comment on "Biotic changes consistent with increased seasonal amplitude of atmospheric CO_2 concentrations." *Journal of Geophysical Research* 93:1747-1748.

Houghton, R., J. Hobbie, J. Melillo, B. Moore, B. Peterson, G. Shaver, and G. Woodwell. 1983. Changes in the carbon content of terrestrial biota and soils between 1860 and 1980: A net release of CO_2 to the atmosphere. *Ecological Monographs* 53:235-262.

IMF (International Monetary Fund). 1988. *International Financial Statistics* XLI(10). IMF, Washington, D.C.

Kahane, A. 1986. *Industrial Electrification: Case Studies of Four Industries.* Berkeley: LBL 22098.

Keepin, W., I. Mintzer, and L. Kristoferson. 1986. Emission of CO_2 into the atmosphere. In Bolin B., B.R. Doos, J. Jager, and R.A. Warrick, eds. *The Greenhouse Effect, Climate Change, and Ecosystems.* Scope 29. John Wiley and Sons, Chichester. 35-91.

Lacis, A., J. Hansen, P. Lee, T. Mitchell, and S. Lebedeff. 1981. Greenhouse effect of trace gases, 1970-1980. *Geophysical Research Letters* 8:1035-1038.

Lovins, A., L. Lovins, F. Krause, and W. Bach. 1981. *Least-Cost Energy: Solving the CO_2 Problems.* Brick House, Andover. 184 pp.

Mintzer, I. 1987. *A Matter of Degrees: The Potential for Controlling the Greenhouse Effect.* World Resources Institute, Washington, D.C. 60 pp.

Mintzer, I. 1988. *Projecting Future Energy Demand in Industrialized Countries: An End-Use Approach.* Prepared for U.S. EPA, Washington, D.C. 40+ pp.

Nordhaus, W.D. & G. Yohe. 1983. Future paths of energy and carbon dioxide emissions. In National Research Council, *Changing Climate.* National Academy Press, Washington, D.C. 87-153.

Oeschger, H., U. Siegenthaler, U. Schotterer, and A. Gugelmann. 1975. A box diffusion model to study the carbon dioxide exchange in nature. Tellus 27:168-192.

OTA (Office of Technology Assessment, U.S. Congress). 1988. *Technology and the American Economic Transition: Choices for the Future.* OTA-TET-283 Washington, DC: U.S. Government Printing Office, May 1988.

Prather, Michael J. 1988. *An Assessment Model for Atmospheric Composition.* NASA Conference Publication XXXX.

Radian Corporation. 1987. *Emissions and Cost Estimates for Globally Significant Anthropogenic Combustion Sources of NO_x, N_2O, CH_4, CO, and CO_2.* Report prepared for U.S. EPA, Research Triangle Park.

Ramanathan, V., R.J. Cicerone, H.B. Singh, and J.T. Kiehl. 1985. Trace gas trends and their potential role in climate change. *Journal of Geophysical Research* 90:5547-5566.

Reilly, J., J. Edmonds, R. Gardner, and A. Brenkeri. 1987. Uncertainty analysis of the IEA/ORAU CO_2 emissions model. *The Energy Journal* 8:1-29.

Rose, D.J., M.M. Miller, and C. Agnew. 1983. *Global Energy Futures and CO_2-Induced Climate Change.* MITEL 83-015. Prepared for National Science Foundation. Massachusetts Institute of Technology, Cambridge.

Rotty, R. 1987a. *Annual Carbon Dioxide Production From Fuels (Updates).* Unpublished Memorandum.

Rotty, R. 1987b. Estimates of seasonal variation in fossil fuel CO_2 emissions. *Tellus* 39B:184-202.

Rotty, R. 1987c. A look at 1983 CO_2 emissions from fossil fuel (with preliminary data for 1984). *Tellus* 39B:203-208.

Ruderman, H., M. Levine, and J. McMahon. 1987. The behavior of the market for energy efficiency in residential appliances including heating and cooling equipment. *The Energy Journal* 8:101-123.

Sanchez, P., and J. Benites. 1987. Low-input cropping for acid soils of the human tropics. *Science* 238:1521-1527.

Schipper, L. 1988. *Approaching Long-Term Energy Demand: Energy Demand for Production or Pleasure?* Presented at the Symposium "Energy Markets and the Future of Energy Demand" in Najing, China in June 1988.

Schurr, S. 1983. *Energy Policy.*

Seidel, S., and D.Keyes. 1983. *Can We Delay a Greenhouse Warming?* U.S. EPA, U.S. Government Printing Office, Washington, DC.

Siegenthaler, U. 1983. Uptake of excess CO_2 by an outcrop-diffusion model of the ocean. *Journal of Geophysical Research* 88(C6):3599-3608.

Smith, J., and D. Tirpak, eds. 1988. *Potential Effects of Global Climate Change on the United States.* U.S. EPA, Washington, D.C., in press.

Strout, A. 1985. *Energy Intensive Materials and the Developing Countries.* ERG Review Paper 053. Massachusetts Institute of Technology, Cambridge.

Strout, Alan. 1988. Structural change in industry over the long run. Presented at: the Lawrence Berkeley Laboratory Workshop, Oakland, CA.

U.S. Bureau of the Census. 1975. *Historical Statistics of the United States: Colonial Times to 1970.* Department of Commerce, Bureau of the Census, Washington, D.C.

U.S. Bureau of the Census. 1987. *World Population Profile: 1987.* U.S. Department of Commerce, Bureau of the Census, Washington, D.C.

U.S. DOE (U.S. Department of Energy). 1987a. *Annual Energy Review.* Energy Information Administration, U.S. DOE. Washington, D.C. 293 pp.

U.S. DOE (U.S. Department of Energy). 1987b. *Long Range Energy Projections to 2010.* Office of Policy, Planning and Analysis, Washington, D.C.

U.S. DOE (U.S. Department of Energy). 1988. *International Energy Outlook 1987.* Energy Information Administration, Washington, D.C.

U.S. EPA (U.S. Environmental Protection Agency). 1987. *Assessing the Risks of Trace Gases That Can Modify the Stratosphere.* Office of Air and Radiation, EPA, Washington, D.C.

U.S. EPA (U.S. Environmental Protection Agency). 1988. *Regulatory Impact Analysis: Protection of Stratospheric Ozone.* Office of Air and Radiation, U.S. EPA, Washington, D.C.

von Hipple, F., and B.G. Levi. 1983. Automotive fuel efficiency: The opportunity and weakness of existing market incentives. *Resources and Conservation* 10:103-124.

Walter, Donald K. 1988. Personal communication to Daniel Lashof. Donald K. Walter, Director. Biofuels and Municipal Waste Technology Division, Department of Energy.

Williams, R.H., E.D. Larson, and M.H. Ross. 1987. Materials, affluence, and industrial use. *The Annual Review of Energy* 12:99-144.

World Bank. 1987. *World Development Report 1987.* Oxford University Press, New York. 285 pp.

World Energy Conference. 1983. *Energy 2000-2020: World Prospects and Regional Stresses; and Oil Substitution: World Outlook to 2020.* London, Grahm and Trotman and Oxford University Press.

Zachariah, K.C., and M.T. Vu. 1988. *World Population Projections, 1987-1988 Edition.* World Bank, Johns Hopkins University Press, Baltimore. 440 pp.

CHAPTER VI

SENSITIVITY ANALYSES

uncertainties surrounding the sources of methane emissions. Using the reference assumptions, this model achieves an atmospheric concentration of 1691 ppb in 1985 compared to the observed value of 1675 ppb. The CH_4 concentrations vary considerably in the sensitivity analyses and range from 1600 ppb to 1780 ppb for alternative chemistry parameters.

Of the three dominant greenhouse gases, the estimates of N_2O concentrations vary the most from historical measurements. The model predicts concentrations of 314 ppb in 1985 compared to 300 to 310 ppb cited in the literature. From 1970 to 1980, the model estimates growth in N_2O concentrations of only 5.5 ppb compared to measurement data that suggests growth of 299 to 310 ppb. One of the possible explanations of these results is that the relative share of emissions of N_2O from anthropogenic sources is larger than estimated in the model. A larger anthropogenic source combined with lower natural emissions or a shorter atmospheric life would be needed to reduce the overall concentrations and obtain the growth in concentrations seen from 1970 to 1985. These results suggest that the model may underestimate future atmospheric concentrations of N_2O.

Finally, the [illegible] shows little deviation from current levels for the short-lived gases, including OH, O_3, and CO. The results for levels in 1870 include higher levels of OH of 14-20% above levels today, [illegible] 25%, lower concentrations of CO by approximately 30%, and [illegible] levels of ozone estimated to be lower by 15%.

ASSUMPTIONS ABOUT TRACE GAS SOURCES AND STRENGTHS

Among the various greenhouse gases there is some uncertainty over the quantity of emissions that can be attributed to specific sources and the ability of these gases to modify the atmosphere. The most critical of these uncertainties are discussed below.

FINDINGS

- The degree of participation by developing countries in policies to limit warming is one of the most important factors affecting equilibrium temperatures in the year 2100. If only industrialized countries adopt policy measures, equilibrium temperatures could increase by 40% or more relative to scenarios with global cooperation. This suggests that, despite uncertainties about future economic growth rates, developing countries will be a significant determinant in the ultimate level of global warming.

- Delaying any response to global warming by OECD and East Bloc Countries until the year 2010 and by developing countries until 2025 might increase the equilibrium warming commitment in 2050 by 30-40%.

- The sensitivity of the climate system to a given increase in greenhouse gases is among the most important causes for uncertainty about the ultimate magnitude of global warming. For most of the analysis in this report, we have assumed that the climate sensitivity to doubling CO_2 is 2.0 to 4.0°C; broadening the range of climate sensitivity to between 1.5 and 5.5°C for a CO_2 doubling causes the estimated range for equilibrium warming in 2050 to become 2.0-7.4°C in the Rapidly Changing World scenario. The impact on realized warming is less: the estimated range for 2050 increases from 1.9-2.8°C to 1.5-3.2°C. This uncertainty has important implications for the timing and stringency of policy responses. Even the lower values, when considered with information on the impacts of global warming, suggests a need for caution about future emissions.

- Uncertainties in biogeochemical feedbacks appear to be the potentially most important reason to suspect that global warming may ultimately be greater than predicted by current general circulation models. Changes in the ocean circulation, methane releases from hydrates, bogs, and rice cultivation and other positive feedbacks could amplify realized warming in 2100 by 20-40% for a climate sensitivity of 2.0-4.0°C. These estimates are speculative, they are based on the fragmentary evidence currently available, and these positive feedbacks may not occur or may be delayed until the later part of the next century, but the potentially large impact on the magnitude of warming suggests that even more drastic policy measures than those considered in the Rapidly Changing World with Stabilizing Policies scenario might be needed.

- Sensitivity analyses with four ocean models for CO_2 uptake suggest that the path of atmospheric concentrations could follow somewhat different trajectories, but very little difference is observed in equilibrium warming for the year 2100. These equilibrium temperatures differ by at most 10% depending on the type of ocean model. More complex ocean circulation models currently in the research stage could broaden or decrease this range in the future.

- Assumptions about the total supply of oil and gas are among the least significant factors affecting global warming in the year 2100. While gas may be desirable as a transition fuel, sensitivity tests that assume very optimistic estimates of oil and gas availability at each price level suggest only small changes in global warming. A larger impact could occur if policy measures were adopted to take advantage of the assumed increases in gas resources.

- The sources of methane are subject to considerable uncertainty. Estimates of some individual emission sources vary by a factor of two to three. Sensitivity tests that consider extreme assumptions about anthropogenic methane emission sources suggest that uncertainties in this budget could cause equilibrium warming commitments in 2100 to vary by about 5%. These results should not be interpreted to mean that methane is not an important greenhouse gas, but simply that uncertainties in the current budget do not greatly affect the ultimate temperatures derived in this report.

- A comparison between current atmospheric concentrations and growth rates for the greenhouse gases and those calculated with the atmospheric composition model, based on estimates of preindustrial concentrations and past emissions, show good agreement. The largest discrepancies are for relatively short-lived gases that have been increasing rapidly in recent years, such as HCFC-22 and carbon tetrachloride.

- Non-greenhouse gases such as NO_x, CO, and Non-methane hydrocarbons (NMHC) affect the lifetimes and concentrations of tropospheric ozone and methane. A comparison of different chemistry models suggests that increases in methane concentrations may vary by approximately a factor of two for similar assumptions about NO_x/CO/NMHC. This range may be attributed to differences in initial budgets and modeling approaches and may ultimately increase or decrease as other models become available.

- The most important determinant of future atmospheric concentrations of methane appears to be the growth rate of methane sources. While NO_x and

CO affect the lifetime of methane, model studies suggest that assumptions about the emissions of these gases are less important than assumptions about the direct emissions of methane. However, considerable research is needed to further our understanding of the chemistry of the atmosphere.

- There is considerable uncertainty about future concentrations of tropospheric ozone and about changes in composition at different altitudes. While model comparisons all suggest that increases in ozone are likely, the effect of these changes in global temperatures is difficult to predict.

- For the major sensitivity analyses presented in this chapter, Table 6-1 summarizes the impact on realized warming and equilibrium warming by 2050 and 2100 (assuming a 3.0°C climate sensitivity). Throughout this chapter results are discussed for 2.0-4.0°C climate sensitivities for the Rapidly Changing World Scenario, with any figures using the midpoint of this range, i.e., a 3.0°C climate sensitivity, unless stated otherwise. Other assumptions would not change the basic findings, only the absolute size of the impacts; these are presented in Appendix C.

TABLE 6-1

Impact of Sensitivity Analyses on Realized Warming and Equilibrium Warming
(degrees Celsius--3.0°C climate sensitivity)

	2050		2100	
	Realized	Equilibrium	Realized	Equilibrium
Rapidly Changing World - No Response (RCW)	2.4°	4.0°	4.7°	7.2°*
Rapid Changing World - Stabilizing Policies (RCWP)	1.5	2.2	1.9	2.5
Sensitivity Case Assumptions				
No Participation by Developing Countries[a]	2.0-2.1	3.1-3.3	3.2-3.7	4.6-5.6
Global Delay in Adopting Policies[b]	2.0	3.1	2.6	3.5
Non-Fossil Technology[c]	2.1-2.3	3.4-3.7	3.6-4.0	5.3-5.9
Fossil Resources				
High Coal Prices[d]	2.1	3.5	3.6	5.3
High Oil Supply[e]	2.4	4.0	4.7	7.1*
High Gas Supply[f]	2.4	4.0	4.7	7.2*
Methane Budget[g]	2.3-2.5	3.9-4.2	4.6-5.0	7.0-7.6*
High CO Emissions		To be added		
N_2O From Fertilizer				
Anhydrous Ammonia[h]	2.4	4.0	4.7	7.2*
N_2O Leaching[i]	2.4	4.0	4.7	7.2*
N_2O From Combustion[j]	2.4	4.0	4.7	7.2*

* Estimates of equilibrium warming commitment greater than 6°C represent extrapolations beyond the range tested in most climate models; this warming may not be fully realized because the strength of some positive feedback mechanisms may decline as the Earth warms.

TABLE 6-1 (continued)

Impact of Sensitivity Analyses on Realized Warming and Equilibrium Warming
(degrees Celsius--3.0°C climate sensitivity)

	2050		2100	
	Realized	Equilibrium	Realized	Equilibrium
CO_2 From Biomass[k]	2.4	4.1	4.8	7.3*
CO_2 Models[l]				
Oeschger et al.[m]	-	3.9	-	6.9
Bolin et al.[n]	-	3.9	-	6.9
Bjorkstrom[o]	-	3.9	-	6.9
Siegenthaler[p]	-	3.9	-	6.3
Unknown Sink[q]	2.2-2.5	3.7-4.2	4.1-4.9	6.2-7.4*
1.5-5.5°C Sensitivity[r]	1.5-3.2	2.0-7.4	2.9-6.6	3.6-13.2*
Heat Diffusion[s]	1.9-2.7	4.0	3.9-5.3	7.2*
Prather Model				
CFC-11 Lifetime[t]	2.4	4.0	4.7	7.2*
Chlorine/Col O_3[u]	2.3	3.8	4.4	6.6
Trop O_3/CH_4[v]	2.4	4.1	4.8	7.3*
OH/NO_x[w]	2.4	4.0-4.1	4.7-4.8	7.1-7.3*
Feedbacks				
Ocean Circulation[x]	3.1	4.1	7.1*	7.6*
Methane[y]	2.7	4.6	5.5	8.4*
CO_2/CH_4/Uptake[z]	3.0	4.9	6.3	8.9*

* Estimates of equilibrium warming commitments greater than 6°C represent extrapolations beyond the range tested in most climate models, and this warming may not be fully realized because the strength of some positive feedback mechanisms may decline as the Earth warms.

TABLE 6-1 -- NOTES

[a] Developing countries were assumed to not participate in climate stabilization policies. The range represents uncertainty in the rate of technological diffusion, i.e., even if developing countries do not participate, they will indirectly benefit from technological improvements as a result of stabilization policies among the developed countries.

[b] Impact if developed countries do not respond to global warming until 2010; developing countries delay to 2025.

[c] These ranges represent modest to optimistic assumptions about future commercial availability of non-fossil technologies, e.g., solar photovoltaics, advanced nuclear power designs, and synthetic fuel production from biomass. Solar photovoltaic costs decline to 4.6 cents/kwh (1985$) by 2020 in the optimistic scenario and by 2050 in the modest assumptions. Nuclear costs decline 0.5% annually with the optimistic assumptions and remain relatively flat in the modest assumptions. The cost of producing and converting biomass to modern fuels reaches $4.00/gigajoule for gas and $6.00 (gigajoule) for liquids by 2020 in the optimistic assumptions and by 2050 in the modest assumptions. The total amount of fuel available from biomass is 210 EJ.

[d] The impact of an escalation in coal prices above the RCW case by about 1% annually from 1985 to 2100.

[e] The impact of an increase in global oil resources to 25,000 EJ, more than double the estimate in the RCW case, assuming proportionate increases in resource availability at each cost level.

[f] The impact of an increase in global natural gas resources to 27,000 EJ, more than 2.5 times the estimate in the RCW case, assuming proportionate increases in resource availability at each cost level.

[g] These ranges represent assumptions about the relative sizes of anthropogenic versus non-anthropogenic sources of methane emissions, thereby affecting growth in emissions over time, i.e., high emission levels (373 Tg CH_4) from anthropogenic activities such as fuel production and landfilling with low emission levels (137 Tg CH_4) from natural processes such as oceans and wetlands, versus low anthropogenic emissions (245 Tg CH_4) with high natural emissions (265 Tg CH_4).

[h] The impact of elevating the emission coefficient for the anhydrous ammonia fertilizer type (the percent of N evolved as N_2O) from 0.5% to 2.0%.

[i] The impact of assuming additional N_2O emissions from fertilizer leaching into surface water and ground water, modeled by increasing all the fertilizer emission coefficients by 1 percentage point.

[j] The impact of higher emission coefficients for N_2O from combustion; assumes that N_2O emissions are about 25% of NO_x emissions and the N_2O emissions from combustion sources in 1985 equaled 2.3 Tg N, over 2 times the level assumed in the RCW case.

[k] The impact of assuming a higher estimate for the amount of carbon initially contained in forest vegetation and soils (roughly a 50-100% increase) and a more rapid rate of change in land-use, resulting in emissions of carbon of 281 Pg from 1980 and 2100 compared to 188 Pg C in the RCW scenario.

TABLE 6-1 -- NOTES (continued)

[l] Realized warming was not calculated in these tests.

[m] This box-diffusion model represents carbon turnover below 75 meters as a purely diffusive process.

[n] This is a 12-compartment regional model which divides the Atlantic and Pacific-Indian Oceans into surface-, intermediate-, deep-, and bottom-water compartments and divides the Arctic and Antarctic Oceans into surface- and deep-water compartments.

[o] This is an advective-diffusive model which divides the ocean into cold and warm compartments; water downwells directly from the cold surface compartment into intermediate and deep layers.

[p] An outcrop-diffusion model that allows direct ventilation of the intermediate and deep oceans in high latitudes by incorporating an outcrop connecting all sublayers to the atmosphere.

[q] These ranges represent the impact of alternative assumptions about the "unknown carbon sink" that absorbs the unaccounted-for carbon in the carbon cycle. Two sensitivities were analyzed: 1) a high case, where the size of the unknown sink increases at the same rate as atmospheric CO_2 levels compared with preindustrial levels; and 2) a low case, where the size decreases to zero exponentially at 2% per year.

[r] Atmospheric response to a doubling of CO_2 was varied from 1.5-5.5°C.

[s] Heat diffusion in the oceans is modeled as a purely diffusive process. To capture some of the uncertainty regarding actual heat uptake, the base case eddy-diffusion coefficient of 0.55×10^{-4} m^2/sec was increased to 2×10^{-4} and decreased to 2×10^{-5} m^2/sec.

[t] The atmospheric lifetime of CFC-11, 65 years in the RCW case, was varied from 55 to 75 years. Increases or decreases in the atmospheric concentration of CFC-11, however, tend to be offset by corresponding decreases or increases in atmospheric concentrations of other trace gases, such as other CFCs and CH_4.

[u] The amount of stratospheric ozone depletion due to chlorine contained in CFCs was increased from a 0.03% to 0.20% decline in total column ozone/$(ppb)^2$ of stratospheric chlorine.

[v] The rate at which tropospheric ozone forms as a result of CH_4 abundance was increased. In the RCW case, this variable for the Northern Hemisphere is a 0.2% change in tropospheric ozone for each percentage change in CH_4 concentration; it was changed to 0.4% in the sensitivity analysis.

[w] Tropospheric OH formation is affected by the level of NO_x emissions. A 0.1% OH change for every 1% change in NO_x emissions for the Northern Hemisphere was assumed in the RCW case; in the sensitivity analysis, a range of 0.05% to 0.2% was evaluated.

[x] For this analysis we assumed that a 2°C increase in realized warming would alter ocean circulation patterns sufficiently to shut off net uptake of CO_2 and heat by the oceans.

TABLE 6-1 -- NOTES (continued)

[y] We assumed that with each 1°C increase in temperature, an additional 110 Tg CH_4 from methane hydrates, 12 Tg CH_4 from bogs, and 7 Tg CH_4 from rice cultivation would be released.

[z] This case illustrates the combined impact of several types of biogeochemical feedbacks: 1) methane emissions from hydrates, bogs, and rice cultivation (see footnote above); 2) increased stability of the thermocline, thereby slowing the rate of heat and CO_2 uptake of the deep ocean by 30% due to less mixing; 3) vegetation albedo, which is a decrease in global albedo as a result of changes in the distribution of terrestrial ecosystems by 0.06% per 1°C warming; 4) disruption of existing ecosystems, resulting in transient reductions in biomass and soil carbon at the rate of 0.5 Pg C per year per 1°C warming; and 5) CO_2 fertilization, which is an increase in the amount of carbon stored in the biosphere in response to higher CO_2 concentrations by 0.3 Pg C per ppm.

INTRODUCTION

The Rapidly Changing World and Slowly Changing World scenarios presented in Chapter V describe two significantly different futures for the global community. Although these two potential paths capture a wide range of uncertainty, they do not represent all possible outcomes. Alternative assumptions are clearly possible for many of the parameters specified in these scenarios; these alternative specifications could alter the timing and magnitude of global climate change described in the Rapidly Changing World and Slowly Changing World scenarios. To understand the importance of these alternative assumptions, this chapter examines how changes in key parameters affect our portrayal of the rate and magnitude of global climate change. These sensitivity analyses include alternative assumptions about: the magnitude and timing of global policies to combat climate change, rates of technological change, trace-gas source strengths and emission coefficients, the carbon cycle, sensitivity of the climate system, atmospheric chemistry, and feedbacks.

The sensitivity analyses discussed in this chapter are generally run relative to the Rapidly Changing World scenario, unless specified otherwise. Overviews of each case are provided to describe the basic results for the reader; more detailed discussion of the sensitivity analyses are provided in Appendix C.

ASSUMPTIONS ABOUT THE MAGNITUDE AND TIMING OF GLOBAL CLIMATE STABILIZATION STRATEGIES

The analyses of the Stabilizing Policy scenarios presented in this Report are based on the assumption that the global community takes immediate, concerted action to contend with the consequences of climate change. Potential actions, which are discussed in Chapters VII - IX, include reducing the amount of energy required to meet the world's increasing needs, developing alternative technologies that do not require the consumption of fossil fuels, halting deforestation, and making changes in agricultural practices, among others. For many reasons, however, the world may not respond to the threat of climate change in a timely fashion. This section explores the consequences of other possibilities, particularly the unwillingness or inability of some countries to participate in climate stabilization programs and the implications of delaying global action until a later date.

No Participation by the Developing Countries

Most of the greenhouse gas emissions currently committing the world to climate change can be traced to activities by the industrialized countries. Although the quantity of emissions generated by developing countries has been increasing, the argument is sometimes made that since the greenhouse problem has been largely caused by the industrialized countries, these countries should be responsible for solving the problem. Also, despite the potential environmental consequences of global climate change, other problems facing the developing countries, such as poverty, inadequate health care, and other apparently more pressing environmental problems may make it difficult for developing countries to commit any resources to climate stabilization policies.

Regardless of the merits of these arguments, for this sensitivity analysis we have assumed that developing countries do not participate in any climate stabilization activities; that is, only developed countries adopt policies to limit global climate change. For this analysis the developing countries include China and Centrally-planned Asian economies, the Middle East, Africa, Latin America, and South/Southeast Asia. We have assumed that industrialized countries (i.e., the U.S., the rest of the OECD countries, and the USSR and Eastern Europe) follow the path assumed in the Rapidly Changing World with Stabilizing Policies (RCWP) scenario, while developing countries follow the path assumed in the Rapidly Changing World No Response (RCW) case, in which the entire global community does not respond to climate change.

Even if developing countries do not participate in global stabilization policies, however, policies adopted by the industrialized countries are likely to lead to technological advancements, altered market conditions, etc., that indirectly reduce emissions in the developing countries as well. For example, advancements by the developed countries in automobile fuel efficiency or fuel supply technologies may be partly adopted by the developing countries, tangentially allowing for some climate stabilization benefits. If the developing countries do not participate, however, they may tend to adopt technological advances more slowly and at a higher cost than if they had participated from the start. This slower rate of technological diffusion could occur for many reasons -- for example, if the industrialized countries take actions that prevent easy access to improved technologies or they are unwilling or unable to make the necessary capital available for investment, or if developing countries decide to invest their limited resources in other areas.

Since we cannot be certain of the direction that non-participation by the developing countries might take, we analyzed two cases to capture the potential range of likely possibilities. In the first case, little technological diffusion was assumed, resulting in a future path of energy consumption and investment trends for developing countries similar to those assumed in the RCW scenario. In the second case, developing countries were assumed to have greater access to the efficiency improvements and technological advances assumed for the RCWP case as a result of policies by the industrialized countries to make these improvements available and extend the credit necessary for investment by the developing countries in these improvements.

In this analysis key assumptions for the developing countries included the following: (1) rates of energy efficiency improvements for all sectors are the same as in the RCW case or midway between the RCW and RCWP case; for example, automobile efficiency levels, which by 2050 in developing countries were 5.9 liters/100 km (40 mpg) in the RCW case and 3.1 liters/100 km (75 mpg) in the RCWP case, were varied from 5.9-4.1 liters/100 km (40-58 mpg); (2) CFCs are not phased out (although compliance with the Montreal Protocol would still occur); (3) agricultural practices that cause methane emissions from rice and enteric fermentation and nitrous oxides from fertilizers do not change or would show modest improvements; (4) deforestation continues as in the RCW case with an exponential decline in forest area; (5) non-fossil energy supply technologies developed by the industrialized countries are available to developing countries at a later date and a higher cost than assumed in the RCWP case; for example, technological diffusion of biomass gasification technology would occur 10 years later than it would in the RCWP case, but feedstock costs would remain high due to a lack of investment by the developing countries in highly productive energy plantations; and (6) no additional incentives are provided for increased use of natural gas.

Without the participation of the developing countries to stabilize the atmosphere, the amount of greenhouse gas emissions will increase substantially: In the analysis considered here, CO_2 emissions are 3-4 Pg C higher than in the RCWP case by 2050 and 6-13 Pg C higher by 2100 (emissions by 2100 are 8.1 to 14.7 Pg C lower than in the RCW case since industrialized countries adopt climate stabilization policies); [1] other greenhouse gas emissions are also higher. These emission increases are sufficient to increase realized warming by 0.4-0.6°C in 2050 compared with the RCWP case and 1.0-2.1°C by 2100 (see Figure 6-1), with equilibrium warming by 2100 up to 1.4-4.1°C higher. Figure 6-1 also shows the results for the Slowly Changing World scenarios. In this scenario, emission increases are sufficient to increase realized warming by 0.3-0.5°C in 2050 compared with the SCWP case and 0.6-1.1°C by 2100, with equilibrium warming by 2100 up to 0.8-2.0°C higher.

The implications of these results are clear: even if the industrialized countries adopt very stringent policies to counteract the effects of climate change, the atmosphere continues to warm at a rapid rate. As a result, unilateral action by the industrialized countries can significantly slow the

[1] Pg = petagram; 1 petagram = 10^{15} grams.

FIGURE 6-1

INCREASE IN REALIZED WARMING
WHEN DEVELOPING COUNTRIES DO NOT PARTICIPATE
(Degrees Celsius; Based on 3.0 Degree Sensitivity)

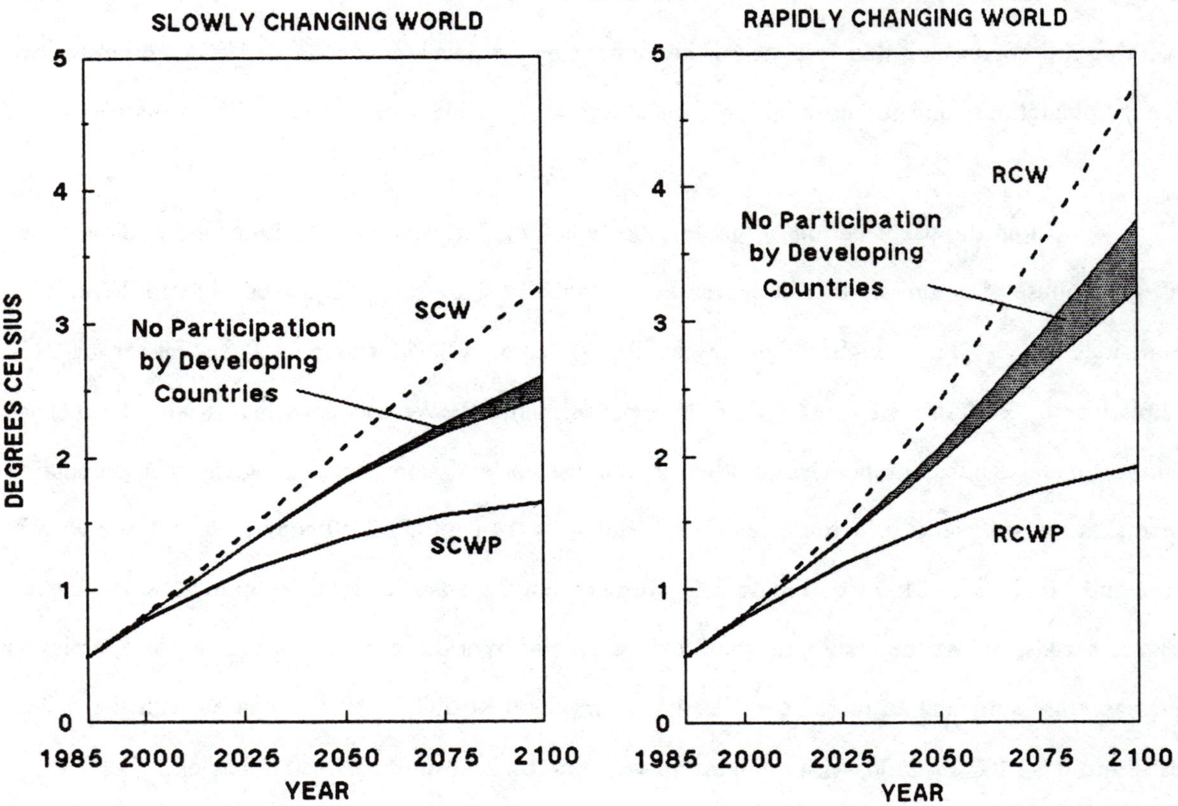

rate and magnitude of climate change, but because of the growing impact that developing countries have on the global climate, without the participation of the developing countries, substantial global warming is unavoidable. Because most of the world's population resides in these countries, their role in climate stabilization becomes increasingly important as the demand for resources to feed and clothe their growing population and improve their standard of living expands.

Delay in Adoption of Policies

For the Stabilizing Policy cases presented in Chapter V it is assumed that the global community takes immediate action to respond to the dangers posed by climate change. For this sensitivity analysis we have assumed that the global community delays any response to the threat of climate change, with developed countries (i.e, the United States, the rest of the OECD countries, the USSR and centrally-planned European economies) delaying action until 2010, and the developing countries delaying action until 2025. Additionally, once regions do initiate action to combat global warming, they do so at a slower rate than assumed in the RCWP case. This slower approach assumes a minimum 25-year delay in attaining the policy goals of the RCWP case; that is, levels of technological improvement, availability of alternative energy supply technologies, etc., will be achieved at least 25 years later. For example, in the RCWP case, automobile efficiency reaches 3.1 liters/100 km (75 mpg) by 2050; in the Delay case industrialized countries reach 3.9 liters/100 km (60 mpg) by 2050, while developing countries reach 4.7 liters/100 km (50 mpg); the rate of energy efficiency improvement for the residential, commercial, and industrial sectors is unchanged from the rates assumed in the RCW case, through 2010 for industrialized countries and 2025 for developing countries. After these years, energy efficiency improvements occur at the same rate assumed in the RCWP case; and the implementation of production and consumption taxes on fossil fuels from the RCWP case was delayed until 2010 for developed countries and 2025 for developing countries.

Delaying the adoption of policies to stabilize the atmosphere significantly increases the Earth's commitment to global warming. With delay by the industrialized countries until 2010 and by the developing countries until 2025, the increase in realized warming compared to that assumed in the RCWP case is 0.4-0.6°C by 2050 and 0.5-0.9°C by 2100; equilibrium warming is 0.6-1.2°C higher by 2050 and 0.7-1.4°C higher by 2100 (based on climate sensitivities of 2.0-4.0°C; see Figure 6-2). Figures 6-2 also shows the results for the Slowly Changing World scenarios. If global delays do occur, the increase in realized warming compared to that assumed in the SCWP case is 0.3-0.5°C by 2050 and 0.3-0.6°C by 2100; equilibrium warming is 0.5-0.9°C higher by 2050 and 0.4-0.8°C higher by 2100 (based on climate sensitivities of 2.0-4.0°C).

ASSUMPTIONS AFFECTING RATES OF TECHNOLOGICAL CHANGE

The extent of global warming will depend on the availability of energy supplies and technologies that minimize dependence on carbon-based fuels, nitrogen-based fertilizers, and other sources of greenhouse gas emissions. The availability of non-fossil fuel technologies and the development of new production methods that significantly increase the supply of natural gas could have an impact on the rate of change in greenhouse gas emissions. Alternative assumptions regarding these factors are presented below.

Availability of Non-Fossil Technologies

Most technologies in use currently rely on fossil fuels to supply their energy needs. In the Rapidly Changing World, fossil-fuel-based technologies continue to dominate throughout the next century: by 2100 fossil fuels still supply over 70% of primary energy needs. However, if non-fossil technologies can be commercialized earlier, the magnitude of global climate change can be reduced because these technologies do not emit the greenhouse gases that cause global warming. To evaluate

FIGURE 6-2

INCREASE IN REALIZED WARMING DUE TO GLOBAL DELAY IN POLICY ADOPTION

(Degrees Celsius; Based on 3.0 Degree Sensitivity)

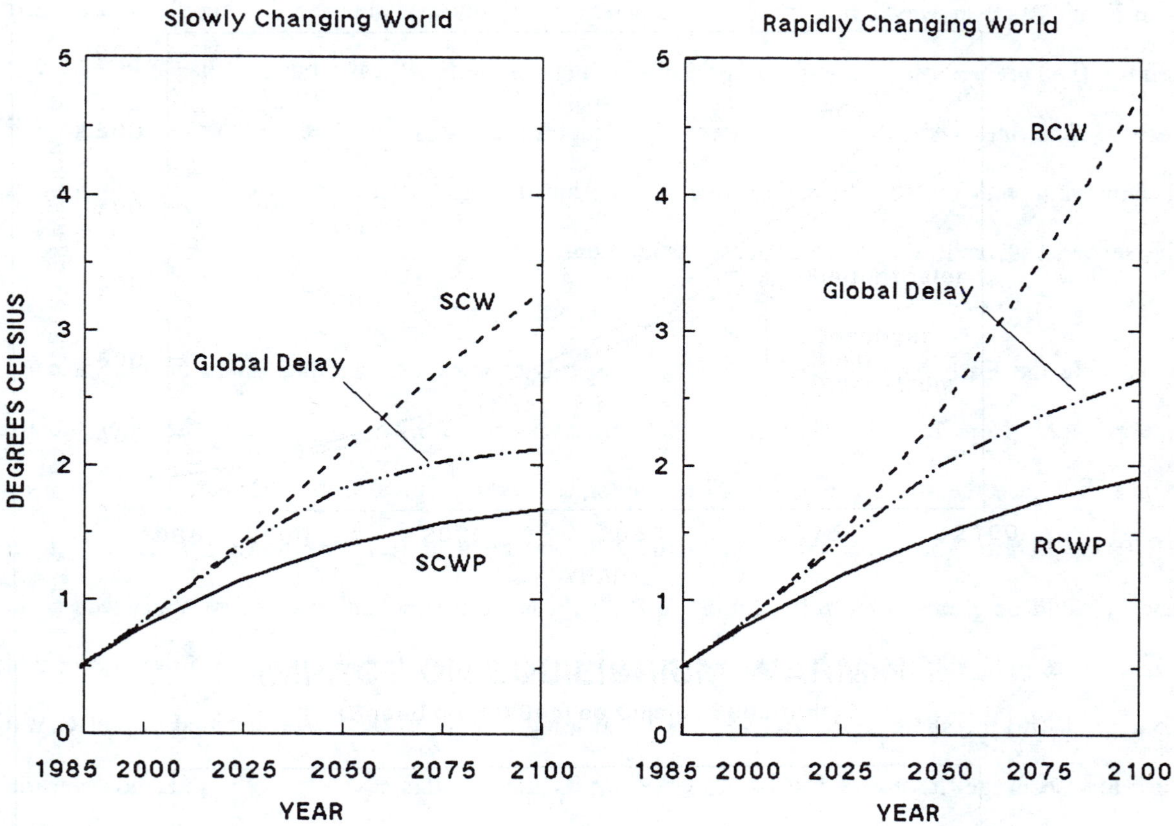

the implications of the availability of non-fossil technologies, two different scenarios were analyzed: (1) an Early Non-Fossil case, in which non-fossil technologies, specifically solar photovoltaics, advanced nuclear power designs, and production of synthetic fuels from biomass, are commercially available by 2000 at a rate faster than that assumed in the RCWP case; and (2) an Intermediate Non-Fossil case, in which non-fossil technologies are widely available by the middle of the next century (i.e., greater use of non-fossil technologies than in the RCW case, but less than in the RCWP case). The intent of these two cases is to capture a range of possible roles for non-fossil technologies, with the first case reflecting very optimistic assumptions on non-fossil availability and the second case reflecting more modest assumptions.

In the Early Non-Fossil case, non-fossil energy sources increase their share of total primary energy supply from 12% in 1985 to about 50% by 2025 and 65% by 2100, while in the Intermediate Non-Fossil case the share for non-fossil technologies increases to 21% by 2025 and about 60% by 2100 (see Figure 6-3a). As shown in Figure 6-3a, in the near term the non-fossil share of total energy could be greater than reflected in the RCWP case if commercial availability is achieved at an earlier date. In this sensitivity analysis, however, the non-fossil share is lower in the long run compared with the share in the RCWP case because other policies that were included in the RCWP case to discourage the use of fossil fuels were not included in this case. In both cases, however, an increased role for non-fossil technologies can affect the amount of global warming. As shown in Figure 6-3b, for the two cases presented here the amount of realized warming compared with the RCW case could be reduced about 0.1-0.3°C by 2050 and 0.6-1.3°C by 2100; equilibrium warming could be reduced about 0.2-0.9°C by 2050 and 0.9-2.5°C by 2100 (based on 2.0-4.0°C climate sensitivities).

FIGURE 6-3

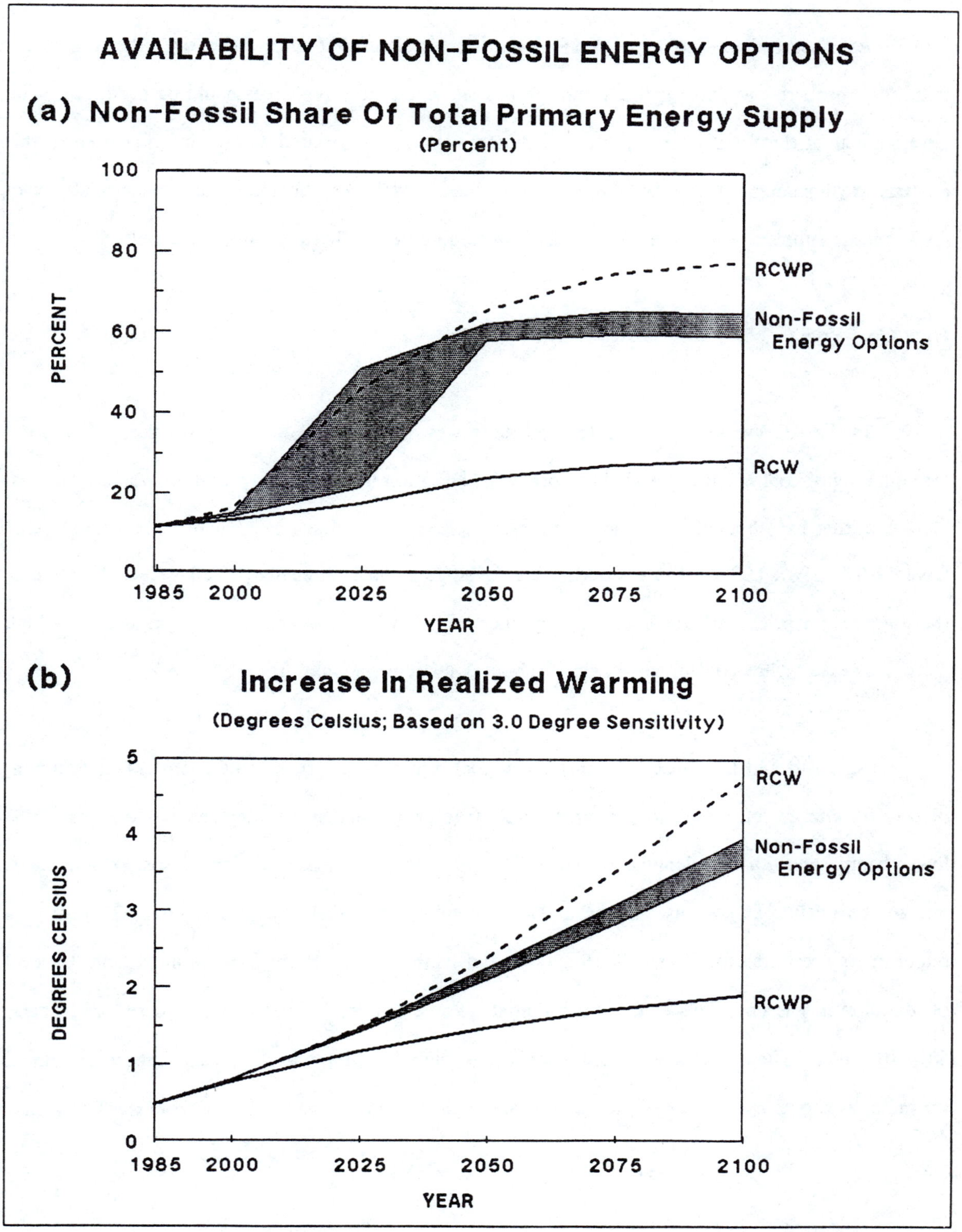
AVAILABILITY OF NON-FOSSIL ENERGY OPTIONS
(a) Non-Fossil Share Of Total Primary Energy Supply
(Percent)
PERCENT
100
80
60
40
20
0
1985
2000
2025
2050
2075
2100
YEAR
RCWP
Non-Fossil
Energy Options
RCW
(b)
Increase In Realized Warming
(Degrees Celsius; Based on 3.0 Degree Sensitivity)
DEGREES CELSIUS
5
4
3
2
1
0
1985
2000
2025
2050
2075
2100
YEAR
RCW
Non-Fossil
Energy Options
RCWP

Cost and Availability of Fossil Fuels

As discussed in Chapters IV and VII, there is significant uncertainty over the amount of fossil-fuel resources available globally and the cost at which these resources could be produced. The development of the fossil energy resource estimates and the associated extraction costs used in this analysis are documented in ICF (1988). Given the uncertainties about the cost and availability of fossil energy supplies, several sensitivity cases were analyzed. These are discussed below.

High Coal Prices

In the RCW case from 1985 to 2050 there was no real escalation in coal prices. Given the vast quantity of coal resources available worldwide, and the rate of productivity improvements in coal extraction that have helped to contain cost increases, coal prices may not escalate in real terms (e.g., from 1949 to 1987, U.S. coal prices declined an average of 0.2% annually [U.S. DOE, 1988]). Since the longer-term price path for coal is highly uncertain, however, we analyzed the impacts of a high price coal case where coal prices escalated about 1 percent annually from 1985 to 2100.

As illustrated in Figure 6-4a, increasing coal prices have a significant impact on the amount of primary energy consumed; for example, by 2100 total primary energy demand is more than 20% lower compared with this demand in the RCW case. Most of this reduction in energy demand is due to the decline in coal use as consumers respond to the escalating prices. Because coal is a major energy resource for electricity production and synthetic fuel production, the impact on the level of greenhouse gas emissions is fairly substantial. For example, CO_2 emissions are reduced nearly 50% by 2100. The reductions in greenhouse gas emissions have a significant impact on global warming, as shown in Figure 6-4b, which indicates a decline in realized warming from the RCW case

FIGURE 6-4

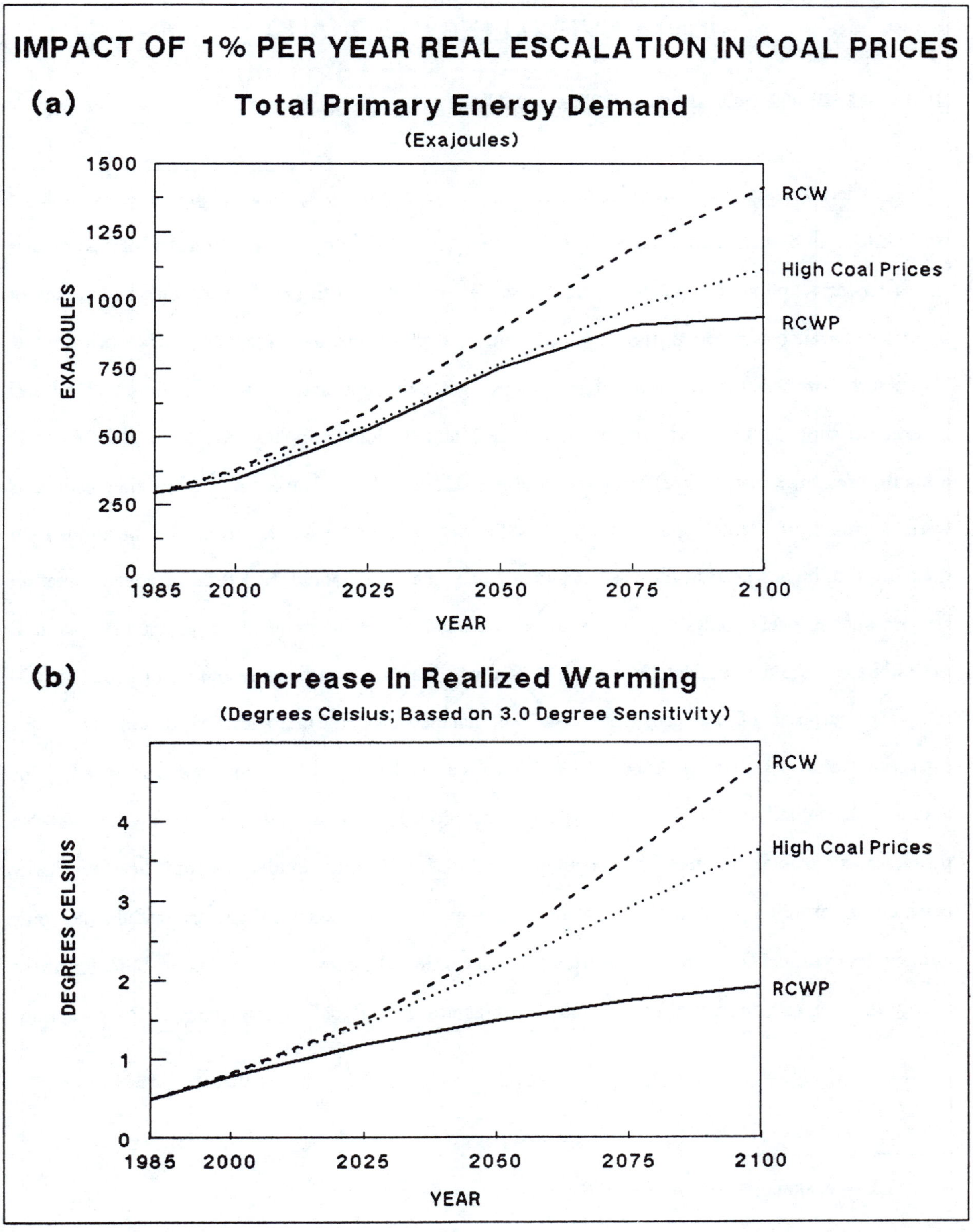
IMPACT OF 1% PER YEAR REAL ESCALATION IN COAL PRICES
(a)
Total Primary Energy Demand
(Exajoules)
EXAJOULES
1500
1250
1000
750
500
250
0
1985
2000
2025
2050
2075
2100
YEAR
RCW
High Coal Prices
RCWP
(b)
Increase In Realized Warming
(Degrees Celsius; Based on 3.0 Degree Sensitivity)
DEGREES CELSIUS
4
3
2
1
0
1985
2000
2025
2050
2075
2100
YEAR
RCW
High Coal Prices
RCWP

of 0.2-0.3°C by 2050 and 0.9-1.3°C by 2100 (assuming 2.0-4.0°C climate sensitivities). The corresponding decrease in equilibrium warming by 2100 is 1.3-2.5°C.

Alternative Oil and Natural Gas Supply Assumptions

There are many uncertainties concerning the amount of oil and natural gas supplies available worldwide. As discussed in Chapter VII, for example, the viability of increased use of natural gas as a near-term option for reducing greenhouse gas emissions critically depends on the amount of natural gas available, its price, the length of time over which adequate supplies can be secured, etc. To explore how sensitive the level of greenhouse gas emissions may be to the amount of oil and natural gas supplies, two sensitivity cases assuming higher global supplies have been analyzed: (1) a high oil resource case and (2) a high natural gas resource case. The higher oil resource estimates were derived from Grossling and Nielsen (1985), who indicated that resources may be more than double the estimates used in the base case analyses (which were about 12,000 EJ of conventional oil resources).[2] For this analysis we assumed conventional oil resources of about 25,000 EJ. Natural gas estimates were derived from Hay et al. (1988), which assumed in-place resources of about 150,000 EJ. For purposes of this sensitivity case, we assumed that technological improvements in gas extraction would permit an additional 10% of in-place resources to be economically recovered. This amount was added to the baseline estimates of proved reserves and economically recoverable resources, for a total resource base of about 27,000 EJ. We must emphasize that these sensitivity cases do not examine policy options that encourage greater oil and natural gas use; rather, they only attempt to examine how current uncertainties concerning the size of the resource base for these energy supplies can directly affect the rate and magnitude of global climate change. Policy options

[2] EJ = exajoule; 1 exajoule = 10^{18} joules.

encouraging greater use of these fuels in conjunction with higher resource estimates would have a substantially different impact.

High Oil Resources. An increase in global oil resources to 25,000 EJ is more than double the resource estimates assumed in the RCW case. These additional resources were assumed to be available at the same economic costs, such that the amount of oil available at any given price was twice the amount assumed in the RCW case. This increase in oil resources had two major impacts: (1) the amount of synthetic production of liquid fuels from coal declined substantially since conventional oil supplies were available at a competitive price to meet this demand; and (2) total demand for energy, mainly oil, increased as consumers responded to the increased availability of oil supplies at the same price (since twice the amount of oil was available at a price equal to that in the RCW case). The net effect of these impacts is a small increase in total primary energy demand (a 4% increase by 2050), a major shift from coal (primarily for synthetic fuel production) to oil, and a decrease in the portion of total primary energy supplied by non-fossil resources since oil is more plentiful and competitive; for example, non-fossil fuels supply about 22% of all energy by 2050 compared with 24% in the RCW case (see Figure 6-5). The net effect of these factors is an increase in CO_2 emissions of 0.4 Pg C by 2050 and 2.2 Pg C by 2100. The decline in coal production, however, lowered methane (CH_4) emissions since the amount of CH_4 emitted during coal mining decreased substantially (e.g., by 2100 CH_4 emissions from fuel production declined from about 360 Tg in the RCW case to 210 Tg), resulting in a modest decline of less than 0.1°C in realized warming by 2100 compared with the RCW case warming (assuming 2.0-4.0°C climate sensitivities).[3]

High Natural Gas Resources. For the high natural gas resource case, natural gas resources were increased from about 10,000 EJ to 27,000 EJ. As in the high oil resource case, higher natural

---

[3] Tg = teragram; 1 teragram = 10^{12} grams.

FIGURE 6-5

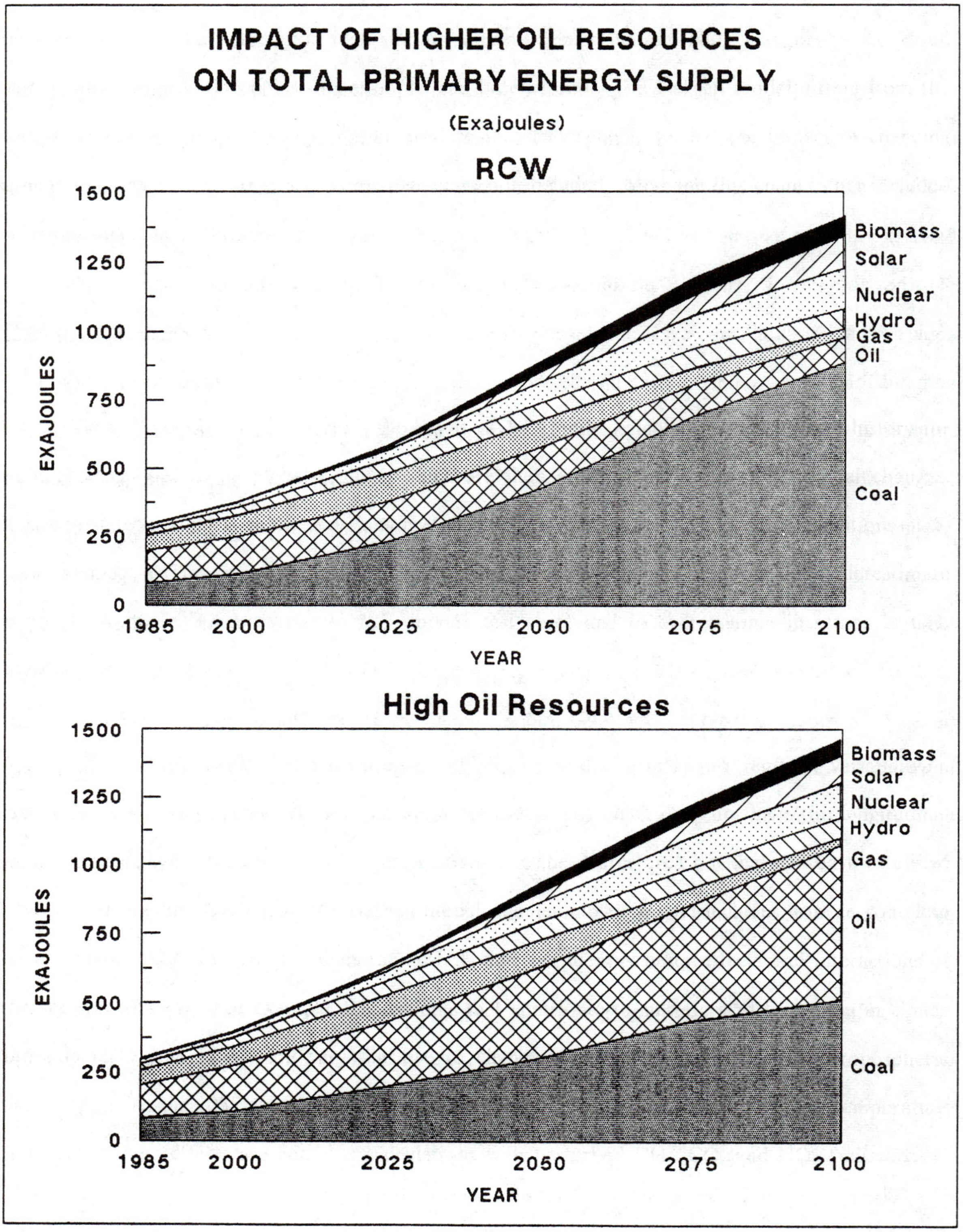
IMPACT OF HIGHER OIL RESOURCES
ON TOTAL PRIMARY ENERGY SUPPLY
(Exajoules)
RCW
1500
1250
1000
750
500
250
0
EXAJOULES
1985
2000
2025
2050
2075
2100
YEAR
Biomass
Solar
Nuclear
Hydro
Gas
Oil
Coal
High Oil Resources
1500
1250
1000
750
500
250
0
EXAJOULES
1985
2000
2025
2050
2075
2100
YEAR
Biomass
Solar
Nuclear
Hydro
Gas
Oil
Coal

gas resource estimates result in two major impacts: (1) an increase in demand for energy, particularly for gas, since natural gas is more plentiful compared with the amount available in the RCW case; and (2) a decline in the conversion of coal to synthetic gas, since natural gas supplies are available to meet the demand.

Overall, by the end of the 21st century the amount of primary energy consumed changes very little from the RCW case. In the near term energy demand increases slightly compared with the RCW case, since natural gas is more plentiful (e.g., by 2025 energy demand is about 2.5% higher compared with the RCW case; see Figure 6-6). However, the total amount of energy required in the long run is less because a greater portion of end-use energy demand is met with natural gas rather than with synthetic gas from coal. This increase in conventional natural gas consumption reduces the total primary energy required to satisfy demand because the decline in synthetic fuel demand from the RCW case reduces the amount of energy required for synthetic fuel conversion, although this impact is small: by 2100 primary energy demand is lower by less than 1%.

The amount of natural gas consumed does increase significantly; for example, in 2050 natural gas consumption increases to 215 EJ compared with 100 EJ in the RCW case. However, the increased availability of natural gas also reduces the portion of energy supplied by non-fossil fuels; for example, by 2050 non-fossil energy sources supply about 20% of total demand compared with 24% in the RCW case. The net impact on CO_2 emissions due to these factors is quite small: no change in emissions by 2050 and a decline of 0.7 Pg C by 2100. The impact on realized and equilibrium warming is negligible (less than 0.1°C).

FIGURE 6-6

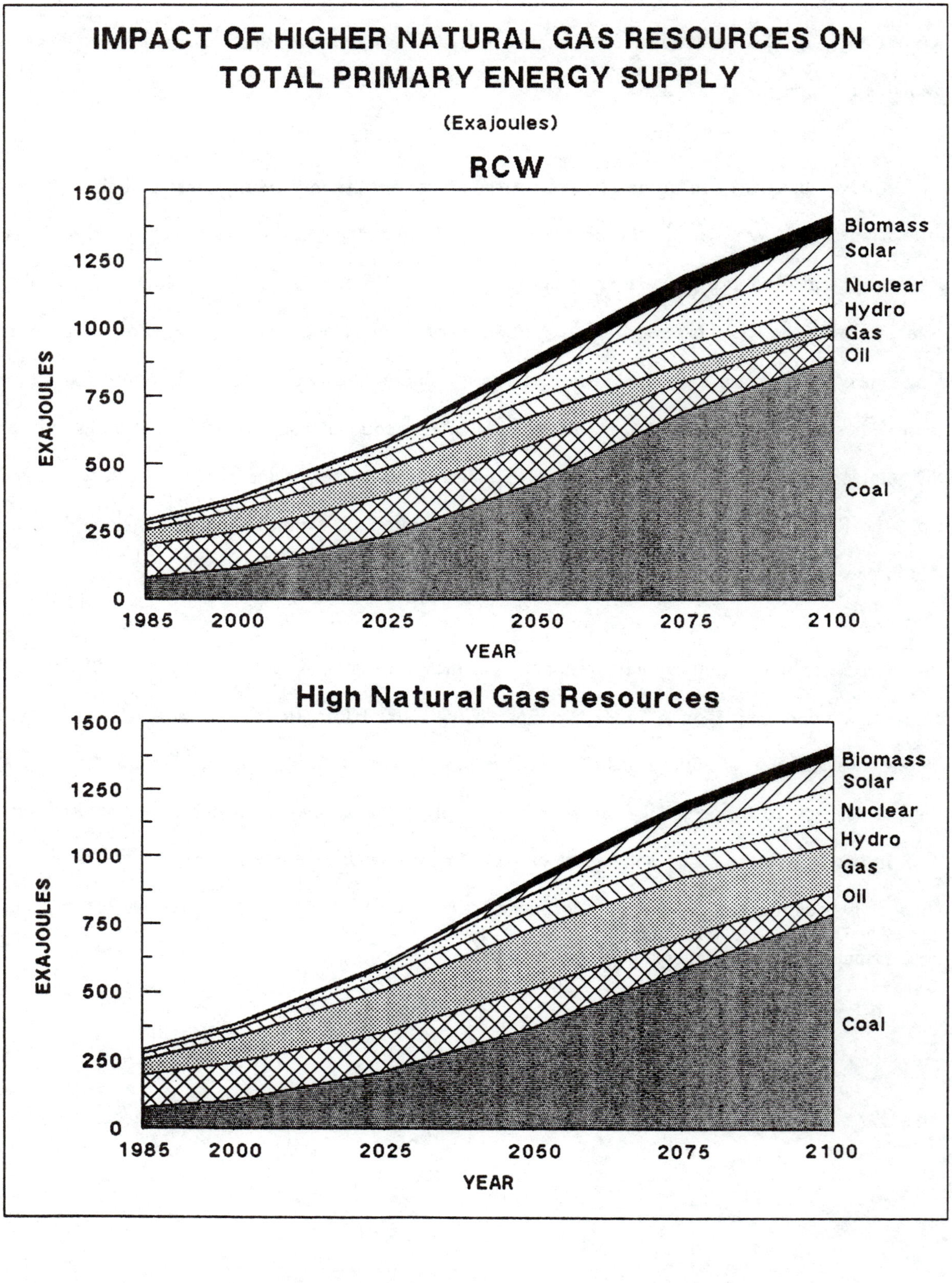
IMPACT OF HIGHER NATURAL GAS RESOURCES ON TOTAL PRIMARY ENERGY SUPPLY
(Exajoules)
RCW
EXAJOULES
0
250
500
750
1000
1250
1500
1985
2000
2025
2050
2075
2100
YEAR
Biomass
Solar
Nuclear
Hydro
Gas
Oil
Coal
High Natural Gas Resources
EXAJOULES
0
250
500
750
1000
1250
1500
1985
2000
2025
2050
2075
2100
YEAR
Biomass
Solar
Nuclear
Hydro
Gas
Oil
Coal

Availability of Methanol-Fueled Vehicles

The transportation sector throughout the world is heavily dependent on petroleum-based fuels. This dependence, particularly on gasoline and diesel fuel, produces substantial quantities of greenhouse gases (see Chapter IV). A variety of non-petroleum-based alternatives are under development, including the use of methanol. There are many potential advantages to using methanol as a transportation fuel rather than gasoline; according to recent research, advanced methanol-fueled vehicles could be 20-40% more energy efficient, emit much lower levels of CO, and reduce non-methane hydrocarbon (NMHC) reactivity up to 95% (Gray, 1987). Methanol's potential to reduce NMHC reactivity could reduce levels of urban ozone, which would improve ambient air quality in urban areas. These reductions could be on the order of about 5-20% of peak ozone levels (DeLuchi et al., 1988). However, it is not clear how reductions in urban ozone levels may translate to reductions in average tropospheric ozone and, therefore, changes in radiative forcing. Current understanding of these atmospheric processes attributes urban ozone changes primarily to NMHC and NO_x flux, while tropospheric ozone changes depend primarily on (in descending order of importance) CH_4, CO, NO_x flux, and NMHC flux (Prather, 1988). Interactions between urban air quality and the rest of the troposphere cannot be evaluated with the aggregate model used here.

Since the ability of methanol to affect tropospheric ozone levels cannot be reliably estimated, we cannot reflect all of the potential advantages of using methanol as a transportation fuel. It is useful to note, however, that in addition to reducing emissions of CO and other gases, methanol can be produced from different types of feedstocks, such as natural gas, coal, or biomass. When biomass is the feedstock, the carbon emitted during the combustion process is recycled from the environment as the biomass is grown. As a result, the net CO_2 emissions are zero when biomass is used. Greenhouse gas emissions from methanol, however, can be greater than those from gasoline if coal is used as the feedstock because additional emissions will be generated during the methanol production process. According to one analysis, methanol production from coal would generate about twice the amount of CO_2-equivalent emissions (based on their radiative effect) compared to gasoline from crude oil, while methanol from natural gas would only be slightly better (about 3%) than petroleum-based fuels (DeLuchi et al. 1988). From a global warming perspective, DeLuchi et al. concluded that only biomass-derived methanol would substantially reduce the amount of radiative forcing from transportation fuels, although as mentioned above, this argument does not incorporate any potential benefits from reductions in urban ozone levels.

ATMOSPHERIC COMPOSITION: COMPARISON OF MODEL RESULTS TO ESTIMATES OF HISTORICAL CONCENTRATIONS

The atmospheric composition model was applied to estimates of historical emissions of trace gases and the results compared to historical data on atmospheric composition. This exercise provides insight on how the model performed under conditions much different from the reference year, 1985, and provided one mechanism to validate the model. The exercise included the development of a single scenario of historical emissions of trace gases and application of the model using different assumptions on climate sensitivity and chemistry parameters in the model.

The scenario of historical emissions of trace gases is based on estimates of natural sources from the Atmospheric Stabilization Framework described in Chapter V, estimates from a study by Darmstadter et al. (1987) on historical emissions from various anthropogenic sources, and estimates of historical CO_2 emissions from Rotty (1987) and Houghton (1988). For natural emission sources, historical emissions were assumed to be constant from 1870 to 1985 at the levels assumed in the scenarios described in Chapter V. The exception is emissions of CH_4 from wetlands, which were assumed to be larger in 1870 by 50% and to decline to current levels due to destruction of wetlands.

The estimates of historical emissions of CFCs and Halons were taken from EPA's Regulatory Impact Analysis on Stratospheric Ozone Protection (U.S. EPA, 1988).

The alternative scenarios of historical atmospheric composition and global warming reflect a range of assumptions concerning the climate sensitivity and the first and second order relationships assumed in the model. Figure 6-7 illustrates the increase in realized warming projected from 1840 to 1985, which ranges from 0.4°C to 0.8°C based on a range of climate sensitivities (from 1.5 to 5.5°C for doubled CO_2). These results compare well with results from Wigley et al. (1986), who estimated a global temperature increase of 0.3-0.7°C in the last century, and Hansen et al. (1988), who estimated a global temperature increase of 0.4-0.8°C during the same period. The model produced estimates of atmospheric concentrations of CO_2, CH_4, N_2O, CO, and CFC-12 within 1.5% and estimates of concentrations of CFC-11 within 3.5% of observed values in 1985. In addition, the pattern of estimated atmospheric concentrations over time conformed well with historical measurements for CO_2, N_2O, and CH_4. Estimates of concentrations of some gases such as CFC-22 varied from the historical measurements to a greater extent, which reflects their more recent introduction and rapid growth in atmospheric concentrations. Table 6-2 summarizes the results for the long-lived gases.

For CO_2, the atmospheric concentration over time matched the Mauna Loa and Ice Core measurements by design through the use of the unknown sink in the model (see **Unknown Sink in Carbon Cycle**). The unknown sink is zero through 1940 and then slowly rises to 1.9 Pg C per year by 1985, which represents about one-third of the estimated anthropogenic emissions.

The estimates of CH_4 concentrations match atmospheric and ice core measurements well, especially given the uncertainties in the emissions estimates and the historical measurements. The model shows somewhat higher than expected growth in the late 19th century, which may reflect the

FIGURE 6-7

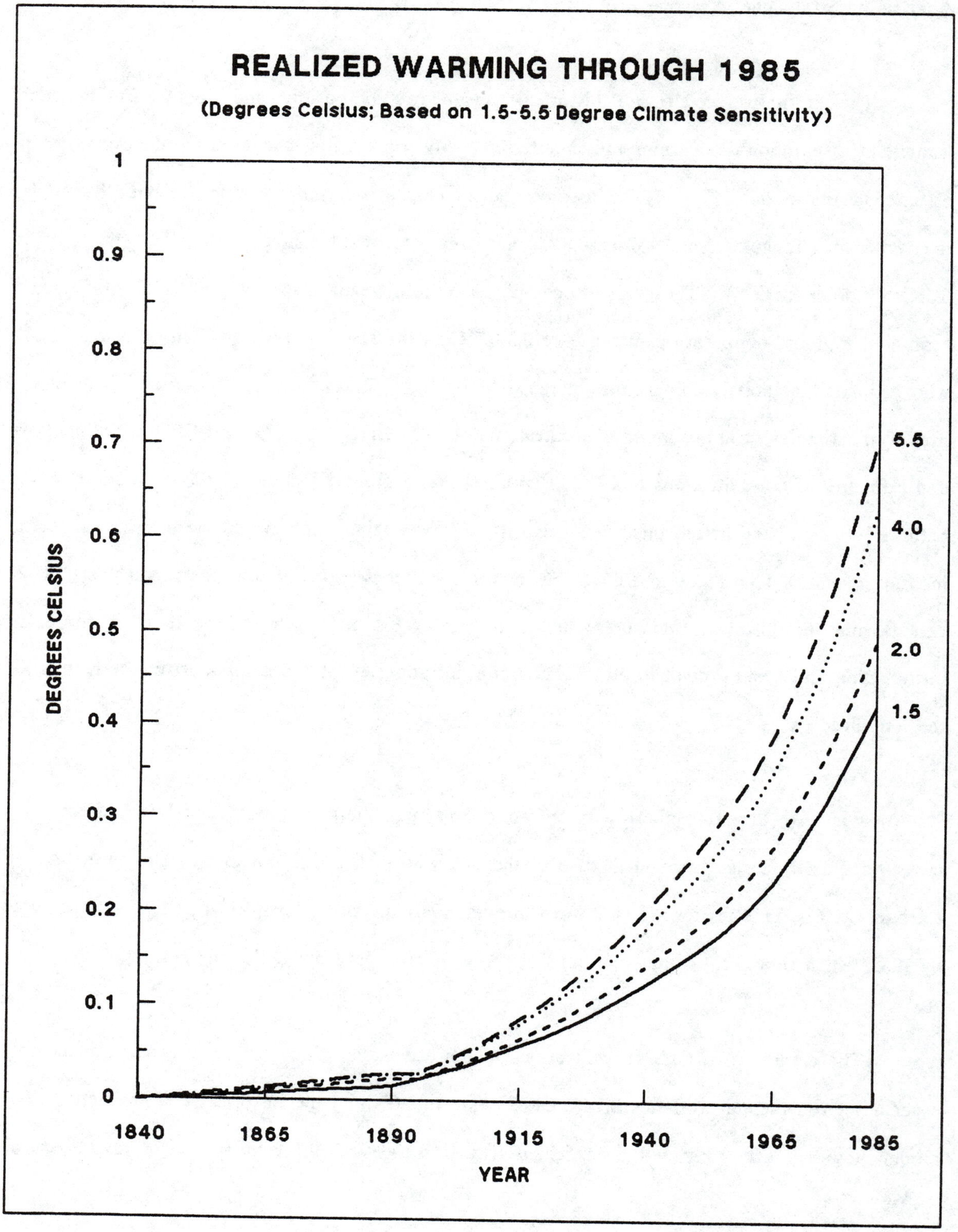
REALIZED WARMING THROUGH 1985
(Degrees Celsius; Based on 1.5-5.5 Degree Climate Sensitivity)
DEGREES CELSIUS
1
0.9
0.8
0.7
0.6
0.5
0.4
0.3
0.2
0.1
0
5.5
4.0
2.0
1.5
1840
1865
1890
1915
1940
1965
1985
YEAR

TABLE 6-2
Comparison of Model Results to Concentrations in 1986

Trace Gas (units)	Model Results	Model Growth Rates	Atmospheric Measurements	Observed Growth Rates
CO_2 (ppm)	346	0.4%	346	0.4%
N_2O (ppb)	314	0.27%	310	0.2-0.3%
CH_4 (ppb)[a]	1650-1750	1%	1675	1%
CFC-11 (ppt)	212-222	4%	226	4%
CFC-12 (ppt)	391	4%	392	4%
HCFC-22 (ppt)	37	14%	100	7%
CCl_4 (ppt)[b]	70	0.6%	121	1.3%
CH_3CCl_3 (ppt)	186	12%	125	6%
Halon 1211 (ppt)	0.4	100%	2	>10%

[a] 1987 value.

[b] 1982 value.

uncertainties surrounding the scenario of historical emissions. Using the reference assumptions, the model achieves an atmospheric concentration of 1671 ppb in 1985 compared to the observed value of 1675 ppb. The CH_4 concentrations vary considerably in the sensitivity analyses and range from 1650 ppb to 1750 ppb for alternative chemistry parameters.

Of the three dominant greenhouse gases, the estimates of N_2O concentrations vary the most from historical measurements. The model predicts concentrations of 314 ppb in 1985 compared to 308 to 310 ppb cited in the literature. From 1979 to 1986, the model estimates growth in N_2O concentrations of 310 to 314 ppb compared to measurement data that suggests growth of 303 to 310 ppb. One of the possible explanations of these results is that the relative share of emissions of N_2O from anthropogenic sources is larger than estimated in the model. A larger anthropogenic source combined with lower natural emissions or a shorter atmospheric life would be needed to reduce the overall concentrations and obtain the growth in concentrations seen from 1979 to 1986. These results suggest that the model may underestimate future atmospheric concentrations of N_2O.

The model "predicts" very little deviation from current levels for the short-lived gases, including OH, O_3, and CO. The results for levels in 1870 include higher levels of OH by 14-26%, lower levels of tropospheric O_3 by 19-29%, lower concentrations of CO by approximately 50%, and increased levels of upper stratospheric ozone by 4.5%.

ASSUMPTIONS ABOUT TRACE-GAS SOURCES AND STRENGTHS

Among the various greenhouse gases there is some uncertainty over the quantity of emissions that can be attributed to specific sources and the ability of these gases to modify the atmosphere. The most critical of these uncertainties are examined below.

Methane Sources

The available evidence on CH_4 indicates that annual production ranges from 400-640 Tg of methane (based on known sources and sinks, its atmospheric lifetime, and current atmospheric concentrations). Within this budget, however, there is much dispute over the size of individual sources. For example, research indicates that current CH_4 emissions from rice paddies could be 60-170 Tg; similarly, estimated emissions from biomass burning range from 50-100 Tg (Cicerone and Oremland, 1989).

To account for these uncertainties, the initial CH_4 budget was varied to construct two cases: (1) a high anthropogenic impact case, where the starting methane budget was biased toward anthropogenic sources by assuming that anthropogenic activities such as fuel production and landfilling caused higher emission levels than assumed in the RCW case, while lower emission estimates were assumed from natural processes such as oceans, wetlands, wildfires, and wild ruminants; and (2) a low anthropogenic impact case, by assuming lower emissions from anthropogenic activities such as fuel production, enteric fermentation, and rice cultivation, with corresponding emission increases from natural processes such as oceans and wetlands. The specific emission assumptions for the starting budget are summarized in Table 6-3.

The alternative starting budgets in Table 6-3 result in different growth paths for CH_4, since emissions from anthropogenic sources increase by different amounts over time. These differences alter the atmospheric concentration of CH_4: by 2100 the atmospheric concentration is about 3500-3700 ppb in the Low Impact case and 5200-5500 ppb in the High Impact case (compared with 4100-4400 ppb in the RCW case). The increase (decrease) in CH_4 also increases (decreases) the amount of tropospheric ozone. The impact on realized warming is summarized in Figure 6-8, which indicates a decline of 0.1-0.2°C by 2100 in the Low Impact case compared with the RCW case and an increase

TABLE 6-3

Low and High Anthropogenic Impact Budgets For Methane
(teragrams/year as of 1985)

Source of Methane	Low Impact	RCW	High Impact
Fuel Production	50	60	95
Enteric Fermentation	70	75	75
Rice Cultivation	60	110	110
Landfills	30	30	58
Oceans	40	15	6
Wetlands	150	115	100
Biomass Burning--Anthropogenic	35	35	35
Biomass Burning--Natural	20	20	15
Wild Ruminants	44	44	10
Other Sources	11	6	6
TOTAL	510	510	510

FIGURE 6-8

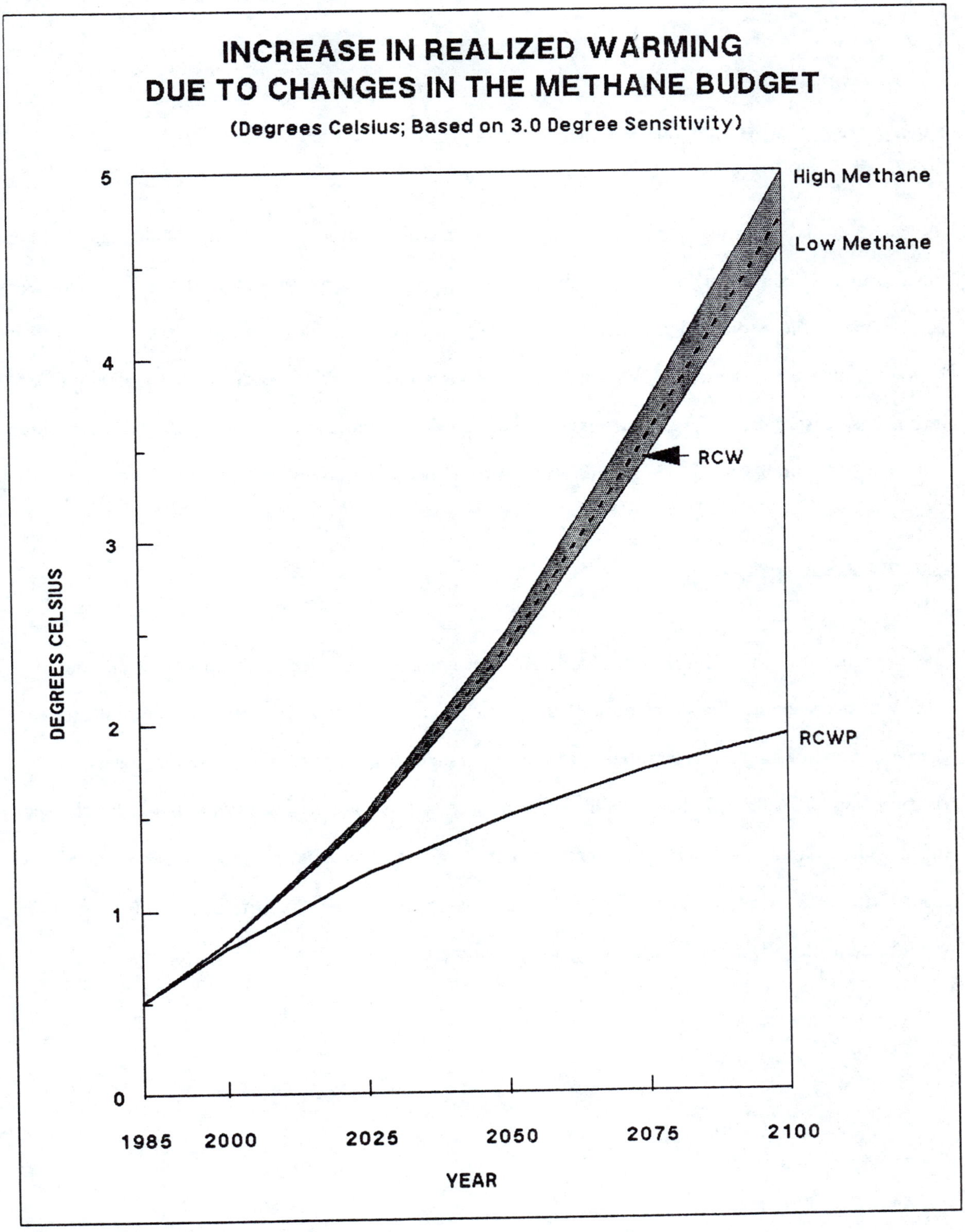
INCREASE IN REALIZED WARMING
DUE TO CHANGES IN THE METHANE BUDGET
(Degrees Celsius; Based on 3.0 Degree Sensitivity)
High Methane
Low Methane
RCW
RCWP
DEGREES CELSIUS
0
1
2
3
4
5
1985
2000
2025
2050
2075
2100
YEAR

of 0.2-0.3°C by 2100 in the High Impact case. The corresponding effects on equilibrium warming by 2100 are a decline of 0.2-0.3°C in the Low Impact case and an increase of 0.3-0.5°C in the High Impact case.

Nitrous Oxide Emissions From Fertilizer

N_2O is naturally produced in soils by microbial processes during denitrification and nitrification. When nitrogen-based fertilizers are applied, N_2O emissions from the soil can increase as a result of the additional nitrogen source. The amount of fertilizer nitrogen evolved as N_2O is highly variable and uncertain. We have used the emission estimates developed by Galbally (1985) in our base cases: 0.5% for anhydrous ammonia, 0.1% for ammonium nitrate, 0.1% for ammonium salts, 0.5% for urea, and 0.05% for nitrates. Alternative assumptions are explored below.

Anhydrous Ammonia

One of the key uncertainties concerns the emission coefficient for anhydrous ammonia. A review of the scientific literature on measurements of N_2O emissions by fertilizer type indicates that the percentage of anhydrous ammonia evolved as N ranges from 0.05-6.84%, with most measurements ranging from 0.5-2.0% (Eichner, 1988). The impact of this uncertainty was evaluated by changing the anhydrous ammonia coefficient from 0.5% to 2.0%. This change increased the amount of N_2O from fertilizer applications by 0.1 Tg of N annually, an increase in 2025 from 0.7 to 0.8 Tg, which was too small to affect the amount of global warming.

N_2O Leaching From Fertilizer

As discussed above, in the RCW case N_2O emissions from fertilizer were based on estimates by Galbally (1985). One N_2O emission pathway not included in this estimate is leaching from the fields into the ground water or surface water due to the application of the fertilizer. The rate of emissions related to leaching is highly uncertain; Conrad et al. (1983) and Kaplan et al. (1978) have suggested that the amount of N_2O evolved due to leaching may be as large as N_2O from the denitrification/nitrification processes in the soil. The impact of leaching on total N_2O emissions and the resulting global warming was analyzed by increasing all of Galbally's emission coefficients by one percentage point. The higher rate of N_2O from fertilizer due to leaching resulted in an increase in emissions of about 1.0-1.5 Tg annually.

Atmospheric N_2O concentrations increase about 20 ppb by 2100 compared with the RCW case (from 403 to 424 ppb; see Figure 6-9). While N_2O concentrations increase when leaching is assumed, the impact on global warming is not as certain. In this case, global warming was slightly reduced (less than 0.1°C) due to the chemical interactions that occur with increased N_2O levels. Specifically, higher N_2O levels in the stratosphere reduce the amount of stratospheric ozone, which in turn allows more ultraviolet (UV) radiation to penetrate to lower elevations. The increased UV radiation increases the amount of CFC destruction, which reduces the contribution of CFCs to global warming. None of these reactions are very strong, since the change in N_2O emissions due to leaching does not have a major effect on atmospheric concentrations, but they are sufficient to counteract the warming effect of higher N_2O concentrations alone.

N_2O Emissions From Combustion

During the combustion process, chemical interactions downstream from the combustion

FIGURE 6-9

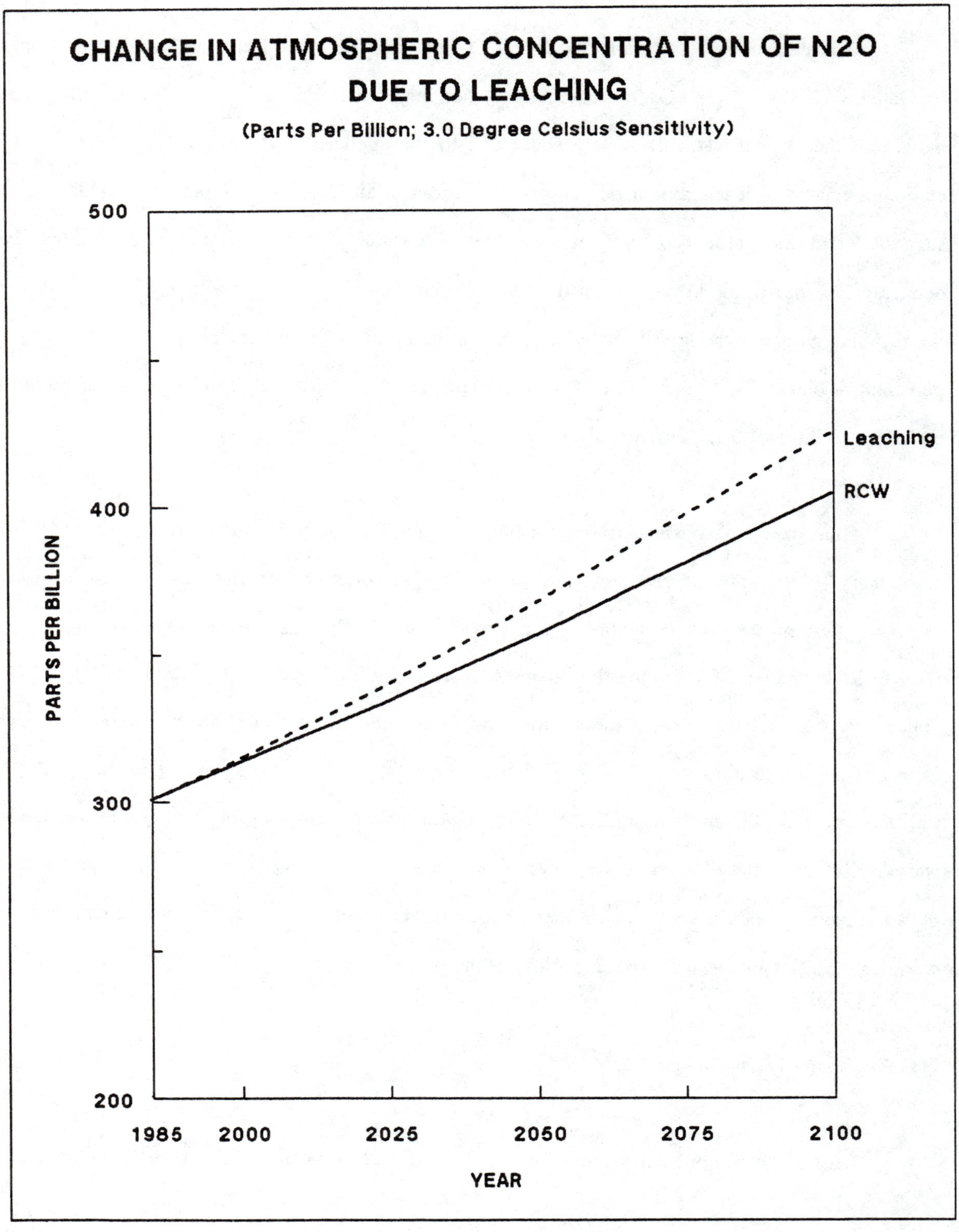
CHANGE IN ATMOSPHERIC CONCENTRATION OF N2O
DUE TO LEACHING
(Parts Per Billion; 3.0 Degree Celsius Sensitivity)
500
400
300
200
PARTS PER BILLION
Leaching
RCW
1985
2000
2025
2050
2075
2100
YEAR

chamber can lead to N_2O formation from nitrogen oxides. The rate of this formation is highly uncertain, although recent evidence indicates that it is likely to be fairly small. In the RCW case these low emission coefficients were assumed (see Chapter II). To ascertain the impact of higher emission coefficients, N_2O coefficients from combustion were increased such that emissions from energy in 1985 were 2.3 Tg N rather than 1.1 Tg N as obtained in the RCW case. The higher N_2O emission levels increased atmospheric concentrations about 50 ppb by 2100 (as shown in Figure 6-10); the resulting impact on global warming was negligible (less than 0.1°C) for the same reasons discussed above under leaching from fertilizer.

UNCERTAINTIES IN THE GLOBAL CARBON CYCLE

The global carbon cycle, which regulates the flow of carbon through the environment, including the atmosphere, biosphere, and hydrosphere, was discussed in Chapters II and III. Uncertainties in the size of the various sources and sinks for carbon and the interactions that govern the flow of carbon increase the difficulty of estimating the impact of anthropogenic activities on global climate. In this section the major uncertainties in the global carbon cycle are evaluated. The first part focuses on the impact of deforestation on CO_2 emissions. The second part discusses the ability of the oceans to absorb CO_2 and heat. Currently, the oceans are the dominant sink for anthropogenic CO_2 emissions, with the mixed layer alone containing about as much carbon as the atmosphere. The oceans' ability to operate as a net sink for carbon and heat is an important component of the global climate system; any changes in this absorption ability could have profound effects on global climate (see Chapter III).

FIGURE 6-10

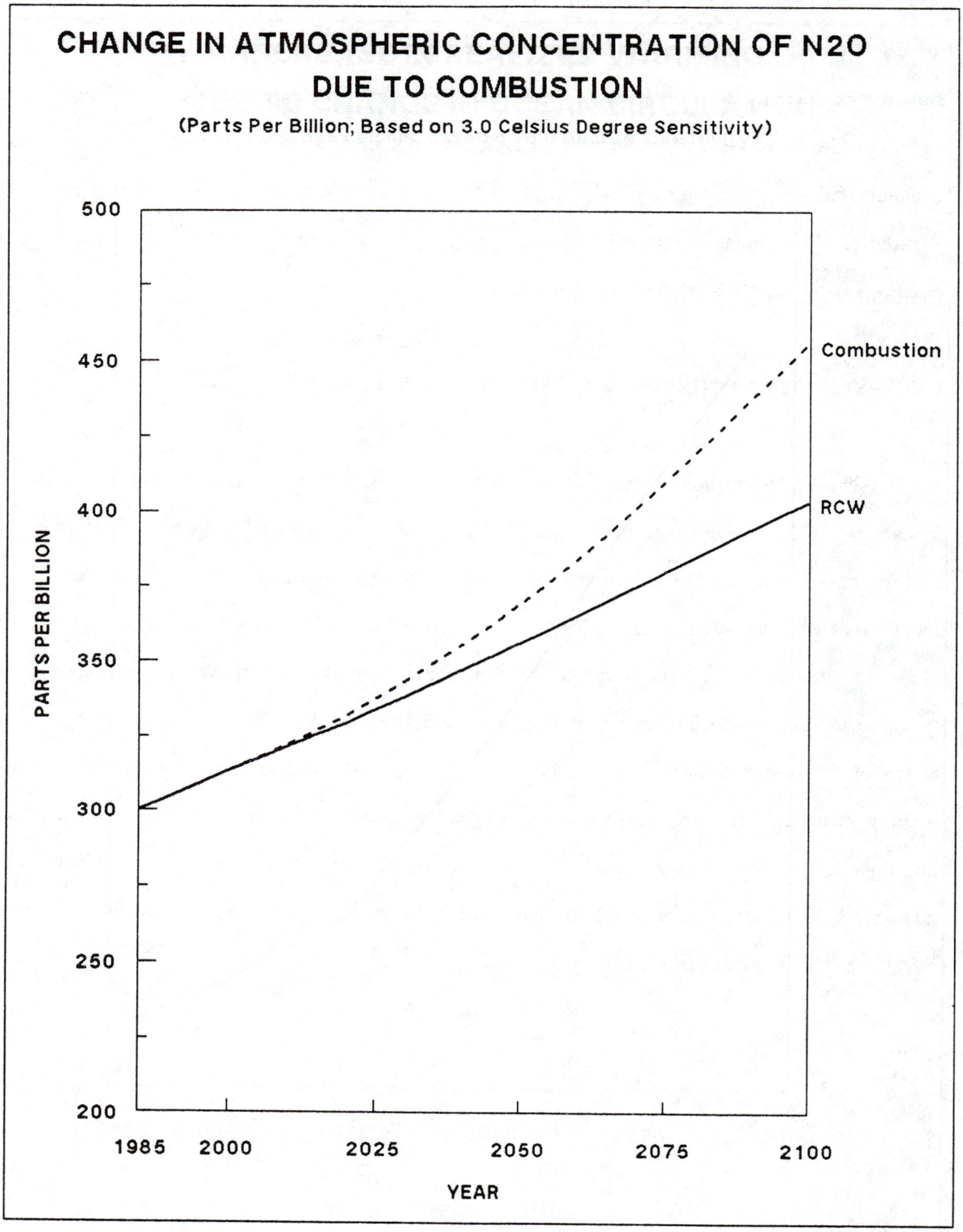
CHANGE IN ATMOSPHERIC CONCENTRATION OF N2O
DUE TO COMBUSTION
(Parts Per Billion; Based on 3.0 Celsius Degree Sensitivity)
500
450
400
350
300
250
200
PARTS PER BILLION
Combustion
RCW
1985
2000
2025
2050
2075
2100
YEAR

Unknown Sink In Carbon Cycle

Atmospheric CO_2 concentrations have changed historically due to an imbalance between the sources and sinks for carbon. If the production of carbon exceeds the ability of the various carbon sinks to absorb it, then the atmospheric CO_2 concentrations will increase (and vice versa). When analyzing the amount of carbon produced from various sources in the past, atmospheric scientists have been unable to balance the carbon cycle. That is, given current estimates of carbon sources, it would appear that atmospheric CO_2 concentrations would have to be higher than currently measured, since all known sinks do not appear to be able to absorb all of the carbon produced. To account for this imbalance, we have assumed the existence of an "unknown sink" that absorbs the unaccounted-for carbon. The size of this unknown sink depends on the assumed magnitude of known sources and sinks--by definition, the unknown sink is simply: sources minus sinks minus atmospheric accumulation.

For our base cases, the size of the unknown sink was kept constant at 1.6 Pg annually based on its calculated value (from the model) for 1975-1985. However, alternative assumptions are plausible. To capture these uncertainties, two sensitivities were analyzed: (1) a high case, where the size of the unknown sink increases at the same rate as atmospheric CO_2 levels compared with preindustrial levels (this increase might occur, e.g., because the size of the unknown sink is related to the fertilization of terrestrial ecosystems by increasing CO_2); and (2) a low case, where the size decreases to zero exponentially at 2% per year (e.g., because the process responsible for the unknown sink has a limited capacity).

When the unknown sink is assumed to increase in proportion to CO_2 concentrations in the RCW case, the amount of carbon absorbed by the unknown sink increases to 11.6 Pg annually by 2100. This rate of carbon absorption results in a decline in CO_2 concentrations relative to the

Rapidly Changing World, which reduces realized warming by 0.1-0.2°C in 2050 and 0.5-0.7°C in 2100; equilibrium warming is reduced in 2050 by 0.2-0.5°C and in 2100 by 0.7-1.4°C (based on 2.0-4.0°C climate sensitivities).

In the low case, that is, when the unknown sink decreases to zero, the estimated impact on warming is significantly lower, since the unknown sink was only 1.6 Pg annually to start. As a result, CO_2 concentrations do increase, but the increase in realized warming is less than 0.1°C in 2050 and 0.1-0.2°C in 2100 (based on 2.0-4.0°C climate sensitivities; see Figure 6-11).

Amount of CO_2 From Deforestation

Estimates of the amount of CO_2 emitted from deforestation activities vary due to different assumptions on the rate of deforestation, the fate of the deforested lands, and the amount of carbon contained in the forest vegetation and soils. In the base cases we used the lower carbon estimates (i.e., lower biomass estimates) given by Houghton (1988); for 1980 the resulting net flux of carbon to the atmosphere was about 0.4 Pg of carbon. Higher estimates of initial biomass have also been analyzed by Houghton (1988); with these estimates the net flux of carbon to the atmosphere in 1980 would have been about 2.2 Pg. These higher biomass estimates are evaluated here for the three deforestation scenarios discussed in Chapter V. The net flux of carbon for each of these scenarios is presented in Figure 6-12.

In the RCW case the rate of CO_2 emissions from deforestation was based on an exponential decline in forest area using the lower biomass assumptions. If the higher biomass estimates are used, the total carbon flux from deforestation from 1980 to 2100 is 281 Pg compared with 118 Pg using the low estimates of carbon stocks (Houghton 1988). Similarly, in the population-based deforestation scenario the total carbon flux to the atmosphere from 1980 to 2100 is about 138 Pg using the lower

FIGURE 6-11

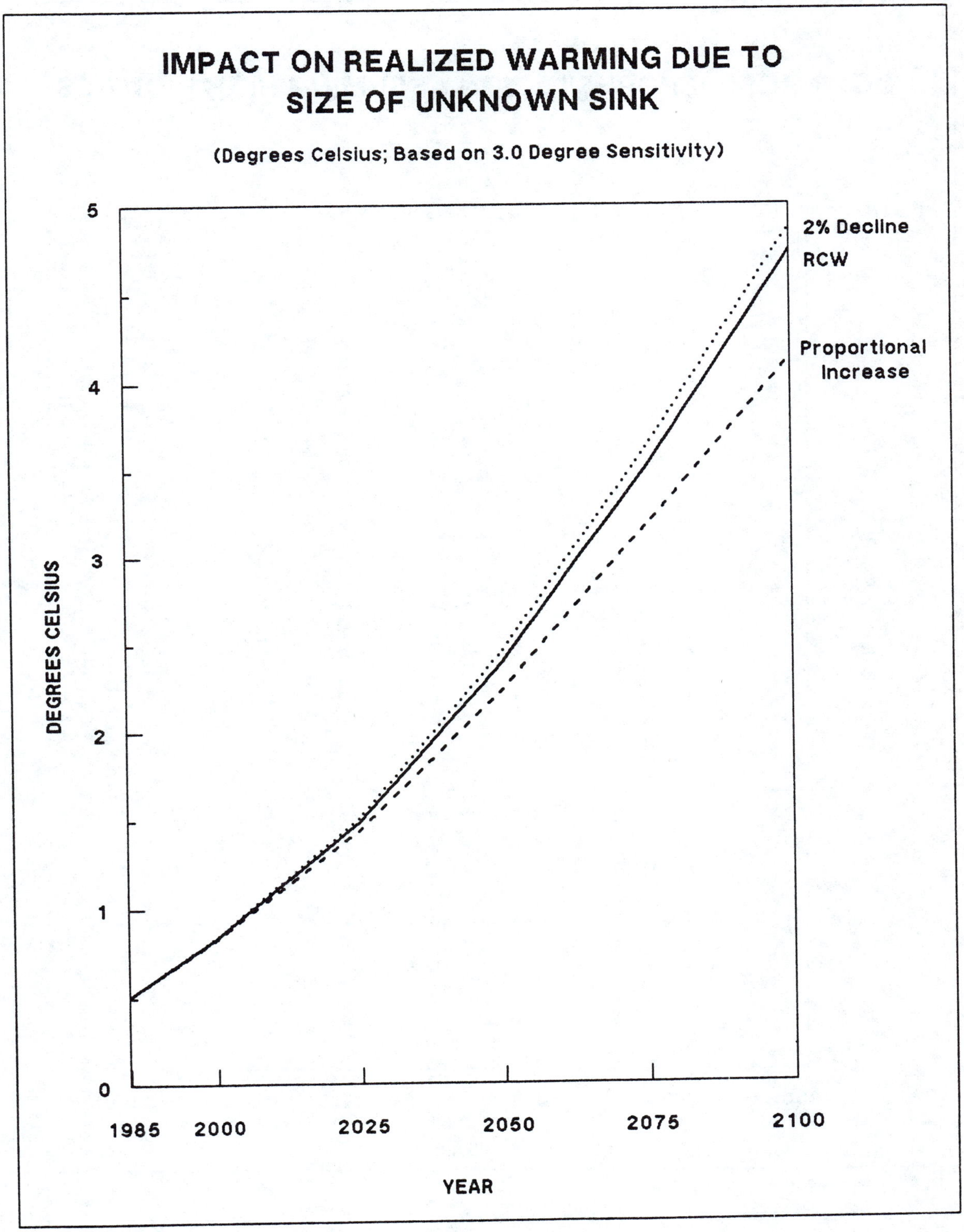
IMPACT ON REALIZED WARMING DUE TO
SIZE OF UNKNOWN SINK
(Degrees Celsius; Based on 3.0 Degree Sensitivity)
DEGREES CELSIUS
0
1
2
3
4
5
1985
2000
2025
2050
2075
2100
YEAR
2% Decline
RCW
Proportional
Increase

FIGURE 6-12

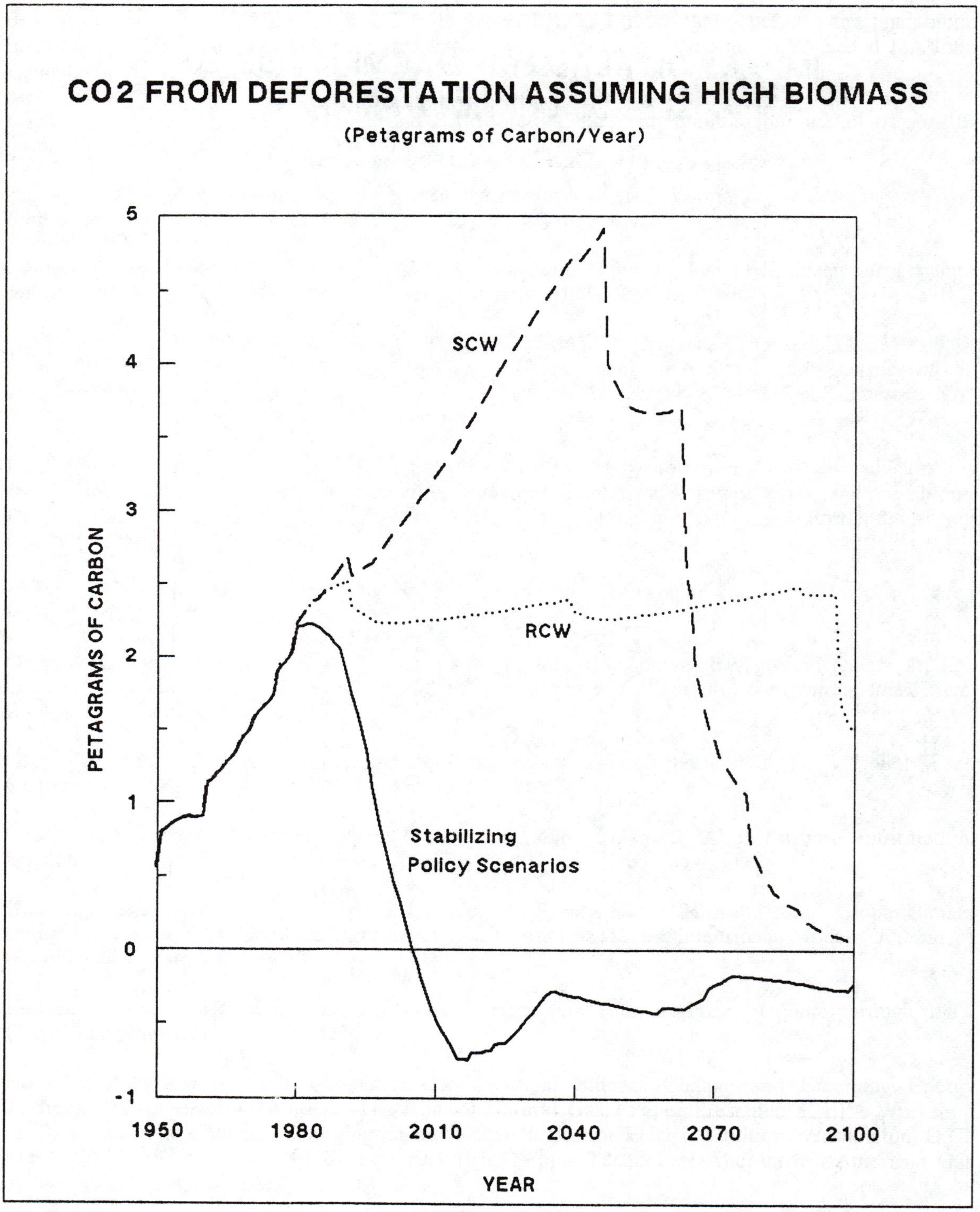
CO2 FROM DEFORESTATION ASSUMING HIGH BIOMASS
(Petagrams of Carbon/Year)
PETAGRAMS OF CARBON
5
4
3
2
1
0
-1
SCW
RCW
Stabilizing
Policy Scenarios
1950
1980
2010
2040
2070
2100
YEAR

biomass estimates and 324 Pg using the higher biomass estimates. In the reforestation scenario, the total accumulation of carbon from the atmosphere was 38 Pg using the lower biomass estimates and 59 Pg using the higher biomass estimates.

Despite the substantial increase in the amount of carbon from deforestation when the higher biomass estimates are used (e.g., by 2050 CO_2 emissions from deforestation are 2.3 Pg compared with 1.0 Pg in the RCW with the lower estimates), the resulting atmospheric concentration of CO_2 is only slightly higher (see Figure 6-13 for the differences in the RCW case, i.e., forest area declines exponentially). This result is due to the larger size of the "unknown carbon sink" in our model when higher deforestation emissions are assumed (see **Unknown Sink** above). In our analysis the increase in the size of the unknown sink was sufficient to absorb some of the additional carbon when the higher biomass estimates are used, assuming that the size of the unknown sink remains constant at its average 1975-1985 value (i.e., 2.6 Pg C with high biomass vs. 1.6 Pg C with low biomass). The additional increase in CO_2 increased realized warming and equilibrium warming less than 0.1°C by 2100 compared with the RCW case warming (assuming 2.0-4.0°C climate sensitivities).

Alternative CO_2 Models of Ocean Chemistry and Circulation

In the RCW case ocean chemistry was represented using a diffusion model of the ocean (the Modified GISS model) based on the model described by Hansen et al. (1988). Several other approaches have also been developed and adopted for the EPA framework by W. Emmanuel and B. Moore. These include:

- Box-Diffusion Model introduced by Oeschger et al. (1975), which represents the turnover of carbon below 75 meters as a purely diffusive process.

FIGURE 6-13

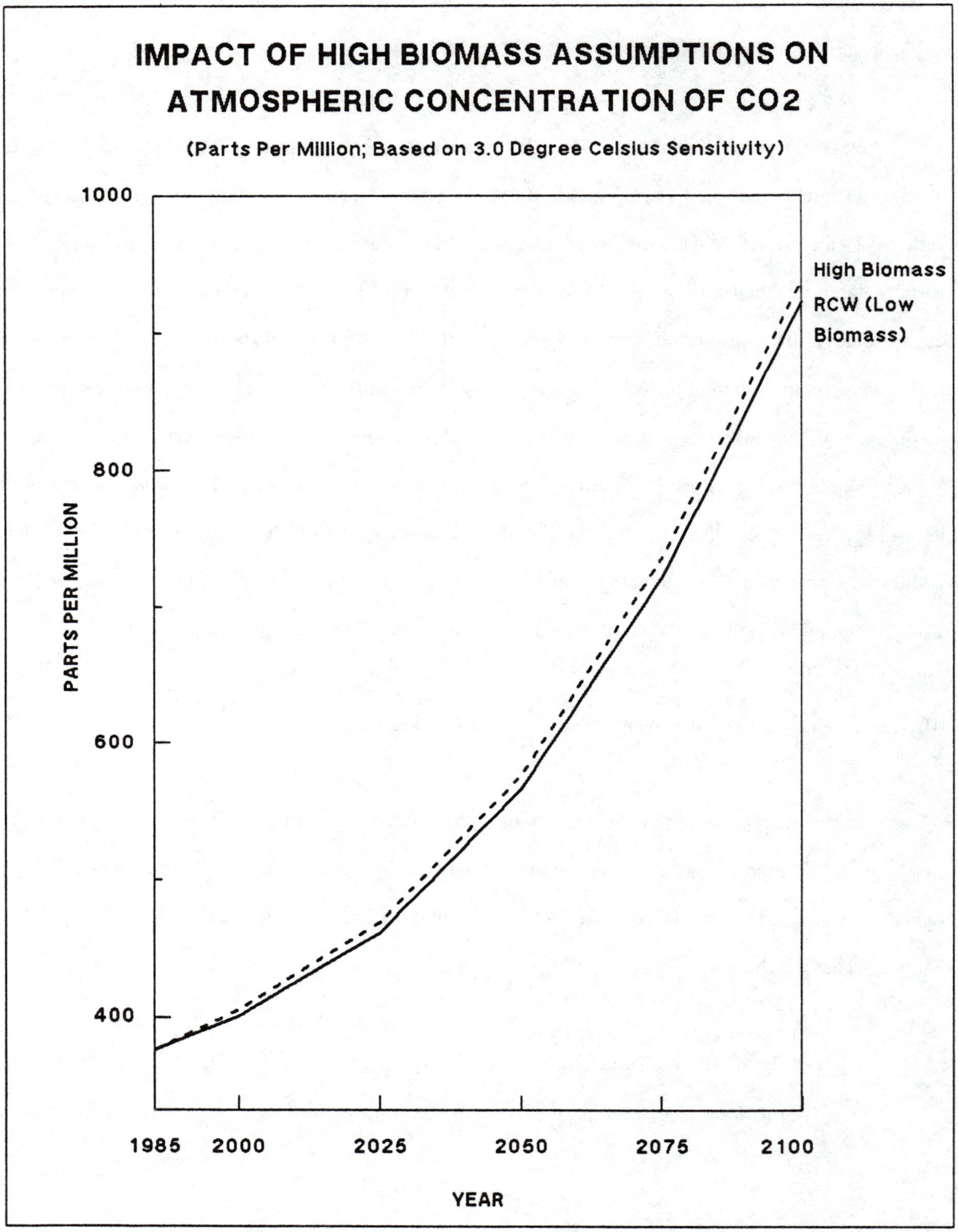
IMPACT OF HIGH BIOMASS ASSUMPTIONS ON ATMOSPHERIC CONCENTRATION OF CO2
(Parts Per Million; Based on 3.0 Degree Celsius Sensitivity)
1000
800
600
400
PARTS PER MILLION
High Biomass
RCW (Low Biomass)
1985
2000
2025
2050
2075
2100
YEAR

- 12-Compartment Regional Model by Bolin et al. (1983), which divides the Atlantic and Pacific-Indian Oceans into surface-, intermediate-, deep-, and bottom-water compartments and divides the Arctic and Antarctic Oceans into surface- and deep-water compartments.

- Advective-Diffusive Model by Bjorkstrom (1979), which divides the surface ocean into cold and warm compartments; water downwells directly from the cold surface compartment into intermediate and deep layers.

- Outcrop-Diffusion Model by Siegenthaler (1983), which allows direct ventilation of the intermediate and deep oceans at high latitudes by incorporating outcrops connecting all sublayers to the atmosphere.

Because each of these models uses a different approach to evaluate ocean chemistry, the resulting impact on atmospheric CO_2 concentrations could vary from one approach to the next. To determine how comparable these models were, the RCW case was evaluated using each model in turn.

The estimates of future CO_2 concentrations from each model are summarized in Figure 6-14a. These results indicate that the Modified GISS model tends to project higher atmospheric CO_2 concentrations than the other models; for example, by 2100 CO_2 concentrations are about 6-7% higher than concentrations estimated by Oeschger et al., Bolin et al., or Bjorkstrom, and about 19% higher than those estimated by Siegenthaler. There are two basic reasons for these differences: (1) The Modified GISS model, unlike the other models, incorporates temperature feedback that alters ocean carbonate chemistry; that is, as the mixed layer of the oceans warms due to atmospheric warming, the amount of carbon that can be absorbed by the oceans decreases; and (2) The Modified

GISS model does not incorporate any heat or CO_2 transfer between the thermocline and the deep ocean (below 1,000 meters); to the extent heat or CO_2 is transported to the ocean depths in the long run, the Modified GISS model understates the oceans' absorption capacity.

Siegenthaler's Outcrop-Diffusion Model estimates lower CO_2 concentrations than any of the other models. This result is anticipated because the Outcrop-Diffusion Model allows CO_2 to be absorbed from the atmosphere to the deep layers rather than diffuse through the intervening layers, so that, in this model, carbon is absorbed more quickly in the oceans than in the other models. By 2100 equilibrium warming using Siegenthaler's model is 0.6-1.2°C lower than the RCW case (see Figure 6-14b for warming estimates from all five models).

ASSUMPTIONS ABOUT CLIMATE SENSITIVITY AND TIMING

Sensitivity of the Climate System

A general benchmark for comparing atmospheric models is their response to a doubling of CO_2 concentrations ($2xCO_2$; see Chapter III). Put simply, this benchmark describes how much warming would be expected once the atmosphere stabilizes following a two-fold increase in CO_2 concentrations. In our analyses we have used the range from 2.0-4.0°C. As discussed in Chapter III, there is a great deal of uncertainty about the strength of internal climate feedbacks, and, in some cases, whether a feedback will be positive or negative. If cloud and surface albedo changes produce large positive feedbacks, as suggested by some analyses, the climate sensitivity could be 5.5°C or greater. On the other hand, these feedbacks could be weak and cloud feedbacks could be negative, resulting in a climate sensitivity as low as 1.5°C. For the sensitivity analysis, therefore, we have evaluated the extent of global warming using 1.5 and 5.5°C as lower and upper bounds, respectively. The impact of these assumptions on realized warming is summarized in Figure 6-15 for the RCW

FIGURE 6-14

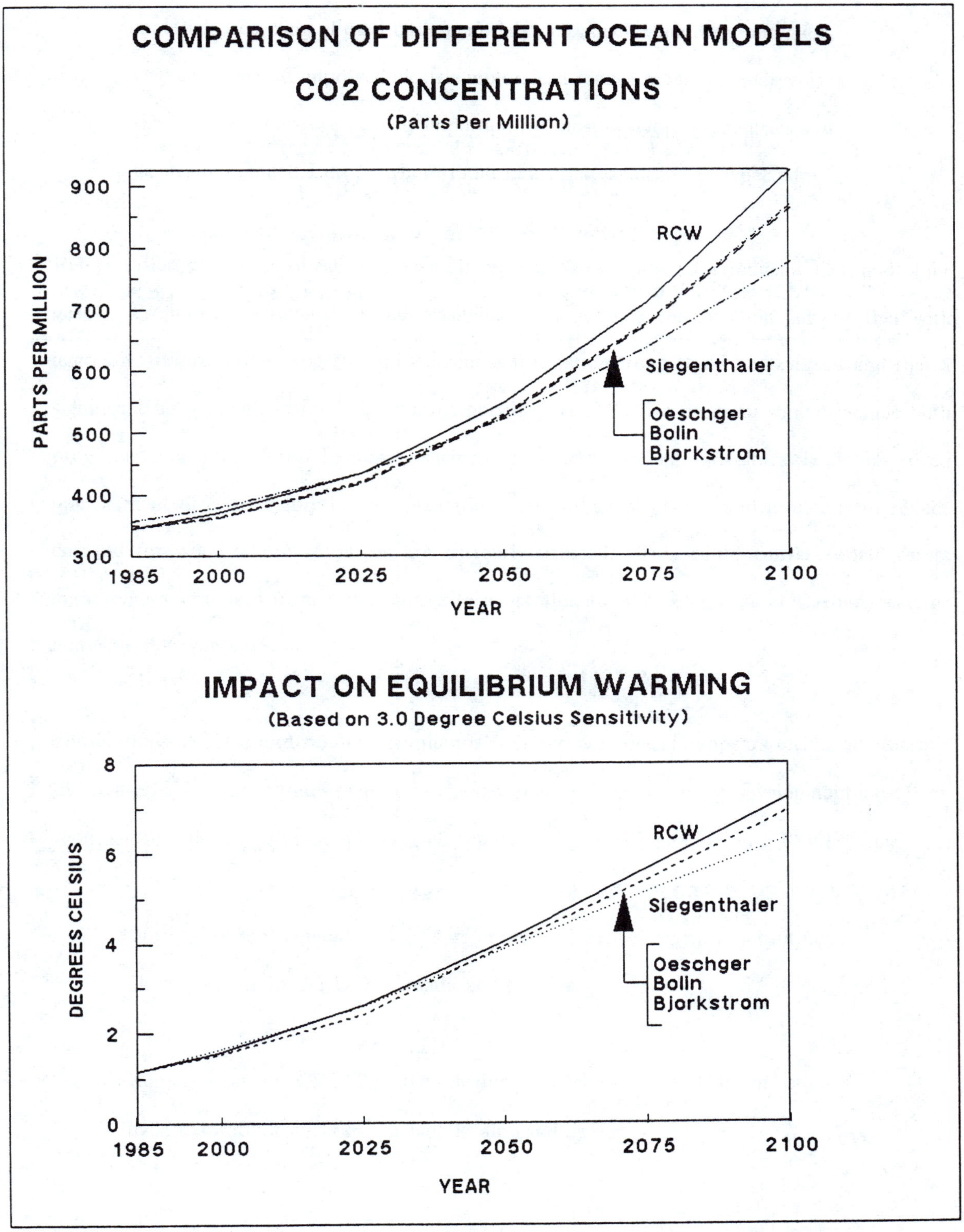
COMPARISON OF DIFFERENT OCEAN MODELS
CO2 CONCENTRATIONS
(Parts Per Million)
PARTS PER MILLION
900
800
700
600
500
400
300
1985
2000
2025
2050
2075
2100
YEAR
RCW
Siegenthaler
Oeschger
Bolin
Bjorkstrom
IMPACT ON EQUILIBRIUM WARMING
(Based on 3.0 Degree Celsius Sensitivity)
DEGREES CELSIUS
8
6
4
2
0
1985
2000
2025
2050
2075
2100
YEAR
RCW
Siegenthaler
Oeschger
Bolin
Bjorkstrom

FIGURE 6-15

IMPACT OF CLIMATE SENSITIVITY ON REALIZED WARMING

(Degrees Celsius; 1.5-5.5 Degree Climate Sensitivity)

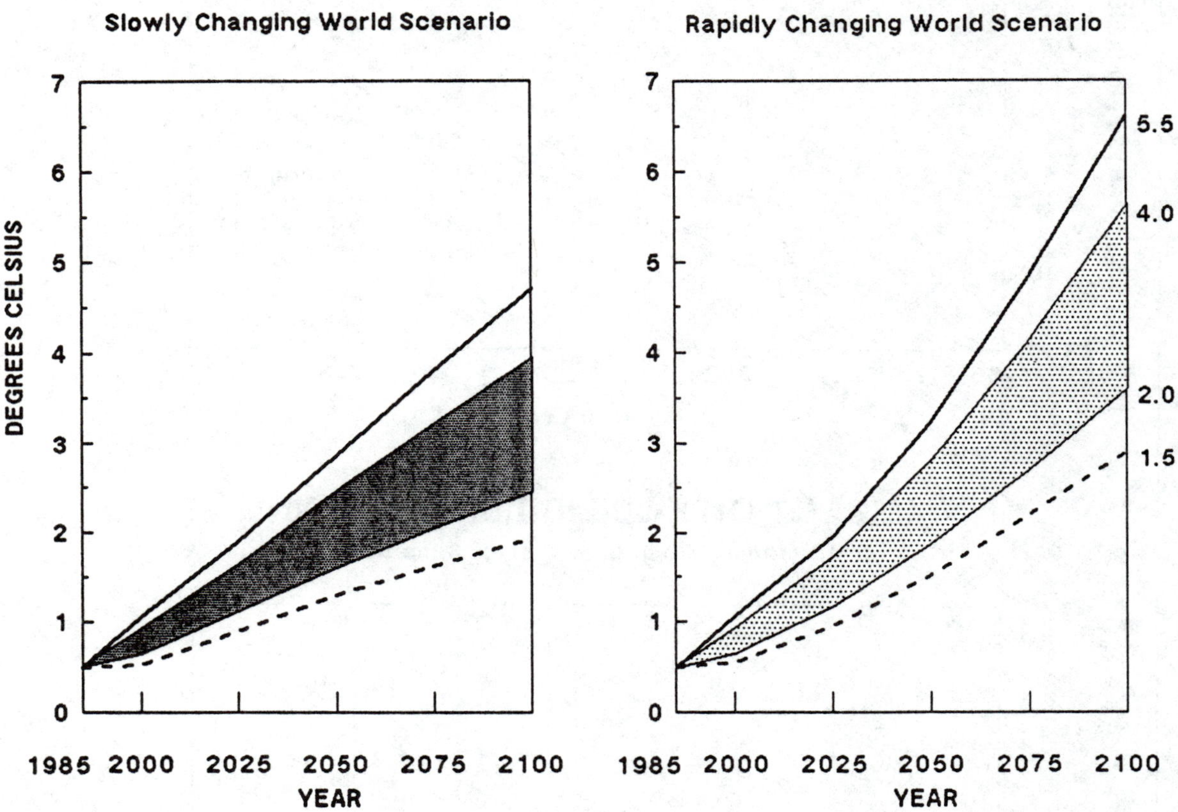

and SCW cases. In the RCW case, the range of realized warming for a 1.5-5.5°C climate sensitivity would be 1.5-3.2°C by 2050 and 2.9-6.6°C by 2100, compared with a range of 1.9-2.8°C by 2050 and 3.6-5.6°C by 2100 when the climate sensitivity is bounded by 2.0-4.0°C. The corresponding values for equilibrium warming for a 1.5-5.5°C climate sensitivity are 2.0-7.4°C by 2050 and 3.6-13.2°C by 2100, compared with 2.7-5.4°C by 2050 and 4.8-9.6°C by 2100 for a 2.0-4.0°C climate sensitivity. In the SCW case, the range of realized warming for a 1.5-5.5°C climate sensitivity would be 1.3-2.9°C by 2050 and 2.0-4.7°C by 2100, compared with a range of 1.6-2.5°C by 2050 and 2.5-4.0°C by 2100 when the range of climate sensitivity is 2.0-4.0°C. The corresponding values for equilibrium warming for a 1.5-5.5°C climate sensitivity are 1.7-6.1°C by 2050 and 2.4-8.6°C by 2100, compared with 2.2-4.5°C by 2050 and 3.1-6.3°C for a 2.0-4.0°C climate sensitivity.

Rate of Heat Diffusion

CO_2 and heat are currently transferred from the atmosphere to the oceans and within the ocean itself as a result of many complex chemical and physical interactions. One of these interactions is the transfer of heat from the mixed layer to the thermocline, thereby delaying global warming. Additionally, changes in ocean mixing and circulation patterns as a result of climate change could alter the capacity of the oceans to absorb heat (see **BIOGEOCHEMICAL FEEDBACKS** for further discussion). The rate at which heat is absorbed only affects the rate of realized warming, not the rate of equilibrium warming, because the oceans cannot absorb heat indefinitely.

In our model the rate at which mixing occurs between the mixed layer and the thermocline is parameterized with an eddy-diffusion coefficient (see Chapter III). The value of the eddy-diffusion coefficient in the base cases was assumed to be 0.55×10^{-4} m^2/sec. For purposes of this sensitivity analysis alternative values of 2×10^{-5} and 2×10^{-4} have been evaluated.

As shown in Figure 6-16 the rate at which the oceans absorb heat can noticeably affect the amount of realized warming by 2100. If the rate of heat absorption is greater than that assumed in the base cases (i.e., if the eddy-diffusion coefficient is 2 x 10^{-4} m^2/sec), realized warming by 2100 would be 0.5-1.2°C less than in the RCW case (assuming 2.0-4.0°C climate sensitivities). For the smaller eddy-diffusion coefficient of 2 x 10^{-5} m^2/sec, realized warming by 2100 would be 0.3-0.9°C higher.

ASSUMPTIONS ABOUT ATMOSPHERIC CHEMISTRY: A COMPARISON OF MODELS

As discussed in Chapters II and III, the chemistry of the future troposphere is one of the uncertainties in the prediction of atmospheric composition. The principal factors contributing to this uncertainty are: (1) the complexity and tremendous natural variability of chemistry in the troposphere, especially regarding oxidant formation; (2) the range of interactions between tropospheric chemistry and radiation perturbed by climate change and changes in stratospheric composition; and (3) the range of uncertainties in future emissions of CH_4, CO, NO_x, and non-methane hydrocarbons (NMHC). This section focuses on the first two aspects of uncertainty in atmospheric composition.

Recognizing the uncertainty in tropospheric chemistry, EPA sponsored a workshop on atmospheric composition to discuss recent modelling efforts among members of the atmospheric sciences community and to construct a parameterized atmospheric chemistry model that would incorporate the latest scientific findings. The end result was the Assessment Model for Atmospheric Composition (AMAC), the model used to obtain the findings discussed in this report. AMAC was developed by Prather of NASA/GISS as a result of the workshop, which was held in January 1988 (see Prather, 1988). To obtain insight into the uncertainties introduced by the physical simplifications made by the AMAC and to ensure results that are comparable to current, more-detailed modeling

FIGURE 6-16

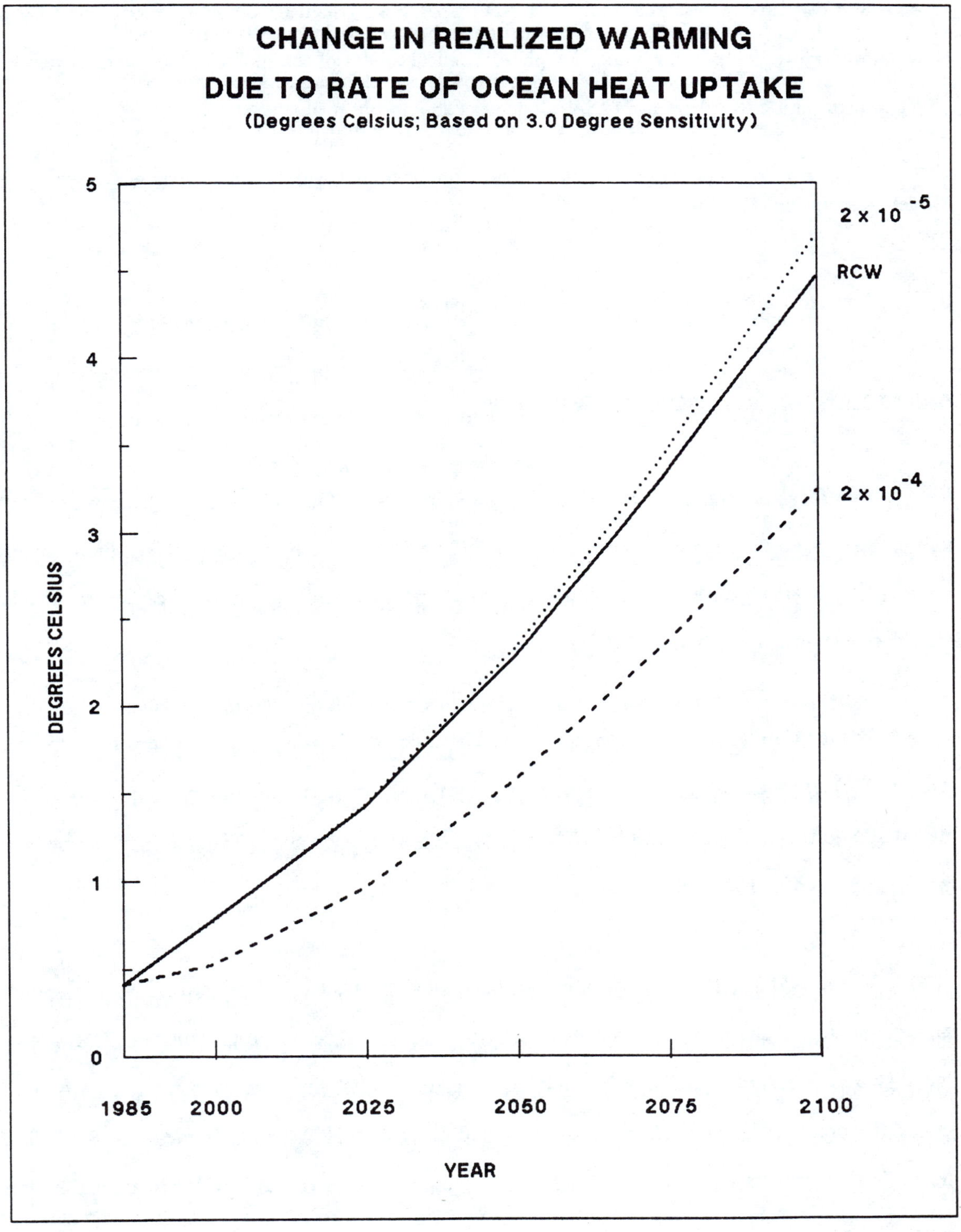
CHANGE IN REALIZED WARMING
DUE TO RATE OF OCEAN HEAT UPTAKE
(Degrees Celsius; Based on 3.0 Degree Sensitivity)
DEGREES CELSIUS
5
4
3
2
1
0
1985
2000
2025
2050
2075
2100
YEAR
2 x 10^-5
RCW
2 x 10^-4

efforts, a set of common scenarios of CH_4, CO, and NO_x emissions were analyzed in the AMAC, as well as in two current research models: a 2-D tropospheric chemistry model developed by Isaksen (Isaksen and Hov, 1987); and a multi-box photochemical model of the global troposphere developed by Thompson and co-workers at NASA/Goddard (Thompson et al., 1988).

Model Descriptions

Each of these models is briefly described below.

Assessment Model for Atmospheric Composition

The focus of interest in tropospheric composition is on O_3, CH_4, and OH, because the two former gases are key greenhouse absorbers and OH (together with ozone) determines the oxidizing capacity of the atmosphere and the abundance of many gases such as methane, carbon monoxide, methyl chloroform, and HCFC-22 (CHF_2Cl).

For the simulation of the troposphere in this model, the Northern and Southern Hemispheres (NH & SH) are treated separately because significant asymmetries are observed in many of the important shorter-lived gases, such as CO and NO_x. These species play a major role in the budgets for CH_4, O_3, and OH in each hemisphere.

In the AMAC, tropospheric OH can be treated as a steady-state variable as it responds immediately to the annual average values of the trace gases. To derive perturbations to OH, a non-linear system is solved equating a "production" term to a "loss" term. OH losses are partitioned among the predicted gases (CH_4, CO), the specified fluxes (NMHC), and self-reactions (OH). The production side of the equation includes a positive response to increases in UV radiation (i.e., loss in column ozone) and in tropospheric H_2O, O_3, and NO_x fluxes. Coefficients for variations in either the production or loss terms with respect to column O_3, tropospheric water vapor, trop-O_3, CO, CH_4, and fluxes of both NMHCs and NO_x are based on results from 1-D and 2-D models (Liu et al., 1987; Thompson and Cicerone, 1986; Isaksen and Hov, 1987; Isaksen et al., 1988). Major sources of uncertainty in calculating OH are the spatial averaging for this highly variable constituent and the nonlinearity in perturbation coefficients, especially with respect to NO_x distribution.

Perturbations to tropospheric ozone affect both tropospheric temperatures and the long-lived source gases controlled by OH. A significant fraction of tropospheric ozone originates in the stratosphere and is destroyed by surface deposition; it is sufficiently short-lived (a few months) that the AMAC calculates ozone perturbations separately for each hemisphere. Changes in tropospheric ozone are associated with perturbations to the total ozone column, and to tropospheric chemical reactions, which are evaluated with sensitivity coefficients, $d\ln(O_3)/d\ln(X)$, ascribed to the precursor gases (Column -O_3, 0.8; CH_4, O.2; CO, O.1; NO_x flux, O.1; NMHC flux, O.1). The coefficients are based on detailed photochemical models for typical tropospheric air parcels (Liu et al., 1987; Thompson et al., 1988), but their uncertainties are large, approximately a factor of 2. Also, the efficiency of O_3 production varies widely with the NO_x levels (Liu et al., 1987), which in turn cannot be adequately characterized throughout the entire troposphere due to their large dynamic range. A similar concern applies to the simplified treatment of non-methane hydrocarbons.

Isaksen Model

The Isaksen model is a 2-D transport model that calculates absolute concentrations for O_3 and OH (and several dozen other trace chemical constituents in the troposphere) as functions of altitude and latitude, as emissions are varied over time (Isaksen and Hov, 1987; Isaksen et al., 1988). Unlike the AMAC, this model resolves latitudinal and altitude distributions, and emission changes are introduced with latitudinal discrimination. The transport of longer-lived constituents can locate key areas of tropospheric ozone and OH change that the AMAC will miss; because a 2-D model resolves altitude, the effects of high-altitude aircraft emissions on NO_x and ozone or cloud perturbations to radiation fields, for example, can be explored. The Isaksen model differs from the AMAC in that the troposphere is not coupled to the stratosphere, so that the impact of changing climate or perturbations on stratospheric ozone are not included. Methane flux changes are included in annual updates of the model.

Thompson et al. Model

The Thompson model couples the result of 1-D model calculations of the time history for eight chemically coherent global regions, which are then averaged to estimate net global changes. A steady-state method is used: emissions are specified in simulations to represent conditions at 5-year intervals. This is somewhat inadequate for lifetime changes, tending to underestimate tropospheric ozone increases (up to 30% in one test case) and to overestimate increases in CH_4 concentrations.

The description of chemically coherent regions offers insight into regional variability, a feature lacking in the version of the Isaksen model used here, which does not have the longitudinal variation needed to treat emissions that are restricted to the source area but that can have extensive effects on ozone and OH. Like the Isaksen model, the Thompson model includes a more complete set of chemical constituents than does the AMAC and can identify other effects and interactions of climate perturbation. For example, the oxidants that contribute to sulfuric acid formation in clouds and rain (HO_2 and H_2O_2) are very sensitive to changes in stratospheric ozone and tropospheric water vapor. At the 2035 year point, stratospheric ozone depletion and climate change (temperature and water vapor) effects are added to calculations with perturbed CH_4, CO, and NO_x emissions.

Results from the Common Scenarios

It is not easy to compare the models because the structure, input, and derived quantities from the three models are not treated comparably. Nevertheless, insights into uncertainties can be obtained by comparing selected results from each model. EPA supplied eight scenarios of alternative estimates of CH_4, CO, and NO_x for evaluation in each model. In this section one of these scenarios is discussed (EPA Scenario #2), which assumes low CH_4, low CO, and high NO_x growth in sources, a rapid growth scenario for CO_2 and N_2O from combustion, and a CFC and halon scenario consistent with the Montreal Protocol. Table 6-4 summarizes the emission estimates for this scenario and compares them to estimates from the Rapidly Changing World (RCW) and Slowly Changing World (SCW) scenarios (Appendix C provides more detail for all eight scenarios). The RCW and SCW cases could not be explicitly included for this model comparison because the development of these cases occurred simultaneously with the model comparison. Table 6-5 summarizes the results for all eight scenarios.

The EPA #2 emission estimates are in the same range as those of the other two cases except for CO emissions. These estimates are much lower than both the RCW and SCW cases and are similar to the Stabilizing Policy cases due to stringent control assumptions on transport vehicle emissions. For the other emissions, the CO_2 emission estimates in EPA #2 fall between the RCW and SCW cases for most of the time periods, approaching the RCW estimates by 2100. The CH_4 and NO_x estimates are similar to those for the SCW case, except that NO_x estimates after 2050 fall between the RCW and SCW estimates.

The AMAC's troposphere is basically a parameterized 2-box model: it reports mean tropospheric values (ppb) for CH_4, and separate perturbations (% change) to OH and O_3 in each hemisphere. For the global average perturbation to OH and O_3, Northern and Southern Hemisphere

TABLE 6-4

Comparison of Emission Estimates For EPA #2, RCW, and SCW cases
(in teragrams, unless indicated otherwise)

	Emissions Estimates by Year					
Trace Gas	1985	2000	2025	2050	2075	2100
CO_2 (Pg C)						
EPA #2	6.3	7.3	10.4	13.7	17.9	25.2
RCW	5.9	7.6	11.5	16.6	22.2	25.5
SCW	5.9	7.2	9.2	9.8	9.7	11.4
CO (as C)						
EPA #2	316	226	194	192	178	192
RCW	502	571	699	895	1050	1207
SCW	502	616	842	859	595	604
CH_4						
EPA #2	500	548	640	721	779	809
RCW	510	577	712	880	1025	1089
SCW	510	569	676	740	773	816
NO_x (as N)						
EPA #2	59	59	64	72	82	104
RCW	53	60	73	92	108	118
SCW	53	59	69	71	65	70

TABLE 6-5

Comparison of Results From Atmospheric Chemistry Models for the Year 2050 Compared to 1985

Model	Test Case Results			
	LOW CH_4		HIGH CH_4	
	Low CO	High CO	Low CO	High CO
Increases in Methane (ppb)				
Low NO_x	EPA#1	EPA#3	EPA#5	EPA#7
Prather	806	1031	1750	2022
Isaksen	400	720	950	1220
Thompson et al.	750	1000	1710	1890
High NO_x	EPA#2	EPA#4	EPA#6	EPA#8
Prather	801	1048	2082	2242
Isaksen	350	550	1010	1200
Thompson et al.	870	1240	2210	2560
Percent Change in CO				
Low NO_x	EPA#1	EPA#3	EPA#5	EPA#7
Prather	16	43	55	82
Isaksen	-13	8	0	17
Thompson et al.	-10	25	16	52
High NO_x	EPA#2	EPA#4	EPA#6	EPA#8
Prather	17	44	70	99
Isaksen	-9	8	8	19
Thompson et al.	-2.4	58	35	97

TABLE 6-5 (continued)

Comparison of Results From Atmospheric Chemistry Models for the Year 2050 Compared to 1985

Model	Test Case Results			
	LOW CH_4		HIGH CH_4	
	Low CO	High CO	Low CO	High CO
Percent Change in OH				
Low NO_x	EPA#1	EPA#3	EPA#5	EPA#7
Prather	-9	-14	-23	-26
Isaksen	5	- 1	1	- 2
Thompson et al.	-9.4	-15.5	-20.1	-27.6
High NO_x	EPA#2	EPA#4	EPA#6	EPA#8
Prather	- 2	- 9	-22	-25
Isaksen	8	4	5	3
Thompson et al.	- 8.9	-17.4	-22.1	-23.9
Percent Change in O_3				
Low NO_x	EPA#1	EPA#3	EPA#5	EPA#7
Prather	1	8	21	27
Isaksen	-1	2	3	5
Thompson et al.	3.7	9.9	13.2	18.3
High NO_x	EPA#2	EPA#4	EPA#6	EPA#8
Prather	5	13	33	39
Isaksen	5	8	10	13
Thompson et al.	10.0	17.5	23.1	29.0

results are averaged with equal weight. In addition to the perturbed species discussed here with the tropospheric chemical models, the AMAC calculated other significant perturbations, such as a 12% decrease in column ozone, a 2°K rise in mean tropospheric temperature, along with a 10% increase in tropospheric water vapor. These perturbations have an impact on tropospheric OH, O_3, CO, and CH_4. Additionally, unlike the other two models, which provide point estimates, the AMAC produces a range of trace-gas scenarios in response to specified uncertainties in the model coefficients (only the mean of each range is provided in Table 6-5).

The Thompson model averages over eight "chemically coherent regions" (Appendix C gives a description of how the EPA scenarios were assigned to the regions). This approach is probably adequate for short-lived species such as OH, and possibly for tropospheric O_3. However, it makes it difficult to interpret CH_4 calculations, which predict different CH_4 concentrations among the boxes, when in fact the long lifetime of CH_4 ensures that it is well mixed throughout the troposphere. The methane results in Table 6-5 have been averaged over the eight regions and scaled to account for the CH_4 lifetime changes occurring in the perturbed atmosphere. Also summarized are percent changes in CO (surface mixing ratios) and OH and O_3 (column-integrated from 0-15 km). The CH_4 and CO changes obtained by the Thompson model are very similar to those obtained with the AMAC. Although not shown in the global averages in Table 6-5, the most useful results of the regional calculations are localized estimates of OH and O_3 changes in each chemically defined region where CO and NO_x growth rates may differ considerably. The differences between areas with controlled emissions (Urban 1) and without controlled emissions (Urban 2) are very striking (Figure 6-17).

The Isaksen model calculates perturbations as a function of latitude, altitude (0-16 km), and time of year. The increase in CH_4 is distributed uniformly throughout the troposphere as expected. There is a problem with the implementation of the EPA #2 scenario in that the CH_4 concentrations

FIGURE 6-17

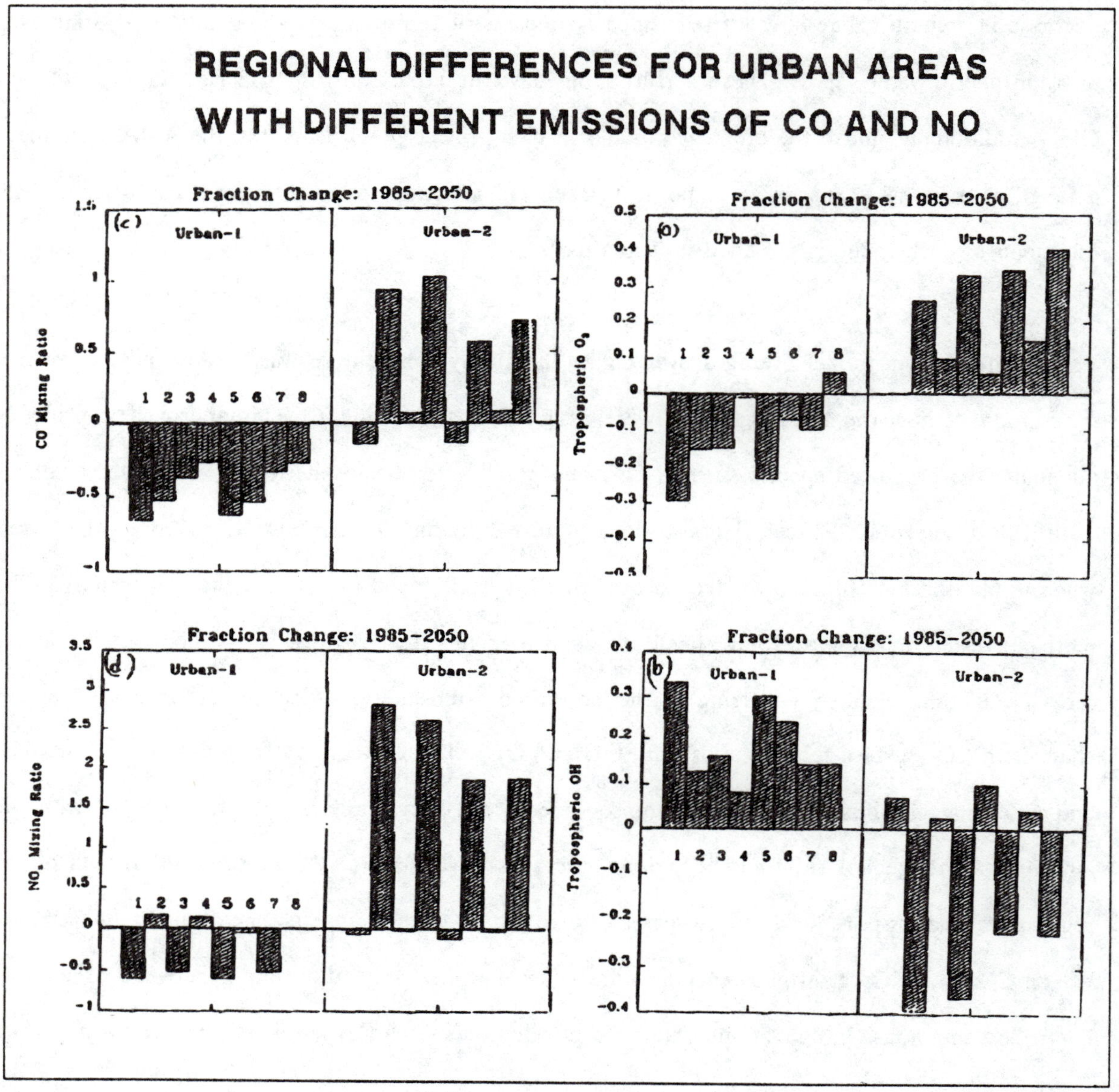

Source: Thompson, 1988.

decline at the beginning of the model integration. This may be due to the low estimate of global CO flux. Initial fluxes were scaled in the AMAC and Thompson models to obtain a steady-state of current concentrations. CH_4 does not recover to its initial concentration for at least 20 years into the scenario, and this is probably the reason for the Isaksen model predicting such a small increase in CH_4. The patterns for OH and O_3 perturbations are distinct (Figure 6-18). The greatest changes in O_3 are below 2 km altitude: there is a large increase between 0° and 35°N and a small decrease centered at 50°N. The spatial pattern of changes in OH are interesting: in the upper troposphere between 12 and 16 km the OH increases by 10-30% in the Northern Hemisphere, whereas throughout most of the Southern Hemisphere OH decreases. Both of these changes may be driven by increases in CH_4. In the dry upper troposphere in the presence of NO_x, CH_4 increases the OH concentration during its atmospheric oxidation, but in the lower troposphere the CH_4 provides merely a sink for OH.

Overall, all three models predict similar increases in tropospheric O_3. The Thompson et al. and AMAC models predict decreases in tropospheric OH, while the Isaksen model reports a globally averaged increase. This discrepancy may be explained by the large increases in OH above 12 km as noted above, something that is also calculated by the Thompson model. However, most of the difference in OH levels seems attributable to the lower CO and CH_4 concentrations calculated by Isaksen compared with the other two models. As shown in Table 6-5 for all eight scenarios, none of the increases in CO by 2050 are more than 15-20% in Isaksen, whereas Thompson et al. and AMAC show CO increases up to 100% (see scenario #8). Some of the CO and OH differences between Isaksen et al. and the other two models are due to the difference in initialization described above, but most of the OH difference may be due to how CO behaves in each model. This may be one of the more prominent uncertainties in predicting future tropospheric composition. CO has a moderate lifetime (typically about a few months) with considerable spatial variability that is not well

FIGURE 6-18

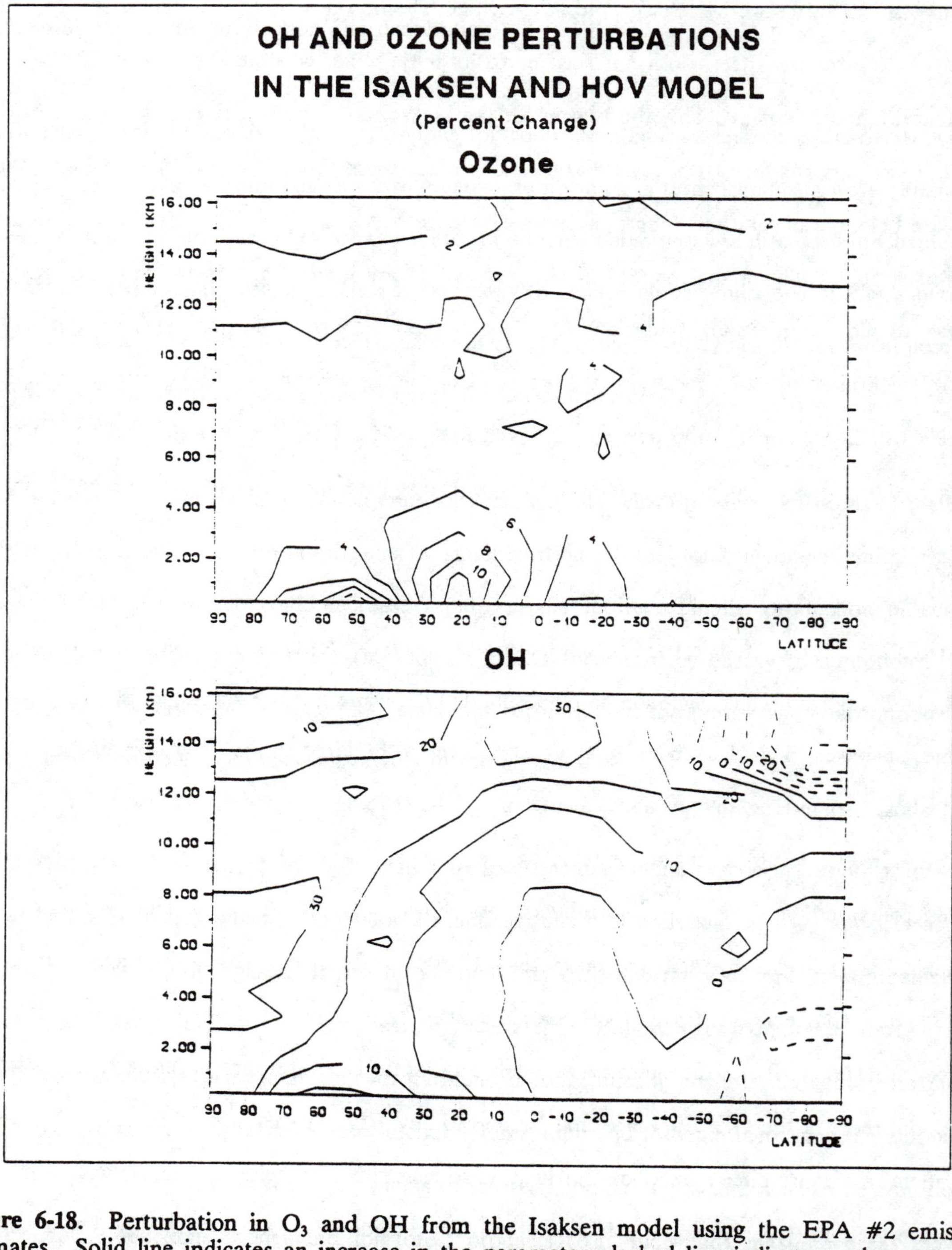

Figure 6-18. Perturbation in O_3 and OH from the Isaksen model using the EPA #2 emission estimates. Solid line indicates an increase in the parameter; dashed line indicates a decrease. O_3 shows large increase between 0° and 35°N; OH shows increases up to 30% in Northern Hemisphere, and decreases in Southern Hemisphere. Source: Isaksen and Hov, 1987.

resolved in any of the models. Perhaps the Isaksen model gives a lower limit to CO and OH changes, and the other two models estimate the largest expected changes.

EVALUATION OF UNCERTAINTIES USING AMAC

Comparing the results of AMAC to other models given identical scenarios provides one approach to evaluating uncertainties related to atmospheric chemistry. Valuable information can also be obtained by testing the robustness of the AMAC results to changes in critical model parameters. This section examines these impacts by varying key parameters within AMAC and then comparing the results to the RCW scenario.

Atmospheric Lifetime of CFC-11

The assumed atmospheric lifetime for CFC-11 in the AMAC for the RCW case was 65 years. Its atmospheric lifetime, however, may range from 55 to 75 years (Prather 1988); these estimates were evaluated to determine the impact on atmospheric chemistry. The changes in atmospheric concentration for CFC-11 are summarized in Figure 6-19, which indicates that concentration levels may vary from about 650 to 810 ppt by 2100. Increases (decreases) in the atmospheric concentration of CFC-11, however, tend to be offset by corresponding decreases (increases) in atmospheric concentrations of other trace gases, such as other CFCs and CH_4. That is, the increase (decrease) in the lifetime of CFC-11 increases (decreases) the amount of stratospheric ozone depletion, which increases (decreases) the amount of UV radiation; these higher (lower) UV levels increase (decrease) the rate of destruction of these other gases. As a result, the impacts on global warming are negligible (less than 0.1°C).

FIGURE 6-19

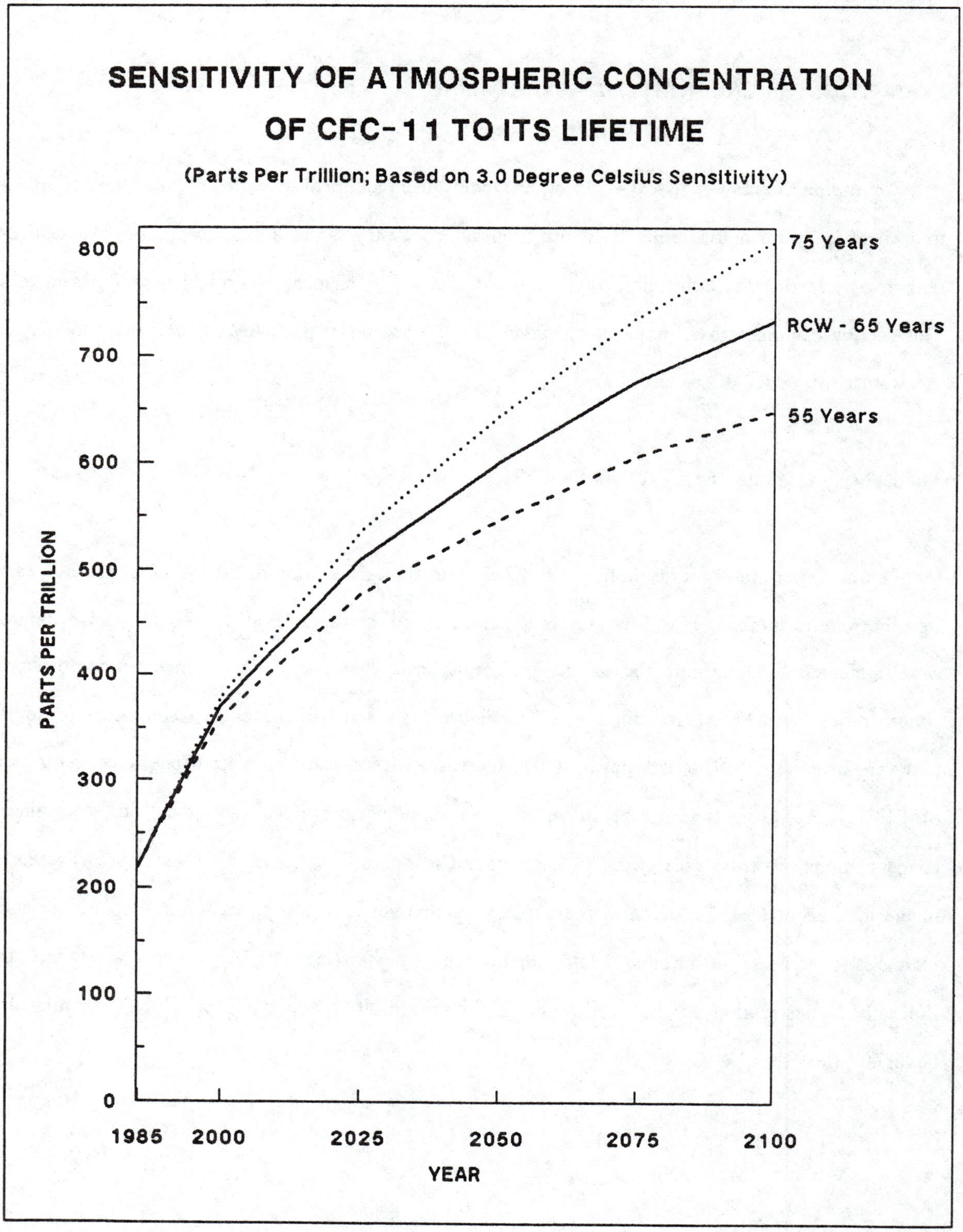
SENSITIVITY OF ATMOSPHERIC CONCENTRATION
OF CFC-11 TO ITS LIFETIME
(Parts Per Trillion; Based on 3.0 Degree Celsius Sensitivity)
75 Years
RCW - 65 Years
55 Years
PARTS PER TRILLION
800
700
600
500
400
300
200
100
0
1985
2000
2025
2050
2075
2100
YEAR

Interaction of Chlorine with Column Ozone

Chlorine in the stratosphere has a negative, non-linear impact on total column ozone. This chemical interaction is one of the primary causes of stratospheric ozone depletion due to the chlorine contained in CFCs; this interaction has been included in the AMAC, however, primarily for its ability to affect the rate of tropospheric ozone formation. In the RCW case this relationship was defined as a 0.03% decline in total column ozone/$(ppb)^2$ of stratospheric chlorine. A higher value, 0.20%, was evaluated, which would increase the rate of column ozone destruction.

With the 0.20% assumption, total column ozone depletion was 45-47% by 2050 (assuming 2.0-4.0°C climate sensitivities) compared with a total column ozone depletion of 16.8% with the lower value (i.e., the -0.03% value used in the RCW case). The increase in total column ozone depletion has a positive feedback on the tropospheric OH levels due to the increase in UV radiation. The resulting OH interactions with other trace gases substantially reduces the atmospheric concentration of CH_4, HCFC-22, methyl chloroform, and methyl chloride, and reduces the rate of tropospheric ozone formation. (The role of O_3 is problematic, O_3 at 10-12 km probably would increase. At this altitude, O_3 probably has the largest greenhouse effect. See Chapter II) These impacts reduce the amount of global warming; as shown in Figure 6-20, the decline in realized warming is 0.1°C by 2050, compared with the RCW case, and 0.3-0.4°C by 2100; the decline in equilibrium warming by 2100 is 0.4-0.8°C (assuming 2.0-4.0°C climate sensitivities).

Sensitivity of Tropospheric Ozone to CH_4 Abundance

Tropospheric ozone formation is affected by the amount of CH_4 present, although the rate at which tropospheric ozone forms as a result of CH_4 abundance is subject to some uncertainty. In the RCW case, this variable for the Northern Hemisphere was assumed to be a 0.2% change in

FIGURE 6-20

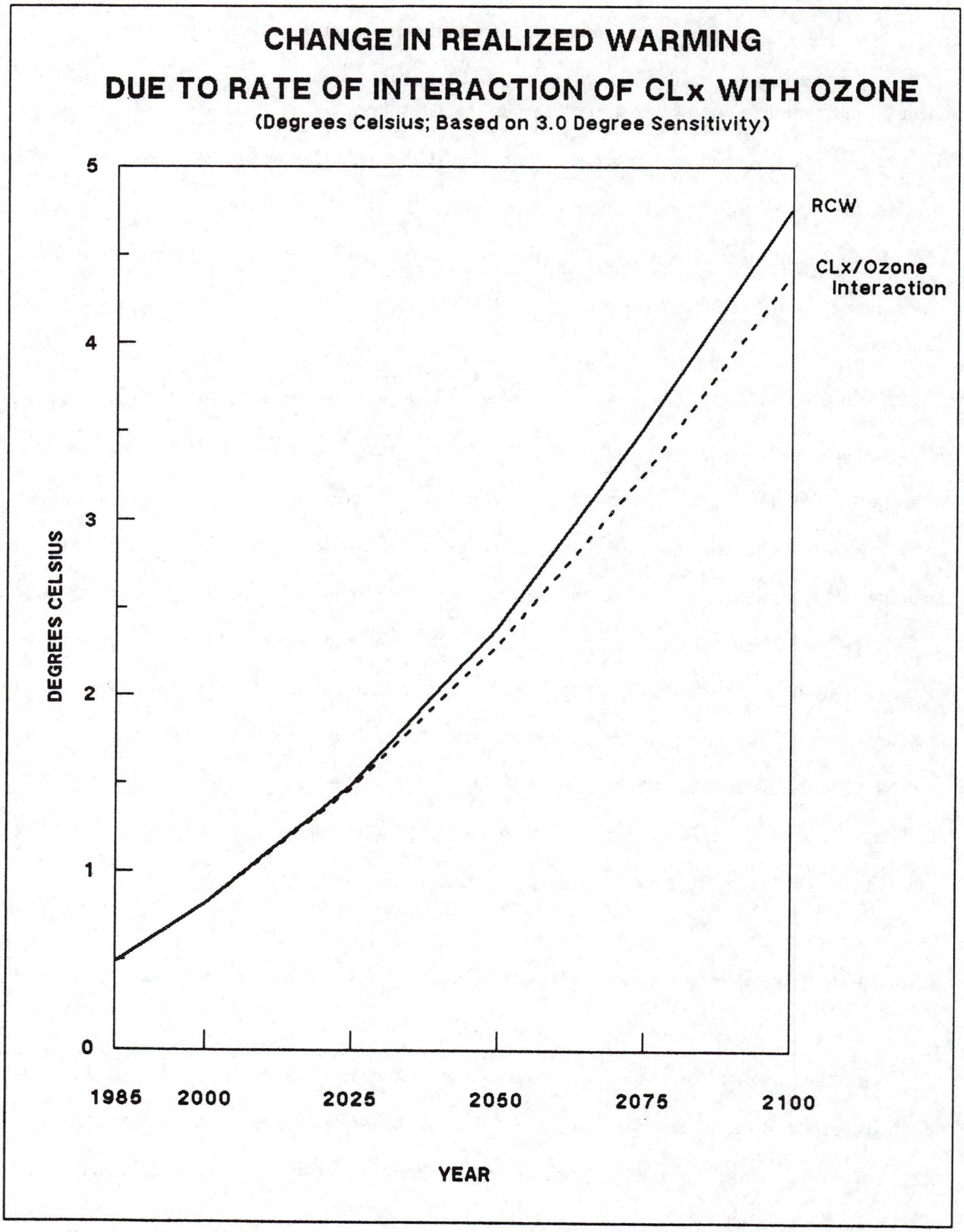
CHANGE IN REALIZED WARMING
DUE TO RATE OF INTERACTION OF CLx WITH OZONE
(Degrees Celsius; Based on 3.0 Degree Sensitivity)
DEGREES CELSIUS
0
1
2
3
4
5
RCW
CLx/Ozone Interaction
1985
2000
2025
2050
2075
2100
YEAR

tropospheric ozone for each percentage change in CH_4 concentration; other evidence suggests that a higher value, 0.4%, is possible (Prather, 1988).

Using this higher value increases the change in tropospheric ozone in 2100 by about 50% over the RCW case (tropospheric ozone increases by about 63% compared with about 43% when a value of 0.2% is assumed). The increase in tropospheric ozone indirectly results in a decrease in CH_4 concentrations since the tropospheric ozone increase also increases OH formation, which destroys CH_4. Due to this partially offsetting effect, the increase in global warming is less than 0.1°C.

Sensitivity of OH to NO_x

Tropospheric OH formation is affected by the level of NO_x emissions, although the rate of OH formation is uncertain. In the RCW case, we assumed a 0.1% OH change for every 1.0% change in NO_x emissions for the Northern Hemisphere. We evaluated a range of uncertainty from 0.05% to 0.2%.

An increase (decrease) in the amount of OH due to a higher (lower) sensitivity to NO_x emissions results in less (more) tropospheric ozone formation as well as lower (higher) levels of CO and CH_4. The higher sensitivity value of 0.2% reduces realized warming about 0.1°C by 2100 compared with the RCW case (assuming 2.0-4.0°C climate sensitivities; equilibrium warming is about 0.2°C lower by 2100), while the lower sensitivity value of 0.05% increases realized warming less than 0.1°C by 2100 (equilibrium warming increases a maximum of 0.1°C by 2100).

BIOGEOCHEMICAL FEEDBACKS

The sensitivity of the climate system to anthropogenic perturbations is determined by a combination of feedbacks that amplify or dampen the direct radiative effects of increasing concentrations of greenhouse gases. Several important internal climate feedbacks, such as those resulting from changes in water vapor, clouds, and sea ice albedo, are included in the estimates of climate sensitivity discussed throughout this Report. There are a number of feedbacks of a biogeochemical origin, however, that may also play an important role in climatic change that were not included in the analyses on which this range is based. Biogeochemical sources of feedback include releases of methane hydrates; changes in ocean chemistry, biology, and circulation; and changes in the albedo of the global vegetation.

Any attempt to quantify the impact of biogeochemical feedbacks is necessarily quite speculative at this time; however, it does appear that they could have an important impact on global climate. For example, Lashof (1989) has estimated that the gain from biogeochemical feedbacks ranges from 0.05-0.29 compared with a 0.17-0.77 gain from internal climate feedbacks. (The gain is defined as the portion of global equilibrium temperature change attributable to the feedback divided by the total global equilibrium temperature when the feedback is included). Some of these key feedbacks were incorporated into the AMAC for these sensitivity cases to determine the magnitude of their impact on global warming.

Ocean Circulation

As mentioned above, the oceans are currently a major sink for heat and CO_2. Concerns have been raised, however, that the basic circulation patterns that allow these processes to continue could be significantly altered as the global climate changes. This possibility is suggested by the rapid rate of atmospheric CO_2 change during past periods of climate change (e.g., see Chapter III). If circulation patterns did change, it is plausible that the oceans would no longer be a net sink for heat and CO_2.

It is not known at what point ocean circulation would be altered. For this analysis we assumed that a 2°C increase in realized warming would alter ocean circulation patterns sufficiently to shut off net uptake of CO_2 and heat by the oceans. This would increase atmospheric CO_2 concentrations from 10-25% by 2100, and would reduce the difference between realized and equilibrium warming as the atmosphere warmed more quickly due to the oceans' inability to continue to act as a heat sink. As shown in Figure 6-21, this feedback is sufficient to increase realized warming up to 1.4°C by 2050 and 1.3-3.5°C by 2100 compared with the warming estimated for the RCW case.

Methane Feedbacks

Increases in global temperature could increase the amount of CH_4 emissions due to several feedback processes: (1) release of methane from hydrates, which are methane compounds contained in continental slope sediments, as increasing temperatures destabilize the formations; (2) additional methane from high-latitude bogs due to longer growing seasons and higher temperatures; and (3) increased rate of methanogenesis from rice cultivation. The amount of CH_4 that could be released from each of these feedback processes, and the rate at which any releases might occur, are highly speculative. For each process we have assumed that the rate of CH_4 release is linearly related to the increase in temperature, with each 1°C increase leading to an additional 110 Tg from methane hydrates, 12 Tg from bogs, and 7 Tg from rice cultivation (Lashof, 1989). These methane feedbacks could have a major impact on atmospheric CH_4 concentrations: by 2100 concentrations would increase to about 7000-8350 ppb, compared with 4150-4400 ppb in the RCW case. As shown in

FIGURE 6-21

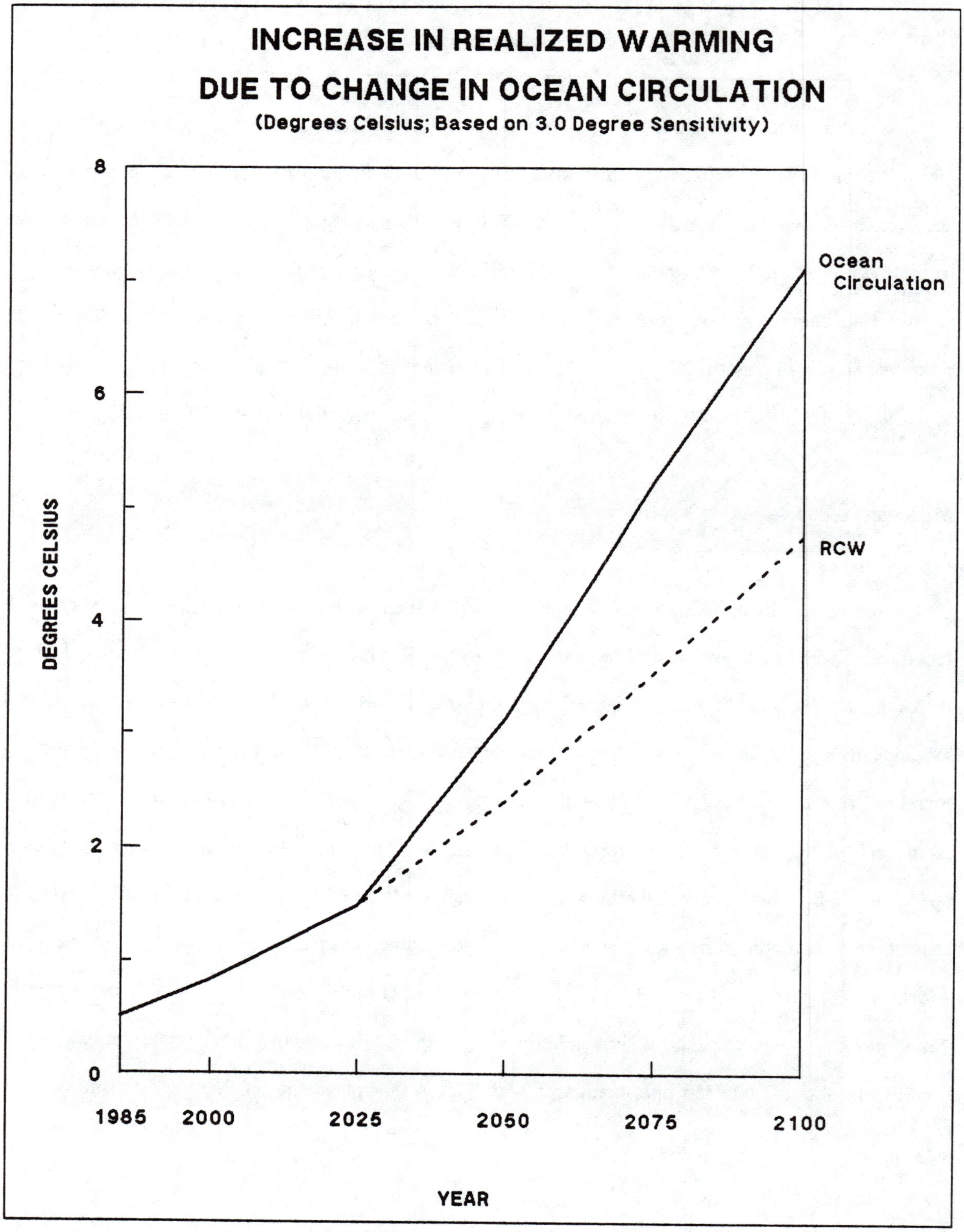
INCREASE IN REALIZED WARMING
DUE TO CHANGE IN OCEAN CIRCULATION
(Degrees Celsius; Based on 3.0 Degree Sensitivity)
DEGREES CELSIUS
0
2
4
6
8
1985
2000
2025
2050
2075
2100
YEAR
Ocean Circulation
RCW

Figure 6-22, this increase in CH_4 would be sufficient to increase realized warming relative to the RCW case about 0.2-0.4°C by 2050 and 0.4-1.0°C by 2100 (assuming 2.0-4.0°C climate sensitivities).

Combined Feedbacks

In addition to the two separate feedbacks discussed above, we analyzed the combined impact of several types of biogeochemical feedbacks. The following specific feedbacks were included: (1) methane from hydrates, bogs, and rice cultivation, as previously discussed; (2) increased stability of the thermocline, thereby slowing the rate of heat and CO_2 uptake by the deep ocean by 30% due to less mixing; (3) vegetation albedo, which is a decrease in global albedo as a result of changes in the distribution of terrestrial ecosystems by 0.06% per 1°C warming; (4) disruption of existing ecosystems, resulting in transient reductions in biomass and soil carbon at the rate of 0.5 Pg C per year per 1°C warming; and (5) CO_2 fertilization, which is an increase in the amount of carbon stored in the biosphere in response to higher CO_2 concentrations at the rate of 0.3 Pg C per ppm. See Lashof (1989) for further discussion.

The combined impact of these feedbacks on realized warming is an increase of 0.4-0.9°C by 2050 and 0.8-2.5°C by 2100 relative to the RCW case (assuming 2.0-4.0°C climate sensitivities; see Figure 6-23); the increase in equilibrium warming is 0.4-1.7°C by 2050 and 0.7-3.2°C by 2100. These preliminary analyses strongly suggest that biogeochemical feedbacks could have a major impact on the rate of climatic change during the next century.

FIGURE 6-22

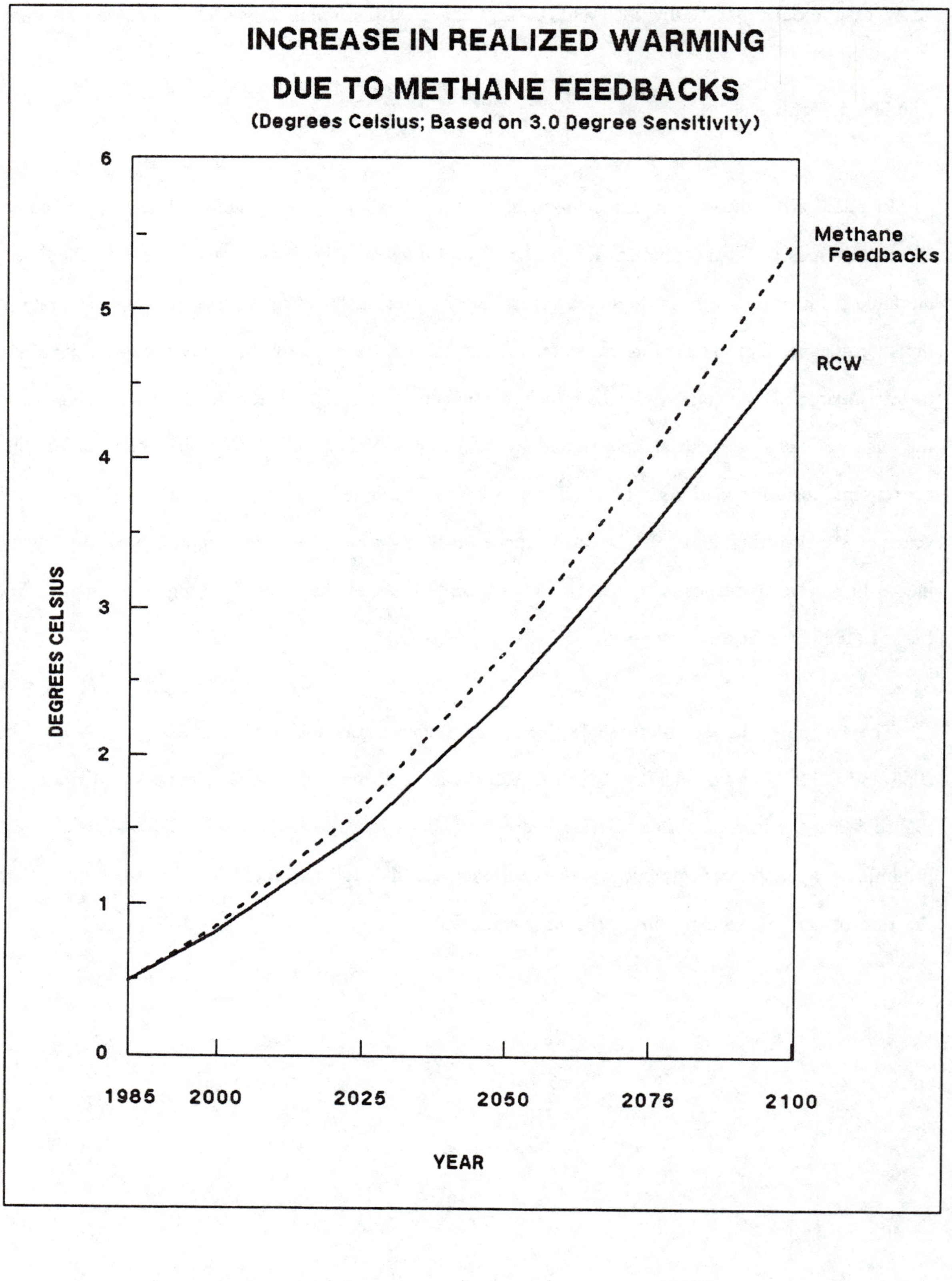
INCREASE IN REALIZED WARMING
DUE TO METHANE FEEDBACKS
(Degrees Celsius; Based on 3.0 Degree Sensitivity)
DEGREES CELSIUS
0
1
2
3
4
5
6
1985
2000
2025
2050
2075
2100
YEAR
Methane Feedbacks
RCW

FIGURE 6-23

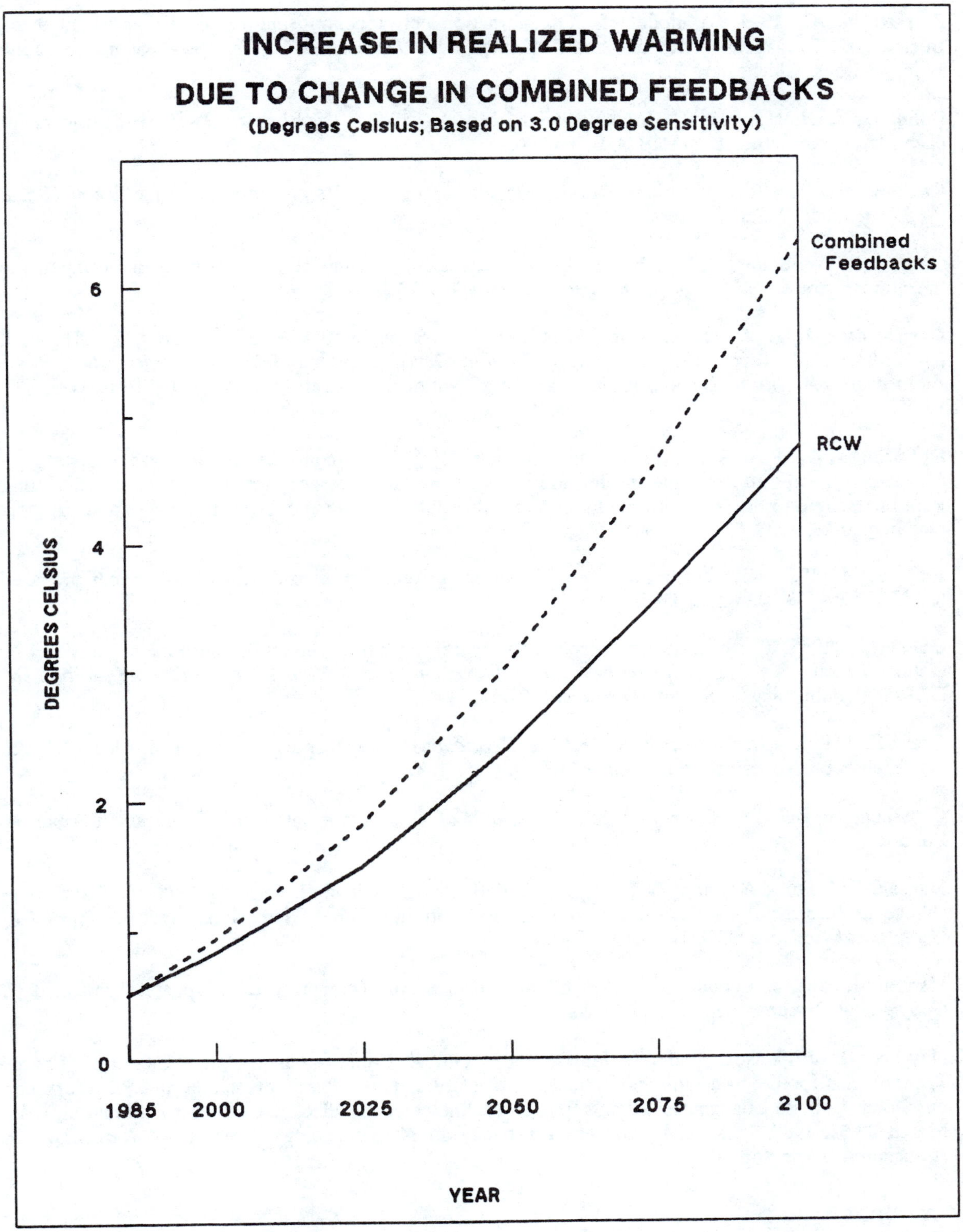
INCREASE IN REALIZED WARMING
DUE TO CHANGE IN COMBINED FEEDBACKS
(Degrees Celsius; Based on 3.0 Degree Sensitivity)
Combined Feedbacks
RCW
DEGREES CELSIUS
6
4
2
0
1985
2000
2025
2050
2075
2100
YEAR

REFERENCES

Bjorkstrom, A. 1979. A model of CO_2 interaction between atmosphere, oceans, and land biota. In Bolin, B., E. Dagens, S. Kempe, and P. Katner, eds. *The Global Carbon Cycle*. Scope 13. John Wiley & Sons, Chichester. 403-457.

Bolin, B., A. Björkström, K. Holmen, and B. Moore. 1983. The simultaneous use of tracers for ocean circulation studies. *Tellus* 35B:206-236.

Cicerone, R.J. and R.S. Oremland. 1989. Biogeochemical aspects of atmospheric methane. *Global Biogeochemical Cycles*, in press.

Conrad, R., W. Seiler, and G. Bunse. 1983. Factors influencing the loss of fertilizer nitrogen into the atmosphere as N_2O. *Geophysical Research Letters* 9:1221-1224.

Darmstadter, J., L. Ayres, R. Ayres, W. Clark, P. Crossan, P. Crutzen, T. Graedel, R. McGill, J. Richards, and J. Tarr. 1987. *Impacts of World Development on Selected Characteristics of the Atmosphere: An Integrative Approach*. Oak Ridge National Laboratory, Oak Ridge, Tennessee, Vol. 1.

DeLuchi, M.A., R.A. Johnston and D. Sperling 1988. Methanol vs. natural gas vehicles: A comparison of resource supply, performance, emissions, fuel storage, safety, costs, and transitions. Paper presented to the International Fuels and Lubricants Meeting and Exposition. Portland, Oregon. October 10-13, 1988.

Eichner, M. 1988. Current knowledge of fertilizer-derived nitrous oxide emissions. Paper prepared for U.S. EPA, Washington, D.C.

Galbally, I. 1985. The emission of nitrogen to the remote atmosphere: Background paper. In J.N. Galloway et al. (eds.). *The Biogeochemical Cycling of Sulfur and Nitrogen in the Remote Atmosphere*. D. Reidel Publishing Company, Dordrecht. 27-53.

Gray, C. 1987. *The Research Behind the Regulations*. Paper presented at The Alcohol Week Conference on New Fuels for Cleaner Air. July 16, 1987.

Grossling, B., and D. Nielsen. 1985. *In Search of Oil*. Financial Times Business Information, London.

Hansen, J., I. Fung, A. Lacis, D. Rind, S. Lebedeff, R. Ruedy, and G. Russell. 1988. Global climate change as forecasts by Goddard Institute of Space Studies Three-Dimensional Model. *Journal of Geophysical Research* 93:9341-9364.

Hansen, J., and S. Lebedeff. 1988. Global Surface Air Temperatures: Update through 1987. *Geophysical Research Letters* 15:323-326.

Hay, N., P. Wilkinson, and W. James. 1988. Global Climate Change and Emerging Energy Technologies for Electric Utilities: The Role of Natural Gas. Paper Presented at EPA Workshop on Global Climate Change: Emerging Energy Technologies for Electric Utilities. Washington, D.C., March 30-31, 1988. Session II: Current Electricity Supply Technology Alternatives. American Gas Association, Arlington, VA.

Houghton, R. 1988. *The Flux of CO_2 Between Atmosphere and Land as a Result of Deforestation and Reforestation from 1850-2100*. The Woods Hole Research Center, Woods Hole. 12 pp.

ICF. 1988. *Documentation of the Demographic, Economic, and Energy Assumptions for the Atmospheric Stabilization Framework*. Fairfax, VA.

Isaksen, I.S.A., and O. Hov. 1987. Calculation of trends in the tropospheric concentration of O_3, OH, CO, CH_4, and NO. *Tellus* 39B:271-285.

Isaksen I., T. Berntsen, and S. Solberg. 1988. Calculated Change in Tropospheric Distributions of Long Lived Primary Trace Species and Secondary Species Resulting from Releases of Pollutants. University of Oslo, Norway, 1988.

Kaplan, W., J. Elkins, C. Kolb, M. McElroy, S. Wofsy, and A. Duran. 1978. Nitrous oxide in fresh water systems: an estimate for the yield of atmospheric N_2O associated with disposal of human waste. *PAGEOPH* 116:423-438.

Lashof, D. 1989. The Dynamic Greenhouse: Feedback Processes That May Influence Future Concentrations of Atmospheric Trace Gases. *Climate Change*, in press.

Liu, S.C., M. Trainer, F.C. Fehsenfeld, D.D. Parrish, E.J. Williams, D.W. Fahley, G. Hubler, and P.C. Murphy. 1987. Ozone production in the rural troposphere and the implications for regional and global ozone distributions. *Journal of Geophysical Research* 92:4194-4207.

Oeschger, H., U. Siegenthaler, U. Schotterer, and A. Gugelmann. 1975. A Box Diffusion Model to study the carbon dioxide exchange in nature. *Tellus* 27:168-192.

Prather, M. 1988. *Conference Publication*. From: Assessment Model for Atmospheric Composition Workshop. NASA, Goddard Institute for Space Studies, New York, January 10-13, 1988.

Rotty, R. 1987. A look at 1983 CO_2 emissions from fossil fuel (with preliminary data for 1984). *Tellus* 39B: 203-208.

Siegenthaler, U. 1983. Uptake of excess CO_2 by an outcrop-diffusion model of the ocean. *Journal of Geophysical Research* 88:3599-3608.

Thompson, A.M., R.W. Stewart, M.A. Owens, and J.A. Herwehe. 1988. Sensitivity of tropospheric oxidants to global chemical and climate change. *Atmospheric Environment*. in press.

Thompson, A.M., and R.J. Cicerone. 1986. Possible perturbations to atmospheric CO, CH_4, and OH. *Journal of Geophysical Research* 91:10853-10864.

U.S. DOE (U.S. Department of Energy). 1988. *Annual Energy Review 1987*. Washington, D.C., 301 pp.

U.S. EPA (U.S. Environmental Protection Agency). 1988. *Regulatory Impact Analysis: Protection of Stratospheric Ozone*. Office of Air and Radiation, EPA, Washington, D.C.

CHAPTER VII

TECHNICAL OPTIONS FOR REDUCING GREENHOUSE GAS EMISSIONS

FINDINGS

- **Many technical changes are feasible at reasonable economic costs, which can reduce sources of greenhouse emissions. No single technology or small set of technical options offers "a solution" to greenhouse gas emissions.** Only by aggregating the effects of many technical opportunities over a long time can significant reductions in greenhouse gas emissions be achieved. This chapter highlights options which appear to be "relatively cost-effective". **Detailed analysis necessary to quantify total costs of the measures assumed in the scenarios has not been conducted.**

- Improvements in End-Use Energy Efficiency provide the best option for reducing CO_2 emissions over the next few decades. Reductions in energy use would also reduce emissions of CH_4, N_2O, NO_x, and CO. Examples of potential efficiency improvements are:

  -- **Transportation** -- 50 mile per gallon automobiles are technically feasible with currently available technology. Further improvements could increase fuel efficiency to more than 80 miles per gallon. In addition, major fuel efficiency improvements in diesel trucks, rail transport and aircraft are possible.

  -- **Residential and Commercial** -- Accelerated improvements in building shells, lighting, space conditioning, and appliances could reduce energy consumption per square foot by 75% below current levels by 2025 for residences and 50% for commercial buildings.

-- **Industrial Energy** - Advanced industrial processes for energy intensive basic materials, recycling of used basic materials - i.e., steel, aluminum, and glass, cogeneration and improved electric motors can reduce industrial energy use significantly. This is especially important for most developing countries and Centrally Planned Europe where rapid industrial expansion is expected.

- **Reforestation** offers one of the most cost-effective technical options for reducing CO_2 and other gases. Preliminary estimates of the feasibility of large-scale reforestation suggest that with aggressive reforestation programs the current deforestation trend could be reversed and that a significant net increase in forest biomass is possible. An effective program should include both programs to increase forest biomass - such as replanting marginal agricultural lands, improved management of existing forests, tree plantations, and urban planting - and programs to reduce demand for wood where resources are currently stressed. **This could convert world forest management practices from a net source to a net sink for 0.7 petagrams of Carbon/year, or more, by the year 2025.**

- **Elimination of CFCs and related compounds** over the next decade appears technically feasible and cost-effective. Substitutes or process changes now available or under development have been identified to reduce or eliminate almost all applications of CFCs and Halons. Key examples are:

-- Worldwide replacement of CFCs as an **aerosol propellent** with substitutes already in use in the U.S., Canada and Sweden.

-- Replacement of CFC-12 with substitutes (such as HFC-134a or other hydrochlorofluorocarbons) in **mobile air conditioners.**

-- Replacement of CFC **solvent use** with aqueous cleaning in the electronics industry.

-- Replacement of **CFC-blown foam insulation** with other materials or use of alternative blowing agents.

- **Several actions may be possible to reduce greenhouse gas emissions from agriculture.** The magnitudes of the agricultural sources as well as the potential effects of control measures are difficult to quantify. Further detailed research and data collection will be required in order to produce credible estimates. Major agricultural activities of interest are:

-- **Methane emissions from livestock** could be reduced through increases in productivity of livestock systems, and use of methanogenesis inhibitors for beef cattle.

-- **Methane emissions from rice production** may be reduced somewhat through productivity increases, and removal of crop residues. In the long term, improvements in varieties of rice, soil amendments and water management practices could decrease methane emissions.

-- **Biomass burning** associated with agriculture produces N_20, CO and CH_4. Changes in agricultural practices - sustainable agriculture, utilization of crop residues, etc. - are technically feasible which could substantially reduce this source.

-- **N_2O emissions from fertilizer use** could probably be reduced through better placement of the fertilizer, nitrification inhibitors and fertilizer coatings, which may also reduce farming costs and agricultural runoff problems.

- **Near-term reductions in greenhouse gas emissions from electricity generation** are possible through:

-- **Efficiency improvements** - Improved fossil electricity generation technology, such as advanced combustion turbines and cogeneration, can increase the efficiency of using these fuels by up to 25%. More efficient transmission and distribution, availability or capacity improvements at existing non-fossil power plants (i.e., nuclear, hydro), and changes in electric utility system operation (i.e., dispatching, wheeling across regions or even internationally) can reduce CO_2 emissions by a few percent per kWh of electricity delivered.

-- **Fuel switching** - Use of more natural gas to displace coal as a fuel for generating electricity could reduce CO_2 emissions by a substantial amount in the near term. The potential of this option is largely dependent on availability and cost of natural gas in the future. Wood, municipal waste, wind, etc. could play a somewhat larger role than they currently play in the near term.

- **Alternative fuels** that do not emit significant amounts of greenhouse gasses could make an important contribution to reducing these emissions in the medium term and could virtually eliminate many categories of emissions over the long term. For widespread use of alternative fuels, important engineering, economic, environmental and social issues must be resolved.

-- **Hydroelectric power** is already making a significant contribution to global energy production. There is significant potential to expand this contribution although environmental and social impacts of large scale projects must be considered carefully.

-- **Biomass energy** is currently being extensively utilized, particularly in developing countries. Current and emerging technologies could vastly improve the efficiency of that use. More advanced technologies, especially for conversion of biomass to gaseous and liquid fuels and electricity could become economically competitive within a decade. Measures to increase biological productivity are also under study. These advances would allow biomass to provide a much larger share of global energy services over the long term, particularly in developing countries. Environmental and social issues of large scale biomass include land use, competition with food production, particulate and organic emissions, etc.

-- **Solar energy** offers a large range of options. Direct use of solar thermal energy either passively or in active systems is already commercially available for water and space heating applications in many regions. These applications could be expanded considerably in the near to medium term. Solar thermal concentrating technologies are being widely tested for power generation or industrial process heat. Solar photovoltaic (PV) cells are economically competitive for some remote power generation needs, especially in developing countries. If current research and testing succeeds in lowering the cost in the next decade, solar electricity generation could play a major role in meeting energy needs in the next century.

-- **Geothermal energy** resources are extensive and widely distributed. Systems are commercially available for electricity generation, and over 7 gigawatts of capacity are currently in operation worldwide. Technologies are improving and being demonstrated rapidly. It may be possible to expand this energy source significantly in future.

-- **Wind energy** systems are currently commercial in some applications and locations. In recent years engineering advances have resulted in reductions in cost and improvements in performance. Assuming this trend continues, wind energy can play a larger role in future energy production.

-- **Nuclear fission** is a technology which is currently widely used and growing in its contribution to global energy supply, due to the completion of power plants ordered during the 1970s. High cost and concerns about safety, nuclear weapons proliferation and radioactive waste disposal have brought new orders to a halt in most countries. It is technically feasible to expand the contribution of this energy source beyond what is currently projected in the future, if these problems are resolved.

- **Emission controls** - Control technologies are currently available to reduce CO from automotive and industrial sources and NO_x produced by power generation at relatively low cost. Other technologies are available which remove larger fractions of these pollutants but at higher cost. Emerging control technologies and combustion technologies with inherently lower NO_x emissions are being tested and could reduce NO_x emissions drastically at lower cost. In a few very limited situations (i.e., enhanced oil recovery) CO_2 recovery from power plant flue gas may be economic.

- **Methane emissions from coal seams, natural gas production and landfills can be reduced.** Currently minor emissions from coal production and landfills are projected to grow in future. Natural gas (primarily methane) is sometimes vented and often flared in conjunction with oil production. Technologies exist for economically recovering this methane and utilizing it for energy production, thereby partially augmenting natural gas supply.

- **Aggressive research programs** may be the most important policy option **for the long run.** Resolution of several key technical issues could vastly expand the economically attractive options for reducing greenhouse gas emissions in the next century. Some important examples are:

  -- **Improved characterization of sources and control options** in several areas would allow better policy and research planning decisions to be made. Sources of N_20 and CH_4 are poorly understood at present. Field measurement and data collection work are needed to improve understanding of the potential role reductions in these emissions could play in an overall climate stabilization strategy. Detailed cost analysis is also needed for most of the technical reduction options identified to support policy decisions in future.

  -- **Solar photovoltaic** technology has been improving rapidly over the last decade. Continuing or accelerating this progress could bring this technology into widespread commercial viability early in the next century.

  -- **Biomass conversion** technologies currently exist which could make substantially greater contributions. Commercial demonstrations of some existing technologies and additional research on advanced biomass conversion technologies and improvements in biomass productivity could greatly expand the role of biomass energy.

-- **Nuclear fission** does not currently appear viable in many countries because of safety, waste disposal, proliferation and cost concerns. Research is underway in several countries to develop and demonstrate "second generation" fission technologies which reduce cost and safety concerns. The establishment of waste disposal plans acceptable to society is also an area of intense study in several countries already committed to nuclear fission. Satisfactory resolution of these problems could expand the role of nuclear fission in future decades.

-- **Energy storage** technology could play a crucial role in integrating intermittent technologies such as solar and wind into energy supply systems. A number of promising concepts are currently under study. Accelerating research and testing to reduce cost and improve performance of storage technologies for electrical energy could greatly expand the potential roles of some alternative energy sources.

-- **Hydrogen energy** systems offer a long term potential for reducing or eliminating CO_2 emissions, if the hydrogen is produced from non-fossil energy inputs. Hydrogen is not a primary energy source but rather an "energy carrier," an intermediate form like electricity. As an energy carrier, it can help resolve some of the energy storage issues with renewables and substitute in some existing fossil fuel applications. Research needs include improved conversion processes using solar, hydroelectric, nuclear, wind, or other renewable energy inputs. Additional concerns include transmission and storage, and applications to transportation, space heating, industrial process and other end uses.

-- **Selected research in energy efficiency** could be helpful in accelerating the rate of improvement. Industrial technology, for example, could be developed to allow major increases in the standards of living in developing countries and Centrally Planned Europe without the enormous increases in CO_2 emissions which accompanied this development in the OECD. A major cooperative research effort to adapt advanced technologies under development in the OECD to the particular constraints and needs of the rapidly industrializing areas could be effective.

-- **Agricultural research** to identify and develop alternative **rice production** systems which reduce the production of methane could play a significant role in a long term solution to the greenhouse problem. Similarly, improvements in productivity and other technological options for reducing methane emissions from **domestic animals** (cattle, sheep, etc.) are possible. Concentrated research in these areas could make a major long term contribution to greenhouse emissions reductions.

INTRODUCTION

This chapter describes the substitute technologies and other means by which greenhouse gas emissions could be reduced relative to the scenarios described in Chapter V. A range of policies that might be used to promote such reductions are described in the next two chapters, which address domestic and international issues.

The preceding chapters discuss the diverse sources and economic activities responsible for greenhouse gas emissions. It should therefore not be surprising to find that there are an equally diverse array of methods for reducing greenhouse gas emissions. The primary means of accomplishing this goal is the **development and use of technologies** that reduce energy requirements (i.e., improve energy efficiency), use less carbon-intensive fuels, or that replace or reduce emissions of other greenhouse gases. In addition to this technological approach, there are also several areas in which **management strategies** are the means of reducing greenhouse gas emissions, particularly with respect to the buildup of gases resulting from some agricultural practices and forest resources.

The Relationship Between Technical Control Options and EPA's Modeling Results

As noted in Chapter V, the single most important determinant of greenhouse gas emissions is the level of energy demand and the combination of sources used to supply that energy. Given the dominance of fossil fuels as a source of greenhouse gas emissions, technologies to reduce use of fossil fuels must play a central role in any effort to stabilize concentrations. Fossil-fuel-based technologies supply over 70 percent of global primary energy needs in the No Response scenarios. A major focus for policies to reduce emissions, as discussed in Chapters VIII and IX, must accordingly be to promote demand-side measures that reduce total energy demand and supply-side measures that promote less carbon-intensive fuels. Technologies to achieve these goals are therefore a major focus of this chapter.

The No Response scenarios assume substantial efficiency gains will occur due to technological innovation and market forces, as discussed in Chapter V. The demand-side measures discussed in this chapter illustrate how this assumed efficiency improvement might occur as well as improvements incorporated into the Stabilizing Policy scenarios and more rapid improvements analyzed in Chapter VIII.

The Economics of Control Options

The uncertainties associated with many of the options described in this chapter preclude detailed assessment of costs. The primary focus of this review is to identify techniques that appear promising although for cost or other reasons, they may not be widely-accepted in today's markets. This is necessary given that global climate change is a long-term problem that may require development of new energy sources and other significant technological breakthroughs. Current prices also may be a misleading basis for assessment since climate change is potentially a major cost not currently reflected in the cost of goods and services; as discussed in Chapter VIII, it may be desirable to rectify the absence of a market price for the risk of climate change through the imposition of carbon fees or other policies, in which case currently more expensive options may become more competitive.

The price of some options is also difficult to evaluate simply because the absence of a market for reducing the risk of climate change has meant relatively little effort toward research and development. The importance of creating a market incentive to improve technology and reduce costs is demonstrated from the rapid recent development of substitutes for CFCs discussed in this chapter. Until such time as it became apparent that environmental regulation would create a market for CFC substitutes, industry reported that there were few feasible options at **any** price. More recently, an intensely-competitive race to commercialize substitutes has begun around the world.

While a detailed economic analysis of the options discussed in this report is impractical, it is worth noting that many of the control options described in this report are economically justified or nearly so today, even based on current costs. This is particularly true in the context of measures to improve energy efficiency, where substantial opportunities for cost saving investments exist despite recent progress, as discussed in Chapter VIII.

The Role of Long- and Short-Term Options

In the time frames considered in this report, long-term options become critical. In order to substantially reduce the concentration of greenhouse gases, new sources of energy supply and dramatic improvements in efficiency will have to assume a significant role. However, there is also much that could be done to reduce greenhouse gas emissions over the next decade by improving energy efficiency, making greater use of natural gas, reducing use of CFCs, and promoting reforestation.

While this current generation of techniques will not be sufficient to stabilize greenhouse gas emissions several decades hence, efforts to adopt such technologies are nevertheless likely to be exceedingly valuable for several reasons. First, reducing the rate of growth in emissions now could make it easier to stabilize concentrations in the future because of the long atmospheric life of these gases. Second, short-term strategies are often intermediate steps toward long-term strategies, e.g., currently-available efficiency improvements will facilitate still greater future efficiency improvements. Finally, the incentives necessary for longer-term strategies, such as emission fees on carbon-intensive fuels, will generally be consistent with short-term strategies.

Over the long term the most important options are advanced non-fossil energy technologies, possibly combined with major breakthroughs in end-use technology that would drastically reduce energy requirements. Also, changes in agricultural and forest management technologies could become important. In addition to the incentives that may flow out of short-term strategies, it is important in the short term to promote national research and development by governments and the private sector toward identifying and advancing promising long-term technologies.

Control Techniques and Worldwide Emissions

As the modeling results make clear, the U.S. is likely to account for a declining share of future greenhouse gas emissions. Stabilizing concentrations will require control options applicable to the needs of other countries, particularly developing countries with very different resources. Some technologies, such as more efficient lighting, are relevant to virtually all parts of the world, but other needs vary considerably. This chapter therefore discusses improved cookstoves, strategies for arresting tropical deforestation, and other options of particular relevance to developing countries. Table 7-1 illustrates some of the promising options for various regions and time horizons.

Organization of this Chapter

Figure 7-1a identifies the "greenhouse gases" and illustrates their estimated percentage contributions to the greenhouse effect in the 1980s. All of these gases are produced through a diverse range of human activities, which we have classified into five broad categories. Energy-related activities have been broken down into two categories: "applications" of energy, that is, energy services, or "end uses," and the production of energy, or energy supply. Other emissions-producing activities are related to industry, which includes the use of chlorofluorocarbons (CFCs), forestry (particularly deforestation), and agriculture. Often a single broad category of activity--fossil energy

TABLE 7-1

Key Technical Options By Region and Time Horizon

Region	Near Term (by 2010)	Long Term (All Regions)
OECD	Energy Efficiency - autos, lighting, space heating	
	CFC Controls	**Alternative Fuels**
	Reforestation	Biomass
	Technology Development	Solar
		Nuclear
Developing Countries	Energy Efficiency - industrial processes, transport	Hydrogen
	Low-Carbon Energy hydroelectricity biomass natural gas	**Agriculture** Rice Production
	Reversing Deforestation	Animals
Centrally Planned Europe	Energy Efficiency - industrial processes, space heating, transport	**Forest Plantations**
	Natural Gas	
	Nonfossil electricity	

FIGURE 7-1

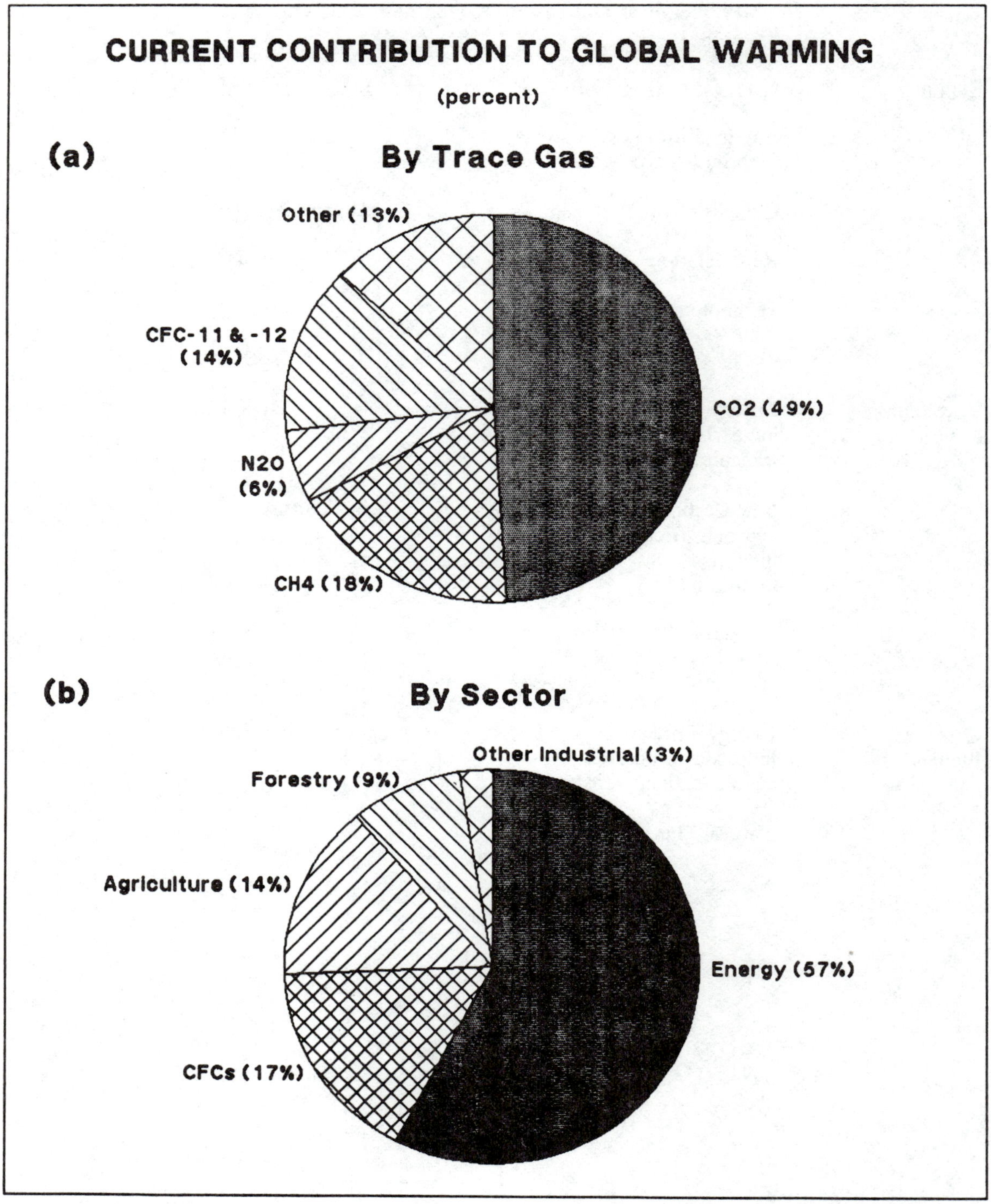
CURRENT CONTRIBUTION TO GLOBAL WARMING
(percent)
(a)
By Trace Gas
Other (13%)
CFC-11 & -12 (14%)
N2O (6%)
CH4 (18%)
CO2 (49%)
(b)
By Sector
Other Industrial (3%)
Forestry (9%)
Agriculture (14%)
CFCs (17%)
Energy (57%)

consumption, deforestation, for example--contributes to several of the gases of concern. Figure 7-1b shows the proportion that the major categories of human activity contribute to global warming.

Because of the enormous amount of information presented in this Chapter (in truth, a separate chapter could be devoted to each major source category discussed here), we have departed somewhat from the format used throughout the rest of this Report to Congress--to make it easier for the reader to follow the discussion. Thus, the remainder of this chapter is divided into five parts.

Energy use causes, in different proportions, emissions of five important gases: carbon dioxide (CO_2), carbon monoxide (CO), methane (CH_4), nitrous oxide (N_20), and nitrogen oxide (NO_x). Energy use is integrally linked with virtually all forms of economic and recreational activity within industrialized countries. In developing countries current biomass energy use contributes to greenhouse gas emissions, and potential increases in commercial energy use could be the largest source of increasing greenhouse gas emissions in the future.

Thus, evaluation of the options for reducing greenhouse gas emissions from energy use must begin with a systematic analysis of all aspects of energy use. Although there are "end-of-pipe" control options for removing some of the relevant emissions from energy use while leaving the basic process intact, the potential impact of such approaches is very small relative to the magnitude of the overall problem. A strategy designed to significantly reduce the warming potential of energy-related activities must incorporate major reductions in fossil energy consumption.

Part One of this chapter reviews the basic applications for which energy is ultimately used and the opportunities for reducing greenhouse emissions at the point of end use. Part Two reviews energy supply and conversion activities and related opportunities for reducing emissions. This includes improvements in efficiency in energy conversion and distribution, and potential for increasing supplies of non-fossil energy sources. Also discussed are reductions in emissions of CH_4 from coal mining and natural gas production and distribution.

In Part Three we discuss technical options for controlling emissions from industrial activities. Non-energy industrial activities contribute to greenhouse warming in three significant ways. First, and most important, industrial activities are the source of all CFC emissions. As discussed in Chapter IV and IX, an international process is already underway to reduce global emissions of CFCs because of their role in depleting the stratospheric ozone layer. This chapter discusses the technical potential for reducing CFCs further than required under the current protocol.

A second source of industry-related emissions are landfills, which produce emissions of methane (CH_4). This source category represents a small portion of the total methane emissions, but emissions from this source could increase rapidly in the future. Finally, production of cement produces CO_2 as a process emission (in addition to CO_2 produced by energy consumption). This again is a small component of total CO_2 emissions currently, but the percentage contribution from cement production has been growing rapidly in recent years.

In Part Four, we discuss options related to forest management practices (resulting in a net annual loss of biomass), which account for a significant share of emissions of CO_2 and CO as well as some portions of other gases. It should be noted that the importance of forestry is greater than its percentage share in Figure 7-1 implies. Forests are the only category that, over time, can be shifted from a source to a major sink for carbon. It is technically possible, though by no means simple, to reverse the long-term trend of global deforestation and to begin increasing the amount of forested lands. There are several components to reforestation strategies that need to be considered. For example, reductions in demand for forest products (e.g., fuelwood) in some areas may be necessary to relieve the pressures that have caused deforestation in recent years.

Agricultural activities are an important source of CH_4, N_2O, NO_x, and CO. The principal activities of interest are rice production, enteric fermentation in domestic animals (primarily cattle, sheep, etc.), fertilizer use, and biomass burning. In the final section of this Chapter, Part Five, we discuss the technical options for reducing emissions resulting from these activities. It is apparent that considerable flexibility exists, particularly in the long term, to alter agricultural practices in the specific categories that constitute the large emitters, but the technical potential is difficult to quantify.

In general, most of the research and analysis of agriculture to date has focused on opportunities for improving productivity. Productivity improvements should automatically lead to some reductions in greenhouse gas emissions per unit of output; however, this relationship is not well-quantified. Additional options for reducing emissions, beyond productivity changes, can be envisioned for each of the specific categories mentioned, though, again, very little work has been done as yet to quantify their potential effects and costs.

Limitations

In this review the information presented, although somewhat detailed, can only begin to illustrate potential technical options that currently appear most promising. The analysis also highlights the uncertainties and need for further study of many options. In some areas, particularly techniques for reducing emissions from agricultural sources, the impact of specific technologies cannot be estimated at present. However, in general, the uncertainties that exist concerning the means of achieving emission reductions are much less, or are likely to be much more easily resolved, than the uncertainties that exist concerning the rate of warming, change in climate, and the ultimate effects of an increase in greenhouse gas concentrations. While some currently emerging technologies may not fulfill current expectations, the diversity of opportunities suggests we can emphasize the development and use of technologies that are relatively less-intensive sources of greenhouse gas emissions if we choose.

The source of most of the uncertainties discussed in this chapter is the difficulty of predicting future technological developments. Many of the options that may have the greatest impact on the buildup of greenhouse gases, such as the development of new engine technology for cars, new designs for nuclear plants, and the use of hydrogen as a substitute for liquid fuels, are long-term possibilities that require substantial further research. It is important to recognize several key limitations:

- The discussion cannot deal exhaustively with the tremendous range of technical options that have been identified for energy efficiency improvements and fuel substitution and for industrial emissions reductions, as well as for forest management and for implementing changes in agricultural practices.

- Because of the limited information available, but even more because of the extensive scope of this study--both in the range of emitting activities and the global diversity of emissions--it was not possible to provide detailed quantification of expected emissions reductions and costs for many of the technical options discussed.

- Much data development and detailed analytical work remains to be done. A more detailed technical assessment has been mandated by the congress. This study is beginning now and will be completed in 1989 by the U.S. Department of Energy and EPA.

- As discussed in Chapter V, scenarios are constrained by current expectations about types of services and economic activities demanded by consumers. The control options presented are intended to produce the same types of goods and services--electricity, lighting, transportation, space heating, etc.--

but with a lower level of greenhouse gas emissions. These technologies and strategies are therefore generally consistent with current lifestyles and consumer preferences. However, some of them are likely to increase or lower costs that could lead to changes in demand. This may result in underestimating the effects of some technical options and strategies on future greenhouse gas emissions.

- The No Response scenarios used in this report assume a considerable amount of efficiency improvement due to technological innovation and market forces. As discussed in Chapter V, the energy use in these scenarios is substantially lower than some other reference projections (e.g., DOE/NEPP; see **CHAPTER V**). If energy use is higher than reflected in the No Response Scenario, then the impact of policies that promote energy efficiency, such as those identified in this chapter, could be much more effective in reducing energy consumption than indicated here.

- Discussions of the potential performance of technical options in this chapter are often based on engineering design calculations, prototype performance, laboratory results, etc. Achievable performance in practice may be less since mass production often requires some engineering compromises relative to laboratory or prototype specifications. Also, performance of technology **in use** often deteriorates somewhat from design or new product performance. On the other hand, currently unforeseen developments may improve performance beyond levels estimated today.

- This chapter identifies and attempts where possible to quantify **technical potentials** for reductions in greenhouse gas emissions. Even where technical options appear economically attractive on a life-cycle basis, there are generally institutional, behavioral, and policy constraints that currently operate to limit their penetration. The portion of identified technical potential that can be achieved in practice is largely a function of the availability and effectiveness of policy options, as discussed in **CHAPTERS VIII and IX.**

PART ONE: ENERGY SERVICES

The services that energy provides (called end-uses), such as lighting and fuel-driven locomotion, are an integral part of human society and, at the same time, the largest category of greenhouse gas emissions. While the production and conversion of primary energy (e.g., coal, oil, gas) is the immediate source of a large portion of energy-related greenhouse gas emissions, it is the applications of this energy to provide specific services that justify this production and conversion. Thus, minimizing the energy required in various end-uses or using non-fossil fuels in specific end-use applications can reduce production- and conversion-related emissions as well.

For convenience, energy services are classified as belonging to one of three major sectors: transportation, residential/commercial, and industrial (including agriculture). Each of these sectors uses energy in distinctly different ways and offers different opportunities for reducing energy use and/or shifting to alternative fuels. Figure 7-2 shows the relative contributions of the three sectors to global energy use as of 1985. Figure 7-2a shows the secondary energy actually consumed at end-use points. Figure 7-2b shows the energy use by sector in primary energy production equivalent terms; that is, the production, conversion, and transmission losses are ascribed to the end-use sectors based on the characteristics of the energy they use. Figure 7-2c shows the proportions that the equivalent primary energy use for these three categories contributes to global greenhouse emissions in the 1980s. The differences are due to the variations in unit contributions of different primary energy sources to greenhouse gas emissions.

There are two time horizons that are useful in discussing technical options. Near-term options refer to technologies currently available or expected to be commercially available by the year 2000. These are the options about which information is available and which could also provide a basis for

FIGURE 7-2

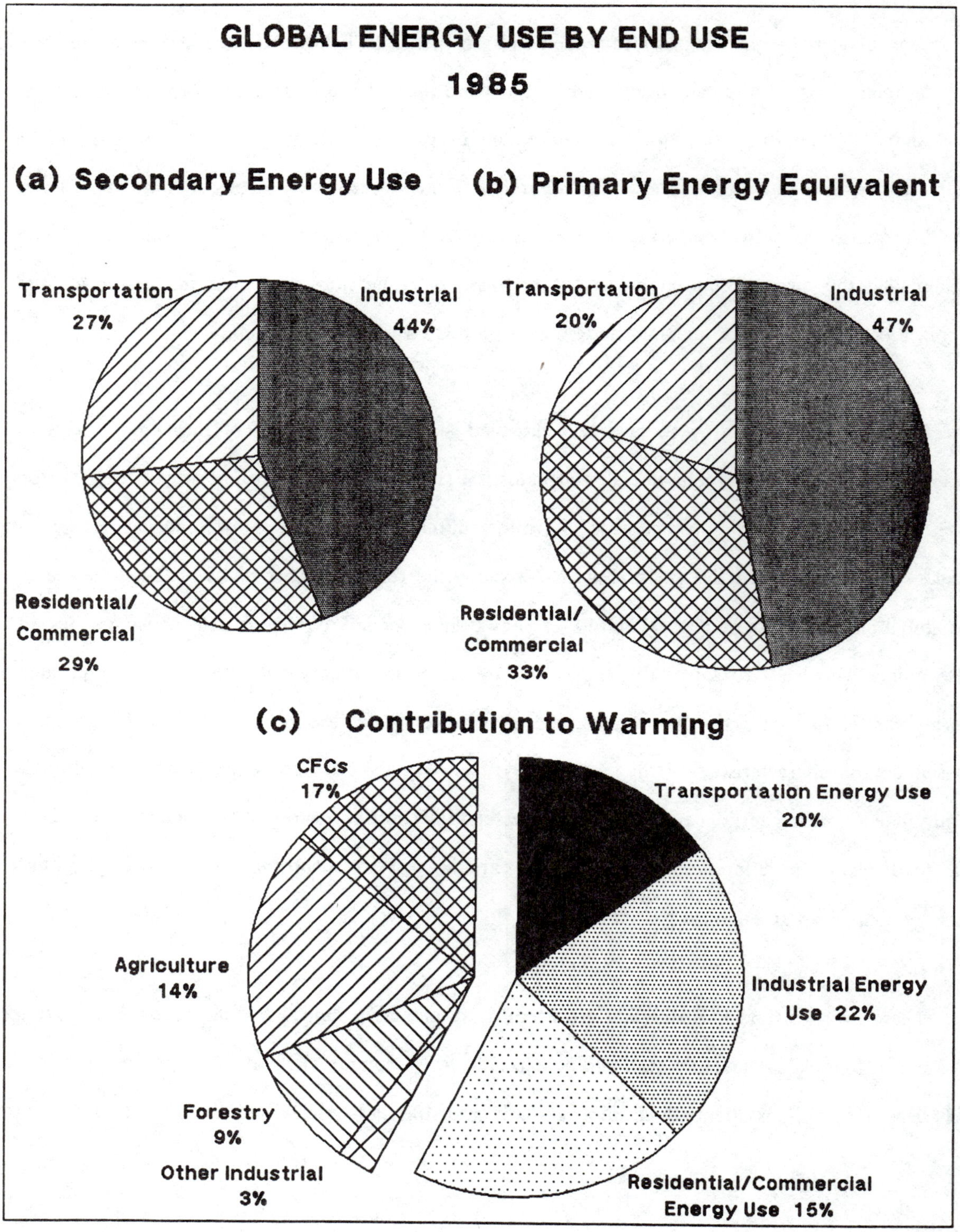
GLOBAL ENERGY USE BY END USE
1985
(a) Secondary Energy Use
Transportation 27%
Industrial 44%
Residential/ Commercial 29%
(b) Primary Energy Equivalent
Transportation 20%
Industrial 47%
Residential/ Commercial 33%
(c) Contribution to Warming
CFCs 17%
Transportation Energy Use 20%
Agriculture 14%
Industrial Energy Use 22%
Forestry 9%
Other Industrial 3%
Residential/Commercial Energy Use 15%

near-term policy action. Long term refers to options that are not expected to be available until after 2000, in some cases well after.

An attempt is made to hypothesize about potential technological developments over the longer term and to discuss the role research could play in accelerating the availability of advanced options. Some of these technologies are speculative and require additional research. They are included in the analysis because they have the potential to play major roles in the long run.

In discussing global energy use patterns, it is important to distinguish between modern and traditional energy forms, particularly in understanding energy use in developing countries. In this discussion modern or commercial energy is used to describe all fuels and energy forms that are priced and sold in energy markets (or, in the case of centrally-planned economies, accounted for and valued in national economic planning). In this category are virtually all of the fossil fuels, which are the major source of greenhouse gases, as well as electricity from all sources. Readily-available energy statistics deal almost exclusively with modern energy and accurately represent energy use patterns in industrialized countries.

However, in developing countries the modern energy transactions alone can give a very misleading picture of energy use. Traditional energy is used to mean fuels such as firewood, agricultural waste, and animal waste that are gathered and used informally without being priced and sold in commercial energy markets. In many developing countries this type of energy accounts for a substantial fraction of the total energy used.

Technical options, especially in the near- to mid-term, vary substantially from region to region and often among individual countries. We focus on the U.S. explicitly, and three blocs of countries: the OECD, developing countries, and the USSR and Eastern Europe. Figure 7-3 shows the pattern

FIGURE 7-3

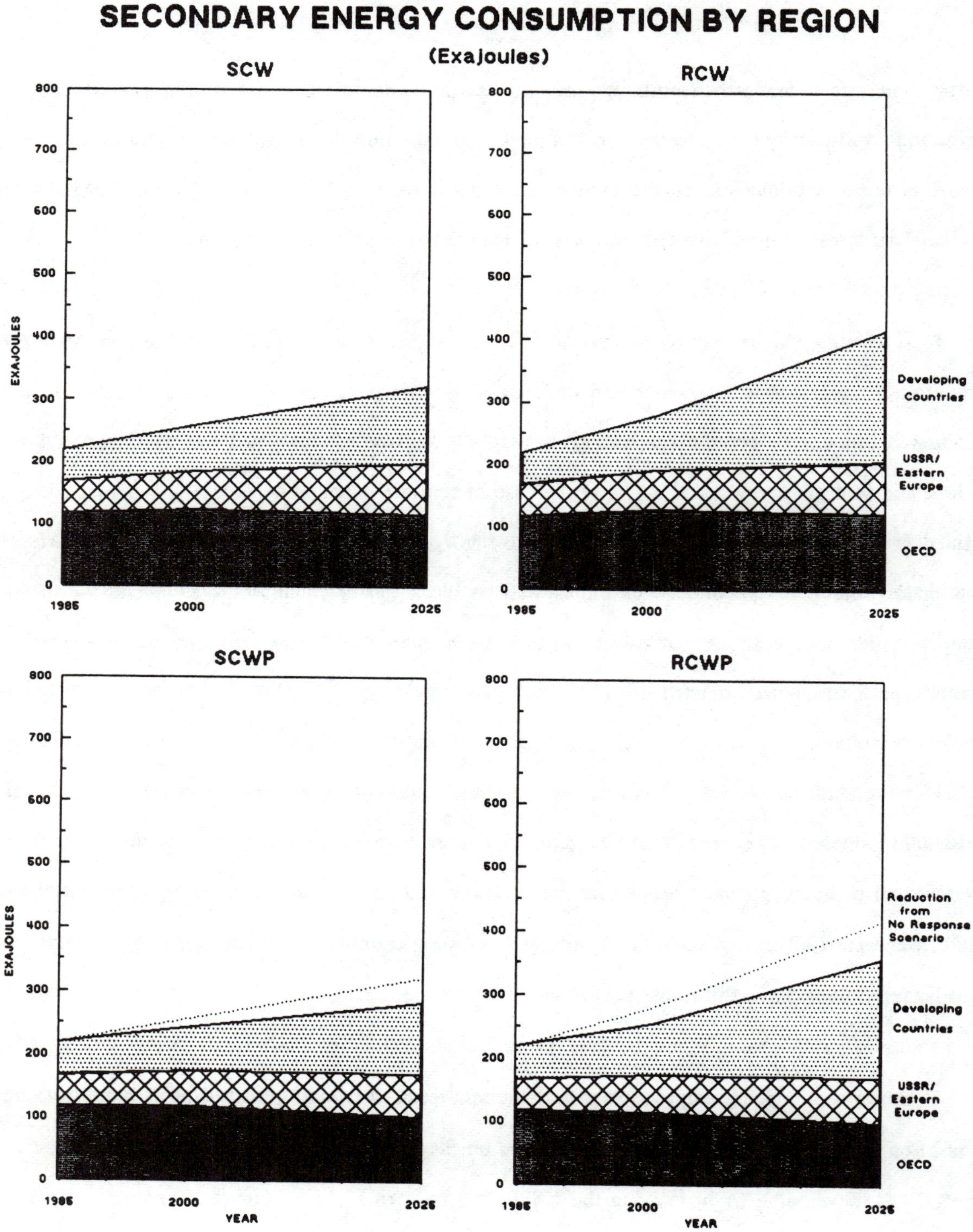
SECONDARY ENERGY CONSUMPTION BY REGION
(Exajoules)
SCW
RCW
SCWP
RCWP
EXAJOULES
YEAR
800
700
600
500
400
300
200
100
0
1985
2000
2025
Developing Countries
USSR/ Eastern Europe
OECD
Reduction from No Response Scenario

of energy consumption for these blocs of countries over time under both the No Response Scenarios (SCW and RCW) and the Stabilizing Policy Scenarios (RCWP and SCWP).

The industrial market economies, the members of the Organization for Economic Cooperation and Development (OECD), have many similarities in terms of economic activities and resources, and hence are discussed as a group. As shown in Figure 7-3, energy use in these countries is relatively high but not expected to grow significantly; rising incomes are devoted increasingly to products that do not require significant energy inputs.

The developing countries are vastly different from industrialized countries both in current levels and types of economic activities and available resources. As shown in Figure 7-3, energy use in the developing countries grows significantly in the scenarios, but there is great uncertainty about the rate of growth. Depending on their rate of development, energy use in these countries increases by a factor of 2.5 to 4 by the year 2025.

The USSR and Eastern Europe share with most developing countries a much greater emphasis on government intervention in economic planning and industrial activities than do OECD countries. On the other hand, these countries have massive and, in many ways, technologically-sophisticated industrial infrastructures that are much more similar to those of the OECD countries than they are to the industrial infrastructures of most developing countries. Energy use in Eastern Bloc countries (and associated greenhouse gas emissions) has been growing rapidly and, as shown in Figure 7-3, is projected to grow in the future. While less is known about the technical potential to reduce emissions in these countries, it is generally believed that very substantial improvements can be made in energy efficiency.

Figure 7-4 shows how secondary energy use by end-use category changes in the No Response scenarios (SCW and RCW) and the two Stabilizing Policy scenarios (SCWP and RCWP). Overall, secondary energy use in 2025 is 13-15% lower in the Stabilizing Policy scenarios; one reason for the limited reductions in energy use is that significant improvements in energy efficiency are assumed to occur in the No Response scenarios. The Rapid Reduction Scenario described in Chapter VIII illustrates some of the additional potential. In that case total secondary energy use was reduced by 20% from the RCW scenario by 2025. As discussed in this section, these are very conservative reductions relative to the technical potential. In both policy scenarios, energy use in transportation decreases the most as a result of the introduction of fuel-efficient, light-duty vehicles and the relatively rapid turnover of the vehicle stock.

Residential/commercial energy savings are smaller, reflecting the longer turnover times. The industrial sector represents the largest current component of energy use and the largest portion of projected growth in the No Response scenarios. Despite this, the reductions in industrial energy consumption in the Stabilizing Policy scenarios are relatively modest, a reflection primarily of the diversity of the industrial sector and the difficulty in designing broad policy assumptions that will affect this sector in the modeling analysis. There are clearly large opportunities for energy savings in the industrial sector, as discussed below.

TRANSPORTATION SECTOR

Transportation currently consumes approximately 27% of global modern energy use. As shown in Figure 7-4, it accounts for 20-30% through 2025 in the No Response Scenarios. Virtually all of the energy used in transportation is derived from oil. In 1985 in OECD countries, energy used for transportation accounted for about 23% of all energy consumed (27% in the U.S.), expressed as

FIGURE 7-4

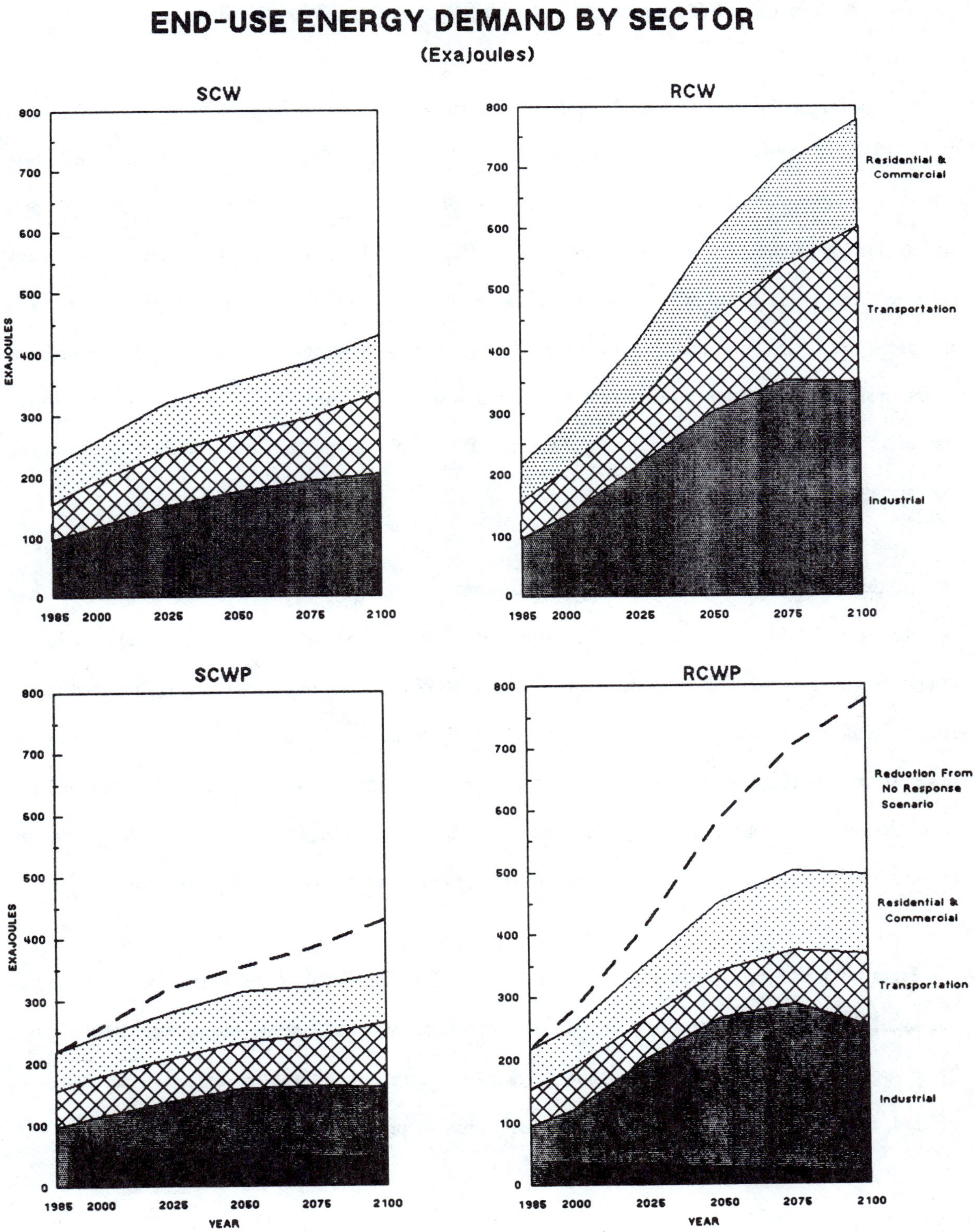
END-USE ENERGY DEMAND BY SECTOR
(Exajoules)
SCW
RCW
SCWP
RCWP
EXAJOULES
YEAR
0
100
200
300
400
500
600
700
800
1985
2000
2025
2050
2075
2100
Residential & Commercial
Transportation
Industrial
Reduction From No Response Scenario

primary energy equivalents. As a share of secondary energy consumption, transportation accounted for about 34% in the OECD, about 97% of which was oil (U.S. DOE, 1987b).

As shown in Figure 7-5, global transportation energy use grows by 50-65% between 1985 and 2025 in the No Response scenarios. In the U.S. and the other OECD countries, in both the Slowly Changing and Rapidly Changing World scenarios, improvements in efficiency as a result of economics and technological innovation slightly more than offset the relatively slow rate of growth in energy services (i.e., miles travelled). The OECD accounts for over 65% of transportation energy use currently, but their share declines in absolute terms and as a percentage of the total. In the last few years, transportation energy efficiency has declined slightly in the OECD. If this trend were to continue, OECD transportation energy use could be higher than indicated in the No Response scenarios.

In the USSR and Eastern Europe, transportation energy use increases rapidly, more than doubling by 2025 in the No Response scenarios. Transportation energy use in the developing countries grows at a roughly equivalent rate in the SCW scenario, but at a much higher rate in the RCW scenario, reaching 3.5 times its current level by 2025. In both the Soviet bloc and the developing countries, freight transport and passenger mass transit (rail, bus) currently account for a much larger share of total energy than in the OECD. The explosive growth in both regions, however, is largely due to rapid expansion in the number and use of light-duty vehicles.

Figure 7-5 also illustrates the impact that Stabilizing Policies could have on transportation energy use in 2025. Overall energy efficiency improves significantly in all regions, and energy use decreases by a large percentage. While total energy demand decreases, alternative fuels are more heavily utilized, thus reducing the amount of fossil fuels used to produce transportation services.

FIGURE 7-5

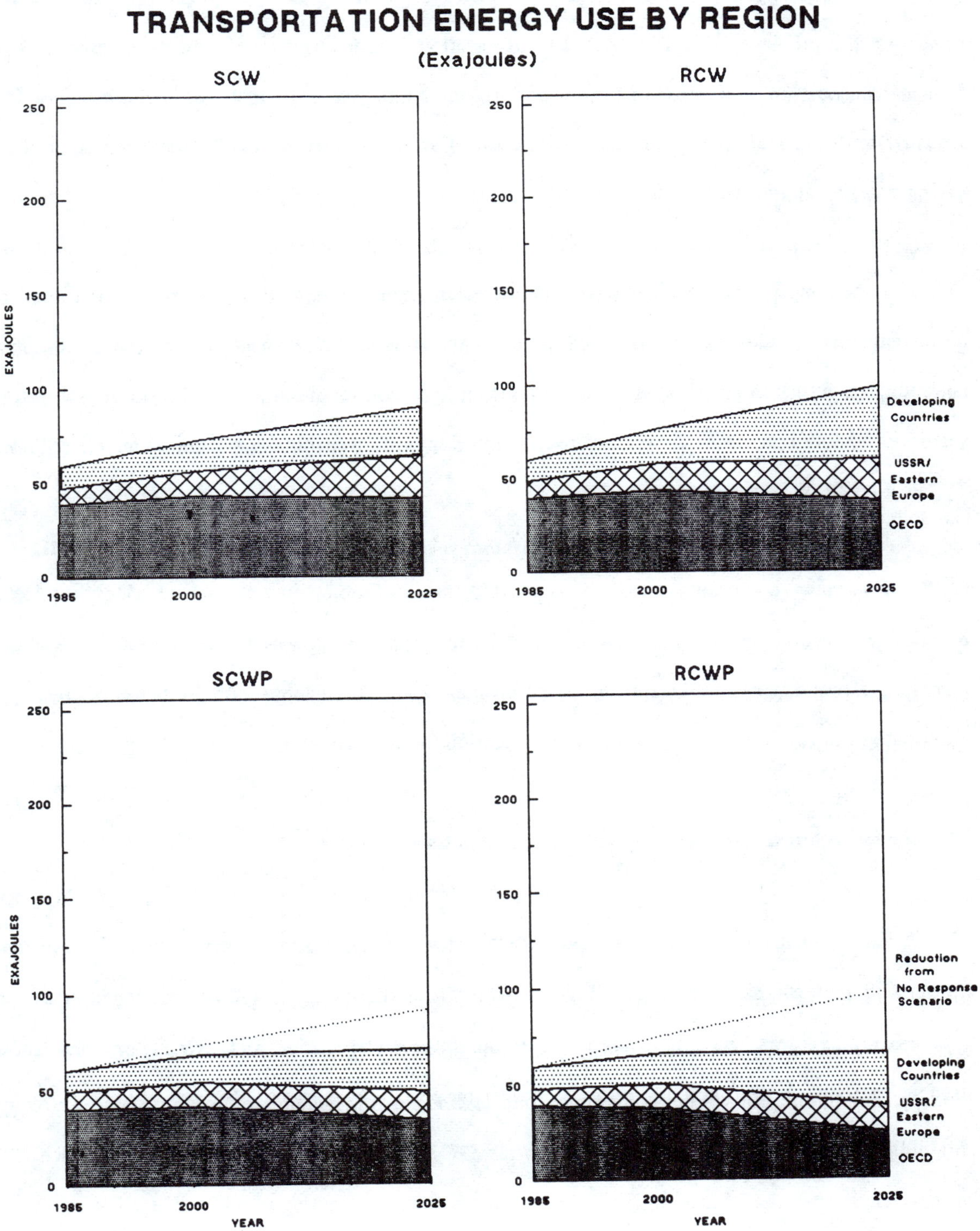
TRANSPORTATION ENERGY USE BY REGION
(Exajoules)
SCW
RCW
SCWP
RCWP
EXAJOULES
YEAR
0
50
100
150
200
250
1985
2000
2025
Developing Countries
USSR/ Eastern Europe
OECD
Reduction from No Response Scenario

In industrialized countries it is likely that the private automobile will continue to be the primary means of transportation in the near future. Fortunately, there are near-term opportunities for improving the efficiency of this mode of transportation. Near-term improvements in efficiency of freight transport and aircraft will also be cost-effective over the next few decades. A few industrialized countries (e.g., Canada, New Zealand) are also pushing ahead with major alternative fuels programs in the near term.

Over the longer term, additional reductions in energy consumption may come from shifting to more efficient modes of transportation and substitutes for transportation (e.g., advanced communication technologies). Also, increasing the use of non-fossil-based fuels in the transportation sector is essential in order to greatly reduce or eliminate greenhouse gas emissions (See **PART TWO: ENERGY SUPPLY**).

In developing countries and the Soviet bloc, transportation currently accounts for a smaller percentage of total energy use. (In some African countries, however, the percentage is much higher.) However, in the future, as economies expand and incomes rise in these regions, the potential exists for explosive growth in transportation energy use.

Near-Term Technical Options: Industrialized Countries

Within the transportation sector, light-duty vehicles, mainly passenger cars, account for the bulk (about 63%) of current energy use (Figure 7-6). Other major contributors are freight transport vehicles (diesel trucks, ships, and railroads), accounting for about 25%, and aircraft, primarily those used in passenger travel, accounting for 12% of transportation energy.

FIGURE 7-6

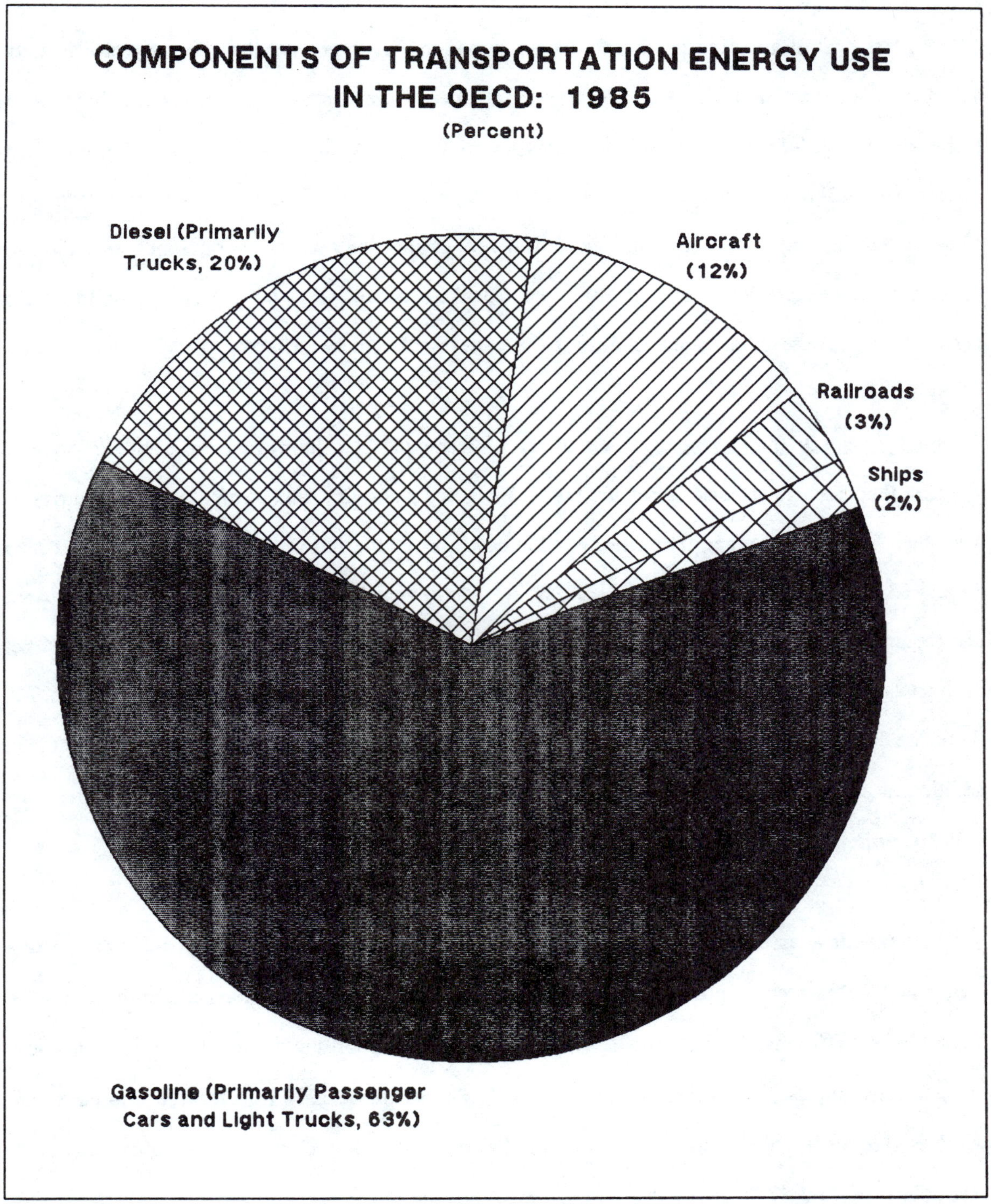

(OECD, 1987)

Light-Duty Vehicles

A great deal of attention has been devoted to options for improving the efficiency of light-duty vehicles in the past decade. Consequently, a number of very promising approaches have been identified and well-documented (see, for example, Bleviss, 1988; Goldemberg et al., 1988, for more extensive discussions of the technical options for improving fuel efficiency of light-duty vehicles). These efficiency improvements must be considered in the context of several other societal and consumer concerns related to light vehicles, such as urban air quality, safety, comfort, and performance. These other goals also affect the patterns of vehicle technology development.

Fuel-Efficiency Improvements. A number of techniques for improving the fuel efficiency of vehicles that use traditional petroleum-based fuels (i.e., gasoline and diesel fuel) are currently available. Although average fuel efficiency for new cars in the industrialized countries is between 25 and 33 mpg (7.7-10 litre/100 km) (IEA, 1987), several vehicles that are roughly twice as efficient are commercially available: the Ford Escort diesel, the Honda City and Civic, and the Chevrolet Sprint all average greater than 50 miles per gallon (5 liters/100 km). As indicated in Table 7-2, there are larger prototype vehicles currently being tested that are substantially more efficient. In addition, a great deal of research on improving fuel efficiency is being conducted by the automobile industry. Improvements are possible in several areas as outlined in Box 7-1.

Box 7-1 provides only a few examples of the many improvements already possible with current technology. Research is proceeding rapidly and will undoubtedly yield further opportunities for improved fuel efficiency in the next decade. It is clear that opportunities exist for major reductions in light-duty fuel use by the end of this century. This is particularly true for the United States, which is still lagging behind most other industrialized countries in the average fuel efficiency of new cars sold (see Table 7-3), partly due to the preference for larger cars in the U.S.

TABLE 7-2

High Fuel Economy Prototype Vehicles

Company	General Motors	British Leyland	Volkswagen	Volkswagen	Volvo	Renault	Renault	Peugeot	Peugeot	Ford	Toyota
Model	TPC (Gasoline)	ECV-3 (Diesel)	Auto 2000 (Diesel)	VW-E80 (Diesel)	LCP2000 (Diesel)	EVE+ (Diesel)	VESTA2 (Gasoline)	VERA+ (Diesel)	ECO 2000 (Gasoline)	____ (Diesel)	AXV (Diesel)
Number of Passengers	2	4-5	4-5	4	2-4	4-5	2-4	4-5	4	4-5	4-5
Aerody-Namic Drag Coefficient	.31	.24-.25	.25	.35	.25-.28	.225	.186	.22	.21	.40	.26
Curb Weight (LB)	1040	1460	1716	1540	1555	1880	1047	1740	990	1875	1430-Target
Maximum Power (HP)	38	72	53	51	52,88	50	27	50	28	40	56
Fuel Economy (MPG)	61 City 74 Hwy	41 City 52 Hwy	63 City 71 Hwy	74 City 99 Hwy	63 City 81 Hwy	63 City 81 Hwy	78 City 107 Hwy	55 City 87 Hwy	70 City 77 Hwy	57 City 92 Hwy	89 City 110 Hwy

Source: Bleviss, 1988.

TABLE 7-3

Actual New Passenger Car Fuel Efficiency
(Gasoline consumption in liters per 100 kilometers)

									Projections/Targets	
	1973	1978	1979	1980	1982	1983	1984	1985	1990	2000
Australia	NA	11.8	11.2	10.1	9.8	9.5	9.5		8.5[1]	
Canada	NA	11.5	11.5	10.3	8.5	8.5	8.4	8.2	7.4	6.8[2]
Denmark	9.0[3]	NA	NA	8.6	NA	7.3	7.0	7.1		
Germany	10.3	9.6	9.4	9.0	8.3	8.0	7.7	7.5		
Italy[4]	8.4	8.3	8.3	8.1	8.3	8.0	NA	7.8	7.4	
Japan	10.4	8.8	8.6	8.3	7.7	7.8	7.8	7.8		
Netherlands	NA	9.2	8.9	8.8	8.5	NA		9.1	8.6	7.8
Sweden	NA	9.3	9.2	9.0	8.6	8.6	8.5	8.5		
United Kingdom	11.0	9.1	9.0	8.7	8.1	7.9	8.8	7.6		
United States	16.6	11.8	11.6	10.0	8.9	9.0	8.9	8.7	8.8	8.2

[1] 1987 target.
[2] 1995.
[3] 1975.
[4] The figures for Italy represent average efficiency of the total car fleet.

Source: IEA, 1987a

BOX 7-1. TECHNOLOGIES FOR AUTOMOTIVE FUEL EFFICIENCY

- Weight Reductions--Many of the most efficient cars substitute non-traditional materials for steel to achieve weight reductions, which contributes to their superior operating efficiency. Much greater weight reduction appears possible by substitution of high-strength steels, aluminum, plastics, ceramics, and composite materials (Bleviss, 1988).

- Aerodynamic and Drag Improvements--In 1979 the average "coefficient of drag" (C_D) for the U.S. was 0.48 and in Europe was 0.44. Currently, some production models such as the Ford Sable and Taurus, Subaru XT Coupe GL-10, and Peugeot 405 achieve a C_D of 0.3 or less. An experimental prototype, the Ford Probe V, has achieved a C_D of 0.137 (Bleviss, 1988). Incorporation of some of the design features currently in prototypes could reduce drag for production vehicles significantly over the next decade. Rolling resistance is being reduced with advanced radial tires. General Motors has recently introduced a "fourth generation" radial tire that reduces rolling resistance by 10-12% from the previous generation. An Austrian company has developed a more advanced tire concept, a liquid-injection-molded (LIM) polyurethane tire. Preliminary tests indicate improvements in rolling resistance as well as tread mileage (Bleviss, 1988).

- Engine and Drive Train Improvements--Several researchers have identified a number of improvements to conventional light vehicle propulsion systems and transmissions that could dramatically increase efficiency (see Bleviss, 1988; von Hippel and Levi 1983; Gray 1983; OTA 1982). One interesting example is the use of continuously variable transmissions (CVT), which eliminate some of the energy losses during shifting and allow the engine to be operated closer to full load at varying speeds. Another possible innovation is an engine-off feature with energy storage capability during idle and coast. In addition, advanced engine designs currently in prototype could be much more fuel efficient than current technology. One example is the adiabatic diesel engine shown in Box 7-2.

Despite the fact that improvements have been identified, it is not clear whether, and at what rate, new technology will be incorporated into automobile designs. Clearly, opportunities for dramatic efficiency improvements exist, but costs associated with these opportunities must be considered. The U.S. Office of Technology Assessment conducted a detailed analysis of the potential for and cost of future automobile fuel efficiency improvements. The study included considerable interaction with automobile manufacturers and other experts on fuel efficiency measures, and concluded that it was technically feasible to achieve average new car fuel efficiency levels in the range of 3.4-4.7 liters/100 km (50-70 miles per gallon) by 2000. It also found that "The consumer costs of fuel efficiency range from values that are easily competitive with today's gasoline prices to values that are considerably higher." (OTA, 1982, p.14).

OTA found that the cost of efficiency improvements varied greatly depending on the actual performance of potential design changes, the success of developing production techniques that can hold down the variable cost increases, and the value consumers place on future fuel savings. With optimistic assumptions, OTA estimated the cost of fuel efficiency measures to be as low as $60-$130/car during the 1985-2000 time period. With alternative assumptions, the cost of efficiency improvements could be as high as $800-$2,300/car.

Von Hippel and Levi (1983) conducted a computer analysis of the cost of introducing a number of specific measures that would improve fuel efficiency, beginning with the 1981 Volkswagen Rabbit diesel. A package of specific improvements that would improve the fuel economy from 5.2 to 3.3 l/100 km (45 to 71 mpg) was estimated to cost about $500 per car.

Goldemberg et al. (1988) and Bleviss (1988) however, suggest that cost estimates toward the lower end of the OTA range are more likely for several reasons. First, a number of other benefits to the consumer would result from some of the efficiency improvements. Alternative materials, for example, also may reduce maintenance costs. There is also some evidence that, contrary to popular expectations, use of more plastics and plastic composites may in some cases increase passenger safety (Bleviss, 1988). The inclusion of the engine-off feature may prolong engine life. If the value of these other benefits is deducted from the cost of fuel-efficiency improvements, more rapid improvements may be cost-effective.

A second reason for lower cost estimates is that the cost of individual improvements may not be additive. Some costs may be offset by the savings from combining measures and integrating a number of related changes into ongoing production process changes. Vehicle manufacturers periodically make major investments in design changes, incorporating style concerns as well as engineering improvements. Chrysler recently conducted a study comparing the costs of producing a conventional steel vehicle and an alternative made principally of composite plastics. Although the composite material is more costly, its use allows a dramatic reduction in the number of parts and hence, assembly costs. The study concluded that the number of parts might be reduced by as much as 75%, and that the overall production costs for the composite vehicle would be only 40% of those for the corresponding steel vehicle (*Automotive News*, 1986).

On the other hand, automobile manufacturers have recently expressed the view that further fuel efficiency improvements may be more difficult and costly to achieve than estimates from the early 1980s suggest (Plotkin, 1989). Concerns raised by the manufacturers include the following:

- most of the cost-effective efficiency measures that are acceptable to consumers have been implemented in the last decade;
- performance of actual production models incorporating design changes for efficiency have fallen short of engineering calculations;
- there are major technical uncertainties and marketability problems associated with many of the fuel efficiency technologies (e.g., advanced diesels, two-stroke engines)
- real trade-offs do exist between further fuel efficiency improvements and environmental (particularly for NO_x emissions) and safety standards.

Alternative Fuels. A number of alternative fuels have been proposed for light-duty vehicles in the past few years. In the U.S. these proposals have been driven primarily by concerns about the effects of emissions on urban air quality and the impact of petroleum-based fuels on energy security. Only recently have analysts focused attention on the greenhouse contributions of alternative fuels. Near-term options of interest include the use of alcohols--both ethanol and methanol, as blends with traditional fuels or as complete substitutes--and direct use of compressed natural gas. The discussion below refers to dedicated alternative-fueled vehicles designed and optimized for the alternative fuel. In the near term, however, "flexible fuel" vehicles may be produced that would be capable of burning one or more of the alternative fuels as well as gasoline. In this case, because vehicles are not optimized, energy efficiency may be less than optimal (and greenhouse gas emissions would be higher).

Methanol, as well as ethanol, can be produced from biomass. If these resources are replaced as they are used, then their combustion should not contribute to the global warming problem in the long run. In the U.S. and globally, however, natural gas would be the most likely feedstock for methanol in the near term. The estimated net contribution to greenhouse gases when gas is used as a feedstock will be roughly equivalent to that from burning gasoline from petroleum. On the other hand, greenhouse gas emissions from the use of coal-based methanol, measured over the entire fuel cycle, are about double those from crude oil-derived gasoline (DeLuchi et al., 1987). A shift from natural gas to coal in the long run could lead to large increases in greenhouse gas emissions.

Compressed natural gas (CNG) as a transport fuel produces fewer CO_2 emissions per unit of energy released than any other fossil fuel, and also seems to be among the cleanest fuels available when considering its emissions of other gases that affect urban air quality (e.g., NO_x, CO, and VOCs, although some questions remain about the level of NO_x emissions from CNG vehicles). However, leaks of natural gas, primarily methane (CH_4), from the production, distribution, and refueling processes, could add to the concentrations of this greenhouse gas. Some researchers estimate that this increase in methane could offset the advantage of lower CO_2 emissions. The degree to which methane releases would increase or could be controlled is highly uncertain.

CNG is currently used as a transport fuel in Canada and New Zealand, among other industrialized countries. In New Zealand, CNG, LPG, and synthetic gasoline (from natural gas) meet half of the total gasoline demand. Other industrialized countries have small alternative fuels programs, mainly based on CNG, LPG, and methanol (Sathaye et al., 1988).

Ethanol is likely to be produced from biomass but is also likely to have difficulty competing economically unless its production is heavily subsidized by government. Additionally, ethanol production by means of current technologies relies on biomass feedstocks such as corn and sugarcane, which are also food crops. This competition with food production raises concerns about the long-run viability of this approach.

In summary, in the near term it appears that the technical potential for industrialized countries to achieve reductions in greenhouse gas emissions from the use of alternative fuels is limited. The CO_2 reductions from alternative fuels in industrialized countries would not be enough to offset projected growth in vehicle miles traveled. However, their use in combination with fuel efficiency improvements may help alleviate other concerns related to urban air quality and energy security without increasing the global warming commitment. As discussed later, in the longer term, alternative fuels derived from renewable sources may play a key role in reducing greenhouse gas emissions in the transportation sector.

Tradeoffs With Other Goals. The governments of many industrialized countries regulate light-duty vehicles to reduce emissions of a number of air pollutants. In addition, consumers value performance attributes other than fuel efficiency, as well as comfort, safety, and cost in their choice of automobiles. To the extent that proposed fuel-efficiency improvements or alternative fuels involve tradeoffs against these other goals, or are perceived to require such sacrifices, they may be more difficult to implement.

Safety is a major concern long associated with light-duty vehicles. Some industrialized countries regulate the manufacture, sale, and maintenance of vehicles to improve safety. In the 1970s and 1980s, U.S. safety standards significantly improved vehicle safety. Some evidence exists of a correlation between size and weight reductions and increases in injury and fatality for currently available vehicles (OTA, 1982). Clearly, effects on safety must be considered in the evaluation of technical alternatives for improving fuel efficiency.

Weight reductions to improve efficiency may in fact reduce the structural strength of vehicles, thus making them less safe; however, it is not true that the use of lighter materials always reduces safety. U.S. government crash tests on the Nova, for example, have consistently shown the car's crash performance to be superior to other vehicles weighing as much as 50% more (Bleviss, 1988). In addition, options for improving automobile safety, such as air bags, automatic seat belts, as well as fundamental design changes, could improve both efficiency and safety if pursued as a national goal by regulatory bodies and manufacturers.

Another major concern associated with automobiles is their emissions. Vehicles emit several pollutants--particulate, volatile organic compounds (VOCs), carbon monoxide (CO), and nitrogen oxides (NO_x)--which contribute to urban air quality problems and indirectly to climate change. Some options for increasing fuel efficiency (and thus reducing greenhouse gas emissions) have the potential to worsen local air pollution problems. For example, diesel engines are more fuel efficient than gasoline-powered automobiles, but diesel engines tend to produce much greater emissions of particulates (many of which are cancer-causing compounds) per mile travelled.

Another complicating factor is related to the way in which vehicle emissions standards are defined. In most industrialized countries pollutant emission requirements for vehicles are applied on a grams-per-mile (or kilometer) basis. One concern about this approach is that it may not encourage development of technologies that would simultaneously improve fuel efficiency and reduce emissions of urban air pollutants, even though such technological options exist (Bleviss, 1988). In addition, as fuel efficiency improves, the marginal cost of driving would decline, assuming fuel prices remain constant, which might induce vehicle operators to drive more than they otherwise would. If this effect is significant (an unresolved empirical question), it would offset some of the expected reductions in greenhouse gas emissions and could also result in a net increase in emissions of conventional urban air pollutants.

Conversely, some of the options for reducing emissions of pollutants can lead to increases in emissions of greenhouse gases. As noted above, switching to methanol fuels could increase greenhouse gas emissions if the methanol is derived from coal. Similarly, use of electric cars could result in large net increases in greenhouse gas emissions if the electricity were generated from fossil fuels.

There are potential solutions to all of the problems illustrated. For example, new diesel cars are much cleaner than earlier models. Mercedes Benz and Volkswagen have now developed emission control devices that make it possible for their diesel cars to meet the strict California particulate standard of 0.2 grams/mile (Bleviss, 1988). In addition, emission standards could be modified (e.g., to grams per gallon in conjunction with higher efficiency standards, or direct regulation of CO_2 emissions) to encourage fuel-efficiency improvements that would also benefit local air quality. Over the long term, it is important that options like methanol-fueled and electric vehicles are promoted in conjunction with non-fossil (or at least non-coal) energy inputs.

Many consumers are concerned about sacrificing size, comfort, or driving performance to achieve major improvements in fuel economy. For several years, however, analysts have demonstrated possible design changes that would reduce fuel consumption while retaining the size and performance of a large automobile (e.g., Forster, 1983).

Some of the prototype high-efficiency vehicles have already demonstrated that size and acceleration do not necessarily have to be sacrificed. The Volvo LCP 2000, with an average fuel efficiency of 3.9 l/100 km (65 mpg), accelerates from 0 to 60 miles per hour in 11 seconds--compared with 13.1 seconds for the average acceleration of the U.S. new-car fleet (1986 models, which averaged only 8.4 l/100 km, or 28 miles per gallon). Likewise, a recently developed Toyota lightweight prototype car is designed to seat 5-6 passengers while achieving 2.9 l/100 km (80 mpg) under urban driving conditions (Bleviss, 1988).

In summary, it appears technologically quite feasible to achieve, for example, a new-car average fuel efficiency of 4.7 l/100 km (50 mpg) in the U.S. by 2000, while maintaining or improving current standards of safety, air pollutant emissions, comfort, and engine performance. This would imply, through vehicle turnover and continuing technical innovation, that fleet average efficiency would reach about 4.7 l/100 km (50 mpg) by 2010. This achievement would require, however, a strong commitment by government and industry. The reductions in energy (and greenhouse gas emissions) from this improvement would be significant.

The U.S. Department of Energy (U.S. DOE, 1987c) has recently projected that automobile vehicle miles travelled (VMT) in the U.S. will increase over 50% by the year 2010 (from 1,315 to

2,032 billion miles). In calculating future transportation energy use, the report also projects that automobiles will average about 8.7 l/100 km (27 mpg) in 2010 (overall average of all operating vehicles, not new-car average). If automobiles averaged 4.7 l/100 km (50 mpg), the energy consumed would be reduced by over 45%, or more than 30 billion gallons of gasoline (3.8 EJ[1]) in 2010--more than a 40% decline from the 1985 consumption level (as opposed to the 7% increase projected by DOE). To achieve these results, it would be necessary for automobiles to perform at the specified fuel efficiency over the life of the vehicle. Currently, there is considerable degradation in fuel efficiency over the life of an automobile. This requires further study and could limit the expected energy savings from fuel economy programs. In addition, the current trend in the U.S. toward light-duty trucks as personal transportation vehicles could offset some of the efficiency gains if it continues in the future. Efficiency improvements in light-duty trucks and/or programs to discourage personal use of these vehicles could produce significant energy savings in the U.S. (The VMT and average mpg assumptions in the DOE analysis are higher than in the No Response scenarios used in this report, but illustrate the potential for improvement). Decreasing fuel use in the U.S. could also support many important national goals such as reducing international trade deficits and improving energy security. Improvements of this magnitude could be made for the OECD as a whole and would undoubtedly also spill over to the non-OECD countries that produce vehicles for sale in the OECD and/or import vehicles or technology from the OECD. Most light-duty vehicles currently in production are derived from designs that originated in the OECD (Bleviss, 1988).

Freight Transport Vehicles

Diesel trucks use about 14% of the oil consumed for transportation in OECD countries (von Hippel and Levi, 1983). Improvements in efficiency in this sector could, therefore, have a noticeable

---

[1] EJ = exajoule; 1 exajoule = 10^{18} joules

effect on total energy use. Current prototype vehicles could reduce energy use per ton mile in the U.S. by as much as 40% for truck transport (*Automotive News*, 1983). Goldemberg et al. (1988) have estimated that further improvements on the order of 50% or better could be achieved with existing technology and at reasonable cost. Box 7-2 describes one of the most promising advanced diesel technologies. OECD truck fuel use for 1985 totaled about 5.3 EJ. A 50% improvement in fuel efficiency over the next several decades seems feasible if aggressively pursued. As truck freight ton miles are projected to grow in the future, this level of improvement could save at least 2.6 EJ by 2010.

Both rail and water transport are much more efficient than trucks on a ton-mile basis. To the extent that shippers could be encouraged to shift freight to these modes in the future, either through price incentives or other policy mechanisms, net energy use for freight transport could be reduced. One interesting approach is being testing by General Motors Corporation (GM). GM has developed a new truck trailer that can also be easily converted to a rail car and connected to a freight engine. This would allow loading of truck trailers at source points, truck hauling to the nearest rail terminal, and conversion to rail without unloading/reloading. Likewise, near the destination, the transition back to truck is simple. GM has estimated that for hauls of over 200 miles, this approach could reduce energy use to 20% of current energy use for an all-truck haul or 50% of a conventional rail shipment (Sobey, 1988).

Improvements in ship fuel efficiency are also possible. Using some of the same diesel engine technologies identified for truck engines, 30-40% improvements in efficiency may be possible over the next several decades. Wind-aided cargo ships may also improve efficiency somewhat in the future.

BOX 7-2. ADIABATIC DIESEL ENGINE TECHNOLOGY

The diesel engine is currently the most efficient power plant used in heavy duty and some light duty vehicles. Little has changed in its basic design over the years. However, motivated by the energy crises of the 1970s, engine designers began trying to improve the fuel-efficiency of diesels even further. One of the most promising results of this research is the adiabatic engine, which combines new structural ceramic materials and turbocharging to increase the effective use of the heat generated during combustion.

- Adiabatic design, which means "without heat loss," increases efficiency by retaining heat in the combustion chamber instead of losing it to exhaust gases and the engine coolant, harnesses the high pressure exhaust gases, and reduces weight and parasitic power losses by eliminating the normal cooling system.

- As shown in the figure below, turbocompounding increases the pressure of gases in the combustion chamber using a turbocharger and then harnesses the extra pressure of the exhaust gases with a turbine connected to the engine crankshaft. Structural ceramics, which are being developed to withstand temperatures and pressures reaching 1000°C and 2000 psi, respectively, will be used to insulate the combustion chamber, allowing greater thermal efficiency.

- Cummins Engines and Adiabatics in the U.S., as well as Japanese automakers, are at the forefront of introducing the adiabatic engine both for heavy duty trucks and passenger vehicles. A Ford Tempo with an adiabatic engine is projected to be able to have a fuel economy of 80 mpg. Along with the consequent reduction in CO_2, additional improvements are expected in hydrocarbon, CO, and NO_x emissions, and particulates are expected to be reduced by as much as 60-80% over current diesel technology.

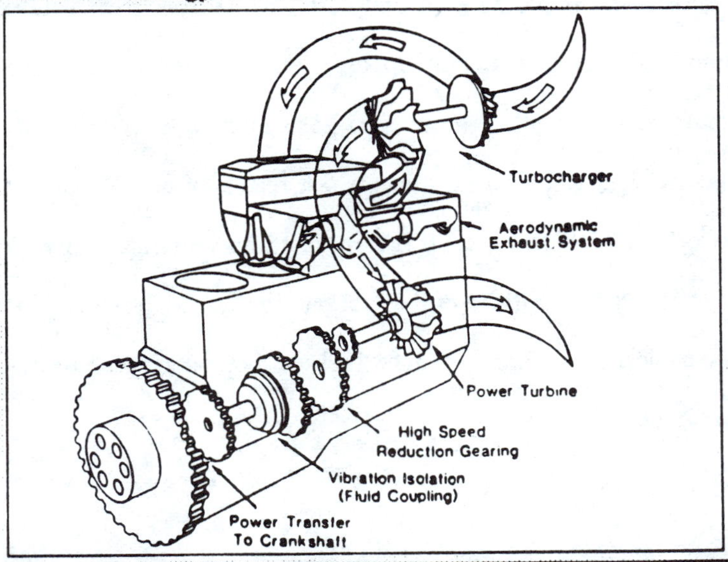

Source:
Kamo, 1987.

Aircraft

Dramatic improvements in fuel efficiency have been achieved in the airline industry over the past decades. From 1970 to 1980 fuel use per passenger mile in the U.S. declined by over 40% (Holcomb et al., 1987). Nevertheless, by 1980 energy still accounted for about 30% of total operating costs for commercial airlines in the U.S. (Goldemberg et al., 1988). Since 1980, energy intensity has continued to decline due to some additions of more efficient aircraft and continued improvements in load factors (revenue passenger miles divided by available seat miles - U.S. DOT, 1988), but at a slower rate (from 1980 to 1984 the improvement was about 4% - Holcomb et al., 1987). The decline in the rate of efficiency improvement is highly correlated with declining energy costs. For example, between 1980 and 1987, nominal fuel prices declined by nearly 40% (U.S. DOE, 1988b), reflecting a drop of more than 50% in real terms. Combined with some efficiency improvements, this has resulted in energy costs being reduced to a much smaller percentage, 10-15%, of total operating costs. Thus, the economic incentive to reduce energy intensity has been greatly reduced.

Because of the historical importance of fuel costs, however, a great deal of research has been conducted and within the industry to identify opportunities for improvements in efficiency. Currently, commercially-available new planes are more than 25% more efficient than the 1980 fleet average (Smith, 1981 for 1980 average, Ropelowski, 1982 for test results of new 757 and 767 models). Already improvements have been identified which, if incorporated into aircraft design, could reduce fuel use per passenger mile to less than one-third of the current average (Maglieri and Dollyhigh, 1982; Smith, 1988). It may be technically possible to reduce fuel use per passenger mile to 50% of the current average by the year 2010.

The rate at which airline passenger miles will increase in the future is, however, a matter of considerable uncertainty. Some industry sources are projecting very high rates of growth in the next few decades, which could offset the projected gain due to improvements in energy per passenger mile. Kavanaugh (1988), for example, projects increases in global jet fuel use of 60 to 120 percent by the year 2025 despite assuming significant substitution of more fuel-efficient aircraft during the same period.

Control of NO_x and CO Emissions from Mobile Sources

The United States and most other OECD countries currently regulate the emission of hydrocarbons (HC), carbon monoxide (CO), nitrogen oxide (NO_x), and particulate matter (PM) on a gram per kilometer basis. Many developing countries have recently adopted some emission control standards as well. The standards vary for the weight class of the vehicle as well as by the type of engine.

International comparisons of emissions standards are difficult because test procedures vary, but it is generally recognized that U.S. standards are among the most stringent in the world. For light-duty gasoline vehicles (which constitute a majority of the U.S. fleet), the U.S. standards are 2.1, 0.25, 0.62 g/km for CO, HC, and NO_x, respectively. Emission standards in most European countries are significantly less stringent than in the U.S. (OECD, 1988). Many developing countries have much less stringent standards, or none at all.

Vehicles sold in the United States control emissions in two steps. The first step is to control the amount of pollutants formed during the combustion process. During combustion, the prime determinant of the amount of carbon monoxide formed is the air/fuel mixture. As the air/fuel ratio increases, CO emissions fall. On the other hand, NO_x emissions increase as the air/fuel ratio rises, a result of the concurrent increase in combustion temperatures, the primary determinant of NO_x formation. Higher combustion temperatures are often associated with increased power. An engine designed for greater power will generally produce higher "engine-out" NO_x emissions than a engine designed for fuel economy. Electronic engine management systems have, however, been able to minimize these tradeoffs, since many of the critical engine parameters (like the air/fuel ratio) can be controlled much more precisely (NAPAP, 1988; OECD, 1988).

The second step is to treat the "engine-out" exhaust after combustion to reduce emissions to the acceptable standard. To control the "engine out" emissions, both catalysts and exhaust gas recirculation (EGR) are used on virtually all new U.S. (and Japanese) passenger cars, usually in conjunction with electronic engine management systems. Since 1981, the primary catalyst system used on U.S.-bound cars has been the three-way converter--named for its ability to reduce emissions of HC, NO_x, and CO, as opposed to just one or two of the gases. Some cars also use a second oxidation catalyst to catch additional HC and CO. Catalysts use a variety of precious metals, including platinum, rhodium, and palladium, to breakdown or reduce the unwanted emissions without causing the metals themselves to react (*Automotive News*, 1988; OECD, 1988; White, 1982). Exhaust gas recirculation reduces NO_x emissions by reinjecting a portion of the inert exhaust into the engine's incoming air. The inert exhaust gas cannot react in the combustion process and reduces peak temperatures during combustion. As a result, less NO_x is formed. Today, all U.S. light-duty vehicles have EGR systems (Husselbee, 1984).

Some tightening of existing U.S. standards has been considered and determined to be technically feasible, although expensive (NAPAP, 1988). More significant improvements in global emissions of NO_x and CO would result from the extension of U.S. standards to the rest of the OECD and ultimately, to the rest of the world. These extensions are assumed over time in the Stabilizing Policy scenarios.

Current technologies for compliance with U.S. emissions standards have added substantially to the cost of new vehicles, although improving technology may reduce the incremental cost in the future (U.S. EPA, 1985, NAPAP, 1987). Also, with existing technologies a modest tradeoff has been documented between emissions control, and fuel efficiency and performance, at the current level of U.S. emissions standards (White, 1982, OECD, 1988). Tradeoffs and cost for tighter standards, however, would depend on the stringency of the standards and the level of technology, as well as the demand for other characteristics like performance.

Near-Term Technical Options: Developing Countries

Transportation energy use is a serious concern in developing countries for several reasons. Worldwide, energy used for transportation is almost exclusively oil. From 1973 to 1986 oil use by developing countries increased by 60%. During this same period, oil use by OECD countries declined by 13%. Recent projections by the Department of Energy (U.S. DOE, 1987c) indicate that the overwhelming majority (86%) of growth in oil consumption in the "free world" (defined by DOE to exclude the centrally-planned economies of Eastern Europe, the Soviet Union, China, Cuba, Kampuchea, North Korea, Laos, Mongolia, and Vietnam) through the year 2010 could come from developing countries. The largest component of the dramatic increases in oil use in developing countries during the last decade is in the transportation sector. For 15 of the largest developing countries, about 50% of the growth in oil consumption in the 1970-1984 period has been in transportation applications (Meyers, 1988). Thus, indications are that energy use for transportation has been growing at a very rapid rate in recent years in developing countries, and this high rate of growth is projected to continue in the future.

Although some developing countries have been successful in implementing programs to reduce fossil-fuel use (or at least oil use) in other sectors, very little attention has been paid to the transportation sector, either by these countries themselves or by international development assistance agencies (with a few notable exceptions, e.g., Brazil and, more recently, several other countries, including the Philippines). There are a number of reasons why it is difficult for developing countries to implement the technical options that have been relatively effective to date in the industrialized countries.

First, information about transportation energy use is limited in many developing countries. It is very difficult to estimate, for example, what portion of the fuel is used in new versus old vehicles, light trucks versus heavy trucks, two- and three-wheeled vehicles, etc. The information that is available, however, suggests other problems.

Road vehicles tend to be kept in service far longer in developing countries and often are used for purposes other than what they were originally designed for. There is some evidence that aging vehicles are often poorly maintained in developing countries. Poor roads also often contribute to increased energy use per vehicle kilometer of travel (VKT). Urban congestion can also have this effect, although in some cases it may also act as a deterrent to increases in VKT. Although heavily utilized, mass-transit systems in developing countries are generally poorly developed and are also affected by poor or congested road systems.

Thus, the approaches that have been successful in slowing the growth of transportation energy use in the industrialized countries over the past 15 years may not be as effective in many developing countries. Industrialized countries have been able to significantly reduce the average consumption of fuel per highway mile by replacing existing vehicles with more efficient newer models and developing or expanding mass transit in urban areas. The effectiveness of both strategies in most developing countries is much more limited because of the slower turnover of vehicles and lack of capital to invest in infrastructure improvements.

On the other hand, it is expected that as developing countries reach a certain per capita income, they will experience a rapid explosion in the demand for personal vehicles that will dramatically increase transportation energy use (Sathaye et al., 1988). The fuel efficiency of new vehicles available during that period can have a significant effect on future transportation energy use.

Those developing countries that do not have domestic oil resources are concerned that increasing their imports of oil will diminish the already-limited foreign exchange available to finance development and hence, limit long-term growth. Thus, there is great interest emerging in limiting oil use for transportation. Efficiency improvements, though often more difficult to achieve than in industrialized countries, clearly are part of the solution. Biomass-based alternative fuels are very important for some developing countries (Brazil, for example) in the near term. Other developing countries (like some industrialized countries) may have natural gas that can be used for transportation or other types of biomass-based options. Generally, the transportation energy solutions that would be attractive to developing countries for the purpose of reducing oil imports - efficiency improvements and alternative fuels - are also beneficial in reducing greenhouse gas emissions. One exception is the use of coal as a rail fuel, which occurs in China and India and results in both lower energy efficiency in rail transport and in more CO_2 per unit of fuel consumed. However, this option may appear attractive to countries concerned primarily with minimizing oil imports.

Fuel-Efficiency Improvements

As individual developing countries reach a certain level of economic activity and per capita income, it is expected that the demand for personal vehicles will "take off," as occurred historically in several more-industrialized countries. Currently, gasoline prices are higher in many developing countries and oil imports frequently make up a large fraction of total imports in developing countries.

Therefore, improvements in the efficiency of new vehicles, both light- and heavy-duty, should be very attractive in developing countries to reduce fuel costs.

Improving Existing Vehicles

Because of capital scarcity, all types of vehicles are kept in service far longer than they are in industrialized countries. Certainly for some developing countries, programs to improve the quality of maintenance or accelerate retirement of older vehicles may be very useful in reducing oil consumption (and greenhouse gas emissions).

For the same reason that aging vehicles are kept in service, new classes of intermediate vehicles, often used as low-cost transport vehicles, have emerged in developing countries. These are generally produced locally, often by small companies or even a single individual. Some alternative vehicles, notably the Chinese tractor converted for passenger transport, are adaptations of vehicles designed for other purposes. India also does a significant amount of road transport in rural areas with tractors. As a means of passenger or freight transport, many of these vehicles are very energy-inefficient. The Chinese tractors, for example, are estimated to use 75% more fuel than would a four-ton truck while carrying only one ton; these tractors account for 27% of the total diesel fuel use in China (World Bank, 1985). Programs to improve the efficiency of these vehicles or to replace them with more efficient alternatives may be very effective in the near term in some developing countries.

Alleviating Congestion and Improving Roads

In rural areas of many developing countries, roads are so poor that traffic must move much more slowly than is customary for intercity travel in industrialized countries. Frequent stops and starts are also a problem in these areas. These conditions inevitably lead to reduced energy efficiency regardless of the quality of the vehicles themselves. A recent study has concluded, however, that poor road conditions do not appear to act as a deterrent to increased vehicle ownership in rural areas of developing countries (Meyers, 1988). Thus, carefully planned highway improvements could result in net reductions in fuel use in some countries. In addition to reducing fuel use in the existing mix of vehicles, better roads may allow "upsizing" of some of the existing traffic to larger trucks and buses that are much more efficient on a passenger- or ton-km basis. Encouraging the use of more efficient modes, such as rail transport, may also be possible in some cases.

In urban areas, congestion is already clearly a major problem in many developing countries (as it is in many industrialized countries) and is likely to become more severe as rapid urbanization continues in many developing countries. Although congestion, as already mentioned, reduces efficiency in fuel use, it may also act as a deterrent to increased vehicle use, promoting the widespread use of more efficient alternatives to personal automobiles, such as motorcycles and mass transit. The degree to which congestion functions as a deterrent to increased transportation energy use is uncertain and needs further investigation.

Thus, it is important to carefully evaluate local conditions in designing improvements to urban road systems, including more extensive roads, but also better road maintenance and better planned roads. It may also be important to combine road improvements with other measures such as mass transit to achieve overall improvements in energy efficiency.

Alternative Modes of Transportation

In addition to encouraging expansion of urban mass transit, developing countries may wish to promote alternatives to highway transport in both rural and inter-city travel. Because major investments may be required to develop or improve highways for these types of travel, it may be more economically attractive to direct this investment into improved rail systems, for example, which move passengers and freight more efficiently. If highway improvement programs are carried out, however, they may be combined with the introduction of efficient bus systems, which would offer an attractive alternative to personal vehicle ownership and use.

Alternative Fuels

Alternative fuels based on locally-available resources may be economically viable and important options in developing countries much sooner than is the case for industrialized countries. In countries like Brazil, which have abundant agricultural land, commercial technologies to convert crops, such as sugarcane or corn, to produce ethanol may make sense (see **PART TWO: ENERGY SUPPLY**, for more detailed discussion of the Brazilian ethanol program). A fuels program based on sustainable biomass production is extremely beneficial in reducing net CO_2 emissions. Most developing countries would have difficulty, however, in diverting significant amounts of biomass, which is currently used for food, to energy purposes. Converting agricultural residues or forest products to fuel may make more sense if the conversion processes can be made economically attractive.

In other countries locally available natural gas may be readily converted to compressed natural gas (CNG), which would reduce emissions of CO_2 as well as many other air pollutants and also reduce oil imports. A recent review of international programs noted that most Asian countries and many Latin American countries that have domestic gas resources are conducting feasibility studies of CNG use and many have pilot programs in place (Sathaye et al., 1988). Where natural gas is currently vented or flared as a by-product of oil production, (e.g., in the Middle East), the availability of cheap local oil discourages investments in natural gas distribution and utilization systems.

A final fuel-switching option that could be helpful in the near term is replacing coal with diesel fuel or electricity in rail transportation systems. Coal use as a rail fuel has virtually disappeared in industrialized countries--primarily because coal-fired rail systems are markedly less efficient than the alternatives. In energy consumed per mile travelled, diesel trains are more than 3.5 times as efficient, and electrified rail is 13 times more efficient (compared on the basis of secondary energy consumed). If electricity is generated from coal, primary energy consumed and CO_2 emissions would be 3 times greater than the end-use energy consumed. Electric rail in this case would be about 4 times more efficient than coal-fired rail in primary energy consumption. However, a few developing countries with abundant coal resources and extensive rail systems--notably India and China--still use coal in rail transport. Although considerably less efficient, this option may be appealing from the perspective of minimizing oil imports. India has a program of gradual replacement underway, which is expected to eliminate coal use in rail systems by the year 2000. A shift to diesel fuel should reduce the CO_2 emissions from this source by a factor of five.

Near-Term Technical Options: Soviet Bloc Countries

In the USSR and Eastern Europe, transportation makes up a much smaller proportion of total energy use than is the case in industrialized countries--primarily because there are many fewer automobiles and trucks. However, that number is growing rapidly: from 1970 to 1980 the number of automobiles in the Soviet Union increased from 1.6 to 6.9 million (a rate of 15%/year), and the number of trucks rose from 3.2 to 5.1 million (4.7%/year).

On the other hand, fairly significant improvements have been made in recent decades in the efficiency of freight transport, primarily transport by rail. From 1960 to 1975, ton-kilometers of freight hauled by all forms of transport in the Soviet Union increased by 276%, while fuel use increased by 2.4% and electricity use increased by 418% (from a very small base). Overall, this represents a significant increase in energy efficiency due primarily to the replacement of coal locomotives by much more efficient diesel and electric engines (Hewett, 1984).

Given the rapid growth in the numbers of automobiles and trucks in the recent past, and in the number projected in the scenarios developed for this report, it appears that an important option for these countries will be to increase the efficiency of new highway vehicles. In addition, because of the large natural gas resources available in the Soviet Union (discussed in **PART TWO: ENERGY SUPPLY**), the feasibility of using compressed natural gas as a vehicle fuel may deserve further investigation.

Summary of Near-Term Technical Potential in the Transportation Sector

With aggressive programs to improve transportation energy efficiency across the board in industrialized countries, it appears that significant overall reductions could be made over time. Improvements in the efficiency of light-duty vehicles and freight transport could result in reductions of energy use on the order of 7 EJ of transportation energy from the No Response scenario levels by 2010 in the OECD. If the same efficient technologies were transferred to developing countries and the Soviet bloc, even larger reductions could be achieved below scenario levels because of the rapid expansion of vehicle stock in those areas. This suggests that the technical potential may exist to reduce energy use by 25 EJ by 2010. This estimate is higher than that assumed in the climate policy scenarios, where the maximum transportation energy reduction is about 32 EJ by the year 2025. Furthermore, it appears that there are fuel-switching options that could be implemented during this time period, which could reduce the greenhouse gases emitted per unit of energy consumed in transportation. The overall potential of these options has not been quantified.

Long-Term Potential in the Transportation Sector

Over the long term, technical possibilities for reducing greenhouse emissions from transportation greatly increase. The range of options runs from further improvements in highway vehicles and expanded use of alternative fuels, to alternative transportation modes, to measures that would reduce the need for transportation. All of these options are somewhat speculative and very sensitive to assumptions about the nature of society in the long term. The discussion here is intended to illustrate possible options rather than to suggest which choices should be made. Considerably more systematic and detailed analysis is required before useful comparisons can be made.

Urban Planning and Mass Transit

A major concern in many urban areas throughout the industrialized countries (and in many developing countries as well) is increasing traffic congestion. When vehicles spend an increasingly greater proportion of their time idling in stop-and-go traffic, they use more fuel and emit more air pollutants per mile travelled. In the near term, solutions to urban congestion problems are extremely difficult. For the long-term, however, alternative approaches to alleviating this problem will tend to incidentally benefit the climate warming problem. Mass transit systems not only reduce highway commuter traffic but the energy used per passenger mile is much lower. The energy intensity of one person commuting alone by car is over 4.5 MJ/km.[2] Average intensities (over all time periods) for bus and rail transit are reported to range from about 2.0-2.5 MJ/passenger-km (Holcomb et al., 1987). At normal commuting times transit systems tend to be much closer to fully-loaded, however, so that the energy intensities per passenger-mile would be even lower. The energy intensity of one person commuting alone by car is over 4.5 MJ/km (7,000 Btu/mile). Average intensities (over all

[2] MJ = megajoule = 10^6 joules.

time periods) for bus and rail transit are reported to range from about 2.0-2.5 MJ/passenger-km (Holcomb et al., 1987). At normal commuting times transit systems tend to be much closer to fully-loaded, however, so that the energy intensities per passenger-mile would be even lower. Similarly, efforts to encourage carpooling can relieve congestion and reduce fuel use at the same time.

Another possibility is shifting to smaller, commuter vehicles. General Motors Corporation is testing a three-wheeled, one- or two- passenger, narrow commuter car, which is essentially more like a covered motorcycle than a traditional automobile. Because the vehicle is much narrower than a normal passenger car, a standard traffic lane could be split in half to double the carrying capacity of existing roads. Because the vehicle is small and aerodynamically-designed, it would also be much more fuel efficient than today's cars, achieving over 100 miles per gallon (Sobey, 1988). Clearly, some safety issues and other complexities in integrating such vehicles into current urban traffic patterns remain to be worked out.

The technical potential exists to design and construct urban areas that are much more energy efficient in terms of their transportation requirements (as well as in their energy requirements for other end uses). By comparing cities whose transportation energy use is very low, relative to global averages, with cities whose energy use in this sector is high, it is possible to identify differences in location patterns, mass transit systems, and other factors that can partially explain the differences in energy demands. In theory, it should be possible to introduce incentives that will cause the cities to developed similarly to the low energy examples over time. This may be especially important in developing countries where populations, especially urban populations, are growing rapidly.

Alternative Fuels

Use of **alcohol fuels** as an alternative to gasoline in highway vehicles is an option that is already receiving considerable interest for a number of reasons, as described earlier. Technologies currently exist for producing ethanol and methanol from various types of biomass. Further research, testing and commercial demonstrations could be helpful in improving performance and lowering cost. In the long run it appears that methanol may be preferable because the biomass feedstock does not necessarily have a food value and therefore is not in direct competition with food production.

Technologies currently exist for operating highway vehicles with **hydrogen fuel.** Such vehicles are not currently viewed as commercial, primarily because of the high cost of hydrogen fuel and difficulties in storing enough hydrogen on board for highway driving. A hydrogen-powered automobile built by Daimler-Benz of West Germany and a hydrogen-powered bus built by the Billings Energy Corporation of Provo, Utah, are examples of currently-operating test vehicles. These two vehicles use metal hydride storage tanks, one promising approach to the storage problem (Ogden and Williams, 1988). In addition, if the fuel-efficiency improvements described above are incorporated, future vehicles may be able to achieve driving ranges comparable to today's vehicles while carrying much less fuel on board. The possibilities for producing hydrogen at competitive costs are also improving with the development of solar photovoltaic technology (see **PART TWO: ENERGY SUPPLY**). Another attractive feature of hydrogen vehicles is that the engines and internal structure required are very similar to what CNG-powered vehicles would require. CNG is already being used in fleet applications in some countries and will be more widely-used in the future. This could provide a market for the initial transition to hydrogen when and if the cost and storage problems are resolved.

Electric-powered vehicles have been discussed for many years as an option for reducing oil use and/or urban pollution. As with hydrogen, the problem of storing enough energy on board to provide a reasonable driving range is an as-yet-unresolved obstacle. In addition, vehicle cost and performance ability comparable to the cost and performance of today's vehicles have not yet been realized. Concerns about higher future electricity costs could also retard penetration of electric cars, even if other problems were resolved.

The source of primary energy for both electric and hydrogen-powered vehicles will determine whether a switch to these fuels increases or decreases greenhouse gas emissions. Use of non-fossil sources of electricity, such as solar or nuclear, would decrease emissions of CO_2 and other greenhouse gases. Conversely, use of coal-fired generating capacity as the primary source of electricity or hydrogen would lead to increases in greenhouse gas emissions.

Expanded Use of Emerging Technologies

Telecommunications may substitute for many transportation services in the future. Teleconferencing is already replacing some types of business travel, although the magnitude of this substitution has not been quantified in energy terms as yet. In the future, as video conferencing equipment improves and is more widely available, this option could become more important, particularly if higher transportation costs and congestion act as incentives. Catalogue shopping, electronic mail, electronic advertising, and electronic banking are other applications of telecommunications that are beginning to replace specific transportation needs. A recent analysis projects that transportation energy use may decline in OECD countries because of these substitutions. As is the case with many energy efficiency improvements, the motivation for this substitution has very little to do with energy use. These substitutions are taking place because consumers and businesses perceive advantages in convenience, time saving, access to wider selections, and cost savings (Schipper, et al., 1989).

Fuel cell technology has many potential applications (described briefly in **PART TWO: ENERGY SUPPLY**). Two appealing characteristics of fuel cells are that cost-effectiveness is not fundamentally a function of size, as with many energy technologies, and that they are virtually pollution-free at the point of use (Jessup, 1988). Because of these characteristics, one possibility is to use small fuel cells to power highway vehicles. Input fuel can be derived from natural gas, coal,. or, ultimately, renewables-based hydrogen. Several different fuel-cell approaches are being researched and tested currently. If they prove economic, the fuel cell could provide very efficient and clean power for mobile sources at some point in the future.

High-speed rail systems are currently in commercial use in Japan and France. They compete well with aircraft or automobiles on a performance basis for some intercity travel. Energy consumed per passenger-kilometer is significantly lower than with automobile or air travel alternatives. As technologies develop in the future (e.g., superconductors), these systems could become more efficient and economically attractive. The primary constraints appears to be cost of constructing the systems and concerns about safety and rights of way.

RESIDENTIAL/COMMERCIAL SECTOR

In the United States residential and commercial energy services consumed 17 EJ in 1985, or about 29% of total secondary energy (36% of equivalent primary energy). For the OECD as a whole, the picture is quite similar, with residential and commercial energy use amounting to about 30% of the total secondary energy. Figure 7-7 shows the distribution of energy use within this category in the U.S. The largest component (more than a third) of residential/commercial energy

FIGURE 7-7

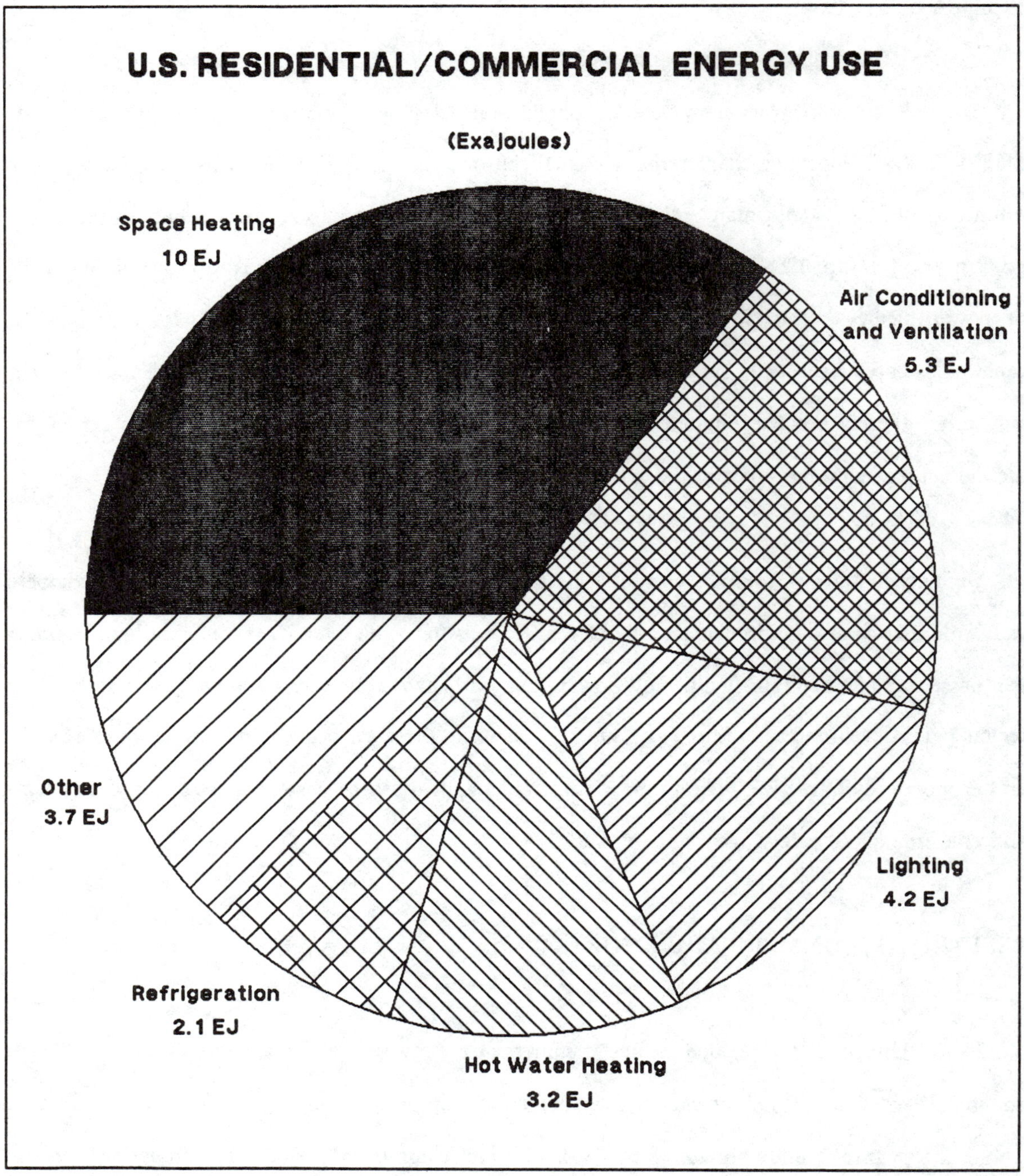

U.S DOE, 1987a.

use is for space heating; combined with air conditioning and ventilation, the overall use of energy for space conditioning accounts to more than half (54%) of all residential and commercial energy use. Lighting accounts for another 15%, hot water heating, 11%, refrigeration, 7%, and the remaining energy (13%) is divided among all of the other appliances and equipment used in residences and commercial establishments (U.S. DOE, 1987b).

As shown in Figure 7-8, global energy use in the residential and commercial sectors in the No Response scenarios grows by 29-65% by the year 2025. The relatively-wide range is apparent in all regions and reflects alternative assumptions about the rate of investment in buildings and in population growth. In addition, a shift toward electricity for a higher percentage of energy use in these sectors results in increases in end-use efficiency, but implies that primary energy required (accounting for losses in electricity generation) is growing more rapidly.

In the scenarios, the percentage of total energy used in the industrialized countries declines significantly over time. Residential/commercial energy use in 2025 for the OECD in the No Response scenarios ranges from slightly higher than today's levels to slightly lower. This reflects the slow rate of population growth and the technological advances that are already increasing energy efficiency in these countries.

In contrast, residential and commercial energy use in developing countries for 2025 ranges from a doubling to more than a tripling of current levels in the No Response scenarios. It is expected that economic growth in these countries in the future will rapidly translate into increasing demands for energy-related amenities in homes and commercial buildings. In the Soviet Bloc increases are on the order of 50-75% for similar reasons.

FIGURE 7-8

RESIDENTIAL/COMMERCIAL ENERGY USE BY REGION

(Exajoules)

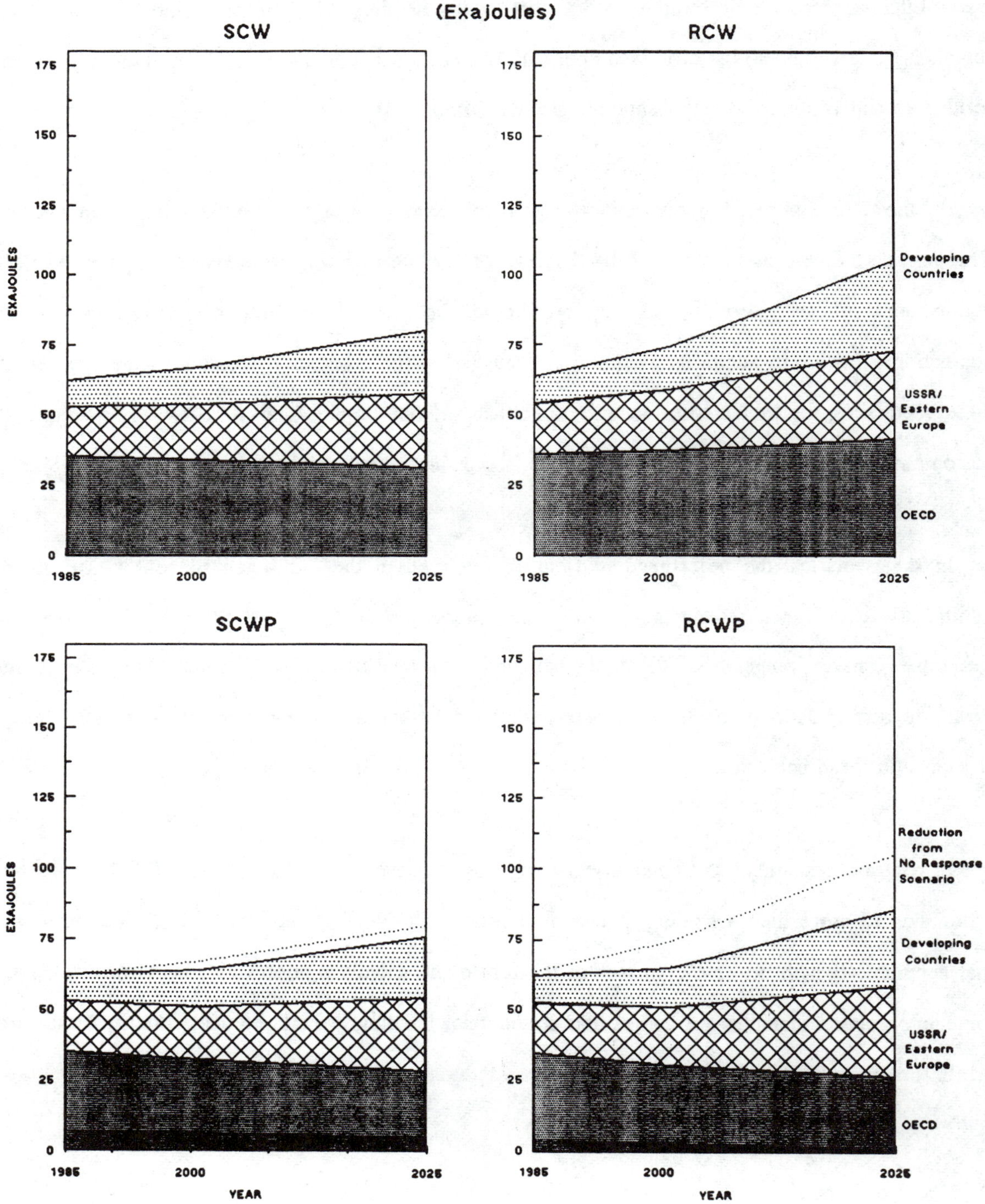

Major technical improvements demonstrated in recent years offer the possibility that new residential and commercial buildings built in the future could require substantially less energy for heating and cooling. Because the stock of buildings turns over so slowly, it is also important to focus on retrofitting, which could reduce air conditioning and heating needs in existing buildings. Dramatic improvements in efficiency are also possible in lighting, particularly in commercial buildings, where lighting may account for a large share of the electrical energy used. Improvements in lighting efficiency also frequently have the added benefit of reducing the amount of waste heat produced by the lighting system and thus reducing the air conditioning requirements as well.

Options for reducing energy use in the residential and commercial sectors are fairly well characterized, at least in the OECD. As discussed below, potentials are very great, but due to the long turnover time of building stock, improvements may have to be phased in over a long time period. Figure 7-8 also indicates reductions of 5-16% in secondary energy use in these sectors by 2025 in the Stabilizing Policy cases. In addition, alternative fuels are introduced, particularly in developing countries, which further reduces the greenhouse impact of energy use. As discussed below, these reductions are small relative to the technical potential for improvements.

Near-Term Technical Options: Industrialized Countries

<u>Improvements in Space Conditioning</u>

Improved efficiency in space conditioning (heating and cooling) can be obtained in several ways. First, the design of new buildings can be altered to improve their insulating qualities, thus reducing losses in heating or cooling. Second, improved technologies can be applied to make existing buildings more weathertight, requiring less energy for heating and cooling. Finally, advanced technologies for heating and cooling equipment can be dramatically more efficient than devices currently in wide use.

A number of very thorough and high-quality reviews of the potential for energy efficiency improvements in buildings have been produced in recent years (see, for example, Hirst et al., 1986; Schipper et al., 1985). The discussion below draws on the extensive published literature to illustrate the technical potential for improvement.

New Residences. The potential for improving energy efficiency in new homes is very significant. Simply by **modifying the building shell to improve its insulating capabilities**, space heating energy requirements can be reduced dramatically. Current new homes in the U.S. require, on average, almost 40% less energy to achieve the same level of heating as the average existing house in the U.S. (See Box 7-3, which illustrates the range of energy requirements for space heating on a per unit of floor space basis). What is even more interesting is that the most efficient new houses built currently are 50% more efficient than the average new home. (It should be pointed out, however, that relatively few of these extremely-efficient homes are being built currently.) Very advanced prototypes and design calculations indicate that it is technically possible to build homes whose heating energy requirements would range from 15 to 20 kJ per square meter per degree day, or 10-12% of the average requirements for today's homes.[3]

In addition to the potential for improving the thermal properties of building shells, equally important advances have been made in developing **high-efficiency equipment** for both space heating and cooling. Already being marketed in the U.S. and in Europe are high-efficiency gas and oil furnaces that are about 95% efficient, compared with the average (about 75% efficient) for new furnaces in the U.S. (Geller 1988).

Substantial energy savings are possible in electrically-heated (and cooled) homes with recent efficient heat pump designs. The most efficient designs on today's market are about one-third more

---

[3] kJ = kilojoule, 1 kilojoule = 10^3 joules.

BOX 7-3 IMPROVING ENERGY EFFICIENCY IN SINGLE FAMILY HOMES

Space Heat Requirements in Single-Family Dwellings
(Kilojoules per square meter per degree day)

United States	
Average, housing stock	160
New (1980) construction in U.S.	100
Mean measured value for 97 houses in Minnesota's Energy Efficient Housing Demonstration Program	51
Mean measured value for 9 houses built in Eugene, Oregon	48
Calculated value for a Northern Energy Home, New York area	15
Sweden	
Average, housing stock	135
Homes built to conform to the 1975 Swedish Building Code	65
Mean measured value for 39 houses built in Skane, Sweden	36
House of Mats Wolgast, in Sweden	18
Calculated value for alternative versions of the prefabricated house sold by Faluhus	83
Version #1	83
Version #2	17

Source: Goldemberg, 1988.

The striking reduction, up to 90%, that is possible between the average home and new "low energy" homes, as illustrated by the figures above, is achieved through the use of state-of-the-art construction and design techniques and technologies; a few of the areas where significant changes have occurred include:

- Building Envelope - Larger wall and ceiling cavities, allowing for significantly more insulation, have been obtained with new construction materials and designs, such as "I-beam" framing members, raising R-values to as high as R-38 in some low-energy homes. Polyethylene vapor/air barriers in external walls reduce infiltration of outside air, one of the major sources of heat loss in most homes. Windows are being triple and even quadruple glazed, and, in some cases, incorporate low-emissivity films and inert gases, such as argon, between the panes to improve their insulating quality.

BOX 7-3 IMPROVING ENERGY EFFICIENCY IN SINGLE FAMILY HOMES (continued)

- Mechanical - These include heating, cooling, and ventilating systems. In addition to high efficiency furnaces, heat pumps and air conditioners, many low-energy homes incorporate air-to-air heat exchangers. These are needed to bring fresh air into the house, since natural infiltration is significantly reduced, but they improve energy efficiency by extracting approximately half the heat from the exhaust air and transferring it to incoming air. The newest development in mechanical ventilation is the heat pump exhaust system, which uses warm exhaust air to provide water heating.

- Active/Passive Solar Design, Thermal Storage - Many different designs have been developed to make better use of solar gain to provide heat. These include use of south-facing windows, greenhouses, atriums, etc., and are often combined with thermal storage systems that store heat collected during peak daylight hours for redistribution when it is needed at other times. Storage systems can also work in the reverse way, collecting cool air during the night and circulating it during warmer hours.

efficient than average. Advanced designs (such as ground-coupled heat pumps) not yet commercially available may provide even more efficient options over the next decade.

With super-insulated shells it may not be necessary to install a central heating system at all. Some of the advanced designs require so little heat input that small electric resistance heaters may be cost-effective in moderate climate areas. The greatly-reduced capital cost for this option may offset the increased cost for the super-insulating features. In very cold regions, the added cost of very efficient gas or oil furnaces would be justified. In regions where cooling is also required, the advanced heat pumps would probably be the most economic choice.

If all options were used in combination, it appears technically quite feasible for advanced building shells and efficient heating and cooling equipment to reduce space conditioning energy requirements in new homes to less than 10 percent of current average use.

Existing Residences. It is expected that the net growth in housing stock will be slow in the future in industrialized countries because of the extensive existing stock and low population growth. In addition, existing housing stocks have very long lifetimes. Therefore, existing stocks will dominate the residential sector for many decades. Thus, it is extremely important to focus on opportunities for reducing energy requirements in existing buildings. Many of the advances that have made possible the enormous potential energy savings for new homes are to some extent applicable to existing homes.

In general, **retrofit improvements** are less effective and more costly than those incorporated in the initial design of a new home. Nonetheless, cost-effective technical options exist for substantially reducing energy requirements of existing homes. Storm windows, added insulation, clock thermostats, and retrofit heating system improvements have been considered conventional conservation measures for several years. Programs to encourage consumers to implement these conservation options have been carried out in a number of areas and have shown considerable success. One study of 40,000 retrofits monitored by U.S. utility companies in the early 1980s showed that energy consumption fell by 25% on average, and homeowners received a 23% return on their investments (Goldman, 1984).

Despite the favorable economics of some retrofit conservation measures, only a small portion of the potential energy savings from conservation retrofits has been realized to date. This is especially true in rental housing where the landlords do not perceive a financial interest in investing in retrofits. Also, low-income families, even if they own their homes, often lack the information or upfront capital to carry out cost-effective conservation measures. If conventional retrofit programs could be extended to larger percentages of existing housing stock, energy savings could be substantial.

Beyond conventional conservation programs, there are now more sophisticated options for improving the retrofit savings. Detailed measurements since the late 1970s have shown that existing homes have obscure defects in their thermal envelopes, leading to very large heat losses. Conventional walk-through energy audits are unlikely to identify these defects, nor would subsequent conservation retrofits correct them. New instrumented analysis procedures developed over the last few years can locate these defects quickly, but these instrumented audits are expensive compared with the standard energy audits now provided by many utilities.

On the other hand, many of the "hidden" defects, once detected, can be easily corrected at small cost. This has led to the development of the "house-doctor" concept as an alternative to traditional audits. For this type of audit a team of technicians conducts an instrumented audit and repairs many of the defects on the spot. One test of this concept showed average immediate energy savings of 19% from one-day "house-doctor" visits. Subsequent conservation retrofitting done at the recommendation of the house doctors increased the average energy reduction to 30%. The average cost of all retrofit measures was $1300 and the average real internal rate of return in fuel savings was 20% (See Goldemberg et al., 1988).

Another approach to improving residential and commercial energy efficiency has been proposed recently by researchers at Lawrence Berkeley Laboratory. Their approach is a reworking of the age-old concept of using shade trees to assist in cooling residential buildings. In their analysis the authors point out that in addition to the direct benefit in terms of reducing air conditioning loads at each house, the indirect effect of planting trees throughout an urban or suburban area can reduce the "heat island" effect, lowering ambient temperatures and further reducing air conditioning loads. In addition, of course, the trees directly remove CO_2 from the atmosphere, although the CO_2 reductions due to reduced cooling loads from well-placed trees are probably much greater than the CO_2 absorption by the trees (see Rosenfeld, 1988; Akbari et al., 1988). Los Angeles has announced a program to do this.

The advanced furnaces and heat pumps available for new homes could also provide significant benefits in the retrofit market. Economics of efficient equipment are generally more favorable as retrofits. Even after a retrofit shell improvement program, including the house-doctor approach, energy use in existing homes will remain well above the best levels achievable for new super-insulated homes. Thus, the added expense of sophisticated advanced heating and cooling equipment will be paid back much faster in energy savings.

In one example, existing homes that had already been visited by house doctors and had received associated shell improvements were evaluated for improved furnaces. Shell improvements were shown to have reduced gas use for space heating to about 70% of the previous requirement. Researchers estimated that retrofit of advanced condensing gas furnaces would further reduce space heating energy use to 44% of the original requirement. The estimated incremental investment (to replace a worn-out furnace with the 95% advanced furnace rather than a conventional 69% model) was estimated to average $1000 and result in fuel savings that correspond to a real rate of return of 15% (Goldemberg et al., 1988).

Commercial Buildings. Like residences, commercial and institutional buildings currently use significant amounts of energy, particularly for space conditioning and lighting. Opportunities for efficiency improvements also appear significant. Commercial buildings in recent years in the U.S. have used about 3.6 EJ of fossil fuels (mostly gas in the U.S. -- other OECD countries continue to heat a significant percentage of buildings with oil) and about 2.6 EJ of electricity (equivalent to about 8 EJ of primary energy) (Rosenfeld and Hafemeister, 1985).

Some progress is already being made in improving energy efficiency in commercial buildings. While surveyed commercial space in the U.S. increased by almost 10% between 1979 and 1983, total energy consumption actually declined by about 4%. Electricity consumption, however, increased in absolute terms as well as in share of total commercial energy use. Thus, the primary energy equivalent of commercial end-use energy consumption has increased (U.S. DOE, 1987a).

While progress has been made, it is evident that commercial energy use in the U.S. could still be reduced significantly with cost-effective efficiency measures. Estimated commercial energy use in the U.S. was about 3.0 gigajoules (GJ) [4] per square meter per year in 1980 (expressed as equivalent primary energy production). This figure is down from 5.7 GJ in 1973, but could still be greatly improved (Flavin and Durning, 1988.) One recent analysis estimated that energy use in standard new commercial buildings could be reduced by more than 50% below the current averages (Rosenfeld and Hafemeister, 1985).

As in the residential sector, traditional conservation measures, such as added insulation and window glazing, reductions in infiltration rates of outside air, passive solar energy concepts, and heat exchange between exhaust and incoming ventilation air are effective although less important. The traditional approach of tree shading, with its indirect heat island and CO_2 removal benefits, as discussed above, can also be applied to commercial buildings.

In addition, some more sophisticated techniques are cost effective for larger commercial buildings. New commercial building are being designed with "smart" energy management systems. These computerized systems monitor outdoor and indoor temperatures, levels of sunlight and location of people in the building. The system can then allocate heating, cooling and ventilation efficiently (Brody, 1987).

---

[4] GJ = gigajoule, 1 gigajoule = 10^9 joules.

Another advanced technique being applied for commercial energy efficiency is thermal storage. In this case, some storage medium, such as a body of water, is used to store heat or cooling when it is readily-available and then release the warm or cool air later when it is needed. This concept has been used in new commercial buildings in Sweden, storing heat energy from people and equipment, and in Nevada to chill water with cool night air and use the chilled water to offset the need for air conditioning during the day (Rosenfeld and Hafemeister, 1985).

Window technology is also improving rapidly. A special "heat mirror" film, which doubles the insulation value of windows, is now commercially available. The film is designed to let light in without allowing heat to escape. Another available technology is to create a vacuum in the space between two panes, creating a "thermos" effect. These and other advanced technologies may allow commercially-available windows in the 1990s to have the same insulating value as ordinary walls (Brody, 1987; Selkowitz, 1985).

As was the case with residences, many of the advances in energy efficiency for new commercial buildings are transferable to retrofit applications to a lesser degree. A study conducted some years ago attempted to estimate the potential energy savings available from retrofits of existing commercial buildings. The conclusion drawn from a survey of several experienced engineers and architects was that a target of a 50% reduction in energy use in U.S. commercial buildings by the year 2000 was reasonable (SERI, 1981).

More recently, Amory Lovins and others at the Rocky Mountain Institute conducted a detailed analysis of the retrofit potential of commercial buildings in the Austin, Texas, area. This study identified potential savings in electrical energy that totaled 73% of the buildings' current energy use (for lighting and other equipment as well as space conditioning). The cost of these measures was estimated to be lower than the operating costs of existing powerplants (Rocky Mountain Institute,

1986), although, there is some question as to what proportion of these savings could be achieved in practice.

Indoor Air Quality

One of the concerns about increasing the energy efficiency of buildings is the increase in indoor air pollution that can result, primarily from efforts to reduce air infiltration. Most homes in the U.S. have air exchange rates, on average, of more than one change per hour, meaning all the air in the house is replaced by outside air once each hour. As new homes are constructed more tightly and some older homes are retrofitted with vapor barriers and better seals around doors and windows, the air exchange rate drops. Without additional ventilation measures, concentrations of several harmful pollutants, including radon gas, formaldehyde, combustion products from tobacco smoke and wood stoves, and asbestos particles, can reach harmful levels.

Control strategies for these forms of indoor air pollution include air cleaning, local ventilation, mechanical ventilation with heat recovery, and exhaust ventilation with heat pump heat recovery. Heat recovery or air-to-air exchange systems have become more popular as homes have become better insulated and more tightly sealed. These can ensure the generally-accepted minimum standard of about 0.5 air changes per hour, while reducing heat loss by using the heated exhaust air to warm the incoming fresh air. Unfortunately, the cost of these systems are high and studies in Canada suggest that their efficiencies may be lower than manufacturers have claimed (Hirst, 1986). This may be due in part to improper maintenance by homeowners. The systems, which usually draw air continuously and can have problems with condensation forming in the heat exchanger, require more routine maintenance than the average homeowner is accustomed to devoting to a major appliance. The use of exhaust ventilation systems connected to an air or water heat pump is a new technological approach that may hold some promise for improving cost-effectiveness and improving some of the maintenance issues. They are being developed primarily in Sweden.

Lighting

Lighting consumes about 20% of U.S. electricity use, most of it in residential and commercial buildings. This end use offers some of the most cost-effective opportunities for saving energy. One study has estimated that 40 large U.S. powerplants could be replaced by simply implementing currently-available, cost-effective lighting efficiency improvements (Rosenfeld and Hafemeister, 1985). Cutting electricity use for lighting in industrialized countries by three quarters has been proposed as a reasonable goal (Flavin and Durning, 1988). Box 7-4 describes several key advanced lighting technologies.

A number of currently commercial measures, implemented in combination, can achieve dramatic energy reductions--over 75% in commercial/institutional settings. These measures include improved controls, reflectors, spacing of lighting, and more efficient bulbs and ballasts. The University of Rhode Island reported reductions of 78% in lighting energy after implementing such a program. The cost of saved energy was calculated to be less than 1 cent per kilowatt-hour (New England Energy Policy Council, 1987). An added benefit of lighting improvements in warm climates is that more efficient lighting reduces waste heat and, therefore, air conditioning loads. The California Energy Commission has estimated that in Fresno, every 100-watt savings in lighting reduces air conditioning energy requirements by 38 watts (Rocky Mountain Institute, 1986).

BOX 7-4 INCREASING LIGHTING EFFICIENCY

Research and development into energy efficient lighting and design features has produced a constant stream of new products and advances. Below are brief descriptions of several of the areas where major advances have taken place.

- Compact Fluorescent Lamps: Compact fluorescent lamps are designed to be screwed into a standard light socket and thus have begun to compete directly with incandescent bulbs. Because they are 60-70% more energy-efficient than incandescent (Hirst, 1986) and are beginning to gain wider acceptance, they represent potentially significant energy savings. If compact fluorescent replaced all incandescent lighting, it is claimed that they could displace 7.5% of total electrical consumption in the U.S. (Lovins and Sardinsky, 1988).

- High Intensity Discharge (HID) Lamps: These are designed primarily for warehouses, factories, street lighting, etc. Three types of HID lamps are currently in use: high pressure sodium, low pressure sodium and metal halide. High pressure sodium and metal halide give approximately 45-60% savings over mercury vapor or fluorescent lighting. Low pressure sodium is somewhat more efficient, but its intense yellow light can be undesirable for many applications. (Hirst et al., 1986).

- Electronic Ballasts: In conventional fluorescent lights the voltage required for operation is provided by an electromechanical ballast which itself consumes a portion of the energy used. New electronic ballasts reduce this additional power consumption by 20-35% (Hirst et al., 1986), and, because of their smaller size, are a key factor in the emergence of compact fluorescent.

- Daylighting: Daylighting is a design approach which enhances the use of natural light either from windows, sidelighting, clerestories, monitors and skylights, or from the use of light pipes or optical fibers to transmit light to the location needed. The use of light colored paints and light shelves helps to distribute the light into the building interior.

- Additional Advances: Several other advances in lighting technology and design deserve mention. One advance is specular ("mirrorlike") reflectors which increase total reflectivity, direct the light in a more optically favorable direction, and maintain their high reflectivity significantly longer. Lighting controls, which include time clocks, scheduling controls, personnel or occupancy sensors, and daylighting sensors, reduce power consumption by turning off lights when they are not needed. Task lighting improves efficiency by directing light onto the specific task area where it is needed most, rather than lighting entire areas.

Appliances

After space conditioning and lighting, remaining energy uses in residential and commercial buildings are largely associated with large appliances. Opportunities for significant energy efficiency improvements in this category have been well-documented. Table 7-4 illustrates some of these opportunities. U.S.-made refrigerators, for example, currently average 1450 kilowatt-hours per year. The best currently commercial model in the U.S. uses about half that much energy. A recent study calculated that efficient new refrigerator freezers that would use about 200 kilowatt-hours per year, or less than 15% of the current average use, could be cost-effectively produced (Goldstein and Miller, 1986). Water heaters also account for a large percentage of appliance energy use. The potential for energy savings is significant through the use of the most efficient technologies and also by switching from electricity to gas. Other energy-intensive appliances also provide opportunities for energy savings. As shown in Table 7-4, the potential exists for advanced technologies that could be produced in the 1990s that are at least 50% more energy efficient than the 1986 average for all major energy-using residential appliances (Geller, 1988).

Near-Term Technical Options: Developing Countries

In developing countries markedly different strategies may be necessary to address residential and commercial energy services. In many developing countries there are distinct modern and traditional sectors. In the modern sector, energy-use patterns are very similar to those in industrial economies (adjusted for climate differences). Commercially-marketed fossil fuels and electricity provide the energy input for a similar mix of energy services: space conditioning, water heating, lighting, and appliances for cooking, refrigeration, entertainment, etc. This modern sector, however, is often smaller than the traditional sector, which exhibits completely different energy-use patterns.

TABLE 7-4

Summary of Energy Consumption and Conservation Potential with Major Residential Equipment

Product	1986 Stock UEC[a]	1986 New UEC[b] (kWh/yr or therms/yr)	1986 Best UEC[c]	Advanced technology for 1990s[d]
Refrigerator	1450	1100	750	300-500
Freezer	1050	750	430	200-300
Central AC	3500	2900	1800	1200-1500
Room AC	900	750	500	300-400
El. water heating	4000	3500	1600	1000-1500
El. range	800	750	700	400-500
El. clothes dryer	1000	900	800	250-500
Gas space heating	730	620	500	300-500
Gas water heating	270	250	200	100-150
Gas range	70	50	40	25-30

[a] Unit energy consumption per typical installation in the 1986 housing stock.

[b] Unit energy consumption for the typical model produced in 1986.

[c] Unit energy consumption for the best model mass-produced in 1986.

[d] Unit energy consumption possible in new models by the mid-1990s if further cost-effective advances in energy efficiency are made.

Source: Geller, 1988.

The energy sources in the traditional sector are largely "noncommercial" biomass, used primarily for cooking and space heating in some colder or high-altitude regions of developing countries. Also, fossil fuels (e.g., kerosene) are frequently used for lighting. (In China, unlike most other developing countries, coal is also used for residential cooking and space heating in the traditional sector.) The task of development projects in these poorer sectors is to vastly increase the level of energy services available for residential and commercial applications. Altruism and development objectives aside, this approach can contribute significantly to solving the climate warming problem because many developing countries have used fuelwood to such an extent that they have become net consumers of forests, and global deforestation is one of the significant causes of increasing greenhouse gas concentrations. The important issue from a climate perspective is to increase energy services without increasing greenhouse gas emissions.

As the developing countries continue to increase their per capita energy use, the implications in terms of energy use and greenhouse gas emissions are enormous. It is technically possible, however, for developing countries to substantially increase per capita energy use without substantially increasing fossil-fuel use. Emissions-reducing strategies similar to those proposed for industrialized countries can be promoted in the modern sectors of the developing countries. However, strategies suitable for the traditional, poorer sectors must be integrated into ongoing development programs if they are to be accepted by the local population. Technical options for reducing greenhouse gas emissions must not only be efficient, they must also be designed to increase energy services to these poorer sectors.

Increasing Efficiency of Fuelwood Use

The primary use of biomass energy in developing countries is in residential cooking, traditionally done in inefficient and smoky conditions. The inefficiency of combustion exacerbates deforestation, and leads to increased time and effort devoted to gathering fuelwood (and fodder), and the smoky combustion results in exposure to significant emissions of health-damaging air pollutants.

Recognition of these problems has focused a great deal of attention on improving the cookstove as a low-cost solution. Existing cookstoves have efficiencies only on the order of 10%. Relatively simple improvements in stove design can in theory reduce wood requirements by 35-70% (Goldemberg, et al., 1987). However, getting people in developing countries to accept and use better-designed stoves has proved difficult for a number of reasons (Miller, et al., 1986). In spite of spirited efforts by a number of groups and generous grants by international aid agencies, traditional cooking stoves and practices have proved surprisingly difficult to dislodge. Traditional stoves come in a bewildering variety of designs and materials and have evolved to suit local fuels and diets. They perform a multitude of functions that were not considered by the early "improved" stove designers and promoters. Failure of the newer designs on some of these dimensions often hampered their acceptance. Current programs represent the third generation of improved designs (Smith, 1987).

There are currently half a dozen examples of successful dissemination efforts. Notable dissemination programs are in place in West Africa, in Kenya and in Karnataka, India (Baldwin et. al., 1985). Successful designs are backed by sound principles of heat transfer (Baldwin, 1987); are targeted to a particular region, generally where cooking fuel is traded; require no substantial behavioral modification from users; and are provided with follow-up support.

Fuel savings with improved stoves, predicted on the basis of laboratory water-boiling efficiency tests, have invariably proved to be over-estimates for field conditions. If fuel savings observed in the laboratory were directly transferable to the field, an improved stove with 40% efficiency would result in a 75% fuel saving when it replaced a traditional stove with 10% efficiency. Yet only a few programs have reported fuel savings, at best on the order of 20%, though greater savings could be possible with better-run programs.

Even with existing limited efficiencies, where fuelwood is traded the payback period is on the order of a few months and therefore economically attractive (Manibog, 1985). As Williams (1985) points out, the adoption of such stoves is a far more cost-effective method of dealing with the fuelwood crisis than any "supply-oriented" solution to the problem that emphasizes growing trees for fuel.

Widespread introduction of improved stoves, while reducing total emissions of oxides of carbon per cooking task, will change the ratio of CO_2:CO emitted. This ratio on a mass basis for traditional stoves is perhaps close to 10:1, whereas for more efficient stoves, this could be reduced to 5:1 reflecting the more complete combustion in traditional stoves. Although CO is not a radiatively interactive gas, it does interact with hydroxyl ions; as a result, its presence affects the concentration of methane and ozone in the troposphere (See **CHAPTER TWO**).

Substituting More Efficient Fuels

In the traditional sectors of many developing countries, substitution of fossil-based end-use technology for traditional biomass use may be desirable as part of a larger strategy even though it may directly increase greenhouse gas emissions to a small degree. Gaseous fuels (natural gas, LPG, etc.) are very attractive relative to fuelwood for several reasons, including their vastly superior convenience and controllability. In addition, a well-designed gas-fueled stove can be five to eight times more efficient than traditional firewood stoves (Goldemberg et al., 1988). Thus, the shift to gaseous cooking fuel can decrease the demand for fuelwood, which can slow the rate of deforestation, or free up vast amounts of fuelwood for use as a feedstock for advanced biomass energy systems, or both. In a few developing countries, such as China, coal is used for cooking and space heating; the advantages discussed above for gaseous fuels also apply as a replacement for coal (with associated reductions in greenhouse gas emissions).

In the traditional sectors of many developing countries, lighting frequently is achieved with very inefficient combustion technologies. In India, for example, it is estimated that 80% of the households illuminate with kerosene lamps. The efficiency of illumination of these lamps is very low. Providing the same level of illumination with incandescent electric bulbs would be 200 times more efficient at the end use (Goldemberg et al., 1988). (Of course, the availability of electricity might actually allow a much greater level of illumination in homes). Thus, even if the electricity production is only 30% efficient, and is coal-fired (worst-case assumptions), the equivalent electric lighting would still produce less net CO_2. If the most efficient current compact fluorescent lighting were used, the benefit would be even greater. Of course, the major constraint to substituting electricity for fossil fuels is the limited availability of electricity in developing countries. Small-scale local generation based on renewable technologies could make major contributions in this situation. These and other options for increasing electricity in developing countries are discussed in **PART TWO: ENERGY SUPPLY.**

Retrofit Efficiency Measures for the Modern Sector

For those residential and commercial segments of developing countries that have similar characteristics to industrialized countries, many of the same retrofit efficiency measures are appropriate. In fact, it is likely that many retrofit measures would be more effective in developing countries. As pointed out in a recent U.S. AID study (1988b), while "industrialized countries have made major strides in using electricity more efficiently over the last decade, few achievements have been made in developing countries in using electricity more efficiently. The opportunities for improvements are tremendous, and the cost is only a fraction of the generation expansion option." In addition, air conditioning requirements are generally heavier in tropical developing countries.

Thus, improvements in building shells and air conditioning equipment could be very effective in reducing electricity use. Similarly, the air conditioning benefit (due to less waste heat) of improved lighting would also be greater in developing countries.

A recent study of Pakistan identified cost-effective efficiency improvements that could reduce commercial sector electricity use by over 30%. Based on commercially available improvements in lighting, air conditioning and fans, and thermal insulation, the study projected national savings of 1800 MW of generating capacity and 18,200 GWh of electricity generation (U.S. AID, 1988b). An analysis in Brazil indicated the potential to reduce electricity use for lighting by 60% in many commercial buildings (Geller, 1984).

New Homes and Commercial Buildings

It is expected that rapid expansion of the residential and commercial building stock will occur in conjunction with economic development in the developing countries over the coming decades. Use of the efficiency options discussed for industrialized countries, adapted to local conditions and objectives, could minimize the increases in energy use associated with this growth. Obviously, since the rate of construction of new building space (and distribution of appliances, etc.) will be much higher in developing countries, the importance and potential impact of efficiency measures will be proportionately greater as well.

Several recent studies have identified significant potential for reductions in energy use in new commercial buildings in developing countries (see, for example, Turiel et al., 1984, Deringer et al., 1987). Careful use of daylighting alone has the potential to reduce energy use by roughly 20% relative to the current building stock in Singapore (Turiel et al., 1984). Other important improvements include efficient lighting systems, external shading, and size and placement of windows.

Near-Term Technical Options: Soviet Bloc Countries

Energy use in buildings in the Soviet Union is dominated by space heating. 25% of the population live in a climate which experiences from 210 to over 300 heating days per year. An additional 40% live in a climate characterized by a 180-280-day heating season (Tarnizhevsky, 1987). Since the early 1970s, the Soviets have made dramatic progress in improving energy efficiency in specific applications, "in large energy uses easily identified and controlled by the planning apparatus" (Hewitt, 1984). The dominant approach advocated by Soviet researchers for space conditioning is consistent with this experience. Centralized, or "district" heating systems, have been the preferred approach to space heating for some time. By 1980, about 70% of residential and commercial heat demand in cities and towns was provided by central systems. Sources of heat for these systems are primarily cogeneration from both fossil-fueled and nuclear powerplants and some waste heat recovery from large industrial process heat uses (e.g., ferrous metal, chemicals, and petrochemicals) (Tarvarshy et al., 1985). Plans for improving energy efficiency include 1) further replacement of local heat sources with district heat from cogeneration, 2) weatherization of buildings, and 3) reductions in heat losses in distribution networks (Tarnizhevsky, 1987).

Electricity consumption in buildings is also growing in the Soviet Union, principally for lighting and household appliances (including some electric ranges). The most significant sources of growth from 1975-1985 were attributed to increasing proliferation of household applications associated with improving living standards. One Soviet researcher estimated that significant electricity savings are achievable through manufacture of more energy-efficient appliances and lighting, optimized use of lighting, elimination of known losses or "wastes" of electricity, etc. (Tarnizhevsky, 1987).

Summary of Near-Term Technical Potential in the Residential/Commercial Sector

In summary, it appears technically feasible with today's technology to reduce space conditioning energy requirements in new homes to 50% of the current average for new homes. Retrofits of existing homes could reduce space conditioning energy use by an average of 25% with the "house-doctor" approach. Reductions of energy use in existing commercial buildings by at least 50% may be technically feasible, and new commercial buildings could easily be 75% more efficient than the average U.S. commercial building. In 1985 the U.S. consumed about 15.3 EJ of energy (primary energy equivalent) in residential and commercial space conditioning (U.S. DOE, 1987a). Retrofits to existing stock could save at least 4 EJ.

Projected estimates for the climate scenarios show residential and commercial energy use in the U.S. and in the OECD as a whole remaining roughly constant through 2025. With rapid widespread penetration of the most efficient new buildings, instead of gradual improvement, the growth in energy consumption from new buildings could be greatly reduced.

Based on potential energy savings of more than 60% for almost all types of appliances, a 50% overall reduction in appliance energy use by the year 2010 is technically feasible. To achieve this reduction would require aggressive policy actions that would ensure that all appliances produced in the next decade be as energy efficient as the best current technology can produce and that would encourage rapid turnover of existing appliances. Current energy use from such appliances is in the range of 7.4-8.4 EJ (expressed as primary energy equivalent).

With aggressive programs to improve energy efficiency in residential and commercial buildings, it appears technically feasible to reduce projected U.S. energy use in this sector by at least 50% in the year 2010. This is roughly equivalent to the reduction assumed in the policy scenarios by 2025.

The technical potential in the OECD as a whole is probably close to that of the U.S. Although more detailed analysis is required, preliminary indications are that the technical potential to reduce projected residential and commercial energy in the much more rapidly growing developing countries and in the Soviet bloc countries is even greater than that estimated for industrialized countries.

Long-Term Potential in the Residential/Commercial Sector

In the long term, the potential for reducing energy use in buildings is considerable. The majority of current per capita energy consumption in buildings could be eliminated over the long run by simply incorporating the best currently-available building and equipment technologies into housing and commercial building stock as it is expanded and replaced over the next 50-100 years. Further efficiency improvements could come from emerging technologies and broader application of existing technologies. In space conditioning, examples include "smart windows" which sense light and adjust opacity to utilize solar heat and light most effectively, and new building materials which may provide better insulating qualities at a reduced cost. Existing technologies for large buildings, such as use of thermal storage and computer controls, could be applied to small buildings and residences as well.

As improved building and equipment technologies are incorporated over time, space conditioning will probably become a much smaller component of total building energy use. Appliances and information technologies (computers, telecommunications) may become more important determinants of residential and commercial energy consumption. Advanced technologies may provide comparable or improved services with less energy. For example, some experimental concepts have been developed for storing food that might greatly reduce the need for refrigeration. As information technology continues to evolve, it may well provide improved energy efficiency as a byproduct, as has been the case in the evolution from vacuum tubes to semiconductors to integrated circuits.

Alternative fuels could also play a more important role in buildings over the long term. Advances in solar photovoltaic technology may make it economic to generate most or all of the needed electricity locally. Hydrogen may become an energy option for building energy needs utilizing or adapting existing infrastructure for distribution of natural gas (See **PART TWO: ENERGY SUPPLY**).

INDUSTRIAL SECTOR

Industrial end uses account for the largest single component of energy use in the industrialized countries--almost 43% of the energy consumed in the OECD in 1985 (in primary energy equivalent terms [U.S. DOE, 1987b]). Actual secondary energy consumption was about 36%. In developing countries, if agriculture is included as part of industry (as it is in this chapter), then the industrial sector generally consumes an even higher percentage of total commercially-traded energy. In developing countries as a whole, industrial energy makes up almost 60% of total modern energy use. In the Soviet bloc, the percentage is slightly under 50% (see **CHAPTER IV**), but in the scenarios this proportion declines over time, accounting for only 36-40% by 2025.

Industrial use is also an area in which impressive efficiency gains have been observed in recent years. In the United States, for example, "end-use energy consumption per constant dollar of industrial output declined by 28% between 1974 and 1984 - reflecting substantial improvements in energy efficiency as well as the relative decline in output from energy-intensive industries in this country" (U.S. DOE, 1987b).

Several researchers have documented the components of changing industrial energy use in developed countries (see Ross, 1984 and 1986; Goldemberg et al., 1987; and Williams et al., 1987). One significant component of declining industrial energy use is a structural shift to products that are inherently less energy-intensive to produce. It is now a well-documented phenomenon that as industrialized countries proceed beyond a certain level of economic development and affluence, their per capita consumption of some of the most energy-intensive industrial products (e.g., cement, steel, and durable goods) declines. Thus, having approached "saturation" in many energy-intensive products, industrialized countries will likely continue to consume less energy per dollar of GDP in the future as incomes continue to rise.

The other major component of declines in energy intensity over the past decade and a half has been actual improvements in the efficiency of production processes. Energy price shocks of the 1970s often affected industrial energy users more than other sectors. As stated by Goldemberg et al. (1988): "Because the cost of providing them energy involves much less unit transport and marketing cost, industrial users are more sensitive than other energy consumers to cost increases at or near the point of energy production." In addition, industrial users, particularly where energy costs are a significant component of product cost, tend to be more aware of and responsive to the return on investments in energy efficiency than are customers in other sectors. Ross (1986) has noted many cases in which industrial managers have pursued aggressive efficiency improvement policies in response to the price signals of the 1970s.

Efficiency improvements to date have largely been in a few industries that are the most energy intensive: petroleum refining, chemicals, cement, metals, pulp and paper, glass, clay, etc. (See Table 7-5). There is reason to believe that efficiency improvements will continue rapidly in the future in the energy-intensive sectors of industry in the developed countries. It is also expected that the structural shifts to less energy-intensive products will also continue in these countries (Williams et al., 1987). Technical options may exist for accelerating these trends and also taking advantage of additional efficiency improvements possible in other industries. Evidence exists that economically-attractive energy conservation investment opportunities exist in industries outside of those few that

TABLE 7-5

Reduction of Energy Intensity In the Basic Materials Industries (1972-1983)

	Percent
Chemicals[b]	31
Steel	18
Aluminum	17
Paper[c]	26
Petroleum refining[d]	10
Energy Weighted Reduction	21

[a] Generally energy per pound of product, unadjusted for environmental and other changes. Purchased electricity accounted for at 10,000 Btu/kwh (2.5 Mcal/kwh).
[b] Not including fuels used as feedstock.
[c] Not including wood-based fuels.
[d] Changes in inputs and outputs and environmental regulations have had a particularly strong impact on petroleum-refining energy. Adjusted for such changes, energy intensity was reduced 26%.

Source: Ross, 1985.

are energy-intensive. For a number of reasons these investment opportunities are often being overlooked (Ross, 1984, 1986).

In contrast to industrialized countries, the developing countries generally are in the midst of, or beginning, a period of rapid expansion of energy- and materials-intensive industries to raise per capita income levels. In addition, industries in these countries currently use energy far less efficiently than do similar industries in industrialized countries. In many cases, this is related to the government's subsidizing of energy prices, lack of access to the most modern technologies, and lack of management skills for identifying and implementing efficient options (Flavin and Durning, 1988). If developing countries industrialize without dramatically improving energy efficiency, the result would be enormous increases in industrial energy use.

The No Response scenarios indicate the potential for extensive growth in industrial energy use by 2025. Figure 7-9 illustrates these results and also shows that the overwhelming majority of growth in this category (81-92%) occurs in developing countries. As these countries develop industrial infrastructure and as their standards of living rise in the future, the energy requirements for the production of basic industrial materials could be enormous.

There is a widening recognition, however, that this type of growth may not be sustainable, and may indeed become self-defeating. Continuing along current energy-intensive industrial development paths may result in increases in energy imports, which absorb foreign exchange credits that are needed for further development. Local environmental concerns and capital scarcity may also constrain the conventional industrial development option. Developing countries and development assistance organizations are now focusing much more attention on the energy consequences of industrial development decisions.

FIGURE 7-9

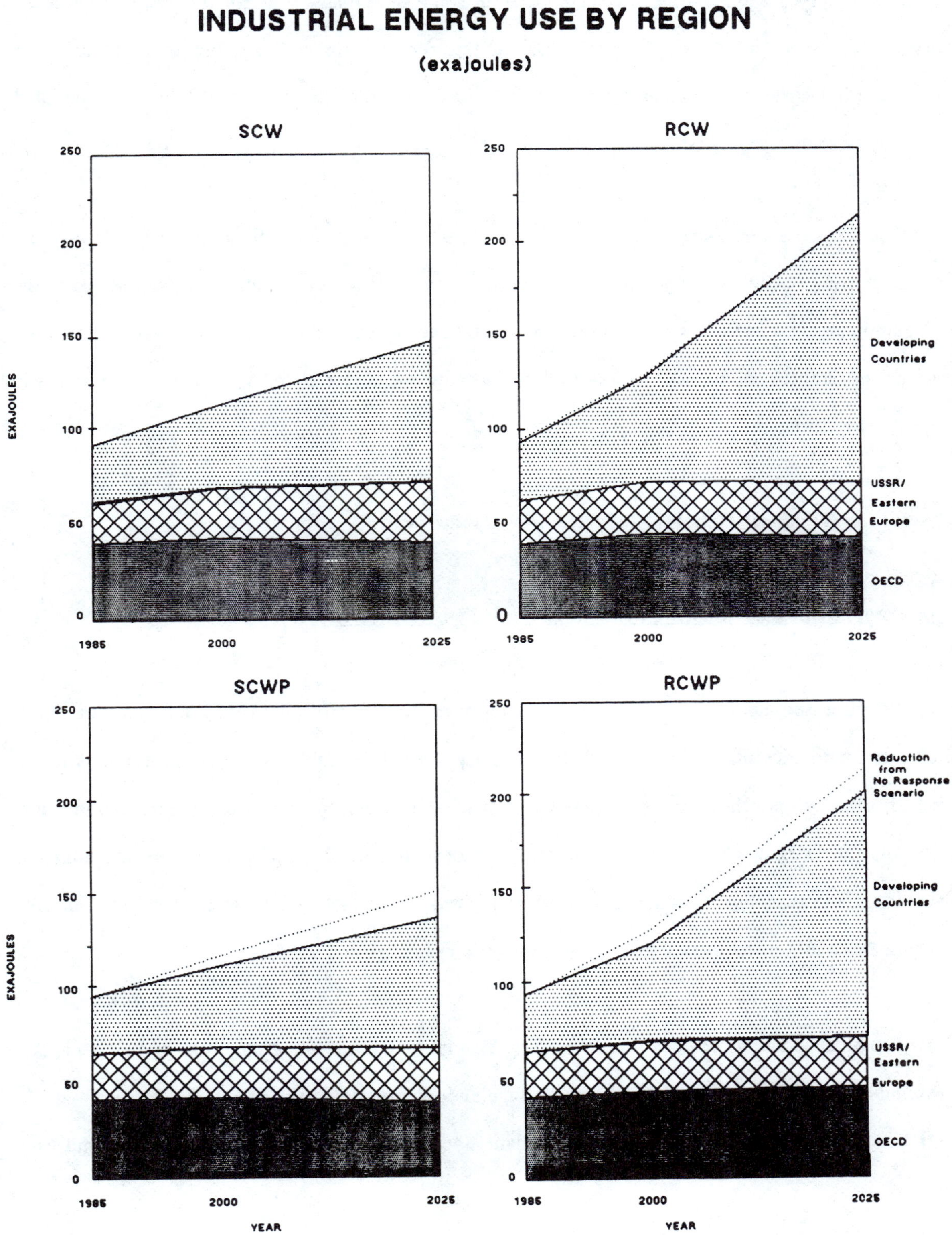
INDUSTRIAL ENERGY USE BY REGION
(exajoules)
SCW
RCW
SCWP
RCWP
EXAJOULES
YEAR
0
50
100
150
200
250
1985
2000
2025
Reduction from No Response Scenario
Developing Countries
USSR/ Eastern Europe
OECD

Technical options may exist for "leapfrogging" from current obsolete energy-inefficient technologies directly to very advanced efficient technologies in some developing countries (although new policy actions will be required; see **CHAPTER IX** of this report). In addition, opportunities exist for designing industrial development based on locally-available alternative fuels, which has been initiated in some developing countries.

These options are poorly defined as yet and clearly require more detailed attention. As a result, the Stabilizing Policy scenarios incorporated in this report are somewhat conservative in terms of the assumptions about efficiency improvements. As shown in Figure 7-9, these scenarios result inefficiency improvements of 6-10% overall in the industrial sector by 2025, most of which occurs in the developing countries.

Near-Term Technical Options: Industrialized Countries

Accelerated Efficiency Improvements in Energy-Intensive Industries

As discussed above, significant improvements in energy efficiency were made in the basic materials industries during the late 1970s and early 1980s (see Table 7-5 for improvements in key industries). Despite these improvements, in most industries the opportunities for further reductions are still quite large. This is true for two basic reasons. Industrial process technology has improved significantly in recent years, and the pace of energy-related technology investment has been relatively slow, particularly in industries that are not growing overall (Ross, 1988).

The steel industry provides one interesting example. For the integrated, or ore-based industry (excluding scrap-based steel making), the average energy intensity in the U.S. in 1983 was about 31.2 GJ per ton (Ross, 1987). The reference plant documented by the International Iron and Steel

Institute (1982) would consume about 19.2 GJ per ton producing roughly the same mix of products. Thus, replacing existing U.S. technology with the best currently commercial, cost-effective technology would produce a 39% savings.

More energy-efficient technologies, once proven commercially, will likely be extremely attractive on the basis of cost-effectiveness, including improvement in overall process cost and environmental impact as well as energy efficiency. Thus, the marketplace can be expected to encourage all producers to adopt advanced technologies in the long run. From the perspective of climate warming, the important concern is whether there are opportunities for accelerating this turnover, either through research initiatives or through programs to promote more rapid capital replacement in selected industries.

Other major energy-consuming industries--petroleum refining, chemicals, pulp and paper, etc.--are also undergoing transformations that will result in continuing declines in energy intensity in the future. Quantifying the potential energy efficiency improvements in each major industry and in various OECD countries will require more detailed analysis than has been carried out to date. Technically-feasible improvements (e.g., assuming that each industry average improves to match the energy efficiency of the best currently-available or emerging technology) could be much greater. With the exception of primary aluminum, other energy-intensive industries have even greater technical opportunities for energy conservation than the steel industry (Ross, 1988).

One important component of the technical potential for reducing energy use in basic industries is materials recycling. Recycling of inorganic wastes, such as bottles and aluminum cans, saves energy and reduces waste streams. Substituting recovered materials for virgin materials to produce steel, aluminum, and glass conserves energy. Depending upon the types of materials recycled, estimates of energy savings can range from an average of 5 GJ per ton of material recycled (Gordon, 1979) to 25 GJ per ton (Stauffer, 1988).[5] Recycling 24 million tons of paper, cardboard, glass, and aluminum (about 16% of our current waste stream) could result in savings of up to 0.6 EJ (Stauffer, 1988).

Unfortunately, economic policies discourage recycling and the use of recycled materials. Differentials in transportation rates favor virgin loads over secondary loads, and favorable tax treatment toward production from virgin materials continues to make recycled materials more expensive to use. If these policies were eliminated or reversed (e.g., creating incentives to boost demand for recycled products), the technical potential for energy savings is quite large. Increased recycling would also have the added benefit of reducing waste.

Aggressive Efficiency Improvements of Other Industries

While there is clearly measurable progress in the energy-intensive, and therefore energy-price-sensitive, industries, less progress has been made in energy efficiency in other industries. Industries for which energy is not a major component of product cost frequently pass up opportunities for investing in energy efficiency with high estimated rates of return (Ross, 1984).

The technical potential for energy savings could be important for widespread penetration of just a few energy-saving measures. The Electric Power Research Institute estimates that 95% of industrial electricity use is represented by three major applications: electromechanical drives or motors, 70%, electrolysis, 15%, and process heat, 10% (Kahane and Squitieri, 1987). Recent studies have estimated that cost-effective replacement of motors and the addition of variable-speed drives could reduce total

[5] PJ = petajoule; 1 petajoule = 10^{15} Joules.

electricity use in motors by as much as 17% in some regions (Geller et al., 1987; Alliance to Save Energy, 1987).

Comparable efficiency improvement measures have also been identified for the other major components of industrial electricity use (Kahane and Squitieri, 1987). In addition, many of the efficiency measures identified earlier for lighting and space conditioning in large commercial buildings are also applicable to industrial users. A recent study examined electricity-saving projects currently being undertaken in automobile manufacturing plants in the U.S. and Europe. It shows that roughly 30% savings from the current cost of purchased electricity are being achieved (Price and Ross, 1988).

Cogeneration

Technologies for cogeneration--production of electricity and heat or steam for other useful purposes from a single combustion source--are described in **PART TWO: ENERGY SUPPLY**. The primary market for cogeneration is in large industrial facilities (although large commercial/institutional applications are important in the Eastern Bloc and are also beginning to be seen in the U.S.). From industry's perspective, cogeneration is one method for improving energy efficiency. The industrial facility benefits, either in the form of electricity used on-site or revenues from sales of electricity to the local utility, from the fuel it would otherwise have consumed solely for industrial production purposes.

Industrial cogeneration has grown rapidly in the U.S. since the 1978 passage of the Public Utilities Regulatory Polices Act (PURPA), which ensures that cogenerators (among others) can sell electricity to utilities at the utilities' avoided cost (the cost that the utility would otherwise have to pay to produce or obtain the electricity). As of 1985, 13 gigawatts of cogeneration capacity were in operation (Edison Electric Institute, 1985). Projects that would yield an additional 47 gigawatts have been registered with the Federal Energy Regulatory Commission (FERC) through October 1987 (FERC, 1988). One company specializing in cogeneration has estimated that cogeneration capacity could reach 100 gigawatts (equal to about 15% of current capacity) by the year 2000 (Naill, 1987).

All of these projects tend to reduce potential greenhouse gas emissions as they result in more efficient use of energy. To the extent that industrial cogeneration projects are based on natural gas or oil--or industrial waste products such as black liquor or bark in pulp and paper--and displace new coal-fired electric generating capacity, the net impact on greenhouse gas emissions (as well as local environmental loadings) could be much greater.

Near-Term Technical Options: Developing Countries

As noted above, developing countries generally are in the early stages of a rapid expansion in producing energy-intensive materials associated with infrastructure development and widespread access to basic consumer durables. If this process proceeds along the path experienced historically by the industrialized countries, the increases in energy consumption and CO_2 emissions will be enormous. In fact, the need to import fossil fuels to support rapid expansion of heavy industry could become self-defeating, operating as a brake on the rate at which some developing countries can industrialize.

Understandably, developing countries and development assistance organization are concerned about alternative approaches to industrial development that would allow developing countries to increase economic activity without having to devote ever-increasing shares of their foreign exchange earnings to financing fossil-fuel imports. Several possible strategies for achieving this goal are very compatible with concerns about long-term greenhouse warming.

Technological Leapfrogging

This phrase has been used by Goldemberg et al., (1987) in discussing industrial development options for both industrialized and developing countries. Theoretically, at least, developing countries could adopt the most efficient process technologies currently available, or even push ahead with currently-experimental technologies as they invest in major expansions of heavy industry necessary to foster economic development. If this strategy were implemented, the projected massive increases in industrial energy use in the developing countries would be significantly reduced.

There are a number of reasons why this may not occur in practice. Because capital (and entrepreneurial experience) is scarce in developing countries, there is a tendency to avoid risky investments. Conventional wisdom suggests that these countries should adopt technologies that are already "mature" in the industrialized countries. This "conventional wisdom" has affected decisions both by developing countries themselves and by relatively conservative leading institutions (banks or assistance organizations) called upon to provide capital. Often, though by no means always, advanced energy-efficient technologies require larger capital investments than older technologies. Also extremely important is that the availability and cost of certain components of production--labor, capital, and natural resources--vary widely between developing countries and industrialized countries, and among individual developing countries. Since many developing countries have difficulty raising the capital required for many investments, some form of international financial arrangements may be necessary to ensure that developing countries adopt energy-efficient technologies as they industrialize (see **CHAPTER IX** for further discussion).

Thus, developing countries will have difficulty basing rapid industrial development on currently mature, but also relatively energy-intensive, technologies available from the industrialized countries. On the other hand, "leapfrogging" to the most advanced technologies now being developed in industrialized countries may be inappropriate in terms of the local availability of labor, capital, and natural resources. What appears to be needed is to identify and develop advanced industrial process technologies that are appropriate to each developing country's individual endowment of resources. To the extent that developing countries move in this direction, energy efficiency, or at least efficiency in fossil-fuel use, should be a major characteristic of the desirable options for most developing countries.

Alternative Fuels

As discussed above, most developing countries are interested in limiting increases in fossil-fuel imports associated with industrialization. For this reason, they will undoubtedly provide a proving ground for development of heavy industry based on alternative fuels.

One option for some developing countries may be to develop potential hydroelectric generation resources and base industrial development on advanced electricity-intensive processes (Opportunities and difficulties in developing hydroelectric generation are discussed in more detail in **PART TWO: ENERGY SUPPLY**). In general, developing countries are relatively rich in biomass resources and in some cases undeveloped hydropower, while they are net importers of fossil fuels. Thus, the tailoring of industrial development strategies to local resources will likely reduce greenhouse gas emissions in addition to achieving other benefits.

For those developing countries with abundant coal resources, most notably China, there may be a tradeoff between energy self-sufficiency goals and climate warming concerns. Even in these situations, promoting the most energy-efficient technologies should be a common goal. When developing countries have both natural gas and coal resources, local environmental and economic concerns, including consideration of the greenhouse phenomenon, should all encourage a near-term emphasis on natural gas.

Retrofit Energy Efficiency Programs

A recent report by the U.S. Agency for International Development (U.S. AID, 1988b) reviewed the electricity supply-and-demand situation in a number of developing countries. The report documented serious concerns in many developing countries about current or projected shortages in electricity. However, it also found that "few achievements have been made in developing countries in using electricity more efficiently. The opportunities are tremendous, and the cost is only a fraction of the generation expansion option." According to the report, over 40% of the electricity use in developing countries is by electric motors in the industrial (including agriculture) sector (U.S. AID, 1988b). One detailed energy analysis in Pakistan identified specific industrial efficiency improvements, including improved controls and lighting, which could reduce industrial energy consumption by more than 20% in 2005 (Miller et al.).

Other recent studies in Kenya and South Korea indicate that efficiency programs have been successful in reducing energy use in heavy industry (Geller, 1986). A detailed analysis of the electricity conservation potential in Brazil indicates that improvements in electric motors and motor controls could reduce industrial electricity consumption by an amount equivalent to 8.7 Gw of new generating capacity (Geller, 1984). Many of the potential industrial energy savings in developing countries are undoubtedly not yet identified because of the relatively little attention that has been given to this issue in the past. However, the fragmentary evidence currently available suggests that these opportunities are much greater in percentage terms than those in industrialized countries and are much cheaper than the incremental cost of increasing energy consumption.

Agricultural Energy Use

On a global scale, agriculture accounts for a small part of "commercial" (excluding traditional biomass) energy use--about 3.5% in 1972-73. However, the percentage in some developing countries may be higher. In addition, the projected transition in developing countries from traditional labor-intensive agricultural practices to modern mechanized practices is expected to increase the use of commercial energy in agriculture significantly in developing countries by the year 2000 (FAO, 1981).

A major goal of most developing countries is to increase productivity in agricultural production either for domestic consumption or for export purposes. If energy conservation is viewed as a constraint to such improvements, it generally will not be viewed as a more important objective. However, productivity increases may be possible through several alternative approaches, with markedly different implications for energy and employment.

Many developing countries are interested in holding down imports of fossil fuels. They may also experience widespread unemployment or underemployment and may seek modernization without displacement of employment. It is important that agricultural modernization be incorporated into an overall development strategy appropriate to each individual country. In this context, energy savings may be achieved in conjunction with other objectives. Goldemberg et al. (1988) point out that some agricultural modernization can occur without such large increases in commercial energy consumption (and associated reductions in labor requirements). Using one type of rice production as an example, they illustrate that many benefits of the "green revolution" in increasing yields per hectare can be achieved with intermediate approaches that do not go as far in substituting mechanical energy for labor.

Expanded agricultural energy needs can also present an attractive opportunity for biomass energy development. FAO (1981) projected that an increase of 17 PJ of oil-equivalent agricultural energy use would be required to double food production in developing countries by 2000. Goldemberg et al. (1988) calculate that this amount of energy in the form of methanol could be produced by thermochemical processes from 40% of the present organic wastes (crop residues, animal manure, and food-processing wastes) in developing countries. Alternatively, feedstocks could come from tree plantations representing land equivalent to 3% of current forest land in developing countries. Other possible areas for efficiency improvements include electromechanical pumping of water for irrigation, more efficient use of existing water resources, and the use of alternative energy supplies such as wind for pumping.

Near-Term Technical Options: Soviet Bloc Countries

In the Soviet Union and Eastern Europe, industrial energy use accounts for nearly fifty percent of secondary energy use. As a share of primary energy equivalent (with electricity conversion losses allocated to end uses of electricity) it is even larger (Mintzer, 1988). These countries have very high energy consumption per unit of GNP. One recent analysis indicates that the current energy intensity of the Soviet economy is "akin to the IEA energy economies of the early 1970s," as shown in Table 7-6 (IEA, 1988). In fact, while energy intensity in OECD countries was declining by over 20% from 1973 to 1986, energy intensity in the USSR actually increased slightly. Industrial energy conservation could, therefore, provide significant improvement in these countries.

A closer look at some specific industries verifies the general impression of energy-inefficiency. The USSR is by far the world's largest producer of steel (WRI and IIED, 1988). Unfortunately, it is also apparently close to last in the world in efficiency. Estimates for 1983 indicate that the USSR used 31 GJ to produce a ton of steel as compared to the Japanese standard of about 19 GJ

TABLE 7-6

Energy Intensities of Selected Economies
(energy/unit of Gross Domestic Product)

	1973	1986
Canada	0.88	0.76
United States	0.76	0.57
IEA Pacific	0.42	0.31
IEA Europe	0.40	0.34
IEA Total	0.56	0.44
Soviet Union	0.99	1.03

Sources: IEA, 1988.

(Chandler, 1986). As shown in Table 7-7, the Soviet Union and many Eastern European countries continue to produce a large percentage of their steel in very inefficient "open hearth" furnaces, which have been virtually eliminated in the OECD. They also do very little recycling, as indicated by the small percentage of steel produced in electric arc furnaces.

One analyst has suggested that Soviet industry has shown some success in improving efficiency with existing technology in their heavy industry (e.g., "housekeeping" measures, refinements of existing technologies), but have failed to assimilate distinctly different, and inherently less energy-intensive technologies. An example is the cement industry, a heavy energy-using sector for which energy use can be greatly reduced by switching from the wet process to a newer dry calcining process. Although the dry process is available in the Soviet Union, it has not been widely utilized (Hewitt, 1984).

There are several reasons for the extraordinarily high energy intensity of Eastern Bloc industry. The Soviet Union (and to a lesser extent Eastern Europe) has enormous energy resources and has historically invested heavily in energy development due to national economic policy rather than demand. Hence, scarcity of energy resources has not provided an incentive to conserve. In addition, the "mark-up" pricing systems used in these countries does not provide a strong incentive for efficiency improvements. In fact, it has been argued that the reward system has actually provided managers with an incentive to consume more energy (Chandler, 1986).

More recently, the value of energy efficiency has been recognized by Soviet leadership. As described by Soviet energy analysts, "the energy economy of the Soviet Union is entering a new period of development. The most economical and favorable located oil and gas resources...are gradually running out." The result is that development of new energy resources is much more expensive and difficult than in the past (Makarov, et al., 1987). One result of rising costs of energy

TABLE 7-7

Innovation in Steel Production Technology
Selected Countries, 1985

Country	Economy Type	"Inefficient" Open Hearth (% of production)	"Recycling" Electric Arc (% of production)
Spain	M	0	61
Italy	M	0	53
South Korea	M	0	31
United Kingdom	M	0	29
Japan	M	0	29
West Germany	M	0	19
Brazil	M	4	25
United States	M	7	33
Romania	C	29	22
China [a]	C	31	19
Yugoslavia	C	34	26
India [a]	C	42	19
Poland	C	42	15
East Germany	C	34	31
Hungary [a]	C	51	13
Soviet Union	C	57	11

M = Market-oriented; C = Centrally planned.

[a] Though this country's agricultural economy is market-oriented, its industry is not.

Source: Chandler, 1986.

development has been the explicit inclusion of energy efficiency measures in central economic and energy planning. Targets have been set that would result in more than a 20% reduction in the energy intensity of the Soviet economy (Makarov, et al., 1987).

The Soviet Union has historically done very well in the utilization of industrial waste heat for electricity generation (cogeneration) and for other heat needs, such as district heating. This trend is continuing. In the Leningrad region, for example, the 1981-1990 energy plan calls for savings of about 2 EJ of previously-projected industrial energy use through a combination of improved industrial technology and expanded use of waste heat for both electricity and other heat needs (Glebov and Kovlenko, 1987).

In addition to efficiency improvements, other options for reducing greenhouse gas emissions from industry in the Eastern Bloc are possible through fuel switching. Although the major resources are generally located far from current industrial centers, the Soviet Union has the world's largest proven recoverable natural gas reserves in the world, amounting to about 39% of the World total (IEA, 1988). In addition, its undeveloped geothermal and hydroelectric resources are significant (as discussed in **PART TWO: ENERGY SUPPLY**). It would be technically possible, therefore, for the USSR to substitute various alternative fuels for current and projected coal use in the industrial sector, although the costs of such substitution have not been well analyzed.

Summary of Near-Term Technical Potential in the Industrial Sector

Some available projections based on trends in structural change and process technology improvement alone suggest that industrial energy use in the OECD countries may not rise significantly over the next 25-40 years (Williams et al., 1987). The No Response scenarios used in this study are consistent with this view, showing roughly constant to slightly increasing industrial energy use for industrialized countries through 2025. Very recent data for 1987 and 1988 in the U.S. show sharp increases in durable goods and basic material production (U.S. DOC, 1988). If these very recent trends continue, industrial energy use in OECD countries would be higher than indicated in the No Response scenarios. Industrial energy consumption in the East Bloc and developing countries increases substantially in the No Response scenarios.

The technical options identified above could reduce industrial energy consumption by 25% below the levels in the No Response scenarios in the 2000-2010 time frame. One recent analysis indicates that cost-effective conservation could result in an absolute decline of about 19% in industrial energy consumption (fuel and purchased electricity) in the U.S. by the year 2010 (Ross, 1988). Changes incorporated into the Stabilizing Policy cases in this document result in small reductions in the U.S. and industrialized countries. These scenarios appear to be well within the cost-effective technical potential; the achievability of these targets within real-world policy constraints is addressed in **CHAPTER VIII.**

Long-Term Potential in the Industrial Sector

Over the long term, technological options for improving efficiency in industrial energy use are very speculative. Several concepts have been advanced by various analysts that may warrant further study to identify potential options for reducing long-term industrial demand for fossil fuels.

<u>Structural Shifts</u>

As discussed above, over the long term, shifts in the structural composition of industrial activity will undoubtedly continue, with significant energy consequences. As discussed by Ross et al. (1987), production of basic (and inherently energy-intensive) materials has tended to decline over time as a share of GNP after an economy achieves a certain level of affluence. Three components of this shift are:

- Substitution of new (and often less energy-intensive) materials;
- Product and production design changes that result in more efficient materials use; and
- Saturation of major markets for material-intensive products, including infrastructure and material-intensive consumer products.

These effects can be expected to continue in the future and to some degree are incorporated in all the scenarios of future energy use presented in this report. With combinations of policies, such as economic incentives and aggressive research and development, these effects can be accelerated considerably, especially in the developing countries where major increases in industrial energy use are expected. It is currently very difficult to identify, much less quantify, the effects of actions to achieve the technical potential. Further study of long-term trends in the structure of industrial activity and specific policy options for influencing these trends is needed.

Advanced Process Technologies

As pointed out by Ross (1985), major reductions in energy intensity in industrial processes can come about through "revolutionary" change in process technology. Typically, process change is not motivated primarily by energy conservation, but large reductions in energy intensity often result due to technological advances. An example is advanced technology for steel production now under development that could result in very large energy savings per unit of output. About 40% of the energy used in iron and steel production is related to shaping and treating starting with liquid steel. Advanced processes utilize controlled solidification, perhaps very rapid, of thin castings near their final shape. When fully developed, this technology should eliminate almost all of the energy use currently associated with rolling and shaping (Ross, 1985).

In the petrochemical industry research is currently underway to identify advanced separation techniques that could eliminate many of the losses inherent in the current distillation process. Gas separation membranes, gas adsorption and liquid mass separating agents are all currently commercial for very specialized petrochemical applications and use considerably less energy than the distillation processes they have replaced (Mix, 1987). Over the long term, wider applications of these or similar technologies may further reduce process energy requirements.

In general, advanced technologies are attractive because of lower total cost, better quality control, reduction in inventories, greater flexibility, etc. as well as improved energy efficiency. It is difficult to identify possible process technology improvements that could become available through research and development over the long term, although it is likely that energy efficiency improvements will be part of these developments.

Another long term area of process technology advancement is the use of biological phenomena in production processes. Although it is difficult to even identify specific applications of biological processes at this time, they appear to offer possible improvements in speed, control, and precision of process technologies (Berg, 1988). As such technologies emerge, they will undoubtedly offer additional opportunities for reductions in fossil energy use. Industrial policies designed to stimulate technological advances and capital turnover will ultimately also lead to reduced energy requirements.

Non-fossil Energy

Although it is likely that energy intensity of industrial activity will continue to decline over the long term, significant levels of energy consumption will no doubt still be required. Opportunities exist for meeting these demands with non-fossil energy sources. Many industrial technology experts (see, for example, Berg, 1988, Schmidt, 1987) have suggested that increasing use of electricity in industrial processes over the long run is likely because of its superior characteristics of controllability and minimal loss or error.

Even though production of electricity involves large losses, this is often offset because the more sophisticated and precise production processes it allows are generally less energy-intensive at the end use. Increasing electrification of industrial processes at the end-use point allows for a wide variety of alternative generation options (as discussed in **PART TWO: ENERGY SUPPLY**), which can reduce greenhouse gas emissions.

In addition, one analyst has suggested that economic competition will cause energy-intensive industries over the long run to locate geographically where isolated resources of cheap and hard-to-transport energy are available. Examples cited include natural gas and hydro locations in Canada, Brazilian hydropower sites, and natural gas locations in the Middle East (Ross, 1985). The same logic could also apply to locations with geothermal and solar resources as those technologies become competitive. Policies could be implemented to discourage fossil fuel use (see **CHAPTERS VII** and **IX**), as well as to encourage movement of heavy industry to locations where renewable resources are economically attractive.

PART TWO: ENERGY SUPPLY

This section discusses technical options for reducing greenhouse gas emissions by (1) utilizing fuels for power generation more efficiently, and (2) altering the types of energy supplies we use. These two goals have one common objective: to supply a similar amount of useful energy services compared to current energy consumption practices, but in such a way that the amount of greenhouse gas emissions are minimized. Options for reducing greenhouse gas emissions in end-use applications, particularly by improving energy efficiency and switching fuels, were discussed in **PART ONE: ENERGY SERVICES.** This section focuses on possible options for improving the delivery of energy services by reducing the losses during energy production and conversion processes. Additionally, the types of energy supplies can be altered by developing sustainable sources of energy that do not emit greenhouse gases.

As discussed in chapters IV and V, fossil fuels currently dominate commercial energy use globally (88 percent of primary energy supplied in 1985). This demand for fossil fuels is expected to continue in the future, e.g., fossil energy production in the No Response scenarios increases by 28-73% by the year 2025 in the Slowly Changing World and Rapidly Changing World scenarios, respectively. Much of this increase in fossil fuel demand will be for electricity generation, with coal as the primary fuel, since it is a relatively inexpensive, abundant resource globally. For example, by 2025 coal use for electricity generation increases by 30% in the SCW scenario and by 130% in the RCW scenario.

In the Stabilizing Policy cases first presented in Chapter V, several measures are assumed to reduce greenhouse gases emissions from the production of electricity. These include improved efficiency in electricity generation, greater use of natural gas to displace coal, recovery of methane from coal seams, and increased use of renewables, e.g., biomass and solar, for producing electricity and synthetic fuels. The contribution that these measures can make to reducing greenhouse gas emissions can be considerable. These options help to reduce the percentage of electric utility primary energy consumption from fossil fuels from over 60% in 1985 to under 40% in 2025 in both the SCWP and RCWP scenarios (see Figure 7-10). The extent to which any specific technological option is emphasized to reduce global warming is a matter for future consideration; as discussed in this section, however, there are many options for altering our current dependence on fossil fuels.

The first part of this section discusses possible options for altering current patterns of fossil fuel use. This is clearly one of the highest near-term priorities, since as discussed in Chapters IV and V, current commercial energy use globally is dominated by fossil fuels. In many cases, however, options exist for reducing emissions from these applications. This is followed by a discussion of possible supply alternatives to fossil fuels, including increased use of biomass, solar resources, additional renewable energy resources, nuclear power, and options for enhancing energy storage and delivery to consumers.

FOSSIL FUELS

As discussed in Chapter IV, fossil fuel consumption is responsible for the vast majority of CO_2 emissions. On an energy-equivalent basis, coal produces the most CO_2 (about 24 kg C/gigajoule), oil about 80 percent the amount of CO_2 compared to coal (about 19 kg C/gigajoule), and natural gas about 55-60 percent the amount of CO_2 (about 14 kg C/gigajoule). Given these rates of CO_2 emissions, major reductions in fossil fuel consumption (or their elimination) would be necessary over the long run to control greenhouse gas emissions. With the current global reliance on fossil fuels, however, the shift away from fossil fuels can not be accomplished easily, even with international agreement to pursue this objective. Steps can be taken now to begin or accelerate the transition to non-fossil energy by minimizing the greenhouse impact of the fossil fuels that are used.

FIGURE 7-10

ELECTRICITY UTILITY DEMAND BY FUEL TYPE

(exajoules)

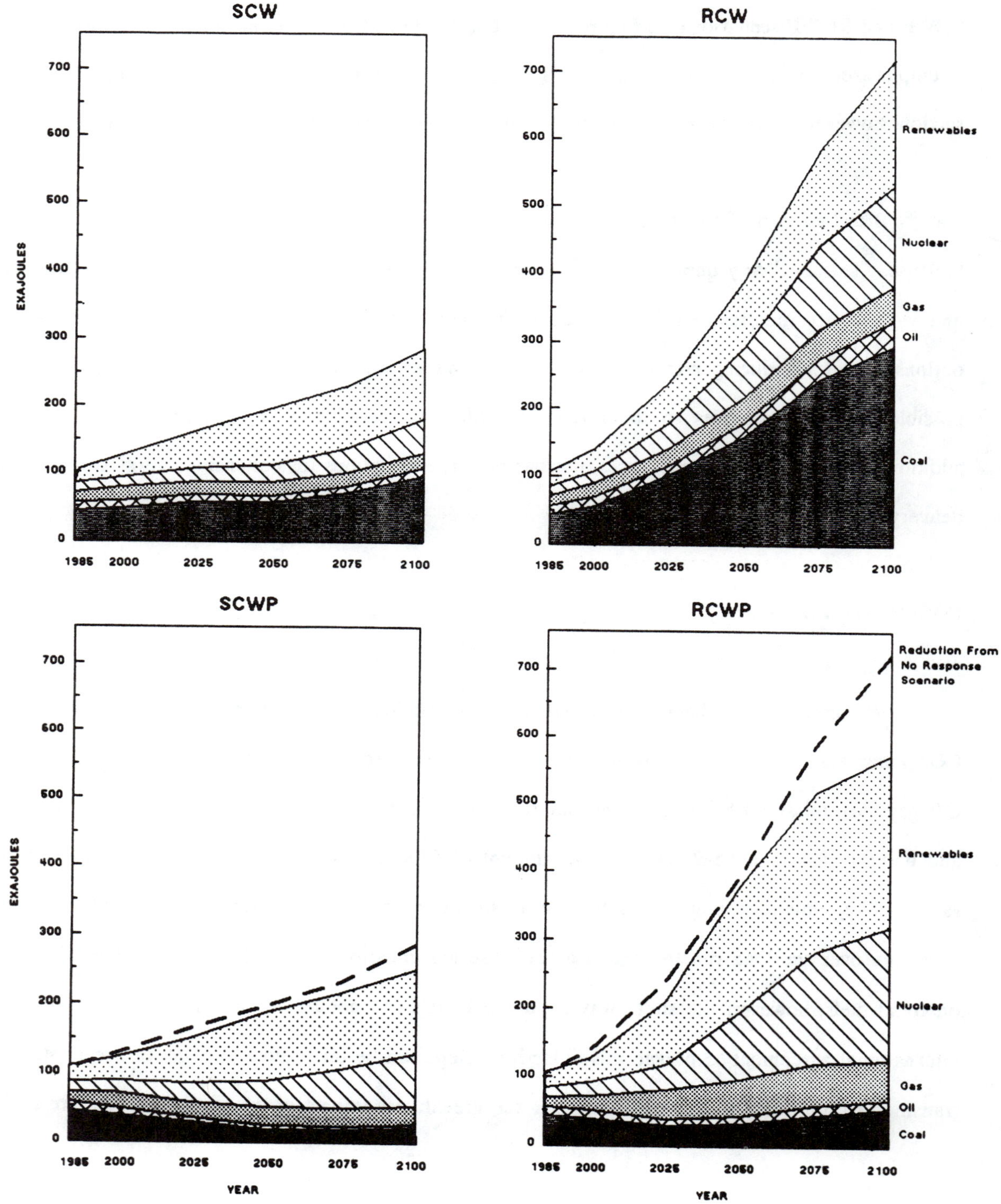

Possible actions include improving the efficiency with which fossil fuels are produced and converted to electricity, switching from more carbon-intensive fuels to less carbon-intensive fuels (e.g., coal to natural gas), and applying various engineering controls to reduce emissions of greenhouse gases during the production and consumption of fossil fuels (e.g., NO_x control, CH_4 recovery).

Since one of the primary uses of fossil fuels currently is the production of electricity, one potential option is to produce electrical power more efficiently, using less energy input to produce electricity. For example, in the U.S. during the 1950s and early 1960s, the efficiency of powerplants consuming fossil fuels increased from 25 to about 32 percent (see Figure 7-11). This improvement has stalled since the early 1960s because of fewer technical improvements in combustion techniques, higher energy consumption by auxiliary equipment used for pollution control (e.g., electrostatic precipitators for particulate removal and scrubbers for SO_2 removal), and increased use of less efficient fuels, such as subbituminous coal and lignite.

Although electricity production is currently one of the primary uses for fossil fuels, other applications have been proposed in order to meet future energy needs. Specifically, as conventional petroleum resources are depleted, it is expected that much of the demand for liquid (oil and natural gas liquids) and gaseous (natural gas) fuels will ultimately be met by synthetic fuel production. Although there is currently little synthetic fuel production in the world, processes have been developed to convert relatively abundant solid energy resources such as coal, oil shale, and tar sands to liquid or gaseous products that could be consumed in the same end-use applications as more conventional resources.

The production of synthetic fuels, however, typically requires the consumption of significant amounts of energy to produce the liquid or gaseous fuels. These conversion processes produce greenhouse gas emissions, particularly CO_2, so that the total emissions per unit of energy are higher

FIGURE 7-11

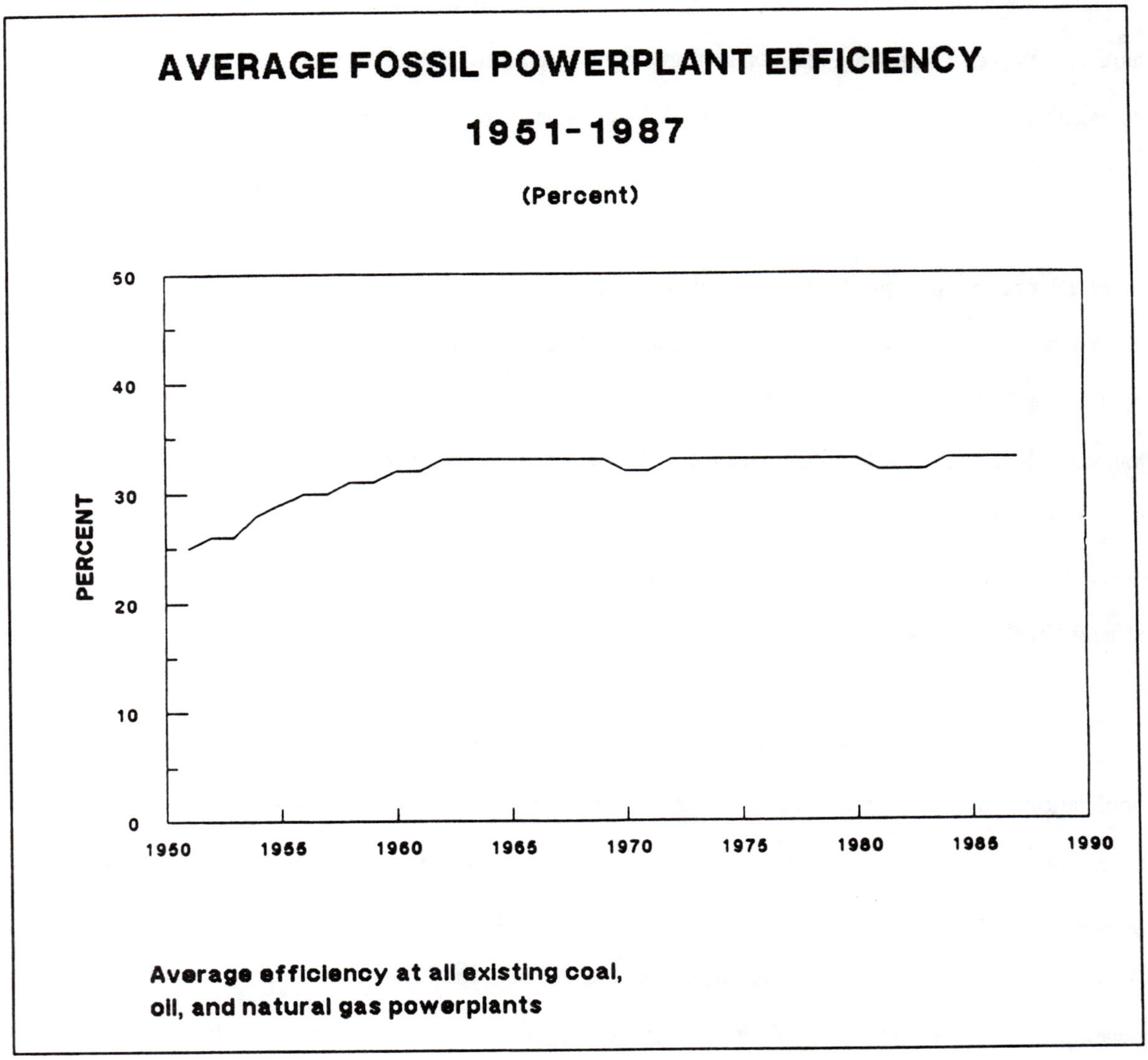
AVERAGE FOSSIL POWERPLANT EFFICIENCY
1951-1987
(Percent)
PERCENT
50
40
30
20
10
0
1950
1955
1960
1965
1970
1975
1980
1985
1990
Average efficiency at all existing coal, oil, and natural gas powerplants

for synthetic fuels than for conventional fossil fuels. For example, the CO_2 emissions from production and consumption of liquid fuels from coal is about 1.8 times the amount from conventional liquid fuels from crude oil (see Chapter IV for further discussion).

The following sections explore possible options for reducing the greenhouse impact of fossil fuels. Using fossil fuels more efficiently is discussed, including refurbishment of existing powerplants, repowering opportunities (including application of clean coal technologies), and cogeneration. Greater use of natural gas is discussed since it produces less CO_2 than oil or coal. Methods for controlling greenhouse gas emissions are also presented.

Refurbishment of Existing Powerplants

Energy use at existing powerplants can be reduced by refurbishing the plant to keep it operating at optimal efficiency. Over time, these efficiencies decline due to wear and various aging processes. For many economic reasons it has become clear over the past decade that utilities in the U.S. are planning to keep existing powerplants in service longer than initially planned (Democker, et al., 1986). It is likely that this trend will occur globally as well to minimize the need to invest in new powerplants. As these powerplants age, however, declining efficiency will become more widespread than in the past.

In response, the U.S. electric power industry has developed new techniques for extending the life of powerplants. The extent of efficiency improvement depends on how badly the specific plant had degraded and the extensiveness of the upgrades. Increases in efficiency in the range of 3-4% appear to be possible at many existing units, with even greater improvements possible in some cases (PEI 1988).

In developing countries, the opportunities for generating efficiency improvements may be much greater. According to a recent study by U.S. AID (1988b), "The majority of thermal powerplants in developing countries operate at lower-than-design capacity and efficiency." Many of these powerplants use more than 13 MJ[1] of fuel to generate a kWh of electricity, compared to typical design heat rates of 9-11 MJ/kWh. It is estimated that a rigorous program of powerplant rehabilitation could improve overall fuel use efficiency for thermal power generation by 10% or more in most developing countries (U.S. AID, 1988b).

Clean Coal Technologies and Repowering

As new powerplants are constructed to meet increasing electricity needs, many of these plants are likely to be fossil-fueled. To the extent that new powerplants use fossil fuels, greenhouse gas emissions can still be reduced by using the most efficient conversion technologies. The U.S. Department of Energy, the Electric Power Research Institute, and many other organizations have been investing significant funding in research, development, and demonstration of "clean coal technologies" designed to allow burning of coal to generate electricity with maximum efficiency and minimum environmental impact (U.S. DOE, 1987e). These technologies offer the potential to significantly reduce the amount of traditional air pollutants such as sulfur dioxide and nitrogen dioxides. However, they may also affect the amount of greenhouse gas emissions, particularly for those technologies that improve the overall efficiency of converting coal to electricity. For example, some of these technologies can improve efficiency 10-25% relative to conventional coal combustion technologies.

---

[1] 1 MJ = 1 Megajoule = 10^6 joules.

Three of these advanced technologies currently in the demonstration phase are atmospheric fluidized bed combustion (AFBC), pressurized fluidized bed combustion (PFBC), and Integrated Gasification/Combined Cycle (IGCC). AFBC is likely to be very similar in efficiency to conventional technology, and therefore not beneficial in reducing greenhouse gas emissions. The PFBC and IGCC systems, however, are projected to increase conversion efficiencies as much as 10-20%, with corresponding reductions in CO_2 emissions per unit of electricity produced (U.S. DOE, 1987e).

Clean coal technologies can be used for newly-constructed powerplants, but also to "repower" existing powerplants. In repowering, the basic combustion components of existing powerplants are replaced with one of the new technologies. Additionally, new components, such as a gas turbine cycle, may also be installed in combination with some refurbished components of the existing powerplant. The result is a hybrid plant with performance very much like that of an efficient new unit.

Cogeneration

Cogeneration is typically the production of both steam and electricity from the same source, with the steam used to meet heating and process requirements at a facility and the electricity used on-site or sold to electricity customers. Because it is more energy-efficient than conventional generating options, it has been encouraged in the U.S. recently by many regulatory and legislative initiatives. For example, the Public Utilities Regulatory Policy Act (PURPA) encouraged cogeneration by establishing a process ensuring that cogenerators with low production costs could sell this cogenerated electrical power to electric utilities.

As discussed earlier **PART ONE: ENERGY SERVICES**, cogeneration has been very popular with large industrial energy users as one approach for reducing their overall energy costs. Most of the approximately 18 GW of currently-operating cogeneration projects in the U.S. and an additional 29 GW under active development fall into this category (Williams, 1988). Similarly, as electric utilities face more competition from industrial cogenerators and independent power producers (i.e., companies not regulated as utilities but do produce power), they may choose to retrofit some existing generating facilities with cogeneration, using the waste heat from electricity production for district heating or industrial process heat applications. Engineering assessments have shown this potential retrofit option to be economically attractive for powerplants burning coal and located close to steam load centers (Hu, et al., 1984).

Natural Gas Substitution

Natural gas (which is primarily methane) has just over half the carbon per unit of energy as coal. In applications such as electricity production where coal is frequently used, switching to natural gas would substantially decrease CO_2 emissions. As discussed previously in **PART ONE: ENERGY SERVICES**, natural gas is currently used in several key end-use applications, particularly a wide variety of industrial energy applications and in residential and commercial space heating. In addition to its end-use potential, natural gas can be used as a fuel for electricity generation. The discussion below focuses on the technical options and advantages of using natural gas as a fuel for electricity generation, and then reviews the issue of how much natural gas is available overall since its role as an end-use fuel and a fuel for electricity generation depends on the amount of natural gas resources available and the cost at which these resources can be supplied.

<u>Natural Gas Use At Existing Powerplants</u>

There are several near-term alternatives for increasing the use of natural gas for electricity generation. One relatively inexpensive option would be to increase the utilization of existing natural gas and oil-fired (most of which can also consume natural gas) powerplants. For example, in 1987 average capacity utilization rates for U.S. oil and gas powerplants were 40% relative to 58% for coal powerplants. Thus, there is some technical potential to increase gas use by increasing the utilization of natural gas-fired powerplants. However, these plants are utilized less because the variable cost of power is higher at most oil and gas plants than at coal, nuclear, and hydro powerplants and oil and gas-fired powerplants can be switched on and off more easily and with less wear on the systems to meet rapid increases in electricity demand. Since electric utilities produce electricity with their least expensive powerplant, policies would have to be adopted to increase utilization of natural gas capacity.

Advanced Gas-Fired Combustion Technologies

An additional option for increasing natural gas use is the construction of new gas-fired combined cycle or combustion turbine powerplants. These powerplants cost significantly less to build than coal powerplants and are typically more energy efficient. They could also be part of a near-term solution since the lead times for plant siting and construction average about two to four years versus six to ten years for coal-fired powerplants. These advanced combustion technologies are not in greater use primarily because of investors' expectations that the costs of natural gas over the operating life of the facility would be substantially higher than other fuel alternatives (such as coal). As a result, despite the lower capital costs of these technologies, electric utilities have often invested in other alternatives because total operating costs of combined cycle or combustion turbine technologies have been perceived to be greater than coal-fired powerplants.

Combustion turbines--Simple/Combined Cycle. Combustion turbines are similar to jet aircraft engines in that fuel is burned in compressed air, with the combustion gases then used to turn a turbine for electricity generation. This process is known as a simple cycle turbine. After the hot gases are used to produce electricity, the exhaust gases can be converted to steam, which can be used to generate additional power in a process known as combined cycling. Most combustion turbines in use are simple cycles, which are used primarily for peak power requirements due to their favorable operating characteristics (primarily their ability to increase power production quickly) and low capital costs, although operating costs are high overall. Combined cycle capital costs are higher and more efficient, and hence, are currently preferred for units expected to be utilized more frequently.

Aeroderivative Combustion Turbines. Recent advances in jet engine (aeroderivative) technology, including new materials and designs that enable combustion to occur at higher temperatures, have made turbines more efficient. Many of these advances are being, or could be, applied to turbines for generating electrical power. For example, existing simple and combined cycle systems have efficiencies of about 32% and 42%, respectively, compared to conventional coal powerplants that have efficiencies of about 31-32%. Recent improvements in aeroderivative technology could significantly improve these efficiencies. For example, one technology that has been recently commercialized is the steam-injected gas turbine (STIG). STIG units take any steam not needed for process heat requirements and inject it back into the combustor for added power and efficiency, e.g., in a simple cycle application the efficiency of the turbine might be 33% with an output of 33 MW, while with full steam injection the efficiency would increase to 40% with an output of 51 MW (Williams, 1988). An improvement in steam injection that has been proposed is the intercooled steam-injected gas turbine (ISTIG). ISTIG cools the compressor bleed air used for turbine blade cooling, allowing a much higher turbine inlet temperature. With this technology the single cycle efficiency cited above would increase from 33% to 47% and output from 33 MW to 110 MW; the estimated capital cost of ISTIG is about $400/kW (Williams, 1988).

Natural Gas Resource Limitations

There is a significant amount of research and debate over the quantity of natural gas that is available. As discussed in Chapter IV, global gas resources are estimated to be significantly smaller than global coal resources (resource estimates by the World Energy Conference indicate that coal resources are about 30 times greater than gas resources). Within the U.S. there is disagreement over the size of this difference. One source has indicated that in the U.S. the natural gas resource base is as large as the coal resource base when one compares the economically recoverable and usable resources; the supply should be adequate for hundreds of years (Hay et al., 1988). On the other hand, based on 1985 consumption levels of about 18.6 EJ, the U.S. DOE has estimated that technically-recoverable U.S. gas resources would last only about 70 years and only about 45 years if limited to supplies that could be marketed for about a maximum of \$5/GJ (see Table 7-8). In contrast, U.S. coal reserves are estimated to be about 350 times greater than 1985 U.S. consumption levels (U.S. DOE 1985; U.S. DOE 1986).

In addition to the potential limit on gas supplies, there are also questions about the cost of additional gas supplies, the location of supplies vis-a-vis the areas of demand, the cost of improved distribution systems, and in the case of international trade in liquified natural gas (LNG), the costs of liquefaction, transportation, and regasification facilities.

These concerns over the cost and availability of natural gas may appear unwarranted given the existence of excess capacity and falling prices in the U.S. gas industry in recent years. However, these market conditions may be temporary. Current excess capacity follows a period in the 1970s when natural gas was in short supply. The recent changes are primarily due to natural gas price deregulation, which allowed prices to increase from previously controlled levels. The price increases had two effects: (1) a decrease in demand as prices increased, and (2) an increase in supplies as the

TABLE 7-8

Total U.S. Gas Reserves And Resources (Exajoules)

	Technically Recoverable Gas[a]	Recoverable Gas By Price[b]	
		<$3/Mcf	$3-5/Mcf
Lower 48 (Conventional)			
Proved Reserved, 12/31/86, Onshore and Offshore	171	171	---
Inferred Reserves/Probable Resources, 12/31/86, Onshore	91	91	---
Inferred Reserved, 12/31/86, Offshore	25	25	---
Extended Reserve Growth in Nonassociated Fields, Onshore	128	60	19
Gas Resources Associated with Oil Reserve Growth[c]	66	32	12
Undiscovered Onshore Resources	236	95	63
Undiscovered Offshore Resources[d]	144	58	30
Subtotal	860	533	125
Lower 48 (Unconventional)			
Gas in Low-Permeability Reservoirs	194	75	53
Coalbed methane	52	9	4
Shale Gas	33	11	5
Subtotal	279	95	63
Alaska			
Reserves	36	8	0
Inferred Reserved (Cook Inlet Area)	3	3	0
Undiscovered, Onshore and Offshore	100	2	2
Subtotal	139	13	2
Total	1,278	640	189

[a] Volumes of gas judged recoverable with existing technology.
[b] Volumes of gas judged recoverable with existing technology at wellhead prices shown (1987$). Mcf represents one thousand cubic feet (about 1.1 GJ), which is a common measure of gas quantity.
[c] Judged at oil prices of <$24/bbl and $24-40/bbl.
[d] Outer Continental Shelf

Source: U.S. DOE 1988.

higher prices of gas prompted increased exploration and development. The duration of the current black market conditions is a matter of much debate, but as the demand and supply for natural gas come into balance, natural gas may no longer be available at current prices.

The relevance of these concerns is not that natural gas cannot play a role in reducing CO_2 emissions; unquestionably, it can. However, the ability of natural gas to replace higher carbon content fuels is a function of the quantity of natural gas available and the cost at which natural gas can be supplied to consumers. That is, even if natural gas is available, its ability to replace other fuels will depend on its cost relative to alternative fuels. Any policies promoting increased use of natural gas need to recognize these factors.

Electric Utility Gas Consumption. In 1985, gas consumption by U.S. electric utilities was about one-fifth of total U.S. gas use, or about 3 EJ. Even though U.S. utilities consume only a minor fraction of natural gas, they can have a large impact on prices because they are frequently the marginal buyer. That is, utilities often have alternative generation options such as oil or coal and can easily switch, unlike many residential or commercial gas users who do not have such flexibility. Increases in electric utility demand for natural gas would affect residential, commercial, and industrial customers by raising gas prices. For example, assuming 13 EJ of consumption, a $1/gigajoule price increase would increase natural gas costs $13 billion per year among all consumers.

To replace a significant amount of electric utility coal consumption would require a very large increase in natural gas consumption. For example, compared to 1985 utility gas consumption of 3 EJ, utility coal consumption was about 15 EJ. Thus, to replace 40% of coal consumption (6 EJ), a 200% increase in utility gas consumption would be necessary, raising U.S. electric utility gas use to unprecedented levels. As a result, any policy to increase natural gas use needs to recognize possible impacts on gas supply and the market price of gas.

Additional Gas Resources

This section discusses two other sources for increasing the available supply of gas to consumers: (1) methane emissions during the production and distribution of natural gas, and (2) methane recovery from coal seams. In each of these cases, increasing the amount of methane available to consumers would often have other positive environmental benefits by reducing the amount of CH_4 and CO_2 emitted to the atmosphere.

Emissions from Natural Gas Flaring, Venting, and Leaking. As discussed in Chapter IV, during the production of oil and natural gas, natural gas[2] may be vented to the atmosphere as CH_4 or flared (producing CO_2). Additional CH_4 emissions are also produced during the refining, transmission, and distribution of natural gas. These emissions can be reduced through more careful production and maintenance procedures or by capturing the gas for on-site use or sale to gas customers.

Gas vented or flared in the U.S. represents around 0.5% of annual domestic production, while gas losses during transmission and distribution represent less than 2% of production. These values are estimated to be much lower than the global average due to a number of factors, including regulations prohibiting the flaring and venting of gas in the U.S. and the existence of a market and infrastructure to transport and sell the gas. In areas in the U.S. where no market for the gas exists (e.g., Alaska), gas produced during oil production activities is reinjected into the reservoirs in order to maintain pressure. Opportunities do exist in the U.S. for reducing gas losses during transmission and distribution and through maintenance and replacement of old, outdated distribution lines.

[2] Natural gas is mostly methane, but the methane content can vary from less than 70% to nearly 100% depending on the source of the gas. We refer to the vented or flared gases as methane, although other trace gases may also be present.

Globally, a larger percentage of natural gas is vented or flared due to fewer regulations governing these releases and a less-developed infrastructure for utilizing the gas. The quantities of gas vented or flared could be reduced through development of an infrastructure to market the gas or through regulations governing these releases.

Coalbed Methane. As discussed in Chapter IV, during coal mining, particularly underground mining, methane trapped in the coal seam is released. Historically, coalbed methane has been viewed as a safety problem during coal mining since methane can accumulate in the coal mine and explode. In the U.S., mining regulations require that underground coal mines be adequately ventilated to prevent this problem. Recently, however, there has been a growing interest in utilizing coalbed methane as a natural gas resource.

Methane extraction from coalbeds, which is being done commercially in a few areas in the U.S., differs from traditional natural gas production in several respects. Perhaps most importantly, the gas production profile differs from conventional gas wells in that maximum output generally occurs two to three years after the wells are in place, compared to immediately afterwards for conventional wells. For a given production site, more wells are drilled to maximize methane flow. Also, because these wells are drilled into available coal seams, they are generally quite shallow (i.e., not more than 3000-4000 feet below ground). While the relatively shallow access helps to reduce drilling costs, more groundwater is encountered, requiring additional efforts to combat this problem.

Due to the relatively new state of the coalbed methane recovery industry, it is difficult to quantify the potential size of this resource. In addition to offering another gas source, coalbed methane recovery could potentially be used to remove methane prior to mining the coal seam. Such recovery would help to ease the problem of methane buildup in coal mines (possibly reducing coal mining costs) and avoid the emissions of CH_4 to the atmosphere that result from current coal-mining operations.

Emission Controls

One technological option for reducing the amount of greenhouse gas emissions is the use of emission control techniques on combustion technologies that generate these emissions. NO_x and CO_2 emission control options for stationary combustion sources, such as electric utility powerplants, are discussed below.

NO_x Controls

Nitrogen oxides (NO_x) are formed during combustion primarily by the combination at high temperatures of nitrogen (N_2) and oxygen (O_2) naturally found in the air and secondarily by the nitrogen that is found in fuels such as coal and oil. Of these two factors, it is the combustion temperature that is the most critical factor affecting the NO_x emission rate. There are a number of currently-available methods for controlling NO_x emissions (based on NAPAP, 1987):

- Low Excess Air (LEA), Overfire Air. These two combustion techniques alter the flow of air during the combustion process. With low excess air the amount of excess combustion air is reduced, thereby lowering emissions up to 15%. With overfire air, some combustion air is redirected to a region above the burners, which can reduce emissions by 30%. Potential drawbacks are incomplete combustion of the fuel, increased smoke, and the extensive plant modifications that may be required.

- Low NO_x burners. This control technique operates within the furnace to limit the mixing of coal and combustion air to create a low-temperature combustion zone. Removal efficiencies approach 45-60%. This technique can be applied to existing and new units, although experience on existing units is quite limited.

- Air and Fuel Staging. When combined, these two controls can achieve 80% removal efficiencies. With air staging, up to 50% of the combustion air is directed above low-NO_x burners. With fuel staging (also known as reburning) additional fuel is burned in a region above the burners to create a fuel-rich combustion zone. Within this zone NO_x is destroyed by reducing conditions that convert NO_x to molecular nitrogen. This technique has been used in full-scale applications abroad, but only in pilot-scale facilities in the U.S.

- Selective Catalytic Reduction (SCR). SCR is a post-combustion control technology that uses a catalyst to reduce NO_x; reductions of 50-80% are possible. Its advantages include relatively simple equipment, no byproduct, and minimal efficiency loss. Its cost-effectiveness, however, depends on catalyst lifetime, which depends primarily on fuel characteristics. SCR has been used abroad, particularly in Japan and West Germany.

CO_2 Controls

Technologies have been developed to remove carbon dioxide from powerplant flue gases and dispose of it in a manner that prevents it from reaching the atmosphere. However, these technologies are unproven and very costly at this time. In one process, carbon dioxide in the flue gas is mixed with water in a solvent solution at temperatures slightly above ambient conditions. The carbon dioxide binds to the reagent and passes to a regenerator chamber where temperatures are elevated. The reverse reaction occurs and carbon dioxide is released, removed, pressurized, and liquified. The reagent is regenerated and reused. The liquid carbon dioxide could then be used for various commercial applications, or pumped to deep ocean locations, deep wells, or salt domes for permanent disposal.

In order to understand the relative costs of this CO_2 removal process, the costs of this system are compared to a conventional sulfur dioxide scrubber in Table 7-9. The carbon dioxide scrubber is 250-350% more costly than the sulfur dioxide scrubber and increases electric power costs 60-80%. The cost of transporting and disposing of the large volume of removed CO_2 has not been adequately assessed, but is likely to substantially increase this estimate. Although some CO_2 could be used for enhanced oil recovery or stored in exhausted oil and gas wells, large-scale disposal would most likely have to be in the ocean. This raises serious environmental concerns that have not been examined.

Emerging Electricity Generation Technologies

There are a number of technological options that could be available in the longer term that would substantially alter the way fossil fuels are utilized. Two of the most-discussed options--fuels cells and magnetohydrodynamics--are discussed below.

Fuel cells

Fuel cells are a new set of technologies now in use in the U.S. space program and expected to eventually be commercially available to the electric power industry. The technology converts fuel energy to electricity using an electrochemical process similar to that employed in chemical batteries.

TABLE 7-9

CO_2 Scrubber Costs Compared To SO_2 Scrubber Costs[a]

	CO_2 Scrubber[b]	SO_2 Scrubber[c]
Capital Cost ($/kw)[d]		
Scrubber	810	220
Pipeline/Disposal	80-710	NA
Variable Operation and Maintenance Costs (mills/kwh)	NA	3.5
Energy Penalty (%)	25	4.5
Capacity Penalty (%)	22	2.5
Fixed Operation and Maintenance Costs ($/kwyr)	NA	10
Total Cost (mills/kwh)[e]	36-47	10.7

a 90 percent removal of both CO_2 and SO_2.

b Steinberg, Cherg, Horn, 1984

c 1987 EPA Interim Acid Rain Base Case Estimates.

d Greenfield Site.

e 65 percent capacity factor; 9 percent capital charge rate; incremental power costs 65 mills per kilowatt-hour; new plant costs of $1,200/kw; fixed O&M for CO_2 scrubber is assumed to be ten dollars per kilowatt-year for comparison purposes only, actual costs could well be higher; 1988 dollars assumed to be worth 42 percent less than 1980 dollars.

Total fuel cell plants may achieve efficiencies of up to 85%. There are several drawbacks to the technology at this time, especially: (1) special fuel requirements, and (2) relatively low reliability.

The fuel cell closest to commercialization is the Phosphoric Acid Fuel Cell (PAFC). This fuel cell converts hydrogen into electricity and water. The hydrogen must be produced, however; considering the conversion losses, overall powerplant efficiencies for large fuel cell plants (e.g., several MW) approach 45%. In theory, the chemical reaction could continue as long as fuel is supplied, but in practice the materials fail after prolonged operation. The current goal is 40,000 hours which, even if achieved, is still well below that of conventional powerplants. Another drawback is that the production of hydrogen from fossil fuels creates CO_2. The section on hydrogen at the end of the chapter discusses this further.

The use of hydrogen in fuel cells suggests the possibility of coupling fuel cells with renewable energy sources. For example, solar powerplants could use any power not delivered to customers to create hydrogen, which could then be used when solar powerplants are not operating or when demand exceeds solar capacity.

Magnetohydrodynamics (MHD)

Magnetohydrodynamics (MHD) is an advanced, efficient generation technology that could use coal as a fuel. In a MHD system, coal is burned at very high temperatures and the hot combustion gases are chemically treated. The gases then pass through a magnetic field created using superconductors, thereby generating power. The gases can also be used in a steam cycle to produce additional power. The MHD system is expected to eventually achieve efficiencies of 60%. In comparison, conventional coal powerplants operate at about 31-32% efficiencies and advanced pressurized fluidized bed combustion (PFBC) and Integrated Gasification/Combined Cycle (IGCC) coal units, now being demonstrated, are expected to achieve 35-37% efficiency.

There are several drawbacks to the MHD technology. It is very capital-intensive, requires superconductors that have only reached the laboratory stage of development, and the high temperature combustion could result in high nitrogen oxide emission rates. Nonetheless, many observers believe MHD systems will eventually be available as these obstacles are overcome, creating an option for very efficient coal combustion.

BIOMASS

Biomass in one form or another continues to be the predominant source of energy for at least half of the world's population. In many countries, such as Nepal, Ethiopia, and Guatemala, over 90% of total energy used comes from biomass (Goldemberg, et al., 1988). Although on a global basis it accounts for only one-seventh of all energy consumed, for over 2 billion people it is close to being the only source of energy.

A number of studies have reported on the alarming rate of global deforestation as the current demand for biomass resources far exceeds the natural rate of regeneration (e.g., IUCN 1980, WRI 1985). It is this difference between growth and use that contributes to net greenhouse gas emissions. Despite this current situation, biomass energy over the long term offers some of the most promising opportunities for displacing large amounts of fossil fuel use.

Although biomass fuels account for about 14% of global primary energy use, they deliver a much smaller fraction of useful energy because of the inefficient nature of their current use. However, technologies exist or are under development to provide many times the current level of energy services from the amount of biomass currently consumed globally. In addition, it is also technically possible to greatly increase the annual increment of biomass available for energy use. Methods for increasing supply through agroforestry, biotechnology, etc., and for reducing extractive use of forest products for non-energy purposes are discussed in **PART FOUR: FORESTS**. In this section various approaches are discussed to reduce biofuel demand through fuel switching, by improving end-use efficiencies, and by increasing conversion efficiencies using upgraded fuels (see Figure 7-12).

Direct Firing of Biomass

The primary uses of biomass in direct combustion applications include use in cookstoves, space heaters, bakeries, brick kilns, and boilers of various sizes. Typical conversion efficiencies range from 5-15%. There is tremendous potential for improving the end-use efficiency in each of these energy conversion processes. In fact, this may be the most cost-effective, immediate option for decreasing the demand for biomass resources (e.g., see discussion on technical options for improving efficiency of biomass use in **PART ONE: ENERGY SERVICES**).

Wood or wood products (bark, sawdust, chips, bagasse) are already used directly as a boiler fuel (especially in agroforestry-based industries with readily-available access to wood supplies). But there are a number of modern technologies that can be used to extract much more useful output. Due to the physical variability of biomass (and the lower density of crop residues), very often some improvements in combustion properties can be achieved just by properly sizing the solid biofuels. The energy requirements for sizing and densification must be weighed against improved combustion, convenience, and ease of transport. While this is not absolutely essential when the application is for heating, some amount of sizing and/or briquetting is critical in more efficient boiler or gasifier designs.

FIGURE 7-12

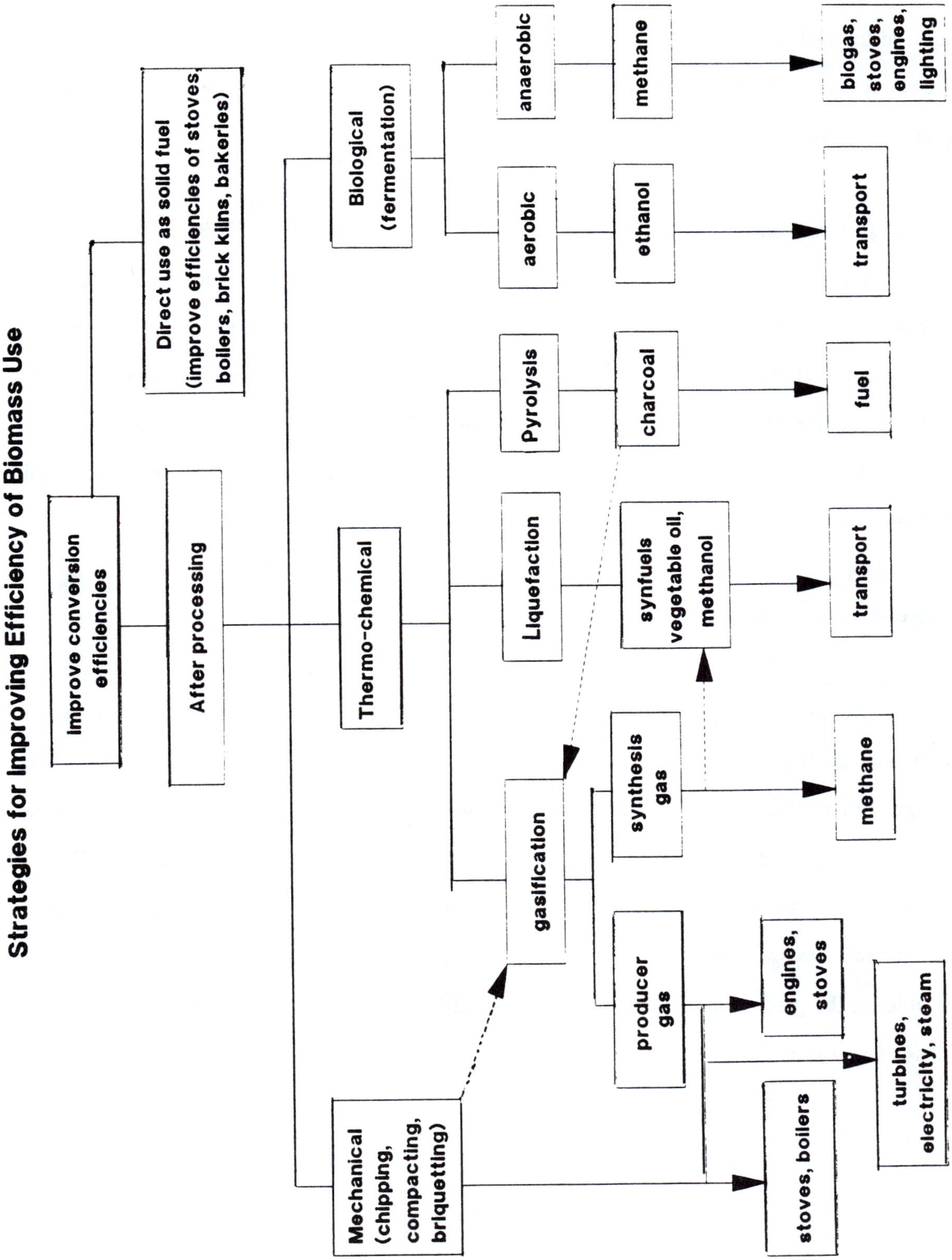
Strategies for Improving Efficiency of Biomass Use
Improve conversion efficiencies
Direct use as solid fuel (improve efficiencies of stoves, boilers, brick kilns, bakeries)
After processing
Biological (fermentation)
Thermo-chemical
Mechanical (chipping, compacting, briquetting)
anaerobic
aerobic
Pyrolysis
Liquefaction
gasification
methane
ethanol
charcoal
synfuels vegetable oil, methanol
synthesis gas
producer gas
biogas, stoves, engines, lighting
transport
fuel
transport
methane
engines, stoves
turbines, electricity, steam
stoves, boilers

Another area in which combustion efficiency can be significantly increased is in the use of advanced burner systems, such as fluidized bed combustors. These use a stream of hot air to suspend the fuel and thus achieve more complete combustion. Some of these technologies can also use certain fuels with high ash or silica content, such as rice husks, that would be harmful to standard combustion systems.

In a few areas, central power stations have been constructed specifically to use wood as a fuel. The facilities that are burning wood for electrical power range in size from 5 to 50 megawatts. Depending on the cost of fuel, conventional wood burning is generally not competitive with conventional fuels due to lower combustion efficiency and greater fuel bulk, which necessitates larger fuel-handling facilities. The localities served by these plants gain various benefits that may offset these inefficiencies. For example, local timber industries may be assisted, and environmental benefits may be obtained from reduced CO_2 emissions.

Charcoal Production

Charcoal is produced by heating wood in the absence of air (also known as pyrolysis). The traditional method of producing charcoal in kilns made of earth, as used for centuries in many regions such as Africa, is very inefficient. Only 15-30% of the energy content of the wood is retained in the charcoal as all of the gases produced during pyrolysis are allowed to escape. Substantial efficiency improvements are possible, e.g., Brazilian kilns built of brick, which produce charcoal for steel manufacture, have achieved overall efficiencies of up to 50%, twice that of traditional methods, by utilizing the gaseous by-products (Miller et al., 1986). Even further improvements are expected (Goldemberg et al., 1987).

Anaerobic Digestion

Anaerobic digestion is a biological process whereby a combustible gas is produced in the absence of oxygen; this gas is a mixture of methane and carbon dioxide similar to the marsh gas produced in swamps and landfills. This technology has the attractive feature of separating the energy content of biomass from its value as a soil conditioner and fertilizer. The energy content is released in a gaseous form (i.e., biogas) that can be utilized at a higher efficiency than the original solid biomass. Biogas is primarily used for cooking and lighting, but can also partially substitute for diesel in engines used for irrigation pumping.

China and India have had extensive biogas programs for more than 15 years (Moulik, 1985). Both programs have concentrated on household systems, with manure and other farm wastes being the most common feedstocks, with over 7 million systems installed, most of these in China. Programs in both countries have had to deal with design, construction, and maintenance problems so that many of the digesters are no longer functioning (Miller, et al., 1986), demonstrating the difficulty of introducing even relatively "simple" low-cost technologies in remote, rural areas of developing countries. Some larger institutional plants have had a higher probability of succeeding than household-size facilities.

Financially, the feasibility of biogas plants depends largely on whether the biogas and biomass residues substitute for fuels or fertilizers that have traditionally been purchased or obtained at zero financial cost. In cases where the biogas or residues do not generate incomes or reduce cash outflows, they are less likely to be viable. However, the nonfinancial benefits of these programs, such as improvements in public health due to lower emissions during cooking or increased mortality of pathogens in the digester, reduced deforestation, less reliance on imported fuels, etc., motivate countries such as China and India to continue subsidies for biogas projects (Gunnerson and Stuckey, 1982).

There are a number of other approaches being tried to make this technology more appealing. Ideally, what is required is a low capital cost technology that permits the use of a wide variety of feedstocks and results in high gas yields over a range of ambient temperatures. In Taiwan, a durable and cheap above-ground bag digester made from bauxite refining wastes has been used with success (Miller et al, 1986). Other promising anaerobic digester designs being explored include the upflow sludge blanket and baffled reactor (Gunnerson and Stuckey, 1982).

Gasification

Many cellulosic materials can be gasified and then either used directly or liquified into methanol. There are two routes for gasifying biomass--one resulting in producer (or wood) gas and the other in higher calorific value synthesis gas. Producer gas is made by heating wood in an almost oxygen-free environment so that the unburned fuel breaks down into gases, ash, and tar. Producer gas was used to propel trucks in World War II when oil was scarce. Synthesis gas is made at higher temperatures with a more pure oxygen source than producer gas, thus eliminating nitrogen, and lends itself to conversion into methanol (see later section on liquid fuels from biomass).

Producer gas can be made using a number of raw materials: wood, charcoal, crop residues, and urban refuse. There are more than 20 companies world-wide selling gasifiers, with Brazil and the Philippines leading in the use of this technology. Gasifiers have been used to power tractors, motor boats, and irrigation pumps and to provide electricity for food processing and other rural needs. They can also replace oil as a boiler fuel.

Electrical Power Applications. Kjellstrom (1985) has shown that for conventional power generation applications, biomass gasifiers compete favorably if the operating times are long, load factors high, and power outputs large. The economics are most favorable for wood-based systems rather than charcoal systems, and could become substantially more attractive if oil prices were to increase or biomass input costs were to decline. One drawback to gasifiers at this time is that current designs do not accept all types of crop residues. Kjellstrom also demonstrated that at 1985 diesel prices it would be uneconomic to run diesel tractors or trucks on producer gas or to use it in industrial applications except for industries with surplus biomass residues. Another technical problem is the need to understand the pollution impacts due to CO emissions and tar condensates from the biomass material.

While there are a number of problems to be resolved before biomass gasifiers can be easily used to replace natural gas in conventional applications, research at Princeton University has shown that integrated gasifier-combustion turbine systems could provide, at small scales (e.g., 5-50 MW), power that is less expensive than power from new hydroelectric, coal, or nuclear plants. (See Box 7-5, Larson, et al., 1987).

Liquid Fuels From Biomass

Biomass can also be used to produce liquid fuels. There are two basic energy types that can be produced with biomass--ethanol and methanol. Each of these fuels is discussed below.

Methanol

Methanol is attractive because current technologies use raw materials grown on lands not required for food production (unlike ethanol). Methanol is produced by producing synthesis gas from

Box 7-5. Biomass-Fired Combustion Turbines

Adaptation of the integrated-gasifier-combustion-turbine technology discussed above (combined cycle or aeroderivative turbines--see **FOSSIL FUELS**) to use biomass presents an attractive option for generating electricity in developing countries (Larson et al., 1987), and could also become important in industrialized countries with advances in biomass plantation productivity (see **FORESTRY**). Indeed, the use of this technology with biomass should be simpler than with coal because biomass is easier to gasify and sulfur removal is unnecessary. The use of aeroderivative turbines (STIG and ISTIG, see **Natural Gas Substitution**) may be particularly attractive because this technology remains quite economic at small scales. While the individual components of such a system have been tested, a commercial demonstration of the integrated package could be very important in increasing investors confidence in this technology.

Biomass-based electricity generation is most economical where the fuel has already been collected for other reasons (e.g., the forest products industry, the cane sugar industry). A prime example is the use of bagasse (i.e., sugarcane residues) at sugar mills and ethanol distilleries. Although these waste products are currently used to provide on-site needs for process steam and often electricity, there has been little interest in increasing efficiency once on-site requirements are satisfied. Indeed, steam and electricity production is often intentionally inefficient because of the desire to consume the wastes. Substituting an ISTIG-based cogeneration system for current steam turbines could increase electrical output by a factor of 20, while still meeting process steam requirements; total system efficiency could exceed 50% (Larson et al., 1987). This could make the cane sugar industry a major electricity producer in developing countries: If this technology were adopted in the 70 sugar-producing countries, total capacity could exceed 50 GW, increasing electricity supplier by 25% in these countries.

Design calculations suggest that the costs would be very competitive with alternatives. Capital costs could be less than $1,000/kW, and electricity could be generated for 3-4 cents/kWh where biomass is available for $2/GJ or less (Larson et al., 1987). (Bagasse in the sugar cane industry is essentially free, implying that the additional costs for producing electricity could be less than 2 cents/kWh.) The barriers to adoption of this technology appear to be primarily institutional -- collaboration between the cane sugar or other biomass producers and the utility industry is required. Furthermore, an integrated biomass-based advanced turbine system needs to be commercially demonstrated, but it is difficult to attract investors for projects considered to entail technological risk in developing countries where this technology is most attractive (see Chapter IX).

biomass, which is then converted to methanol. While the technology for conversion of synthesis gas to methanol is well-established, converting biomass to synthesis gas is a major technical and economic challenge (Williams, 1985). Although this part has been demonstrated (e.g., in West Germany and Brazil), the use of biomass is still not as economic as methanol derived from fossil sources such as natural gas and coal. In the U.S. for wood costing $20/dry ton, the cost of methanol, which burns more efficiently than gasoline, would be $14/GJ, which is equivalent to a gasoline price of $11/GJ (or about $1.50/gallon). As a result, the production of methanol from biomass should be regarded as a promising long-term opportunity if alternative fuel prices rise and/or technical improvements reduce cost. (Williams, 1985).

Ethanol

Biomass can also be used to produce ethanol. Ethanol is useful both as an automotive fuel and as a feedstock for the production of ethylene and other chemicals. Because of certain inherent requirements, as discussed below, its applicability is limited to a few countries. However, to the extent it can replace petroleum products, net emissions of CO_2 will be zero since ethanol from biomass is produced on a renewable basis.

Today, Brazil has the most extensive ethanol development program in the world. The purpose is to provide alternative transportation fuels, based largely on sugarcane-derived ethanol (Sathaye et al., 1988). Although Brazil has succeeded in reducing its oil imports, its program has received mixed reviews, largely because the program has depended on substantial subsidies to ensure its success. Similar efforts in other countries, such as Kenya, have been less successful, usually due to a lack of low-cost biomass or industrial expertise such as that available in Brazil (Miller, et al., 1986). Despite these problems, alcohol production may offer future benefits, as a number of other feedstocks (e.g., grain, sorghum, corn) may provide cheaper ethanol (Brown, 1980) and as technologies to produce ethanol from wood are developed.

Other

One technically-feasible option with biomass-derived liquid fuels is to partially substitute coconut, palm, and other vegetable oils with diesel fuel. These oils can be grown on plantations with high yields, but their worth is usually greater as food than as fuel. The Philippines briefly experimented with a mixture of coconut oil and diesel until changes in relative prices made the approach uneconomical (Miller et al., 1986).

SOLAR ENERGY

Solar energy technologies, as used in this section, refer to technologies that collect, concentrate, and convert solar radiation into useful energy. Technologies for converting solar energy into useable energy offer some of the greatest long-term opportunities for replacing fossil fuels. Within this category, however, there are several types of technologies. Some solar energy applications for the residential and commercial sectors were discussed in **PART ONE: ENERGY SERVICES**. This section will focus on solar thermal and solar photovoltaic options.

Solar thermal systems that are more sophisticated than the residential and commercial systems described earlier are being developed that can concentrate solar radiation to produce higher temperatures. These higher temperatures could be used to make steam, electricity, or power for other industrial process heat applications. The most sophisticated solar conversion approaches are photovoltaic technologies that convert solar radiation directly into electricity. These techniques have received considerable research attention over the last 10-15 years and have progressed considerably as a result. The potential of both photovoltaic and solar thermal technologies, which produce power directly only during daylight hours, will be even greater as more efficient storage technologies increase the cost-effectiveness of storing excess power generated during the daytime for use at other times, as discussed further in **STORAGE TECHNOLOGIES.**

Solar Thermal

Solar thermal technology is a promising area of solar research since it can provide thermal energy at temperatures from 150-1700°C for heat or electricity applications, at almost any scale, and with conversion efficiencies as high as 31.6% for electricity and 80% for heat (IEA, 1987). Solar thermal concentrating systems have been extensively demonstrated in recent years for both industrial heat and electricity generation. 670 MW of electric generating capacity are already in place or under construction in seven countries (Shea, 1988). In the U.S., solar thermal capacity is projected to increase to 550 MW in the next 5 years based on announced industry plans (U.S. DOE, 1987a). The U.S. Department of Energy has estimated recently that a further cost reduction by a factor of 1.5 to 2 would make these technologies competitive with conventional electric generating technologies. DOE has also identified several key technical improvements currently being researched that could achieve the needed cost improvements, leading to economic competitiveness by the mid-1990s (U.S. DOE, 1987a). In recent years most research and development has focused on three thermal technologies--parabolic troughs, parabolic dishes, and central receivers. These technologies, along with solar ponds, are discussed below.

Parabolic Troughs

Parabolic troughs are often referred to as line focus systems because they use one axis only to track the sun; each concentrator (collector) has its own receiver, with parabolic troughs often connected to form a series of collectors (see Figure 7-13). Troughs operate at lower temperatures than most other technologies, e.g., up to 400°C, making them most suitable for industrial process heat

FIGURE 7-13

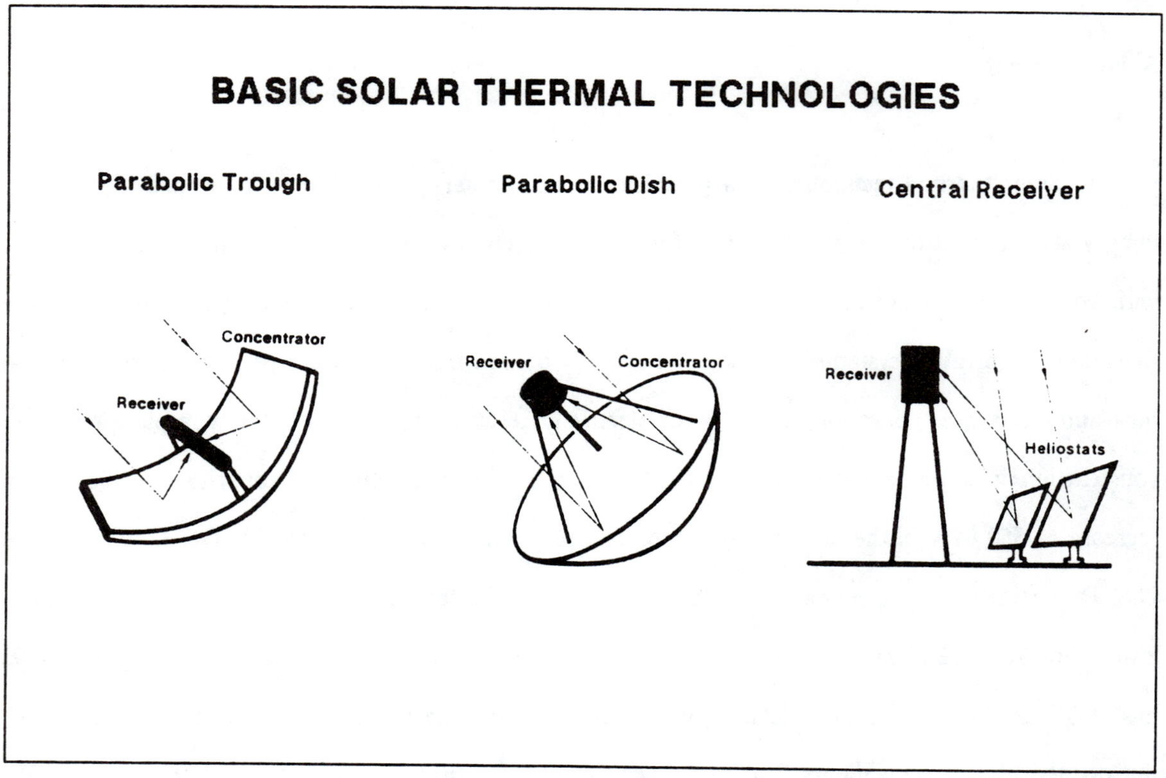

Source: IEA, 1987

applications. The thermal efficiencies of these units have improved significantly in recent years, reaching 65-70%, with availabilities exceeding 95% (IEA, 1987).

Parabolic Dishes

Parabolic dishes differ from troughs in that they are point focus systems, i.e., a two-axis tracking system is employed to follow the sun (see Figure 7-13). The higher concentration ratios allowed by this design (from one hundred to several thousand suns) produces temperatures approaching 1700°C, making electricity conversion possible. This technology currently holds many efficiency records, including the highest conversion efficiency (31.6%). Several of the most ambitious development projects are in the U.S., with installed capacity at some projects approaching 5 MW. DOE estimates total system cost at $3,400/kW, although cost reductions of 40% are considered feasible (IEA 1987).

Central Receivers

Central scale receivers are the largest scale thermal technology. The receiver is typically mounted on a tower surrounded by a large field of nearly flat tracking mirrors called heliostats (see Figure 7-13). The heliostats focus the solar energy on the receiver, achieving working fluid temperatures of 1500°C or higher (IEA 1987). Several countries are studying central receivers, with the largest plant built in southern California (10 MW). This plant has exceeded peak design output by 20%, operated at night from storage, and achieved overall plant efficiency of 13% (IEA 1987). Heliostats account for 40-50% of total costs, and consequently are the focus of most cost reduction efforts; recent developments indicate that cost reductions of 50% could be achieved in one or two years (IEA 1987).

Solar Ponds

Solar ponds collect and store heat in large bodies of water and operate at much lower temperatures than other solar thermal technologies (85-100°C). In a typical design, a salt gradient below the surface acts as an insulating barrier by trapping incoming solar radiation. Large heat exchangers are used to extract the thermal energy, which can be used for many purposes, including seasonal heat storage for space heating, low temperature process heat applications, or even electricity production. Research into this technology has achieved thermal efficiencies of 15%, with electrical conversion efficiencies of about 1-2% (IEA 1987). Despite these low efficiencies, solar ponds can be attractive since capital costs are very low.

Solar photovoltaic

Solar photovoltaic (PV) technologies convert solar radiation to electricity (DC, or direct current) without moving parts or thermal energy sources. Photovoltaic technology was initially developed in the 1950s; these first systems were about 100 times more expensive than conventional generation technologies, but major improvements have reduced this differential to 3-4 times current energy costs. This differential is even lower when compared to current replacement costs. Figure 7-14 shows the dramatic progress that has been made since 1975 in reducing the costs of electricity from photovoltaic systems.

The principal drawback with photovoltaics is their high capital costs. The current goal is to reduce costs to about 6 cents/kwh (which is comparable to the best conventional powerplant costs), at least for conditions in the southwestern United States where insolation is high. The U.S. DOE is projecting that its cost goal ($0.06/kwh) may be achieved in the 1990s and that photovoltaic systems could provide up to 1.5 GW of generation capacity by 2005 (U.S. DOE, 1987a). Researchers at

FIGURE 7-14

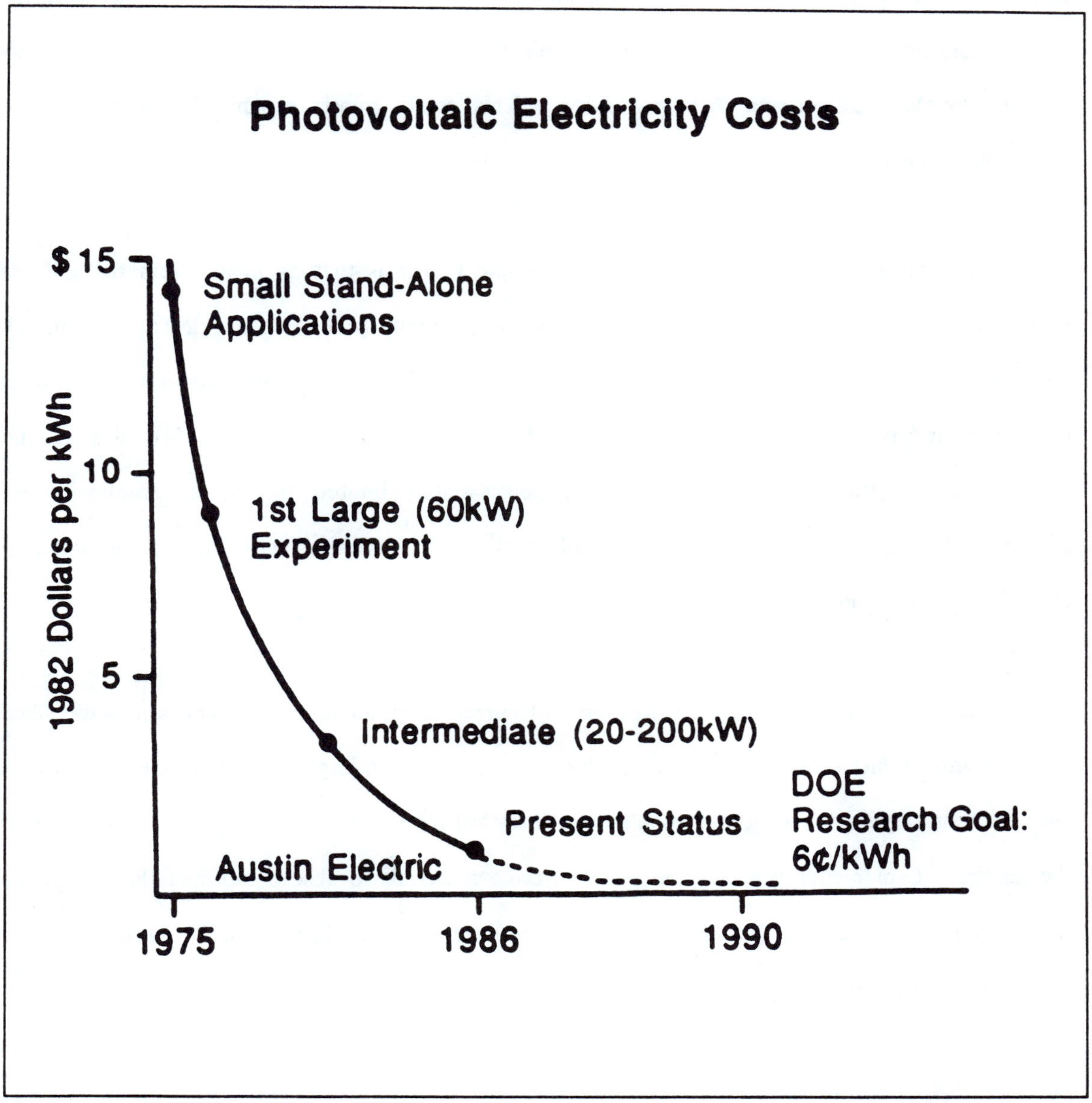
Photovoltaic Electricity Costs
$15
10
5
1982 Dollars per kWh
Small Stand-Alone Applications
1st Large (60kW) Experiment
Intermediate (20-200kW)
Present Status
Austin Electric
DOE Research Goal: 6¢/kWh
1975
1986
1990

Princeton University are also projecting major technological breakthroughs in amorphous silicon solar cell technology. They have projected that capital costs (almost the entire cost for PV) for these cells may decline by over 80% by the year 2000, while the conversion efficiency of photovoltaic arrays roughly doubles (Ogden and Williams, 1988). This could greatly expand the contribution of photovoltaic electricity generation over the next several decades.

The basic principle behind solar photovoltaic power is that as light enters the PV cell, electrons are freed from the semi-conductor materials, thereby generating an electric current. To generate substantial amounts of power with PV, individual cells (which produce about 1 watt of direct current electric power) are combined into a weatherproof-unit called a module. Modules can then be connected together into an array, with the power output limited only by the amount of area available. The modular nature of PV arrays allows them to be built in increments to conform more closely to power requirements as demand for electricity grows.

The principal photovoltaic technologies for current commercial power applications use silicon for the semi-conductor materials. Several efforts are underway to improve silicon-based photovoltaic cells, including optical tracking that orients the cell toward the sun, concentrating devices that increase the amount of solar energy hitting the plate, layering materials to absorb more of the energy, and using amorphous thin film techniques to lower production costs. The major PV technologies are briefly discussed below.

Crystalline Cells

Crystalline silicon cells are the earliest and most-established PV technologies. The most popular material has been single-crystal silicon, which had 90% of the global market in 1980 but only 44% (10.8 MW) by 1985 (IEA 1987). Single-crystal silicon PV cells are relatively efficient (production efficiencies are 12-14%; laboratory efficiencies up to 21%), but have lost market share primarily due to the high cost of manufacturing. This involves several energy-intensive stages to produce the large silicon crystal ingots from which wafers are cut (which destroys about half the crystal ingot), then final preparation and assembly.

In response to some of the problems encountered with single-crystal technology, other crystalline technologies have been developed. Polycrystalline silicon cells use a casting process that is less expensive but only slightly less efficient than single-crystal silicon, e.g., commercially-produced cells have a conversion efficiency of 11-12%, with laboratory efficiencies of 18% (IEA 1987). Polycrystalline silicon cells captured about 20% of the global market in 1985. Another alternative is polycrystalline silicon ribbon, which avoids the need to produce slices (or wafers) by producing sheets or ribbons of polycrystalline silicon. The advantages of this technology are the potential for high-speed production and less material waste, although its efficiency is somewhat lower, e.g., production efficiencies are 10-13% and laboratory efficiencies are 17% (IEA 1987). This technology has been commercialized only recently.

Thin-Film Technologies

As an alternative to crystalline technologies, researchers have focused on thin films. Thin-film solar cells can be produced less expensively using less material and automated production techniques. There are many semi-conductor thin-film materials under investigation, including amorphous silicon, copper indium diselenide, gallium arsenide, and cadmium telluride.

Amorphous Silicon. Amorphous silicon cells have received the most attention of the thin-film technologies, supplying 35% of the global market in 1985 (IEA 1987). The vast majority of this was for use in the consumer market, especially calculators. Because this material does not possess the natural photovoltaic properties of crystalline silicon, efficiency levels have been a problem--laboratory efficiencies are about 13%. Additionally, amorphous silicon suffers from a light-induced power degradation problem which can cause a 22-30% loss of power output over time. Thus, large-scale projects can currently expect to achieve about 5% efficiency over the life of the project. Research into this problem continues, with amorphous silicon thin-film cells considered one of the best potential technologies for power applications as efficiencies improve and production costs are lowered.

Other Semiconductor Materials. Other materials are being investigated that do not utilize silicon, including copper indium diselenide, gallium arsenide, and cadmium telluride. Copper indium diselenide is attractive due to its high efficiency (up to 12% currently) and stability when exposed to sunlight over extended periods. Gallium arsenide technologies have achieved the highest efficiency of any PV material (29%, Poole, 1988), but are also some of the most costly. Research continues on production processes, such as electroplating, that would significantly reduce current production costs.

Multi-Junction Technologies

Multi-junction, or tandem, technologies combine the characteristics of two or more different PV technologies to take advantage of different light absorption characteristics, thereby increasing total cell efficiency (e.g., two-junction devices can achieve efficiencies of 18-35%; U.S. DOE, 1987c). This technique has been used to produce the most efficient solar cell to date--a 31% efficient cell that combined a single-crystal gallium arsenide cell and a single-crystal silicon cell (Poole, 1988). Multi-junction devices are also thought to be a critical component of solar concentrators, which are optical systems designed to improve PV output by increasing the amount of sunlight striking a cell by ten to a thousand times. In combination these technologies may help to achieve in practice the 25-30%

efficiency range considered critical for utility applications (U.S. DOE, 1987c). Multi-junction thin film technologies are also expected to become more important as multi-layering increases system costs proportionately less than the resultant increase in efficiency.

ADDITIONAL PRIMARY RENEWABLE ENERGY OPTIONS

There are opportunities for increased use of additional renewable energy sources, such as hydroelectric, wind, geothermal, and ocean energy. Many of these resources are utilized today (often to supply electricity) and, with continued research and development, have the potential to make important contributions to meeting future energy needs without increasing emissions of greenhouse gases.

Hydroelectric Power

Hydro power currently provides the largest share of renewable electricity generation in the U.S. and globally. In the U.S., hydroelectric capacity accounted for 10-14% of total electricity generation for the years 1983-1987 (U.S. DOE, 1988). Globally, hydroelectric generation accounts for about 20% of total electricity and 7% of primary energy production (United Nations, 1988). The technical potential exists to expand the contribution of hydro at least by a factor of two by the year 2025.

Industrialized Countries

Traditional large-scale hydropower projects are no longer a significant option for most industrialized countries. Many of the most attractive sites have already been exploited in these countries and remaining potential sites are often highly valued in their natural state for recreational, wilderness or ecological purposes. The U.S. has more operational large hydroelectric capacity than any other country in the world. However, new sites are no longer being developed, and the U.S. Bureau of Reclamation, the federal agency historically responsible for large dam-building projects in the West, announced in 1987 that its mandate to develop new water supplies has virtually expired and that it would be contracting in size and shifting its focus to other activities over the next decade (Shabecoff, 1987).

One exception among industrialized countries is the potential for greatly expanding hydroelectric generating capacity in Canada. The U.S. Department of Energy recently identified potential hydro sites in Canada which, if developed, could more that double peak hydro generating capacity in Canada (from about 55 GW to over 127 GW). This level of expansion would not be required to meet projected Canadian demand for many years. However, excess capacity developed in Canada could be used to generate power for transmission to the U.S. where it could compete favorably with other generation options in many regions. The DOE analysis estimates that U.S. imports of power from Canada could more than double by 2010 if the potential hydro sites discussed above were developed (U.S. DOE, 1987d). In lieu of this hydro development, additional generating capacity in the U.S. during this period would predominantly be fossil-fueled. Thus, displacing U.S. capacity additions with Canadian hydro development would reduce CO_2 emissions. However, any hydro development raises potential bilateral and environmental issues with Canada that would need to be resolved.

Expansion of small-scale hydropower (e.g., less than 30 MW) could be an option in many industrialized countries. In the U.S. up to 10 GW of potential capacity additions in this category have been identified (U.S. DOE, 1987a). Other OECD countries, such as Canada and West Germany, are evaluating the potential for expanding small-scale hydroelectric generation (Shea, 1988).

In the USSR and Eastern Europe, there appears to be significant technical potential for expanding both large- and small-scale hydro use. In 1981, hydroelectric generation provided about 13% of total electricity in the USSR. Soviet researchers estimated that this was only about 19% of the country's potential large hydro capacity. Much of the untapped potential is in Siberia, requiring potentially costly long distance transmission to reach major load centers (Hewitt, 1984). In Eastern Europe there are indications that opportunities for large hydro projects still exist as well. Between 1980 and 1985 Romania increased its hydroelectric capacity by 2.5 GW, which was more than a 70% increase in total capacity (World Bank, 1984).

Opportunities also exist for expanding small-scale hydro in the USSR and Eastern Europe. Poland, for example, has recently initiated a program to rehabilitate 640 small dams that had fallen into disrepair and return them to electric generation (Shea, 1988).

Developing Countries

In developing countries the potential for large-scale and small-scale hydro development is very large. For example, by 1980 it is estimated that North America and Europe had developed 59% and 36% of their large hydropower potential, while in contrast, Asia, Africa, and Latin America were estimated to have developed only 5-9% of potential resources. In fact, most large-scale hydro development in the 1980s has taken place in the developing countries.

Between 1980 and 1985, hydro capacity additions in twelve developing countries totaled over 38 GW (over 8% of total generating capacity in developing countries). Several developing countries, notably Brazil, China, and India, have ambitious large hydro development programs planned for the future. Total capacity additions in developing countries could exceed 200 GW by 1995 if these plans are implemented (World Bank, 1984).

There is some question, however, if these plans will be fully implemented. A recent study by the U.S. Agency for International Development (U.S. AID, 1988b) points out that many developing countries are finding it increasingly difficult to finance capital-intensive power projects, especially hydro, which have construction periods of 10 years or more. In addition, the AID and others have pointed out that concerns about ecological and land use impacts such as submergence of forested areas as well as resettlement impacts of large hydroelectric projects may slow such development in the future.

As in industrialized countries, there is also significant potential for small-scale hydro potential in developing countries. At least 28 developing countries already have active programs for developing small-scale hydropower (World Bank, 1984). Equipment for small-scale hydro generation is now manufactured in a number of developing countries, resulting in designs that are both less costly and more suitable to local conditions (U.S. AID, 1988a).

Wind Energy

Wind-powered turbines were first connected to electric power systems in 1941. Currently, there are several technologies available, primarily horizontal axis wind turbines and vertical axis wind turbines. Under optimal conditions, these systems can produce power at 10 to 15 cents per kilowatt-hour, or about two to three times more expensively than conventional fossil powerplants. The goal is to reduce these costs such that in areas with wind resources, these systems can be used economically. In particular, wind energy systems may be very suitable for remote sites where the cost of conventional generation technologies may also be quite high.

Considerable improvement in the performance and economics of wind electric generation was achieved in the early 1980s. In California, electricity generated from "wind farms" increased from 10 terajoule-hours to 16 petajoule-hours (primary energy equivalent) while capital costs fell from $3,100 to $1,250/kw between 1981 and 1987 (Shea, 1988). The Department of Energy reports that the cost of electricity from wind turbines has fallen from about 30 cents/kwh to 10-15 cents/kwh in the 1980s. DOE projects that this improvement will continue and that wind energy could provide as much as 0.7 EJ of electricity in the U.S. by 2005 (DOE, 1987a).

Considerable international attention is now being paid to wind energy. Wind farms are being installed in Denmark, the Netherlands, Great Britain, Greece, and Spain (IEA, 1987). Other nations that have announced plans for expanded wind energy development include China, Australia, Belgium, Israel, Italy, the Soviet Union, and West Germany (Shea, 1988).

Geothermal energy

Geothermal energy is thermal energy stored in rocks and fluids within the earth. It has been estimated that approximately 10% of the world's land mass contains accessible geothermal resources that could theoretically provide hundreds of thousands of megawatts of energy for many decades (IEA, 1987b). As indicated in Table 7-10, geothermal resources suitable for generating electricity are extensive and geographically widespread. From a global warming context, several countries with the most extensive geothermal potential, e.g., the U.S., China, and the USSR, are also currently heavily dependent on coal consumption. Geothermal energy may, in these countries, provide one option for displacing coal as a source of baseload electricity generation and industrial heat. Significant geothermal resources also exist, and are being developed, in several Pacific Rim countries, where economic growth rates, and thus demand for additional energy, is expected to be high.

While certain types of geothermal resources, specifically hydrothermal and geopressured reservoirs, are not strictly renewable on a human time scale, resources are so extensive that with

TABLE 7-10

Estimates of Worldwide Geothermal Electric Power Capacity Potential
(in Megawatts)

Country	MW	Country	MW
Argentina	19,950	Kenya	79,450
Bolivia	63,100	Korea (N. & S.)	79,450
Cameroon	15,150	Mexico	257,050
Canada	446,700	Morocco	19,950
Chile	30,200	New Guinea	30,900
China	537,050	New Zealand	30,900
Columbia	77,650	Nicaragua	33,900
Costa Rica	12,600	Peru	302,000
Ecuador	100,000	Philippines	67,600
El Salvador	5,000	Portugal	1,000
Ethiopia	154,900	Saudi Arabia	15,850
Greece	8,900	Soviet Union	239,900
Guadeloupe	387	Spain	5,900
Honduras	12,600	Taiwan	8,150
Iceland	22,900	Tanzania	6,200
India	15,200	Turkey	87,100
Indonesia	436,500	U.S.	501,200
Iran	75,850	Venezuela	39,800
Italy	33,900	Vietnam	37,150
Japan	79,450		

Source: U.S. DOE, 1985c.

careful phasing and reservoir management, geothermal energy can make a significant long-term contribution to global energy needs. Unlike renewable energy sources such as solar radiation and wind, geothermal resources are available on a constant basis, making them suitable for baseload electricity generation and industrial applications without the storage problems associated with intermittent energy sources.

There are several types of geothermal resources that require somewhat different approaches for exploitation. Hydrothermal resources contain hot water and/or steam trapped in fractured or porous rock at accessible depths (e.g., 100-4500 meters). These are the most commonly used resources currently, and the only resources currently commercially exploited. Technology for exploiting these resources involves sinking wells, extracting the hot fluids, and using the steam or hot water for electricity generation or direct heat applications. Steam can be used directly in steam electric generating turbines. Hot water resources are either "flashed" to produce steam, or used to vaporize another working fluid which in turn drives a turbine.

Another type of geothermal resource of long-term interest is hot dry rock. These potential resources are widely distributed around the world, but more difficult to exploit than hydrothermal resources. To extract heat from hot dry rock, it is necessary to inject liquid into one well and withdraw it through another well after it has absorbed heat. Considerable research is underway to improve technologies for extraction of energy from hot dry rock. A geothermal resource still at the conceptual stage of development is magma which is liquefied or partially-liquefied rock. Potential resources may be greater than any other geothermal resource and the very high temperatures available suggest that power could be economically produced; however, development of the necessary technologies is seen as a major challenge.

Geothermal energy is currently used in several countries for direct heat and electricity generation. Table 7-11 shows the extent of direct heat use in 1984. Table 7-12 indicates the installed capacity for geothermal electricity generation by country in 1985. Although they represent only a small fraction of total energy use, these facilities have demonstrated the commercial viability of geothermal technologies. In the U.S. geothermal systems currently produce electricity at a cost that is competitive with coal and nuclear plants, and the average geothermal unit today is available on-line more than 95% of the time (U.S. DOE, 1987a). The Department of Energy projects that U.S. geothermal electric capacity will more than double by 1995 to about 4.7 GW.

Geothermal energy may also play an even more important role in some developing countries. Eight developing countries currently have installed geothermal electric capacity, and about 50 more have been identified as having potential for geothermal development (U.S. AID, 1988a). In the 1980s, the costs of geothermal technology have come down considerably, and small-scale, ready-to-install generators (1-1.5 MW) have also been developed and proven reliable. These and other developments should make geothermal electricity more attractive to many developing countries (U.S. AID, 1988a).

Ocean Energy

There are several types of potential ocean energy sources, including thermal gradients and waves. Research is underway in many countries, including the U.S., to develop technologies that can exploit these resources. In one technology, cold water located deep in the ocean is used to condense a working gas such as freon or ammonia. The liquid then is reconverted to gas by warm surface waters and used to drive a turbine and generate electricity. Current costs for such a system are roughly three times higher than commercial alternatives, and significant technological uncertainties remain regarding system components and operation in an ocean environment.

TABLE 7-11

Capacity of Direct Use Geothermal Plants
In Operation - 1984
(For countries having capacity above 100 MW)

Country	Power MW	Energy GWh	Load %
China	393	1945	56
France	300	788	30
Hungary	1001	2615	30
Iceland	889	5517	71
Italy	288	1365	54
Japan	2686	6805	29
New Zealand	215	1484	79
Romania	251	987	45
Soviet Union	402	1056	30
Turkey	166	423	29
United States	339	390	13
Other	142	582	47
Total	7072	23957	39[1]

[1] Based on total thermal power and energy.

Source: IEA, 1987.

TABLE 7-12

Geothermal Powerplants On-Line As Of 1985

Country	No. Units	Type(s)[a]	MW
United States	56	DS,1F,2F,B	2022.11
Philippines	21	1F	894.0
Mexico	16	1F,2F	645.0[b]
Italy	43	DS,1F	519.2[b]
Japan	9	DS,1F,2F	215.1
New Zealand	10	2F	167.2
El Salvador	3	1F,2F	95.0
Kenya	3	1F	45.0
Iceland	5	1F,2F	39.0
Nicaragua	1	1F	35.0
Indonesia	3	DS,1F	32.25
Turkey	2	1F	20.6
China	12	1F,1F,B	14.32[b]
Soviet Union	1	F	11.0
France (Guadeloupe)	1	2F	4.2
Portugal (Azores)	1	1F	3.0
Greece (Milos)	1	1F	2.0[b]
Totals	188		4763.98

[a] DS = dry steam; 1F,2F = 1- and 2-flash steam, B = binary.
[b] Includes plants under construction and scheduled for completion in 1985.

Source: IEA, 1987.

Another technology would exploit wave energy using a device in which wave motion would compress trapped air. The compressed air would then be used to drive a turbine and generate electricity. This system is also currently unproven and expensive. Further, the output of a wave-based system would fluctuate with wave conditions, which would not necessarily match well with electricity load requirements. As with many other renewable technologies, economic and reliable energy storage would be necessary to make this option viable.

NUCLEAR POWER

This section discusses the potential role for nuclear power to meet future energy needs. From the perspective of global warming, nuclear power technologies are attractive in that they do not emit greenhouse gases. As will be discussed, however, there are other problems that beset the nuclear power industry. The first part of this section discusses fission technologies; nuclear fission is the technology currently used in operating nuclear powerplants. One of the key attributes of this technology is its need for fissionable, radioactive material in order to operate. The last part of this section discusses fusion technology, which is a longer term nuclear option currently in the research and development stage. Unlike fission technologies, fusion technologies would not require large inventories of radioactive materials such as uranium and plutonium for their operation.

Nuclear Fission

Nuclear fission technology is an important source of electricity in many regions of the world. For example, in the U.S. nuclear plants provided about 17% of total electricity generated in 1987. This total is projected to increase throughout the remainder of this century as about 20 GW of new nuclear powerplants, which are currently under construction, are completed. However, the prospects for further capacity additions are clouded. In the words of the U.S. Department of Energy, "No additional orders for nuclear powerplants are likely in this country until the conditions of the past few years are changed" (U.S. DOE, 1987a). Moreover, between 2000 and 2015, about 40% of existing nuclear powerplant capacity will have to retire unless current operating licenses are extended beyond their expiration dates.

The situation is somewhat similar in many other industrialized countries. The International Energy Agency reports that nuclear energy was the fastest growing fuel for electricity generation in the OECD countries between 1985 and 1987. However, the report continues: "It should be kept in mind, however, that present additions to nuclear capacity come from stations authorized in the second relatively intense "cycle" of nuclear plant construction activity in the 1970s. Compared with some 239 GW of nuclear capacity operating in the OECD countries at the end of 1987, around 52 GW are under construction" (IEA, 1988). In addition, planned nuclear electricity production in the 1980s in the USSR has been consistently behind schedule due to construction delays. By 1986, nuclear generation was providing about 10% of electricity in the USSR. In the wake of the Chernobyl disaster of April 1986, the nuclear program in the USSR is experiencing further delays and future contributions are difficult to project (IEA, 1988).

Nuclear power has been beset by a series of problems. Plant capital costs have increased so dramatically that new nuclear powerplants are no longer considered economical (see Figure 7-15). Powerplant lead times (i.e., the time for project initiation to completion) are greater than ten years, increasing project risk. Additionally, in many countries such as the U.S., the regulatory environment is generally unfavorable toward large, long-term capital investments and nuclear powerplants in particular (U.S. DOE, 1987a).

Public opposition to nuclear powerplants is strong, in part due to concerns about safety in the aftermath of the accidents at Three Mile Island and Chernobyl. In addition to plant safety, waste

FIGURE 7-15

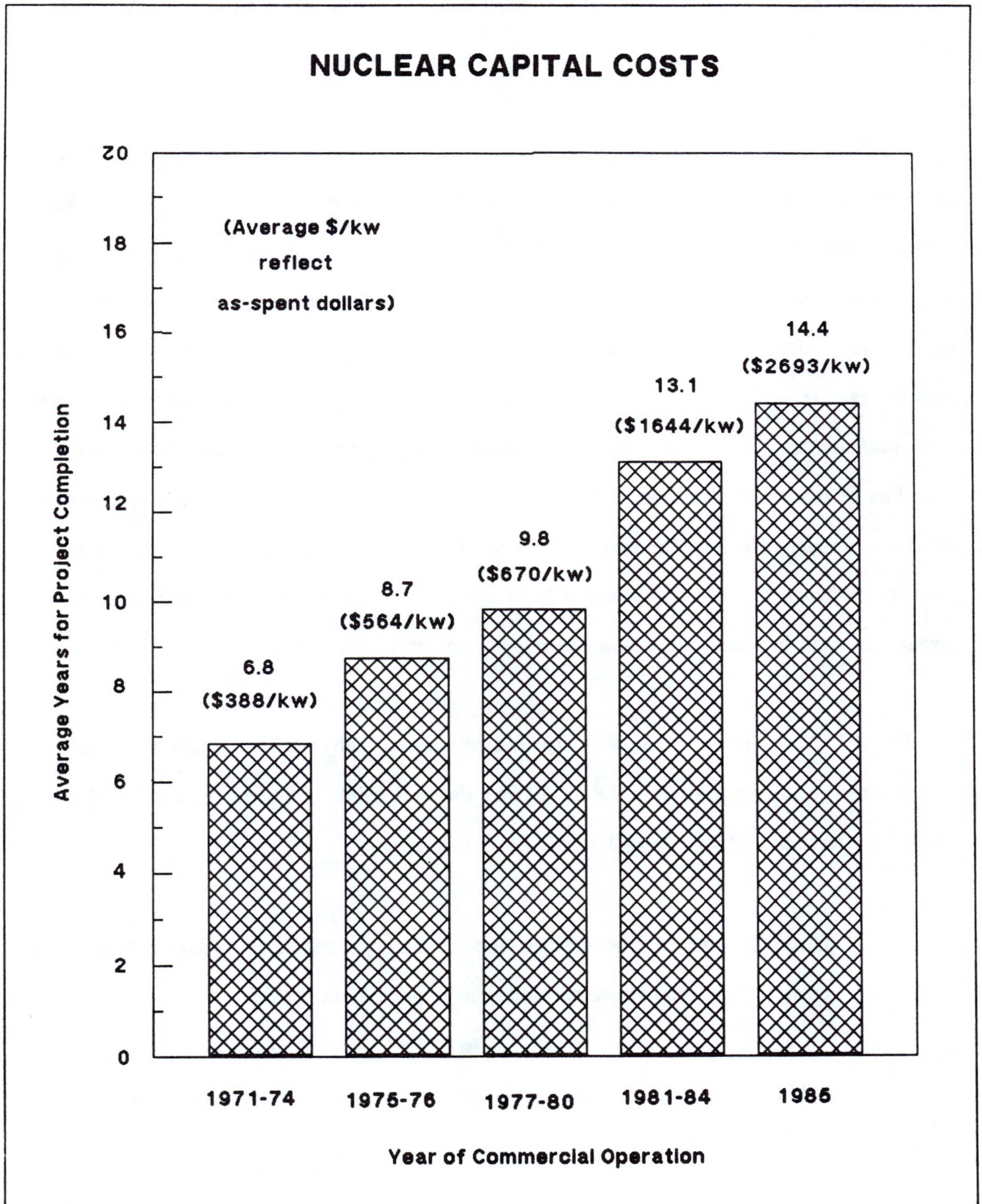

U.S. DOE, 1987a.

disposal is an issue of major concern. Although several countries utilizing nuclear power are working on the problem, none have identified an acceptable solution to the problems of long-term nuclear waste disposal.

Finally, the nuclear weapons proliferation issue must be addressed if nuclear power is to become publicly acceptable. A present day 1000 MW nuclear plant produces some 141 kg of fissionable plutonium annually in its spent fuel. For comparison, it takes less than 10 kg of fissionable plutonium to make a nuclear weapon. Currently, this spent, intensely radioactive fuel is stored at plant sites. If nuclear power were to be greatly expanded in the future, limited supplies of uranium worldwide would rapidly force a shift from today's "once through" nuclear cycles (i.e., the nuclear fuel is only used once) to fuel cycles involving the reprocessing of spent fuel and the recycling of recovered plutonium for use as fuel in present reactor types and a new generation of plutonium breeder reactors. The amount of fissionable material required would pose a formidable institutional challenge to the world community to safeguard relatively large quantities of plutonium that would circulate in worldwide commerce (Ogden and Williams, 1988).

In the U.S., the Department of Energy has recognized many of these potential obstacles. In its 1987 Energy Security report (U.S. DOE, 1987a), the U.S. DOE identified three basic obstacles that must be overcome before new orders will be forthcoming:

- Given current reactor designs and regulatory processes, nuclear power is not economically competitive with alternative generation options.

- Public concerns about reactor safety make licensing of new plants difficult.

- Currently unresolved questions about the viability of long-term waste disposal options have become a significant barrier to expansion of nuclear power.

DOE has initiated efforts to deal with all of these problems. First, it has underway a research program aimed at the development of improved reactor designs. Program goals include new designs for "enhanced safety, increased simplicity, and improved reliability." To meet these goals, the new designs incorporate innovative concepts of passive safety that hopefully would ensure that any equipment or operator failure would cause the plant to shut down automatically. A second important attribute of the advanced designs is standardization of designs, which could reduce cost, shorten the permitting process, and improve reliability and safety. Third, the new designs are for smaller, modular units, which should make them more compatible with the utility industry's needs for smaller capacity addition. Another design concept is that the portion of the plant that is actually within the nuclear containment vessel can be minimized and modularized, thereby allowing the balance of the plant to be built to standard construction specifications. An example of this approach is Sweden's ASEA-ATOM's PIUS reactor, which places the steam generator and primary pumps outside the containment vessel (EIA, 1988). More stringent safety standards for nuclear powerplants could then focus on the smaller nuclear component. Ultimately, the nuclear portion may be designed to be fabricated at the manufacturing plant and transported in its entirety to the powerplant site for insertion into the remainder of the plant (Griffith, 1988).

These principles are being applied to several basic reactor types:

- Advanced Light Water Reactor (ALWR). Light water reactor technology is currently the type most widely in use in the U.S. The hope is to develop a standardized design incorporating the lessons of recent experience.

- Modular High Temperature Gas Reactor (HTGR). This is an advanced concept that has not been demonstrated yet but is being designed to provide a next generation nuclear capability that goes further toward meeting the design criteria mentioned above than the ALWR.

- Advanced Liquid Metal Reactor (LMR). For the longer term, designs are being developed that would incorporate the design goals stated above into a potential breeder reactor option.

Finally, on the waste disposal issue, DOE has announced that it will continue to implement the long-term Geologic Repository Program for disposal of high-level wastes while also constructing a Monitored Retrievable Storage (MRS) Facility (U.S. DOE, 1987a). Thus, DOE is attempting to deal with many of the constraints affecting the nuclear power industry. It is difficult to predict at this point how effective these programs will be. In addition, the DOE report does not mention the weapons proliferation problem that is also perceived as a long-term constraint by many observers.

Nuclear Fusion

Nuclear fusion, like nuclear fission, is an attractive power generation technology from a global warming perspective because it does not generate greenhouse gas emissions. Fusion power has two key advantages over fission power: (1) It uses secure and inexhaustible fuels: lithium and deuterium are obtainable from seawater, and (2) it does not create large inventories of radioactive wastes.

However, fusion reactor technology is only in the early stages of research and development; it is not expected to be a viable technology until after 2025. The costs of development for this technology are expected to be high. To hasten fusion development and to defray the costs that need to be borne by any single country, international cooperative R&D agreements are likely to be signed in the next few years.

ELECTRICAL SYSTEM OPERATION IMPROVEMENTS

One possible option for reducing the amount of greenhouse gas emissions from electricity is to improve the efficiency of transmission, distribution, and storage of electrical power. Each of these areas is discussed below.

Transmission and Distribution

Electrical energy losses from transmission and distribution are normally in the 5-10% range in industrialized countries. When generation is primarily from fossil fuels, as in the U.S., programs to reduce efficiency losses associated with electricity transmission and distribution may provide one option for modest reductions in greenhouse gas emissions. In developing countries, however, it is very likely that significant improvements are technically possible. Many developing countries experience losses of over 20%, even 30%. Relatively inexpensive and straightforward technological solutions exist for upgrading utility transmission and distribution systems to recover a major portion of these losses. On the other hand, in many countries it is estimated that half of the "controllable" losses are due to theft (U.S. AID, 1988b); despite many reasons for reducing these thefts, it is not clear whether a straightforward solution to these electricity thefts exists nor would this mean a significant reduction in greenhouse gas emissions, except to the extent demand is reduced as consumers are charged for the power they use.

Electric utility companies normally operate their systems and interconnections with other systems in such a way as to minimize generation costs, subject to a number of other constraints.

Thus, powerplants with the lowest variable operating (including fuel) costs would be operated more of the time, and power available from other utility systems would be purchased when its cost is lower than the incremental cost of generating additional power within the system.

Within this framework, there may be some flexibility to shift more of the overall load to non-fossil or natural gas-fired capacity without greatly increasing overall generating costs. Although difficult to achieve institutionally, such alternative dispatching options may be technically feasible and cost-effective in the near term. This could include "wheeling" power from region to region to take maximum advantage of non-fossil generating capacity. On its largest scale, this type of strategy could result in international electricity transfers from countries with less carbon-intensive generating capacity to countries with more carbon-intensive available options. For example, expanded Canadian power imports to the U.S. based on hydroelectric generation would be one such possibility.

Superconductors

Superconductors offer no resistance to electrical flow. Recently, breakthroughs in superconductivity research has increased the prospects that this technology could be applied to power production. Until recently, superconductivity could only be achieved in extremely cold environments (e.g., more than 200°C below zero). The superconductivity effect can now be achieved at much higher temperatures, although they are still well below ambient conditions. In the long term, superconductivity could significantly improve the transmission of power. Conventional transmission systems lose about 8% of the power they conduct; superconductors could greatly reduce, if not eliminate, these losses. Superconductivity could also be beneficial for power production. For example, several technologies use electromagnets, including the MHD system discussed above. During the power production process electromagnets are typically the source of some lost power.

Superconductors could reduce the energy lost in generating power by reducing these losses associated with electromagnets. Also, as discussed below, superconductivity could be useful for energy storage.

Storage Technologies

There are a variety of technologies currently available or under development for energy storage. Storage systems can perform several tasks, including:

- Load-leveling, which would allow inexpensive baseload electricity to be stored during periods of low demand and released during periods in which the marginal cost of electricity is high, e.g., powerplants could be run during the night and the power stored to meet peak requirements in the afternoon. One caveat is that storage may allow baseload coal to substitute for natural gas, increasing greenhouse gas emissions, unless renewable energy sources are used.

- Spinning-reserve, in which the energy would be used to backup failed generating systems.

- System-regulation, i.e., to balance a utility's constantly-shifting generation and load requirements.

The development of adequate energy storage systems could be particularly crucial for the competitiveness of many renewable energy technologies which can only produce power when the resource is available, e.g., during daylight hours for solar or when the wind is blowing for wind energy systems. To enhance their competitiveness, renewable technologies could be used in tandem with storage systems that would allow power to be generated whenever available and then stored until needed.

Types of Storage Technologies

There are many different types of energy storage systems, including pumped storage, batteries, thermal, compressed air, and superconducting magnetic energy storage. Several major energy storage systems are discussed below.

Pumped Storage. Pumped storage is a hydroelectric power storage option that has been used recently by the electric utility industry to meet electricity requirements during periods of peak demand. With this system water is pumped from a lower storage reservoir to an upper storage reservoir during off-peak hours, using electricity from a baseload powerplant (which is frequently coal-fired or nuclear). During peak demand hours, the water is released to the lower reservoir much like it would be at a typical hydroelectric dam. This system essentially stores power from the baseload powerplant when it is not needed for use during peak demand periods when the baseload plant is already committed 100% to meeting the peak power requirements.

While there are numerous pumped storage plants in the U.S., their efficiency is low and additional siting is likely to be difficult if they involve large, above-ground reservoirs. When underground reservoirs are used, the systems are only economical in very large facilities (e.g., 1000 MW).

Batteries. Batteries are attractive primarily for their flexibility; because they are modular, plants can be constructed quickly on an as-needed basis. Recent research and development has focused on advanced lead-acid batteries and zinc-chloride batteries. Lead-acid battery technology has been used for decades, e.g., in automobiles, although their use in utility applications may be limited by their capital costs. Due primarily to the lower cost of the construction materials, the zinc-chloride battery is expected to be less expensive than the lead-acid battery in the longer term; possibly less than $500/kW compared to $600/kW or higher for lead-acid batteries (OTA 1985). Plans to test both types of batteries on a commercial scale are currently being planned. Other battery technologies that could be available in the longer term include zinc-bromide, sodium-sulfur, iron-chromium, and lithium-iron sulfide batteries.

Compressed Air Storage. Compressed air energy storage (CAES) uses off-peak electricity to store energy in the form of compressed air in an underground cavern such as salt reservoirs, hard rock reservoirs, or aquifers. The compressed air is used in tandem with natural gas or oil in a modified gas turbine where the compressed air is used in lieu of a conventional compressor in the turbine cycle, thereby allowing the turbine efficiency to increase up to three times its normal efficiency (OTA 1985). CAES is dependent on geological characteristics that can be found in about 3/4 of the U.S. and uses technology that is well-advanced. However, no CAES facility has yet been built in the U.S.; there is a facility that has been operating since 1978 in West Germany. There are two sizes proposed for CAES plants; a mini-CAES (about 50 MW) costing about $392/kW and a maxi-CAES (about 220 MW) costing about $515/kW; construction lead times are estimated to be 4 to 8 years (U.S. OTA, 1985).

Superconducting Magnetic Energy Storage. Superconducting magnetic energy storage (SMES) would function like more conventional storage technologies, but would be able to store energy with an efficiency of approximately 95% compared to 75% for pumped storage or 65% for battery storage (Schlabach 1988). At this time SMES is clearly a long-term technology pending further improvements in basic superconductivity design.

HYDROGEN

One concept for reducing CO_2 emissions is the long-term phaseout of a carbon-based economy and the adoption of one utilizing hydrogen and electricity as complementary energy currencies. Hydrogen would serve as an "energy carrier" since electricity or some other source of power is required to produce it. Hydrogen is attractive for this role because it is nonpolluting, portable, and relatively safe. Hydrogen could be produced in a number of ways, including chemical processes beginning with coal or natural gas, and electrolysis using electricity from the full range of potential electricity production options. Of course, production chemically from fossil fuels or the use of fossil-based electricity would not provide significant long-term reductions in greenhouse gas emissions. However, these technologies are currently available and could play a role in a transition to greater use of hydrogen as an energy source (Harvey, 1988).

One technology under development holds some promise for allowing hydrogen to be generated from fossil fuels without CO_2 emissions. The process, known as hydrocarb, produces hydrogen and carbon black through a process of gasification and distillation. When coal is used as the feedstock, approximately 20% of the energy value is converted into hydrogen. The carbon black produced could be disposed in mines or other disposal facilities. Biomass can be used as the feedstock, although it is a less efficient source of hydrogen (Grohse and Steinberg, 1987).

While large-scale hydrogen use would have to be considered a long-term option, this fuel could begin to make contributions in the near term. Researchers at Princeton University have suggested that use of hydrogen as a transportation fuel in urban areas may be its first significant role in replacing traditional fuels. This is because existing transport fuels are high-priced premium fuels such that the economics would be more favorable, and because urban air quality problems are already forcing many cities to look for alternatives to gasoline and diesel fuel in the transportation sector.

They further suggest that recent and projected improvements in the economics of amorphous silicon photovoltaic cells may make production of hydrogen fuel for transportation use from solar PV electricity cost competitive in some areas of the U.S. before the end of this century (Ogden and Williams, 1988).

Another attractive feature of hydrogen that would be helpful in a long-term transition away from fossil fuels is that it can substitute relatively easily for natural gas in many applications. For example, natural gas space heating could be replaced by hydrogen over the long term, natural gas pipelines could be used to transport hydrogen, and hydrogen is the basic fuel for fuel cells. Storage of hydrogen could utilize salt mines, aquifers, and depleted oil and gas fields for large needs and liquid hydrogen and metal hydrides for small applications.

In a long-term hydrogen economy, non-fossil energy could be provided by a variety of renewable sources, with conversion to hydrogen accomplished by the electrolysis of water. Conversion efficiencies in producing hydrogen from renewable energy exceed 80% and efficiencies for conversion back to energy in fuel cells range from 58-70%.

PART THREE: INDUSTRY

Figure 7-16 illustrates the overall current contribution of industrial processes (excluding energy) to the greenhouse warming problem. By far the largest component is the production and ultimate release to the atmosphere of chlorofluorocarbons (CFCs), halons, and chlorocarbons. Other industrial processes are relatively minor (but growing) contributors: CO_2 is emitted from cement manufacture and methane (CH_4) is produced by solid waste landfills. In addition, industrial process emissions of carbon monoxide contribute to atmospheric chemistry, which indirectly affects the concentration of tropospheric ozone and methane.

CFCs AND RELATED COMPOUNDS

As a result of the Montreal Protocol (discussed in Chapter IX), emissions of the most important CFCs will be capped in the near future and will be reduced to half of 1986 levels by 1998. Halons will be frozen at 1986 levels beginning in 1992. Other related compounds, not covered by the protocol, are projected to grow in the No Response scenarios, partly because some of these compounds will be used as substitutes for the regulated compounds.

Some of the substitute compounds affect greenhouse warming but generally to a much smaller degree than do the controlled substances. In addition, most of the unregulated compounds have much shorter atmospheric lifetimes, which further decreases their impact on the greenhouse problem.

In the No Response scenarios, we have assumed general compliance with the Montreal Protocol (100% in U.S., 94 % in other industrialized countries, and 65% in developing countries). Figure 7-17 shows projections for future emissions for the regulated compounds, as well as the

FIGURE 7-16

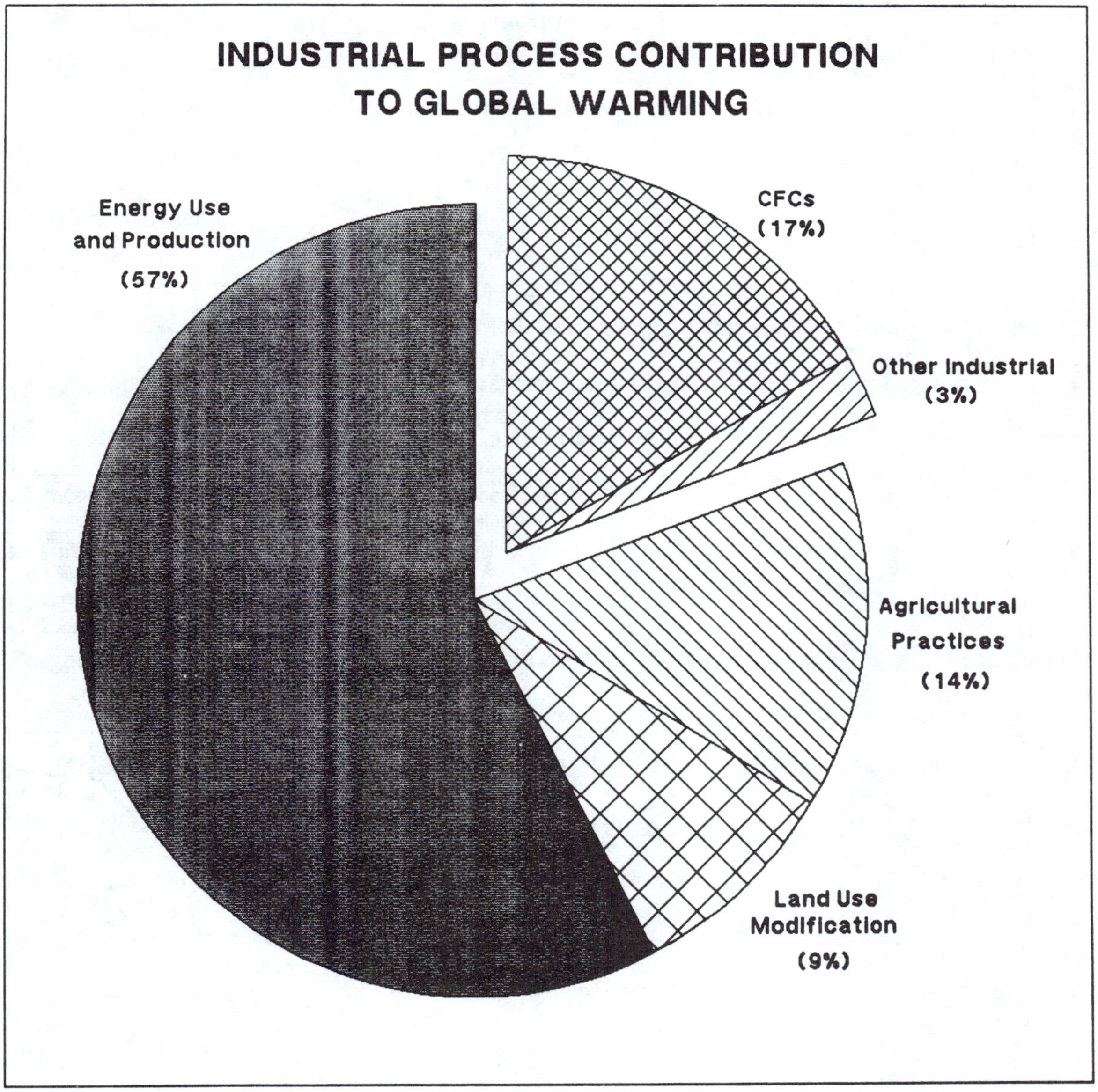
INDUSTRIAL PROCESS CONTRIBUTION
TO GLOBAL WARMING
Energy Use
and Production
(57%)
CFCs
(17%)
Other Industrial
(3%)
Agricultural
Practices
(14%)
Land Use
Modification
(9%)

FIGURE 7-17

EMISSIONS OF MAJOR CFCs

(Gigagrams)

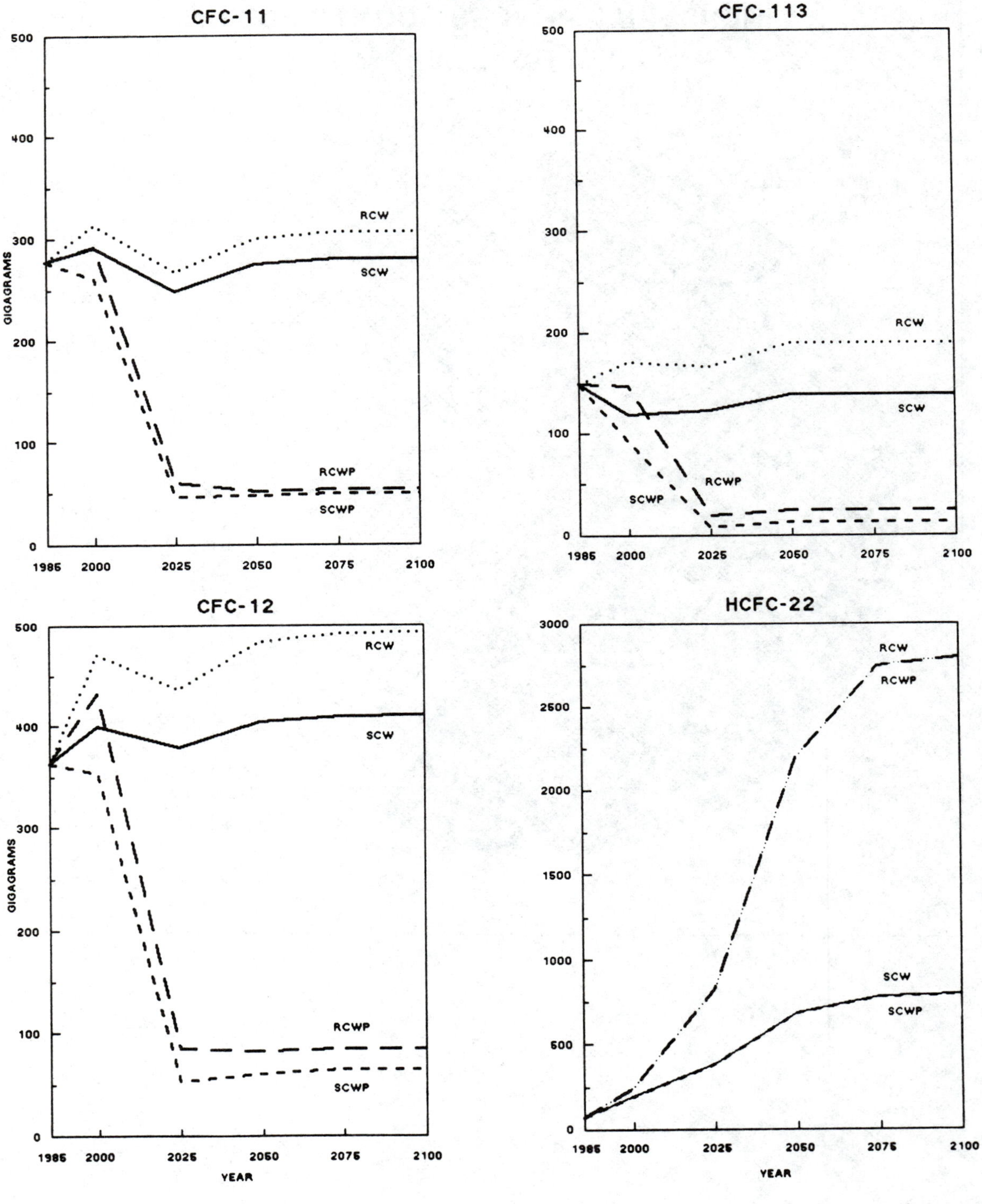

growing global emissions of unregulated substitute compounds, which have some radiative forcing impact.

A considerable amount of recent analysis evaluates the potential of further reducing emissions of CFCs and related compounds (see, for example, Hoffman and Gibbs, 1988, or Makhijani et al., 1988). As shown in Figure 7-17, the SCWP and RCWP scenarios incorporate a complete phase-out of the major CFCs and Halons by 2003 and a cap on global emissions of methyl chloroform at 1986 levels (100% participation in industrialized countries and 85% in developing countries). This phase-out schedule is feasible given that substitutes and alternative technologies now being developed and tested are expected to become available over the next decade -- as a result of the considerable research currently underway in response to the Montreal Protocol.

Use and emissions of CFCs could be reduced by three possible mechanisms:

- Chemical substitution - switching from production processes in which CFCs are used to those in which other chemicals are used; for example, using FC-134a or blends of other non-fully halogenated HCFCs instead of CFC-12 in mobile air conditioning.

- Engineering controls - in the near term, switching to production technologies that use fewer CFCs per unit of output, such as recycling equipment that collects and recycles CFC emissions during the production of electronics.

- Product Substitution - switching from CFC-using products to other products; for example, replacing CFC-based foam egg cartons with paper-based egg cartons.

This chapter surveys existing and future product substitutes, engineering controls, and chemical substitutes that can reduce use and emissions of CFCs and halons. The information presented is based on a detailed series of industry studies performed by EPA for use in its *Regulatory Impact Analysis* (U.S. EPA, 1987) for stratospheric ozone protection. Unless otherwise noted, information in this section is drawn from U.S. EPA (1988b).

Technical Options For Reducing Emissions

Chemical Substitutes

Several chemical substitutes that have physical properties (e.g., boiling point) similar to those of CFCs either do not contain chlorine or have short atmospheric lifetimes that reduce their potential to deplete stratospheric ozone.

FC-134a and blends of non-fully-halogenated HCFCs appear to be the most promising chemical substitutes for refrigeration and air conditioning applications, including commercial chillers and mobile air conditioning, and along with HCFC-141b and HCFC-142b appear to be likely alternatives to CFC-11 and CFC-12 in production of polystyrene sheet and polystyrene boardstock. Several major chemical producers have begun limited production of FC-134a. An international consortium of chemical producers has been formed to undertake toxicity testing of FC-134a and other chemical substitutes.

HCFC-141b and **HCFC-123** are expected to become commercially available in the early 1990s and could replace CFC-11 currently being used in manufacturing slabstock flexible polyurethane foam and rigid polyurethane foam. In blends with HCFC-22, **HCFC-142b** appears to be an option for replacing the remaining "essential" CFC use in aerosols.

HCFC-22 is currently used in residential air conditioners and might be used in commercial chillers. It could substitute for CFC-11 and CFC-12 as a leak-testing agent in several refrigeration applications. Mixtures of HCFC-22 and other compounds could be used in mobile air conditioners.

HCFC-22 has already been accepted as a substitute for CFC-blown polystyrene sheet products. It was recently approved by the Food and Drug Administration as an alternative blowing agent for use in food packaging. The Foodservice and Packaging Institute, in concert with several environmental groups, recently announced an industry-wide program to eliminate within one year the use of CFC-11 and CFC-12 use in food service packaging by substituting HCFC-22.

A major manufacturer of extruded polystyrene boardstock recently announced that it will substitute HCFC-22 and other partially-halogenated CFCs in its manufacturing processes beginning in 1989.

Blends of HCFC-22 with **dimethyl ether** and HCFC-142b can be used to replace the remaining "essential" CFC use in aerosols. **Hydrocarbons** have largely replaced CFCs as aerosol propellants in the United States. Other nations can reduce their use of CFCs in aerosol propellants by reformulating aerosol products.

Ethylene oxide (EO) is currently blended with CFC-12 for use in sterilization of medical equipment and instruments. Reductions in CFC-12 use could be achieved by using pure EO, a blend of CO_2/EO, radiation, or the use of CFC substitutes.

Terpene-based solvents can be used instead of CFC-113 to clean electronic components. One major manufacturer expects to replace one-third of all its CFC use in electronics manufacturing with terpene-based solvents. **Aqueous cleaning** can also reduce CFC-113 use in many applications for cleaning electronic components.

Engineering Controls

Recovery and recycling can reduce CFC emissions from several applications. Recovery and recycling during servicing of air conditioning and refrigeration equipment, such as commercial chillers and mobile air conditioning units, can achieve large reductions in CFC emissions.

Carbon adsorption units could be installed to capture CFC emissions during manufacture of slabstock flexible polyurethane foam. Simple housekeeping improvements and process modifications such as automatic hoists and covers, carbon adsorption, and reclamation and recycling can reduce CFC-113 emissions during solvent cleaning.

Alternative leak-testing agents can reduce halon emissions during discharge testing of total flooding fire-extinguishing systems.

Improved system design can reduce CFC emissions. Attractive techniques in mobile air conditioning include design improvements, such as the use of more refrigerant-tight hose materials, shorter hoses, and improved compressor seals and fittings.

Alternative processes can be used to produce final products without using CFCs. For example, the CFC-blown foam process can be modified to eliminate use of CFC-11, but the foam will be slightly denser as a result. Nearly all uses of molded flexible foam could be converted to water-blown formulations.

Training could reduce unnecessary discharges of CFCs during air conditioning servicing. Adoption of new training procedures, such as use of simulators, could reduce halon emissions during military training exercises.

Product Substitutes

Research efforts are exploring ways to replace the use of CFC-blown insulation in refrigerator walls by insulating vacuum panels. A prototype vacuum panel refrigerator, which could achieve sizeable gains in insulating properties, is currently being built.

CFC-blown slabstock flexible polyurethane foams can be replaced by other products. Fiberfill materials, cotton batting, latex foams, and built-up cushioning that contains springs may be suitable substitutes, but they are more expensive and lack the durability of flexible polyurethane.

Some product substitutes are available for rigid polyurethane foam products. Many alternative products are currently available for use as sheathing or roof insulating materials. Expanded polystyrene foam beadboard, fiberglass, fiberboard, and gypsum, for example, could be used instead of polyurethane foam, as they were 30 years ago when polyurethane foams were not yet manufactured. In some cases, wall and roofing insulation can be made thicker to achieve the same insulating capacity as at present, but the use of foam blown with chemical substances is likely to continue where it offers the advantages of reduced labor costs and smaller bulk and meets a building's energy efficiency requirements.

Some product substitutes are available for poured or sprayed foam, depending on the specific use. In applications such as packaging and flotation, product substitutes are numerous. Combinations of plastic and non-plastic materials can provide equivalent degrees of cushioning, shock resistance, and water resistance. At present, however, no other insulation materials have equivalent ability to be poured or sprayed, nor can other materials of the same thickness insulate as well as rigid polyurethane foams.

Polystyrene sheet competes with many other disposable packaging and single service products, including paper, cardboard, solid plastic, metal foils, and laminar composites of foil, plastic film and paper. Any of these substitutes could eliminate CFC use in food packaging applications.

Current and possible future alternatives to foam boardstock as insulation include a host of product substitutes, including fiberglass board, perlite, expanded polystyrene, fiberboard, cellular glass, insulating concrete, rock wool, vermiculite, gypsum, plywood, foil-faced laminated board, and insulating brick. Greater thicknesses of these alternatives may be required to provide equal energy efficiency.

Pre-sterilized, disposable products are one possible alternative that will enable hospitals to reduce their dependence on CFC-12 for sterilization.

Pump sprays and roll-ons are product substitutes that can replace CFC-propelled personal care aerosols.

Summary of Technical Potential

A recent detailed report examined the available and emerging options for reducing CFCs, as well as halons and chlorocarbons, which are of potential concern for both stratospheric ozone depletion and greenhouse warming (Makhijani et al., 1988). These authors assumed that both CFCs and halons as well as carbon tetrachloride and methyl chloroform would be phased out by the year 2000. The report asserted that technical options are currently available for eliminating all but two of the important applications; however, it is not certain that all of the substitutes proposed are actually available or will perform as hoped.

Given the rapid pace of current technical innovation, however, it is not unreasonable to expect that all sources of industrial emissions could be eliminated over the next decade. However, even the less stringent CFC, halon and methyl chloroform controls assumed in the Stabilizing Policy scenarios were very effective in reducing climate warming commitment. The recent detailed analysis indicates that even more significant reductions are technically possible.

METHANE EMISSIONS FROM LANDFILLS

As indicated in Figure 7-18, methane emissions from municipal landfills currently are a very small component of global methane emissions but are expected to increase in the future. In the No Response scenarios, landfill emissions are projected to increase from 3% of global methane emissions to 7-8.5% by 2025.

Over 1.6 million tons of municipal solid waste is generated on the planet each day (adapted from Bingemer and Crutzen, 1987). Approximately 80% of this volume is disposed on land in

FIGURE 7-18

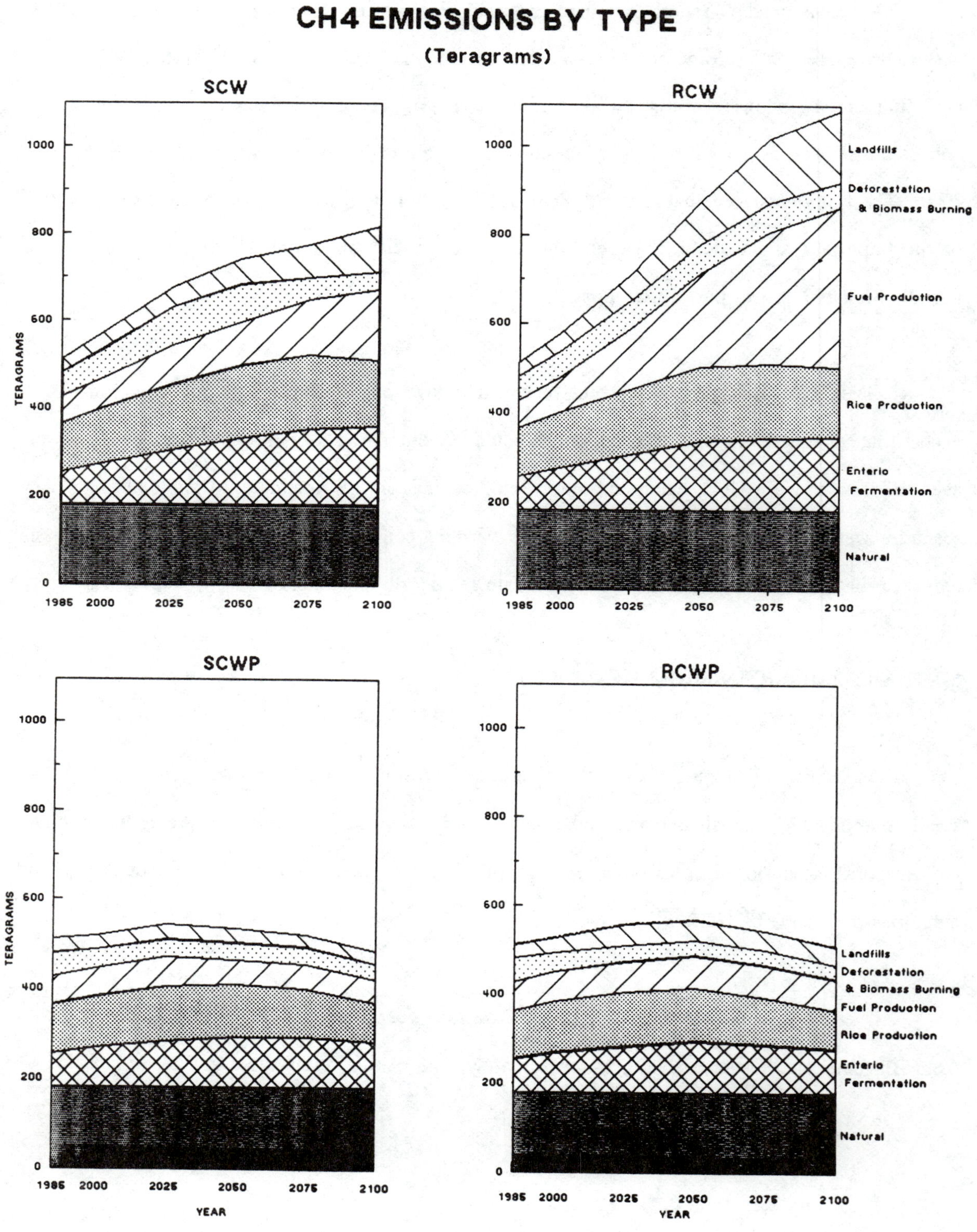
CH4 EMISSIONS BY TYPE
(Teragrams)
SCW
RCW
SCWP
RCWP
TERAGRAMS
YEAR
0
200
400
600
800
1000
1985
2000
2025
2050
2075
2100
Landfills
Deforestation
& Biomass Burning
Fuel Production
Rice Production
Enteric
Fermentation
Natural

landfills or open dumps. Anaerobic decomposition of municipal and industrial solid wastes in landfills results in the generation of 30-70 Tg of methane each year (Bingemer and Crutzen, 1987).

The amount of gas produced by a given site is a function of the amount and composition of landfilled wastes. The rate of gas production is a function of the age of material in the landfill, oxygen and moisture concentrations, pH, and the presence of nutrients. Methane production is highest when the organic content of the refuse is high, the wastes are relatively new, and there is adequate moisture available. The decomposition process only occurs in an environment that is oxygen-free and has a moderate pH. There is a one- to two-year lag period between landfilling of wastes and the beginning of gas generation. Methane production occurs once all available oxygen has been consumed and the environment becomes anaerobic. Food and garden wastes generally decompose over a 1- to 5-year time frame, while paper wastes could require 5 to 20 years to decompose (Bingemer and Crutzen, 1987). These factors and the active life of a landfill affect the duration of methane production. It can take anywhere from 10 years to over 100 years for a landfill to produce significant amounts of methane (Wilkey et al., 1982).

Estimates place the rate of methane production between 1000 and 7000 cubic feet per ton (31-218 m^3/mt) of municipal solid waste deposited (Wilkey et al., 1982). The United States generated approximately 148 million tons of municipal solid waste in 1988 (U.S. EPA, 1988c), which, using the rates above, would produce between 2.9 and 20.7 Tg of methane. Using a methane production rate suggested by Bingemer and Crutzen (1987), the same amount of solid waste would produce an estimated 7 Tg of methane.[1]

[1] This estimate assumes a carbon content of 22% (Bingemer and Crutzen, 1987), that 90% of municipal solid waste generated is landfilled, and a conservative methane production efficiency of 0.25 ton of methane per ton of carbon.

Methane Recovery

Landfill gas, which is typically comprised of approximately 50% methane, 50% carbon dioxide, and some trace constituents of volatile organic compounds, can be recovered and used as fuel. The gas is a medium-Btu fuel (approximately 500 Btu/standard cubic foot), which can be used directly in boilers or for space heating, in gas turbines to generate electricity, or can be processed to high-Btu pipeline quality gas (Zimmerman et al., 1983). The gas is purified prior to use: for medium Btu-gas, processing requires removal of particulates and water; for high-Btu gas, carbon dioxide and most trace components must also be removed.

Landfill gas can present an environmental hazard because of its high combustibility and ability to migrate through soil. Methane is flammable in concentrations between 5 and 15% by volume in air at ordinary temperatures. Methane can rise vertically or can migrate horizontally out of a landfill. Methane migrates easily through porous soils, drainage corridors and other open areas, often to significant distances. Migrating methane gas has caused explosions and flash fires, resulting in property damage and death. Malodors and vegetative damage have also been attributed to migrating landfill gas.

Methane control systems, required under certain circumstances by the Resource Conservation and Recovery Act (RCRA), can help to mitigate malodors, gas hazards, and vegetative stress. RCRA requires that concentrations not exceed a lower explosive limit of 5% methane (10% landfill gas) at the landfill boundary. Controls include impermeable barriers, induced exhaust systems, evacuating and venting or flaring (burning) of the gas, and recovery for use as an energy source. Of these controls, only flaring and recovery of the gas reduce the amount of methane emitted into the atmosphere.

Of the 6,584 municipal solid waste landfills in operation in the U.S., 1539 (16.6%) employ some methane recovery or mitigation system such as venting or flaring. Only 123 of these sites (1.9% of total) recover methane for energy use (U.S. EPA, 1988c). New regulations proposed by EPA under the Clean Air Act would require collection and control of landfill gas at both new and existing landfills. These regulations have the potential to significantly reduce the emission of methane from landfills in the U.S. (U.S. EPA, 1988a).

Estimates place the quantity of gas generated by sanitary landfills in the United States at 1% of the nation's annual energy needs, or approximately 5% of current natural gas utilization (Escor, 1982). The recovery efficiency of methane from landfills can range between 60 and 90% of the gas produced, depending upon the quality and design of the gas recovery system, spacing of recovery wells, and landfill covering.

The economic viability of methane recovery at a landfill depends upon the landfill's size, location, proximity to potential users, current competing energy costs, and regulations governing the site. The capital costs of recovery projects are about $1000/kw. Suggested minimum requirements for recovery include an in-place refuse tonnage of 2 million tons, a disposal rate of 150 tons per day, an average refuse depth of 40 ft, a surface area of 40 acres, and two years of remaining active fill life (EMCOM Associates and Gas Recovery Systems, Inc., 1981).

It is probable that fewer than 1000 landfills in the U.S. are of sufficient size to meet these criteria. Sanitary landfills in developed countries hold the best potential for economical recovery of methane. Currently, there is little potential for methane recovery from open dumps in the developing world; if the practice of sanitary landfilling is adopted, the prospect of methane recovery will improve.

Over 408 (about 5%) of the municipal solid waste landfills in the U.S. receive more than 500 tons of municipal solid waste per day (U.S. EPA, 1988c). Collectively, these landfills receive over 75 million tons of municipal solid waste each year (over 50% of the total generated in the U.S.). If methane recovery were implemented only on the largest 5% of landfills in the U.S., an estimated 2.2-3.3 Tg of methane could be recovered.

Constraints on the economic feasibility of recovery projects have hampered further adoption of this technology. Under current market conditions, projects are not economically viable unless there is a suitable gas user within 2-3 miles of the site, or the electricity generation can be tied into an electricity grid. Current regulations governing many sites also discourage the recovery of methane. Some state regulations subject resource recovery projects to unlimited liability for any potential contamination problems at a landfill - a significant disincentive to recovery.

Various techniques can be used to enhance gas production and yield from a landfill, including controlled addition of moisture and nutrients (usually in the form of landfill leachate or landfill gas condensate), bacterial seeding, and pH control. The addition of leachate and condensate is only permitted at landfills where there is a liner. Further research in these areas could increase the economic viability of methane recovery.

Recycling and Resource Recovery

Recycling and resource recovery hold the potential to affect emissions of greenhouse gases through both waste reduction and reducing energy demand. Source separation and waste-stream reduction have many benefits for the municipality and the environment. A reduced waste stream means less refuse going to the landfill, an increasingly limited resource. In addition, the energy savings from recycling can be significant as discussed in **PART ONE: ENERGY SERVICES.**

Separating organics from the waste stream, such as paper and food, lawn, and garden wastes, can achieve many benefits, including reduced production of methane. Reducing organics in the landfill results in less methane production from that source. Organics that are separated and composted do not produce methane if the composting includes aeration to keep the process aerobic.

Within the industrialized countries of the OECD, garden and park wastes make up about 12-18% of the municipal solid waste stream and account for about 10-14% of its organic content, while food wastes make up between 20 and 50% of the stream and account for over 20% of the organic content. Within developing countries, garden and park wastes are insignificant, while food wastes account for between 40 and 80% of the waste stream and over 70% of the organic content (adapted from Bingemer and Crutzen, 1987). Given these high proportions of food, garden, and park wastes, the potential for reducing methane production through aerobic composting could be significant.

CO_2 Emissions From Cement Production

Carbon dioxide emissions from cement production originate from two sources: 1) as a chemical by-product of the manufacturing process, and 2) as a by-product of fossil fuel combustion used for kiln firing and plant electricity. Energy consumption emissions are discussed in the end-use section of this chapter. The CO_2 emissions that result as a chemical by-product occur during the firing process, when the raw materials (cement rock, limestone, clay, and shale) are exposed to progressively higher temperatures in a kiln. It is during calcination, which occurs at approximately 900 to 1000°C, that the limestone ($CaCO_3$) is converted to lime (CaO) and CO_2, and the CO_2 is released. For every million tons of cement produced, approximately 0.137 Tg C as CO_2 is emitted as a result of this chemical process (Rotty, 1987). For comparison, approximately 0.165 Tg C per million tons of cement produced result from energy consumption.

Although cement manufacture currently accounts for only a small percentage (approximately 2%, not including the energy consumption emissions) of the global anthropogenic source of CO_2, emissions associated with this industry have increased rapidly over the last few decades and can be expected to continue to grow in the future as demand for cement grows. Between 1950 and 1985, cement production and associated CO_2 emissions grew at an average annual rate of 6%. Regional production growth rates have varied during this period due to economic fluctuation in the construction industry (cement's primary market) and shifts in international competition between the cement-producing countries. Today, the USSR, China, Japan, and the U.S. account for 43% of the world's cement production.

Since CO_2 is an inherent product of cement manufacture, the only way to slow the rate of growth emissions is to limit the amount of cement required, i.e., reduce demand through more efficient use of cement and/or through substitution with other materials (e.g., steel and glass). Increases in efficiency can be achieved through both material and fabrication improvements, e.g., through the use of pre-stressed and steel-reinforced concrete products. Substitution of cement with other materials such as steel, glass or wood also would slow the growth in cement demand, although substitution with such energy-intensive materials as steel may result in greater net CO_2 emissions. In fact, in some applications cement products have been used in lieu of other materials, such as steel. Improved efficiency has already occurred in much of the developed world due to improved engineering design in construction. Also, most basic infrastructure has been built in the developed world, so demand may slow somewhat in the future. This is not likely to be the case in the developing world, however, where demand will probably continue to grow more rapidly than GNP, particularly since cement is an inexpensive building material.

PART FOUR: FORESTRY

FORESTS AND CARBON EMISSIONS

Forests, which store 20-100 times more carbon per unit area than croplands, play a critical role in the terrestrial carbon cycle (Houghton, 1988a). Active forest management to maintain high amounts of standing biomass, to reduce tropical deforestation, and to aggressively reforest surplus agricultural or degraded lands, offers significant potential for slowing atmospheric buildup of CO_2, N_2O, and CH_4.

The ecological diversity and geographic range of vegetation communities determine the degree of carbon sequestering by forests and the rate of carbon emissions due to disturbances of forests. Forests cover about one-third of the Earth's land, or 4 billion hectares, of which about 42% is in developed countries (mostly temperate) and 58% is in developing countries (mostly tropical) (FAO/WRI/World Bank/UNDP, 1987).[1] The carbon content of tropical moist forests (with closed canopies, like Amazonian rain forest) averages 155-160 t C/ha of standing biomass in Latin America and Asia and ranges up to 187 t C/ha in Africa.[2] The carbon content of dry tropical forests (open forests with grassy or herbal ground cover, as in African savannas) average 27 t C/ha in Latin America and Asia and 63 t C/ha in Africa (S. Brown, 1988b).

Anthropogenic alterations of forest ecosystems now account for emissions of atmospheric CO_2 equal to about 10-30% of total emissions from combustion of fossil fuels, as carbon stored in vegetation and soils is released by clearing, fire, or decay (Houghton, 1988a). Recent estimates of

---

[1] hectare = 1 ha = 2.471 acres

[2] t C = tons of carbon.

annual net carbon flux from deforestation range from 0.4 to 2.6 Pg C/yr for 1980, primarily due to land use change in the tropics (Detwiler and Hall, 1988a; Houghton et al., 1987).[3] Brazil, Indonesia, and Colombia are the largest of the top ten producers of net carbon release from tropical deforestation; these ten combined account for about half of the CO_2 emitted due to land use change (Houghton et al., 1987).

Uncertainties still exist in determining carbon storage in and emissions from swidden (i.e., shifting, or slash and burn cycle) agriculture versus sedentary (permanent) agriculture, including agroforestry systems. Neither do we have reliable estimates of biomass, carbon content, and trace gas emissions in tropical forests (both standing and those being cleared) and carbon in disturbed tropical soils (about one-third of carbon flux from deforestation) (Houghton et al., 1988). Permanent conversion of natural forest to pasture or cropland results in net loss of carbon stored both in standing biomass and soils; the amount lost is dependent on the biomass productivity and soil carbon storage rates of the former versus the new land use. Cyclical harvest of forest for timber or fuelwood similarly releases carbon from slash (nonmarketable tree parts: branches, leaves) burned or left to decay on-site, and from timber milled into non-durable wood and paper products (and wastes like sawdust and scrap) that are soon burned or discarded.

Forests in temperate regions are essentially now in balance in terms of carbon cycling, with annual incremental growth rates roughly equal to rates of timber harvest and deforestation for urban growth and other land uses. Consequently, temperate forests do not currently contribute significantly to the increase in atmospheric CO_2 (Houghton et al., 1987). However, they now cover much smaller areas than in the past, and have historically contributed heavily to global carbon emissions, as forests were cleared in Europe and North America for agricultural production.

---

[3] Pg = petagram. 1 Pg = 10^{15} grams

Widespread reforestation programs that could expand temperate forests into former forest ranges and reduce net carbon emissions are discussed below. Trees newly planted in urban areas would alleviate the greenhouse problem in two ways: first by reducing the need for air conditioning and hence electricity, and second by increasing the uptake of carbon in biomass growth (Akbari et al., 1988).

DEFORESTATION

Each year, at least 11.3 million (and perhaps as high as 15 million, see below) hectares of forest are cleared in the tropics, an area larger than Austria or Tennessee (Lanly, 1982; IIED and WRI, 1987). The rate of deforestation -- combined with the escalating growth in demand for forest products -- is such that while 33 tropical countries are currently net exporters of wood products, this number may decline to fewer than 10 by the end of the century (Repetto, 1988). If this trend could be halted and reversed, tropical forests could serve as a vast carbon sink, reducing global CO_2 levels.

Figure 7-19 illustrates the movement of tropical forest lands among different stages of the deforestation cycle (an approach used by several researchers, e.g., Houghton et al., 1985; Lugo, 1988), and the conversion stages that the response options discussed below would address.

The underlying causes of deforestation vary widely by ecosystem and region, and are often complex, involving the interplay of historic, biological, economic, and political factors at both macro (national and transnational) and micro (household and village) levels. A recent international conference on the state of the world's tropical forests (Bellagio, 1987) concluded that:

FIGURE 7-19

MOVEMENT OF TROPICAL FOREST LANDS AMONG STAGES OF DEFORESTATION AND POTENTIAL TECHNICAL RESPONSE OPTIONS

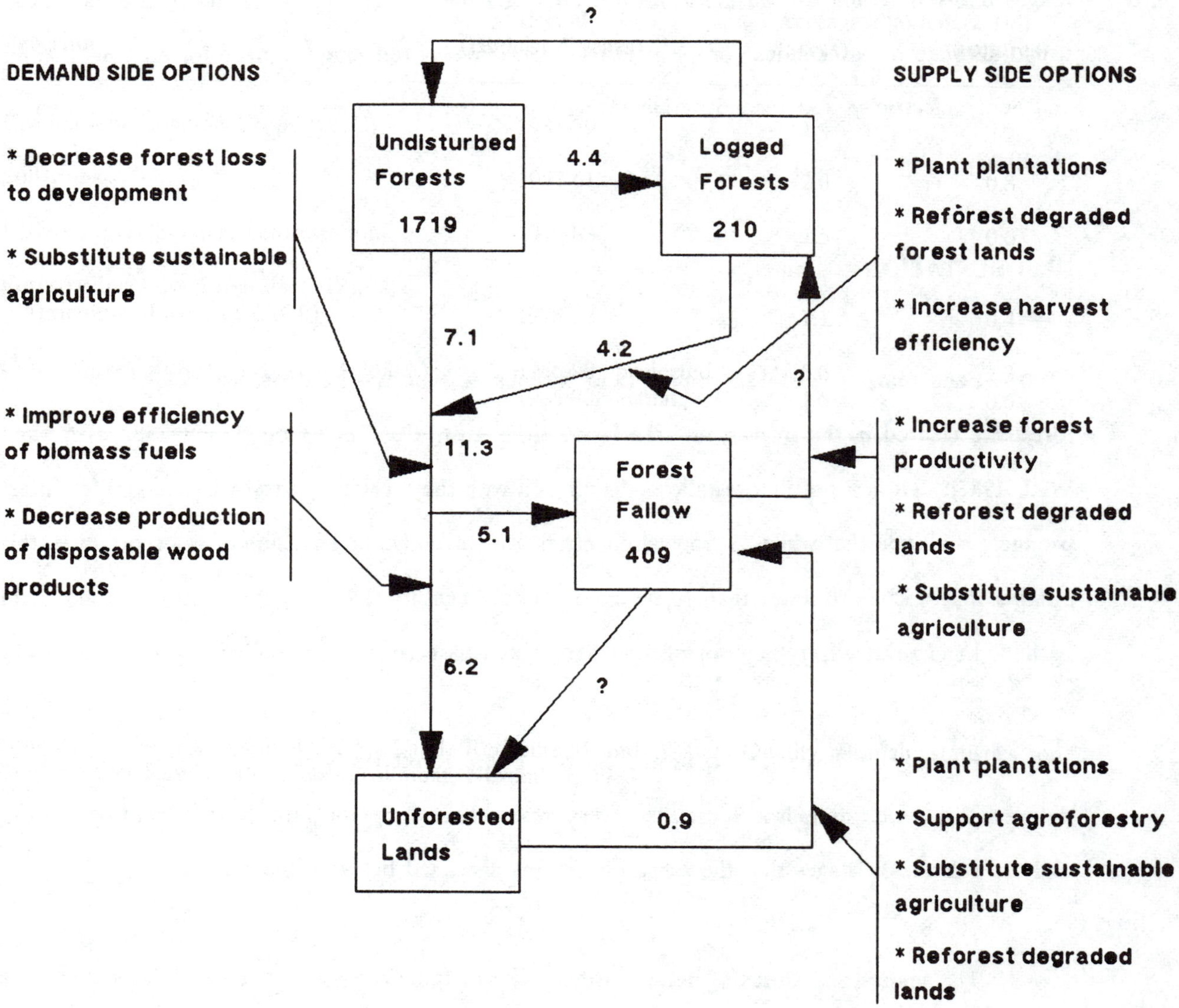

Figure 7-19. Pathways of conversion of tropical closed and open forest lands, and where technical response options discussed here would intervene to slow conversion. Data were derived from FAO (1981) and from Lanly (1982), and are expressed in millions of hectares. Numbers inside boxes represent total area in category in 1980. Numbers on lines tipped with arrowheads represent annual rates of conversion. Data include both closed forests (complete canopy) and open forests (incomplete canopy and grass herbaceous layer). Source: Pathway data modified from Lugo, 1988. Houghton et al. (1985) offer similar conversion data.

> The causes of deforestation are well known. They include population pressure for agricultural land, the demand for industrial timber production and export, and inappropriate government policies regarding land tenure, economic incentives, forest settlement, and other population issues.

The predominate causes of deforestation vary by region. In tropical Africa and in South and Southeast Asia, rapid population growth appears to be the critical factor affecting deforestation. The majority of the population practices agriculture, and most of the increases in agricultural production necessary to sustain high birth rates have come from increases in the area under cultivation through deforestation. Seventy percent of Africa's deforestation stems from swidden (shifting) agriculture. Logging activities, both commercial and individual, in Malaysia, Indonesia, and the Philippines provide access to partially cleared forest lands that facilitates further clearing of forest for agriculture (Houghton, 1988a).

Rural populations rely on wood as the major source of energy, another important cause of deforestation. More than a billion people are currently affected globally by fuelwood shortages that cause them to spend much of their time searching increasing distances for wood to cook meals and heat homes. The fuelwood deficit in arid and semi-arid regions of the world in 1980 affected 29.3 million people, and totaled 13.1 million m^3 of wood. In the dry topics of Africa, the current annual rate of fuelwood consumption exceeds the annual rate of additions to supply through mean annual increment (growth of trees) by the following margins:

Sahelian countries (total):	30%
Sudan:	70%
Northern Nigeria:	75%
Ethiopia:	150%
Niger:	200%

This fuelwood gap between consumption and supply is anticipated to grow to 130% for the Sahelian countries overall by 2000, with single-country forecasts as high as 620% for Niger (Anderson and Fishwick, 1984).

The Amazon region in Brazil is experiencing one of the highest rates of tropical deforestation in the world (IIED and WRI, 1987; Fearnside, 1987). As a consequence, Amazonia is emitting greenhouse gases at rates and quantities high enough to affect global CO_2 and climate cycles. Yet due to the nature of centralized government policies regarding forests and development, the vast scale of its forest resources, and international pressures, Brazil offers high potential for slowing deforestation through proactive adjustments in government, commercial, and colonizing forest use and development practices. The problem is complex, however, as shown in Figure 7-20 which illustrates the complexity of deforestation pressures and consequences in Amazonia, and the implied difficulty of devising technical control solutions.

Brazil has 357 million hectares (ha) of closed tropical forest -- 30% of the global total, and three times as much as Indonesia, which is second to Brazil in its extent of forest area. Only 10% of the Atlantic coastal forest is left, mostly in southern Bahia; the rest has been lost due to logging and plantation clearing. The rate of deforestation in Rondonia doubled between 1976 and 1981, while the population increased 15% per year (Woodwell et al., 1986).

The situation in Brazil is changing rapidly. Analysis conducted at the Brazilian Space Research Center found that forest fires covered 20 million ha (77,000 square miles, or 1.5 times the area of New York state) during 1987, of which 8 million ha were virgin forest (Setzer and Pereira, 1988). This observation has forced reevaluation of standard mid-1980s estimates (e.g., Lanly, 1982) of 11.3 million ha deforested for the entire globe's closed and open tropical forests (Houghton, 1988b), and could raise estimates, perhaps as high as 15 million ha/yr for the late 1980s. The

FIGURE 7-20

POPULATION GROWTH, ROAD BUILDING, AND DEFORESTATION IN AMAZONIA

A. Growth of population and deforested area in the state of Rondônia.

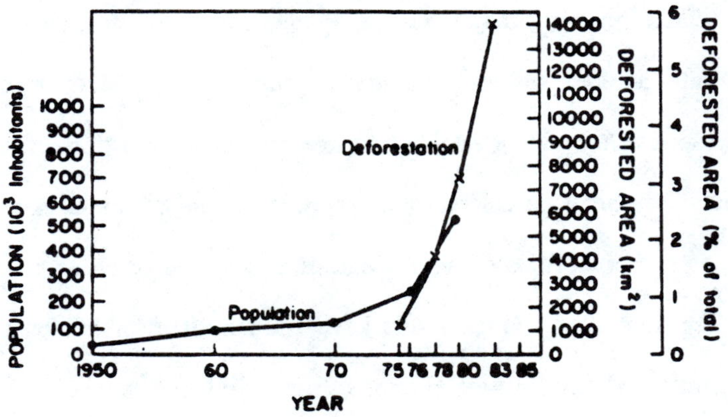

B. Causal loop diagram of the relationship between road building and deforestation.

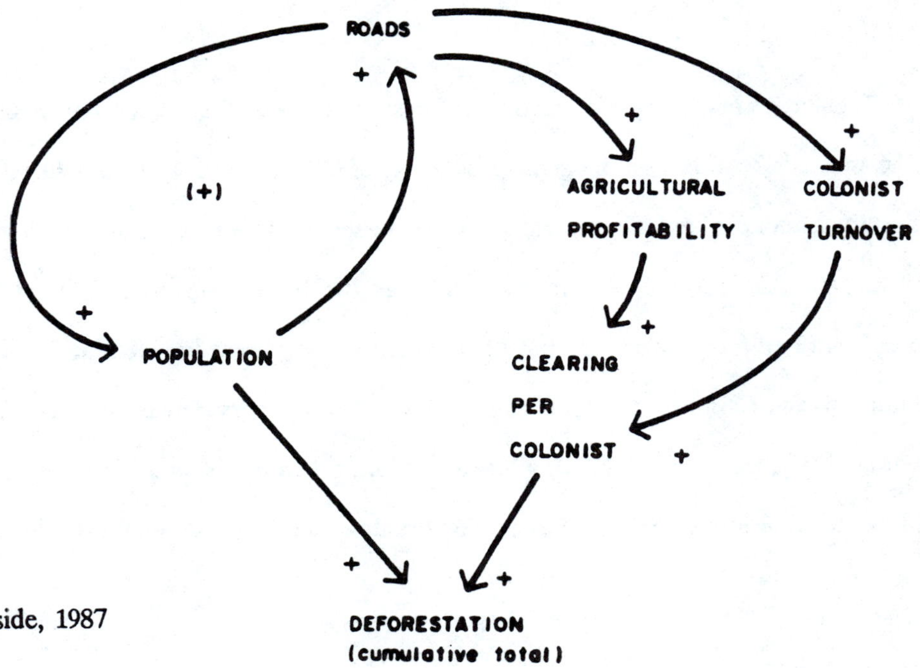

Source: Fearnside, 1987

emissions from these fires contribute roughly 10% of total global emissions of CO_2 (Fearnside, 1985). If the Brazilian Amazon were completely cleared, 11 Pg of carbon would be released immediately, augmented by a continuing gradual release that would elevate the total to 62 Pg (Fearnside, 1985). Thus, deforestation in Brazil poses serious global consequences for climate change as well as the much discussed loss of species diversity.

In the Brazilian Amazon, major factors driving the loss of forests include land speculation, inflation, negative-interest and special crop loans, production of beef for export, and population redistribution in response to high growth rates and the mechanization of agriculture in southern Brazil. The Amazon has been perceived by entrepreneurs and planners as an undeveloped frontier capable of producing vast quantities of timber and minerals, and of absorbing underemployed urban populations from the southeast. Once roads have been built into the forest, land is worth more cleared than forested, and the profits obtained from land speculation are reinvested in further clearing (Fearnside, 1986, 1987; Maguire and Brown, 1986).

TECHNICAL CONTROL OPTIONS

Technical control options involving forestry can sequester carbon through the growth of woody plants, can reduce anthropogenic production of CO_2, and can complement other strategies for reducing the buildup of greenhouse gases. Forestry sector strategies for responding to the threat of global warming (see Table 7-13) basically fall under two broad categories: (1) Reduce the demand for forest land and forest products, and (2) Increase the supply of forested land and forest products. In addition to the obvious benefit, from the climate change perspective, of increasing the supply of forested land (i.e., trees absorb CO_2), afforestation has a number of valuable ecological and economic benefits worthwhile on their own merits. For example, more forests may provide jobs in the forest

TABLE 7-13
Major Forestry Sector Strategies
for Stabilizing Climate Change

Reduce Demand for Forest Land and Forest Products

- Substitute sustainable, sedentary agricultural technologies for swidden (shifting) agriculture.
- Decrease consumption of forest for cash crops and development projects, through environmental planning and management.
- Improve the efficiency of biomass (fuelwood) combustion in cookstoves and industrial uses.
- Decrease the production of disposable forest products (e.g., paper, disposable chopsticks) by substituting durable wood or other goods, and by recycling wood products.

Increase Supply of Forested Land and Forest Products

- Improve forest productivity on existing forests, through management and biotechnology on managed and plantation forests.
- Increase harvest efficiency in forests, by harvesting more species with methods that damage fewer standing trees and use more of total biomass.
- Establish plantations on surplus cropland and urban lands in industrialized temperate zones, to produce high biomass and/or fast-growth species to fix carbon.
- Reforest degraded forests and establish plantations and agroforestry projects in the tropics, using both fast-growth and high-biomass species on short rotations for biomass and timber.

products industry, maintenance of biodiversity, watershed protection, non-point pollution reduction, and areas for recreation.

In the RCW scenario, a low rate of replanting is assumed and the area of tropical forests decreases exponentially from 1980 to 2100. (This assumption is the same as in Houghton, 1988b.) By contrast, the policy scenario (RCWP) assumes a rapid decline in deforestation (which ceases by 2015), and a linear increase in the establishment of plantations in the tropics. Implementation of the control options discussed below may make it possible to achieve the RCWP scenario, which results in a net CO_2 uptake of up to 0.7 Pg C/yr in 2025.

Compared with the annual crop cycles of agriculture, rotation time in forestry (the time necessary for one cycle of a forest crop to be planted, grown, and harvested) is slow -- typically on the order of 25-80 years in temperate zones and 8-50 years in the tropics (where growth rates are higher). Essentially, therefore, forestry climate strategies can create a net CO_2 sink for a fixed period, albeit long, since trees eventually die or are cut and release carbon through decay or burning. Harvesting on rotation (with aggressive replanting) allows storage of carbon in durable products like construction beams, crates, or fences, until they decay, and regeneration of new biomass at high rates of growth and carbon fixation.

All strategies listed in Table 7-13 could be pursued simultaneously. Some, however, are better suited to industrialized countries and some are more appropriate for developing countries. Strategies for maximizing both biomass growth rates and volume of standing stock are needed. Species that produce high volumes of biomass (e.g., Douglas fir in the Northwest) usually grow slowly (e.g., 80-100 years to mature) and are most useful in industrialized countries, whose conditions favor long-term forest protection and intensive management. Developing countries especially need to maximize biomass growth rates, to restore degraded and desertifying lands, and to produce fuelwood and timber. Developing countries face rapid forest depletion and high population growth rates, and have limited institutional capability to guarantee long forest rotation times in the face of these realities. However, developing countries should also plant high-value, long-rotation hardwood stands, and protect existing old-growth forests from cutting and burning.

All potential strategies, whether categorized as demand reduction or supply enhancement, to the extent feasible should be:

- sustainable over time, without deteriorating the natural resource base or introducing ecological changes (i.e., pests),
- capable of addressing the direct and indirect causes of forest loss by providing viable alternatives to current land use patterns,
- economically attractive (low-cost and offering income commensurate to present land uses),
- capable of providing an equivalent spectrum of forest products and jobs, at rates of return to labor (or time) and capital comparable to current forest use patterns,
- socially integrative or adaptive, building on local needs and tradition,
- technologically simple and durable, and
- readily adaptable to changing economic, political, social, and ecological realities (e.g., civil war, drought, and resource-driven population shifts).

In addition, these response options need to address the full range of causes of, and stages in, the rapid deforestation occurring in the tropics, including:

- logging and clearing by colonizers in closed canopy forests,
- fuelwood harvesting and swidden (shifting) agriculture in closed and open canopy forests,
- poor resource management of both undisturbed (virgin) and secondary (disturbed or fallow) forests,
- low reforestation planting and success rates on lands degraded by human resource use patterns (e.g., upland forests cut for timber and fuelwood and then overgrazed by goat and sheep herders).

See Table 7-14 for a summary of potential technical options for implementing forestry strategies to reduce demand and to increase supply.

Because of the long rotation times for forest growth, technical options will need to integrate short-term educational, harvesting, and research work with longer-term adaptations in forest planting, management, and product use.

Reduce Demand for Forest Land and Products

Both tropical and temperate forest lands are in high demand to provide a range of alternative land uses and forest products. Forests are consumed in areas of rapid population growth worldwide as villages and cities expand, transportation corridors are built to connect them, and additional arable land is sought for food production. Other major forces contributing to deforestation in the tropics, in particular, include swidden (shifting) agriculture; large-scale economic development projects (often financed by multilateral banks); cattle ranches and palm oil, timber, and rubber plantations; and fuelwood demand.

Based on population growth, projected demand for fuelwood by the year 2000 will require the creation of 20-25 x 10^6 ha (or perhaps as high as 50 million ha) of new closed forest plantations for fuelwood, at a cost of $50 billion, a rate 10-20 times current planting tallies (Nambiar, 1984; FAO, 1981; Lundgren and van Gelder, 1984). An additional 200 x 10^6 ha of croplands will be needed (FAO, 1978) just to maintain the already inadequate 1980 levels of per capita food supply.

Currently, more than ten hectares are lost to each one that is planted, based on the ratio of global deforestation to tree planting (Lanly, 1982). Models of forest product demand from 1980-

TABLE 7-14
Potential Forestry Strategies and Technical Options to Slow Climate Change

Strategy	Technical Options	Regions Most Effective In
Reduce Demand for Forest Land and Forest Products		
Substitute sustainable, integrated forest and natural resource management	• Natural forest management systems • Extractive reserves • Biosphere reserves	• Tropical moist and dry forests • Temperate forests (public or private)
Substitute sustainable agriculture	• Small-scale agroforestry • Financial incentives for swidden colonists to shift to sustainable practices • Technical aid in soil and crop selection • Plan into development projects by banks, state agencies	• Tropical moist & dry forests with strong central governments (Brazil, Colombia, Malaysia, Indonesia)
Decrease forest consumption for development	• Assistance from multilateral banks and state agencies contingent on planning. Loans contingent on minimal forest loss • Mitigation of loss by 2:1 protection of forest	• Strong central governments and banks with ability to plan and manage (Brazil, Costa Rica, India, China, Mexico)
Improve efficiency of biomass combustion	• Widely distribute efficient cookstoves • Incentives for industrial cogeneration	• Areas with inefficient current stoves, good extension, and difficult-access or expensive fuelwood (Nepal, India, Sahel, Haiti)
Decrease production of disposable forest products	• Substitute durable wood products for disposables • Establish recycling programs for wood	• Areas with cheap substitutes for wood & developed markets (industrialized areas of developing countries, Japan, U.S.)

TABLE 7-14 (Continued)
Potential Forestry Strategies and Technical Options to Slow Climate Change

Strategy	Technical Options	Regions Most Effective In
Increase Supply of Forested Land and Forest Products		
Increase forest productivity • Manage temperate forests • Apply natural forest management in tropics • Increase plantation productivity	• Intensify timber stand improvement all forests • Apply fertilizers & biotechnology to plantations • Extend natural forest management practices	• Developed countries and industrialized developing countries, with extension capability
Increase forest harvest efficiency	• Increase number of species harvested • Decrease damage to standing trees • Use harvest slash & mill scraps	• Any area, if marketed, and countries with large or multinational logging concessions (Brazil, Malaysia)
Plant trees on crop & urban lands in temperate zone • Expand tree planting programs • Reforest croplands • Reforest urban areas • Reforest highway corridors	• Reforest surplus cropland • Plant fast-growth plantations • Plant trees near buildings, highways, rivers	• Developed countries • US: Southeast, North Central states, West Coast
Reforest degraded forest lands and plant plantations in tropics	• Incentives for agroforestry • Establish extension farms & workers for agroforestry • Require commercial & village loggers to replant • Organize village tree planting • Mobilize youth and religious groups to plant trees	• Tropics, where rainfall and soils are adequate

2020 project tropical forest removals (harvest for timber) to double between 1980 and 2000, and then plummet to 72% of their 1980 level by 2020 (WRI et al., 1988; Grainger, 1987).

Option 1: Slow Deforestation by Introducing Sustainable Forest Use Systems

Net deforestation in an area results when demand for timber, fuelwood, or forested land exceeds the local supply and the productivity rates of forest lands allowed to regenerate do not keep place with the harvesting of products. One key method of reducing demand for virgin or secondary forest is to introduce sustainable natural resource management techniques at the village and regional levels that provide a stream of forest benefits but minimize cutting of natural forest. McNeely (1988) and McNeely and Miller (1984) offer theory and case studies illustrating the economics of non-consumptive, integrated natural resource management.

Natural forest management (NFM) applies silvicultural techniques to allow smaller sustainable harvests of natural forests, instead of traditional clear-cuts of large tracts to maximize short-term profits. NFM may increase forest productivity, and provides a wide range of non-wood products with high economic returns (e.g., nuts, herbal medicines, nature tourism operations; see Gradwohl and Greenberg, 1988). Extractive reserves are a newly evolved example of NFM in which economic products like nuts and rubber are extracted from forest reserves in Brazil. They maintain standing forest while providing jobs and wages to organized rubber tappers and nut collectors otherwise dependent upon income from logging (Schwartzman and Allegretti, 1987).

The Malaysian Uniform System is a 60-year NFM rotation technique developed to rapidly regenerate harvested dipterocarp forest (lowland closed forest dominated by trees of the Dipterocarpaceae family) in peninsular Malaysia. Another method is the Celos Silvicultural System, practiced on long-term research plots in Suriname, which uses carefully planned logging trails and winches to reduce damage to standing trees during harvest from 25% down to about 12%. Numerous small areas are cut on 20-year rotations, rather than single huge tracts, and 3 improvement thinnings of non-target tree species and vines are made each rotation.

The Palcazu Development Project in Peru, funded by the U.S. Agency for International Development, has devised a system of active forest management that harvests thin swaths of forest 20-50 meters wide on 30-40 year rotations. Old-growth forest left surrounding the strips naturally provides seed dispersal after harvest, as in natural tree-gap regeneration processes, and maintains biological diversity lost in logging operations. Potential net profits after the wood is processed could be as high as $3500/ha worked, according to estimates by researchers (Hartshorn et. al, 1987).

NFM systems could be widely introduced via forest extension programs, bilateral and multilateral rural development projects, and integrated management of protected reserves and adjacent lands (e.g., the Biosphere Reserve concept of multiple-purpose protected areas combining preservation, research, and economic use zones [McNeely and Miller, 1984]). NFM offers a promising vehicle for maintaining high-biomass standing forest, slowing deforestation, and allowing the 54 million ha of forests already logged (WRI and IIED, 1988) to regenerate.

Option 2: Substitute Sustainable Agriculture for Swidden Forest Practices

Sustainable agricultural systems are those that rely on biological recycling of chemical nutrients (in soils) and energy, and on naturally occurring mechanisms for protecting crops from pests, to produce annual harvests that can be sustained in perpetuity (Dover and Talbot, 1987). Generally such systems use low levels of agricultural technology (i.e., minimal agricultural pesticides or fertilizers or improved seeds, and few conservation measures).

Sustainable agriculture offers three major types of benefits for the carbon cycle:

- increased supply of woody biomass fixing carbon in natural forests and in forest-crop systems.
- reduced demand for natural forest wood products, and
- reduced demand for new land cut from primary or secondary forest for swidden agriculture (by substituting higher-nutrient sedentary systems on permanent plots).

Other non-carbon-cycle benefits of sustainable agricultural systems include the preservation of biological diversity, low soil erosion, maintenance of the hydrological cycle and soil moisture, and recreation and tourism.

Swidden (or shifting) agricultural methods involve cutting and, usually, burning forest patches, to plant crops that are harvested for 1-7 years, and then abandoning and leaving fallow the patches for about 7-14 years as new patches are cut and farmed. About 41 million ha of tropical primary and secondary forest are burned per year (see Chapter IV). Tropical forests store up to 90% of a plot's nutrients in woody plants (compared with temperate forests, where only 3% are stored in plants and 97% in soils), some of which are released by burning (Keay, 1978). Swidden systems persist throughout the world, especially in remote and hill districts, and during times of individual or regional economic stress. On Negros Island in the Philippines, the number of swidden farmers rose by 80% in only two years in the mid-1980s because of declines in the sugarcane industry that forced underemployed workers into swidden agriculture to grow food. Ecologists predicted that a major rain forest there could be destroyed by the year 2000 (Dover and Talbot, 1987).

Sustainable agriculture and soil management systems can be introduced as a substitute for swidden systems that require destruction of virgin or secondary forest. For every hectare farmers put into such methods, 5-10 hectares of tropical rainforest may be saved from destruction to store carbon, conserve hydrologic cycles, and retain biological diversity, according to Sanchez (1988). Table 7-15 shows the equivalent area of forests needed for traditional swidden practices for every one hectare of forest land needed for more resource-intensive sustainable uses with higher productivity rates. Agroforestry is the combination of agricultural and forestry techniques to produce woody plants on the same parcels as food or commodity crops or animals, with a mutually beneficial synergism. It

TABLE 7-15

Comparison of Land Required for Sustainable Swidden Versus Agricultural Practices

Sustainable Agricultural Practices	Number of Swidden Hectares Required for Every One Hectare of Sustainable
Flooded rice	11.0
Low-input cropping (transitional)	4.6
High-input cropping	8.8
Legume-based pastures	10.5
Agroforestry systems	not determined

Source: Derived from 17-year ongoing research by North Carolina State team at Yurimaguas, Peru, in tropical moist lowland forest (Sanchez, 1988, and Sanchez and Benites, 1987).

offers one of the most promising approaches to providing both fuelwood and food needs, while reducing environmental externalities associated with monocultural row cropping (pesticide use, pest population surges, high irrigation requirements). Interest in agroforestry has surged since the late 1970s, and international research and project lending is underway in approximately 100 developing nations.

Agroforestry systems derive from traditional forest farming practices of many indigenous peoples and are sustainable over long rotations, large acreages, and low population densities. Innovative research programs should build on these local methods, where feasible. The Lacandon Maya Indians living in rainforest in Chiapas, Mexico, practice a multiple-layer cropping system utilizing up to 75 species in 1-hectare plots that produce crops for 5-7 years. As soil fertility wanes, the Lacandon plant tree crops (cacao, citrus, rubber, avocado) that provide valuable products as the plots regenerate with secondary forest. This forest is cut again when the trees overgrow the managed species. It is estimated that only 10 hectares of rain forest are consumed per farmer throughout his career with this method of sustained agroforestry. In contrast, immigrant colonists consume two or three times as much forest area (Nations and Komer, 1983).

Three major classes of agroforestry are being practiced in Asia, Africa, and Latin America:

1. agro-silvicultural (including fuelwood trees, hedgerow or "alley" intercropping, Taungya forestry, shelterbelts and windbreaks),
2. silvo-pastoral (fodder production, living fences, trees in pastures), and
3. agro-silvo-pastoral (woody hedges for grazing by livestock and producing mulch, home gardens with woody plants).

Other systems include traditional swidden, and aquaforestry (silviculture in mangrove swamps and fish ponds). One example of a model agroforestry farm in Rwanda is illustrated in Figure 7-21, incorporating trees and bushes in erosion-control strips, hedges, nitrogen-fixing trees in fields, and cash and fodder crops (Dover and Talbot, 1987). (Further information on the array of systems is

FIGURE 7-21

MODEL AGROFORESTRY FARM LAYOUT, RWANDA

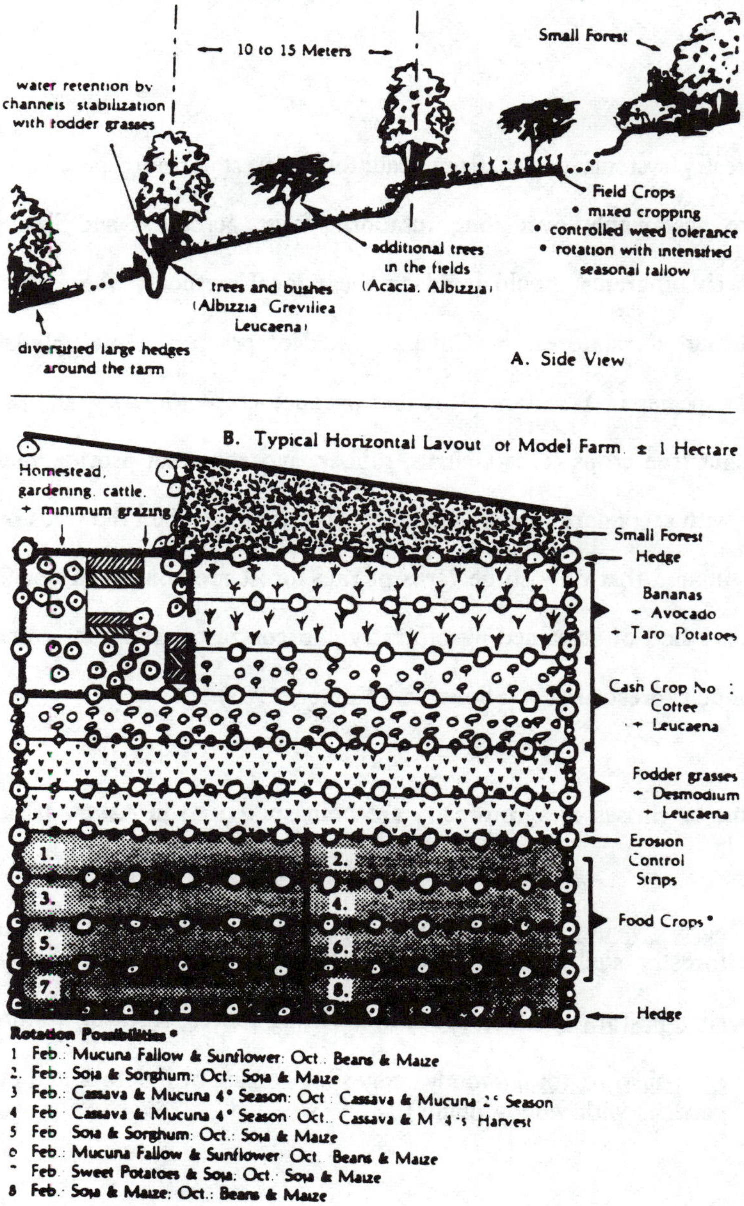

Source: Dover and Talbot, 1987, derived from Behmel and Neumann, 1981

available in Winterbottom and Hazelwood, 1987; Dover and Talbot, 1987; Winterbottom et al., in press; and OTA, 1984).

Integrated crop-forest systems not only reduce demand for forest, but also retain carbon stored in tropical soils (otherwise emitted after disturbance), and in woody biomass, although at rates well below mature forest. A typical agroforestry system in steep uplands with poor soils in Himachal Pradesh, India, is stocked with 20.5 trees/ha, which produce a yield of 2.0 m^3/ha/yr of wood, or 0.8 t C/ha/yr, plus the potential savings of roughly a 5:1 ratio of ha of virgin forest retained intact per ha converted to permanent cultivation. A more intensive stocking rate of 322 trees/ha in home gardens in Surakarta, Indonesia, yields 7.3 m^3/ha/yr wood, or 1.9 t C/ha/yr. Data from Indonesia and Tanzania indicate that 200-300 trees are sufficient to provide wood-production needs of a typical household (Lundgren and van Gelder, 1984).

Reliable estimates of global acreage with potential for substitution of agroforestry for swidden agriculture, and net carbon benefits, are not presently available. Preliminary calculations for an integrated agroforestry and plantation project in Guatemala designed for an American utility (AES) to offset its CO_2 emissions are offered by Trexler (in press) and outlined in Table 7-23. An overview of potential carbon cycle and biomass productivity benefits from a range of agroforestry systems is presented in Table 7-16.

Agroforestry projects considered successful from a climate change perspective would require suitable environmental conditions (soils and rainfall), and human population densities and institutions adequate to encourage multi-year resource management. Either overcrowding or pervasive poverty will shift the focus to short-term survival. Social and economic factors likely to promote success include clear and relatively equitable land tenure for farmers; local decision-making, and political systems that at least tolerate medium-term investment by villagers of various classes; developed and accessible markets for crop and forest products; and adequate protection of agroforestry systems

TABLE 7-16

Potential Carbon Fixation and Biomass Production Benefits from Agroforestry Systems

Type of System	Location	Trees Per Hectare	Productivity (t C/ha/yr)	Species Used	Products Produced
Natural forest management and crops	Guesselbodi forest, Niger		0.8	native shrubs (*Combretum micranthum, Guiera senegalensis*)	wood, mulch, crops, gums, fodder, medicines
Steep uplands, poor soils system	Himachal Pradesh, India	20.5	0.8		fuelwood, fodder, crops
Alley cropping	IITA, Nigeria		0.9-3	nitrogen-fixing shrubs (*Glicidia, Leucaena, Calli-andra, Sesbania*)	maize in alleys between hedgerows cut for mulch and stakes
Home gardens	Surakarta, Indonesia	322	1.9		fruit, fodder, mulch fuelwood
CARE Agrofor-estry and Re-source Conser-vation Project	Guatemala	400	4.7	conifers (high-lands), hard-woods (lowlands). 20 species in total	fuelwood, fodder, crops

Source: Winterbottom and Hazelwood, 1987, and WRI et al., 1988 (Niger, Nigeria); Lungren and van Gelder, 1984 (India, Indonesia); Trexler, in press, and WRI, 1988 (Guatemala).

from grazing livestock, villagers, civil strife, and rapid economic changes (Winterbottom and Hazelwood, 1987; IIED and WRI, 1987).

Constraints to wide diffusion of agroforestry include the difficulty of technology transfer to remote populations often holding traditional values that do not encourage innovation; the need for systems tailored to specific site conditions; capital requirements to purchase and maintain seedling nurseries and to fund extension efforts and research; and the vast scale of implementation necessary to slow forest degradation (Lundgren and van Gelder, 1984). The long-term sustainability of new agroforestry systems has not been fully demonstrated in many first-generation projects, which often still rely upon high levels of fertilizer and labor.

Option 3: Reduce Demand For Other Land Uses That Have Deforestation As A Byproduct

Large development projects, especially those planned in the tropics by transnational corporations and multilateral development banks, consume huge tracts of forest for roads, hydroelectric project reservoirs, and new communities. The Mahaweli project in Sri Lanka will destroy 260,000 ha of tropical moist forest in order to generate 466 MW of electric power and flood control and irrigation benefits. The Narmada Valley Project in Madhya Pradeshi, India, may inundate 350,000 ha of teak and bamboo forest (Kalpavriksh, 1985).

Pressure on banks and governments to reduce adverse impacts on tropical forests has mounted since 1980 and spawned new development planning methods that include protection of forest tracts, in order to offset consumptive use of other forested lands (Rich, 1989; Gradwohl and Greenberg, 1988). In 1986, the World Bank issued a 6-element new policy on wildlands to guide planning of Bank development projects. This policy states preference for choosing already degraded (e.g., logged over) and the least-valuable lands for development purposes. It requires compensation for wildlands converted by Bank projects, in the form of an added wildlands management and preservation component (Goodland, 1988; World Bank, 1986). India is experimenting with compensatory reforestation, at a 2:1 replacement ratio, for forest cut for hydro projects, although problems still remain with such mitigation approaches -- including management responsibility over long time frames, and potential productivity rates of new compensatory forests.

Option 4: Increase Conversion Efficiencies Of Technologies Using Fuelwood

Fuelwood demand from tropical forests accounts for significant deforestation. Wood supplies over 90% of total energy use in Burkino Faso, Malawi, Tanzania, and Nepal; 50% in Indonesia; 25% in China; and 20% in Brazil (Brown et al., 1988a). Annual average fuelwood consumption for agricultural and industrial uses in Tanzania from 1979-80, for example, consumed 1.9 x 10^6 m^3 of fuelwood, releasing 0.5 x 10^6 t C, to cure tobacco, smoke fish, dry tea, fire pottery, and burn bricks (Mwandosya and Luhanga, 1985).

The introduction of more efficient cookstoves and industrial technologies could reduce wood requirements by 25-70% at very low cost (Goldemberg et al., 1987; Brown et al., 1988a), as discussed in the energy section of this chapter. The most successful strategy for reducing fuelwood-related deforestation in the long run may be the substitution of kerosene, gas, and electricity for fuelwood, and widespread distribution of cogeneration technologies to produce higher benefits from fuelwood use.

Option 5: Decrease Production of Disposable Forest Products

Forests harvested in developing countries or managed in industrialized nations to generate wood products that are consumed and then burned or buried in landfills in the short term (e.g., newsprint, paper goods, and fast-food packaging) contribute greenhouse gas emissions. Work has begun on introducing technologies and programs to replace consumable forest products with durable goods that are used repeatedly and/or recycled, avoiding these emissions and providing carbon storage. Two major control options are discussed below.

Substitute durable wood or non-wood products for high-volume disposable uses of wood

Current storage of durable wood products has been estimated at about 10 x 10^9 m^3 of solid wood (2.6 Pg of carbon), or roughly 25% of world industrial harvest over the past 35 years (Sedjo and Solomon, in press).

The global forest industry has been stagnating for the past 15 years, as real prices have decreased, growth in consumption has shrunk, and competition on world markets has accelerated from developing countries. FAO (1986a) and Kuusela (1987), however, project annual growth rates of wood-based panels for 1985-95 at about 2.5-4.0%, down from 6.9% in 1963-75, but along with printing and writing paper the most quickly rising rate. Total world production of principal forest products for 1978-82 averaged 805.5 x 10^6 t/yr, or about 0.4 Pg C/yr, and is projected to rise to a mean estimate of 1333 x 10^6 t/yr (0.66 Pg C) by 2000.

Accelerated harvest and storage of wood products could provide a technical response option for reducing demand for products from virgin forest, and increasing supply from managed forests. For example, if production of all wood products was increased by 30% on average above the projected growth from 1982-2000, and this increase occurred through increased use of durable wood products, then production would rise to 1733 x 10^6 t/yr, storing an additional 0.4 Pg C over the 18-year period. Potential storage over the next 50 years to 2030 has not been calculated, but might reach a total of more than 1 Pg C.

Examples of potential product shifts from consumables to durables include eliminating the use of disposable chopsticks in Japan and elsewhere, in favor of permanent wood or plastic ones; replacing single-use wooden crates for shipping, fruit and pallets with metal or plastic ones; and widely installing wood paneling in homes and commercial structures. To achieve net greenhouse gas emission benefits, production processes for durables must minimize generation of gaseous byproducts through use of energy conservation measures noted earlier in this chapter.

Expand recycling programs for forest products

Global production of newsprint, paperboard, and other paper averaged 334 million t/yr from 1978-82, with growth rates to 2000 anticipated to hover around 3% per annum (FAO, 1986a,b).

In the U.S., consumption of all paper products in 1988 totalled 79.8 million tons. Post-consumer recycling of paper in the U.S. now provides 28% of domestic production of paper and paperboard, and totaled about 22.3 million tons in 1986 -- virtually twice the amount recycled in 1970 (U.S. EPA, 1988). Current obstacles to enhanced recycling rates include market development, regulatory, and financial issues (Ruston, 1988).

Recycling of paper products in the U.S. could be pursued as a climate change response option. If, for example, recycling rose to 80% (with 10% diverted to (stored in) durable books or construction, and 10% disposed of), then the difference between the number of tons burned or buried in landfills in 1986 at 28% recycling (about 50 million tons) and at an 80% rate (about 16 million tons) would equal 34 million tons/yr. Methane production from landfills (see Landfills section, this chapter) and carbon emissions from incineration would decline. Paper products formerly treated as consumables would become converted to essentially durable recycled products, thereby increasing the net total stock of carbon (assuming that forests previously used to grow wood for paper remain undisturbed).

Increase Supply of Forested Land and Forest Products

Option 1: Increase Forest Productivity: Manage Temperate Forests For Higher Yields

Modern forestry management techniques applied to commercial, state, and large private forest lands offer the greatest potential for large-scale increases in productivity.

In the U.S., new forest area growing at average rates would not be sufficient to offset our current annual production of 1.3 Pg C. Per capita annual carbon production for 237 million Americans is about 5 t C/capita. U.S. commercial forests (those producing greater than 1.4 m^3/ha/yr and not set aside in parks) totalled 195.3 x 10^6 ha in 1977, with a net average growth rate of 3.15 m^3/ha/yr, or 0.82 t C/ha/yr. Thus, 6 ha (15 ac) of forest would be required to sequester each person's fossil fuel emissions. For 237 million people this would require 1.5 x 10^9 ha of average forest -- a tract 50% larger than the 0.9 x 10^9 ha land area of the country (Marland, 1988).

However, intensive forest management to increase biological productivity or economic returns on forest land may offer a partial solution. The U.S. Forest Service (USFS, 1982) estimates that if current commercial forests simply were fully stocked (i.e., they were universally managed to grow the tree densities and volumes of mature stands), their net annual growth could be increased by about 65%, which would sequester 0.1 Pg C/yr. This full stocking option is appealing, since many areas not presently growing forests are sites so poor that even intensive management or aggressive planting are likely to provide only negligible net annual growth (USFS, 1982). Forest Service estimates also indicate that forest owners could go further by managing forests to take advantage of economically feasible opportunities that would offer a 4% annual return on investment. These management options could produce an additional 18 billion cubic feet of annual forest growth, equal to 0.16 Pg C (327 x 10^6 t wood) (Hagenstein, 1988; Moll, 1988).

Intensive management techniques that improve productivity include site-specific species selection (through provenance trials with seeds), thinning and release cuttings, control of spacing among trees, weed and pest control, fire suppression, fertilization, irrigation, and planting genetically improved seedlings. All options requiring significant labor tend to be prohibitively expensive if implemented on large scales; however, access to volunteer labor from youth or citizen groups might make these options more feasible.

The economics of fertilization on large tracts vary; for many species and sites, the costs of chemical fertilizers exceed growth stimulated. Yet according to Ford (1984), "fertilization is the most important single treatment that the forest manager can apply during the growth of the crop to accelerate growth." More than half of the loblolly pine plantations in the Southeast would show value-added growth from fertilization, according to one observer (Binkley, 1986).

Obstacles to use of fertilizers include trees' low nutrient recovery rates, due to leaching and microbial activity; the nutrient status of the site; and whether slash is removed from the site during harvest (tropical forests store 90% of available nutrients in biomass and only 10% in soils, compared with about 30% for temperate forest biomass) (Marland, 1988; Ballard, 1984). Nitrogen-fixing legume species, like black locust and honey locust in the U.S. and *Leucaena* and *Calliandra* in the tropics, offer the advantages of supplying their own nutrient requirements, growing well in the depleted soils of degraded lands, and cutting fertilizer costs.

Other constraints to intensive management include the need for very short rotations to maintain high growth rates (and associated labor costs and need to sequester large volumes of harvested wood); pest and genetic diversity problems associated with monocultural stands; costs of energy and labor for plantations; the tradeoff between maintaining large volumes of standing biomass and fast growth rates; and fire management costs.

Option 2: Increase Forest Productivity: Improve Natural Forest Management of Tropical Little-Disturbed And Secondary Forests

Natural tropical moist forests produce annual wood increments higher then managed forests on average (Table 7-17), due to the latter's higher harvest offtake volumes, minimal replanting success, and reduced biomass in regenerating forests. From a carbon-cycle perspective, this offers an argument in favor of a two-prong forest management strategy managing virgin and secondary forests as sustainable high-biomass sinks, and managing fast-growth plantations for provision of

TABLE 7-17

Natural and Managed Tropical Moist Forest Yields
(mean annual wood increment, metric tons per hectare)

Continent	**Unmanaged Forest**[1]	**Managed Forest**
Africa	2.3-5.8	0.6-1.3
Latin America	0.9-1.9	0.6-2.0
Asia	3.7-4.5	1.3-2.6

[1] Unmanaged forests are stands in balance, where annual increment roughly equals forest losses from natural causes.

Source: IIED and WRI, 1987 (from Wadsworth, 1983).

forest products. Natural forest management (NFM) techniques (discussed under demand option 1, above) can generate products and services that sustain indigenous and village populations otherwise engaged in forest felling, and serve as high-biomass carbon sinks. Natural forests comprise 83% of the tropical forest under intensive management, on 35 million ha; only 7.07 million ha of intensively managed tropical forests have been planted or artificially regenerated (IIED and WRI, 1987). Hence, NFM, while requiring considerable investments of labor, may present the most viable long-run option for forestry in the vast majority of tropical forest lands.

Multiple-use management employs timber harvesting, replanting, stand improvement (release cutting), and forest protection to confer benefits from timber sales, recreation, and flood control. Fully 16% of tropical moist forest species have non-timber economic benefits, according to one recent survey by IUCN (IIED and WRI, 1987). Minor forest products like rattan, latex, resins, medicinal plants, and bamboo contributed $150 million to Indonesia's economy in 1982 (Repetto, 1988).

Option 3: Increase Forest Productivity: Plantation Forests

Plantation biomass productivity can be improved by three types of actions:

1. silvicultural practices that yield biomass gains, especially in industrialized country forests;
2. lengthened and stabilized land tenure for commercial and community forestry projects in developing countries, to encourage forest management for multiple (rather than single) rotations and the ensuing environmental benefits; and
3. biotechnology advancements utilizing genetics and seed selection.

This discussion focuses primarily on biotechnology and genetic potential, as silvicultural management is addressed below, and land tenure considerations reside in the realm of policy.

Plantations managed to grow a mix of short, medium, and long-rotation species, if site conditions allow, are most likely to provide the continuous stream of forest products and income necessary to meet timber and fuelwood demand in developing country villages. Similarly, mixed-stand (multiple species) plantations in temperate zones may reduce ecological problems (e.g., pest infestations) and timber market disruptions, although mixed-stand plantations may require more intensive management and harvest techniques.

Applied genetics or tree improvement may produce greater increases in yields of biomass than improved silvicultural methods. Short-rotation intensive culture (SRIC) efforts apply intensive agronomic practices to growing selected and/or genetically improved hardwood species in plantations, to achieve maximal productivity rates at competitive costs. The Department of Energy's Short-Rotation Woody Crops Program, begun in 1977, has conducted field trials to boost productivity and reduce costs of woody species managed under SRIC as an energy source. Its target is to achieve average productivity of 20 dry t/ha/yr (10 t C/ha) of biomass at a cost of $2.25/GJ on optimal plots by 1995, and a competitive technology 5 years later.

By 1987, productivity reached 13 dry t/ha/yr (6.5 t C/ha), at a cost of $55/dry t, or $3.25/GJ (ranging from $2.90 to $5.10 delivered), on SCS site class I-III soils (i.e, largely fertile and flat). Planting densities ranged from 2500-4000 trees/ha, on coppice rotations of 5-8 years. Ongoing research is focused on 4 species (silver maple, sweetgum, American sycamore, and black locust) and one genus (*Populus*, including cottonwood, poplar, and aspen). Collectively, average SRIC growth rates have reached about 9.5 dry t/ha, or 4.7 t C/ha. Rates as high as 12-14 t C/ha have been documented for exotic or hybrid trees (Ranney et. al., 1987).

Other species with high potential for plantation biomass production include wood grass (*Populus*), grown at densities of about 1700 trees/ha for 4-year rotations, and 25,000/ha for 2-year rotations (Shen, 1988). Kenaf, an African annual crop closely related to hibiscus, grows to 6 meters in 120-150 days after seeding, and can produce 3-5 times more pulp for paper than trees, on an annual basis, potentially freeing forests for biomass production and carbon storage. Field trials in Texas by the Department of Agriculture have found that kenaf grows well without pesticides in the

Cotton Belt and with irrigation in the drier Southern states, and requires less chemical input than wood to produce and whiten pulp. Ninety percent of its original weight can be converted to usable fiber (Brody, 1988).

Working with tropical species in Espirito Santo, Brazil, Aracruz pulp company has produced eucalyptus hybrids with 30% increased height and 80% improved diameter at breast height (dbh) over parents at 4 years, through selection of parent tree seeds, breeding for desired characteristics, and planting into specific microsites. Average yields of 70 m^3/ha/yr (18.2 t C/ha/yr) for 14 million trees per year grown from rooted cuttings of 54 species have been achieved, and yields of 100 m^3 (26 t C) or greater are projected. Stands are managed for bleached pulp and charcoal for steel making. The trees reach 20 meters in height in less than three years (OTA, 1984; IIED and WRI, 1987). Other trials with eucalyptus from 32 sites in 18 countries have generated productivity gains of several hundred percent simply by matching optimal seed characteristics with site soil and microenvironmental conditions (Palmberg, 1981).

Pine breeding for straightness, reduced forking, and drought resistance, coordinated by seed cooperatives in Latin American and Europe, are showing significant improvements. Other tree-improvement methods include seed orchards relying on grafting and rooted cuttings that produce clones of trees, but these methods are limited to only a few species. Tissue culture (manipulating sprouts from germinating seeds) is still in its infancy in forestry, although the Weyerhaeuser Corporation in the U.S. Northwest plans soon to produce 100,000 clones of Douglas fir per year. Research on producing cell culture embryoids (microscopic infant trees) is underway in major developed countries on temperate species, and early work on teak and Caribbean pine is encouraging (IIED and WRI, 1987).

Table 7-18 surveys productivity increases from intensive management and applied genetics for selected species in the U.S. and in the tropics.

TABLE 7-18
Productivity Increases Attributable to Intensive Plantation Management

Management Technique	Maximum Mean Annual Yield (Tons Carbon/ha/yr)
Douglas Fir in Washington	
Natural stands	2.8
Silvicultural treatments	
Plantation establishment	3.6
Nitrogen Fertilization	4.4
First-generation genetics	4.9
TARGET	12.5
Loblolly Pine in North Carolina Pocosins	
Natural stands	1.8
Silvicultural treatments	
Drain and plant	3.5
Bedding	4.3
Preplant phosphorus	5.3
Nitrogen fertilization	5.9
First- and second-generation genetics	7.2
TARGET	15.0
SRIC Hardwoods, Various Sites in U.S.	
Short rotation, genetics, site preparation, fertilizer, coppicing	6.5
TARGET for year 1996	10.0
Energy Crop Plantations, Temperate Zone	
Intensive management, mixed species;	
TARGET for year 2025	24.7
Eucalyptus Hybrids in Espirito Santo, Brazil	
Seed selection, breeding, microsite planting	18.2
TARGET	26.0

Source: Based on Farnum et al., 1983, and Marland, 1988 (Douglas fir and loblolly pine); Ranney et al., 1987 (SRIC hardwoods); Walter, 1988 (energy crop plantations); WRI et al., 1988 (Eucalyptus).

The implications for carbon fixing are clear; a 30% rise in biomass production allows 30% less land in forest -- or a 30% offtake of fuelwood or timber, either managed or poached -- to achieve the same results. Drought-resistant strains would encourage forest-sector and donor community investment in arid and degraded land projects.

The potential benefits of productivity increases on plantation forests currently are limited by the slow pace of research and field trials on promising species and varying site conditions. Eighty-seven percent of plantation forestry in the tropics focuses on species of only 3 types: pine, teak, and eucalyptus (Vietmeyer, 1986). Obstacles to expanding plantations are reviewed below.

Option 4: Improve Forest Harvesting Efficiency

Commercial forest management, especially in tropical forests with extremely high species diversity per hectare, has targeted harvesting on only about 5% of species. Reasons for this high-grading -- selective cutting of high-value trees -- include tradition, lack of demonstrated uses and educated markets for other species, and the availability of virgin stands open to resource "mining" without costly management and with government support. As a result, fully 85% of total wood produced from tropical natural forests in 1970 went unused, left as slash or wasted at the mill (Goldemberg et al., 1987).

A survey of Malaysian and Indonesian logged forests recently found that 45-75% of standing trees had been injured (Gillis, in press). In southeast Amazonia, remote sensing study has revealed that while 90,000 km^2 had been clearcut by 1985, three times that area (266,000 km^2) had been seriously damaged by logging and colonization (WRI et al., 1988; Malingreau and Tucker, 1988). Improvements in forest management would reduce waste of non-target species damaged during logging (e.g., the Celos Silvicultural System, discussed under option 1 above).

Efficient harvesting would require less virgin and mature secondary forest to be cut (Mergen and Vincent, 1987). Malaysian government policy has raised the number of commercial tree species for harvest from 100 in the mid-1960s to over 600 today (IIED and WRI, 1987). The harvest and marketing of under-used species and size classes throughout the tropics, encouraged by government regulations and forestry company practices, could reduce tree losses from harvest and improve stand yields more than silvicultural innovations could, especially in secondary forests (OTA, 1984).

Option 5: Expand Current Tree Planting Programs in the Temperate Zone

Forested area in temperate and boreal zones is considered roughly constant in most studies but may be declining in this decade (Houghton, 1988a). However, temperate forests were much larger historically and could be expanded. Some European countries have increased their net forested area: France was 14% forest in 1789; today, 25% of France is forested (Brown et al., 1988a).

According to the Forest Service report for FY 1987, total U.S. tree planting by Federal and State agencies revegetated 1.2 million hectares (3 million acres) in 1987 (USFS, 1987). Table 7-19 lists the five major tree-planting programs in the U.S. since 1935 and the number of acres planted in their highest 5-year period.

If current programs planted 1.2 million ha with existing financial and programmatic incentives geared to replacement levels of planting, then additional enticements, on the order of $220-345/ha ($90-140/ac), would probably stimulate tree planting on hundreds of thousands of hectares. The Conservation Reserve Program of USDA (see Option 2, below) paid an average of $219/ha (average rental payment of $125/ha ($50/acre) plus half of establishment costs at an average of $94/ha), to plant 648,000 ha of trees (1.6×10^6 ac) from 1986 to mid-1988. Youth groups could be mobilized to plant trees annually on Arbor Day or during weekend or summer work camps.

TABLE 7-19
Summary of Major Tree Planting Programs in the U.S.

Program	Period (Highest 5 years)	Acreage Planted ($x 10^6$) (acres)	(hectares)
Civilian Conservation Crops (CCC)	1935-39	1.4	0.6
Soil Bank	1957-61	2.0	0.8
USDA-Forest Service Reforestation	1979-83	1.5	0.6
Forestry Incentive Program (FIP), Agriculture Incentive (ACP)	1978-82	1.1	0.4
Conservation Reserve Program (USDA)	1986-90 projected	5.6	2.3
	1986-88 actual	1.6	0.6

Sources: Conrad, 1986; USDA, Land Retirement and Water Quality Branch statistics on Conservation Reserve Program enrollment, November, 1988.

Large-scale reforestation by individuals, companies, and/or government programs has been proposed as a possibility in temperate zones (see Table 7-21). However, this would be far less cost effective (by perhaps a factor of 3-10) than in the tropics because of higher land costs and slower tree growth. Recent, more targeted proposals for tree planting in the U.S. focus on activation of croplands considered surplus during periods of diminished exports and high costs for farm support programs.

Option 6: Reforest Surplus Agricultural Lands

The Conservation Reserve Program (CRP) administered by USDA has been proposed as the quickest, most cost-effective way to stimulate tree planting at the scale necessary to partially offset CO_2 emissions. Planting tree carbon sinks may be comparatively cheaper than other current CO_2-limiting options, for example, planting short-rotation biomass energy plantations, investing in energy conservation measures, or scrubbing CO_2 from industrial emissions (Dudek, 1988a,b).

The CRP was established by the conservation title of the Food Security Act of 1985 to retire highly erodible cropland, reduce production and boost prices of surplus food commodities, and reduce Treasury outlays. Participating landowners contract to retire cropland for 10 years. They are reimbursed for 50% of the costs of planting the necessary vegetative cover, and receive an annual rental payment.

Participation in the tree-planting program was targeted at 12.5% of projected total retired acreage (16.2-18.2 million ha, or 40-45 million acres by 1990), or about 2.3 million ha (5.6 million acres). However, tree-planting enrollment totals only 0.40 million ha (of 10.5 million ha enrolled in CRP by November 1988), mostly in Southern states already producing plantations of loblolly pine in favorable soils and climate. Planting rates have been low due to low bid prices, farmers' reluctance to lose base acreage in Federal crop support programs, and inadequate support for tree planting by extension offices. Better financial incentives (i.e., higher bid prices, higher share of planting costs paid) may greatly increase tree acreage enrollment.

For example, all new CO_2 emissions from fossil-fuel electricity plants projected for 1987-96 could be offset by planting trees (see Table 7-20). Estimates of planned increases in fossil-fuel electric generating capacity for 1987-96 total 25,223 MW, producing 45.5 Tg C (equivalent to 166.7 million tons CO_2) (NAERC, 1987).

The uptake of CO_2 varies by species. Silver maple, for example, a relatively inefficient species, absorbs 5 t C/ha/yr under optimal conditions, while American sycamore absorbs as high as 7.5 t C/ha/yr (Steinbeck and Brown, 1976; Marland, 1988). Depending upon species chosen, it would take between 4.5-13 million ha of short-rotation (4-5 years) monocultural (single-species) plantations on good sites to offset the planned 25,223 MW of additional fossil-fuel electricity.

Mixed stands of numerous tree species are more desirable to prevent the ecological problems associated with monocultural stands -- increased pest populations, low species diversity, vulnerability of even-aged stands. The acreage requirements for mixed stands to offset 45.5 Tg C could rise to 13 million ha (32 million acres), or about 70% of the total enrollment target for the 10-year CRP program. Further analysis is necessary to take into account availability of productive soils, variations in actual site and mixed-stand productivity rates.

USDA's share of costs for establishing trees on CRP acres averages $94 per hectare, plus rental payments averaging $125 per ha per year for the 10 years of the current CRP. If the full costs of the planting are assumed to be $370/ha, and fertilizer costs are $62/ha, then the mid-range estimate of 9.1 million ha of trees added to the CRP would cost $3.9 billion to establish ($3.1 billion if current average USDA cost-share expenditures continued) (see Table 7-20). Rental payments by USDA or utilities, estimated to rise to about $250/ha/yr, would need to continue over the 50-year life of the electricity plants whose emissions would be offset.

TABLE 7-20
Estimates of CRP Program Acreage Necessary to Offset CO_2 Production from New Fossil Fuel-Fired Electric Plants, 1987-96, by Tree Species or Forest Type

Tree Species/Forest Type	Carbon Fixing Rate Used (t C/ha/yr)	Land Requirements for Offset (x 10^6) Hectares	Acres
Average growth US commercial forests 1977 (USFS, 1982)	0.82	55.5	137.1
Average growth US commercial forests fully stocked (USFS, 1982)	1.35	33.7	83.2
Estimate for large mixed stands of moderate-growth species	3.5	13.0	32.1
Silver maple (Ranney et al., 1987)	5.0	9.1	22.5
SRIC program average productivity by 1987 (Ranney et al., 1987)	6.5	7.0	17.3
American sycamore (Marland, 1988)	7.5	6.1	15.1
SRIC program target for 1996 (Ranney et al., 1987)	10.0	4.5	11.1
SRIC program, highest documented from exotics (Ranney et. al., 1987)	13.0	3.5	8.6
Loblolly pine, target after genetics (Farnum et al., 1983)	15.0	3.0	7.4

Notes:
1. New fossil fuel-fired electric utility plant emissions are assumed to be 166.7 million tons CO_2 total for period 1987-96 (or 45.5 Tg C) (Dudek, 1988a; NAERC, 1987). 1 metric ton CO_2 = 0.27 ton carbon.

2. 1 ton biomass = 0.50 ton carbon. 1 hectare = 2.47 acres. Units are expressed as metric (not English) tons/acre for comparison.

3. Rotations for all estimates are assumed to be 4-5 years.

4. The carbon fixation rates for silver maple and American sycamore differ from those cited in Dudek (1988a), following personal communication with Dudek.

Some lands not presently producing timber or crops might be available for afforestation. In 1976, the U.S. had 70 x 10^6 ha (173 million acres) of land currently not productive -- although used for recreation or other purposes -- with rainfall greater to 50 cm/yr that may offer suitable (although not optimal) habitat for tree planting (Fraser et al., 1976). Any additional lands required would have to be diverted to forestry from other competing land uses.

The stream of direct and indirect benefits that would accrue from afforestation, including timber harvest, reduced soil erosion and nonpoint source water pollution, and increases in recreation use and wildlife, has not been considered yet, and would tend to reduce costs attributed to climate change programs if utilities or farmers managed forested land for multiple uses. Quantification of these potential benefits is needed.

CO_2 uptake rates might be increased through biotechnology improvements to perhaps 10 t C/ha/yr within 10-15 years (Ranney et al., 1987). Higher uptake rates would reduce both offset acreage requirements and costs. If economic hardwood species like American sycamore (used in flooring and millwork) and silver maple (for furniture and box and crate production) or softwoods (for lumber) are harvested for durable products, carbon storage continues after harvest.

Option 7: Reforest Urban Areas

Urban areas, which currently contain 75% of the U.S. population, on 28 million ha, are increasing by 0.53 million ha (1.3 million ac) per year (USDA, 1982). A study of urban forests in 20 cities found that for every four trees removed, only one tree is planted, and for a third of the cities, only one in eight trees lost is replanted (Moll, 1987). The total number of trees in Chicago dropped by 43,853 between 1979 and 1986; trees planted per year declined from 24,675 in 1979 to 9,380 in 1980 (NASF/USFS, 1988; Open Lands Project, 1987).

An urban tree is about 15 times more valuable than a forest tree in terms of reducing CO_2 emissions. Trees break up urban "heat islands" by providing shade, which reduces cooling loads (air conditioning) in warm weather by reducing building's heat gain, and cuts heating loads in cool weather by slowing evaporative cooling and increasing wind shielding. Three strategically placed trees per house can cut home air conditioning energy needs by 10-50% (Akbari et al., 1988). Trees planted as windbreaks around houses and buildings also can reduce winter heating energy use by 10-50% (Robinette, 1977). Thus urban trees both sequester CO_2 and reduce consumption of fossil fuels, making strategic planting around buildings a small but efficient response option (Meier and Friesen, 1987). The American Forestry Association (AFA) recently conducted a survey of urban forest needs, and launched Project ReLeaf to plant 100 million trees in city streets, parks, and rural areas. AFA estimates savings of 40 billion kilowatt-hours of energy from these new trees (based on Akbari et al., 1988), which would provide a carbon cycle benefit equivalent to 4.9 Tg C annually. This benefit would accrue from a combination of absorption of CO_2 and reduced emissions from electricity generation (Sampson, 1988).

Option 8: Afforestation for Highway Corridors

Highway corridors offer significant opportunities for tree planting, along 6.2 million kilometers (km) (3.9 million miles) of roads in the U.S. In 1985, 11.9 million ha (29.5 million acres) of land (totalling 1.3% of the contiguous U.S.) were in use as highways, including right-of-ways and buffer strips -- 9.9 million ha in rural areas, and 2.0 million in urban areas (calculations based on average municipal right-of-way as 50 feet, reported in U.S. EPA, 1987; data from U.S. DOT, 1985). The North-Central states have 4.2 million ha (10.3 million acres) in roads, the South has 3.9 million ha, and the Northeast, 1.1 million ha, all regions with generally good site characteristics for tree planting. If, for example, an additional 10% of the 9.2 million acres of interstate, state, and local highway corridors in these regions were planted with trees, 0.9 million ha would be available -- about 10% of the roughly 9 million ha in trees necessary to offset new CO_2 emissions from power plants from 1987-96 (see above). At average costs of establishment and fertilization of $432/ha, total cost would approach \$390 million. If 20% of highway corridors were planted, 1.8 million ha (4.52 million acres) would be produced, at a cost of \$777 million.

Option 9: Reforest Tropical Countries

Numerous estimates have been made of tree planting desirable for economic, social, and environmental reasons unrelated to climate warming. One authority concludes that "at least 100 million hectares of tree planting worldwide appears necessary to restore and maintain the productivity of soil and water resources" (an area equivalent to the size of Egypt) (Brown et al., 1988a). Major tree-planting programs are being promoted in many parts of the world, partly in response to a 1985 international initiative, the Tropical Forestry Action Plan, jointly sponsored by the Food and Agriculture Organization (FAO), the United Nations Development Programme (UNDP), The World Bank, and the World Resources Institute (WRI) (see Chapter 9). In response to the Plan, global funding for forestry by multilateral development banks and bilateral agencies is expected to rise from \$600 million in 1984 to \$1 billion in 1988. In a parallel effort, China doubled tree planting to 8 million hectares by 1985, and planted 3.3 million ha of seedlings in 1986 alone (Houghton, 1988a, quoting FAO data), although the survival rates initially hovered around 30%. China has set a goal of 20% forest cover by the year 2000 (up from 12.7% forest in 1978) (Brown et al., 1988a).

Carbon benefits achieved from afforestation are not likely to be realized solely to slow global warming. Instead, social forestry projects designed to integrate provision of human needs with economic incentives and environmental stabilization -- with carbon-reduction goals piggybacked on top -- provide the most feasible approach.

Forest plantation planting in the tropics to date has focused on establishing commercial hardwoods (722,000 ha/yr) (WRI et al., 1988), and on providing fuelwood (550,000 ha/yr) (Brown et al., 1988a). The gap between fuelwood supply and demand in the rapidly expanding developing countries could reach 1 billion m^3 by the year 2000 (Marland, 1988). World Bank tallies suggest that

55 million ha of high-yielding fuelwood plantations will need to be established by 2000, to close this projected deficit -- fully 2.7 million ha/yr, or 5 times current fuelwood rates. Thus, allowing for overlap, Postel and Heise (Brown et. al., 1988a) calculate that a total of about 110 million ha of planting is necessary both to restore degraded lands and to provide fuel requirements in 2000, which would sequester approximately 0.7 Pg C/yr for the roughly 40-year life of the forest.

Several crude estimates have explored the possibility of very large reforestation efforts in tropical regions to provide a sink for fossil fuel emissions. For example, Sedjo and Solomon (in press) have proposed that the current annual atmospheric net increase in carbon (approximately 2.9 Pg C) could be sequestered for about 30 years in approximately 465 million hectares of plantation forests, at a cost possibly as low as $186 billion in the tropics or $372 billion in the temperate zone, a large but not inconceivable sum. This area would also produce as much as 4.7 billion cubic meters of industrial wood annually, three times the current annual harvest. The opportunity cost to society to offset carbon production from global annual deforestation of 11.3×10^6 ha in the tropics can be calculated from Sedjo and Solomon's replacement cost figures, and equals $400 per hectare -- or $4.5 billion per year, a sum the world community should be willing to pay for forest preservation in order to avoid paying carbon offset replanting costs. Myers (1988; see also Booth, 1988) has suggested that 300×10^6 ha of plantation eucalyptus or pine -- a landmass the size of Zaire -- absorbing about 10 t C/ha/yr could offset the 2.9 Pg C accumulating in the atmosphere annually. Dyson and Marland (1976) and Marland (1988) suggested that 700×10^6 ha of land -- an area about the size of Australia, or equal to the total global forest cleared since agriculture began (Matthews, 1983) -- in short-rotation American sycamore, fixing carbon at a rate of 7.5 t/ha/yr, would be required to offset total global production of 5 Pg C per year.

All of these estimates suffer from their application of some of the highest growth rates observed to vast tracts of highly differentiated site conditions. Also, global reforestation estimates thus far have focused on the potential for using forest growth to completely offset total or net global carbon emissions. New estimates are needed that consider more feasible offset goals and growth rates. More complete analyses, based on better field estimates of carbon fixation rates for a range of mixed-species stands and agrosilvicultural systems, are underway to identify available acreage (with adequate soils) for potential planting programs in specific countries -- to generate inductive assessments of the potential of this approach, rather than the deductive approaches utilized thus far. Table 7-21 gives a summary of preliminary estimates of forest acreage required to offset global CO_2 emissions.

Is there adequate idle land to seriously entertain the notion of massive tree planting? Houghton (1988b) has roughly calculated for tropical Asia and Africa availability of tropical land climatically and edaphically suitable for forest growth (i.e., climate and soils previously supported forest). Crude measures of land formerly in forest but presently degraded or in pasture, and not in use for crops or development, suggest that about 100 x 10^6 ha are available for reforestation in South and Southeast Asia (excluding arid lands in India and Pakistan). For tropical Africa, ratios of land once forested and land currently in "other land" categories in FAO estimates (FAO, 1987) are less reliable, but create a range of 20-150 x 10^6 ha available. Human-initiated fires set to create and maintain savanna for cattle grazing and other uses hypothetically could be suppressed (although difficult to manage in practice), potentially allowing another 191 x 10^6 ha of savanna to revert to closed forest along the northern savanna and in western Africa, providing an upper limit of 340 x 10^6 ha with potential for reforestation (Houghton, 1988b; FAO/UNEP, 1981).

Reforesting Degraded Lands. Restoration of lands formerly forested but degraded by anthropogenic practices -- logging, overgrazing, swidden and inappropriate agricultural methods -- could increase reforestation and carbon fixation rates. Forest fallow in swidden agriculture, often with low volumes of standing biomass, totals 1 billion ha globally, with another 1 billion in some form of degradation. Desertification (the reduction of biological productivity, primarily from human activities, usually in dry forests or rangelands) has moderately or severely affected 1.98 billion ha globally, especially in the African Sahel, Southern Africa, Southern Asia, and China and Mongolia (WRI, 1987; OTA, 1984). Sedjo and Solomon (in press) stress the low purchase price of degraded lands, while

TABLE 7-21
Estimates of Forest Acreage Required to Offset Various CO_2 Emissions Goals

Estimate	Location	Carbon Sequester Rate Assumed (t C/ha/yr)	Offset Goal	Hectares ($x10^6$)	Total Planting Cost ($\$ \times 10^9$)	Average Cost/ha ($)
Dyson and Marland (1976)	Temperate Zone	7.5	5 Pg C (total annual fossil fuel use)	700		
Marland (1988)	Tropics	9.6	5 Pg C (total annual fossil fuel use)	500		
Myers (1988)[1]	Tropics	10.0	2.9 Pg (net annual increase C)	300	120	398
Sedjo and Solomon (in press)	Tropics or Temperate	6.2	2.9 Pg (net annual increase C)	465	186 tropics 372 temperate	395 800-1400[2]
Woodwell (1987)	Tropics	5.0	1-2 Pg (net annual increase C from tropics)	200-400		
Postel and Heise (Brown et al., 1988a)	Tropics	6.8	0.7 Pg (C benefits from new fuelwood plantations and restored forests)	110		
Dudek (1988a,b)	U.S.	10-12	0.05 Pg (1987-96 new electric plant C)	3.4-4.5	1.6-1.9[3]	432[3]
EPA (this study)	U.S.	3.5-10	0.05 Pg (1987-96 new electric plant C)	4.5-13	1.9-5.6[3]	432[3]

Notes:

1. See also Booth, 1988.
2. Includes $400/ha land purchase cost.
3. Assumes no cost-sharing of establishment costs, and no annual rental payments (see text).

recognizing that low-productivity soils on degraded sites are likely to greatly increase acreage requirements for a given offset goal. Grainger (1988) concluded that 758 million ha of degraded tropical land (including 203 million ha forested in the past) could be restocked with forest.

The substitution of sustainable resource use practices and active reforestation on these lands can reduce further loss of woody biomass and increase rates of carbon fixing and fuelwood production. The price of establishing plantations on degraded grasslands in Indonesia has hovered around $400 per ha (JICA, 1986). For example, Houghton (1988b) estimates that by replacing swidden cultivation with permanent, low-input agriculture, about 365 x 10^6 ha of fallow land would be available for reforestation during the period 1990-2015. Managed reforestation of arid lands has been successful in some sites where natural mulch or commercial fertilizers were applied to seedlings, and native species well-adapted to local pest and environmental conditions were planted -- supplemented by cautious use of fast-growth or nitrogen-fixing leguminous exotic species, like *Leucaena*, *Pinus*, *Acacia*, and *Eucalyptus*, which tend to be more susceptible to pest invasions. Research and field tests on leguminous trees inoculated with *Rhizobium* fungi, which produce nitrogen-fixing nodules on legume roots, have shown that damaged tropical soils depleted of essential micorrhizal fungi can be replanted effectively and inexpensively (less than 1 cent per tree) (Janos, 1980).

Subtropical dry forests in the Western Hemisphere have been reduced by 98%. Botanist Daniel Janzen has organized an ambitious plan to restore dry forest cut for agriculture and manage habitat fragments to expand Guanacaste National Park in Costa Rica (Jansen, 1988a,b). The plan calls for establishing a local and international environmental educational and research program; suppressing human fire activities, cattle ranching, and agricultural clearing; purchasing intact remnant dry forest habitat adjacent to moist forest tracts or protected areas to provide seed sources; developing a management plan stressing species diversity and zoning for habitat use, including provision of economic opportunities. A closed-canopy dry forest with significant representation of its previous fauna is expected to evolve within 10-50 years. Expansion of this innovative approach throughout the dry tropics may be feasible, if stable land tenure and managed use can be attained.

In Nyabisindu, Rwanda, hillsides denuded by intensive swidden agriculture featuring rapid deforestation, soil erosion, and overgrazing, are being restored to productivity by strips of densely planted trees positioned across steep slopes to catch soil and create terraces for crops, and to produce fruit and fuelwood. The restoration project stocks these hedges from a tree nursery producing 5 million seedlings per year for farm fruit trees, shade trees lining roads and hilltop woodlots. A typical farm family can produce 25-50% more fuelwood than it consumes with this mixed crop-tree system (Dover and Talbot, 1987).

Obstacles to Large-Scale Reforestation in Industrialized Countries

Economic and institutional obstacles to widespread reforestation center on the high costs of site preparation, planting, forest management, and necessary financial incentives to private landowners. However, 1.2 million ha of trees were planted in 1987 without strong incentives (USFS, 1986). U.S. state foresters maintain that financial incentives on the order of $125-250/ha ($50-100/ac) would be sufficient to bolster reforestation of harvested woodland and surplus croplands in most states. With about 3 years' lead time, existing tree nurseries could accelerate production of seedlings enough to plant 3-10 times current acreage per year (NASF, 1988).

Ecological drawbacks to massive reforestation schemes include the low levels of genetic variability that characterize vast tracts of monocultural stands, and their reduced resistance to infestations by pests (e.g., gypsy moth, pine bark beetles) that can lead to widespread forest decline and mortality. Large-scale plantations may strain surface and ground water resources in areas already experiencing overdrafts of and escalating demand on aquifers (e.g., the Ogallala aquifer in the southern Great Plains, and Southeast coastal plain ground water) (Los Alamos National Laboratory, 1987).

Air pollution -- acid precipitation, ozone and other photochemical oxidants -- is already affecting the health of forests in the U.S., Europe, and China. Decline in at least seven coniferous and 4 broadleaf important species in European forests, and at least 8 important species in North American forests have been documented since 1979; 52% of West German and 36% of Swiss forests were in decline by 1986. In the Southeast U.S., natural-stand diameter growth rates for yellow and loblolly pines have declined 30-50% in the past 30 years (WRI and IIED, 1986; IIED and WRI, 1987).

Air pollution impacts may be exacerbated by the combined effect of warming and increased ultra-violet radiation due to ozone depletion. The long-term process of climate change is likely to complicate the task of actively expanding net forest area, a topic explored in Smith and Tirpak (1989). The increase of carbon dioxide (through the effect of CO_2 fertilization) may spur tree growth rates, but the net result of climate change on forest growth under changing temperature and precipitation conditions is difficult to predict. Early studies suggest that Southern bottomlands hardwood, oak, and pine forests, and Northern conifer forests are likely to advance northward in a doubled CO_2 scenario (as temperatures increase). The natural rate of forest migration -- about 100 to perhaps 400 km per 1000 years -- appears unlikely to keep up with the likely rate of forest decline (Davis and Zabinski, in press; Shands and Hoffman, 1987; Smith and Tirpak, 1989). Technical methods of speeding seed dispersal to favorable soils (e.g., seeding from aircraft) are not complex but may be costly, with potentially low seedling survival rates (WWF, no date).

Forests under the stress of climate change may have difficulty maintaining current productivity rates, let alone increased rates and expanded geographic range, as envisioned under reforestation policies. Forests migrating north in response to rising temperatures will tend to encounter poorer soils, slow natural seed dispersal methods, stresses on ecosystems, competition from other land use sectors and ecosystems also responding to climate change, and either reduced or increased precipitation and water supply conditions, depending on the region. While global boreal forests are likely to increase in total area and tree density (Shugart and Urban, no date), they probably will contain, on average, lower standing biomass volumes (and carbon fixation rates) than current boreal forests. Sedjo and Solomon (in press) estimate a net loss of 24 Pg C storage in global forests under a 2xCO_2 scenario, largely due to declines in total biomass in boreal forests.

Climatic change therefore may affect the viability of reforestation strategies as a mitigation measure. Further research is needed.

Obstacles to Reforestation in Developing Countries

Tropical reforestation schemes face many obstacles. Pest management in monocultural plantations (e.g., rats, fungus, nematodes for *Leucaena* in the Philippines), and loss of genetic diversity that might allow adaptation to pest or viral infestations have proved major deterrents to large-scale plantation projects in the Philippines, at Jari, Brazil, and elsewhere. Other issues include rapid removal of soil nutrients by fast-growth species, reduced growth rates at altitudes as low as 450 meters and in frost belts for some species, and sheet erosion. Practical limitations of plantations include the large areas of land necessary, limited transportation infrastructure to move biomass to users, plantation security (fencing to prevent poaching of trees in northern Nigeria in 1976 cost $160-200 per ha), costs of establishment, and limited availability of skilled technicians (Villavicencio, 1983).

Pastures cut from tropical forests have been invaded by resilient monocultural grasses like *Imperata cylindrica* that must be suppressed before trees can survive. Research into recolonization of large, highly degraded pastures in the Amazon (Uhl, 1988) suggests that three factors inhibit reforestation: few seeds of forest species are being dispersed into pastures, because 85% of tree seeds are transported by animal pollinators that do not frequent pastures; most seeds dispersed to pastures are eaten by predators; and moisture and energy conditions near the ground surface in pastures are radically different from forest conditions.

Economic and population pressures in many regions make net afforestation difficult to achieve. China, for example, is experiencing rapid economic development that has led to housing construction that consumed 195 million cubic meters of wood (0.05 Pg C) from 1981-85, equivalent to the total annual growth of all China's forests (Brown et al., 1988a).

Other environmental stresses on extant forests -- including seasonal climatic variations, and the litany of stresses addressed for industrialized nations -- reduce their ability to meet current and projected demand for forest products, let alone newly supply large increases in productivity and plantation acreage. Persistent drought already plays a prominent role in the migration of environmental refugees from traditional agricultural areas such as the Sahel, forcing relocation in areas of marginal dry forest exposed to new pressures (Houghton, 1988a; Brown, 1988).

Summary of Forestry Technical Control Options

Slowing tropical deforestation, and rapidly expanding temperate and tropical reforestation, may offer two of the most cost-effective policy responses to increasing CO_2 emissions. However, only rudimentary estimates of the feasibility, costs, and consequences of large-scale reforestation have been performed. Table 7-22 summarizes these options. All calculations are preliminary estimates of planting costs only. Total costs of these measures are not estimated, and would vary greatly depending on land costs and management costs. These estimates are provided to give a general sense of the relative costs of these options. Few viable, global-scale plans to slow forest loss in the tropics have been advanced; most are policy options rather than technical fixes, and are discussed in chapters 8 and 9. Replacement of swidden agriculture with permanent low-input, sustainable agricultural systems offers particular promise.

Integrated natural resource management and social forestry projects -- designed to provide the full spectrum of forest and food products in demand; forest protection from swidden agriculture, logging, and fire; economic opportunities and reasonable rates of return; and cooperative management

TABLE 7-22
Comparison of Selected Forest Sector Control Options: Preliminary Estimates

	No. ha Required ($x10^6$)	Planting Cost/ha ($)	Total Planting Cost ($x10^6$ $)	Fixation Rate (t C/ha)	Carbon Offset Goal (Pg C/Sequestered)	Planting Cost/Pg Carbon ($x10^9$ $)
A. DOMESTIC U.S. OPTIONS						
Plant trees to offset annual CO_2 emissions US	240.0	432	104,000	5	1.2 Pg C/yr	10
Plant 100 million urban trees	0.0		100	5	0.052 Pg C/yr	0.2
Reforest 20% of highway corridors	1.8	432	778	5	0.009 Pg C/yr	10
Plant trees to offset new US electric plant CO_2 (1987-96)	9.1	432	3,930	5	0.045 Pg C/yr	10
Implement current forest economic opportunities at 4% ROR	58.3		0		0.164 Pg C/yr	0
B. INTERNATIONAL FORESTRY OPTIONS (TROPICS)						
Plant trees to offset new US electric plant CO_2 (1987-96) in LDCs	6.0	400	2,400	7.5	0.045 Pg C/yr	6
Plant trees in Guatemala to offset CO_2 emissions of 180 MW electric plant in US	0.1		14[1]		0.16 Pg C total over 40-year period	0.9

TABLE 7-22 (continued)
Comparison of Selected Forest Sector Control Options: Preliminary Estimates

	No. ha Required ($x10^6$)	Planting Cost/ha ($)	Total Planting Cost ($x10^6$ $)	Fixation Rate (t C/ha)	Carbon Offset Goal (Pg C/Sequestered)	Planting Cost/Pg Carbon ($x10^9$ $)
Substitute sustainable agriculture for all swidden agriculture	41					
Restore degraded lands and provide fuelwood for year 2000 and reduce tropical C release by 40%	110	400	44,000	7.5	0.7 Pg C/yr over 40 yrs.	6
Reforest tropical lands to offset global annual net increase in C	465	400	186,000	6.2	2.9 Pg C/yr	8
Replant tropical forest cut or burned each year	11.3/yr	400	4500/yr		1.0 Pg C/yr	4.5

1. This number is a total project cost including forest protection, plantation establishment, and agroforestry (see Table 7-4.11).

Note: All calculations made from estimates and references listed in this section; and all calculations are preliminary estimates of planting costs only, without management, harvest, rental or purchase costs included.

by local people and resource professionals -- are likely to be the most successful in addressing climate change effects and offset goals. Examples of two such projects are displayed in Table 7-23. These two projects were proposed to offset the lifetime CO_2 emissions of a 180 MW coal-fired electric plant planned by Applied Energy Services for Uncasville, Connecticut, by planting and managing forests in the tropics.

Our RCWP scenario envisions a "high reforestation" case (Houghton, 1988a,b), in which deforestation is gradually reduced to zero by 2025. This decline is projected through three major assumptions. First, 85% of shifting agriculture is replaced with sedentary cultivation, freeing 14.6 x 10^6 ha of fallow lands for reforestation each year, totalling 365 x 10^6 ha from 1990-2015. Second, 10 x 106 ha of pastures in Latin America are abandoned each year from 1990-2000, providing 100 million ha for reforestation; 5 million ha of grasslands are abandoned and reforested in Asia each year from 1990-2010, producing 100 x 10^6 ha; and 5 million ha of degraded land or savanna in Africa is reforested each year from 1990-2050, freeing another 300 million ha. Third, establishment of tree plantations (on 40-year rotations) grows linearly from about 1 x 10^6 ha in 1980 to about 5 x 10^6 ha in 2100. Over the 120-year period, 320 million ha are planted.

The total accumulation of carbon in terrestrial ecosystems from these three assumptions rises to about 80 Pg (assuming low biomass volumes per ha) to 183 Pg (using high-biomass assumptions). Peak carbon <u>sequestering</u> rates reach between 0.7 and 3.5 Pg/yr (low and high biomass assumptions), reversing the current <u>emission</u> of 0.4-2.6 Pg/yr from land use changes in the tropics.

Obstacles to slowing deforestation and to planting and maintaining reforested lands in the tropics include degraded soils, limited research on appropriate species and systems, limitations in institutional capabilities, perverse government incentives, and soaring population growth rates.

TABLE 7-23
Overview of Two Social Forestry Projects Proposed to Offset CO_2 Emissions of a 180-MW Electric Plant in Connecticut

Forest Attribute	CARE/WRI/Guatemala	WWF/Costa Rica
Total area of project	101,000 ha	122,000 ha
Protected in forest reserves	19,740 ha	72,000 ha
Logged or managed forests	38,000 ha	
Newly established woodlots (plantations)	13,140 ha	12,000 ha
Agroforestry lands	68,350 ha	
Carbon sequestered over 40-year life of plant	15.8 Tg C	21.8 Tg C
Cost estimate (cash and in-kind)	$14 million	$9.6 million

Note: Offset goal = 0.39 Tg C/yr.

Sources: WRI, 1988b, and Trexler, in press (Guatemala); WWF, 1988 (Costa Rica).

PART FIVE: AGRICULTURE

Agriculture contributes to the emission of greenhouse gases through three primary means: rice cultivation, nitrogenous fertilizer use, and enteric fermentation in domestic animals. Estimates place the annual contribution of rice cultivation and domestic animals at approximately 20 and 15%, respectively, of global methane production. The use of nitrogenous fertilizers is estimated to account for between 1 and 17% of the current global source of N_2O (See Chapter II - Greenhouse Gas Trends). Figure 7-22 illustrates the net effect of these agricultural sources on current greenhouse warming.

Both the magnitude of agricultural source emissions themselves and the potential effectiveness and costs of possible reduction measures are very uncertain. While considerable research has been done on the agricultural activities of interest, relatively little attention has been focused on agriculture-related emissions of greenhouse gases and how various changes in agricultural practices affect these emissions.

In the No Response scenarios (SCW and RCW), as shown in Figure 7-23, emissions from all three categories of agricultural practices are expected to increase over 1985 levels: Global CH_4 emissions from rice and enteric fermentation increase about 35% and 65%, respectively, by 2025; and N_2O emissions from fertilizer use are projected to increase 133% by 2025. By 2100, emissions from rice, enteric fermentation, and nitrogenous fertilizer increase by approximately 40%, 125%, and 175%, respectively.

In the Stabilizing Policy cases (SCWP, RCWP) it is assumed that growth in emissions from all three categories can be reduced somewhat, as shown in Figure 7-23: By 2100, methane emissions from rice are reduced about 20% from 1985 levels. Methane emissions from enteric fermentation

FIGURE 7-22

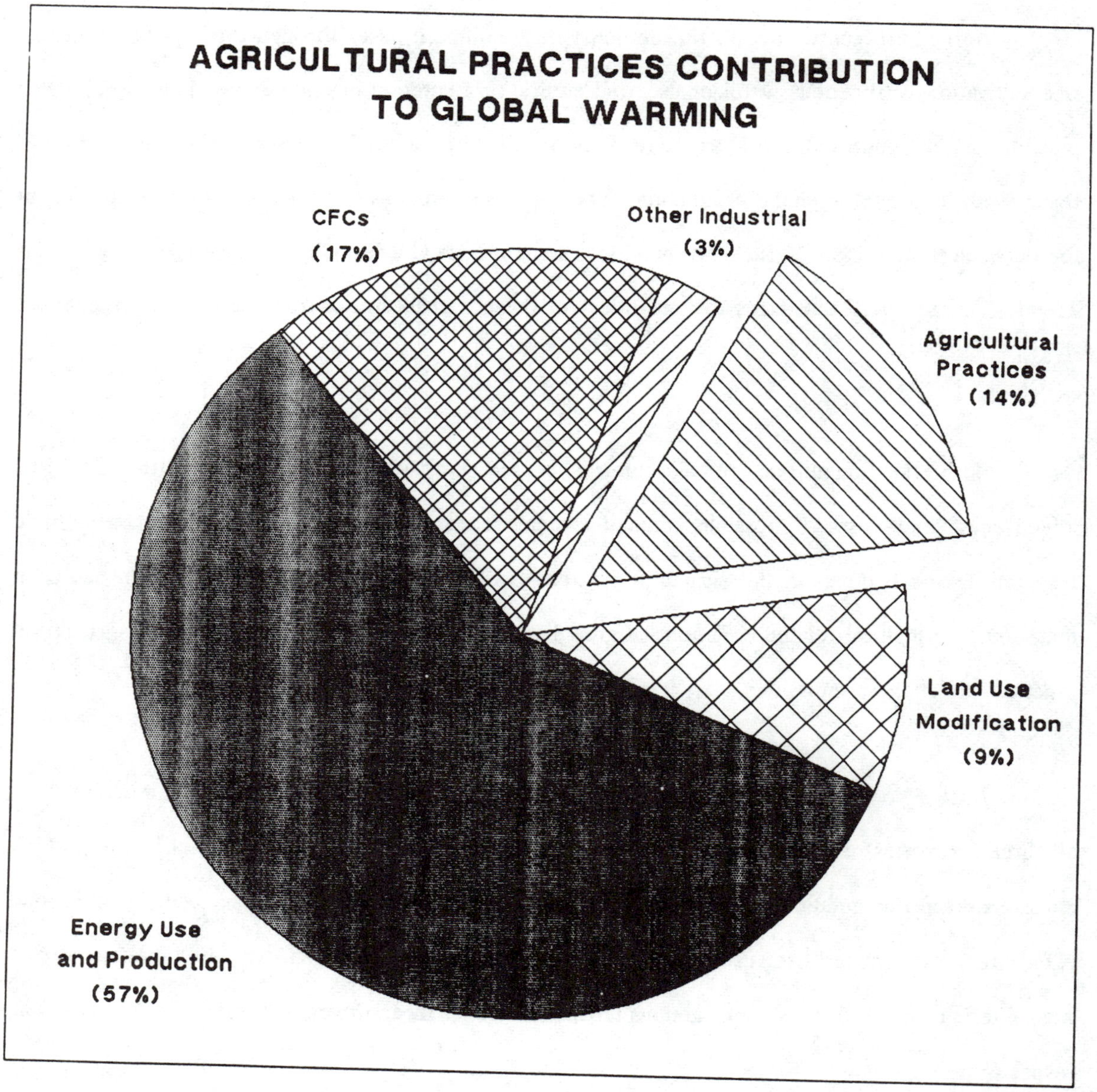
AGRICULTURAL PRACTICES CONTRIBUTION
TO GLOBAL WARMING
CFCs
(17%)
Other Industrial
(3%)
Agricultural
Practices
(14%)
Land Use
Modification
(9%)
Energy Use
and Production
(57%)

FIGURE 7-23

TRACE GAS EMISSIONS FROM AGRICULTURAL ACTIVITIES

(Teragrams CH4 and N2O)

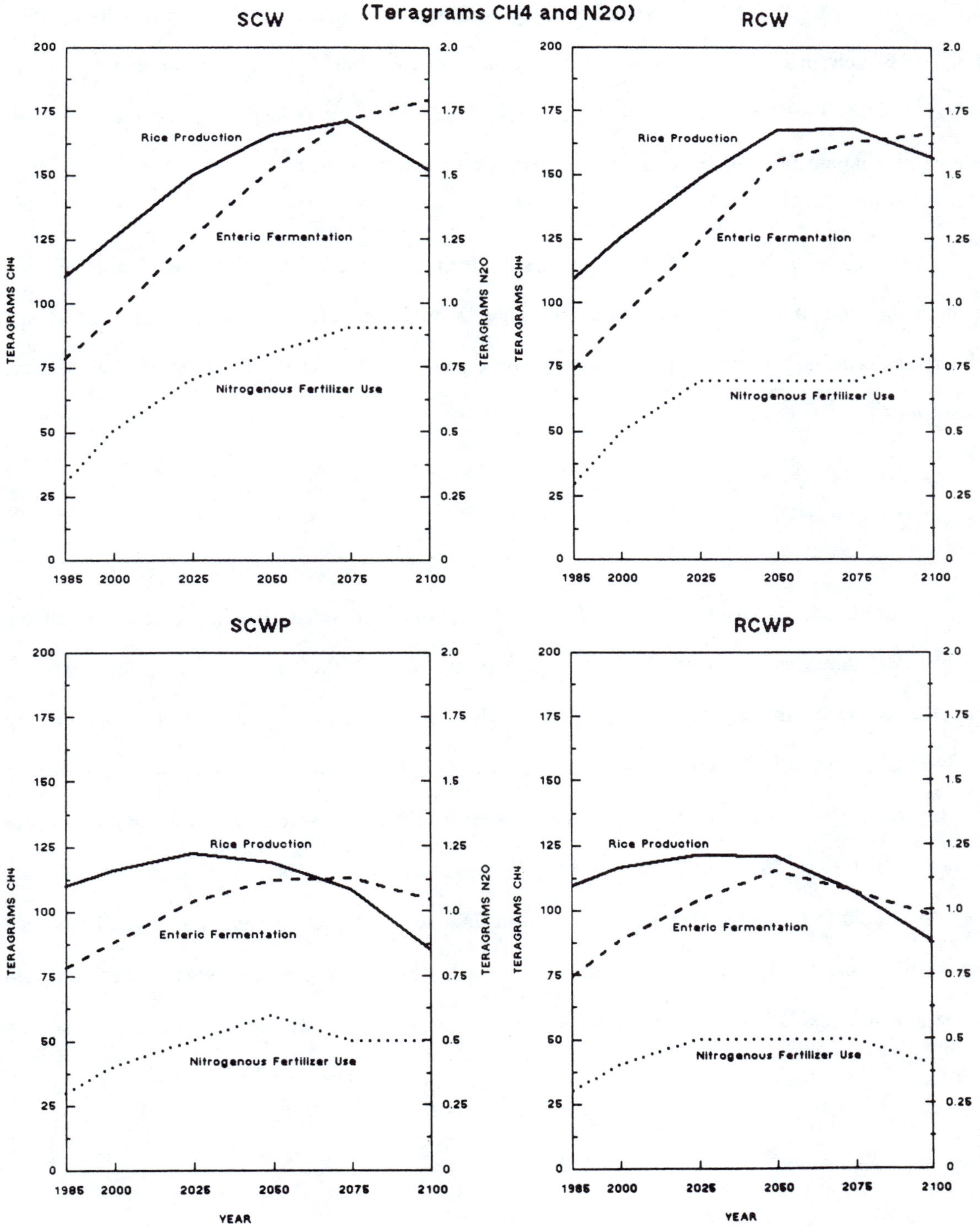

fermentation and N_2O emissions from fertilizer usage increase by 2100 in the policy cases, but at reduced rates than in the non-policy cases. Methane from enteric fermentation increases by about 32% between 1985 and 2100 in the policy cases, and N_2O from fertilizer use increases by 35-65%. It appears likely that reductions in these ranges are possible, but it is not possible at this time to precisely define the technical measures necessary to achieve these reductions, their costs, or their other (non-climate related) benefits of these emission-reducing policies.

For each of the three major trace-gas-producing agricultural practices, we discuss existing technologies and management practices, emerging technologies, and areas that require additional research. Technical options for reducing greenhouse gas emissions are discussed under each of these subsections.

RICE CULTIVATION

Methane is produced by anaerobic decomposition in flooded rice fields. Some methane reaches the atmosphere through ebullition (bubbling up through the water column), but most of it (about 95%) passes through the rice plants themselves, which act as conduits. Methane production is affected by the particular growth phase of the rice plant, temperature, irrigation practice, fertilizer usage, presence of organic matter, rice species under cultivation, and number and duration of rice harvests (Fung et al., 1988). A limited number of measurements have been performed in California, Italy, and Spain to evaluate methane production in flooded rice fields. No measurements have been published for the major rice producing areas of Asia however, where environmental conditions and cultivation practices differ significantly from those in these more temperate regions (Cicerone and Shetter, 1981).

Rice cultivation is estimated to contribute between 60 and 170 Tg methane per year currently (Fung et al., 1988), and with rice-harvested area increasing between 0.5 and 1.0% per year (IRRI, 1986), rice cultivation will continue to be a significant source of methane emissions.

Rice cultivation practices vary widely. With over 60,000 varieties of rice, there is great variation in water requirements, fertilizer response, pest and disease resistance, growing season, plant height, and yield potential. Traditional rice varieties are generally tall plants with a low grain-to-straw ratio that have a good resistance to endemic weeds and pests and a high tolerance for moisture, including flooded monsoon conditions. Cultivation of traditional varieties of rice does not involve heavy fertilization because these plants tend to lodge (fall over) at high levels of fertilization. The highest most stable rice yields are achieved with irrigation.

The research of the 1960s, which led to the Green Revolution, produced a high-yielding rice cultivar that is short, stiff-stemmed, and very fertilizer responsive. These modern, high-yielding varieties also have a short growing time, which allows for multiple plantings during the year. The first of these modern varieties, IR8, established a maximum yield potential of 10 mt/ha under ideal conditions, and reduced the growing time by a month to 130 days (Barker et al., 1985.)

Existing Technologies and Management Practices

No technology currently available can inhibit the production of methane in rice paddies, but cultivation practices and plant variety affect the amount of methane produced.

Rice Production System. The nature of the rice production system has a substantial effect on the amount of methane produced. Wet, paddy rice produces methane, while dry, upland rice does not. Wetland rice comprises about 87% of rice area worldwide. Of global rice area, about

53% is irrigated, 22.6% is shallow rainfed, 8.2% is deepwater, and 3.4% is tidal wetland. The remaining 13% is dry upland rice (Dalrymple, 1986). The majority of upland rice area is in Africa and Latin America. Rice paddies are inundated to varying depths and for different lengths of time depending on production system used, which affects the rate of CH_4 generation. How these rates differ has yet to be quantified.

Organic Matter. The kind and volume of organic matter added to the rice paddy has been shown to affect methane production. Laboratory experiments have shown that adding organic matter leads to an early peak in methane production and an overall increase equivalent to 2-5% of the added organic matter (Delwiche, 1988).

Organic fertilizers are often used in rice cultivation in Asia. Sources include animal manures, composted garbage, night soil (human feces), and plant residues. After harvest, rice plant residue is often incorporated into the soil as a source of organic material--a practice that appears to increase methane emissions. In addition, whether and to what extent the introduction of manure from domestic animals during plowing and harrowing affects methane production, are issues that must be examined.

Crop Residues. Crop residues can be burnt, buried, incorporated into the rice paddy, or used for some other activity. Burning the residue releases carbon dioxide. Buried residue partially decomposes and produces methane, but less than the amount produced by incorporating the residue into the paddy.

In southern India and Sri Lanka, there is a preference for intermediate height varieties of rice, which produce more straw for fodder and fuel (Barker et al., 1985). Finding additional uses for rice straw, such as for fiber or building materials, is necessary in order to reduce the incentive for plowing crop residue back into the field.

High-Yield Varieties. The shift to high-yield varieties of rice has helped to reduce the amount of methane produced per unit of rice. Modern, short-stemmed varieties have a grain-to-straw ratio that is about 50% higher than traditional varieties, which means less organic material (straw) is left to decompose (assuming it is not burned). In addition, the shorter growing season of high-yield varieties results in a reduction in methane emissions. However, the shorter growing season also allows multiple plantings during the course of a year. Overall methane emissions could increase as a result of multiple-planting, particularly if more organic material is incorporated into the paddies to support the greater number of plantings

High-yield varieties of rice are largely limited to irrigated areas where yield response is the highest. These modern varieties show the best fertilizer response and overall yield in irrigated areas during the dry season, when water levels are best controlled and solar energy is at its peak. The yield response is lower and highly variable under the summer monsoon conditions when flooding, pests, and diseases are most common. Some farmers in India and Bangladesh cultivate modern varieties in the dry season and traditional varieties in the wet season (Barker et al., 1985).

The uncertainty of adequate moisture in many areas, and flooded conditions in others, discourages the use of modern fertilizer-responsive varieties in non-irrigated production regimes. But, in some countries, the use of modern varieties has expanded well beyond the irrigated zones. In the Philippines, modern varieties are grown extensively in rainfed lowlands. In Burma, modern deepwater and upland varieties now cover an area three times that which is irrigated (Dalrymple, 1986; see Table 7-24).

TABLE 7-24
Water Regime and Modern Variety Adoption for Rice Production in Selected Asian Countries

		Percent rice area by water regime							
		Irrigated		Rainfed				
COUNTRY	Total rice area (000 ha)	Season Dry	Wet	Shallow (b)	Deep Water (c)	Floating (d)	Dryland	% Cropped in Modern Varieties
Bangladesh	10190	10	2	42	26	11	9	25
Burma	4751	2	15	43	21	4	15	53
China (e)	34288	(f)	95	5	-	-	-	20 (g)
India	39515	7	32	28	12	6	15	53
Indonesia	8913	30	51	6	3	2	8	64 (g)
Korea Rep	1229	(f)	100	-	-	-	-	34
Malaysia	680	29	29	29	-	-	13	44
Nepal	1261	-	23	54	15	4	4	36 (g)
Pakistan	1999	-	100	-	-	-	-	46
Philippines	3535	17	26	34	11	-	12	85
Sri Lanka	818	27	40	27	-	-	6	89
Thailand	8679	6	14	53	12	4	11	13

Source: IRRI, 1986.

a A dash (-) indicates magnitude zero.
b 0-30 cm.
c 30-100 cm.
d More than 100 cm.
e Allocations based on impressions rather than data.
f All rice grown during the summer months and shown under wet season.
g Number of MV adoption in disagreement with "Development and Spread of High-yielding Rice Varieties in Developing Countries." AID numbers: for China - 91%; Indonesia - 82%; Nepal - 53%.

Further dissemination of the short, stiff-stemmed, fertilizer-responsive modern varieties of rice has potential for decreasing emissions of methane. World rice yield is currently at about one-half of genetic potential, which is as much as 14-16 mt/ha (Mikkelsen, 1988). There is considerable room for efficiency improvements in rice production, which would lead to a relative decrease in methane production.

Fertilizer Use. Widespread adoption of modern varieties has been somewhat impeded because of the capital required for new seed and fertilizer. Switching to modern varieties has resulted in an increase in the use of both **organic and chemical fertilizers**. Research indicates that a significant increase in production could be achieved through more efficient application of chemical fertilizer. Asia currently has a fertilizer-use efficiency of between 30 and 40%. Direct placement of fertilizer into the soil when rice is transplanted could double fertilizer-use efficiency and would increase yield (Mikkelsen, 1988).

The increase in fertilizer use and intermittent flooding of rice paddies both lead to an increase in nitrous oxide production through denitrification. (See **USE OF NITROGENOUS FERTILIZER**.)

Emerging Technologies

Current research at the International Rice Research Institute (IRRI) and other rice research centers is focused on the development of varieties better suited to a wide range of environmental conditions (Dalrymple, 1986). The development of high-yield varieties that can withstand the drought conditions of upland and shallow rainfed systems, as well as the flooded conditions associated with lowland and tidal rice cultivation, would increase the dissemination of high-yielding varieties in these areas.

Greater emphasis is being placed on efforts to understand the complexities of the farming system by conducting research in farmer's fields and encouraging farmer involvement. This so-called farming-systems style of research combines the knowledge of researchers with the direct experience of farmers (Barker et al., 1985). Research that focuses on developing cultivars that consistently produce a good yield under a wide range of conditions, rather than a high yield under ideal conditions, holds the most promise for improving food supply stability and decreasing methane emissions from rice cultivation in Asia.

There are likely to be further shifts to irrigated rice production in the future; however, improving the potential of modern varieties to perform well under rainfed conditions would reduce the incentive to shift to irrigated production systems, which are large producers of methane.

Research Needs and Economic Considerations

Before a comprehensive strategy to reduce methane production in rice can be developed, research in several areas is needed. In particular, to estimate the amount of methane produced per unit of rice we need to quantify the amount of methane produced from different cultivars, under various cultivation practices, particularly under different water management regimes. There is also a need for experiments in Asia, where the majority of the world's rice is grown and no data is currently available.

Rice is the cornerstone of the Asian economy. Many Asian countries are striving for self-sufficiency and have protected internal markets against the price fluctuations of the international market. Some countries have initiated price floors and others price ceilings. Price supports in countries such as the U.S. and Japan cause an increase in the production of rice.

USE OF NITROGENOUS FERTILIZER

Denitrification and nitrification are the primary processes that lead to the evolution of N_2O from soils fertilized with nitrogenous fertilizers. In well-aerated soils, nitrification is the primary process producing N_2O (Breitenbeck et al., 1980). Denitrification is prevalent in poorly drained, wet soils; rice cultivation is the largest agricultural contributor to denitrification losses (Hauck, 1988). Leaching of fertilizer into ground water and surface water is an additional source of nitrous oxide: between 5 and 30% of fertilizer leaves the soil system via leaching or runoff (Breitenbeck, 1988). Researchers need to derive a more precise estimate of nitrous oxide from this source.

Fertilizer-derived emissions of N_2O are estimated to be 0.14-2.4 Tg N annually (Fung et al., 1988), based on global consumption of 70.5 Tg of nitrogenous fertilizer in 1984/1985 (FAO, 1987). Nitrogenous fertilizer use is increasing at an estimated 1.3% per year in industrialized countries and 4.1% per year in developing countries (World Bank, 1988). By 2050, global fertilizer consumption is estimated to increase by a factor of 3.5 over the 1990 level (Frohberg et al., 1988).

Anthropogenic factors affecting the fertilizer-derived emissions of N_2O include the type and amount of fertilizer applied, application technique, timing of application, tillage practices, use of chemicals, irrigation practices, vegetation type, and residual nitrogen in the soil. Natural factors such as temperature, precipitation, and organic matter content and pH of soil also affect N_2O emissions (Fung, 1988).

N_2O emissions are difficult to estimate because of the complexity and variability of fertilized soil systems. Estimates suggest that fertilizer-derived emissions of N_2O are highest for anhydrous ammonia (between 1 and 5% of N applied), followed by urea (0.5% of N applied), ammonium nitrate, ammonium sulfate, and ammonium phosphate (0.1% of N applied), and nitrogen solutions

(0.05% of N applied) (Fung, 1988). Anhydrous ammonia is only used extensively in the United States, where it comprises about 38% of nitrogenous fertilizer consumption. Urea is used extensively in Asia and South America, where it accounts for 69% and 58%, respectively, of nitrogenous fertilizer consumption.

Existing Technologies and Management Practices

The wide variety of agricultural systems and fertilizer management practices produce very different quantities of N_2O emissions. These emissions can be reduced by improving the efficiency of fertilizer use, which can be achieved through changes in management, such as better placement in the soil, or in technology, for example, introducing nitrification inhibitors and fertilizer coatings, which both improve the efficiency of fertilizer applied and reduce the amount required.

Adoption of more efficient fertilizer management practices and technologies by farmers has been slow, however, particularly in developing countries. Traditional agricultural practices have proven to be difficult to dislodge. Immediate benefits of alternative practices must be made apparent or incentives provided in order to achieve widespread adoption of these more efficient practices and technologies.

Management Practices That Affect N_2O Production

Type of Fertilizer. Nitrous oxide emissions vary by one to two orders of magnitude between nitrogen solutions, urea, and anhydrous ammonia. The feasibility of using fertilizers with lower N_2O emission rates for different cropping situations needs to be examined.

Rate and Timing of Fertilizer Application. Increased fertilizer application results in increased emissions of N_2O. Adjusting fertilizer application in accordance with plant requirements would improve fertilizer-use efficiency. Currently, fertilizer subsidies and pricing encourages a higher than optimum level of fertilizer use.

The timing of fertilizer application is likely to affect the evolution of N_2O from the soil. Limited studies on the subject suggest that emissions from fertilizer applied in the fall exceed those from fertilizer applied in the spring (Bremner et al, 1981).

Placement of Fertilizer. Proper (deep) placement of fertilizer can improve fertilizer efficiency by curbing nitrogen losses and thus N_2O emissions. Broadcasting and hand placement of fertilizer results in higher nitrogen losses than does deep placement. Deep placement is particularly important in flooded fields where fertilizer is used inefficiently. In Asia, only about one third of the nitrogen applied benefits rice crops. Placement of fertilizer into the reduced zone, which prevents microbial action, can double fertilizer-use efficiency (Stangel, 1988). This technique improves efficiency regardless of the water regime (Eriksen et al., 1985).

Rice can either be grown from seeds or transplanted. Transplanted rice may reduce denitrification losses by allowing more efficient fertilizer application. A simple tool that allows for proper fertilizer placement at the time of transplanting greatly improves fertilizer efficiency, however, this tool has not been widely adopted in Asia (Stangel, 1988). Rice farms in Asia average 3 hectares or less and are very labor-intensive operations. Consequently, there has been little incentive to develop and disseminate fertilizer technology.

Water Management. Intermittent flooding of rice paddies increases nitrogen loss and the creation of N_2O (Eriksen et al., 1985; Olmeda and Abruna, 1986).

Tillage Practices and Herbicide Use. Measurements suggest that denitrification activity and N_2O emissions are higher under no-till systems than under conventionally tilled systems (Groffman et al., 1987). This increase could be the result of fertilizer placement or the increased use of herbicides associated with no-till systems. Preliminary observations indicate that application of post-emergent herbicides leads to significant, but short-lived, increases in N_2O emissions (Breitenbeck, 1988). The influence of tillage practices and herbicide use on N_2O emissions merit further study.

Legumes as a Nitrogen Source. Few studies have been done to determine the role of nitrogen-fixing crops in the emission of N_2O. N_2O emissions from legumes have been shown to be similar to those from fertilized crop systems (Groffman et al., 1987) and to emissions from fallow unfertilized soil (Blackmer et al., 1982). Under no-till cropping systems, there seems to be a greater potential for N_2O emissions when legumes are used as the nitrogen source (Groffman et al., 1987). Although the use of legumes as a nitrogen source is unlikely to reduce emissions of N_2O from agriculture, better quantification of this source is needed.

Technologies that Improve Fertilization Efficiency

Nitrification Inhibitors. Nitrification and urease inhibitors are fertilizer additives that can increase efficiency by decreasing nitrogen loss through volatilization. Nitrification inhibitors can increase fertilizer efficiency by 30% (Stangel, 1988).

Reduced Release Rate. Techniques that limit fertilizer availability, such as slow-release or timed-release fertilizers improve nitrogen efficiency by reducing the amount of nitrogen available at any time for loss from the soil system.

Coatings. Limiting or retarding water solubility through supergranulation or by coating a fertilizer pellet with sulfur can double fertilizer efficiency (Stangel, 1988).

Emerging Technologies

No breakthrough in chemical fertilizer technology is anticipated in the near future, but there are likely to be improvements in fertilizer production and application efficiency. The development of a subsurface fertilizer application method for no-till, for example, could have a significant effect on N_2O emissions from that source.

Advances in biotechnology are likely to affect N_2O emissions from agriculture. Engineering of crop varieties that are more resistant to weeds could reduce herbicide use, and in turn, decrease the emission of N_2O.

Nitrogen-fixing cereal crops, which could be available by 2000, would decrease the use of fertilizer on those crops. The potential for reduced N_2O emissions needs to be examined (OTA, 1986).

Technological advances that increase crop yield will help reduce the amount of land and other inputs needed to produce the goods required to support the population. Improved yield and other efficiency improvements hold potential for reducing N_2O emissions.

Research Needs and Economic Considerations

Many uncertainties surround the factors influencing N_2O emissions from fertilizer. Cultivation practices and environmental practices that affect the fertilizer-derived emissions of N_2O have not been quantified. Additional research in the following areas is needed:

- N_2O emissions by fertilizer type;
- the effect of crop type on emissions;
- the effect of water management practices on emissions;
- N_2O release from drainage and ground water;
- N_2O emissions from legumes and manure;
- the effect of tillage, herbicide use, and alternative cropping systems on N_2O emissions;
- how the rate of fertilizer application affects emissions;
- the effect on emissions of different fertilizer management practices, including fertilizer placement; and
- the contribution of tropical agriculture to N_2O emissions.

Most fertilizer use in Asia is for rice cultivation. Rice is a significant source of both methane and N_2O. The production of the two gases under different management regimes needs to be explored. There is room for significant efficiency improvement in this area.

Fertilizer subsidies, target prices and loan levels encourage a higher level of fertilizer use than would otherwise occur. Such policies also reduce the attractiveness of more efficient fertilizer application and use. Increased fertilizer prices and the threat of shortages in fossil-fuel-based fertilizers should increase interest in fertilizer efficiency in the future.

ENTERIC FERMENTATION IN DOMESTIC ANIMALS

Livestock play a vital role within the agricultural sector, producing meat, dairy products and fiber. Ruminant animals, including cattle, dairy cows, buffalo, sheep, goats, and camels have the capability of converting roughages into usable nutrients through microbial fermentation. This unique capability allows them to convert cellulosic plant material to milk, meat, and fiber that would otherwise be unavailable for human consumption. Some non-ruminant animals (e.g., horses and pigs) also produce methane, although in much smaller quantities. Currently, ruminants contribute approximately 97% of the annual methane emissions from domestic animals; non-ruminants contribute approximately 3%.

Protein of animal origin has the highest biological value of all sources of protein. Within the United States, animal products supply 53% of all foods consumed, including 69% of the protein, 40% of the energy, 80% of the calcium, and 36% of the iron in our diets (English et al., 1984). In developing countries, there is a continued and growing demand for animal proteins.

About 24% of the world's land area is in permanent pasture. Rangeland and pastureland combined account for over 50% of total land area in the world. Within Africa and Oceania (Australia and New Zealand), rangeland and pastureland make up 65 and 75% of the land area, respectively (IIED and WRI, 1987). These areas are typically too steep, arid, rocky, or cold to be suitable for crops, but provide forage for about 3 billion head of cattle, buffalo, sheep and goats (FAO, 1986).

The type and function of livestock systems vary considerably around the globe. Within developed economies and countries in which meat is a major export commodity, livestock are intensively managed for either meat or dairy production. Ranches within these economies are typically large and use management techniques such as selective breeding, rangeland management, feed enhancement, and the use of antibiotics to increase cattle production.

Livestock within these systems are typically larger animals, fed at least part of the time on a high- quality grain diet. Fodder crops are grown specifically for feeding cattle. Most beef cattle are fed a high- quality, low-cellulose diet in a feedlot for several months prior to slaughter. These animals have a much higher yield of meat or dairy product than similar animals in developing countries. Table 7-25 shows average meat production per animal in several regions of the world.

Within less developed economies, livestock are usually more integrated into the whole agricultural system, frequently as a part of a crop/livestock or pasture-based system. These animals are not always managed to maximize reproductive efficiency or slaughtered at the most efficient stage of life. They are frequently maintained as scavengers and live on a low-quality forage diet.

Livestock serve as a buffer within agricultural systems by helping to stabilize cash flows and food supplies. Livestock production can serve to level out the effect of climate variability and the seasonality of rainfall in crop/livestock systems. Ruminant livestock can salvage energy and nitrogen from what otherwise could have been complete crop failures. In a grain crop failure, for example, vegetative matter suitable for livestock consumption is still produced. Livestock are convenient disposal systems for crop residues and provide a sink for surplus and damaged crops not suitable for human consumption (Raun, 1981).

In Tanzania and other Central African countries, tribal groups regard an increase in animal numbers as the best insurance against economic and social risk. The animals afford protection against the uncertainties of climate and destruction of crops by pests (Winrock, 1977). Livestock ownership enhances social status in many developing countries.

TABLE 7-25
Average Meat Yield per Animal
(kilograms)

Region	Beef & Veal	Mutton & Goat
North America	87.7	11.1
Europe & USSR	59.7	6.6
Oceania	45.8	5.7
Latin America	29.4	2.7
Near East	17.3	4.4
Far East & China	8.2	4.0
Africa	13.6	3.5
World Average	31.2	4.7

Source: Stoddart et al., 1975.

In addition to the production of meat, dairy products, and fiber, livestock are a source of fertilizer, fuel, and provide most of the tractor power in many developing countries. Animals are used for plowing, threshing, and to provide power for irrigation. Draft animals can increase farm output several times by increasing the area that a farm family can cultivate. Within Africa and the Far East, over 80% and 90%, respectively, of all draft power is provided by animals (Raun, 1981).

Animal manure is used by about 40% of the world's farmers to enhance soil fertility. Dried dung can also be used as fuel. It is estimated that over 200 million tons of manure are used each year as fuel in developing countries (Winrock, 1977).

Methane Emissions from Livestock

Enteric fermentation, the digestive process that makes livestock so useful, causes them to emit methane into the atmosphere. The 1272 million cattle, 1140 million sheep, 460 million goats, 126 million buffalo (FAO, 1985), and assorted other domesticated animals contribute an estimated 65-100 Tg of methane to the atmosphere annually. Of this amount, approximately 57% comes from beef cattle, 19% is from dairy cattle, and 9% is from sheep (Lerner et al., 1988). With the global cattle population increasing at 1.2% per year (see Chapter IV), livestock will continue to be a significant source of methane.

Livestock cannot metabolize methane. It is lost energy, which neither contributes to the animal's maintenance, nor production of a product. Estimates of gross energy intake lost as methane for cattle have ranged from 3.5% (Johnson, 1988) to 8.3% (Blaxter and Wainman, 1964).

Methane emissions from livestock are affected by differences in quantity and quality of feed, body weight, age, energy expenditure and enteric ecology. All other thing being equal, emissions are higher for animals that are heavier, have higher feeding levels and high-cellulose forage diets, and are ruminants. Emissions are also higher for work animals because some of the additional feed required to supply this energy is converted to methane (Lerner et al., 1988). Manure is an additional, unquantified source of methane (Winrock, 1977).

Although we know that animal type, feed type, and management practice affect the amount of methane generated by an animal, few modifications to current practices are certain to reduce methane production.

Existing Technologies and Management Practices

Livestock System Productivity. Intensively managed, high-productivity livestock systems--those with fewer, larger, more productive animals, produce more animal product per unit of methane (Moe and Tyrell, 1979). In these highly managed livestock systems, as the feeding level is increased, the level of methane production increases, but the energy loss in relation to gross energy intake decreases (Thorbek, 1980). An animal fed at maintenance level, however, loses a larger percentage of its gross energy intake to methanogenesis than does an animal fed at a higher level, and has no increase in animal product.

The vast differences between livestock systems in developed and developing countries make differences in productivity difficult to compare. Additional benefits, such as the work output of draft animals, must be considered, but, intensively managed, high-productivity livestock systems are the most efficient in terms of animal product yield.

Animal Type. Animal type and size affect methane production. Sheep produce less methane per animal than cattle over a shorter time period. Overall, sheep produce about 20% less methane per unit of animal product than do cattle.[1]

Diet. At low levels of feed intake, at or slightly above maintenance, a high-cellulose forage diet results in lower methane production per unit of feed than a high-nitrogen grain diet. Conversely, at high levels of feed intake, approaching three times maintenance, a high-quality grain diet results in lower methane production per unit of feed (Blaxter and Clapperton, 1965).

This information suggests that current feeding practices are reasonably good with regard to minimizing methane generation. In developing countries, where livestock are living near maintenance levels, they subsist on a high-cellulose forage diet. In developed countries, where animals consume several times the maintenance level, animals are fed a high-quality grain diet. The analysis of Blaxter and Clapperton (1965) suggests that this situation is optimal. But, further research is needed in this area, as no measurements of methane generation have been done on livestock within farm conditions of developing countries.

Feed Additives. Feed additives, which increase feed use efficiency and reduce methane generation by modifying rumen fermentation, are currently available for beef cattle. These feed additives, known as ionophores, are fed to most feedlot cattle to improve the efficiency of beef production during finishing for market. Ionophores reduce the amount of methane produced by a ruminant by improving digestive efficiency (less is produced) and by causing the animal to eat less, which leaves less food in the gut to ferment. Ionophores improve the efficiency of beef production

---

[1] This estimate assumed methane production rates of 132 and 17 l/d (adapted from Johnson, 1988); weight gains of 1.05 and .22 kg/d (English et al., 1984); slaughter weights of 310 and 42 kg; and dressing percentages of .6 and .5 for cattle and sheep, respectively.

by 6-8%, while reducing methane production by approximately 5%, for animals on a forage diet, and by approximately 20%, for animals fed a high-grain diet (Johnson, 1988).

Ionophores have been shown to lose effectiveness over time, when the bacteria in the gut develops a tolerance to them (Johnson, 1974). A feeding program that alternates different ionophores can improve overall efficiency by impeding bacterial adaptation (Hubbert et al., 1987).

The use of ionophores is predominately limited to feedlot cattle because of the difficulties associated with administering a drug to range cattle. Range cattle can be fed ionophores using a mineral block lick, but few range cattle receive this drug.

Feed additives are not currently available for dairy cattle, and are unlikely to become available because of problems with efficiency and chemical residues in the milk. In developing countries, where cattle are generally used for milk and meat production, ionophores cannot be used.

Methane from Manure. Microbial decomposition in manure generates methane under anaerobic conditions, but the amount produced from this source is not well quantified. Disposing of manure in lagoons and flooded fields creates anaerobic conditions conducive to the generation of methane.

Energy from manure can be captured in the form of methane in biogas plants. Manure produced by ruminants, particularly cattle and buffalo, is an ideal substrate for anaerobic fermentation in biogas plants. Methane from manure has a value of 5 kcal per cubic meter (about 70% of the energy value of natural gas). In the U.S., the dung from about 40 cows could provide all of the non-mobile fuel for a farm family, including electricity. Disposal of manure in biogas plants can reduce methane generation, while still providing a fertilizer source from the residue (Winrock, 1977).

Emerging Technologies

Efficiency improvements and advances in biotechnology have resulted in considerable productivity increases for beef cattle, dairy cattle, and sheep. Since 1940, dairy production in the U.S. has remained relatively constant while the herd size has been cut in half (English et al., 1984). Efficiency improvements should continue to produce gains in productivity, along with associated reductions in methane emissions.

Bovine growth hormone, a bio-engineered imitation of a naturally occurring protein in cattle, is expected to increase productivity in dairy cattle. This drug, which has not yet been approved by the U.S. Food and Drug Administration, is likely to decrease methane production per unit of milk (Tyrell, 1988).

The development of a methanogenesis inhibitor for dairy cattle is unlikely in the near future. Improving the ionophore delivery system for range cattle could increase use of that additive.

There is also some potential for reducing methane production by inhibiting fungus in the rumen of cattle. These fungal interactions need to be more thoroughly investigated before fungal inhibitors can be developed.

Through longer-term research in biotechnology, there exists the potential to develop microbial species capable of converting hydrogen to a useful hydrogen sink, rather than to methane. Acetic acid is such a hydrogen sink, which is useful as an energy source in cattle.

Animal scientists have long been working on the problems of improving feed-use efficiency and livestock system productivity and, thereby, the problem of methane generation. This research should continue to reduce methane generation per unit of animal product within highly managed systems.

Research Needs and Economic Considerations

Further research is needed on livestock in developing countries in order to devise a strategy to reduce methane generation. For example, estimates of methane generation are needed for a range of livestock (including draft animals) on a variety of diets. Also, the potential for using antibiotics, such as ionophores, on cattle needs to be explored.

We also need to quantify methane generation from manure under a range of management and disposal options.

Commodity programs in the U.S. encourage a larger cattle population than there otherwise would be. Unnaturally low grain prices, which are a result of deficiency payments, are, in effect, a subsidy to beef and dairy production. Beef producers are also protected by a tariff, which in recent years has been converted to a variable levy (Schuh, 1988). These policies result in a larger cattle herd and increased production of methane.

REFERENCES

PART ONE: ENERGY SERVICES

Akbari, H., J. Huang, P. Martien, L. Rainer, A. Rosenfeld, and H. Taha. 1988. The impact of summer heat islands on cooling energy consumption and global CO_2 concentration. In American Council for an Energy Efficient Economy. *Proceedings of the 1988 ACEEE. Summer Study on Energy Efficiency in Buildings.* August.

Alliance to Save Energy. 1987. *The GPU Project: Findings and Implications.* Washington, D.C.

Automotive News. 1983. Cat's 3306B Makes it Big in the Real World. November 7.

Automotive News. 1986. Chrysler Genesis Project Studies Composite Vehicles. May 5, p. 36.

Automotive News. 1988. Ford Develops Catalyst With No Platinum. December 19.

Berg, C. 1988. Industrial Processess in the Long Run. in S. Mevers. ed. *Summary of Presentations at a Workshop on Energy Efficiency and Structural Change: Implications for the Greenhouse Problem.* Oakland, CA. May 1-3, Lawrence Berkeley Laboratory, Berkeley, CA.

Bleviss, D. 1988. *The New Oil Crisis and Fuel Economy Technologies: Preparing the Light Transportation Industry for the 1990's.* Quorum Press, New York, N.Y.

Brody, H. 1987. Energy-Wise Buildings. *High Technology.* February.

Chandler, W.U. 1986. *The Changing Role of the Market in National Economies.* Worldwatch Institute, Washington, D.C. September.

Crandall, R.W., H.K. Gruenspecht, T.E. Keeler, and L.B. Lave. 1986. *Regulating the Automobile.* The Brookings Institution, Washington, D.C.

Deluchi, M.A., R.A. Johnston, and D. Sperling. 1987. *Transportation Fuels and the Greenhouse Effect.* University Wide Energy Research Group, University of California, Berkeley, CA. December.

Deringer, J.J., J.F. Busch, J. Hall, K.S. Kannon, M.D. Levine, A.C. Ayub, and I. Turiel. 1987. *Energy and Economic Analysis in Support of Energy Conservation Standards for New Commercial Buildings in Malaysia.* Lawrence Berkeley Laboratory, Berkeley, CA. April.

Edison Electric Institute. 1985. *1985 Capacity & Generation.* EEI, Washington, D.C.

FERC (Federal Energy Regulatory Commission). 1988. *The Qualifying Facilities Report.* FERC, Washington, D.C. January.

Flavin, C., and A. Durning. 1988. Raising fuel efficiency. In *State of the World 1988: A Worldwatch Institute Report on Progress Toward a Sustainable Society.* W.W. Norton and Co., New York, N.Y.

(FAO) Food and Agriculture Organization. 1981. *Agriculture: Toward 2000.* FAO, Rome.

Forster, H.J. 1983. Big Car, Too, Can be Light on Fuel. *Resources and Conservation* 10:85-102.

Geller, H.S. 1984. *The Potential for Electricity Conservation in Brazil.* Companhia Energeticer de Sao Poalo, Sao Poalo, Brazil.

Geller, H.S. 1986. *End-Use Electricity Conservation: Options for Developing Countries.* Energy Department Paper No: 32. World Bank, Washington, D.C.

Geller, H.S., M.R. Ledbetter, E.L. Miller, and P.M. Miller. 1987. *Acid Rain and Electricity Conservation.* The Energy Conservation Coalition and the American Council for an Energy Efficient Economy, Washington, D.C.

Geller, H.S. 1988. *Residential Equipment Efficiency: 1988 Update.* American Council for an Energy Efficient Economy, Washington. D.C.

Glebov, I.A. and V.P. Kovalenko. 1987. Strategies for Conservation of Fuel and Energy Resources in the Leningrad Region. *Energy* 12:(10/11)949-952.

Goldemberg, J., T.B. Johansson, A.K.N. Reddy, and R.H. Williams. 1988. *Energy for a Sustainable World.* Wiley Eastern Limited, New Dehli, India.

Goldemberg, J., T.B. Johansson, A.K.N. Reddy, and R.H. Williams. 1987. *Energy for a Sustainable World.* World Resources Institute, Washington, D.C.

Goldman, C. 1984. Measured energy savings from residential retrofits: updated results from the BECA-A projects. In American Council for an Energy Efficient Economy (ACEEE). *Proceedings of the 1984 ACEEE Summer Study on Energy Efficientcy in Buildings, Vol B.* ACEEE, Washington, D.C.

Goldstein, D.B., and P. Miller. 1986. Developing cost curves for conserved energy in new refrigerators and freezers. In American Council for an Energy Efficient Economy. *Proceedings of the 1986 Summer Study on Energy Efficiency in Buildings, Vol. 1.* ACEEE, Washington, D.C.

Gray, C. 1983. U.S. light vehicles - some exciting news for the 1990's. *Resources and Conservation* 10:65-84.

Hewett, E.A. 1984. *Energy, Economics and Foreign Policy in the Soviet Union.* The Brookings Institution, Washington, D.C.

Hirst, E., J. Clinton, H. Geller, and W. Kroner. 1986. *Energy Efficiency in Buildings: Progress and Promise.* American Council for an Energy Efficient Economy, Washington, D.C.

Holcomb, M.C., S.D. Floyd, and S.L. Cagle. 1987. *Transportation Energy Data Book: Edition 9.* Oak Ridge National Laboratory Report No: ORNL-6325, Oak Ridge, TN.

Husselbee, W.L. 1984. *Automotive Emission Control.* Reston Publishing Co., Reston, Virginia.

IEA (International Energy Agency). 1987. *Energy Conservation in IEA Countries.* Organization for Economic Cooperation and Development, Paris.

IEA (International Energy Agency). 1988. *Energy in Non-OECD Countries: Selected Topics 1988.* Organization for Economic Cooperation and Development, Paris.

Jessup, P.S. 1988. *Advanced Power Technologies: Their Potential Contribution to Stabilizing CO_2 Emissions.*

Kahane, A., and R. Squitieri. 1987. Electricity use in manufacturing. *Annual Review of Energy* 12: pg #'s?

Kamo, R. 1987. Adiabatic diesel-Engine Technology in Future Transportation. *Energy* 12:(10/11)1073-1080.

Kavanaugh, M. 1988. End-use efficiency and NOx emissions in aviation. In S. Meyers, ed. *Summary of Presentations at a Workshop on Energy Efficiency and Structural Change: Implications for the Greenhouse Problem*. Oakland CA. May 1-3, 1988. Lawrence Berkeley Laboratory, Berkeley, CA.

Lovins, A., and R. Sardinsky. 1988. *The State of the Art: Lighting*. Competitek, The Rocky Mountain Institute, Colorado.

Maglieri, D.J., and S.M. Dollyhigh. 1982. We have just begun to create efficient transport aircraft. *Astronautics and Aeronautics* (Feb.), p.20.

Makarov, A.A., A.A. Beschinsky and A.G. Vigdorchik. 1987. Basic Principles of Energy Policy in the U.S.S.R. *Energy* 12:(10/11)945-948.

Meyers, S. 1988. *Transportation in the LDC's: A Major Area of Growth in World Oil Demand*. Lawrence Berkeley Laboratory, Berkeley, CA. March.

Miller, A.S., I.M. Mintzer, and S.H. Hoagland. 1986. *Growing Power: Bioenergy for Development and Industry*. World Resources Institute, Washington, D.C. April.

Miller, P.M., H.S. Geller and A.T. de Almeida. Undated. *Improving Energy Efficiency of Electrical Equipment in Pakistan*. American Council for an Energy Efficient Economy. Washington, D.C.

Mintzer, I.M. 1988. *Projecting Future Energy Demand in Industrialized Countries: An End-use Oriented Approach*. Prepared for U.S. EPA, Washington, D.C. 40+ pp.

Mix, T.W. 1987. Advanced Seperation Techniques for Petrochemicals. *Energy* 12:(10/11).

Naill, R. 1987. Cogeneration and Small Power Production. Presentation to the Energy Policy Forum, Airlie Va. June 16.

National Acid Precipitation Assessment Program. 1987. *Interim Assessment: The Causes and Effects of Acid Deposition*. Washington, D.C.

National Energy Conservation Center. 1986. *Potential for Energy Efficiency Improvements in the Commercial and Industrial Sectors*. Publ. city?

New England Energy Policy Council. 1987. *Power to Spare: A Plan for Increasing New England's Competitiveness Through Energy Efficiency*. Energy Policy Council, Boston, MA.

OTA (Office of Technology Assessment). 1982. *Increased Automobile Fuel Efficiency*. OTA, Washington, D.C.

OECD (Organization for Economic Co-operation and Development). 1988. *Transport and the Environment*. Paris.

Plotkin, S.E. 1989. U.S. Office of Technology Assessment, Personal communication, January 13.

Price, A. and M. Ross. 1988. *Reducing Electricity Costs in Automotive Assembly, Stamping and Parts Plants.* Unpublished.

Rocky Mountain Institute. 1986. *Advanced Electricity-Saving Technologies and the South Texas Project.* Old Snowmass, CO.

Ropelowski, R.R. 1982. 757 Key to Route Flexibility. *Aviation Week and Space Technology.* August 30, p.36.

Rosenfeld, A.H. 1988. Urban trees and light-colored surfaces: inexpensive and effective strategies for reducing air conditioning energy consumption. In Meyers S., ed. *Summary of Presentations at a Workshop on Energy Efficiency and Structural Change: Implications for the Greenhouse Problem.* Oakland CA, May 1-3, 1988. Lawrence Berkely Laboratory, Berkely, CA.

Rosenfeld, A.H. and D. Hafemeister. 1985. Energy conservation in large buildings. In D. Hafemeister et al. (eds). *Energy Sources: Conservation and Renewables.* Conference Proceedings No. 135. American Institute of Physics, New York.

Rosenfeld, A.H. and D. Hafemeister. 1988. Energy-efficient Buildings. *Scientific American* 258:(4).

Ross, M. 1985. Industrial Energy Conservation. In Hafemeister, D., H. Kelly and B. Levi. *Energy Sources: Conservation and Renewables.* American Institute of Physics, New York.

Ross, M. 1986. *Current Major Issues in Industrial Energy Use.* Prepared for the Office of Policy Integration, U.S. Department of Energy, Washington, D.C.

Ross, M. 1987. Industrial Energy Conservation and the Steel Industry of the United States. *Energy* 12:(10/11).

Ross, M. E.D. Larson and R.H. Williams. 1987. Energy Demand and Materials Flows in the Economy. *Energy* 12:(10/11).

Sathaye, J.A., B. Atkinson and S. Meyers. 1988. *Alternative Fuels Assessment: The International Experience.* Lawrence Berkeley Laboratory, Berkeley, CA. March.

Sathaye, J.A., A.N. Ketoff, L.J. Schipper, and S.M. Lele. 1988. *An End-use Approach to Development of Long-term Energy Demand Scenarios for Developing Countries.* Lawrence Berkeley Laboratory, Berkeley, CA. September Draft.

Schipper, L., S. Meyers, and H. Kelly. 1985. *Coming in from the Cold: Energy-Wise Housing in Sweden.* Seven Locks Press, Washington, D.C.

Schipper, L., et al. 1989. Energy Use and Lifestyle: A Matter of Time? *Annual Review of Energy.* Annual Reviews Inc., Palo Alto,. CA.

Smith, B.A. 1988. Douglas Initiates Marketing Efforts to Promote MD-90 Propfan Program. *Aviation Week and Space Technology.* January 25.

Smith, J.B. 1981. *Trends in Energy Use and Fuel Efficiencies in the U.S. Commercial Airline Industry.* U.S. Department of Energy, Washington, D.C.

SERI (Solar Energy Research Institute). 1981. *A New Prosperity: Building a Sustainable Energy Future.* The SERI Solar/Conservation Study. Brickhouse, Andover, MA.

Sobey, A.J. 1988. Energy use in transportation: 2000 and beyond. In Meyers S., ed. *Summary of Presentations at a Workshop on Energy Efficiency and Structural Change: Implications for the Greenhouse Problem.* Oakland CA, May 1-3, 1988. Lawrence Berkeley Laboratory, Berkeley, CA.

Stauffer, R.F. 1988. *Energy Savings from Recycling.* National Appropriation Technology Assistance Service Paper.

Turiel, I., R. Curtis and M.D. Levine. 1984. Parametric Energy Analysis in Support of Singapore Energy Conservation Standards for Commercial Buildings. K. H. Olson and W.W. Ching eds. Proceedings of the Asean Conference on Energy Conservation in Buildings. Singapore.

U.S. AID (Agency for International Development). 1988a. *New Directions for A.I.D. Renewable Energy Activities.* Office of Energy, Bureau for Science and Technology, Washington D.C. February.

U.S. AID (Agency for International Development). 1988b. *Power Shortages in Developing Countries: Magnitude, Impacts, Solution, and the Role of the Private Sector.* U.S. AID, Washington, D.C. March.

U.S. DOC (Department of Commerce). 1988. *Survey of Current Business* 68(8), Bureau of Economic Analysis, Washington, D.C. August.

U.S. DOE (Department of Energy). 1987a. *Annual Review of Energy 1986.* Energy Information Administration, U.S. DOE, Washington D.C.

U.S. DOE (Department of Energy). 1987b. *Energy Security: A Report to the President of The United States.* U.S. Government Printing Office, Washington, D.C.

U.S. DOE (Department of Energy). 1987c. *Long Range Energy Projections to 2010.* Office of Policy, Planning and Analysis, U.S. DOE, Washington, D.C. September.

U.S. DOE (Department of Energy). 1988. *Annual Energy Review 1987.* Energy Information Administration Report No: DOE/EIA-0384(87), Washington, D.C.

U.S. DOT (Department of Transportation). 1988. *Federal Aviation Administration Aviation Forecasts for Fiscal years 1988 through 1999.* February.

U.S. EPA (Environmental Protection Agency). 1979.

U.S. EPA (Environmental Protection Agency). 1985. *Regulatory Impact Analysis: Oxides of Nitrogen Polutant-Specific Study and Analysis of Comments.* Office of Mobile Sources.

von Hippel, F., and B. V. Levi. 1983. Automotive Fuel Efficiency: the Opportunity and Weakness of Existing Market Incentives. *Resources and Conservation* 10:103-124.

White, L.J. 1982. *The Regulation of Air Pollutant Emissions from Motor Vehicles.* American Enterprise Institute for Public Policy Research. Washington, D.C.

Williams. 1988.

Williams, R.H., E.D. Larson, and M. Ross. 1987. Materials, Affluence, and Industrial Energy Use. *Annual Review of Energy* 12:99-144.

Wohlgemuth, J. 1988. Photovoltaic technology. In *Workshop on Global Climate Change: Emerging Energy Technologies for Electric Utilities*. Alliance Technolgies Inc., Bedford, MA. September.

The World Bank. 1985. China: The Transport Sector, Annex 6 to *China: Long-Term Development Issues and Options*. Publisher, Publisher's City.

WRI and IIED (World Resources Institute and International Institute for Environment and Development). 1988. *World Resources 1988-89*. Basic Books, Inc. New York.

PART TWO: ENERGY SUPPLY

Baldwin, S.F. (1987). Biomass Stoves: Engineering Design, Development and Dissemination. V.I.T.A., Arlington, VA.

Baldwin, S.F., H. Geller, G. Dutt, and N.H. Ravindranath. 1985. Improved Woodburning Stoves: Signs of Success. Ambio 14(4-5):280-7.

Bormann F.H., B.T. Bormann, and K.R. Smith. 1988. Earth to Hearth: A microcomputer for comparing biofuel systems. Submitted for publication.

Brown L.R. 1980. Food or Fuel: New competition for world's cropland. Worldwatch Paper 35, Washington, D.C.

Clarke, J. F. The U.S. Magnetic Fusion Energy Program. Presented at the Workshop on Global Climate Change: Emerging Energy Technologies for Electric Utilities. September, 1988.

DeMocker, J.M., J.M. Greenwald, and P.F. Schwengels. 1986. Extended Lifetimes for Coal-Fired Power Plants: Implications for Air Quality Analysis and Environmental Policy. Paper No: 28.3, Air Pollution Control Association, Annual Meeting, Minneapolis, MN, June 22-27.

EPRI, *Utility Turbopower for the 1990s*. EPRI Journal, April/May 1988

Goldemberg J., T.B. Johansson, A.K.N. Reddy, and R.H. Williams. 1988. Energy for a Sustainable World. Wiley-Eastern, New Delhi.

Grohse, E. and M. Steinberg. 1987, *Economical Clean Carbon and Gaseens Fuels from Coal and other Cabonaceous Raw Materials*. Brookhaven National Laboratory, November. (BNL 40485).

Gunnerson C.G., and D.C. Stuckey. 1986. Anaerobic Digestion: Principles and Practices for Biogas Systems. Technical Paper No. 49, World Bank, Washington,D.C.

Harvey, L.D. 1988. Potential role of electric power utilities in hydrogen economy. Paper presented at Workshop on Global Climate Change: Emergency Energy Technologies for Electric Utilities. March 30-31, 1988. American Institute of Architects, Washington, D.C.

Hay, N. P. Wilkinson, and W. James. 1988. Global climate change and emerging energy technologies for electrical utilities: The role of natural gas. Paper presented at Workshop on Global Climate Change: Current Electricity Supply Alternatives. March 30-31. American Gas Association, Washington, D.C.

Hewitt, E.A. 1984. Energy Economics and Foreign Policy in the Soviet Union. The Brookings Institution, Washington, D.C.

Hu, D., I. Oliker and F. Silaghy. 1984. Power Plant Modification for Cogeneration. Fossil Plant Life Extension Workshop, Washington D.C., June 12-15, Electric Power Research Institute.

IEA (International Energy Agency). 1987b. *Renewable Sources of Energy*. Organization for Economic Cooperation and Development, Paris.

IUCN (International Union for Conservation of Nature and Natural Resources). 1980. World Conservation Strategy for Sustainable World. Gland, Switzerland.

Jessup, P.S. *Advanced Power Technologies: Their Potential Contribution to Stabilizing CO_2 Emissions*. March 1988

Kjellstrom, B. 1985. Biomass Gasifiers for energy supply to Agriculture and small industry. Ambio 14(4-5):267-74.

Larson E.D., and R.H. Williams. 1988. Biomass-fired Steam-injected Gas Turbine Cogeneration. Proc. 1988 ASME Cogen-Turbo Symposium, Montreux, Switzerland, Aug 30-Sep 1.

Larson E.D., J.M. Ogden, and R.H. Williams. 1987. Steam-injected Gas-Turbine Cogeneration for the cane sugar industry. PU/CEES Report No 217, Princeton.

Manibog, F.R. 1984. Improved Cooking Stoves in Developing Countries: Problems and Opportunities. Ann. Rev. Energy 9:199-227.

Miller A.S., I.M. Mintzer, and S.H. Hoagland. 1986. Growing Power: Bioenergy for Development and Industry. WRI Study No 5, Washington.

NAPAP (National Acid Precipitation Assessment Program). 1987. *Interim Assessment: The Causes and Effects of Acidic Deposition*. Volume IV. NAPAC, Washington D.C.

New York Times, October 25, 1988, [article on superconductors - need complete reference] page D23

OECD (Organization for Economic Cooperation and Development). 1988. Transport and the Environment. Paris.

PEI Associates, Inc. 1988. Supply-Side Conservation Techniques for Tennessee Valley Authority and American Electric Power Coal-Fired Boilers. Cincinnati, OH. Draft Report. October.

Poole, R. 1988. Solar cells turn 30. *Science*. 241:900-901.

Ramakrishna J. 1988. Doctoral Dissertation, Dep of Geography, Univ of Hawaii, Honolulu.

Reid W.V., J.N. Barnes, and B. Blackwelder. 1988. Bankrolling Success: A portfolio of sustainable development projects. Environment Policy Institute, Washington.

Sathaye J., B. Atkinson, and S. Meyers. 1988. Alternative Fuels Assessment: the international experience. LBL mimeo, March, Berkeley.

Schlabach. T.D. 1988. Prospects for high-T_0 super conductors in large-scale power utility applications. Paper presented at Workshop on Global Climate Change: Emerging Energy Techniques for Electric Utilities. March 30-31, 1988. American Institute of Architects, Washington, D.C.

Shabecoff, P. 1987. U.S. Bureau for Water Projects Shifts Focus to Conservation. *New York Times*, October 2.

Smith K.R. 1987. The Dialectics of Improved Stoves. presented at 2nd Int. Workshop on Stove Dissemination, Antigua, Guatemala.

Smith K.R. 1987b. *Biofuels, Air Pollution and Health*. Plenum Press, New York.

Smith K.R., A.L. Aggarwal, and R.H. Dave. 1983. Air Pollution and Rural Biomass Fuels in Developing Countries. *Atmos. Envir.* 17:2343-62.

Steinberg, Cheng, and Horn, Brookhaven National Laboratory, *A Systems Study for the Removal, Recovery and Disposal of Carbon Dioxide From Fossil Fuel Power Plants in the U.S.*, 1984

United Nations. 1988. *Energy Statistics Yearbook*. United Nations, New York.

U.S. AID (Agency for International Development). 1988a. *New Directions for A.I.D. Renewable Energy Activities*. Washington, D.C. February.

U.S. AID (Agency for International Development). 1988b. *Power Shortages in Developing Countries: Magnitude, Impacts, Solutions, and the Role of the Private Sector*. Washington, D.C. March.

U.S. DOE (Department of Energy). 1985a. *Federal Ocean Energy Technology Program*.

U.S. DOE (Department of Energy). 1985b. *Five Year Research Plan 1985-1990, Wind Energy Technology: Generating Power from the Wind*.

U.S. DOE (Department of Energy). 1985c. *U.S. Geothermal Technology: Equipment and Services for Worldwide Application*. U.S. DOE, Washington, D.C.

U.S. DOE (U.S. Department of Energy). 1986. *International Energy Annual 1985*. U.S. DOE, Washington, D.C.

U.S. DOE (Department of Energy). 1987a. *Energy Security, A Report to the President of the United States*.

U.S. DOE (Department of Energy). 1987c. *Five Year Research Plan 1987-1991 National Photovoltaics Program*.

U.S. DOE (Department of Energy). 1987d. *Northern Lights: The Economic and Practical Potential of Imported Power from Canada*. Office of Policy, Planning and Analysis, Report No: DOE/PE-0079, Washington, D.C. December.

U.S. DOE (Department of Energy). 1987e. *America's Clean Coal Commitment*. Office of Fossil Energy Report FE/DOE-0083, Washington D.C. February.

U.S. DOE (Department of Energy). Draft 1987. *U.S. Geothermal Energy Program, Five Year Research Plan, 1988-1992.*

U.S. DOE (Department of Energy). 1988. *Electric Power Annual 1987.* Energy Information Administration Report No: DOE/EIA-0348(87), Washigton, D.C. September.

U.S. OTA (Office of Technical Assessment). 1985. *New Electric Power Technologies.* U.S. OTA, Washington, D.C. 329 pp.

Williams, R.H. 1988. Are Runaway Energy Capital Costs a Constraint on Development? Presented at Int. Sem. on the New Era in the World Economy, Sao Paulo, Aug 31-Sep 2.

Williams, R.H. 1985. Potential Roles for Bioenergy in an Energy Efficient World. *Ambio* 14(4-5): 201-9.

Williams, R.H., and E.D. Larson. 1988. Aeroderivative Turbines for Stationary Power. *Annual Review of Energy* 13:

Williams, R.H. Draft 1988. *Advanced Gas Turbines for Stationary Power.*

World Bank. 1980. *Alcohol Production from Biomass in Developing Countries.* Washington, D.C.

World Bank. 1984. *A Survey of the Future Role of Hydroelectric Power in 100 Developing Countries.* Washington, D.C.

WRI (World Resources Institutes). 1985. *Tropical Forests, A Plan for Action.* Washington, D.C.

Wuebbles D.J., and J. Edmonds. 1988. *A Primer on Greenhouse Gases.* U.S. DOE/NBB-0083, March, Washington, D.C.

PART THREE: INDUSTRY

Bingemer, H., and P. Crutzen. 1987. The production of methane from solid wastes. *Journal of Geophysical Research* 92:2189.

Emcon Associates and Gas Recovery Systems. 1981. *Landfill Gas: An Analysis of Options.* Prepared for New York State Energy Office.

Escor, Inc. 1982. *Landfill Methane Recovery Part II: Gas Characterization.* Prepared for Gas Research Institute and Argonne National Laboratory, Department of Energy.

Gordon, J.G. 1979. *Assessment of the Impact of Resource Recovery on the Environment.* U.S. Environmental Protection Agency, Washington D.C.

Hoffman, J.S., and M.J. Gibbs. 1988. *Future Concentrations of Chlorine and Bromine.* U.S. Environmental Protection Agency Report No 400/1-88-005, Washington D.C.

Jansen, G.R. 1988. The Economics of Landfill Gas Projects. Paper from Governmental Refuse Collection and Disposal Association (GRCDA) Symposium. March, 1988.

Makhijani, A., A. Makhijani, and A. Bickel. 1988. *Saving Our Skins: Technical Potential and Policies for the Elimination of Ozone-Depleting Chlorine Compounds*. Environmental Policy Institute and the Institute for Energy and Environmental Research. Washington, D.C.

Pollock, C. 1987. Realizing Recycling's Potential. In *State of the World 1987*. W.W. Norton and Company, New York, pp. 101-121.

Rotty, R.M. 1987. A look at 1983 CO_2 emissions from fossil fuels (with preliminary data for 1984). *Tellus*. 39B:203-208.

Stauffer, R.F. 1988. *Energy Savings from Recycling*. National Appropriation Technology Assistance Service Paper.

U.S. EPA (U.S. Environmental Protection Agency). 1987. *Regulatory Impact Analysis: Protection of Stratospheric Ozone*. Office of Air and Radiation, Washington, D.C.

U.S. EPA (U.S. Environmental Protection Agency). 1988a. *Air Emissions from Municipal Solid Waste Landfills - Background Information for Proposed Standards and Guidelines*. Draft EIS. Washington, D.C.

U.S. EPA (U.S. Environmental Protection Agency). 1988b. *How Industry is Reducing Dependence on Ozone-Depleting Chemicals*. Washington, D.C.

U.S. EPA (U.S. Environmental Protection Agency). 1988c. *Report to Congress: Solid Waste Disposal in the United States*. Washington D.C.

Wilkey, M.L., R.E. Zimmerman, H.R. Isaacson. 1982. *Methane from Landfills: Preliminary Assessment Workbook*. Argonne National Laboratory, Department of Energy.

Zimmerman, R.E., G.R. Lytwynyshyn, and M.L. Wilkey. 1983. *Landfill Gas Recovery: A Technology Status Report*. Argonne National Laboratory, Department of Energy.

PART FOUR: FORESTRY

Akbari, H., J. Huang, P. Martien, L. Rainier, A. Rosenfeld, and H. Taba. 1988. The impact of summer heat islands on cooling energy consumption and CO_2 emissions. Paper presented at ACEEE Summer Study on Energy Efficiency in Buildings, Asilomar, Cal., August.

Anderson, D., and R. Fishwick. 1984. *Fuelwood Consumption and Deforestation in African Countries*. World Bank, Staff Working Papers, Number 704. The World Bank, Washington, D.C. 52 pp.

Ballard, R. 1984. Fertilization of plantations. In G. Bowen, and E. Nambiar, eds. *Nutrition in Plantation Forests*. Academic Press, London.

Bellagio, 1987. Statement of the Bellagio Strategy Meeting on Tropical Forests. Paper presented at the Bellagio Strategy Meeting on Tropical Forests, Bellagio, Italy, July 1-2, 1987.

Behmel, F., and I. Neumann. 1981. An example for agro-forestry in tropical mountain areas. Presented to the Workshop on Agro-Forestry in the African Humid Tropics, Ibadan, Nigeria.

Binkley, D. 1986. *Forest Nutrition Management*. J. Wiley and Sons, New York.

Booth, W. 1988. Johnny appleseed and the greenhouse. *Science* 242:19-20.

Brody, J. 1988. Scientists eye ancient African plant as better source of pulp for paper. *New York Times* December 13, 1988.

Brown, L. 1987. *State of the World 1987.* Norton, New York. 268 pp.

Brown, L. 1988a. *State of the World 1988.* Norton, New York.

Brown, S. 1988b. The global carbon cycle: Letter in response to Detweiler and Hall, 1988. *Science* 241:1739.

Conrad. 1986. The conservation reserve: Tree-planting windfall or tilting at windmills? *American Forests* September.

Davis, M. and C. Zabinski. in press. Rates of dispersal of North American trees: Implications for response to climatic warming. Forthcoming in *Proceedings of the Conference on Consequences of the Greenhouse Effect for Biological Diversity.* Washington, D.C., October 4-6, 1988, World Wildlife Fund-U.S.

Detwiler, R., and C. Hall. 1988a. Tropical forests and the global carbon cycle. *Science* 239:42-47.

Detwiler, R., and C. Hall. 1988b. The global carbon cycle: Response to letters. *Science* 241:1738-1739.

DOT (U.S. Department of Transportation). 1985. *Highway Statistics.* Federal Highway Administration, Department of Transportation, Washington, D.C.

Dover, M., and L. Talbot. 1987. *To Feed the Earth: Agro-Ecology for Sustainable Development.* World Resources Institute, Washington, D.C.

Dudek, D. 1988a. Offsetting new CO_2 emissions. unpublished paper, Environmental Defense Fund, New York.

Dudek, D. 1988b. Testimony before the Senate Committee on Energy and Natural Resources, September 19, 1988.

Dyson, F., and G. Marland. 1979. Technical fixes for the climatic effects of CO_2. In W. Elliott and L. Machta, eds. *Workshop on the Global Effects of Carbon Dioxide from Fossil Fuels.* U.S. Department of Energy, Washington D.C., CONF-7703585, 111-118. Proceedings of workshop at Miami Beach, Florida, March 7-11, 1977, held by DOE.

Farnum, P., R. Timmis, and J. Kulp. 1983. Biotechnology of forest yield. *Science* 219:694-702.

FAO (Food and Agriculture Organization). 1978. [[cropld nded to maintain alrdy low levels food supply. see St of Wld??]

FAO (Food and Agriculture Organization). 1981. *Tropical Forest Resources Assessment Project: Forest Resources of Tropical America, Technical Report 1; Forest Resources of Tropical Asia, Technical Report 2; Forest Resources of Tropical Africa, Technical Report 3.* Food and Agriculture Organization, Rome, UN32/6.1301-78-04.

FAO (Food and Agriculture Organization). 1986a. *Forest World Outlook Projections 1985-2000.* Forestry Paper 73. Rome.

FAO (Food and Agriculture Organization). 1986b. *The Outlook for Pulp and Paper to 1995.* FAO, Rome.

FAO (Food and Agriculture Organization). 1985. *Tropical Forestry Action Plan.* FAO, Rome.

FAO/WRI/World Bank/UNEP (Food and Agriculture Organization/World Resources Institute/World Bank/United Nations Environment Programme). 1987. *Tropical Forestry Action Plan.* Rome.

Fearnside, P.M. 1985. Brazil's Amazon forest and the global carbon problem. *Interciencia* 10:179-186.

Fearnside, P.M. 1986. Spatial concentration of deforestation in the Brazilian Amazon. *Ambio* 15:74-81.

Fearnside, P.M. 1987. Causes of deforestation in the Brazilian Amazon. In R. Dickinson, ed. *The Geophysiology of Amazonia.* John Wiley and Sons, New York.

Ford, E. 1984. The dynamics of plantation growth. In G. Bowen and E. Nambiar, eds. *Nutrition in Plantation Forests.* Academic Press, London. 17-52.

Fraser, M, J. Henry, and C. Vail. 1976. Design, operation and economics of the energy plantation. *Proceedings of Symposium, Clean Fuels from Biomass, Sewage, Urban Refuse, Agricultural Waste.* Institute of Gas Technology, Chicago. 371-95.

Gillis, M. in press. Malaysia: Public policies and the tropical forest. In R. Repetto, and M. Gillis, eds. *Public Policy and the Misuse of Forest Resources*, Cambridge University Press, Cambridge, U.K.

Goldemberg, J., T. Johansson, A. Reddy, and R. Williams. 1987. *Energy for Sustainable Development.* World Resources Institute, Washington, D.C.

Goodland, R. 1988. A major new opportunity to finance the preservation of biodiversity. In Wilson, E., ed. 1988. *Biodiversity.* National Academy Press, Washington, D.C.

Gradwohl, J., and R. Greenberg. 1988. *Saving the Tropical Forests.* Smithsonian Institution/Earthscan Books, Washington, D.C.

Grainger, A. 1987. TROPFORM: A model of future tropical hardwood supplies. In *Proceedings of CINTRAFOR Symposium on Forest Sector and Trade Models.* College of Forest Resources, University of Washington, Seattle. 205-12.

Grainger, A. 1988. Estimating areas of degraded tropical lands requiring replenishment of forest cover. *International Tree Crops Journal* 5:(1-2).

Hagenstein, P. 1988. Forests over the long haul. Paper presented at Conference on Natural Resources for the 21st Century, sponsored by American Forestry Association, Washington, D.C., November 14-17.

Hartshorn, G.S., R. Simeone, and J.A. Tosi. 1987. Sustained yield management of tropical forests: A synopsis of the Palcazu development project in the Central Selva of the Peruvian Amazon. English version of paper published in Spanish in Figueroa, J., F. Wadsworth, and S. Branham, eds. *Management of the Forests of Tropical America: Prospects and Technologies.* Institute of Tropical Forestry, Rio Piedras, Puerto Rico. 235-243.

Houghton, R. 1988a. The flux of CO_2 between atmosphere and land as result of deforestation and reforestation from 1850 to 2100. Unpublished paper prepared for Environmental Protection Agency.

Houghton, R. 1988b. Tropical lands available for reforestation. Unpublished paper and personal correspondence, submitted to Environmental Protection Agency.

Houghton, R., R. Boone, J. Melillo, C. Palm, G. Woodwell, N. Myers, B. Moore, and D. Skole. 1985. Net flux of carbon dioxide from tropical forests in 1980. *Nature* 316:(6029)617-20.

Houghton, R., R. Boone, J. Fruci, J. Hobbie, J. Melillo, C. Palm, B. Peterson, G. Shaver, G. Woodwell, B. Moore, D. Skole, and N. Myers. 1987. The flux of carbon from terrestrial ecosystems to the atmosphere in 1980 due to changes in land use: Geographic distribution of the global flux. *Tellus* 39B:122-39.

Houghton, R. 1988. The global carbon cycle: Letters in response to Detwiler and Hall, 1988. *Science* 241:1736.

IIED and WRI (International Institute for Environment and Development and World Resources Institute) 1987. *World Resources Report 1987*. International Institute for Environment and Development and World Resources Institute, Washington, D.C.

Janos, D. 1980. Mycorrhizae influence tropical succession. *Biotropica* 12 (supplement).

Jansen, D. 1988a. Tropical ecological and biocultural restoration. *Science* 239, January 15.

Jansen, D. 1988b. Tropical dry forests. In Wilson, E.O., ed. 1988. *Biodiversity*. National Academy Press, Washington, D.C.

JICA (Japan International Cooperative Agency). 1986. Technical guidance for afforestation on the grassland in Benmekat South Sumatra. ATA-186, April.

Kalpavriksh. 1985. The Narmada Valley Project--development or destruction? *The Ecologist* 15:(5-6).

Keay, R. 1978. Temperate and tropical: Some comparisons and contrasts. In J. Hawkes, ed. *Conservation and Agriculture*. Allanheld, Osmun and Co., Montclair, N.J. 243-48.

Kuusela. 1987. World Wood Supply. Ambio

Kuusela, K. 1987. Forest products -- world situation. *Ambio* 16:80-85.

Lanly, J. 1982. *Tropical Forest Resources*. FAO Forestry Paper No. 30. U.N. Food and Agriculture Organization, Rome.

Lashof, D. 1988. Technical background on reforestation as a CO_2 sink. Unpublished memorandum from Environmental Protection Agency to World Resources Institute, Washington, D.C., August 5.

Los Alamos National Laboratory. 1987. Reforestation is draining the water supply. *Los Alamos National Laboratory Newsletter* 7:31, August 14.

Lugo, A. 1988. Estimating reductions in the diversity of tropical forest species. In Wilson, E., ed. 1988. *Biodiversity*. National Academy Press, Washington, D.C.

Lundgren, and van Gelder. 1984. Agroforestry paper

Maguire, A., and J. Brown, eds. 1986. *Bordering on Trouble: Resources and Politics in Latin America.* Adler and Adler, Inc., Bethesda, MD, 448 pp.

Malingreau, J-P., and C. Tucker. 1988. Large-scale deforestation in the southeastern Amazon basin of Brazil. *Ambio* 17:(1).

Marland, G. 1988. The prospect of solving the CO_2 problem through global reforestation. U.S. Department of Energy, Office of Energy Research, Washington D.C. DOE/NBB-0082, 66 pp.

Matthews, E. 1983. Global vegetation and land use: New high-resolution data base for climate studies. *Journal of Climate and Applied Meteorology* 22:474-87.

Meier, A., and J. Friesen. 1987. Strategic planting. *Energy Auditor and Retrofiter* July/August.

Mergen, F., and J. Vincent. 1987. *Natural Management of Tropical Moist Forests: Silvicultural and Management Prospects of Sustained Utilization.* Yale School of Forestry and Environmental Studies, New Haven.

McNeely, J. 1988. *Economics and Biological Diversity: Developing and Using Economic Incentives to Conserve Biological Resources.* International Union for Conservation of Nature and Natural Resources, Gland, Switzerland. 236 pp.

McNeely, J., and K. Miller. 1984. *National Parks Conservation and Development: The Role of Protected Areas in Sustaining Society.* Smithsonian Institution Press, Washington, D.C..

Miller, A., I. Mintzer, and S. Hoagland. 1986. *Growing Power: Bioenergy for Developing and Industrialized Countries.* World Resources Institute, Washington, D.C..

Moll, G. 1987. The state of our city forests. *American Forests.* May/June

Moll, G. 1988. Testimony before Senate Committee on Environment and Public Works, Washington, D.C., September.

Mwandosya, M., and M. Luhanga. 1985. Energy use patterns in Tanzania. *Ambio* 14:(4-5)237-41.

Myers, N. 1988. Tropical forests and climate. Unpublished essay dated September 28, and personal communication, Washington, D.C., October 4, 1988.

NAERC (North American Electric Reliability Council). 1987. *Electricity Supply and Demand for 1987-1996.* North American Electric Reliability Council, Princeton, NJ.

Nambiar, E. 1984. Plantation forests: Their scope and a perspective on plantation nutrition. In G. Bowen, and E. Nambiar, eds. *Nutrition in Plantation Forests.* Academic Press, London.

NASF (National Association of State Foresters). 1988. Personal communication with board of directors of NASF, annual board meeting, Washington, D.C., December.

NASF/USFS (National Association of State Foresters/U.S. Forest Service). 1988. *Forestry: A Community Tradition.* National Association of State Foresters/U.S. Forest Service, Washington, D.C., brochure.

Nations, J., and D. Komer. 1983. Central America's tropical rainforests: Positive steps for survival. *Ambio* 12:(5).

OECD (Organization for Economic Co-Operation and Development). 1987. *Energy Statistics, 1970-1985*. Volume 1. OECD, Paris.

OTA (Office of Technology Assessment, U.S. Congress). 1984. *Technologies to Sustain Tropical Forest Resources*. OTA-F-214, Washington, D.C., Government Printing Office.

Open Lands Project. 1987. Untitled article on urban forest replanting. *Open Lands Newsletter* XI:(2).

Palmberg, C. 1981 A vital fuelwood gene pool is in danger. *Unasylva* 33. (133)

Ranney, J., L. Wright, and P. Layton. 1987. Hardwood energy crops: The technology of intensive culture. *Journal of Forestry* September:17-28.

Repetto, R. 1988. *The Forest for the Trees? Government Policies and the Misuse of Forest Resources* (WRI, 1988)

Rich, B. 1989. The politics of tropical deforestation in Latin America: The role of the public international financial institutions. Forthcoming in S. Hecht and J. Nations, eds. *Social Dynamics of Deforestation: Processes and Alternatives*. Cornell University Press.

Robinette, G. 1977. *Landscape Planning for Energy Conservation*. Environmental Design Press, prepared by American Society of Landscape Architecture Foundation.

Ruston, J. 1988. Developing markets for recycled materials. Unpublished paper, Environmental Defense Fund, New York, 49 pp.

Sampson, R. 1988. ReLeaf for global warming. *American Forests* November/December:9-14.

Sanchez, P. 1988. Deforestation reduction initiative: An imperative for world sustainability in the twenty-first century. Paper presented at the Bureau of Science and Technology, U.S. Agency for International Development, Washington, D.C., July 22, 1988.

Sanchez, P., and J. Benites. 1987. Low-input cropping for acid soils of the humid tropics. *Science* 238:1521-1527.

Schwartzman, S. and M. Allegretti. 1987. *Extractive Reserves: A Sustainable Development Alternative for Amazonia*. Report to World Wildlife Fund-U.S.

Sedjo, R. 1988. The global carbon cycle: Letter in response to Detwiler and Hall, 1988. *Science* 241:1737.

Sedjo, R., and A. Solomon. in press. Climate and forests. Paper prepared for Workshop on Controlling and Adapting to Greenhouse Forcing, held by Environmental Protection Agency and National Academy of Sciences, Washington, D.C., June 14-15, 1988. Forthcoming in N. Rosenberg, W. Easterly, P. Crosson, and J. Darmstaedter, eds. *Greenhouse Warming: Abatement and Adaptation*. Resources for the Future, Washington, D.C.

Setzer, A., and M. Pereira. 1988. Amazon biomass burnings in 1987 and their tropospheric emissions. Unpublished manuscript, Brazilian Space Research Center, Sao Jose dos Campos, Brazil.

Shands, W., and J. Hoffman. 1987. *The Greenhouse Effect, Climate change, and U.S. Forests*. The Conservation Foundation, Washington, D.C. 304 pp.

Shen, S. 1988. Biological engineering for sustainable biomass production. In Wilson, E., ed. 1988. *Biodiversity*. National Academy Press, Washington, D.C.

Shugart, H., and D.L. Urban. no date. Using Computer Models to Project Ecosystem Responses. In WWF. Proceedings of the Conference on Consequences of the Greenhouse Effect for Biological Diversity. Washington, D.C., October 4-6, 1988. in press.

Smith, J., and D. Tirpak, eds. 1989. *The Potential Effects of Global Climate Change on the United States*. Environmental Protection Agency, Report to Congress, Washington, D.C.

Steinbeck, K., and C. Brown. 1976. Yield and utilization of hardwood fiber grown on short rotations. In *Applied Polymer Symposium* no. 28, John Wiley and Sons, New York, 393-401.

Stowe, R. 1987. United States foreign policy and the conservation of natural resources: The case of tropical deforestation. *Natural Resources* 27:55-101.

Trexler, M. in press. *Social Forestry as a Response to Global Warming: A Preliminary Analysis and Case Study*.

Uhl, C. 1988. Restoration of degraded lands in the Amazon basin. In Wilson, E., ed. *Biodiversity*. National Academy Press, Washington, D.C. 326-32.

USDA (U.S. Department of Agriculture). 1982. *National Resources Inventory*. Soil Conservation Service, USDA, Washington, D.C., Statistical Bulletin 686.

U.S. EPA (U.S. Environmental Protection Agency). 1987. The transportation sector. unpublished report, Environmental Resources Inventory Project, Office of Policy Analysis, Washington D.C.

U.S. EPA (U.S. Environmental Protection Agency). 1988. The solid waste dilemma: an agenda for action. Office of Solid Waste, Washington, D.C., 70 pp. plus 3 appendices.

USFS (U.S. Forest Service, Department of Agriculture). 1982. *An Analysis of the Timber Situation in the United States, 1952-2030*. Forest Resource Report No. 23, Washington, D.C.

Vietmeyer, N. 1986. Lesser-known plants of potential use in agriculture and forestry. *Science* 232:1379-1384.

Villavicencio, V. 1983. The environmental aspects of Ipil-Ipil plantations. *Renewable Sources of Energy* 1:(1)41-65.

Wadsworth, F.H. 1983. Production of useable wood from tropical forests. In F.B. Golley, ed. *Tropical Rain Forest Ecosystems: Structure and Function*. Elsevier Scientific Publishing Company, New York. 281.

Walter, D.K. 1988. Letter to Daniel A. Lashof (U.S. Environmental Protection Agency) re: future potential worldwide bioenergy production. U.S. Department of Energy, Washington, D.C. May 26, 1988.

Winterbottom, R., et al. in press. Draft manuscript on agroforestry systems in preparation by study group on agroforestry. American Association for the Advancement of Science, David Pimentel, chair.

Winterbottom, R., and P. Hazelwood. 1987. Agroforestry and sustainable development: Making the connection. *Ambio* 16:(2-3).

Woodwell, G., R. Houghton, T. Stone, and A. Park. 1986. Changes in the area of forests in Rondonia, Amazon Basin, measured by satellite imagery. In Trabalka, J.R. and D.E. Reichle, eds. *The Changing Carbon Cycle: A Global Analysis*. Springer-Verlag. New York. 242-257.

World Bank. 1986. *The World Bank's Operational Policy on Wildlands: Their Protection and Management in Economic Development*. The World Bank, Washington, D.C. 21 pp.

Woodwell, G. 1987. The warming of the industrialized middle latitudes. Paper presented at the Workshop on Developing Policies for Responding to Future Climate Change, Villach, Austria.

Woodwell, G. 1988. The global carbon cycle: Letter in response to Detwiler and Hall, 1988. *Science* 241:1736-1737.

WRI and IIED (World Resources Institute and International Institute for Environment and Development). 1986. *World Resources Report 1986*. World Resources Institute and International Institute for Environment and Development, Washington, D.C.

WRI and IIED (World Resources Institute and International Institute for Environment and Development). 1988. *World Resources Report 1988-89*. World Resources Institute, International Institute for Environment and U.N. Environment Programme, Washington, D.C.

WRI (World Resources Institute). 1988. Power company to fund reforestation to offset carbon dioxide emissions, slow greenhouse effect. press release and press conference, October 11, Washington, D.C.

WWF (World Wildlife Fund-U.S.). 1988. The Forestry Fund: An endowment for forest protection, management, and reforestation in Costa Rica. Proposal by Conservation Foundation/World Wildlife Fund (Washington, D.C.) and Fundacion Neotropica (San Jose, Costa Rica), May 1988, 18 pp. plus 2 annexes.

WWF (World Wildlife Fund-U.S.). *Proceedings of the Conference on Consequences of the Greenhouse Effect for Biological Diversity*. Washington, D.C., October 4-6, 1988. no date.

PART FIVE: AGRICULTURE

Anderson, I., and J. Levine. 1986. Relative rates of nitric oxide and nitrous oxide production by nitrifers, dentrifers, and nitrate respirers. *Applied and Environmental Microbiology* 86:938-945.

Barker, R., R. Herdt, and B. Rose. 1985. *The Rice Economy of Asia*. Resources for the Future, Washington D.C.

Bauchop, T. 1967. Inhibition of rumen methanogenesis by methane analogues. *Journal of Bacteriology* 94:171-175.

Benz, D., and D. Johnson. 1982. The Effect of Monensin on Energy Partitions by Forage Fed Steers. Paper from Colorado State University. Fort Collins, Colorado.

Blackmer, A.M., S.G. Robbins, and J.M. Bremner. 1982. Diurnal variability in rate of emission of nitrous oxide from soils. *Soil Science Society of America Journal* 46:937-942.

Blaxter, K. and J. Czerkawski. 1966. Modifications of the methane production of the sheep by supplementation of its diet. *Journal of the Science of Food and Agriculture* 17:417-421.

Blaxter, K.L., and F.W. Wainman. 1964. Utilization of food by sheep and cattle. *Journal of Agricultural Science* 57:419.

Blaxter, K.L., and J.L. Clapperton. 1965. Prediction of the amount of methane produced by ruminants. *British Journal of Nutrition* 19:511.

Breitenbeck, G.A. 1988. Presentation at EPA's Workshop on Agriculture and Climate Change. Washington DC, February, 1988, and subsequent correspondence.

Breitenbeck, G.A., A.M. Blackmer, and J.M. Bremner. 1980. Effects of different nitrogen fertilizers on emission of nitrous oxide from soil. *Geophysical Research Letters* 7:85-88.

Breitenbeck, G.A., and J.M. Bremner. 1986a. Effects of various nitrogen fertilizers on emission of nitrous oxide from soils. *Biology and Fertility of Soils* 2:195-199.

Breitenbeck, G.A., and J.M. Bremner. 1986b. Effects of rate and depth of fertilizer application on emission of nitrous oxide from soil fertilized with anhydrous ammonia. *Biology and Fertility of Soils* 2:201-204.

Bremner, J.M., and A.M. Blackmer. 1978. Nitrous oxide: Emission from soils during nitrification of fertilizer nitrogen. *Science* 199:295-296.

Bremner, J. M., and A. M. Blackmer. 1980. *Natural and Fertilizer-Induced Emissions of Nitrous Oxide from Soils*. Air Pollution Control Association, Montreal, Quebec, p.1-13.

Bremner, J.M., G.A. Breitenbeck, and A.M. Blackmer. 1981. Effect of anhydrous ammonia fertilization on emission of nitrous oxide from the soil. *Journal of Environmental Quality* 10:77-80.

Cicerone, R.J., and J.D. Shetter. 1981. Sources of atmospheric methane: Measurements in rice paddies and a discussion. *Journal of Geophysical Research* 86:7203-7209.

Crutzen, P., I. Aselmann, and W. Seiler. 1986. Methane production by domestic animals, wild ruminants, other herbivorous fauna, and humans. *Tellus* 38B:271-284.

Dalrymple, D. 1986. *Development and Spread of High-Yielding Rice Varieties in Developing Countries*. Bureau for Science and Technology, Agency for International Development, Washington, D.C.

Delwiche C. 1988. Methane emission rates. Presented at U.S. EPA Workshop on Agriculture and Climate Change, Washington, D.C., February 29-March 1, 1988.

Eichner, M. 1988. *Current Knowledge of Fertilizer-Derived Nitrous Oxide Emissions*. Prepared for U.S. EPA, Washington, D.C.

Eliassen, K., and W. Zawadzki. 1985. Effects of monensin on polyamine formation in rumen liquid of small ruminants. *Zentralblatt for Veterinarmedizin (Journal of Veterinary Medicine)* 32:356-367.

English, B., J. Maetzold, B. Holding, and E. Heady. 1984. Future Agricultural Technology and Resource Conservation. Proceedings of the RCA Symposium: Future Agricultural Technology and Resource Conservation, Dec. 5-9, 1982, Washington, D.C. The Iowa State University Press, Ames, Iowa.

Eriksen A., M. Kjeldby, and S. Nilsen. 1985. The effect of intermittent flooding on the growth and yield of wetland rice and nitrogen-loss mechanism with surface applied and deep placed urea. *Plant and Soil* 84:387-401.

FAO (Food and Agricultural Organization of the United Nations). 1986. *1985 FAO Production Yearbook* Vol. 39. FAO, Rome.

FAO (Food and Agriculture Organization of the United Nations). 1987. *1986 FAO Fertilizer Yearbook.* Vol 36, FAO, Rome. 124 pp.

Frohberg, K., and P.R. Vande Kamp. 1988. Results of Eight Policy Scenarios for Reducing Agricultural Sources of Trace Gas Emissions. Center for Agricultural and Rural Development, Iowa State University. Report prepared for U.S. EPA, Washington, D.C.

Fuller, J., and D. Johnson. 1981. Monensin and lasalocid effects on fermentation in vitro. *Journal of Animal Science* 53:1574-1580.

Fung, I. 1988. Agriculture Emission Coefficients Estimates. Presented at U.S. EPA Workshop on Agriculture and Climate Change, Washington, DC, February 29-March 1, 1988, and subsequent correspondence.

Fung, I., E. Matthews, and J. Learner. 1988. Trace Gas Emissions Associated with Agricultural Activities. Prepared for the EPA Workshop on Agriculture and Climate Change, Washington, D.C. February 1988.

Garrett, W.H. 1982. Influence of monensin on the efficiency of energy utilization by cattle. In Ekern, A., and F. Sundstol, ed. Energy metabolism of farm animals. *European Association for Animal Production publication* 29:104.

Groffman, P., P. Hendrix, and D. Crossley. 1987. Nitrogen dynamics in conventional and no-tillage agroecosystems with inorganic fertilizer or legume nitrogen inputs. *Plant and Soil* 97:315-332.

Hauck, R. 1988. Projections of Fertilizer Use. Presented at U.S. EPA Workshop on Agriculture and Climate Change, Washington, D.C., February 29-March 1, 1988, and subsequent correspondence.

Hubbert, M., M. Branine, M.L. Galyean, G.T. Lofgreen, and D.P. Garcia. 1987. *Influence of Alternative Feeding of Monensin and Lassaload on Performance of Feedlot Heifers-Preliminary Data.* Clayton Livestock Research Center Progress Report No. 47, Clayton, New Mexico.

IIED and WRI (International Institute for Environment and Development and World Resources Institute). 1987. *World Resource Report - 1987.* Basic Books, New York.

IRRI (International Rice Research Institute). 1986. *World Rice Statistics 1985.* IRRI, Manila, Philippines.

Johnson, D.E. 1974. Adaptational responses in nitrogen and energy balance of lambs fed a methane inhibitor. *Journal of Animal Science* 38:154-157.

Johnson, D.E. 1988. Livestock Emissions Estimates. Presented at U.S. EPA Workshop on Agriculture and Climate Change, Washington, D.C., February 29-March 1, 1988, and subsequent correspondence.

Leedle, J., and R. Greening. 1988. Postprandial changes in methanogenic and acidogenic bacteria in the rumens of steers fed high- of low-forage diets once daily. *Applied and Environmental Microbiology* 54:(2) 502-506.

Lerner, J., E. Matthews, and I. Fung. 1988. Methane emission from animals: A global high-resolution database. *Global Biogeochemical Cycles* 2:139-156.

Mikkelsen, D. 1988. Technological Options for Limiting Emissions. Presented at U.S. EPA Workshop on Agriculture and Climate Change, Washington, D.C., February 29-March 1, 1988.

Moe, P. and H. Tyrell. 1979. Methane production in dairy cows. *Journal of Dairy Science* 62:1583-1586.

Olmeda, R., and F. Abruna. 1986. Four nitrogen levels and three water management systems on rice yield and nitrogen recovery. *The Journal of Agriculture of the University of Puerto Rico* 70:197-205.

OTA (Office of Technology Assessment). 1986. *Technology, Public Policy, and the Changing Structure of American Agriculture*. OTA, United States Congress, Washington, D.C.

Patrick, W. 1982. Nitrogen Transformations in Submerged Soils. *Nitrogen in Agricultural Soils*. Agronomy Monograph no. 22. Madison, Wisconsin.

Raun, N. 1981. Livestock as a buffer against climate change. In Winrock. *Food and Climate Review 1980-81*. Food and Climate Forum. Aspen, CO.

Rumpler, W., D. Johnson, and B. Bates. 1986. The effect of high dietary cation concentration on methanogenesis by steers fed diets with and without ionophores. *Journal of Animal Science*. 62:1737-1741.

Schuh, G.E. 1988. Agricultural Policies for Climate Changes Induced by Greenhouse Gases. Report prepared for U.S. EPA, Washington, D.C.

Smith, C., M. Wright, and W. Patrick. 1983. The effect of soil redox potential and pH on the reduction and production of nitrous oxide. *Journal of Environmental Quality* 12:186-188.

Stangel, P. 1988. Technological Options Affecting Emissions. Presented at U.S. EPA Workshop on Agriculture and Climate Change, Washington, DC, February 29-March 1, 1988, and subsequent telephone conversations.

Stoddart, L., A. Smith, and T. Box. 1975. *Rangeland Management*. McGraw-Hill, New York.

Thorbek, G. 1980. *Commission on Animal Nutrition, Factors Influencing Energy Losses During Metabolism. Evaluation of Energy Systems. Energy Losses in Methane*. National Institute of Animal Science, Copenhagen.

Tyrell, H. 1988. Livestock Emissions Estimates. Presented at U.S. EPA Workshop on Agriculture and Climate Change, Washington, DC, February 29-March 1, 1988.

U.S. EPA (U.S. Environmental Protection Agency). 1988. Meeting Summary of Workshop on Agriculture and Climate Change. A report prepared for EPA by J,T&A, Inc. Washington, D.C.

Wedegaertner, T., and D. Johnson. 1983. Monensin effects on digestibility, methanogenesis and heat increment of a cracked corn-silage diet fed to steers. *Journal of Animal Science* 57:168-177.

Whitelaw, F., J. Eadie, L. Bruce, and W. Shand. 1984. Methane formation in faunted and ciliate-free cattle and its relationship with rumen volatile fatty acid proportions. *British Journal of Nutrition* 52:261-275.

Winrock. 1977. *Ruminant Products: More than Meat and Milk.* Windrock International Livestock Research and Training Center. Morrilton, Arkansas.

World Bank. 1988. *World Bank, FAO, UNIDO Fertilizer Working Group Nitrogen Supply, Demand and Balances for 1986/87 to 1992/93.* Washington, D.C.

CHAPTER VIII

POLICY OPTIONS

FINDINGS

- If limiting U.S. and global emissions of greenhouse gases is desired, government action will be necessary. Market prices of energy from fossil fuels, products made with CFCs, forest and agricultural products and other commodities responsible for greenhouse gas emissions do not reflect the costs associated with the risk of climate change. As a result, increases in population and economic activity will cause emissions to grow in the absence of countervailing government policies to modify and/or supplement market signals.

- Policy preferences will vary among nations. However, nations can choose among many complementary policy options to reduce emissions consistent with governmental systems and other societal needs.

- There is an important distinction between short-term and long-term policy options. In the short-term, the most effective means of reducing emissions is through strategies that rely on pricing and regulation. In the long-term, policies to increase research and development of new technologies and to enhance markets through information programs, government purchases, and other means could also make a major contribution.

- The most direct means of allowing markets to incorporate the risk of climatic change is to assure that the prices of fossil fuels and other sources of greenhouse gases reflect their full social costs. It may be necessary to impose emission fees on these sources according to their relative contribution to global warming in order to accomplish this goal. This would also raise revenues that could finance other programs. The acceptability of such fees will vary among countries, but would be enhanced if equitably structured.

- Regulatory programs would be a necessary complement when pricing strategies are not effective or produce undesirable impacts. In the U.S., greenhouse gas emissions are influenced by existing federal regulatory programs to control air pollution, increase energy efficiency, and recycle solid waste. Reducing greenhouse gas emissions could be incorporated into the goals of these programs. New programs could focus directly on reducing greenhouse gas emissions through requirements such as emissions offsets (e.g., tree-planting), performance standards, or marketable permits.

- Over the long-term, other policies may be needed to reduce emissions and can complement pricing and regulatory strategies. Other policy options include redirecting research and development priorities in favor of technologies that could reduce greenhouse gas emissions, information programs to build understanding of the problems and solutions, and the selective use of government procurement to promote markets for technological alternatives.

- State and local government policies in such areas as utility regulation, building codes, waste management, transportation planning, and urban forestry could make an important contribution to reducing greenhouse gas emissions.

- Voluntary private efforts to reduce greenhouse gas emissions have already provided valuable precedents for wider action and could play a larger role in the future.

- Government policy is already exerting considerable influence on the rate of growth in greenhouse gases. Policies adopted or under consideration to promote greater use of coal, reduce required improvements in automobile efficiency, subsidize electricity consumption, and similar measures may significantly accelerate the rate of U.S. greenhouse gas emissions. A modeling exercise shows that a combination of policies that increase greenhouse gas emissions globally could substantially increase the rate of climatic change.

- Policies to create incentives to improve energy efficiency, promote renewable energy technologies, encourage tree planting, and other strategies for reducing emissions could substantially reduce emissions. A modeling exercise combining aggressive use of these policies shows that such measures could reduce the rate of warming to levels that may greatly enhance the potential for adaptation by natural and economic systems.

- Near-term actions to reduce greenhouse gas emissions will be necessary if it is decided to limit the concentration of greenhouse gases and facilitate future reductions. Atmospheric concentrations of most greenhouse gases will decline much more slowly than emissions. Delay would allow time to increase knowledge of risks and refine the choice of policies but could substantially increase the cost and reduce the effectiveness of policy responses. Time will be needed to agree on and implement policy responses since rapid changes in patterns of energy use and the industrial infrastructure responsible for emissions are likely to be disruptive and expensive.

- Most of the policies necessary to limit the buildup of greenhouse gases, such as improving energy efficiency and reversing deforestation, would produce other substantial short-term benefits. Policies to limit emissions can therefore be implemented consistent with other important economic and environmental goals.

INTRODUCTION

If the government desires to stabilize the concentration of greenhouse gases, short-term actions must be considered. Market forces do not reflect the risks of climate change and therefore, barring government intervention, emissions--and the risks of climate change--can be expected to grow. The costs of taking action may increase with time; it takes many years to develop new technologies, implement policies and replace the existing capital stock. The chemistry of the atmosphere also has long lags so that even large reductions in emissions are likely to cause atmospheric concentrations of greenhouse gases to decline very slowly or even increase. The earth would continue to warm and the climate to change for decades after atmospheric concentrations of greenhouse gases were stabilized.

Near-term government action to reduce emissions may consist entirely of policies that will contribute other substantial short-term benefits. The justification for policies that reduce greenhouse gas emissions may, therefore, be much greater than it would appear from a traditional comparison of costs and benefits. Finally, decisions not to take action to reduce emissions could also facilitate policies that result in increased emissions since such policies are presumably less likely if they contravene the purpose of other programs. This is a significant risk as we discuss later in this chapter; several recently proposed policies have the potential to significantly increase greenhouse gas emissions.

The government may pursue many policy alternatives to reduce greenhouse gas concentrations. In the short-term, economic incentives and regulatory strategies will be most effective, but in the long-run other policies will complement and support these approaches. Strategies to reduce greenhouse gas emissions can contribute to other goals, including economic growth, enhanced energy security, and a cleaner environment.

The preceding chapter shows the wide range of measures available to limit greenhouse gas emissions. Many of these measures are nearly cost-competitive or are expected to be soon. Is government intervention to reduce emissions therefore necessary? The analysis done for this report strongly suggests that without such action emissions will increase significantly. Without government intervention, there may be insufficient market incentive to develop and adopt the measures identified in Chapter VII.

In the absence of government intervention, market prices will not reflect the cost to society of activities that are detrimental to the environment. Accordingly, we will not see a reduction in consumer demand of goods and services that increase environmental risk.

The obstacles to bringing about consumer response to environmental dangers are particularly challenging for global problems like ozone depletion and the greenhouse effect. In this situation, there is the danger of what is commonly termed the "tragedy of the commons," the ecological destruction that can occur from uncontrolled use of shared resources like the atmosphere. There is probably no country for which reductions in global warming provide an adequate economic incentive to reduce greenhouse gas emissions unilaterally, even though such action could yield substantial global benefits. From any one country's viewpoint, the costs of controlling emissions may exceed the benefits since, without international agreement, reductions achieved by one nation may be offset by another. Therefore, even though the entire world may be better off as a result of efforts to lower emissions, new economic incentives are necessary to lead the market to a socially efficient outcome.

The remainder of this chapter describes the range of domestic and international policies that could be adopted to reduce emissions of greenhouse gases, and also discusses the results of modeling analyses of government actions that could reduce or increase such emissions.

There are many possible policy responses to reduce greenhouse gas emissions; none is likely to be totally effective without other supporting programs. However, as previously mentioned, consumers are not likely to reduce their demand for products and services that produce emissions of greenhouse gases unless the cost begins to exceed the benefits derived. In addition, other policies will be much less effective in the absence of economic incentives. In the short term, economic incentives and regulatory programs are likely to be the most effective policies. In the long term, research and development, information and educational activities, government procurement, and other policies can make a substantial contribution.

The policies discussed here are most applicable to the United States. However, many of the ideas reviewed are likely to be potentially suitable to other countries. Policies most suitable for developing countries are discussed in Chapter IX. However, some programs developed in the U.S., such as energy conservation programs financed by electric utilities, may have even greater relevance for developing countries, which have higher interest costs and a greater need for new generating capacity.

INTERNALIZING THE COST OF CLIMATE CHANGE RISKS

One of the most potentially effective ways to promote energy efficiency and other strategies to reduce greenhouse gas emissions is to increase the cost of activities responsible for emissions to reflect the risks of climate change. The government could increase the price of fossil fuels and other sources of greenhouse gases by imposing charges or fees and at the same time reducing the price of desirable alternatives by providing direct or indirect subsidies. Prices of some relevant commodities, particularly electricity, are already regulated and could be adjusted to promote reduced emissions. Such policies have already been adopted to varying degrees in some states to promote other economic and environmental goals. However, the limitations of economic approaches should also be recognized; other types of policies may be more effective in redressing some market failures, and these can complement policies based on adjusting prices.

Evidence of Market Response to Economic Incentives: Energy Pricing

Proper pricing of energy to reflect the reality of the global warming threat could substantially reduce emissions. It was long assumed (based on historical data) that there was a constant relationship between the rate of growth in energy use and the rate of growth of GNP (Weinberg, 1988). The historical constancy of the relationship was often explained as a consequence of technology and stable or declining fuel prices. The chief flaw in this argument was its failure to anticipate the effects of rising real energy prices. During the century preceding 1973, the relative price of energy had fallen. However, between 1973 and 1985 the price of energy rose, relative to other goods and services, and in accordance with basic economic theory; growth in energy demand slowed, and by some measures declined significantly.

Prior to 1973, U.S. energy demand grew at a rate only slightly less than the rate of growth in real GNP. Had this trend continued, U.S. energy demands would nearly have doubled between 1973 and 1985. Had pre-1972 energy-use trends continued, by 1984 the United States would have been consuming almost 40 percent more energy than it actually did (U.S. DOE, 1987c). But efficiency improvements induced by significant oil price increases enabled the U.S. to hold energy use to 1973 levels while expanding the economy by 40 percent. By one estimate, energy efficiency improvements now save the U.S. economy $160 billion annually.[1]

---

[1] The $160 billion figure is derived by first calculating what total expenditures on energy in 1985 would have been had energy intensities remained constant at 1973 levels and 1985 price levels prevailed, and then subtracting actual expenditures on energy in 1985. For further discussion, see Chandler et al. (1988) and U.S. DOE (1987c).

U.S. energy efficiency improved markedly between 1973 and 1985 while energy prices rose markedly (Figure 8-1). However, efficiency improvements in the U.S. and many other industrialized countries have leveled off in recent years. Whereas energy intensity in the nations of the European Economic Community declined 20 percent from 1973 to 1982, the corresponding figure for 1982 to 1986 is 2.4 percent (EEC, 1988). One important factor is that oil prices have fallen continuously since 1981. Oil prices dropped precipitously in 1986: World oil prices fell from about $25 per barrel in January 1986 to about $10 per barrel in July 1986. Prices have since fluctuated in the $15 per-barrel range. Adjusted for inflation, gasoline prices in the first half of 1988 were 27 percent less than in 1985 and 48 percent less than in 1980 (Geller, 1988). The rapid decline in world energy prices has slowed the rate of improvement in energy efficiency. This trend is a predictable response to pricing and may continue for some time given Department of Energy forecasts for relatively constant or even declining energy prices in the short term (EIA, 1988).

The performance of other countries suggests that higher energy prices are conducive to greater efficiency. Japan, France, and West Germany, much more dependent on imported oil than the U.S., currently use much less energy per unit of output than does the United States (see Table 8-1). Many developing countries are much less energy efficient than the U.S., in part because they subsidize energy prices.

In recent years, the U.S. has debated and rejected higher oil import fees and gasoline taxes as an instrument of energy policy primarily out of concern that such policies would have significant regional and national macroeconomic impacts (U.S. DOE, 1987; U.S. DOE, 1988). However, without such policies the incentive to purchase more efficient vehicles has declined and the trend has been toward a renewed emphasis on more powerful, less efficient engines (Geller, 1988; Bleviss, 1988).

FIGURE 8-1

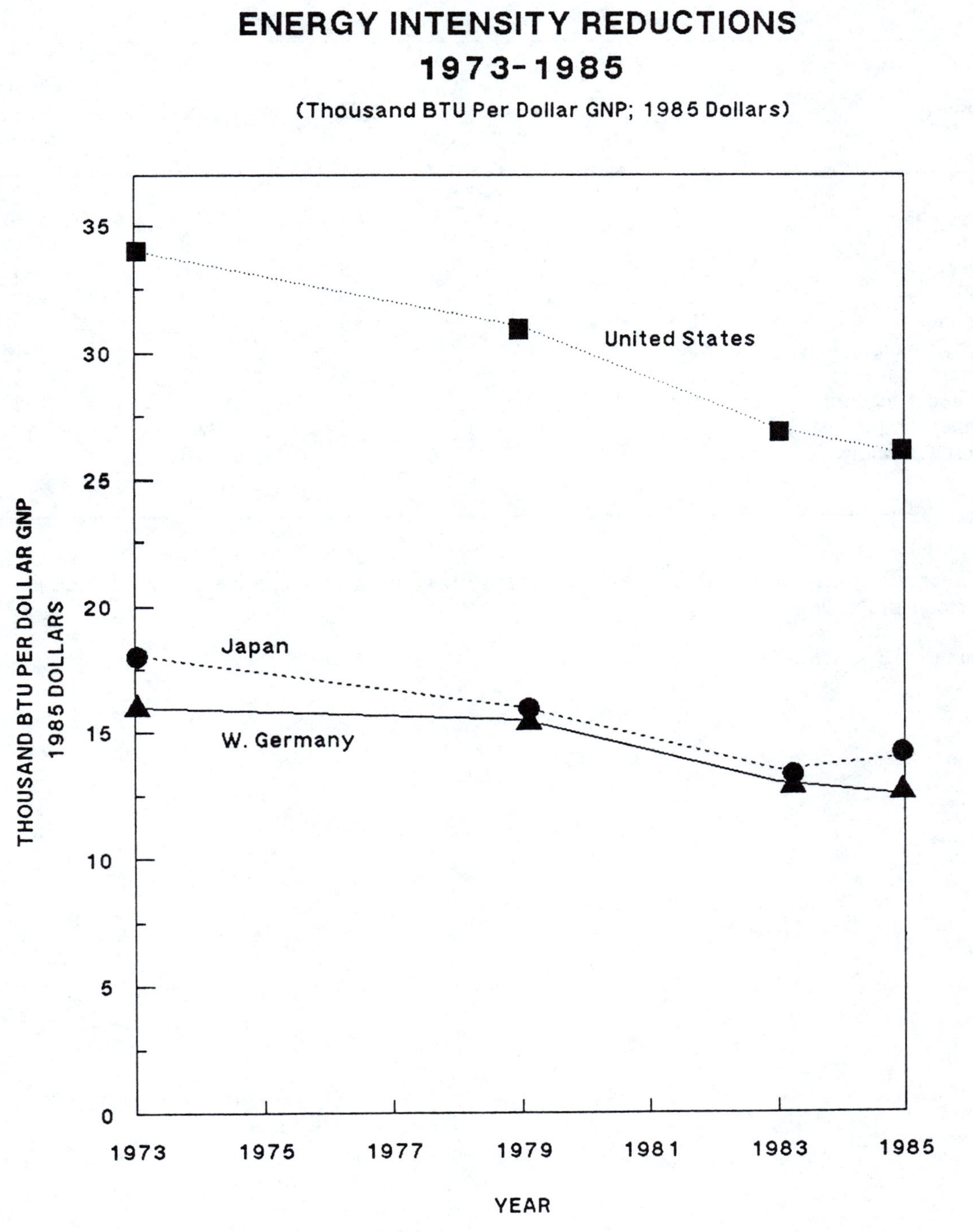

Source: Chandler, et al., 1988

TABLE 8-1

Energy Intensity of Selected National Economies, 1973-85

Country	1973	1979	1983	1985	Change, 1973-85 (percent)
		(megajoules per 1980 dollar of GNP)			
Australia	21.6	23.0	22.1	20.3	-6
Canada	38.3	38.8	36.5	36.0	-6
Greece [a]	17.1	18.5	18.9	19.8	+16
Italy	18.5	17.1	15.3	14.9	-19
Japan	18.9	16.7	13.5	13.1	-31
Netherlands	19.8	18.9	15.8	16.2	-18
Turkey	28.4	24.2	25.7	25.2	-11
United Kingdom	19.8	18.0	15.8	15.8	-20
United States	35.6	32.9	28.8	27.5	-23
West Germany	17.1	16.2	14.0	14.0	-18

[a] Energy intensity increased as a result of a move toward energy-intensive industries such as metal processing

Source: IEA, 1987.

Existing federal law gives an additional economic incentive for efficiency through the so-called "gas-guzzler" tax. For 1986 and later models, the law imposes a $500 tax on any car with a fuel economy rating less than 22.5 mpg and a $3,850 tax on models with ratings less than 12.5 mpg. It was originally proposed that the revenues collected be given to purchasers of highly efficient cars in the form of a rebate; however, this proposal was rejected because at the time the beneficiaries would have been almost exclusively buyers of imported cars (see Bleviss, 1988).

Gas-guzzler fees could be used as a potentially more effective and acceptable substitute for gasoline taxes if set equal to the amount that would be collected from a gas tax over the expected lifetime of the automobile. For example, a 50 cents per gallon tax would become a $1,000 excise tax on a 25 mpg car but only $333 on a 75 mpg car. An advantage of this strategy is that, unlike gasoline taxes, it takes advantage of the first-cost sensitivity of consumers in a non-regressive way, taxing new inefficient vehicles the most.[2] The structure of the charge could vary with vehicle class so as not to unduly favor small cars over larger models. The effectiveness of such a system could be enhanced by rebating some of the revenue to buyers of the most efficient models, a concept that now may be more acceptable since several American models rate among the most efficient.

The market for electricity also demonstrates the influence of prices and competition on growth in demand. Until about 1970, the efficiency of new plants was improving and prices of electricity generation were therefore declining, contributing to steady growth in the demand for electricity. From about 1970 through 1982, electricity prices rose mainly as a result of higher fuel costs and interest rates, construction delays, the cost of environmental controls, and the end to

---

[2] First-cost sensitivity refers to the tendency, common to many consumers, to prefer a product that has a low initial cost to a higher-priced alternative which might be much more cost efficient over the long term. In other words, the upfront costs to consumers tend to dominate the purchase decision more than a conventional economic analysis would predict. The use of a hefty excise tax may affect consumer decisions more acutely than would the gasoline expenses for operation of an inefficient vehicle, which occur in smaller amounts and at intervals rather than all at once.

significant improvements in generating efficiency. In the last five years prices have been stable or declining as fuel costs stabilized and the amount of new capacity being added slowed in response to the decline in demand (Figure 8-2). The Department of Energy projects continued stable prices until the mid-1990s, followed by modest price increases (EIA, 1987c).

Financial Mechanisms to Promote Energy Efficiency

Since the mid-1970s, the U.S. has accumulated a substantial body of experience with a variety of financial incentives for energy efficiency. Most of these programs have been carried out by state and local governments and utility companies, usually without much publicity but often quite successfully.[3] These efforts suggest the possibilities for more widespread programs in the future.

One of the most popular forms of utility programs is rebates for high-efficiency appliances. A recent survey revealed 59 such programs among U.S. utilities in all parts of the country (Berman et al., 1987; ACEEE, 1988). The majority of these programs are only for residential customers, but more than 20 programs also provide incentives to commercial and industrial customers. Rebates are structured in many different ways; most go only to the purchaser, but some are also offered to retailers. Program coverage varies with differences in demand across the country; some summer-peaking utilities promote both high-efficiency air conditioners to reduce summer demand and heat pumps to increase demand in winter. Rebate amounts frequently vary with the size and efficiency of the appliance, sometimes according to a sliding scale to encourage maximum efficiency. In general, rebate programs appear to be highly cost-effective relative to the costs of new generating capacity. For 33 utilities reporting such data, the median cost of peak demand reduction was $200 per kW saved. (For an example of one unique utility conservation program, see Box 8-1).

---

[3] The results of these programs have been periodically reported and evaluated by the Electric Power Research Institute, the American Council for an Energy Efficient Economy, and others. See, for example, Berman et al. (1987) and ACEEE (1988).

FIGURE 8-2

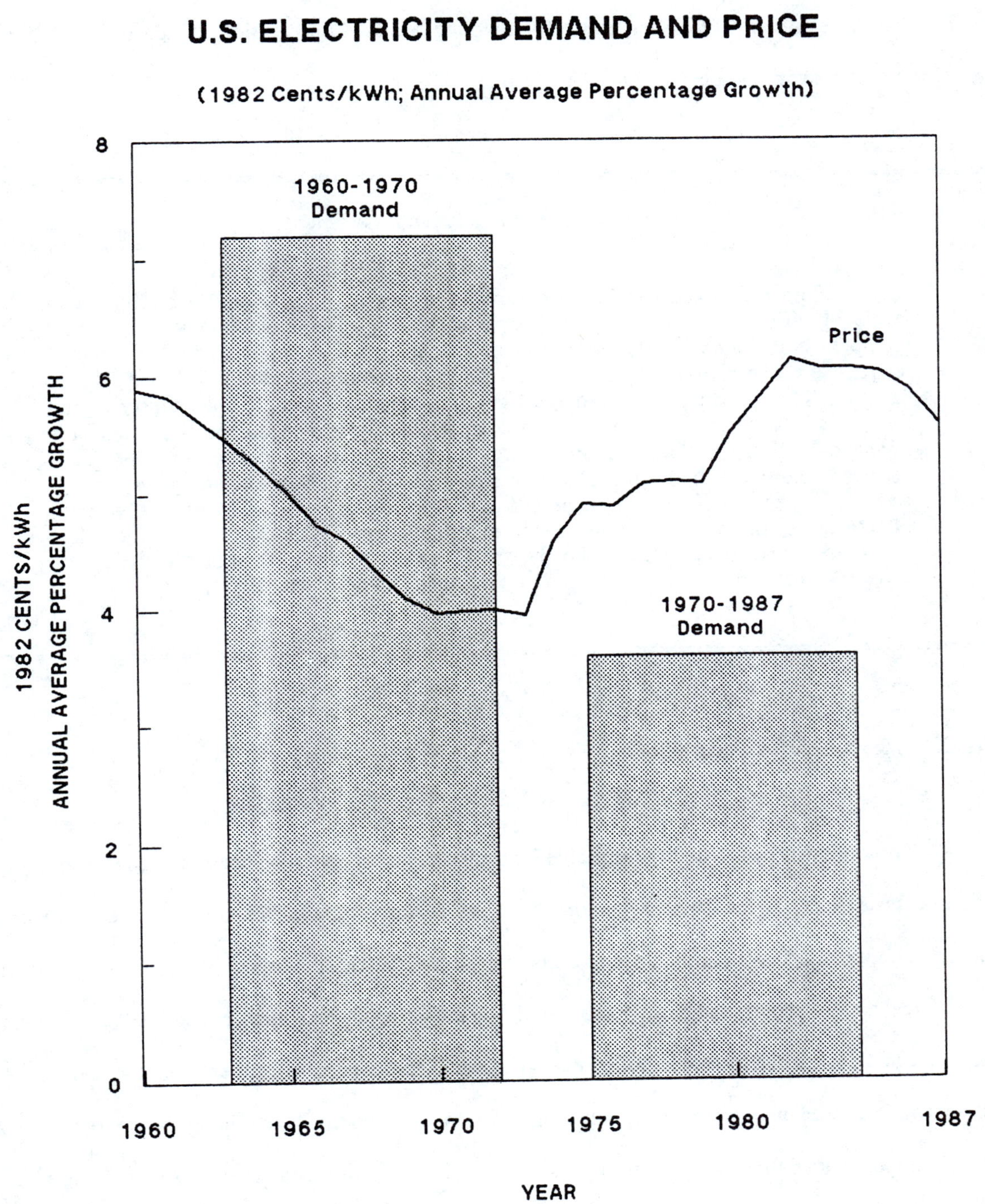

Source: U.S. DOE, 1988.

Some programs demonstrate the potential for giving economic incentives for energy efficiency to builders. For example, the Bonneville Power Administration's "Super Good Cents" program gives builders a **cash grant** of $1,000 per house for meeting Model Conservation Standards developed by the Northwest Power Planning Council (Randolph, 1988a).

Box 8-1. The Hood River Experiment

One recent conservation program illustrates the possibility of achieving high levels of efficiency and broad public participation. The Pacific Power & Light Company and Bonneville Power Administration financed what was arguably the most aggressive conservation program in the U.S. in the small community of Hood River, Oregon. The sponsors offered to audit and implement all cost-effective conservation measures in electrically-heated households. In two years, 85 percent of the eligible households accepted improvements, and most of the remainder of the populace received an audit or had previously improved their homes. Space heating levels were reduced to about half the level expected in pre-project forecasts. However, the cost-effectiveness of the program became difficult to evaluate when electricity consumption dropped sharply shortly before the program went into effect due to price increases in excess of 40 percent (Cavanaugh and Hirst, 1987).

Creating Markets for Conservation

The opportunity for highly profitable investments in energy efficiency has not gone unnoticed by some businesses. If consumers are unresponsive to opportunities for profitable investments in conservation, businesses theoretically could make such investments and share the savings, particularly since a technically expert company could avoid some of the uncertainties facing the typical energy user. Such firms, often referred to as energy service companies, did emerge in the early 1980s. These firms offered a one-stop shopping approach to conservation, combining an energy audit to identify opportunities for savings with installation and financing in exchange for a share of the savings for some fixed period. The owner is not required to assume any risk and is guaranteed a bill no more than would have been the case without the improvements. Federal and State governments are testing shared-saving programs to reduce capital spending requirements as discussed in **STATE AND LOCAL EFFORTS** below.

In practice, the energy service concept has proven to be valuable but not a panacea (Weedall et al., 1986). The negotiation and administration of contract terms can be lengthy and expensive. Relatively large savings are necessary to justify the overhead costs, meaning that privately funded energy service companies have generally been most interested in larger projects, such as commercial buildings. One means of expanding the reach of the energy service companies has been the organization of such services by non-profit groups, which can sometimes operate on a lower margin because they do not pay taxes or return a profit and may also be able to obtain below-market-rate financing. On the other hand, the range of permissible activities for such organizations is also usually limited by charter and funding source. Recently, hybrids containing elements of non-profit and for-profit organizations have begun to appear, with possibilities for achieving the benefits of both (Kerwin and Jolin, 1988).

Another innovative approach to reducing the cost of energy service contracting is to pool enough buildings to achieve economies of scale. The Alliance to Save Energy developed a model for this concept after a demonstration program with eight agencies organized by the United Way of Wilkes-Barre, Pennsylvania. However, the project was difficult to implement and the potential for replication remains to be seen (Prindle and Reid, 1988).

Many other financing mechanisms have been tested or proposed (Lovins and Shepard, 1988). Several utilities offer generic rebates for any demonstrated savings; in one area in Manhattan with exceptionally high cost of service the utility offers commercial customers rebates of $500 per kW peak reduction. Another possibility is sliding-scale hookup fees for new buildings, based on the cost of the generating capacity necessary to supply the building's electricity needs (the converse of subsidies for more efficient buildings) (Rosenfeld and Hafemeister, 1988; also see Box 8-2).

Box 8-2. Creating Markets Through Demand Side Bidding

Another emerging approach to promoting markets for efficiency is the demand-side bidding concept. Bidding programs for new capacity have been developed in many states as a means of obtaining low-cost supply additions, a concept that may receive further encouragement from proposed rules now being considered by the Federal Energy Regulatory Commission (FERC, 1988a). FERC has invited comment on the possibility of addressing demand-side bidding, and several states have already implemented such programs. Energy service companies have some potential here since they can achieve some economies of scale. However, large commercial and industrial firms may want to bid themselves rather than share savings (Cole et al., 1988). In 1987 Central Maine Power requested bids for investments producing demand reductions from large industrial customers. The utility offered to pay up to 50 percent of the project cost or a rebate sufficient to reduce the payback to two years, whichever is less; four contracts were produced for a demand reduction of over 11 MW at a cost of $2.5 million.

Limits to Price-Oriented Policies

Economic incentives could play a critical role in reducing greenhouse gas emissions. Nevertheless, there are often practical and political limits to reliance on price-oriented policies. Demand may be very inelastic due to lack of information, the absence of alternatives in the near term, or shortages of capital. Price increases may also have socially undesirable, disproportionate impacts on particular groups or regions. Price increases may also be politically unpopular relative to other policies. Thus, the price necessary to induce change might be so high as to be inequitable, economically impractical, and politically unattractive. Empirical studies illustrate some situations in which increased energy prices have not resulted in nearly the degree of energy efficiency that would be expected from economic calculations. One study at the Energy Analysis Program at the Lawrence

Berkeley Laboratory examined the market for more energy efficient appliances over the period 1972-1980 (Ruderman et al., 1987). The authors found that consumers demand payback periods of two years or less for investing in more efficient household appliances. Except for air conditioners, the implicit discount rates corresponding to these payback periods are much higher than real interest rates or the discount rates commonly used in life-cycle cost analysis of consumer choice (see Table 8-2). The study suggests several factors are responsible for this substantial underinvestment in energy efficiency: lack of information; lack of access to capital to pay for the added first cost; the relatively small savings; a substantial number of purchases of some appliances by builders or landlords who do not pay operating costs; and manufacturers' decisions to limit high-efficiency features to top-of-the-line models that yield higher profit margins.

Studies of small commercial customers and larger industrial customers indicate they also make greater demands of efficiency investments than of alternatives that are considered to be more within their main line of business (Komor and Katzev, 1988; Lovins and Shepard, 1988; Cavanaugh, 1988; Alliance to Save Energy, 1987). Managers of small enterprises may never see the electric and gas bill, have no knowledge of the rate structure for their building, and are largely unaware of which appliances are responsible for most of their bill. Larger companies may demand payback periods of six months to two years for conservation investments, while simultaneously investing in government bonds at an 11 percent interest rate. Utility companies also typically apply investment criteria to new powerplants that are much less demanding than consumer investments to save an equivalent amount of energy (Cavanaugh, 1986).

Some authorities predict that market forces alone are unlikely to prompt continued demand for automobile efficiency improvements due to the declining share of operating costs attributable to fuel as efficiency improves (Figure 8-3). According to an analysis by the Department of Energy, a typical automobile costs over $10,000 and consumes about 370 gallons a year (U.S. DOE, 1988; see

TABLE 8-2

Payback Period in Years for Appliances, 1972-1980

Appliance	1972	1978	1980
Gas central space heater	2.98	2.38	2.21
Oil central space heater	2.33	1.70	1.18
Room air conditioner	5.11	4.77	5.25
Central air conditioner	4.96	4.16	5.18
Electric water heater	0.48	0.41	0.41
Gas water heater	1.50	1.07	0.98
Refrigerator	1.35	1.45	1.69
Freezer	0.60	0.67	0.72

Source: Ruderman et al., 1987.

FIGURE 8-3

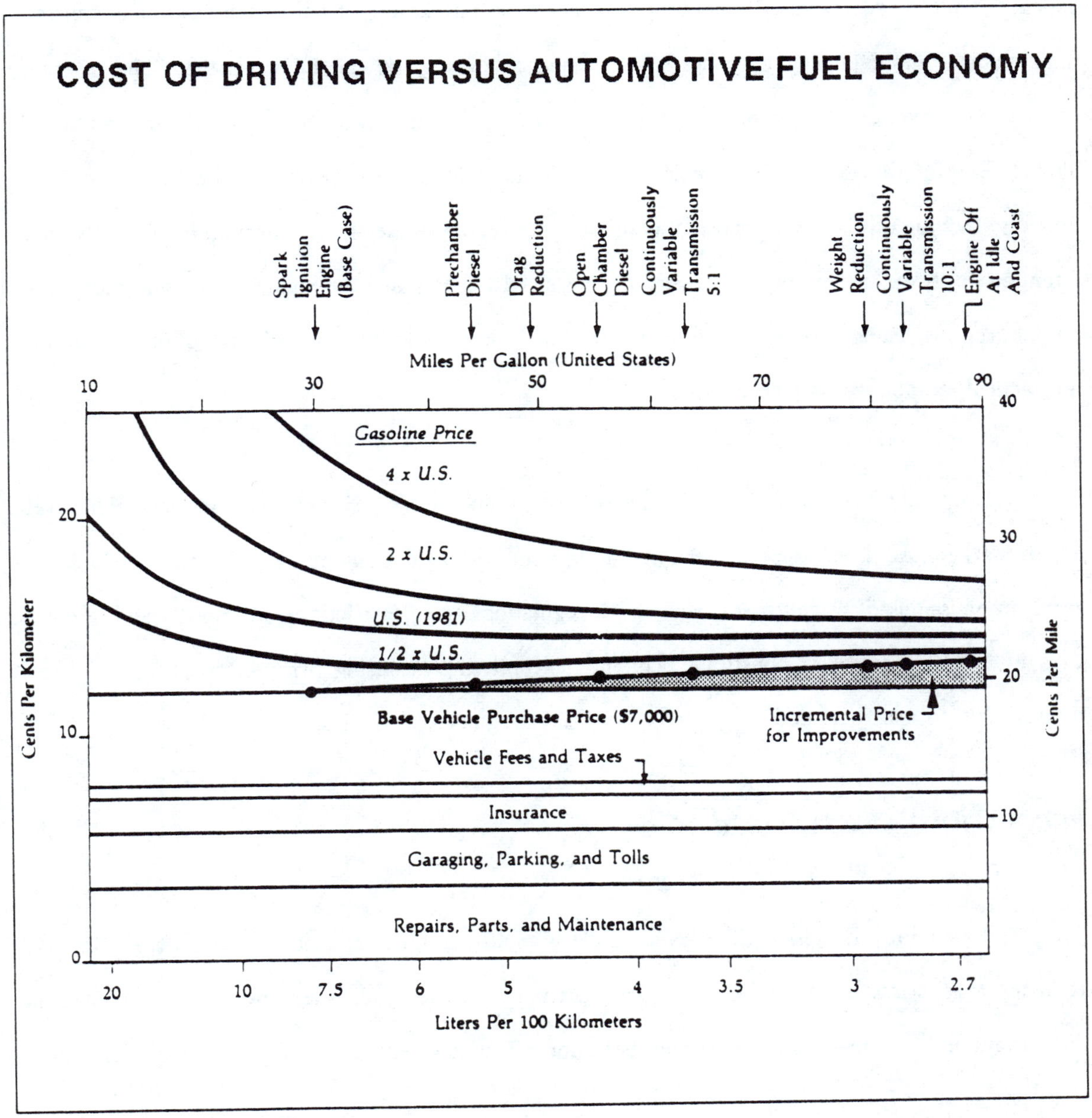

Figure 8.3 The indicated energy performance is based on computer simulations of an automobile having various fuel economy improvements added in the sequence shown at the top of the graph. The base car is a 1981 Volkswagen Rabbit (gas version).

The figure shows that the reduced operating costs associated with various fuel economy improvements are roughly offset by the increased capital costs of these improvements over a wide range of fuel economy.

Source: Goldemberg et al., 1987.

also, Bleviss, 1988; Goldemberg et al., 1987). At $1.10 per gallon, this implies that annual fuel expenses typically represent only about 4 percent of the purchase cost and much less than insurance, maintenance, and financing expenses. The result, according to U.S. DOE, is that "The desire to reduce fuel expenses, even with higher gasoline prices, will not motivate consumers to significantly alter their preferences for automobiles" (U.S. DOE, 1988).[4] Precise evaluation of this effect is difficult; the high visibility of posted gasoline prices may cause some consumers to focus more attention on this cost than on other operating costs. Moreover, at some higher price consumers could be induced to change their preferences, but this price might be so high as to be inequitable, impractical, or politically unacceptable.[5]

Economic incentives alone will not serve to reduce trace-gas emissions to acceptable levels. When markets are unresponsive, there may be a need for regulations on activities and products that result in emissions of trace gases. Many such regulations (e.g., restrictions on imports and domestic uses of CFCs) themselves constitute economic incentives by causing price increases that make it unprofitable to use regulated products or engage in regulated activities.

REGULATIONS AND STANDARDS

Regulation of energy, CFCs, and forestry could complement pricing and other policies for reducing greenhouse gas emissions. In brief, pricing strategies and other government policies may not always induce changes in consumer behavior effectively, either because there is some major

---

[4] One domestic automotive company reportedly estimates that consumers will seek a 1 mpg improvement in fuel efficiency for every 20 cent increase in gasoline prices (Bleviss, 1988).

[5] One expert has noted that "No aspect of America's energy price system is more peculiar, in comparison with other industrial countries, than the absence of a substantial national sales tax on gasoline" (Nivola, 1986).

market failure or because of an unwillingness to set prices high enough to bring about the desired change in demand.

Like pricing and economic incentive policies, government regulations should be viewed as simply one tool for promoting the development and adoption of technologies and behavioral changes necessary to reduce greenhouse gas emissions. They are likely to be most effective (and perhaps most acceptable) when they are targeted to specific groups, are no more restrictive than necessary, and are used to complement economic incentives, research and development programs, information programs, and the other approaches outlined in this chapter.

In this section we will consider possible regulatory strategies for existing regulatory programs, which have been adopted for reasons unrelated to climate change, that could be strengthened to either restrict emissions of greenhouse gases or encourage actions to reduce emissions, as well as concepts for new regulatory programs specifically targeted to reduce emissions.

Existing Regulations that Restrict Greenhouse Gas Emissions

The Federal government already has extensive regulations and regulatory programs that limit greenhouse gas emissions, such as its air pollution control laws, restrictions on the use of chlorofluorocarbons, and regulation of investments and rates charged by utilities. There are also energy efficiency standards for automobiles, appliances, and fluorescent lamp ballasts. These programs could be modified to further reduce greenhouse gas emissions.

Regulation of Chlorofluorocarbons

Chlorofluorocarbons (CFCs) are being regulated because of concern about their impact on stratospheric ozone (see discussion in Chapters IV and VII). The potential for introducing substitutes for CFCs to reduce the risks of ozone depletion and global warming has been known for many years. For example, the DuPont Company stated in 1980 that such substitutes could be produced, but the cost would be much higher (DuPont, 1980). Without regulation, however, there was no market for these alternative chemicals and they were not produced. The Montreal Protocol on Substances That Deplete the Ozone Layer, signed in September 1987, provides a framework for global reductions in CFC emissions:

- The use of CFCs-11, -12, -113, -114 and -115 is to be frozen at 1986 levels starting in approximately mid-1989, reduced to 80 percent of 1986 levels in 1993, and reduced to 50 percent of 1986 levels in 1998. The reduction from 80 percent to 50 percent will take place unless the parties participating in the Protocol vote otherwise.

- The use of Halons -1211, -1301, and -2402 is to be frozen at 1986 levels starting in approximately 1992.

- Beginning in 1990, and at least every four years thereafter, the parties will assess the control measures in light of the current data available. Based on these assessments the parties may adjust the control levels and substances covered by the Protocol.

- Each party shall ban the import of the controlled substances (bulk CFCs and halons) from any State not party to the Protocol beginning one year after the Protocol takes effect. The parties shall, in addition, develop a list of products that contain the controlled substances, which will be subject to the same trade restrictions. The feasibility of restricting trade in products manufactured with the controlled substances shall also be assessed.

- Developing countries with low levels of use per capita are permitted to delay their compliance with the Protocol for up to 10 years. The parties also agree to assist developing countries to make expeditious use of environmentally safe alternative substances and technologies.

On August 1, 1988, EPA announced a comprehensive regulatory program for these chemicals that will reduce production and consumption in three phases leading to a 50 percent cut by July 1, 1998 (*Federal Register*, August 12, 1988). The U.S. regulations become effective upon international ratification of the Protocol. In September 1988 EPA Administrator Lee Thomas stated that recent scientific evidence makes it necessary to consider a complete phase-out of these chemicals. The regulatory approach adopted by EPA restricts CFCs by granting limited production rights to the five U.S. companies who manufactured them in 1986, and by restricting imports. This is expected to cause an increase in price and an increasing incentive for CFC users to find substitutes.

EPA was concerned that the added profits for CFC producers could create an incentive to delay the introduction of substitute chemicals. EPA estimated that CFC producers could earn additional profits between $1.8 billion and $7.2 billion by the year 2000. The Agency therefore requested comments on the merit and legality of taxes on CFCs to remove the added profit and potential incentive to delay.

Another concern is that the price rise alone may not be an effective means of reducing demand for some uses of CFCs for which substitutes may be available, but which have a small impact on total product cost. EPA has indicated it will monitor this possibility and will, if necessary, consider product-specific regulations.

Energy Efficiency Standards

Congress adopted minimum energy efficiency standards in 1987 for refrigerators, water heaters, air conditioners, furnaces, and other appliances. The standards take effect on different dates for each product in recognition of differences in product planning and production needs. The law defines standards for different classes and categories of each appliance and does not prevent new features that may increase total energy consumption (see Table 8-3). By one estimate, the standards will reduce peak electricity demand in the year 2000 by an amount equivalent to the output of 22 large powerplants (Geller, 1986). The appliance standards were adopted on the basis of a consensus

TABLE 8-3

Appliance Efficiency Improvements Required by Law

Product	Average Of Those In Use	New Model Average	Best Commercial Model	Estimated Cost-Effective Potential [1]	Potential Savings [2]
		(kilowatt-hours per year)			(percent)
Refrigerator	1,500	1,100	750	200-400	87
Central Air Conditioner	3,600	2,900	1,800	900-1,200	75
Electric Water Heater	4,000	3,500	1,600	1,000-1,500	75
Electric Range	800	750	700	400-500	50
		(therms per year)			
Gas Furnace	730	620	480	300-480	59
Gas Water Heater	270	250	200	100-150	63
Gas Range	70	50	40	25-30	64

Source: Geller, 1986.

[1] Potential efficiency by mid-nineties of further cost-effective improvements already under study are made.

[2] Percent reduction in energy consumption from average of those in use to best cost-effective potential.

supported by industry and environmental and consumer groups, demonstrating that a cooperative approach to establish achievable energy efficiency goals is possible. Appliance manufacturers were most interested in preemption of State regulation, while utilities and environmental groups were motivated by energy savings.

The energy efficiency standard for new U.S. automobiles as of 1988 is a corporate average fuel economy (CAFE) of 26 mpg. In September 1988, the Department of Transportation set CAFE standards at 26.5 mpg for the 1989 model year, down from a scheduled increase of 27.5 mpg. Automakers are allowed to meet this requirement by offsetting less efficient (usually larger, more powerful) models with more efficient ones; they may also offset their failure to meet the target in one year with credits gained in prior years. However, mpg ratings of domestic models may not be averaged with those of foreign imports. New-car fuel economy has improved markedly since CAFE was adopted (well over 50 percent since 1973), but its value has been vigorously debated. Critics argue that the gains have been primarily a response to higher prices and that the regulations have imposed substantial costs on manufacturers and consumers. They argue that fuel efficiency is improving due to market pressures and technological innovation, but that if more rapid improvement is desired, it could be promoted more efficiently by higher gasoline taxes to promote demand for more efficient vehicles and encourage turnover of older, inefficient cars (U.S. DOE, 1988; Crandall, et. al. 1986).

However, there is no assurance that market forces alone will produce substantial further improvement in vehicle mileage ratings so long as oil prices are stable or declining in real terms. Higher gasoline taxes would help, but may have to be increased sharply to spur demand for efficiency improvements beyond 30 mpg. Exclusively demand-oriented strategies can leave manufacturers with considerable uncertainty about future markets (Bleviss, 1988). Regulations adopted as a supplement to tax increases could give industry a clear target, reducing the need for hedging strategies that dilute efforts and increase investment costs (Bleviss, 1988).

The CAFE approach also could be improved to both increase incentives for efficiency improvements and meet some industry objections (McNutt and Patterson, 1986). The fleet average concept unduly penalized larger cars, when technologies could allow improvements in all sizes and classes. It required annual improvements, when improvements are made in steps over longer time periods, and then sought to provide flexibility with credits that encouraged a search for administrative exemption rather than long-term improvement. Lower standards were set for light trucks, which further undercut the program. Finally, violations of standards are punishable by court-imposed civil penalties, when equivalent economic incentives could be achieved more acceptably by imposing fees proportional to the noncompliance (Bleviss, 1988).

Air Pollution Regulations

Some greenhouse gases are already regulated as air pollutants because of their effects on human health and welfare. Under the Clean Air Act, EPA sets uniform ambient standards for emissions of carbon monoxide, ozone, nitrogen oxides, and other pollutants and uniform technology-based standards for major new sources of these pollutants. However, States retain responsibility for compliance with the ambient standards, which may require allocating greater responsibility on some sources than on others. States are also free to set more stringent requirements on stationary sources.

The regulatory strategy for control of air pollutants to stabilize greenhouse gas concentrations is not identical to that for achieving compliance with ambient air quality standards. In the latter case, the objective is to stay below threshold levels that effect human health or welfare. However, total, as well as peak, emissions affect the buildup of greenhouse gases, which implies a potentially different regulatory approach. An area barely in compliance with ambient air quality standards at all times may contribute more to greenhouse gas buildup than an area with occasional air quality violations but substantial periods of much cleaner air.

Substantial progress has been made in reducing total emissions of regulated pollutants since the Clean Air Act was adopted in 1970 (U.S. EPA, 1987). For example, carbon monoxide emissions from automobiles dropped about 35 percent from 1970 to 1985 despite a 58 percent increase in vehicle miles traveled (Figure 8-4). Emissions of volatile organic compounds from highway vehicles decreased 48 percent in the same period. However, emissions of NO_x during this period were relatively unchanged.

Waste Management

The problem of waste management illustrates another area in which policies to reduce greenhouse gas emissions may complement other environmental objectives. As discussed in Chapter IV, decomposition of solid wastes is a growing source of methane emissions. In the U.S., governments at all levels have accelerated efforts to promote recycling and reduction of solid wastes for both economic and environmental reasons.

State and local governments have historically been primarily responsible for solid waste disposal. The cost of landfills and public opposition to burning waste has generated substantial interest in recycling and waste reduction. Numerous states have recently adopted stringent programs to reduce the waste stream by a minimum of 25 percent. How this will be accomplished has yet to be made completely clear, but economic incentives will play a major role in some States. Florida adopted an advance disposal fee on every container sold at retail that fails to achieve a 50 percent

FIGURE 8-4

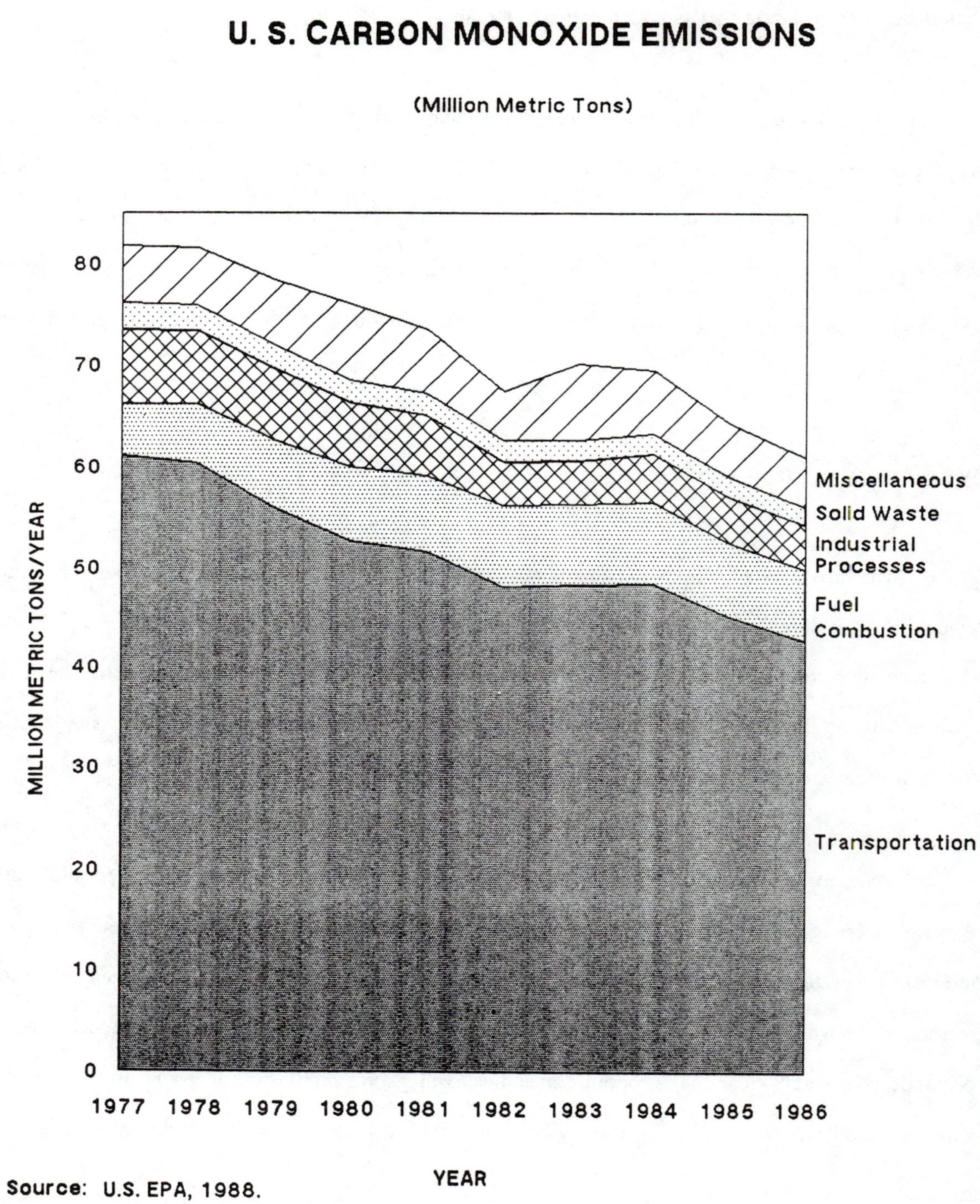

Source: U.S. EPA, 1988.

recycling rate; the fee increases if targets are not met (Carlson, 1988b). California similarly adopted a system of deposit fees with increases scheduled to take effect if goals are not met.

Federal law (RCRA Sec. 4010(c)) assigns EPA the responsibility for ensuring the environmental safety of sanitary landfills. As discussed in Chapter VII, EPA proposed minimum standards on August 1, 1988 that would require gas monitoring stations to detect methane and to plan for its removal, and proposed regulations under the Clean Air Act would require collection of landfill gas at both new and existing landfills. The Agency has also supported a national recycling goal of 25 percent by 1992.

Another source of methane releases discussed in Chapter VII is coal mining. Interest in coalbed methane recovery is increasing for economic reasons, but a this time there are no specific policies to promote such efforts.

Utility Regulation

Regulatory policies also can exert a significant influence on the price and demand for electricity. Electric companies are regulated monopolies whose rates and investments are regulated by State public utility commissions--except for wholesale transactions, which are reviewed by the Federal Energy Regulatory Commission. The structure as well as the overall level of electricity rates is regulated and both influence demand (see Kahn, 1988). For example, some States provide relatively low "lifeline" rates for the initial increment of residential demand to meet the basic energy needs of all customers. The consequence is usually an inverted rate structure, with rates increasing with greater use. Even though rates may be initially set to produce the same total revenue, the result is some reduction in demand because consumers recognize the higher rates associated with greater use. For example, Detroit Edison estimated that a lifeline rate structure imposed in 1981 resulted in reducing demand about 3 percent (Kahn, 1988).

Where the production of electricity to meet peak power needs is more carbon-intensive (e.g., where the marginal fuel is oil and the base fuel has a large percentage of nuclear power), incentives to reduce peak demand and levelize loads may be economically justified and a means of supporting greenhouse gas reductions. With time-of-use metering, utilities can charge rates that more accurately reflect costs, potentially promoting energy efficiency as well as cost savings. This is possible because the costs of generating electricity are not uniform throughout the year; when demand peaks, costs are usually highest because the utility must rely on its most costly units to meet the increased demand. Some utilities in the U.S. and in Europe have tested rate structures that give customers incentives to use electricity at times when costs are low, and conversely, disincentives when costs are high. For example, a recent three-year New York experiment gave customers a one-cent-per-kWh reduction during hours of normal demand, but above a certain threshold level, customers were charged a rate of 40 cents per kWh, several times higher than the usual rate. On average, customers saved 5 to 10 percent annually, reduced their peak period loads over 40 percent relative to the average customer, and expressed strong support for the continuation of the program (Cole and Rizutto, 1988). The metering costs are currently about $500 per house, but the overall program is still expected to produce substantial net benefits. Moreover, it is expected meter costs will drop substantially with large production. (For an example of utility disincentives, see Box 8-3.)

Another significant issue in utility regulation is the creation of competitive markets for the generation of electricity. Federal and State policy has encouraged this concept since the enactment of the Public Utility Regulatory Policies Act (PURPA) in 1978. This law requires utilities to buy power from cogenerators and small renewable energy projects and to pay an amount equal to the

Box 8-3. Disincentives to Utility Conservation

The revenues earned by electric and gas utilities are governed by accounting rules devised to allow utilities to recover operating costs and earn a fair rate of return. These rules were not intended to favor particular types of investments but recent analysis by the Maine and California public utility commissions indicates that they may in fact favor investments in additional supply as opposed to efficiency improvements, even when the latter is much less expensive (Moskovitz, 1988; Cavanaugh, 1988). So long as retail rates exceed short-term production costs, these studies show that, under traditional regulatory practices, utilities will always earn greater profits by selling more electricity, even if the cost of conservation is zero. Allowing utilities to profit from conservation by allowing an equivalent return on such investments does not rectify the situation. Instead, the economic incentive is to do the most costly conservation investments (to achieve the greatest return) with the least impact (to maintain sales). California and Maine have adopted procedures designed to correct these disincentives as one means of encouraging utility interest in demand-side investments.

costs they save as a result. Since the law's adoption, thousands of projects have been developed around the country, despite the absence of a need for new generating capacity in many areas (FERC, 1988c).[6] Federal Energy Regulatory Commission (FERC) compilations indicate cumulative applications from projects equivalent to more than 40 nuclear powerplants (see Table 8-4). FERC is currently reviewing proposed rules to expand the circumstances and type of projects eligible to participate in competitive bidding arrangements (FERC, 1988a, 1988b). FERC has also asked for comment on the possibility of allowing demand-side reduction efforts to compete with additions to capacity.

[6] FERC publishes periodic compilations of "qualifying facilities," projects that have formally requested status as entitled to the benefits of PURPA. However, the list is only an approximation of actual activity, since not all projects file such applications; some file after they have already begun operation, and some applications are made for projects that are never completed.

TABLE 8-4

Cogeneration Facilities (No. Filings), [No. of Facilities] and Capacity in Kw

	New		Existing		Both		Total		Qualified Facilities and Initial Capacity	
Coal	(183)	10,419,659	(6)	461,600	(12)	356,272	(201)	11,237,531	[178]	10,948,083
Natural Gas	(1,264)	25,894,776	(41)	1,888,457	(9)	694,671	(1,314)	28,477,904	[1,269]	26,880,678
Biomass	(122)	1,617,390	(18)	570,497	(9)	615,000	(149)	2,802,887	[138]	2,698,802
Waste	(51)	2,432,239	(2)	40,875	(1)	51,000	(54)	2,524,114	[50]	2,450,114
F06/F02	(47)	324,921	(8)	323,700	(1)	5,350	(56)	653,971	[56]	653,971
Other	(38)	1,258,288	(3)	56,500	(1)	62,250	(42)	1,377,038	[39]	1,311,968
Total Cogeneration	(1,705)	41,947,273	(78)	3,341,629	(33)	1,784,543	(1,816)	47,073,445	[1,730]	44,943,616

NOTES:

1. Under §292.202 of the Commission's rules, "Cogeneration facility means equipment used to produce electric energy and forms of useful thermal energy (such as heat or steam), used for industrial, commercial, heating, or cooling purposes, through the sequential use of energy." Topping-cycle facilities first apply the energy input to electric power production; bottoming-cycle facilities first apply the energy input to a useful thermal application. All but 9 filings in this table are topping cycle facilities. Bottoming-cycle and partially bottoming-cycle facilities are listed below:

QF80-6-000	New	Petro Coke	75,000
QF82-100-000	New	Coal	10,000
QF82-207-000	New	Coal/Waste	125,000
QF83-220-000	New	NG, FO	6,000
QF84-134-000	New	Petro Coke	27,300
QF84-192-000	Ex.	NG	910
QF85-572-000	New	Waste/Petro Coke	49,900
QF86-23-000	New	NG	2,700
QF86-800-000	New	Hydrocarbon Coke	3,300

2. "Other" includes cogeneration facilities using nuclear or solar energy sources.

Source: FERC, 1988c

Existing Regulations that Encourage Emissions Reductions

In contrast with the mandatory regulations discussed above, some regulatory programs stress positive incentives for activities to reduce greenhouse gas concentrations. Such programs could be expanded to encourage greenhouse gas reductions.

A prime example of regulatory incentives for actions that reduce greenhouse gas concentrations is the Conservation Reserve Program (CRP) created by the Food Security Act of 1985. The CRP gives up to 50 percent of the cost of establishing approved conservation practices to farmers who agree to take highly erodible cropland out of production for contracted periods of 10 to 15 years. Under Section 1232(c), at least one-eighth of the number of acres placed in the reserve between 1986 and 1990 are to be devoted to trees. As of the sixth signup, ending February 19, 1988, more than 1.5 million acres had been committed to trees--less than a third of the Congressional goal--at a cost to the government of roughly $58 million (Dudek, 1988).

In recognition of the CRP's broader potential for achieving environmental protection, the program was modified in the sixth signup to allow the inclusion of 66- to 99-foot-wide strips of cropland along streams and waterbodies, irrespective of the erosion rate.[7] Tree planting is one allowed use of the land. These so-called "filter strips" provide a buffer zone that absorbs pollutants that would otherwise go into lakes and streams. The Conservation Reserve program offers a possible model and precedent for further cooperation between EPA and the Forest Service to identify opportunities where tree planting may serve multiple-agency goals.

[7] Joint Memorandum of the Soil Conservation Service, EPA, Fish and Wildlife Service, and Forest Service on the Conservation Reserve Program Filter Strip Initiatives, April 29, 1988.

Because the CRP program provides such diverse benefits, proposals have been made to expand its scope and coverage. In some States, particularly in the Southeast, aggressive tree planting could become a significant means of offsetting the buildup of carbon dioxide. According to a recent study, most eastern States, with the exception of Florida, appear to have enough erodible acreage to provide the acres needed to offset new CO_2 emissions from electricity production if planted with trees (Table 8-5). Roughly 11-22 million acres of new trees, or about a fourth to one-half of the current CRP goal, would offset projected new fossil-fueled generating plants for the 1987-1996 period if planted with optimum species (Dudek, 1988). By this estimate, the costs are modest relative to other environmental requirements associated with operating powerplants, on the order of 0.5 cents per kilowatt-hour.

New strategies to promote tree planting may avoid some of the limitations and complications created by the CRP program, which must identify farmers willing to withdraw farmland and which restricts the land for a period less than the time many trees require to reach maturity. Utility companies and large industrial coal users may also wish to sponsor tree planting. The Forest Service reports that southeastern States--where most timber industry investments have occurred in the past two decades--must replant and manage their dwindling forests to allow its timber industry to keep growing (USFS, 1987).

The Food and Security Act has another provision that demonstrates the potential use of regulations to promote environmental goals. Under Section 1221, any farmer who produces an agricultural commodity on a wetland converted to agricultural use after the effective date of the Act may become ineligible for many forms of USDA financial assistance. The restriction applies to all of the farmer's land, not only the converted wetland area. The impact of the program obviously depends on the value of the financial assistance, which in turn varies with commodity prices, but the precedent is important (see Tripp and Dudek, 1986).

TABLE 8-5

Erodible Acreage Available to Offset CO_2 Emissions From Electricity Production

State	Megawatts (MW)	CO_2 Emission (10^3 tons)	Erodible Acres (10^3 acres)	Offset Acreage (10^3 acres)	Percent (%)
Alabama	1,557	10,458.4	2,558.8	697.2 - 2,091.6	27.2 - 81.7
Arizona	710	4,900.3	187.1	326.7 - 980.1	>100
Colorado	899	6,204.7	6,480.1	413.6 - 1,240.8	6.4 - 19.2
Florida	2,678	15,136.3	670.2	1,009.1 - 3,027.3	>100
Georgia	1,641	11,325.9	3,072.7	755.1 - 2,265.3	24.6 - 73.8
Iowa	50	201.2	17,833.7	13.4 - 40.2	0.1 - 0.3
Indiana	1,600	10,179.6	5,244.2	678.6 - 2,035.8	12.9 - 38.7
Kentucky	1,095	7,557.5	2,412.7	503.8 - 1,511.4	20.9 - 62.7
Louisiana	540	3,727.0	2,171.0	248.5 - 745.5	11.4 - 34.2
Maryland	753	4,870.3	602.7	324.7 - 974.1	53.9 - >100
Minnesota	1,251	8,634.2	10,102.0	575.6 - 1,726.8	5.7 - 17.1
Missouri	1,005	6,792.4	7,223.7	452.8 - 1,358.4	6.3 - 18.9
Mississippi	250	1,725.5	3,009.9	115.0 - 345.0	3.8 - 11.4
North Carolina	690	4,762.3	2,476.4	317.5 - 952.5	12.8 - 38.4
Nevada	1,865	12,469.0	79.1	831.3 - 2,493.9	>100
Ohio	1,300	8,972.4	3,799.8	598.2 - 1,794.6	15.7 - 47.1
Oklahoma	470	2,754.6	3,795.2	183.6 - 550.8	4.8 - 14.4
Texas	5,515	38,063.5	20,417.2	2,537.6 - 7,612.8	12.4 - 37.2
Utah	814	5,577.8	392.6	371.9 - 1,115.7	94.7 - >100
Wisconsin	60	414.1	4,015.6	27.6 - 82.8	0.7 - 2.1
Total	24,743	164,727	96,544.7	10,981.8 - 32,945.4	11.4 - 34.2

Source: Dudek 1988.

Another example of regulatory incentives for actions to reduce greenhouse gas emissions is higher rates of return in some States for utility investments in efficiency or renewable energy. Utilities are also sometimes allowed to choose whether to amortize their investment and receive a return or recover their expenses in the first year. Some States have also adopted revenue adjustment provisions to prevent reductions in utility profits as a consequence of utility-sponsored conservation programs (Cavanaugh, 1988; Moskovitz, 1988).

There are some practical questions to be answered before attempting a large-scale program to offset carbon emissions with trees, particularly if implemented on an international basis. For example, there may be difficulty counting net "new" trees in a developing country undergoing deforestation; there must be assurances that the trees are properly cared for and not cut down prematurely. Similarly, counting net energy reductions may be difficult since a factory may have intended to make improvements in any case.

Despite these concerns, the offset concept offers some advantages that make it worthy of further consideration. First, it can be targeted to achieve greenhouse gas reductions, unlike energy regulations, which may not always reduce carbon-intensive fossil fuels. Second, it is consistent with efficiency and innovation by allowing each developer the flexibility to seek out the least expensive means of reducing his emissions. Third, it can be implemented in a way that promotes international cooperation by allowing reductions to be achieved in developing countries where they can be accomplished most cheaply and also contribute to development. Eligibility for offsets could be limited to countries that join in an international greenhouse agreement as an incentive for participation.

RESEARCH AND DEVELOPMENT

Further research and development (R&D) will be necessary to bring about widespread use of many of the technologies and strategies reviewed in Chapter VII. As discussed below, in some areas substantial programs are already in place, while in others existing priorities may not be entirely consistent with the objective of stabilizing greenhouse gas emissions.

Research programs can serve several different purposes. In some cases, such as finding more efficient methods for producing steel and photovoltaics, basic research is needed. In other areas technologies are nearly commercially ready but their introduction could be accelerated through testing and demonstration. Research is important to improve policies and programs as well as hardware; for example, the U.S. has acquired a great deal of experience with energy conservation programs over the last decade that has yet to be fully evaluated and summarized for use by public and private authorities.[8]

The importance of government support for R&D is widely accepted; industry often has only a weak incentive to pay for research that will be of widespread benefit, or that is long term and high risk (U.S. DOE, 1987c). However, government can create incentives for greater industry efforts. Special efforts are also needed to develop technologies suitable for use in developing countries where local needs and resources may dictate very different solutions.

[8] One excellent source of information on technologies, programs, and policies on energy efficiency in buildings is the biennial proceedings of the summer study program organized by the American Council for an Energy Efficient Economy (ACEEE, 1988).

Energy Research and Development

The energy R&D budget is important from a greenhouse perspective in both its total amount and its relative priorities. Research to promote the more efficient use of all forms of energy is desirable from the standpoint of greenhouse gas emissions, as is research to reduce the cost of renewable and nuclear energy relative to fossil fuels. Since FY 1981 annual appropriations for energy R&D have declined from $3.4 billion to roughly $1.4 billion in FY 1988. An examination of R&D expenditures in other western industrialized countries suggests that their research budgets have tended to fluctuate less than those of the U.S.; as of 1986 the U.S. ranked last among OECD member countries in the percentage of GNP devoted to energy R&D. However, U.S. expenditures still ranked first in absolute terms (Table 8-6).

Priorities among energy sources have also shifted during the 1980s as research on renewables and conservation has declined more than research in other areas (see Figure 8-5). The Clean Coal Program is by far the largest energy R&D initiative this decade; President Reagan requested $525 million for FY 1989 and $1.775 billion for FY 1990-1992. The intent is in part to improve the efficiency of coal use, but equally important to expand the market for coal here and abroad--an objective that would unavoidably increase carbon emissions relative to other fuels used with equivalent efficiency (U.S. DOE, 1987c; see also, National Coal Council, 1987).

The analysis presented in earlier chapters of this report suggests that research priorities may have to be redirected in order to stabilize greenhouse gas emissions. Several prime objectives are to reduce the demand for fossil fuels for transportation, reduce the energy intensity of basic materials production, particularly steel and cement, and improve the efficiency of biomass use for energy. The prospects for research to improve light vehicle fuel economy are discussed in more detail below.

TABLE 8-6

Government Efficiency Research and Development Budgets in OECD Member Countries, 1986

Country	Efficiency Budget	Total R&D Budget	Efficiency as Percent of Total
Japan	78	2,311	3
United States	275	2,261	12
Italy	48	761	6
West Germany	21	566	4
United Kingdom	43	378	11
Canada	34	336	10
Sweden	29	79	37
Greece	0	15	0
Denmark	5	14	36
Total OECD[1]	622	7,133	9

[1] Total includes minor additional expenditures. Excludes France.

Source: IEA, 1987.

FIGURE 8-5

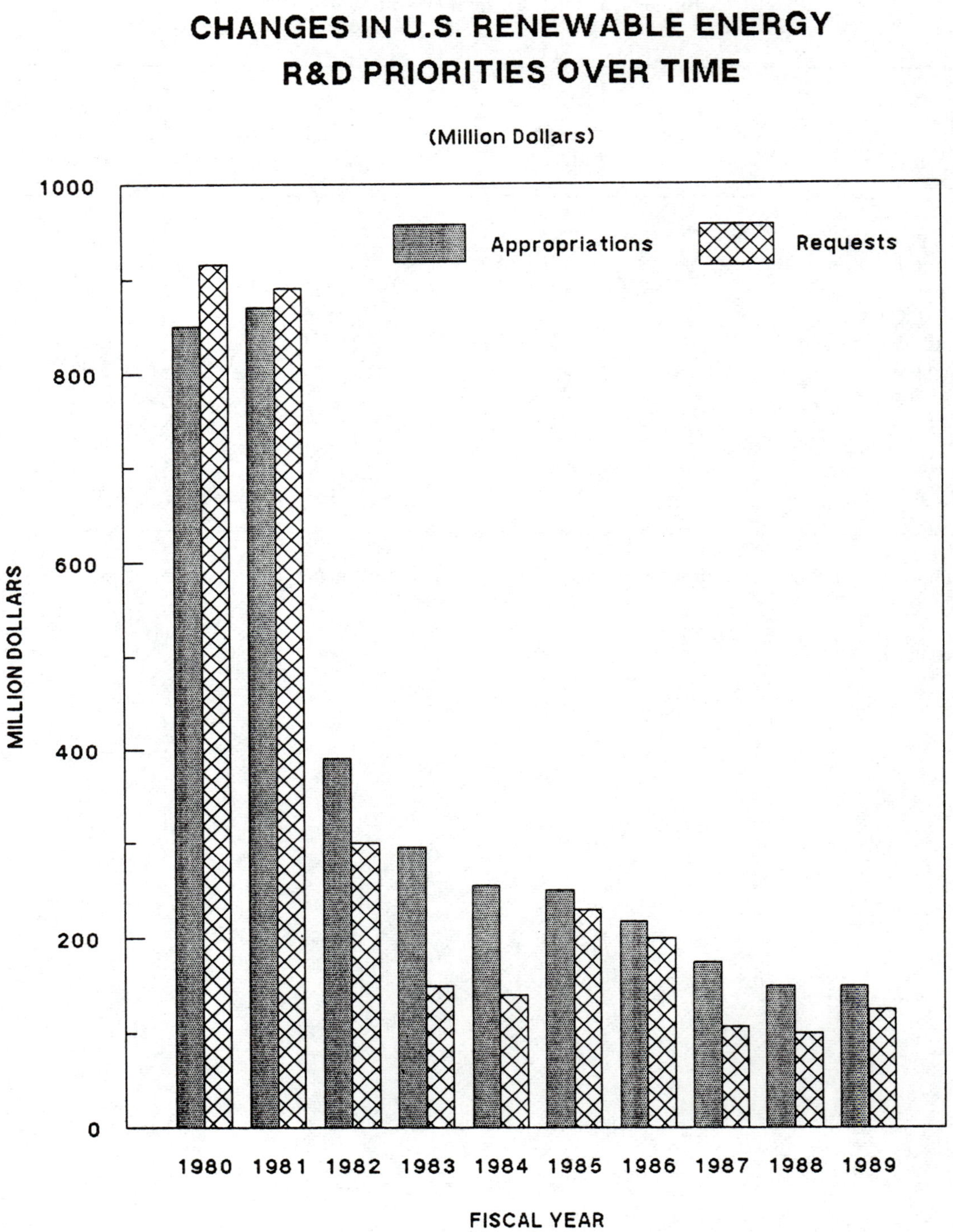

Source: U.S. DOE, 1987a.

Based on projections supplied by automobile manufacturers, the Department of Energy recently estimated that U.S. fuel economy will rise modestly from 27.1 mpg in model year 1988 to 32 mpg in 1995 (U.S. DOE, 1988). U.S. DOE recommended against any action to spur the development or introduction of more efficient vehicles, instead suggesting the use of alternative fuels and increased domestic production of petroleum (U.S. DOE 1988; 1987c). However, these alternatives could exacerbate the greenhouse problem, particularly if alternative fuels were coal-derived (see Box 8-4).

Total federal spending on R&D is not the only measure of success. Some studies suggest that the variations in funding levels and priorities in U.S. R&D efforts have reduced their effectiveness relative to more steady and long-term programs such as those in Japan (Flavin, 1988; Chandler et al., 1988). Federal R&D support may no longer be as important for technologies reaching commercial competitiveness, such as photovoltaics; low energy prices have become a more critical obstacle (Carlson, 1988a). The largest improvements in energy efficiency often result from changes in processes or products that serve multiple purposes; reduced energy costs may be only a secondary consideration (OTA, 1983). Narrowly-focused energy conservation programs may not effectively address this objective. One alternative is to establish research centers for energy-intensive industrial processes. Such centers could research multiple improvements on a cooperative basis with industry and academia, an approach currently incorporated in the Combustion Research Program operated by Sandia National Laboratories (Chandler et al., 1988).

Federal programs are also not the only large source of energy R&D support. Several states, notably New York, California, and North Carolina, have created State agencies to support energy R&D (see **STATE and LOCAL EFFORTS**, below). Private sector support is also large; the utility-supported Electric Power Research Institute and Gas Research Institute both have annual budgets

Box 8-4. Light Vehicle Fuel Economy R&D

Prototype vehicles designed by Toyota and Volvo have already demonstrated that it is technically possible to produce full-sized vehicles with fuel economy of 70 mpg or better, as discussed in Chapter VII. Recent developments in materials technology and other areas suggest that even this level will be greatly exceeded in the near future. The problem therefore is less one of basic science than it is lack of incentives for product development to reduce costs. The relatively low price of oil has diminished incentives to invest in improved efficiency unless it is associated with other features considered more marketable. This is true both in Japan and Western Europe, as well as in the U.S., since the price of oil has declined even more rapidly in these countries due to the relative decline in the value of the dollar.

Barring another large rise in oil prices or some significant government policy intervention, it seems unlikely to expect that consumers will demand, and that manufacturers will produce, vehicles with much higher fuel economy. Indeed, there is some evidence that fuel economy (particularly in U.S.-made vehicles) will stagnate or even decline in the near term due to market demand for larger models and increased acceleration and performance capabilities (Bleviss, 1988).

in excess of $100 million, and some of these organizations give high priority to research on efficiency improvements.

Another important consideration is the allocation between basic science and more applied research. The former has received highest priority but demonstration and technology transfer programs can accelerate acceptance of innovative ideas that have been proven on a small scale but not yet widely adopted. The Department of Energy now supports some technology transfer activities for conservation (U.S. DOE, 1987c) and full-scale demonstration of clean coal technologies to provide "proof-of-concept" experience (*Clean Coal Synfuels Letters*, August 26, 1988, pp.3-4; U.S. DOE, 1987c). Some of the technologies with greatest potential to reduce greenhouse gases, such as advanced gas turbines (see Chapter VII), also could benefit from full-scale demonstration and evaluation (Williams, 1988).

Energy R&D should also include examination of policy and program issues (IEA, 1987). One very critical question is why so many consumers often fail to invest in energy conservation measures despite very high rates of return; this behavior appears to be irrational in economic terms (Kempton and Neiman, 1987; Aronson and Yates, 1985). Consumers also respond differently to loans, rebates, and other subsidies even though they have roughly equal cost to the government.

Another research need for policy purposes is a more detailed data base on international sources of greenhouse gas emissions. The compilation of such data should be an initial goal of the recently formed Intergovernmental Panel on Climatic Change.

Global Forestry Research & Development

The benefits of trees for storage of carbon are widely accepted, but the elements of a large-scale forestry program to stabilize greenhouse gas emissions have yet to be fully described (see discussion in Chapter VII). Such a program would seek to maximize global vegetation and storage of carbon. The steps necessary to accomplish this objective may not be entirely consistent with the emphasis and goals of existing governmental and commercial forestry research programs. The preservation of tropical forests, for example, is a critical international environmental problem with considerable impact on global climate change. However, this effort may not be the most cost-effective way to increase net global forest cover, because the underlying causes are often closely related to deeply-rooted social ills that require very long-term solutions. Large-scale plantation forestry is one alternative, but the economic criteria used in commercial forestry may also have little relationship to maximizing carbon storage.

Traditional forestry research priorities have not emphasized conservation. Recent reviews of forestry research in tropical countries conclude that most projects have focused on the creation of large-scale industrial plantations and other activities to promote the improved industrial use of timber (FAO, WRI, World Bank, and UNDP, 1987a; FAO, 1988). According to the Tropical Forest Action Plan prepared by an international task force, forestry has not received the large-scale, targeted research support devoted to agricultural study of plant breeding and the development of improved crop varieties. The Plan proposed to begin responding to these needs with a five-year, $1 billion research program.

Efforts to date indicate that forestry research can produce markedly higher yields through genetic improvement and better management practices. For example, a Brazilian paper company was able to double yields from 33 to 70 cubic meters per hectare per year through genetic selection and cloning (WRI, World Bank, and UNDP, 1985). Research and demonstration can also help promote sustainable management practices; an AID-funded village woodlot project in Thailand shows that planting fuelwood species can be profitable and environmentally protective (U.S. AID, 1987).

More research on managing public lands, pricing public resources, and other policy aspects of tree planting is also needed. For example, government policies in both the developing and industrialized world have been a major source of pressure on tropical forests, but this relationship has not been thoroughly studied. Excessive consumption of forest products has also sometimes been encouraged by underpricing of public resources in some developing countries (see **CHAPTER IX**).

Research to Eliminate Emissions of CFCs

Industry is now making substantial effort to reduce or eliminate emissions of CFCs as discussed above (U.S. EPA, 1988; Cogan, 1988). However, existing regulatory incentives may not be adequate to assure that all users of CFCs will make a serious effort to find substitutes. The demand for some uses of CFCs, such as automotive air conditioning, is highly inelastic because the cost is very low relative to the total cost of the product and there are no obvious short-term alternatives (*Federal Register*, August 12, 1988). EPA is considering regulatory changes to address this problem, but also is attempting to bolster industry interest in alternatives through a program of cooperative research on promising technologies (U.S. EPA, 1988; Claussen, 1988). Such efforts could be expanded as part of an effort to phase out all CFC emissions as soon as possible and could serve as a model for programs to reduce emissions of other greenhouse gases.

INFORMATION AND TECHNICAL ASSISTANCE PROGRAMS

The government can facilitate the development and adoption of new technologies and strategies through many forms of information and technical assistance programs. These efforts complement research, pricing, and other policies by making consumers more aware of the value of energy conservation and therefore more likely to respond to investment opportunities. Information programs can serve to improve consumer understanding of the significance of energy costs, which are often underestimated because they occur over time. For example, many consumers do not know the relative energy cost of home appliances or that the cost of operating a refrigerator over its lifetime will be greater than the first cost. Information coming from the government is also frequently perceived to be more credible than similar information coming from utility companies or other private firms (Kempton and Neiman, 1987).

Information and technical assistance programs take a variety of forms to serve a range of specialized purposes. Within the Department of Energy, R&D results are disseminated to potential users through "technology transfer" programs (U.S. DOE, 1987c). One element of this program is the National Awards for Energy Innovation, which annually recognizes outstanding achievements in conserving and producing energy.

U.S. DOE also operates several energy information services for different audiences (U.S. DOE, 1987b). The Conservation and Renewable Energy Inquiry and Referral Service (CAREIRS) serves the general public through a toll-free telephone number and refers technical questions to one of several hundred laboratories and expert agencies. In FY 1987, CAREIRS responded to more than 40,000 inquiries. General information on energy production and consumption is also available through the National Energy Information Center and the Solar Technical Information Program. More specialized assistance is provided by the National Appropriate Technology Assistance Service (NATAS), which will help evaluate new technologies and suggest approaches to commercialization.

The Federal government has also provided consumers detailed information on the comparative energy cost of new cars and appliances through mandatory labeling requirements. These programs improve market forces by making it easier for consumers to make decisions about the value of more expensive but more efficient models.

There is no standardized or widely-accepted system for communicating energy cost information for homes and buildings, which may be partly responsible for the slow rate of improvement in this sector (Chandler et al., 1988; Hirst et al., 1986, Alliance to Save Energy, 1988). For example, such a system could be used by lenders to evaluate the expected energy cost of a home as a factor in loan amounts, creating an incentive to improve energy efficiency. Such a system is used on a limited basis now by the Federal National Mortgage Association (U.S. DOE, 1987c). Another model is in use by 12 banks in Seattle (Hirst et al., 1986).

Several Federal information programs directed to small-scale energy use have been gradually curtailed but still operate in many states. One is the Residential and Commercial Conservation Program, which requires electric and gas utilities to offer home energy audits to their customers for a minimal charge. The program was not promoted effectively in most parts of the country and participation has typically been very low, although a few states achieved notably better success (Hirst et al., 1986). The Energy Extension Service provides a small amount of Federal support for State energy information programs and technical assistance programs targeted to individuals and small energy consumers.

Another approach to improving consumer awareness of energy costs is through the introduction of technology that continually reports electricity costs. Advances in microelectronics and communications now make it possible to provide such information for a cost that could be offset by savings achieved through better energy management and shifts in time of use, while at the same time more closely tracking marginal costs (Peddie and Bulleit, 1985; Rosenfeld, 1985).

Federally-supported programs have also helped support development of computer models and other analytical methods for evaluating the energy use from new residential and commercial buildings. These analytical tools can be used by designers to lower the energy use of buildings and as a basis for calculating energy use for purposes of minimum standards and incentive programs (Vine and Harris, 1988). Computer models of building energy use were a key product of the Federal Building Energy Performance Standards. The Standards began as a mandatory Federal regulatory program in 1976 but Congress subsequently amended the law to require only voluntary guidelines. The standards have yet to be released in final form, but much has been learned in the process and interim products have been used by industry and government. The Federal government could facilitate use of the final rules by providing additional technical assistance such as materials on compliance methods (U.S. DOE, 1987c, Chandler et al., 1988).

The integration of research, technical assistance, and public information has a long and productive tradition in Federal agriculture programs. The Federal budget for such activities is about \$3 billion annually, and as of 1984 there were over 3,500 specialist extension agents and over 11,000 county agents (OTA, 1986). As discussed in Chapters IV and VII, modifications in some farming practices, such as selection of crop varieties and fertilizers, water use, and disposal of crop residues may inhibit greenhouse gas emissions, although more research is needed to establish the efficacy of such changes. As our understanding of agricultural sources of greenhouse gas emissions improves, extension activities could be used to teach farmers how to reduce their emissions.

Information and technical assistance is also an important function of bilateral aid, as discussed in **U.S. Bilateral Assistance Programs** (Chapter IX).

CONSERVATION EFFORTS BY FEDERAL AGENCIES

The Federal government is the single largest consumer of energy in the United States, accounting for 2.5 percent of total energy consumption (U.S. DOE, 1987a).[9] The annual energy budget for over 500,000 Federal buildings and facilities is about $4 billion, plus another $2 billion for Federally-assisted housing. A Federal Energy Management Program was established 12 years ago to provide leadership in reducing these costs and has achieved some success. According to U.S. DOE, cumulative energy savings over the past 12 years are nearly 1.6 Quads, equivalent to about $6.5 billion in savings (U.S. DOE, 1987a; see Table 8-7). In FY 1985 a 10-year performance target for improving the energy efficiency of Federal buildings ended, having attained a 16.6 percent reduction in energy per square foot relative to a 20 percent target.

The government could use its buying power to test and demonstrate energy conservation and alternative energy sources. Government procurement could help demonstrate new products, acting as an incentive for manufacturers worried about the lack of a market and consumers worried about being first-of-a-kind purchasers (Goldemberg, 1987). The Department of Energy expressed support

---

[9] Expenditures on energy in FY 1987 were over $8 billion, or 0.8 percent of the Federal budget.

TABLE 8-7

Federal Energy Expenditures and Cost Avoidance
FY 1975-FY 1987

	Annual Energy Cost ($ Million)	Annual Energy Use Reduction Rel. to FY 1975 (BBTU)	Average Annual Energy Cost ($/MBtu)	Annual Energy Cost Avoidance ($Million)
Buildings and Facilities				
1975	1,812.453	---	2.107	---
1976	1,888.673	58,080.4	2.355	136.779
1977	2,162.395	51,847.4	2.675	138.692
1978	2,286.054	66,013.3	2.879	190.052
1979	2,636.361	77,885.9	3.370	262.475
1980	3,168.399	97,947.3	4.157	407.167
1981	3,713.200	99,466.9	4.882	485.597
1982	3,804.965	92,207.3	4.955	456.887
1983	3,863.248	98,559.4	5.073	499.992
1984	3,919.356	64,706.3	4.928	318.873
1985	4,054.799	27,131.1	4.868	132.074
1986	3,828.608	50,877.7	4.731	240.702
1987	3,941.605	6,465.6	4.617	29.852
Total				3,299.142
General Operations				
1975	2,697.494	---	2.578	---
1976	2,310.695	129,389.8	2.520	326.062
1977	2,594.787	115,884.7	2.789	323.202
1978	2,710.483	139,408.6	2.989	416.692
1979	3,705.588	112,222.9	3.967	445.188
1980	6,195.928	94,910.6	6.513	618.153
1981	8,289.325	33,191.0	8.182	271.569
1982	9,828.685	6,669.7	9.454	63.055
1983	8,743.907	9,260.1	8.432	78.081
1984	7,728.081	18,506.1	7.258	134.317
1985	6,336.706	49,557.8	6.358	315.088
1986	4,006.669	63,407.1	4.077	258.511
1987	4,348.108	26,606.8	4.264	113.451
Total				3,363.369
All Energy Total				6,662.511

Note: This table incorporates revisions to previously published energy consumption and cost data submitted to DOE by Federal agencies. Energy costs for FY 1975 - 1981 are estimated, based on data provided by the Defense Fuel Supply Center and the DOE Energy Information Administration (EIA). Energy costs for FY 1982 - 1987 are based on annual reports submitted to DOE by Federal agencies.

Source: U.S. DOE, 1987a.

for this role in the 1987 *Energy Security* report to the President: "The Federal Government should lead by example in testing and adopting cost-effective technologies that use energy more efficiently, especially those that minimize future reliance on insecure supplies of oil" (U.S. DOE, 1987c).

According to the U.S. DOE, a major obstacle to Federal conservation efforts was recently addressed through a change in Federal law. Conservation investments, although cost-effective, may require a substantial initial investment in order to achieve energy and cost savings over the long term (U.S. DOE, 1987a). Agencies may now legally "share the savings" from investments in conservation through contracts for up to 23 years with companies that supply the necessary equipment or services in lieu of paying the full capital cost upfront. Several states already have such programs, which can serve as demonstration programs for other levels of government and the private sector and may facilitate future Federal efficiency investments. Texas has implemented an energy performance audit program for 18 major universities enforced by a potential 10 percent withholding of administrative funds. The audit includes a review of the energy management program, including procedures for tracking and monitoring, use of audit activities, capital outlays, and professional training programs. The entire program has been highly cost-effective and in addition served to increase upper management awareness of the energy management opportunities under their control (Verdict, 1988).

STATE AND LOCAL EFFORTS

State and local governments can make an important contribution to the reduction of greenhouse gas emissions. Much environmental regulation and energy policy are to a considerable degree State and local responsibilities. For example, State public utility commissions oversee decisions about the need for new generating capacity and the choice of fuel and can exercise their discretion to promote or discourage particular fuels to promote environmental objectives (ABA, 1980; Randolph, 1988b).[10]

A growing number of States have adopted programs to reduce costs and growth in electricity demand through comprehensive planning and conservation efforts supported by electric utilities. There is evidence that the potential impact of such programs is very large and could allow substantial displacement of fossil-fuel generating capacity in the future. At least 10 States have statutory requirements and policies requiring that utilities examine efficiency investment opportunities as part of a "least-cost" plan (Machado and Piltz, 1988). However, in recent years some utilities have reduced their commitment to conservation and renewed efforts to promote new demand for electricity in order to reduce the costs of excess capacity.

One of the most comprehensive and thoroughly evaluated programs is in the Pacific Northwest. The Northwest Power Planning Council, a regional planning agency created by Congress in 1980, must by law evaluate conservation as a potential resource comparable to generation, and may not support new plants until after first undertaking less expensive demand-side measures. In cooperation with the Bonneville Power Administration, the region spent over $800 million on conservation between 1979 and October 1987. The Council concluded that its conservation programs had achieved energy savings at costs ranging between 1.9 and 2.9 cents per kilowatt-hour, much less than the cost of generation. The Council has undertaken some conservation programs despite a power surplus because of the potential "lost-opportunity" resource if, for example, new buildings are constructed without cost-effective conservation measures. Savings from improving the efficiency of new buildings alone are estimated at $700 million over 20 years (NPPC, 1988).

---

[10] For example, Washington offers utilities a higher rate of return on investments in renewable energy systems, while Texas requires that renewable energy be considered first when a utility seeks to add new capacity and also requires retail wheeling (use of transmission lines) for renewable energy generators under 10 MW whose owners wish to sell power to non-utilities.

A recent Michigan study focused on opportunities to improve efficiency in the residential sector (Krause et al., 1988). The study, done by the Lawrence Berkeley Laboratory, examined the technical and economic potential for electricity conservation. Some of the key findings were that 3400 GWh/year, or 680 MW of baseload equivalent, can be reliably saved by 2005. This is about 29 percent of the forecasted demand for this date and is about two-thirds of the technical potential (Figure 8-6). The result implies a steady decline in overall residential electricity demand of about 1 percent per year over the next 20 years. Most of the projected savings could be purchased at a cost of less than 3 cents per kWh assuming utilities pay for the full extra first costs of consumer investments; the average cost is about 1.1 cent/kWh, in contrast with short-run marginal costs in Michigan of about 3 cents/kWh and much higher costs for new capacity. The greatest savings are from improving the efficiency of lighting, water heating, and refrigerators. The net present value of implementing these savings over 20 years at a 7 percent discount rate would be $545 million.

Some States have undertaken impressive programs to develop alternative energy technologies suited to their climate and energy needs. The North Carolina Alternative Energy Corporation (AEC), for example, is funded by voluntary contributions from electric utilities recovered from ratepayers (Harris and Kearney, 1988). The AEC has contributed almost $6 million to projects that test or demonstrate either conservation, load management, or renewable energy technologies. New York and California are among the other States with substantial energy research programs.

Local governments also undertake many relevant activities. One traditional local government function is to establish building codes and land use regulations. Some local governments have implemented stringent energy conservation requirements for new housing. For example, Tacoma, Washington, estimates that compliance with its code adds only $2,000 to the cost of a new home but saves electricity at a cost equivalent to 2 cents/kWh (much less than any new generation) and offsets

FIGURE 8-6

COST OF POTENTIAL RESIDENTIAL ELECTRICITY CONSERVATION IN MICHIGAN BY 2000*

(1985 cents/kWh)

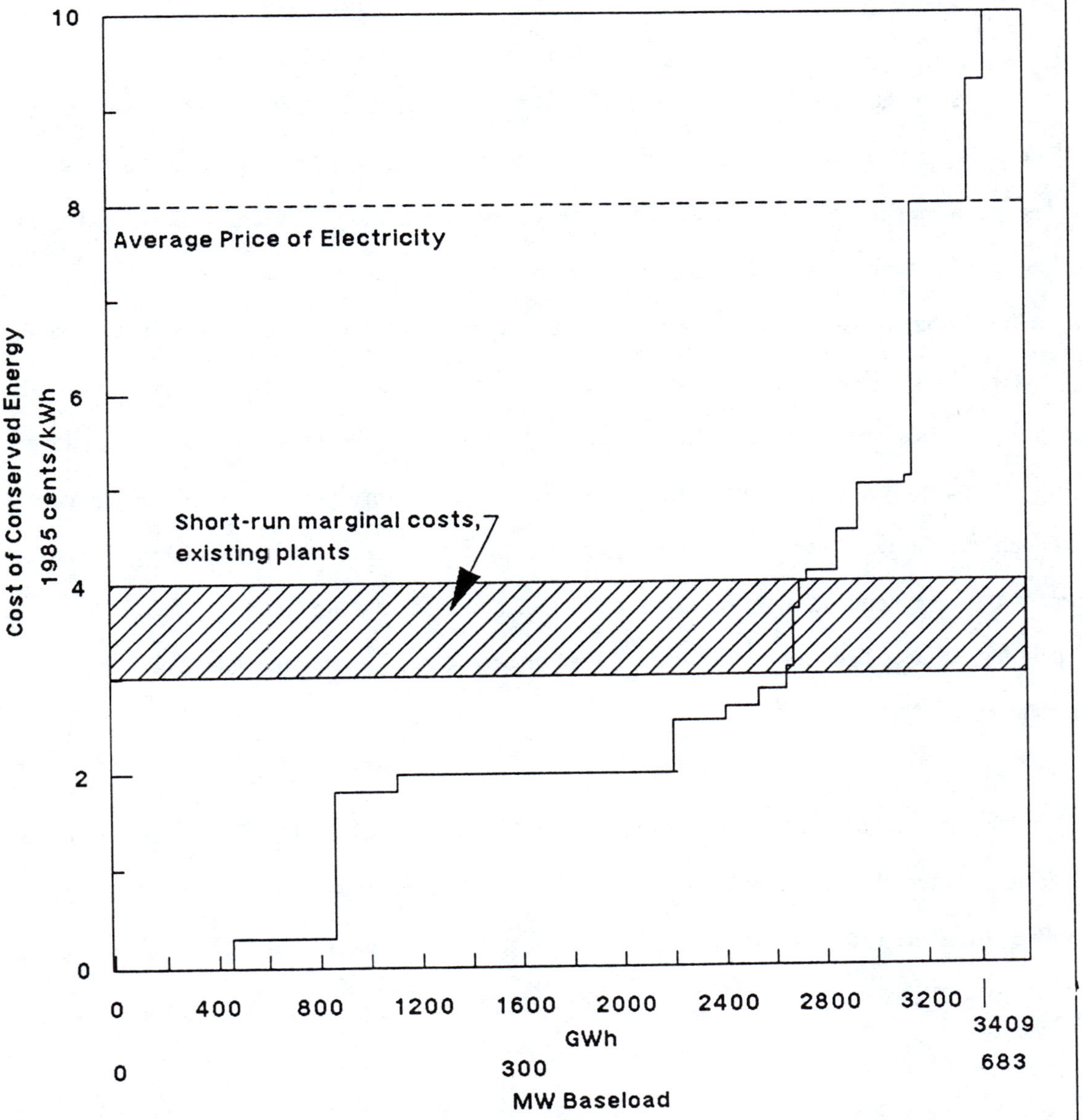

* Based on analysis of territories served by Consumers Power and Detroit Edison companies and assuming a 7 percent discount rate.

Source: LBL, !988.

the need for 1 MW of generation (Randolph, 1988a). Seattle, Santa Monica, and several other cities have worked out cooperative arrangements with their utilities in which the utility provides audits or other services on a reimbursable basis--sometimes at a profit (Randolph, 1988a). Two California cities, Davis and Berkeley, require compliance with minimum residential energy standards as a condition for the sale of a home (Randolph, 1988a).

The State and local role in limiting greenhouse gas emissions is not limited to energy regulation. Other state and local authority that could be exercised to lower greenhouse gas emissions includes management of landfills and regulation of existing stationary sources of air pollution. Another source of large, short-term opportunities may be tree planting programs. As discussed in Chapter VII, urban tree planting can reduce local temperatures, thus reducing summer energy needs for air conditioning, while simultaneously storing carbon dioxide for a relatively modest cost (Akbari et al., 1988). Cities have undertaken large-scale tree planting programs to improve air quality, lower summer temperatures, and beautify neighborhoods. In Los Angeles, a nonprofit group called TreePeople organized a successful effort to plant a million trees prior to the 1984 Olympic games. A new initiative announced by Mayor Bradley in October 1988 calls for planting five million trees and painting surfaces light colors to save 500 MW of peak power, or the equivalent of a large new coal plant (LBL, 1988).

Some States have also taken steps to help reduce emissions of CFCs. For example, Massachusetts recently fined a foam manufacturer for failing to recover CFCs, and obtained an agreement from the company that it will reduce emissions in the future (*New York Times*, August 25, 1988, p. A-21).

PRIVATE SECTOR EFFORTS

Because of the global nature of the greenhouse problem and the lack of direct economic incentive for solutions, much of the impetus for solutions will have to come from governments. Nevertheless, private corporations, non-governmental organizations, and individuals can make important contributions without waiting for government direction. Indeed, there are already several good examples of private initiatives that will contribute to reducing greenhouse gas emissions.

There have also been a number of important actions by private companies to reduce emissions of CFCs in advance of any government mandate. In September 1986--a year before the Montreal Protocol--the DuPont Company announced that they could produce chemical replacements for ozone-depleting CFCs within 5 years if governments provided proper regulatory incentives to support the new market. The substitutes either do not contain chlorine, the chemical that threatens the ozone layer, or they contain hydrogen, resulting in a much shorter and therefore less dangerous atmospheric lifetime. As evidence of the risk to the ozone layer mounted, other companies also announced support for efforts to reduce CFC emissions. The food packaging industry, for example, voluntarily agreed to substitute food packaging made with HCFC-22 rather than with CFC-12; HCFC-22 has an ozone-depletion and greenhouse potential roughly one-twentieth that of CFC-12, as discussed in Chapter III.

Since March 1988, when the NASA Ozone Trends Panel announced its conclusion that global ozone depletion has been detected, several major CFC producers and users have stated their support for an orderly phase-out of all production of CFCs by the turn of the century. Some companies have restricted sales to existing markets, stopped development work on new applications for regulated products, and committed not to increase production capacity nor sell technology to others.

It is often the private non-governmental organizations that are uniquely capable of promoting grassroots development and alleviating poverty (World Bank, 1984). The World Bank has supported the involvement of private companies in the design and implementation of projects, particularly projects on social forestry, agroforestry, and the environment. AID and the Peace Corps have given a high priority to involving private voluntary organizations in community forestry projects undertaken through the Food for Peace Program (Joyce and Burwell, 1985). The Tropical Forest Action Plan also includes strong support for involving private organizations.

Applied Energy Services (AES) of Arlington, Virginia, a private company involved in cogeneration projects, recently hired a non-profit organization, the International Institute for Environment and Development (IIED), to assist in identifying potential reforestation projects capable of providing a carbon sink equal to the emissions from a new 180 MW coal-fired plant in Connecticut. This approach reflects the offset concept discussed above. One problem in making this exchange on a voluntary basis was that the company had difficulty obtaining financing despite the small cost of the trees relative to the total project.[11]

Many of the strategies necessary to reduce greenhouse gas emissions require changes in technologies or policies that can only be accomplished by large corporations and governments. However, there are some important exceptions, particularly tree planting. Reviving the tradition of Arbor Day could provide a very effective symbol of individual responsibility for greenhouse gas emissions; one hectare of well-managed Douglas fir trees can absorb the average American's lifetime

---

[11] The program will cost the company $2 million for an endowment to finance a 10-year tree planting program by CARE, the Guatemalan forest service, and the Peace Corps to support 40,000 small farmers in planting 52 million trees in plantations and agroforestry systems. Over the 40-year life of the plant, 15 million tons of carbon will be fixed at a total cost of $14 million cash and contributed labor. AES is considering providing similar offsetting forestry projects for nine other plants. The costs are expected to be only about 0.1 cents per kWh because the projects will take place in developing countries and some labor will be contributed. CARE officials informally estimate that the project could be replicated perhaps 100 times given current institutional capability and land availability (WRI, 1988).

per capita emissions of CO_2 (about 400 tons, or 5 tons C/year x 80 years).[12] The cost of planting that many trees on a large scale is difficult to estimate, but $1500 is a reasonable approximation--a large but not impossible lifetime investment.[13]

As these examples suggest, the government can help foster private initiatives. In some cases regulatory obstacles impede voluntary efforts and government can help remove them; in other cases the government may be able to help bring parties together and provide technical information or assistance. The important point is that government can help promote voluntary efforts to reduce greenhouse gas emissions.

COMPLEMENTARY STRATEGIES TO REDUCE GREENHOUSE GAS EMISSIONS

The policies described in this chapter are likely to be most effective when used in combination and, wherever possible, linked to the attainment of other national goals. Pricing and regulatory strategies are the most effective policies in the short term, however, the other policies reviewed in this section can complement and enhance pricing and regulatory approaches. In the long run, government-supported R&D, information programs, and other market-enhancing activities can make a significant contribution. Because government action to reduce greenhouse gas emissions will often be closely related to other national policies, programs must also be carefully crafted to meet several goals simultaneously (IEA, 1987).

Improving the efficiency of the U.S. economy is closely related to concerns about the country's economic competitiveness. While the U.S. has reduced its energy intensity a great deal, our

---

[12] Other species would require more land but with research it may be possible to do substantially better.

[13] A more detailed discussion of the feasibility of large-scale tree planting is provided in Chapter VII.

principle economic competitors have done even better and can offer more efficient technologies in many areas. Per capita energy consumption in the U.S. remains more than double that of Europe or Japan; Japanese industry uses half as much energy per dollar of value added as industry in the U.S although spending about as much due to higher energy prices (Zimmerman and Reid, 1988; Flavin and Durning, 1988). Efficiency investments would make available billions of dollars in capital for other investments (Rosenfeld and Hafemeister, 1988). Oil imports are also a growing negative influence on the U.S. balance of trade, accounting for roughly a fourth of the merchandise trade deficit--more than any other single item (Chandler et al., 1988).

Policies for reducing greenhouse gas emissions may also complement efforts to address concerns raised by recent growth in oil imports and potential risks to U.S. energy security (U.S. DOE, 1987c). Oil imports have increased from 27 percent of U.S. supply in 1985 to about 37 percent in 1988, and current forecasts indicate imports could exceed 50 percent by the mid-1990s (U.S. DOE, 1987c). By comparison, imports accounted for 35 percent of supply before the 1973 embargo and 43 percent before the 1979 embargo.

As discussed in Chapter IV, oil consumption for transportation is one of the largest sources of U.S. greenhouse gas emissions, as well as a major contributor to projected increases in global emissions. Transportation also accounts for almost two-thirds of U.S. oil consumption. Further improvements in auto efficiency could therefore contribute to both policy objectives (Mackenzie, 1988).

Complementary strategies are very important in the building sector, where the effectiveness of pricing strategies is limited by extreme first-cost sensitivity and the stringency of regulation is likely to reflect considerable compromise. Studies show that selective financial incentives in combination with information and training programs can be highly effective in promoting innovation not covered by standards, creating incentives for early adoption of standards, and enhancing compliance with standards. The result can be performance that is substantially better than compliance with minimum standards would produce (Vine and Harris, 1988).

Complementary strategies can also work well to achieve the goal of reducing carbon emissions from automobiles. To achieve this goal, which is clearly consistent with efforts to promote higher average fuel economy, consumers must be able to evaluate the relative fuel efficiency of automobiles; Federal fuel efficiency tests, gas mileage guides, and other information programs promote this goal.[14] Manufacturers have rightfully noted that consumer interest in fuel efficiency has declined with lower gasoline prices; higher gasoline taxes, gas guzzler fees, or some other form of economic incentive is necessary to stimulate demand for efficiency.

Better management of solid wastes can reduce methane emissions. States with the most effective programs often combine some or all of the following policies: regulations on landfills, deposits on recyclable materials, tax incentives for recycling companies, and government procurement policies that promote purchase of recycled materials (Shea, 1988).

Research and development programs can assist on the supply side by accelerating the development and testing of new technologies. Procurement programs can provide an initial market and extensive testing for concepts prior to wider marketing. Minimum fuel economy standards fill a different role by providing clear targets and reducing market uncertainty, but regulations must be carefully structured to allow adequate time for adjustment, to avoid conflicts with other socially desirable features, and to preserve fair competition.

---

[14] The Department of Energy publishes *Gas Mileage Guides*, which list the fuel economy of all new car models; however, the number of guides distributed has been reduced from 10 million to 3 million annually.

There is some concern that other national policies for reducing automobile oil consumption and improving urban air quality may lead to increased carbon emissions--for example, if methanol is produced from coal in large quantities (as discussed in Chapter VII), particularly if vehicles using such fuels are permitted to be less efficient as provided in recent Federal legislation.

Federal and state governments have also been considering air pollution regulations to control problems other than global warming that could unintentionally exacerbate global warming. Most prominent of these areas is the ongoing debate over acid rain legislation. Acid rain is caused in part by sulfur dioxide and nitrogen oxide emissions from electric utility powerplants. Some of the methods to limit these emissions could result in greater CO_2 emissions. For example, the use of scrubbers to control sulfur dioxide emissions from coal-fired or oil-fired powerplants would increase CO_2 emissions because the additional fuel required to operate the scrubbers would produce additional CO_2 emissions. However, strategies that rely on natural gas or demand-side reductions would reduce carbon emissions (Centolella et. al., 1988). Thus, an important priority for acid rain mitigation strategies should be to consider the global warming impacts of alternative acid rain programs.

Just as strategies to reduce greenhouse emissions must be applied in combination in order to be maximally effective, so too must we work together at all levels of government, as well as in the private sector.

Many short-term actions to reduce greenhouse gas concentrations could be implemented today on the basis of existing legislative authority. Indeed, some policies directed toward this objective are already under consideration. For example, agencies could make greater use of the environmental impact statement process to consider the extent to which their major projects will contribute to or be effected by climate change. The President's Council on Environmental Quality has done an analysis of projects that might warrant discussion of climate change issues and may issue guidelines for other agencies. EPA, which is legally required to review draft impact statements, has used its authority in at least one instance to comment on the additional greenhouse gas emissions that might result from actions proposed by the Federal Energy Regulatory Commission.

Much more could be done to improve the federal government's energy management within existing legislative authority, beginning with much more disaggregated reporting of energy use by agencies. The laws creating the Federal Energy Management Program require an examination of opportunities for energy-saving alternatives in federal buildings. Such analysis is to be done on the basis of life-cycle costs, reflecting marginal fuel costs and a 7 percent real discount rate--terms intended to favor conservation, as compared with using average costs and a higher discount rate. (Energy Security Act, PL 96-294, Section 405). The Department of Energy proposed implementing regulations in 1980, but final regulations were never completed.

Other short-term possibilities follow from some of the recent and pending agency decisions discussed above, such as ongoing consideration of rules to promote demand-side bidding by the FERC and reconsideration of CAFE standards by the Department of Transportation.

IMPLICATIONS OF POLICY CHOICES AND TIMING

The potential cost of government action to reduce greenhouse gas emissions may be much less than it would appear if such action serves other important economic or environmental objectives. Most of the measures proposed to reduce emissions are already of substantial public interest, for example, policies that promote energy efficiency, reductions in use of CFCs, efforts to halt deforestation, and other desirable social policies--so that the threat of a global warming is often simply another reason to implement such policies. The incremental cost of taking actions to limit global warming today may therefore be modest.

The technical feasibility of implementing various measures that have been proposed for limiting global warming has already been addressed in Chapter VII. However, since the threat of global warming has so far not been an important consideration in government policy-making, it is important to note the other benefits of these measures. For example, serious consideration is being given to eliminating emissions of CFCs because of their impact on the ozone layer. In addition, a substantial number of states are promoting investments in energy-efficient technology in order to reduce energy costs and to meet the need for new generating capacity. Also, Federal and state authorities are promoting waste reduction and recycling as alternatives to land disposal because of the high cost and environmental risks associated with traditional disposal methods. Finally, a large and growing international effort has been organized to slow tropical deforestation because of its impact on economic development and the quality of life in many developing countries.

For several reasons, near-term action may be necessary if it is desired to stabilize greenhouse gas concentrations. For political and economic reasons, actions cannot be immediately implemented once it is agreed they are needed, and for reasons having to do with atmospheric chemistry, concentrations of greenhouse gases--and the attendant risks--will decline only gradually even after actions are implemented.

Policy development and implementation can be a lengthy process, particularly at the international level. Roughly a decade was required for the process that led to international agreement to reduce emissions of CFCs embodied in the Montreal Protocol, and it will take another decade to implement the agreed-upon reductions. An agreement to reduce emissions of greenhouse gases associated with the use of fossil fuels and with deforestation could take much longer to achieve: the activities responsible for CO_2 emissions are more economically valuable, the distribution of emissions is greater, the responsible countries have diverse economies and less shared interests, and the difficulty of implementing alternatives may be greater.

Implementing new technologies can also often be time-consuming. Most emissions result from fundamental activities that support the global economy (transportation, space conditioning, industrial production, land clearing, etc.); therefore, rapid introduction of new technologies for some uses could be very disruptive. Once a technology is cost-effective, it may take years before it achieves a large market share, and decades more for the existing capital stock to be replaced. The historical pattern of gradual displacement of old energy sources by new is shown in Figure 8-7. Government policy may be able to significantly accelerate this transition, although it is debatable whether such efforts have been very successful in the past.

New end-use technology with greater efficiency can replace existing technologies more quickly: it takes 5-10 years to develop new automobile models, and the existing fleet is largely replaced over 8-12 years; major home appliances and space heating and conditioning systems are in use for 10-20 years, and industrial equipment, for 20-40 years. Buildings may be used for 40-100 years or more, however, and the reduction in energy requirements that can be achieved (per unit) as a result of remodeling is more limited than what can be achieved in newly-constructed buildings. Thus, depending on the sector, it traditionally takes roughly 20-50 years to substantially alter the technological base of industrial societies, and thus the cost of reducing emissions could rise dramatically as the time allowed for achieving these reductions is decreased. While the rate of change can be higher in rapidly developing countries, once the industrial infrastructure is built, it will normally be many years before it is replaced. This process will be naturally accelerated if new technologies become sufficiently attractive, and government policies can encourage more rapid retirement of existing buildings and equipment. Nevertheless, such efforts are likely to be more expensive and technically difficult.

FIGURE 8-7

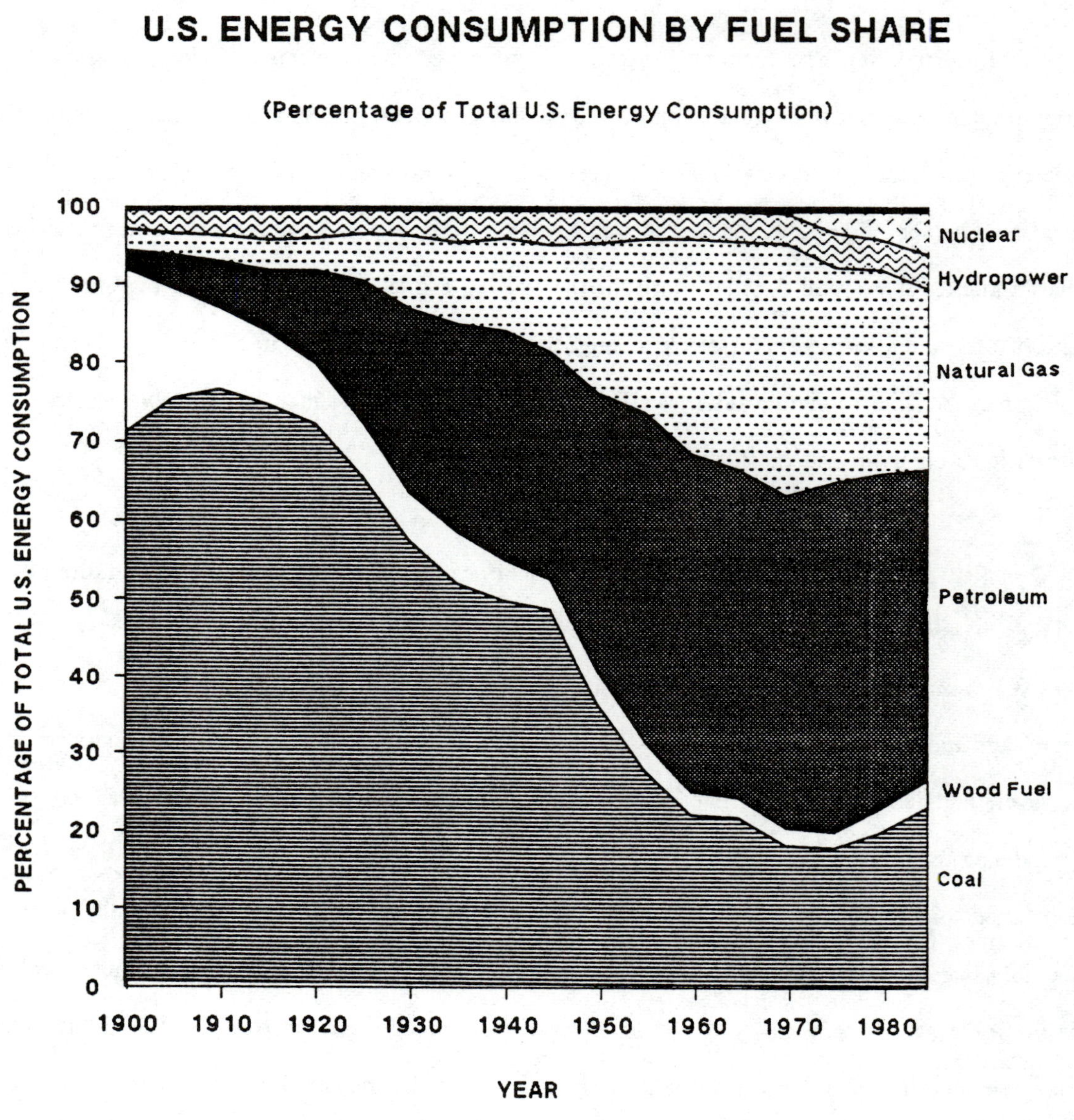

Source: U.S. DOE, 1987a.

Once greenhouse gases have entered the atmosphere, they continue to affect climate for decades. Even if all anthropogenic emissions of carbon dioxide could be suddenly eliminated, it may take more than a century for the oceans to absorb enough carbon to reduce the atmospheric concentration of CO_2 even half way toward its preindustrial value (see discussion in Chapter II). With continued emissions, the time required to reduce excess concentrations by the same percentage increases further. For CFCs and N_2O, it would be more than 50 years before their excess concentrations declined by half even if all anthropogenic emissions were eliminated. Only methane responds relatively quickly--the excess concentration would fall by 50 percent in less than a decade after anthropogenic emissions were eliminated (Figure 8-8).

The climate response also lags behind the radiative forcing imposed by changes in atmospheric composition (Chapter III). Due to the heat capacity of the ocean, the global average temperature will rise more slowly than if climate were continuously in equilibrium with the changing composition of the atmosphere. The climate will also cool more slowly if trace-gas trends are reversed. Consequently, any strategy that involves responding to climate impacts is very risky: Once climate change occurs even draconian measures would not reverse the process for decades.

SENSITIVITY TESTS OF THE EFFECT OF ALTERNATIVE POLICIES ON GREENHOUSE GAS EMISSIONS: RISK TRADE-OFFS

A series of sensitivity tests were run using the EPA's Atmospheric Stabilization Framework to test the relative effect of a range of policies that might increase or moderate greenhouse gas emissions. These tests involved changing key parameter values in the Rapidly Changing World No Response scenario (RCW) and the Rapidly Changing World with Stabilizing Policy scenario (RCWP) to represent more aggressive government intervention. The following section summarizes the results of those sensitivity tests and concludes with some suggestions concerning possible future targets for policy intervention. (For a detailed description of the assumptions and outcomes incorporated into

FIGURE 8-8

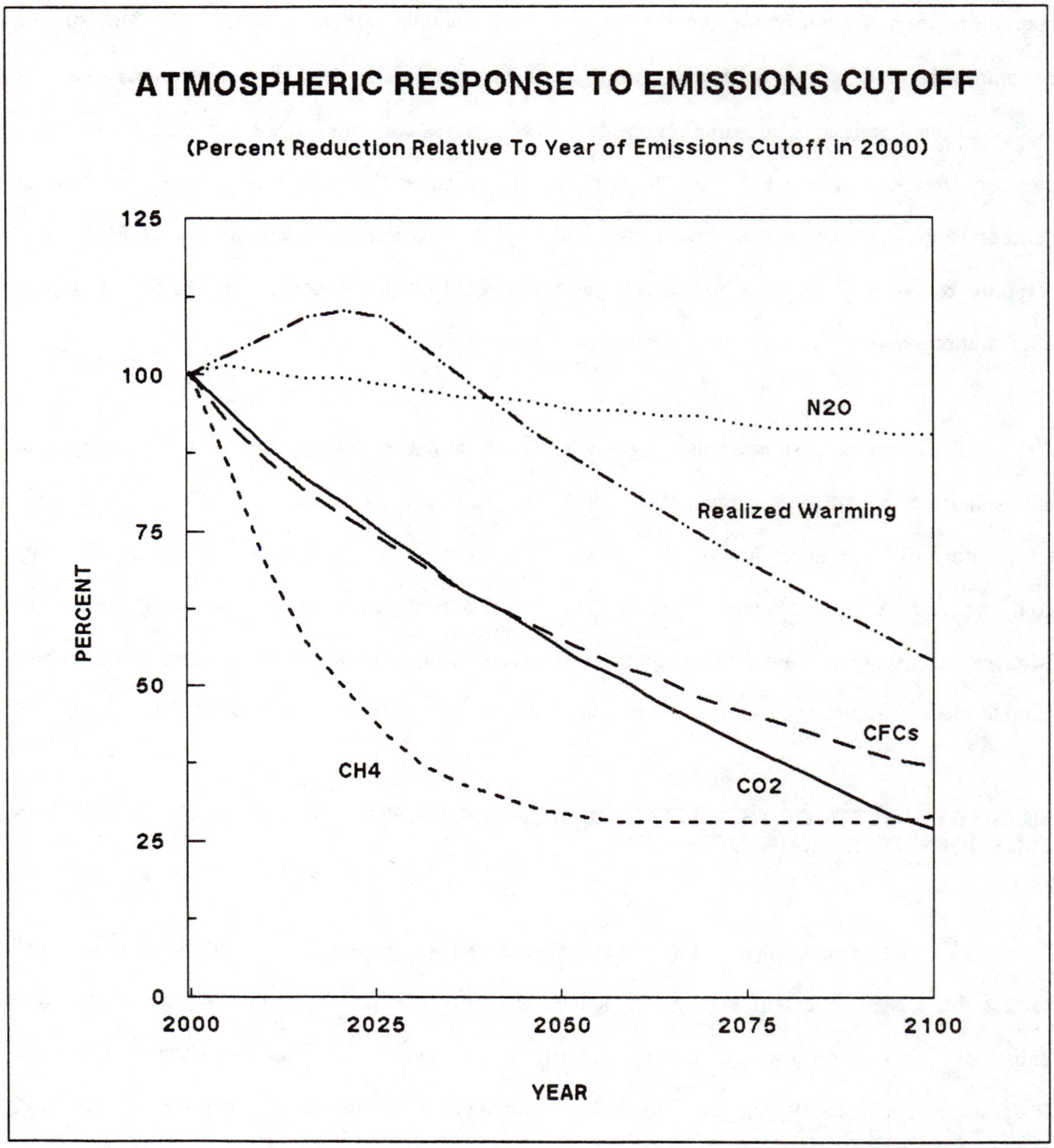

Figure 8-8. Change in atmospheric concentration and temperature following a cutoff in anthropogenic emissions of greenhouse gases. 100% is defined as the year 2000 concentration or temperature and 0% as the preindustrial concentration or temperature. Total methane emissions are reduced to 380 Tg per year, which may represent only a 50% reduction in anthropogenic emissions.

the RCW and the RCWP scenarios, the reader is referred to Chapter V and Appendix B of this report.)

To measure the impact of a range of specific policy goals in the RCW scenario, two sets of sensitivity tests were developed. In each case, the effects of the policy alternatives were first evaluated individually and then in combination to measure the impact of implementing all the identified measures simultaneously. The results were evaluated in terms of the degree to which the projected global warming was amplified or reduced relative to the reference scenarios.

Policies That Increase Greenhouse Gas Emissions

Several recent policy proposals indicate the potential for government actions that will or may promote increased emissions of greenhouse gases. U.S. energy policy currently seeks to increase coal production and use to reduce dependence on imported fuels and boost employment; the Department of Energy has made numerous suggestions concerning various policies to increase the role of coal in relative and absolute terms (U.S. DOE, 1987; National Coal Council, 1987; see Figure 8-9). Recent initiatives in utility regulation and alternative fuels have an ambiguous impact and the Department of Transportation has also decided not to increase automobile fuel efficiency requirements beyond 26.5 mpg.

Improving the efficiency of coal combustion in so-called "clean coal" technologies may reduce greenhouse gas emissions relative to the current generation of coal-burning plants. Over the long run, however, more efficient coal-burning technologies may increase greenhouse gas emissions by making coal economically attractive relative to other fuels. (This proposition is tested in the modeling analysis presented below.) Numerous policy proposals have also been made to increase U.S. coal exports in order to improve the balance of trade. A recent proposal by the U.S. DOE coal advisory

FIGURE 8-9

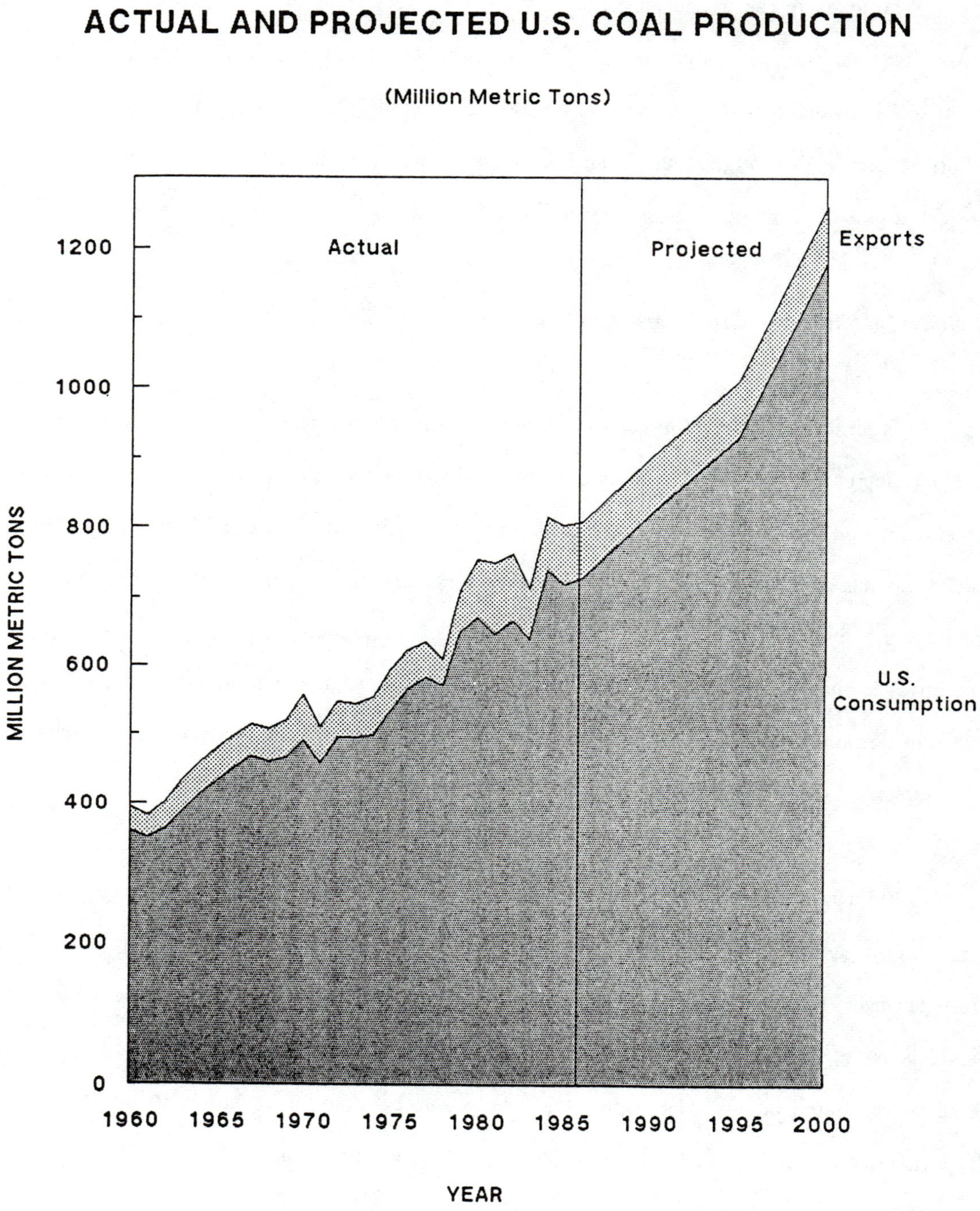

Source: U.S. DOE, 1987a.

committee would link exports of clean-coal technology to an agreement to purchase U.S. coal policy that might slow the adoption of more efficient technology for burning less expensive domestic coal in some developing countries like China (National Coal Council, 1987).

The need to more carefully consider the potential impact of government decisions on greenhouse warming is evident from analyses of two recent policies with ambiguous impact on greenhouse warming. The Methanol and Alternative Fuels Promotion Act enacted this year creates incentives for auto manufacturers to produce vehicles powered by methanol, ethanol, and substitutes for gasoline. This program was adopted to lessen dependence on imported oil and improve urban air quality. However, during Congressional debates concern was expressed that if methanol were produced in large quantities from coal, the result would be a significant increase in greenhouse gas emissions. Congress therefore included provision for study of this relationship. (The potential effect of accelerated synfuels development is presented below.)

Another example of a policy with ambiguous but potentially significant effects on greenhouse gas emissions is rule changes proposed by the Federal Energy Regulatory Commission (FERC) to facilitate non-utility power production. The draft environmental impact statement (DEIS) on these rules concluded that coal-fired technologies have so far played a limited role in the development of independent power projects relative to resource recovery, hydro power, and natural gas. As a result of the FERC proposals, coal could assume a much larger role in the future because of proposed elimination of requirements for cogeneration incompatible with the most economic coal technologies and because larger firms with the resources necessary to undertake large-scale projects that increase the attractiveness of coal technologies may find the power market more attractive. However, alternative assumptions imply natural gas will grow much more than coal (FERC, 1988b).

A set of tests applied to the RCW case measures the effect of policy choices that cause future emissions to grow. In this "Accelerated Emissions" case, eight key parameters in the model, representing domestic and international policies that increase emissions, were varied. These parameters were chosen as proxies for currently-proposed policies (e.g., accelerated development of synfuels) or as possible consequences of government inaction or failure (e.g., high CFCs and deforestation).

The tested policy alternatives included the following:

- A High CFC Case, which assumed a low level of participation in and compliance with the Montreal Protocol.
- A Cheap Coal Case, which assumed that advances in the technology of coal extraction and transport rapidly reduce the market price of coal at the burner tip. In the RCW scenario, coal supply productivity was assumed to improve at a rate of approximately 0.5 percent per year. In this case, it is assumed to improve at a rate of 1 percent per year.
- A Cheap Synfuels Case, which assumed that the price of synthetic oil and gas could be reduced by 50 percent and commercialization rapidly accelerated relative to the RCW case. This case assumes that the minimal production price for synfuels can be achieved in 20 years rather than the 30 years assumed in the RCW case.
- A High Oil Price Case, which assumed that OPEC (or some other political entity) could control production levels and thus raise the border price of oil and gas. To simulate this effect, oil and gas resources were shifted to higher points on the regional supply curves. In addition, extraction costs for oil in each resource grade were increased relative to the assumptions in the RCW case.

- A Slow Improvement Case, which assumed that technical gains in the engineering efficiency of energy use occurred only half as rapidly as had been assumed in the RCW case. In the RCW case, it was assumed that efficiency improved at rates of approximately 1-2 percent per year. In the Slow Improvement case, the assumed rates were reduced to only 0.5-1.0 percent per year.
- A Rapid Deforestation Case, which assumed that rates of deforestation increased at a rate equal to the rate of growth in population.
- A High-Cost Solar Case, which assumed that solar energy was so expensive that its price precluded the possibility of making any significant contribution to global energy supply.
- A High-Cost Nuclear Case, which assumed that the cost of electricity from fission electric systems became so high that their contribution to global energy supply was permanently limited. In this case, an environmental tax of $20/GJ in 1975$ (i.e., $40/GJ in 1985$) was imposed on the price of electricity supplied by nuclear powerplants.
- A Combination Case, in which all of these policy strategies were combined in one scenario.

Figure 8-10 summarizes the results of these sensitivity tests as compared with the RCW case. The results are illustrated in terms of the incremental effect of each policy strategy on the warming commitment in the RCW scenario in 2050 and 2100. Figure 8-10 indicates that the measures that amplify the warming to the greatest extent are those that (1) reduce the rate of efficiency improvement (Case 5), (2) reduce the cost of synfuels (Case 3), and (3) increase the assumed rate of growth in CFC production and use (Case 1).

FIGURE 8-10

ACCELERATED EMISSIONS CASES: PERCENT INCREASE IN EQUILIBRIUM WARMING COMMITMENT

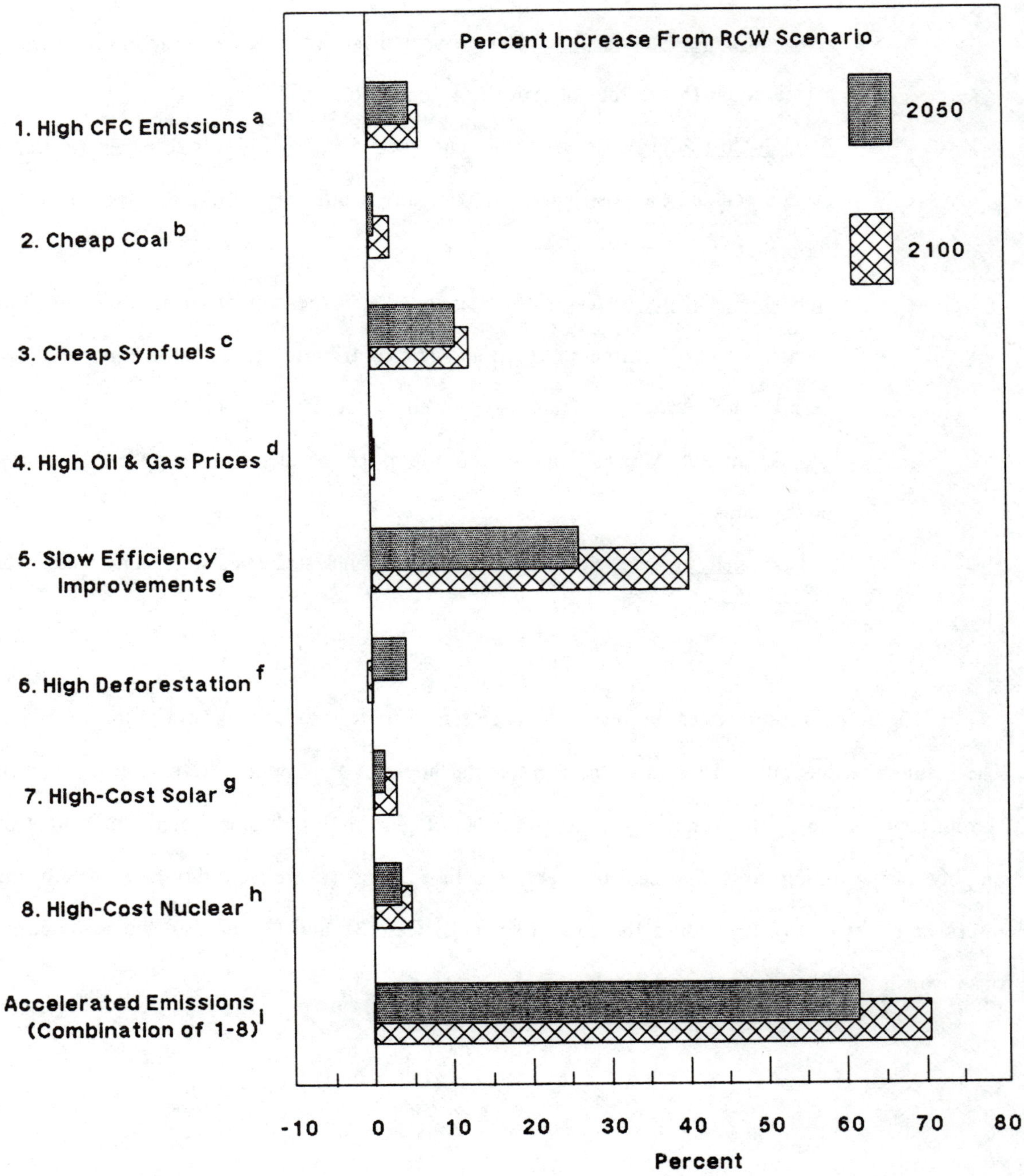

FIGURE 8-10 -- NOTES

Impact Of Accelerated Emissions Policies On Global Warming

[a] Assumes a low level of participation in and compliance with the Montreal Protocol. The assumptions used in this case are similar to those used in the "Low Case" analysis described in the EPA's Regulatory Impact Assessment report.

[b] Assumes that advances in the technology of coal extraction and transport rapidly reduce the market price of coal at the burner tip. In the RCW scenario, the economic efficiency of coal supply is assumed to improve at a rate of approximately 0.5% per year. In this case, it is assumed to improve at a rate of 1% per year.

[c] Assumes that the price of synthetic oil and gas could be reduced by 50% and commercialization rapidly accelerated relative to the RCW case. This case assumes that the minimal production price for synfuels can be achieved in 20 years rather than the 30 years assumed in the RCW case.

[d] Assumes that OPEC (or some other political entity) could control production levels and thus raise the border price of oil and gas. To simulate this effect, oil and gas resources were shifted to higher points on the regional supply curves. In addition, extraction costs for oil in each resource grade were increased relative to the assumptions in the RCW case.

[e] Assumes that technical gains in the engineering efficiency of energy use occurs only half as rapidly as assumed in the RCW case. In the RCW case it is assumed that efficiency improves at rates of approximately 1-2% per year. In the Slow Improvement case the assumed rates were reduced to only 0.5-1.0% per year. The lower rate of improvement is similar to the assumptions in recent projections for the Department of Energy's National Energy Policy Plan.

[f] Assumes annual deforestation increases at a rate equal to the rate of growth in population.

[g] Assumes that solar energy remains so expensive that its price precludes the possibility of making any significant contribution to global energy supply.

[h] Assumes that the cost of electricity from fission electric systems becomes so high that their contribution to global energy supply is permanently limited. In this case, an environmental tax of about $40 (1985$) per gigajoule (GJ) on the price of electricity supplied by nuclear powerplants was phased in by 2050.

[i] All of the above assumptions were combined in one scenario. The result is not equal to the sum of the warming in the RCW and the eight individual cases because of interactions among the assumptions.

Policies Designed to Reduce Greenhouse Gas Emissions

The second set of tests, called the Rapid Reduction scenario, examines the effect of measures that might be imposed to supplement those already analyzed in the RCWP scenario. The assumptions made in this case imply more aggressive response to the risks associated with rapid climate change and the near-term adoption of strategies that rapidly reduce the rate of emissions growth. To parallel the accelerated emissions case, eight separate policy strategies were also tested individually and in combination in this exercise by varying key parameters in the model.

The tested policy alternatives included the following:

- A Production Fee Case, in which fees are imposed on the production of fossil fuels in proportion to the CO_2 emission potential. In this case, fees of $8.50/GJ were imposed on unconventional oil production, $5.70/GJ on coal, $2.30/GJ on oil, and $1.10/GJ on natural gas. These fee levels are specified in 1985$ and are phased in over the period between 1985 and 2050.[15]

[15] The revenue implications of these fees (in billions of U.S. dollars) are as follows:

Year	US	Global
	RCWP	
2000	$30	$134
2025	$46	$338
	Additional Production Fee	
2000	$54	$293
2025	$73	$605
	Consumption Fee	
2000	$59	$255
2025	$70	$507

- A Consumption Fee Case, in which a percentage fee, proportional to the carbon content of the fuel, was levied on fuel use. Consumption fees were also imposed in the RCWP case. In this sensitivity case, the fee on coal consumption was increased from 28 percent of the price to 40 percent; the fee on oil use was increased from 20 percent to 30 percent; the fee on natural gas use was increased from 13 percent to 20 percent; the fee on electricity use was increased from 0 to 5 percent. These fees were phased in and fully applied by 2025.
- A High-mpg Car Case, in which global fleet-average auto efficiencies reach 65 mpg in 2025 and 100 mpg in 2050.
- A High Efficiency Buildings Case, in which the rate of technical efficiency improvement in the residential and commercial sectors improves substantially beyond that assumed in the RCWP case. In this case, the rate of efficiency improvement in the residential and commercial sectors is increased so that a net gain in efficiency of 50 percent relative to the RCWP case is achieved in all regions.
- A High Efficiency Power Case, in which, by 2050, average powerplant conversion efficiency improves by 50 percent relative to the RCWP case. In this scenario, the design efficiencies of all types of generating plants improve rapidly. For example, by 2025, oil-fired generating stations achieve an average conversion efficiency roughly equivalent to that achieved by combined-cycle units today.
- A High Biomass Case, in which the availability of commercial biomass was doubled relative to the assumptions in the RCWP case. In this case, biomass availability was assumed to follow the trajectory outlined in the U.S. DOE Biofuels Program. Conversion costs were assumed to fall by half relative to the assumptions in the RCWP scenario.
- A No Coal Case, in which environmental fees of approximately $20/GJ in 1985$ drove coal out of the bulk power markets by 2050;

- A Rapid Reforestation Case, in which deforestation is halted by 2000 and the biota become a net sink for CO_2 at a rate of about 1 Pg C per year by 2025, about twice the level of carbon storage assumed in the RCWP case.
- A Do It All Case, in which all of the above measures are implemented simultaneously. The impact is much less than the sum of the individual components because many of the measures are not additive.

Figure 8-11 summarizes the results of the second set of sensitivity tests as compared with the RCWP case. The results are illustrated in terms of the effect of each policy strategy in reducing the warming commitment from the RCWP scenario in 2050 and 2100 relative to the RCW scenario (for consistency with the Accelerated Emissions case). Figure 8-11 indicates that the measures that reduce the warming to the greatest extent are those that (1) drive coal out of the power markets (Case 7), (2) impose stiff fees on the production of fossil fuels (Case 1), and (3) increase the assumed level of biomass availability (Case 6).

Conclusions From the Sensitivity Tests

This analysis demonstrates that policy choices can significantly affect the timing and magnitude of future global warming. Comparing the results in Figures 8-10 and 8-11 suggests, however, that the effects of policy choices that increase the rate of growth in greenhouse gas emissions could be much larger than the effects of policies that accelerate reductions in future emissions rates. In other words, it may be easier for government policy to worsen the problem than to ameliorate it. If the policies evaluated here are representative of the range of relevant choices, policies that increase emissions may make the situation much worse than current trends suggest, and may have large effects very quickly.

FIGURE 8-11

RAPID REDUCTION STRATEGIES:
ADDITIONAL DECREASE IN EQUILIBRIUM WARMING COMMITMENT

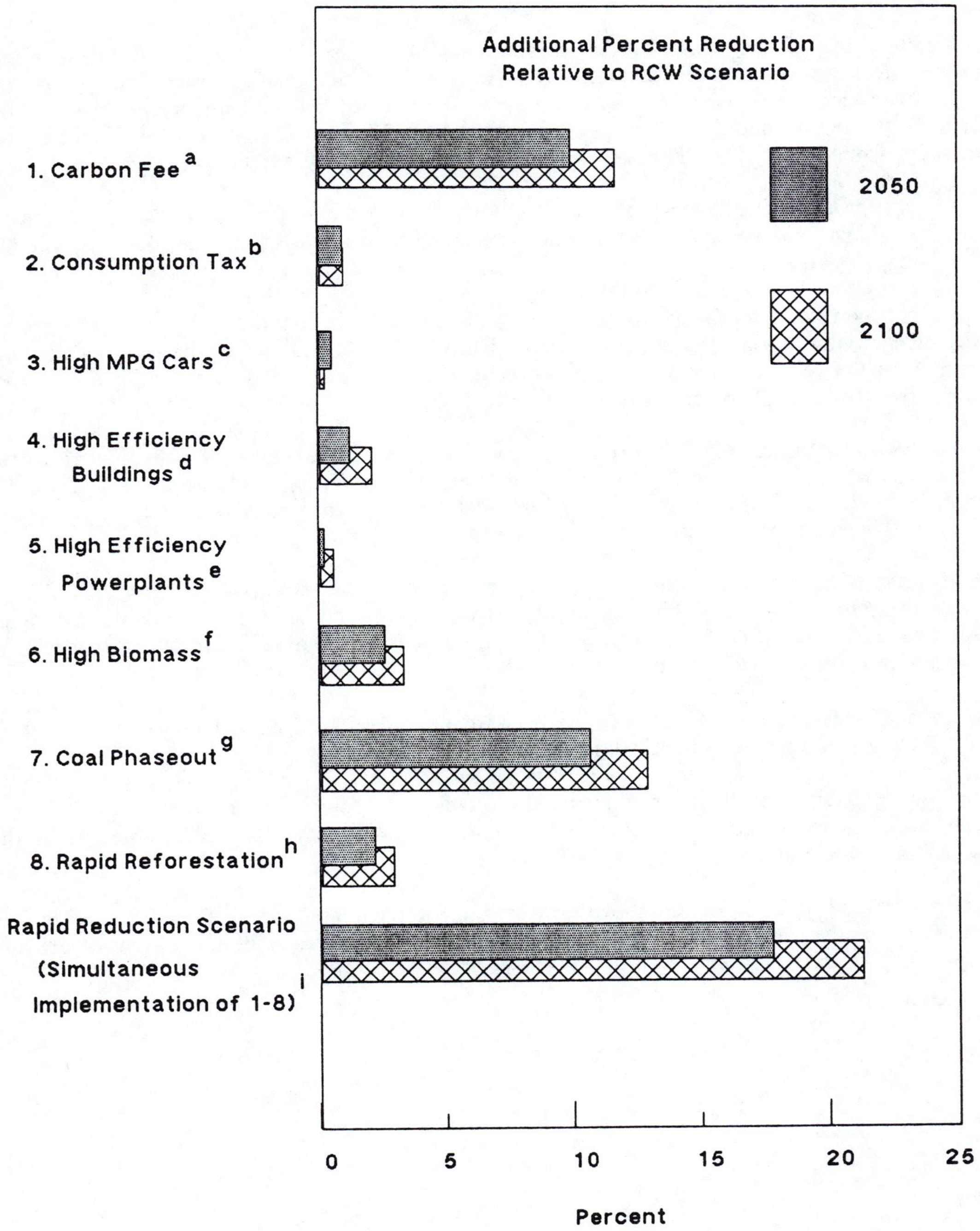

FIGURE 8-11 -- NOTES

Impact Of Rapid Reduction Policies On Global Warming

[a] High carbon emissions fees are imposed on the production of fossil fuels in proportion to the CO_2 emissions potential. In this case, fees of $8.50/GJ were imposed on unconventional oil production, $5.70/GJ on coal, $2.30/GJ on oil, and $1.10/GJ on natural gas. These fee levels are specified in 1985$ and are phased in over the period between 1985 and 2050.

[b] A percentage fee, proportional to the carbon content of the fuel, was levied on fuel use. Consumption fees were also imposed in the RCWP case. In this sensitivity case, the fee on coal consumption was increased from 28% of the price to 40%; the fee on oil use was increased from 20% to 30%; the fee on natural gas use was increased from 13% to 20%; the fee on electricity use was increased from 0% to 5%. These fees were phased in and fully applied by 2025.

[c] Assumes that the average efficiency of new cars in the U.S. reaches 50 mpg (4.7 liters/100 km) in 2000 and that global fleet-average auto efficiencies reach 65 mpg in 2025 (3.6 liters/100 km) and 100 mpg (2.4 liters/100 km) in 2050.

[d] Assumes that the rate of technical efficiency improvement in the residential and commercial sectors improves substantially beyond that assumed in the RCWP case. In this case, the rate of efficiency improvement in the residential and commercial sectors is increased so that a net gain in efficiency of 50% relative to the RCWP case is achieved in all regions.

[e] Assumes that, by 2050, average powerplant conversion efficiency improves by 50% relative to the RCWP case. In this the design efficiencies of all types of generating plants improve significantly. For example, by 2025, oil-fired generating stations achieve an average conversion efficiency roughly equivalent to that achieved by combined-cycle units today.

[f] The availability of commercial biomass was doubled relative to the assumptions in the RCWP case. In this case the rate of increase in biomass productivity is assumed to be at the high end of the range suggested by the U.S. DOE Biofuels Program. Conversion costs were assumed to fall by half relative to the assumptions in the RCWP case.

[g] Environmental fees of about $20/GJ (in 1985$) are phased in by 2050. This has the effect of gradually making coal uncompetitive in utility markets.

[h] A rapid rate of global reforestation is assumed. In this case deforestation is halted by 2000 and the biota become a net sink for CO_2 at a rate of about 1 Pg C per year by 2025, about twice the level of carbon storage assumed in the RCWP case.

[i] Impact on warming when all of the above measures are implemented simultaneously. The impact is much less than the sum of the individual components because many of the measures are not additive.

Of the options available to slow the rate of emissions growth, the most promising near-term strategies are measures to increase the efficiency of energy use. As noted earlier, these measures are improving rapidly and provide other substantial economic, social, and environmental benefits.

The warming estimates for all of the combination scenarios considered in this report are summarized in Table 8-8. Only the most aggressive policy case lowers realized warming below a tenth of a degree celsius per decade. Some experts have suggested this rate of change represents a maximum consistent with acceptable adaptation for many species of plants and animals (Woodwell, 1987).

TABLE 8-8

Scenario Results For Realized And Equilibrium Warming

Realized Warming - 2°C Sensitivity	1985	2000	2025	2050	2075	2100
Accelerated Emissions	0.5	0.7	1.5	2.8	4.5	6.3
RCW	0.5	0.7	1.2	1.9	2.7	3.6
SCW	0.5	0.7	1.1	1.6	2.0	2.5
RCWP	0.5	0.6	0.9	1.1	1.3	1.4
SCWP	0.5	0.6	0.9	1.0	1.2	1.2
Rapid Reduction	0.5	0.6	0.8	0.8	0.8	0.7

Realized Warming - 4°C Sensitivity	1985	2000	2025	2050	2075	2100
Accelerated Emissions	0.5	1.0	2.1	4.1	6.8	9.6
RCW	0.5	0.9	1.7	2.8	4.1	5.6
SCW	0.5	0.9	1.7	2.5	3.2	4.0
RCWP	0.5	0.9	1.4	1.8	2.1	2.3
SCWP	0.5	0.9	1.3	1.7	1.9	2.0
Rapid Reduction	0.5	0.9	1.3	1.4	1.3	1.2

Equilibrium Warming Commitment - 2°C Sensitivity	1985	2000	2025	2050	2075	2100
Accelerated Emissions	0.7	1.1	2.4	4.3	6.4	8.3
RCW	0.7	1.1	1.7	2.7	3.8	4.8
SCW	0.7	1.0	1.6	2.2	2.7	3.1
RCWP	0.7	1.0	1.3	1.5	1.6	1.7
SCWP	0.7	1.0	1.2	1.3	1.4	1.4
Rapid Reduction	0.7	1.0	1.1	1.0	.9	.7

Equilibrium Warming Commitment - 4°C Sensitivity	1985	2000	2025	2050	2075	2100
Accelerated Emissions	1.5	2.2	4.7	8.7	12.4	16.5
RCW	1.5	2.1	3.5	5.4	7.5	9.6
SCW	1.5	2.1	3.3	4.5	5.4	6.3
RCWP	1.5	2.0	2.5	2.9	3.2	3.3
SCWP	1.5	1.9	2.4	2.7	2.8	2.8
Rapid Reduction	1.5	1.9	2.1	2.0	1.7	1.3

REFERENCES

ABA (American Bar Association). 1980. *The Need for Power and the Choice of Technologies: State Decisions on Electric Power Facilities*. ABA Special Committee on Energy Law, Washington, D.C.

ACEEE (American Council for an Energy Efficiency Economy). 1988. *Proceedings of the 1988 ACEEE Summer Study on Energy Efficiency in Buildings* (11 volumes).

Akbari et al. 1988. The Impact of Summer Heat Islands on Cooling Energy Consumption and CO_2 Emissions. In *National and Regional Programs: Proceedings of the 1988 ACEEE Summer Study on Energy Efficiency in Buildings*.

Alliance to Save Energy. 1987. *Industrial Decision-Making Interviews: Findings and Recommendations*. Paper for Michigan Electricity Options Study.

Aronson, E., and S. Yates. 1985. Social Psychological Aspects of Energy Conservation. In Hafemeister, D., H. Kelly, B. Levi, eds., *Energy Sources: Conservation and Renewables*.

Berman, E., M. Cooper, and H. Geller. 1987. *A Compendium of Utility-Sponsored Energy Efficiency Rebate Programs* (EPRI EM-5579).

Bleviss, D. 1988. *The New Oil Crisis and Fuel Economy Technologies: Preparing the Light Transportation Industry for the 1990s*. Quorum, New York.

Carlson, D. 1988a. *Low-Cost Power from Thin-Film Photovoltaics*. Solarex, Newtown, PA.

Carlson, E. 1988b. *Florida Readies Broad Assault on Garbage*. Wall Street Journal, July 20, 1988, p. 25,50

Cavanaugh, R. 1986. Least-Cost Planning Imperatives for Electric Utilities and Their Regulators. *Harvard Environmental Law Review*. 10:299-344.

Cavanaugh, R., and E. Hirst. 1987. The Nation's Conservation Capital. *Amicus Journal* (Summer): 38-42.

Cavanaugh, R. 1988. (journal title?) *Yale Journal on Regulation* 5:331-366.

Centolella, P., F. Pultz, and E. Miller. 1988. Policy Implications of Relying Energy Conservation as an Acid Rain Control Strategy. In *National and Regional Conservation Programs: Proceedings of the 1988 ACEEE Summer Study on Energy Efficiency in Buildings*. ACEEE, Washington, D.C.

Chandler, W., H. Geller, and M. Ledbetter. 1988. *Energy Efficiency: A New Agenda*. ACEEE, Washington, D.C.

Claussen, E . 1988. Moving Forward Together. *Environmental Forum* 5(3): 14-16.

Cogan, D. 1988. *Stones in a Glass House: CFCs and Ozone Depletion*. Investor Responsibility Research Center, Washington, D.C.

Cole, J., and J. Rizutto. 1988. Residential Applications of Real Time Pricing. In ACEEE, *Utility and Private Sector Conservation Programs: Proceedings of the 1988 ACEEE Summer Study*. Washington, D.C.

Crandall, R., H. Gruenspecht, T. Keeler, and L. Lave. 1986. *Regulating the Automobile*. Brookings, Washington, D.C.

Dudek, D. 1988. Testimony before the Senate Committee on Energy and Natural Resources, September 19, 1988. Environmental Defense Fund, New York.

EEC (Commission of the European Communities), 1988. *The Main Findings of the Commission's Review of Member States' Energy Policies, The 1995 Community Energy Objectives*, COM(88) 174, Brussels.

FAO, WRI, World Bank, and UNDP (Food and Agricultural Organization of the United Nations, World Resources Institute, and U.N. Development Programme). 1987. *The Tropical Forestry Action Plan*.

FAO, WRI, World Bank, and UNDP (Food and Agricultural Organization of the United Nations, World Resources Institute, and U.N. Development Programme). 1987. *The Tropical Forestry Action Plan: Investment Priorities and Policy Reforms for Sustainable Forest Land Use*. Keynote paper prepared for the Bellagio Strategy Meeting on Tropical Forests. July 1-2, 1987.

FERC (Federal Energy Regulatory Commission). 1988a. Proposed Regulations Governing Bidding Programs (RM88-5-000), March 16, FERC, Washington, D.C.

FERC (Federal Energy Regulatory Commission). 1988b. Proposed Regulations Governing Independent Power Producers (RM88-4-000), March 16. FERC, Washington, D.C.

FERC (Federal Energy Regulatory Commission). 1988c. The Qualifying Facilities Report, (January 1). FERC, Washington, D.C.

Federal Register. August 12, 1988.

Flavin, C. 1988. Creating a Sustainable Energy Future. In Starke, L., ed., *State of the World 1988*. New York: W.W. Norton 22-40.

Forster, M. 1986. Two Cheers for the Agreement. *Environmental Policy and Law* 16(5): 142-143.

Geller, H. 1986a. Energy-Efficient Appliances; Performance Issues and Policy Options. *Technology and Society Magazine* March.

Geller, H. 1986b. Energy and Economic Savings from National Appliance Efficiency Standards. ACEEE, Washington, D.C.

Geller, H. 1988. Recent Growth in U.S. Energy Demand: A Preliminary Examination. ACEEE, Washington, D.C.

Goldemberg, J., T. Johansson, A. Reddy, and R. Williams. 1987. *Energy for Development*. World Resources Institute, Washington.

Goldemberg, J., T. Johannson, A. Reddy, and R. Williams, 1988. *Energy for a Sustainable World*. Wiley Eastern Limited, New Delhi, India.

Harris, L., and A. Kearney. 1988. Case Study of the North Carolina Alternative Energy Corporation. In *National and Regional Conservation Programs: Proceedings of the 1988 ACEEE Summer Study on Energy Efficiency in Buildings*. ACEEE, Washington, D.C.

Hirst, E., J. Clinton, H. Geller, and W. Kroner, 1986. *Energy Efficiency in Buildings: Progress and Promise*.

IEA (International Energy Agency). 1987. *Energy Conservation in IEA Countries*. OECD/IEA, Paris, 259 pp.

IIED and WRI (International Institute for Environment and Development and World Resources Institute). 1987. *World Resources 1987*. Basic Books, Inc., New York, 369pp.

Kahn, E. 1988. *Electricity Utility Planning and Regulation*. ACEEE, Washington, D.C.

Kempton, W., and M. Neiman, eds. 1987. *Energy Efficiency: Perspectives on Individual Behavior*. ACEEE, Washington, D.C.

Kerwin, J., and S. Jolin. 1988. Hybrids: New Generation of ESCOs? In ACEEE, *Community and Local Conservation Programs: Proceedings of the 1988 ACEEE Summer Study*. Washington, D.C.

Komor, P., and R. Katzev. 1988. Policies for Encouraging Energy Conservation in Small Commercial Buildings. In *Proceedings from the International Symposium: Energy Options for the Year 2000* (Volume 1). University of Delaware.

Krause, F. 1988. *Analysis of Michigan's Demand-Side Electricity Resources in the Residential Sector.* (LBL #2305). Berkeley, CA: Lawrence Berkeley Laboratory.

Lovins, A., and M. Shepard. 1988. Implementation Paper #1: Financial Electric End-Use Efficiency. Rocky Mountain Institute. Old Snowmass, Colorado.

Machado, S., and R. Piltz. 1988. *Reducing the Rate of Global Warming: The State's Role*. Renew America, Washington, D.C.

MacKenzie, J., 1988. *Breathing Easier: Taking Action on Climate Change, Air Pollution, and Energy Insecurity*. World Resources Institute, Washington, D.C.

Miller, A., I. Mintzer, and S. Hoagland. 1985. *Growing Power: Bioenergy for Development and Industry*. World Resources Institute, Washington, D.C.

Moskovitz, D. 1988. Will Least-Cost Planning Work Without Significant Utility Reform? Paper presented to the Least-Cost Planning Seminar of the National Association of Regulatory Utility Commissioners, Aspen, Colorado, April 12, 1988 (revised June 10, 1988).

McNutt, B., and P. Patterson, 1986. *Cafe Standards - Is a Change in Form Needed?* Paper presented at the Society for Automotive Engineers, September 22-25, 1986, Dearborn, Michigan.

National Coal Council. 1987. *Improving International Competitiveness of U.S. Coal and Coal Technologies*. National Coal Council, Washington, D.C.

Nivola, P. 1986. *The Politics of Conservation*. Washington, D.C.: Brookings Institution.

Novick, S., ed. 1987. *The Law of Environmental Protection.* City?____________________: Clark Boardman.

NPPC (Northwest Power Planning Council). 1988.

OTA (Office of Technology Assessment). 1983. *Industrial Energy Use.* OTA, Washington, D.C.

OTA (Office of Technology Assessment). 1986. *Technology, Public Policy, and the Changing Structure of American Agriculture.* OTA, Washington, D.C.

Peddie, R., and D. Bulleit. 1985. Managing Electricity Demand Through Dynamic Pricing. In D. Hafemeister et al., eds. *Energy Sources: Conservation and Renewables,* City?____________ : American Institute of Physics.

Prindle, W., and M. Reid. 1988. "Pooled" Performance Contracting for Nonprofit Agencies. In *Community and Local Government Conservation Programs: Proceedings of the 1988 ACEEE Summer Study.* ACEEE, Washington, D.C.

Randolph, J. 1988a. The Limits of Local Energy Programs: The Experience of U.S. Communities in the 1980s. In *Proceedings from the International Symposium: Energy Options for the Year 2000.* (University of Delaware)

Randolph, J. 1988b. Waste to Energy Development and PURPA. In *Proceedings from the International Symposium Energy Options for the Year 2000.* (Volume 1).

Renner, M. 1988. *Rethinking the Role of the Automobile,* (Worldwatch Paper #84).

Repetto, R. 1988. *The Forest for the Trees? Government Policies and the Misuse of Forest Resources.* World Resources Institute, Washington, D.C.

Rosenfeld, A. 1985. Managing Electricity Demand Through Dynamic Pricing. In *Energy Sources: Conservation and Renewables,* D. Hafemeister et al., eds. American Institute of Physics. Washington, D.C. .

Rosenfeld, A., 1988. "Energy Conservation, Competition and National Security," *Strategic Planning and Energy Management Journal,* 8:1, 5-30.

Rosenfeld, A., and H. Hafemeister, 1988. "Energy-efficient Buildings," *Scientific American,* 258:4, 78-85.

Ruderman, H., M. Levine, and J. McMahon. 1987. The Behavior of the Market for Energy Efficiency in Residential Appliances Including Heating and Cooling Equipment. *The Energy Journal* 8(1): 101-124.

Shea, C. 1988. Building A Market for Recyclables. *World-Watch,* 1:3, 12-18.

Tripp, J., and D. Dudek. 1986. The Swampbuster Provisions of the Food Security Act of 1985: Stronger Wetland Conservation If Properly Implemented and Enforced. *Environmental Law Reporter* 16:10120-10123.

U.S. DOE (U.S. Department of Energy). 1987a. *Annual Report on Federal Government Energy Management, FY 1987.* U.S. DOE, Washington, D.C.

U.S. DOE (U.S. Department of Energy). 1987b. *Eleventh Annual Report to Congress on Federal Energy Conservation Programs: FY 1987*. U.S. DOE, Washington, D.C.

U.S. DOE (U.S. Department of Energy). 1987c. *Energy Security: A Report to the President of the United States*. U.S. DOE, Washington, D.C.

U.S. DOE (U.S. Department of Energy). 1988. *The Role of Fuel Efficiency in Decreasing Oil Dependence*. (Analysis Memorandum). Office of Policy Analysis, U.S. DOE, Washington, D.C.

U.S. EPA (U.S. Environmental Protection Agency). 1987. *National Air Pollutant Emission Estimates: 1940-1985*. (January). U.S. EPA, Washington, D.C.: EPA.

U.S. EPA (U.S. Environmental Protection Agency). 1988. How Industry is Reducing Dependence on Ozone-Depleting Chemicals. EPA Office of Air and Radiation, Washington, D.C.

Weinberg, A. 1988. Energy Policy in an Age of Uncertainty. *Issues in Science and Technology*. 5:81-85.

Williams, R., E. Larson, and M. Ross. 1987. Materials, Affluence, and Industrial Energy Use. *Annual Review of Energy* 12:99-144.

Wolf, E. 1986. Avoiding a Mass Extinction of Species. In Starke, L., ed., *State of the World 1988*. W.W. Norton, New York, 101-117.

Woodwell, G. 1988. *Biotic Implications of Climate Change: Look to Forests and Soils*. Preparing for Climate Change, Proceedings of the First North American Conference on Preparing for Climate Change: A Cooperative Approach, October 27-29, 1987, Washington, D.C.

WRI (World Resources Institute), World Bank, and UNDP (United Nations Development Programme). 1985. *Tropical Forests: A Call for Action*. WRI, Washington, D.C.

Verdict, M, 1988. "The New Energy Audit Tool: Performance Audits for Energy Management." In *National and Regional Conservation Programs: Proceedings of the 1988 ACEEE Summer STudy on Energy Efficiency in Buildings*. ACEEE, Washington, D.C.

Vine, E., and J. Harris. 1988. The Experience of Energy Conservation Programs With New Commercial Buildings. In *National and Regional Conservation Programs: Proceedings of the 1988 ACEEE Summer Study on Energy Efficiency in Buildings*. ACEEE, Washington, D.C.

USFS (United States Forestry Service). 1987. *The South's Fourth Forest*. Washington, D.C., USFS.

Zimmerman, M., and M. Reid. 1988. "Fact Sheet: Energy and Efficiency in Industry and Commerce," Alliance to Save Energy, Washington, D.C.

CHAPTER IX

INTERNATIONAL COOPERATION TO REDUCE GREENHOUSE GAS EMISSIONS

FINDINGS

- Most of the expected growth in greenhouse gas emissions is from other countries, particularly developing countries and the Eastern Bloc. Efforts to promote international solutions are therefore essential to achieve global reductions in greenhouse gas emissions.

- Technological solutions and policy strategies for the U.S. and other western industrialized nations may not be equally applicable in other parts of the world. Differences in economic development, energy resources, and economic systems must be addressed in devising international strategies to reduce greenhouse gas emissions.

- U.S. leadership has historically made important contributions to other important international environmental agreements, such as the Montreal Protocol on Substances that Deplete the Ozone Layer and the Tropical Forest Action Plan. In the future, U.S. leadership could promote international cooperation to reduce emissions worldwide. Reductions in national emissions may encourage similar actions by other nations and could serve as a valuable demonstration that reductions are feasible.

- Developing countries could reduce the expected increase in greenhouse gas emissions consistent with economic development and other environmental and social goals. Energy efficiency improvements are already essential to reduce capital requirements for the power sector, and efforts to halt tropical deforestation will provide many long-run economic and environmental benefits. The U.S. can promote desirable changes in energy and environmental policy in developing countries through judicious use of its bilateral aid programs and its influence on loans extended by multilateral development banks.

- The Soviet Union and Eastern Bloc nations are and will continue to be important contributors to greenhouse gas emissions. The absence of market pricing has hindered efforts to reduce the energy intensity of these economies but their governments have shown increasing interest in curbing energy costs.

- There have already been some important first steps toward building a framework for international cooperation to reduce the risks of climate change. The November 1988 meeting sponsored by UNEP and WMO initiated a process that includes an intergovernmental committee chaired by the United States to discuss policy responses.

INTRODUCTION

The greenhouse problem requires international strategies to promote global cooperation. An important first step has been taken in this direction under the auspices of the United Nations Environment Programme (UNEP) and the World Meteorological Organization (WMO). In 1988 these organizations helped organize an Intergovernmental Panel on Climate Change to review the science, impacts, and response strategies associated with climate change. The U.S., Soviet Union, China, Japan, and many other leading nations agreed to participate. Three panels were organized to consider scientific issues (chaired by the United Kingdom), effects of climate change (chaired by the Soviet Union), and policy responses (chaired by the U.S.). The first meetings are planned for early 1989.

There is an important relationship between the domestic policies discussed in Chapter VIII and the evolution of international cooperation that is the focus of this Chapter. U.S. actions can promote further international cooperation and complementary strategies by other countries. Consideration of the domestic policies reviewed in this report demonstrates that the U.S. no longer views the problem as only a scientific concern. Further analysis and consideration of such policies may also convince other nations of the seriousness of the risks and the need for action. U.S. leadership in the United Nations and other international forums can have a major impact on the evolution of international understanding and ultimately agreements, as illustrated by the Montreal Protocol on Substances that Deplete the Ozone Layer.

Special efforts will be necessary to limit the growth in greenhouse gas emissions from developing countries while addressing their need for energy and economic growth. Solutions to the problems of climate change may be linked to other development needs, such as capital shortages and increased recognition of local environmental problems. U.S. bilateral assistance programs and participation in multi-lateral development banks (MDBs) provide an opportunity to promote policies that reduce greenhouse gas emissions in developing countries consistent with their development needs. Programs should be designed to increase developing countries' stakes in contributing to this global effort, for example, through debt swaps, afforestation programs, and technology transfer agreements that are linked to reductions in greenhouse gas emissions.

The need for international cooperation has already been recognized, and some important first steps have been taken to establish a framework for international cooperation on scientific aspects of the global warming and to discuss policy responses.

THE CONTEXT FOR POLICIES INFLUENCING GREENHOUSE GAS EMISSIONS IN DEVELOPING COUNTRIES

Because of faster growth rates and greater needs for basic materials, the developing countries will contribute an increasing share of greenhouse gas emissions (Figure 9-1). Any global effort to reduce emissions will therefore have to take into account the very different needs, resources, and other constraints in the developing countries, particularly their need to grow economically in order to be able to meet basic human needs despite their limited capital resources for meeting development objectives. These issues will have to be addressed in the international forums that are being created specifically to deal with the greenhouse problem. However, the U.S. can also address these concerns in the short term by recognizing the link between emissions of greenhouse gases and the investment choices encouraged by bilateral aid and lending by multilateral development banks. Before we discuss these options, however, it is useful to examine the different context within which developing countries operate to get an idea of the energy and environmental issues they face and the priority of their concerns.

FIGURE 9-1

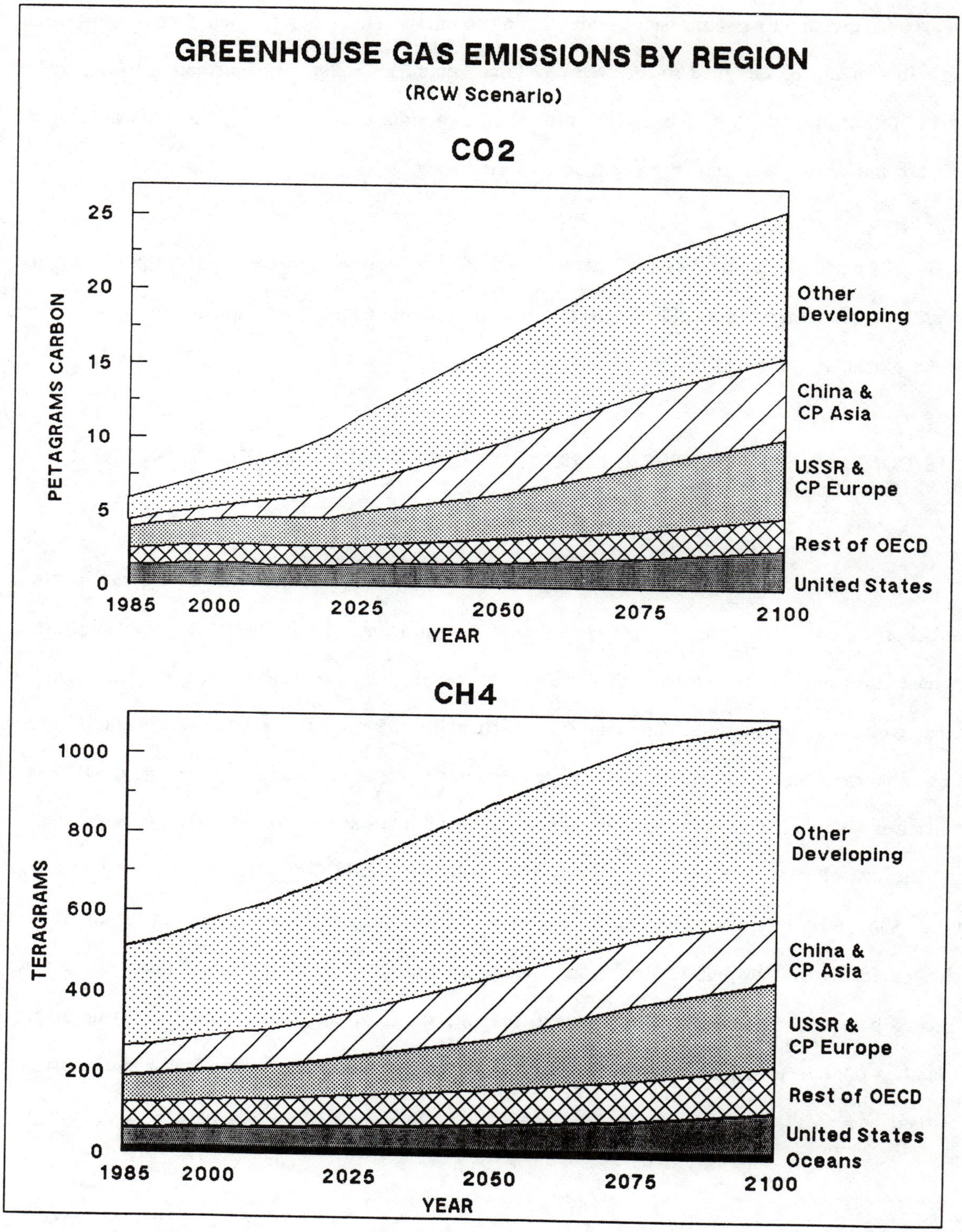
GREENHOUSE GAS EMISSIONS BY REGION
(RCW Scenario)
CO2
PETAGRAMS CARBON
25
20
15
10
5
0
1985
2000
2025
2050
2075
2100
YEAR
Other Developing
China & CP Asia
USSR & CP Europe
Rest of OECD
United States
CH4
TERAGRAMS
1000
800
600
400
200
0
1985
2000
2025
2050
2075
2100
YEAR
Other Developing
China & CP Asia
USSR & CP Europe
Rest of OECD
United States
Oceans

The context for policy formation in developing countries often differs substantially from that in the industrialized countries, reflecting differences in political and economic systems, resources, and societal needs. Some of the major issues associated with the formulation and implementation of energy and related environmental policies in developing countries include socioeconomic equity, financial viability, institutional structure and management of enterprises, and the burgeoning awareness of environmental concerns. Our understanding of these issues is essential to developing emissions reduction policies that are relevant to developing countries, that take into account the conditions existing in these countries and the issues that are of most concern to them.

Economic Development and Energy Use

Developing countries are a much more diverse group than the member countries of the OECD. They range from some of the poorest nations, such as Bangladesh and Ethiopia, to some of the richest ones, such as Saudi Arabia and Singapore. Their annual income per capita varies from \$150 to \$7000 (World Bank, 1987). The group includes oil and natural gas exporters and importers at different levels of per capita income. Their economies vary from a centrally-planned one like China to more market-oriented ones like Brazil and South Korea. Each country has a different mix of institutions engaged in the supply of energy, reflecting different degrees of governmental control and foreign involvement (Table 9-1).

Equity concerns play a strong role in the pricing of fuels. Preferential electricity tariffs for residential customers are often accompanied by even lower tariffs for agricultural ones. (This is not unlike the situation in some developed countries where low "lifeline rates" are charged for small amounts of electricity consumption.) On the other hand, gasoline prices in developing countries are typically set high relative to world market prices, since gasoline is used in cars that are primarily owned by the rich. Petroleum taxes are an important instrument of social policy that can be adjusted

TABLE 9-1

1985 Population and Energy Use Data from Selected Countries

Country	Population[a] (Millions)	GDP (1985 U.S.$ Billions)	GDP/CAP (1985 U.S.$)	ELECTRICITY: Sales Per Capita (KWH)	ELECTRICITY: Growth Rate (1970-85) (%)	MODERN FUELS[b]: Per Capita (GJ)	MODERN FUELS[b]: Growth Rate (1970-85) (%)	Biomass Use Per Capita (GJ)	DOMINANT FUEL: Total	DOMINANT FUEL: %[c]
China	1,045.3	265.53	254	365	6.7	21.25	5.2	8.83	Coal	79
India	750.9	177.10	236	166	7.2	7.91	5.5	5.79	Coal	56
Brazil	134.6	155.45	1,155	1,283	10.2	35.13	7.3	16.16	Hydro	46
Indonesia	165.2	86.20	522	74	14.3	8.57	11.2	9.28	Oil	71
Korea	41.1	86.80	2,112	1,234	13.4	57.03	8.9	2.07	Oil	49
Nigeria	92.0	48.70	529	68	10.7	7.13	16.4	10.92	Oil	80
Mexico	75.6	146.87	1,943	929	7.9	62.72	7.2	5.28	Oil	58
Egypt	47.2	30.06	637	471	9.9	23.68	8.7	NA	Oil	72
Kenya	20.4	5.02	246	107	10.9	2.31	-3.3	NA	Oil	93
U.S.	239.3	3,946.60	16,492	9,719	3.5	325.76	0.7	11.46	Oil	42

[a] based on several sources

[b] modern fuels expressed as fossil fuel equivalent assuming conversion efficiency of 10,000 Btu/kwh

[c] dominant fuel as a percentage of modern fuels (excluding most biomass)

Source: Lawrence Berkeley Laboratory, August, 1988.

to cushion the impact of energy price increases, or to maintain higher prices when international prices decline.

The price of coal in India and China, the two major developing countries that rely heavily on coal, is subsidized with subsidies varying by sector. In China, since the coal price is too low, the government provides subsidies varying from 10 to 27 yuan/ton (or $2.7 to $7.2) to coal producers in order to encourage them to produce more coal (Dadi, 1988). Similarly, despite continual increases in the price of coal in India, Coal India Limited (CIL), an Indian government enterprise that produces over 90 percent of the country's coal, incurred losses equivalent to more than $8 per ton (Hindu Survey, 1988).

Industrial energy use forms a major segment of overall energy use in the developing countries. Because the industries are generally less modern than their counterparts in the developed countries, they tend to be less energy-efficient as well (Table 9-2). According to AID, developing countries can typically save 5-15 percent of commercial fuel through low cost measures and up to 25 percent through cost-effective retrofits (U.S. AID, 1988). A recent detailed study of the potential for cost-effective electricity conservation in Brazil documented potential savings equal to more than a fourth of currently-forecasted needs (Geller et al., 1988; see Table 9-3). The management of government-controlled industries is often inefficient, as a comparison of similar products produced by government-controlled and private-sector industries, even in the same country, will show.

Given adequate capital and technical knowledge, industries in a developing country can operate just as efficiently as comparable ones in a developed country (Schipper, 198?). For example, multinational companies operating in developing countries often use about the same amount of energy per unit of output as the company's comparable plants in the developed countries.

TABLE 9-2

Efficiency of Energy Use in Developing Countries: 1984-85

Country/Region	Energy Use/ Unit of GDP[a]	Average Annual Growth Rate (%)
North America[b,c]		
Canada	0.80	-0.5
United States	0.61	-2.2
Average	0.62	-2.1
Oceania[b,c]		
New Zealand	0.50	+1.8
Australia	0.45	-0.5
Japan	0.29	-3.1
Average	0.32	-2.5
Europe[b,c]		
Luxembourg	0.65	-4.9
Turkey	0.56	-1.0
Portugal	0.49	+1.5
Greece	0.44	+1.2
Ireland	0.44	-1.4
Norway	0.40	-1.4
Sweden	0.40	-0.6
Belgium	0.36	-2.2
Netherlands	0.36	-1.7
United Kingdom	0.35	-2.0
Austria	0.33	-1.2
Italy	0.33	-1.8
Spain	0.32	+0.3
Germany	0.31	-1.7
Denmark	0.27	-1.7
Switzerland	0.25	+0.3
Average	0.34	-1.4

TABLE 9-2 (continued)

Efficiency of Energy Use in Developing Countries: 1984-85

Country/Region	Energy Use/ Unit of GDP[a]	Average Annual Growth Rate (%)
Asia[d,e]		
People's Republic of China	1.40	-1.3
India	0.79	+1.4
Pakistan	0.64	+4.2
Taiwan	0.62	+0.2
Thailand	0.38	-0.8
Malaysia	0.36	+0.3
Indonesia	0.35	+3.3
Philippines	0.35	-2.7
Bangladesh	0.27	NA
Average	0.97	+0.5
Latin America[d,e]		
Venezuela	1.40	+4.6
Brazil	0.68	+2.1
Mexico	0.56	+2.2
Argentina	0.29	+1.8
Average	0.57	+2.7
West Africa[d,f]		
Senegal	0.49	+3.6
Morocco	0.27	0.0
Nigeria	0.18	+9.4
Cote d'Ivoire	0.13	+2.8
Average	0.20	+6.5

[a] Gross domestic product metric tons of oil equivalent per $1,000 U.S. (constant 1980 dollars).
[b] 1985 data.
[c] Average annual growth rate for 1973-85.
[d] 1984 data.
[e] Average annual growth rate for 1973-84.
[f] Average annual growth rate for 1977-84.

NA = Not available.

Sources: Sathaye, et al., 1987; OECD/IEA, 1987.

TABLE 9-3

Potential for Electricity Conservation in Brazil

End-Use Area	Current forecast (TWh)	Savings potential (%)	Savings potential (TWh)
Industrial motors	164.8	20	33.0
Domestic refrigerators	24.7	60	14.8
Domestic lighting	16.5	50	8.2
Commercial motors	28.0	20	5.6
Commercial lighting	25.0	60	15.0
Street lighting	16.8	40	6.7
Total	275.8	--	83.3

Source: Geller et al., 1988.

The energy sector is owned and operated by government-controlled corporations in the poorer developing countries. These corporations often have little incentive to invest in reducing their energy costs because they are protected from competition and rewarded more by measures of production than efficiency of service. This is compounded by regulated prices that are often not high enough to pay for the companies' expenses and non-payment or delayed payment for fuels purchased from other government companies.

Oil Imports, Capital Shortages, and Energy Efficiency

Most developing countries are largely dependent on imports for commercial fuels. During periods of high oil prices, import costs have resulted in serious hardship in these countries. In some countries oil import costs exceeded 25 percent of export earnings in 1984 and much more than that in earlier years (Table 9-4). Future increases in world oil prices may have an even more adverse impact because of the rapid growth in the transportation sector in some countries.

A shortage of capital for large development projects is pervasive in developing country economies and increasingly a constraint on the energy sector. Energy investments in developing countries require a high percentage of available capital (Table 9-5). The World Bank and more recently AID have reviewed the magnitude of energy shortfalls in developing countries and its implications for economic development (World Bank, 1983; U.S. AID, 1988; see Table 9-6). AID estimates that in a current trends scenario, AID-assisted countries would need to spend over $2.6 trillion for the power sector by the year 2008 to meet projected needs. This is an average of over $125 billion per year, compared with the estimated $50 to $60 billion currently being spent annually. Since current expenditures already consume a fourth or more of development budgets, this is potentially a serious constraint on economic development. Aggressive conservation efforts could reduce capital needs by 40-60 percent (Williams, 1988; World Bank, 1983).

TABLE 9-4

Net Oil Imports and Their Relation to Export Earnings for Eight Developing Countries, 1973-1984

	1973	1974	1977	1979	1981	1983	1984
	Net Oil Imports (million U.S. dollars, current prices)						
Kenya	1	27	57	113	316	208	219
Zambia	11	30	53	72	63	274	454
Thailand	173	510	806	1,150	2,170	1,740	1,480
Korea	276	967	1,930	3,100	6,380	5,580	5,770
Philippines	166	570	859	1,120	2,080	1,740	1,470
Brazil	986	3,230	4,200	6,920	11,720	8,890	7,470
Argentina	83	328	338	351	302	-	-
Jamaica	71	193	242	309	490	-	-
India	308	1,170	1,750	3,067	-	-	-
Bangladesh	-	92	172	247	509	286	314
Tanzania	47	153	102	174	306	175	156
	Imports as Percentage of Export Earnings						
Kenya	0.1	4.1	4.8	10.2	26.9	21.2	20.3
Zambia	2.2	5.1	9.5	8.2	7.8	20.8	21.4
Thailand	11.1	20.9	23.1	21.6	30.9	27.3	20.0
Korea	8.6	21.7	19.2	20.6	30.0	22.8	19.7
Philippines	8.8	20.9	27.5	24.4	36.8	35.4	27.8
Brazil	15.9	40.7	34.7	45.4	50.4	40.6	27.7
Argentina	2.5	8.3	6.0	4.5	3.3	-	-
Jamaica	18.1	27.3	32.4	37.7	50.3	-	-
India	10.6	29.7	27.5	39.3	-	-	-
Bangladesh	-	26.5	36.1	37.4	64.6	39.4	33.6
Tanzania	12.8	38.0	20.2	34.8	52.7	47.0	42.3

Source: IMF, 1985 in Goldemberg et al., 1987.

TABLE 9-5

Annual Investment in Energy Supply As a Percent of Annual Total Public Investment (Early 1980s)

Under 20%	20-30%	30-40%	Over 40%
Egypt	Botswana	Ecuador	Argentina
Ethiopia	China	India	Brazil
Ghana	Costa Rica	Pakistan	Columbia
Nigeria	Liberia	Philippines	Korea
Sudan	Nepal	Turkey	Mexico

Source: Munasinghe and Saunders, 1986.

TABLE 9-6

World Bank Estimate of Capital Requirements for Commercial Energy In Developing Countries, 1982-1992

	Low Income	Middle Income		All Countries
		Oil Importers	Oil Exporters	
Total Required Capital				
Electricity	17.6	35.9	13.1	66.6
Oil and Gas	12.1	16.7	40.0	68.8
Coal	5.6	2.8	0.6	9.0
Total	35.3	55.4	53.7	144.4
Foreign Exchange Requirements				
Electricity	3.6	11.4	7.2	22.2
Oil and Gas	4.9	5.9	25.4	36.2
Coal	1.1	1.0	0.3	2.4
Total	9.6	18.3	32.9	71.2[a]

[a] Includes $10.4 billion for refineries that is not included in country group or individual fuel totals.

Source: The World Bank, 1983.

The added cost of more efficient products is often cited as an obstacle to efforts to improve efficiency in capital-short developing countries. However, the high cost of capital may favor investments in efficiency relative to investments in long lead-time supply projects, such as construction of new powerplants. Because efficiency investments pay-off much more quickly, their economic advantage increases with interest rates (Geller, 1987). However, this comparison is not visible to consumers who are not paying the marginal cost of new energy supplies, and energy and utility companies may have no interest in efficiency for institutional reasons (Goldemberg et al., 1987).

Greenhouse Gas Emissions and Technology Transfer

The transfer of state-of-the-art technology will be necessary to significantly reduce commercial energy use in developing countries. However, there are many obstacles to the development and dissemination of such technologies from industrialized countries to potential competitors in developing countries.

Even when state-of-the-art technology is made available, developing country governments or manufacturers may be reluctant to accept it. Such technology will carry a higher capital cost and may produce products of higher quality at a higher price. Heavily-indebted developing countries may not be in a position to secure the additional capital needed for state-of-the-art technology. Developing country markets tend to be more price-sensitive given the lower discretionary income enjoyed by consumers in these countries. A program to reduce the cost of capital for more efficient plants and equipment, and another to induce consumers to purchase products with higher first-costs yet lower lifecycle costs are essential to encourage developing country consumers to acquire and use state-of-the-art technology.

Most developing countries need technologies that take advantage of abundant but unskilled labor and that minimize the need for capital, and many have more biomass resources than fossil fuels (Goldemberg et al., 1987; Williams, 1988). Improved versions of some low-cost technologies no longer used in the industrialized countries, such as wood-burning cookstoves, are therefore a high priority. Developing countries are still rapidly increasing their consumption of basic materials, whereas such consumption is being replaced by high value-added fabrication and finishing activities in industrialized countries (Williams et al., 1987). This implies substantial differences in research priorities, but there are few institutions and much less money devoted to meeting their needs. The tendency, reinforced by the risk-averse policies of multilateral banks, is to make do with what is tried and true in the industrialized nations.

STRATEGIES FOR REDUCING GREENHOUSE GAS EMISSIONS

Studies of future energy use in developing countries indicate that the industrial sector and power generation will retain their large share of total energy demand (Sathaye et al., 1988). However, the mix of fuels is less certain. A slower growth rate would mean continued use of traditional, and therefore, biomass fuels. On the other hand, faster economic growth would reduce this burden, but it would add to that caused by transportation, as individuals with higher income would demand greater mobility and consequent higher consumption of petroleum products. Strategies to reduce CO_2 emissions will therefore depend on the rate of economic growth in the developing countries.

Energy-pricing reform is essential to promote efficient use of fuels; existing subsidies and price controls are major barriers to investments in energy conservation. However, the short-term result may not be to reduce CO_2 emissions since many countries have substantial unmet needs that would utilize any supplies made available by efficiency improvements. Coal is also the least expensive fuel other than biomass in most developing countries, and market pricing may lead to increasing the share of energy from this carbon-intensive source.

AID, the World Bank, and other development agencies have actively promoted energy-pricing reform in developing countries. However, a recent AID analysis concludes that such efforts have often failed due to fears that reforms will be economically and politically destabilizing. However, effective pricing reform does not necessarily require radical reorganization of economies. Hungary, for example, has achieved substantial improvements in agricultural productivity by creating economic incentives while maintaining a primary role for cooperatives (Chandler, 1986). Further study may help identify ways to overcome this problem, but a realistic expectation may be gradual price increases and structural reform to reduce future interference in energy markets (U.S. AID, 1988).

Improvements in energy efficiency in developing countries may not lead to emission reductions in absolute terms because the energy made available may meet unmet needs and pent up demands. This is consistent with efforts to promote economic growth but implies that even highly successful conservation programs may not avoid substantial emissions growth in these countries. However, over the long run the result will be less emissions than would result from continued inefficient use of energy. Pricing reform similarly may not always lead to reduced emissions of greenhouse gases. For example, removing subsidies for kerosene and LPG may push some consumers back to inefficient use of traditional fuels (Leach, 1987, 1988). This may promote deforestation and a net increase in CO_2 emissions if the only alternative is wood from freshly-felled trees. Continued subsidies for modern cooking fuels for some transition period might be desirable in order to reduce CO_2 emissions.

Several financial strategies can promote improved industrial efficiency by increasing the capital available for desired investments. One traditional financial approach is to allocate capital for designated purposes, such as improving energy efficiency, sometimes at subsidized rates. Power sector loans also could be tied to stringent conservation targets involving utilities in promoting conservation as in some industrialized countries. Governments also sometimes deliberately seek to increase competition within the industrial sector, sometimes inviting foreign collaboration, as a way of forcing companies to increase capital spending.

As capital becomes expensive, governments have begun to tap into private capital markets by floating bonds in targeted funds (e.g., the Korean, Taiwanese, or Indian funds introduced in international markets). Domestic markets are being tapped as well in similar ways. Governments could target private capital for proven technologies and government funds for investment in novel, yet economic, energy conservation schemes specially targeted to reduce CO_2 emissions. Industrialized countries may help by inducing banks to reduce the interest charged for investment targeted for CO_2 emissions abatement strategies.

Another possible innovative approach is for governments to allow the release of so-called "black money" funds locked away by private individuals to evade taxes, if directed to investments for CO_2 emissions abatement. In some countries black money may account for 30-40 percent of total GDP (*Economist*, August, 1988). Placing this money in circulation would mean that governments would have to forego or reduce taxes on these funds, in effect conceding that they would not collect them anyway without substantial increase in enforcement.

International Lending and Bilateral Aid

The U.S. can help developing countries reduce their greenhouse gas emissions through its foreign aid programs and contributions to the World Bank and other multilateral development banks (MDBs). Although all such programs (U.S. and foreign) address only a small percentage of total investment in developing countries, they can exert disproportionate influence because they leverage much greater amounts of funds and certify the financial merit of particular technologies and projects.

U.S. Bilateral Assistance Programs

Most U.S. non-military bilateral assistance is administered by the Agency for International Development (AID), including energy- and forestry-related assistance. AID has attempted to improve its sensitivity to environmental concerns in recent years and estimates that it now spends over $100 million for activities aimed at conserving natural resources (U.S. AID, 1987). The Agency has expanded the number of environmental professionals and established an Office of Forestry, Environment, and Natural Resources.

Most AID support for energy is provided through regional and national programs, although a few projects are funded through a central Office of Energy. Some coordination is also encouraged by an Agency Sector Council for Energy and Natural Resources. Energy-related funding for the past five years has averaged slightly less than $200 million per year; the FY 1986 budget included $254 million for energy projects in 23 countries, of which $180 million was spent for electric power (U.S. AID, 1988). This is about 4 percent of the total AID budget, much less than the share allocated to energy by the World Bank and many other bilateral aid programs (Gray et al., 1988).

AID has funded some analysis of promising projects for biomass fuels that indicate the possibilities for redirecting some bilateral assistance to more effectively promote rational energy planning and development and adoption of improved technology. The Multi-Agency Working Group on Power Sector Innovation, coordinated by AID's Office of Energy, was organized in 1987 to promote cooperation among international institutions involved in power sector development. AID has proposed using this group and the International Development Assistance Committee to focus greater attention on the energy/environment relationship (U.S. AID, 1988). Proposals have also been made to expand the scope of the Consultative Group on International Agricultural Research, a highly successful international consortium of lenders for improving crop yields and production in developing countries, to encompass agroforestry, bioenergy, and tropical ecology.

Current AID priorities are to identify projects that utilize indigenous energy resources and that have the greatest potential for replication. AID will attempt to broker technical and financial assistance for promising projects, emphasizing private sector participation. AID's funding priorities are also shifting toward the poorest nations, which generally implies countries with lower energy growth rates - although sometimes high rates of deforestation (Gray et al., 1987). AID is proposing to give greater attention to the environmental implications of the energy sector

AID has steadily upgraded its commitment to forestry in recent years, from about $20 million in 1979 to $56.2 million for 146 projects in FY 1987 (U.S. AID, 1987; Stowe, 1987; Table 9-7). Agroforestry has been emphasized along with training and institution building. A large amount of support has been given to tree planting and forestry-related activities through the Food for Peace Program (PL 480); the program is responsible for direct tree planting on an estimated 1.5 million hectares in 53 countries during the early 1980s (Joyce and Burwell, 1985) and for 38 tree-planting projects in 23 countries in FY 1987, mostly in Africa (U.S. AID, 1988). Combining bilateral and food-aid assistance, AID's tropical forestry programs exceeded $82 million in FY 1987. AID has also been a strong supporter of the Tropical Forest Action Plan.

If the U.S. seeks to promote reductions in greenhouse gas emissions in developing countries, AID would logically play a major role. Section 106 of the Foreign Assistance Act already authorizes AID programs to promote renewable energy and improvements in energy efficiency. Several proposals have been made for greatly expanded efforts in this area. A working group of the Atlantic

TABLE 9-7

U.S. AID Forestry Expenditures by Region

Region	Number of Countries With Projects	Number of Projects Active in FY 1987	Number of New Starts	Number Completed	LOP Forestry Obligations (in $1,000)	FY 1987 Forestry Obligations (in $1,000)
Africa	23	45	3	3	95,150	13,960
Asia/Near East	11	39	1	6	273,212	17,337
Latin America/ Caribbean	12	46	3	8	140,241	17,398
Central Bureaus	NA	16	0	0	78,103	7,488
Totals	46	146	7	17	$586,706	$56,183

Note: Many forestry projects are components of larger natural resource and agricultural projects. To determine the forestry component, a percentage of the total LOP funding was estimated for significant forestry activities based on judgments made by A.I.D. staff and contractors. This percentage was then applied to each year's obligations to arrive at annual figures. Projects can receive funding obligations at any time during the life of the project.

Source: IDEA Inc., 1988.

Council of the United States and the Member Committee of the U.S. World Energy Conference recommended expanding AID's energy assistance programs to promote greater private-sector investment. They suggest using the revolving loan fund and grant program administered by the Bureau for Private Enterprise to make energy loans; its current priority is agri-business (Gray et al., 1988).

Another recent study by the American Council for an Energy-Efficiency Economy proposes that AID's energy budget be increased to $50-$100 million per year. The authors would redirect aid programs away from support of specific projects in favor of building the capabilities of individuals and institutions within developing nations. They conclude that increasing private-sector involvement should be a high priority, noting that U.S. companies could be encouraged to market energy-efficient technologies with the assistance of the Overseas Private Investment Corporation, the Export-Import Bank, and the Trade and Development Program (Chandler et al., 1988).

AID is not the only U.S. agency involved in bilateral assistance. For example, both the Department of Agriculture and the Peace Corps support international cooperative forestry programs. The former has an Office of International Cooperation and Development exclusively devoted to international activities and also funds some research (Stowe, 1987).

<u>Policies and Programs of Multilateral Development Banks</u>

The U.S. contributes over $1 billion to MDBs annually, an amount that leverages many times as much actual borrowing through co-financing and other arrangements. Much of the activity of these banks is directly or indirectly related to greenhouse gas emissions through loans for energy projects, forestry, and agriculture. U.S. influence on these institutions is substantial, though they are not controlled by the U.S. government. Voting on each bank's board is through a board of directors whose membership is proportional to the size of contributions, giving the U.S. _____ percent in the World Bank. (U.S. recommendations on specific loans, however, have been outvoted.) The U.S. is the largest contributor to the Bank, and the Bank president has traditionally been an American. Congress has directed use of the U.S. "voice and vote" to promote several selected policies, such as human rights concerns and the use of "light capital technologies" (22 U.S. Code Sec. 262f and 262d).

The MDBs are able to influence the policies of developing countries to some degree because loans and aid tend to be offered on relatively favorable terms and bank support can help countries obtain credit from lenders. MDB lending has taken on even greater significance as the debt burden and capital requirements of developing countries have grown enormously in recent years. This influence has been used in ways that have had both positive and negative impacts on the energy and forestry sectors. For example, the World Bank has been a strong advocate of energy-pricing reforms (World Bank, 1983; Goldemberg et al., 1988). However, studies have also documented numerous MDB projects that lead directly and indirectly to deforestation, including roads, dams, tree crop plantations, and agricultural settlements (Repetto, 1988).

About one-fourth of World Bank lending goes to energy-related projects, or nearly $4 billion. The rate for other MDBs ranges from 9 percent by the African Development Bank to 34 percent for the Asian Development Bank (Gray et al., 1988). Traditionally, a majority of this funding has gone for development of very large power projects (Table 9-8). By one estimate, over 90 percent of multilateral and bilateral energy assistance has been for large systems for the generation, transmission, and distribution of electricity; new and renewable energy sources have received about 3 percent and end-use efficiency measures less than 1 percent of total energy-related loans (Goldemberg et al., 1987; see Table 9-9).

TABLE 9-8

World Bank Energy Sector Loans in 1987
(Million U.S. Dollars)

	Eastern & Southern Africa	Western Africa	East Asia & Pacific	South Asia	Europe, Middle East & N. Africa	Latin America & Caribbean	Total
Oil/gas/coal	20.0	15.0	0.0	548.0	0.0	104.4	687.4
Hydroelectric	63.0	6.3	684.8	1,312.0	527.0	423.8	3,016.9
Total	83.0	21.3	684.8	1,860.0	527.0	528.2	3,704.3

Source: World Bank, 1987.

TABLE 9-9

Expenditures of Multilateral and Bilateral Aid Agencies in the Energy Area (millions of current dollars)

	Conventional Power Generation (Hydro, Nuclear, Thermal), Transmission; Distribution; Power Sector Studies	Fossil Fuels Recovery (includes Studies and Training)	New and Renewables (includes Geothermal, Fuelwood)	Technical Assistance, Energy Planning, Other	Total Energy Aid
MULTILATERAL AID					
World Bank (FY 1972-December 1978)	5,210	305	170	-	5,686
Inter-American Development Bank (FY 1972-FY 1978)	2,596	158	4	-	2,758
Asian Development Bank (FY 1972-FY 1978)	1,183	21	0	-	1,204
European Development Fund (to May 1978)	141	-	9	-	150
U.N. Development Programme (to Jan. 1979)	72	23	29	13	137
U.N. Center for Natural Resources, Energy and Transport (to Jan. 1979)	3	5	4	5	17
Subtotal	9,205	512	216	18	9,952

TABLE 9-9 (continued)

Expenditures of Multilateral and Bilateral Aid Agencies in the Energy Area (millions of current dollars)

	Conventional Power Generation (Hydro, Nuclear, Thermal), Transmission; Distribution; Power Sector Studies	Fossil Fuels Recovery (includes Studies and Training)	New and Renewables (includes Geothermal, Fuelwood)	Technical Assistance, Energy Planning, Other	Total Energy Aid
BILATERAL AID					
French Aid (1976-1979)	229	16	30	5	280
Canadian International Development Agency (1978-1979, 1979-1980)	88	0	2	1	91
German Aid (1970-present)	1,925	41	81	48	2,095
Kuwait Fund (FY 1973-FY 1978)	437	99	1	-	536
Netherlands-Dutch Development Cooperation (1970-present)	119	71	7	2	198
U.K. Overseas Devel. Admin. (1973-present)	146	1	3	-	149
U.S. AID (FY 1978-FY 1980)	403	2	96	46	546
Grand Total	12,719	757	437	121	14,033
Percentage in Each Sector	91	5	3	1	100

Source: Goldemberg et al., 1987.

The World Bank has had a major role in financing energy sector expansion in the developing countries that are growing most rapidly and relying most heavily on fossil fuels, including China, India, and Pakistan. Coal resources in these countries are often characterized by low heating values and the generation and distribution of electricity is typically much less efficient than in the industrialized countries. Bank projects, therefore, represent a large share of carbon emissions from developing countries.

In 1987 the World Bank announced a general commitment to upgrade its support for environmental analysis and programs (World Bank, 1987). The Bank has created a new environmental department, increased environmental staff, and begun a process of preparing environmental assessments on about 30 developing countries. Special attention is being given to identifying regional environmental projects in Africa. Support for tropical forest conservation and development has also been made a high priority; funds will increase from $152 million in FY 1987 to $350 million in FY 1989.

Some power sector loans have important conservation components; for example, a recent loan to Zimbabwe includes $44 million to upgrade generating equipment, rehabilitate transmission lines, and train workers to increase total output (*World Bank News*, January 14, 1988). The Bank has used its influence to support energy-pricing reforms and the elimination of subsidies and has funded a small number of conservation programs through its Energy Sector Management Assistance Program (Table 9-10). Some industrialization loans may also result in substantial improvements in energy efficiency, although such improvements are not a primary purpose for these loans.

The Bank plans to maintain energy loans as a relatively constant percentage of its total lending (Gray et al., 1988). However, recognizing that capital needs for energy in developing countries are expected to grow much faster than available funding, the Bank plans to increase its analytical, policy, and technical advisory roles and to become more of a catalyst for funds from

TABLE 9-10

World Bank Energy Conservation Projects:
Energy Sector Management Assistance Program (ESMAP)
Energy Efficiency Initiatives

Industrial Energy Efficiency

- Program Design and Institutional Support (Senegal, Ghana)
- Energy Audits (Syria)
- Training of Local Staff (Tanzania)

Power Efficiency (Many Countries)

- Design of Programs to Reduce Technical and Non-Technical Losses
- Pre-Feasibility Studies for Life Extension and Rehabilitation Projects
- Utility Organization Studies

Electricity Savings in Buildings

- Energy Efficient Building Code (Jamaica)
- Appliance Labeling (Jamaica)

Household Energy Savings

- Household Energy Strategy Studies
- Improved Cookstoves Projects
- Improved Charcoaling

Energy-Environmental Studies

- Energy Supply Options to Steel Industry in Carajas Region (Brazil)

Source: WRI, 1988a

non-Bank sources. The Bank, like AID, plans to give more attention to private power generation and other innovative means of financing new power sources.

A Bank representative cited several obstacles to increased support for conservation at an EPA workshop (WRI, 1988a). One is that the Bank traditionally has found it difficult to fund relatively small projects; ways must be found to package them as an element of a larger loan. (This was done in the case of a large power loan to Brazil, which included $2 million for analysis of conservation opportunities.) Implementation of projects that require actions by many individuals is similarly perceived as difficult. Some of the most inefficient industries may be judged uneconomic and therefore inappropriate candidates for loans. Finally, the recipient governments have to be interested and must possess the necessary technical skills to implement conservation programs.

Another problem is the availability of proven technology suited to the needs of developing countries. As discussed above, the technological needs of developing countries differ from the industrialized countries in many areas. However, lending criteria tend to discourage the search for new or innovative technologies by restricting financing to practices that have been fully proven in practice (Daffern, 1987). This reflects an understandable desire to minimize risks, but the effect is often to finance equipment that is not the most efficient. Promising alternatives that would utilize biomass and other local resources are also neglected because the technologies are not widely used in the industrialized countries (Williams, 1988).

New Directions

The possibility of redirecting bilateral aid and Bank loans to further energy conservation and other strategies to reduce greenhouse gas emissions was discussed at the EPA workshop on developing country issues. One proposed remedy is a shift in Bank energy sector lending to general energy programs as opposed to specific projects as a way of facilitating more funding of demand-side efforts (Goldemberg et al., 1988). This proposal may be difficult to implement but may receive more attention as capital constraints lead to greater interest in conservation.

A recent study of innovative financing mechanisms by the World Resources Institute proposes that the MDBs place greater emphasis on promoting policy reforms in conjunction with loan agreements to protect natural resources. The authors suggest that the international development agencies identify and analyze the effects of tax, tariff, credit and pricing policies, as well as the terms and administration of concession agreements, on the use of resources (Repetto, 1988, p. 41).

A possible strategy for creating an economic incentive for protection of tropical forests involves payment of annual fees for custodian services--protection from squatters and illegal development--to affected nations in proportion to the areas under protection (Rubinoff, 1985). Agreements to protect 100 million hectares, or roughly 10 percent of the world's remaining moist tropical forest, might cost on the order of three billion dollars. The status of reserves would be monitored and payments adjusted accordingly. In this way tropical countries would be paid for some management costs, reflecting some of the benefits thereby provided to the rest of the world.

Section 119 of the Foreign Assistance Act requires AID to prepare environmental assessments of its major actions, including effects on the global environment.[1] AID is also specifically directed to consider the impacts of its programs on tropical forests (22 CFR § 2151(a), (p); see also Stowe, 1987, p. 73-74).

---

[1] AID procedures for preparing environmental assessments are published in the Code of Federal Regulations, Volume 22, Part 216.

The importance of these issues is suggested by increasing Congressional interest (Stowe, 1987; Rich, 1985, p. 681). Section 539 of Public Law 99-591 directs the Secretary of the Treasury to instruct U.S. Executive Directors in each of the multilateral development banks to take a number of steps to support environmental reform measures and also requires the Treasury Department to report on progress toward these objectives (U.S. Treasury Department, 1988). The Department's most recent report provides mixed reviews of the MDBs' responsiveness to environmental concerns. As already noted, the World Bank announced a higher priority for environmental concerns in 1987, but the report states that "The Bank has not moved effectively in the area of energy conservation."

REDUCING GREENHOUSE GAS EMISSIONS IN EASTERN BLOC NATIONS

The Soviet Union is the third largest source of carbon dioxide emissions and together with Eastern European nations is likely to remain a significant contributor to global greenhouse gas emissions in the years to come (Figure 9-1). The energy policies of these countries will therefore have an important influence on the greenhouse problem.

The Soviet Union has enormous energy resources, although much of it, including half the world's accessible coal, is in relatively remote areas of Siberia. Soviet oil production is roughly equal to the total production of all Middle Eastern countries combined and the Soviets also rank first in gas production and third in coal production. Coal was the dominant energy source until the late 1950s but declined to less than 30 percent in 1977 with the growth in oil production (Table 9-11). Future plans call for increasing reliance on natural gas and nuclear power and renewed growth in coal use; large coal-burning powerplants and coal slurry pipelines are now under construction. Other Eastern European nations, particularly Poland, East Germany, and Czechoslovakia, are even more dependent on coal (WRI, 1988).

TABLE 9-11

Energy Use in the Soviet Union and Eastern Bloc
(petajoules)

	Liquid Fuels[a]			Solid Fuels			Gaseous Fuels[b]			Electricity[c]			Total		
	1970	1980	1985	1970	1980	1985	1970	1980	1985	1970	1980	1985	1970	1980	1985
Eastern Europe															
Bulgaria	346	575	562	570	615	672	17	148	216	7	49	71	940	1,387	1,521
Czechoslovakia	371	668	567	1,792	1,861	1,878	79	298	330	25	40	71	2,267	2,867	2,846
East Germany	483	690	632	2,602	2,523	2,856	22	302	288	8	54	53	3,065	3,569	3,829
Hungary	223	424	392	496	374	373	121	364	393	13	27	63	853	1,189	1,221
Poland	301	610	512	2,878	4,230	4,170	222	363	363	7	11	6	3,408	5,214	5,051
Romania	406	688	760	410	689	813	971	1,507	1,599	1	47	51	1,788	2,931	3,223
Subtotal	2,080	3,655	3,425	8,748	10,292	10,762	1,432	2,982	3,189	61	228	315	12,321	17,157	17,691
U.S.R.R.	9,283	14,770	14,450	12,933	14,440	14,596	6,611	13,209	19,761	442	811	1,257	29,270	43,229	50,065
Total	11,363	18,425	17,875	21,681	24,732	25,358	8,043	16,191	22,950	503	1,039	1,572	41,591	60,386	67,756

Notes:

a All petroleum products and natural gas liquids.
b Natural gas, manufactured gas, coke-oven gas, and blast furnace gas.
c Hydro, nuclear, and geothermal power.

Source: WRI, 1988.

The energy intensity of the Soviet economy changed relatively little between 1970 and 1985, in contrast with the substantial reductions achieved during the same period in the U.S. and other industrialized nations. The government has become increasingly interested in energy conservation because of the growing capital and fuel costs of energy production. More than 20 percent of national capital investments and 14 percent of primary energy resources are consumed by the energy supply system (NAS, 1987). Oil is also a valuable export commodity, and known reserves may be exhausted at current production rates by the year 2000. Siberian reserves may be large but production costs will be much higher (WRI, 1988).

According to a recent report by a Soviet energy expert, improvements in energy efficiency continue to prove elusive contrary to official progress reports (Martynenko, 1988). This is because industry does not pay market prices for energy and therefore has little economic incentive to conserve. However, the government is gradually increasing fuel costs within the Soviet economy and recent economic reforms may give industry greater incentive to conserve (NAS, 1987).

Energy pricing and planning in Eastern Europe is complicated by trade agreements with the USSR. These countries have provided labor for construction on Soviet oil and gas projects in return for options to purchase oil at a price equal to the average price paid for Soviet oil exports over the previous five years--a rewarding arrangement when prices were rising but unattractive at recent low oil prices (WRI, 1988).

The Soviet Union and United States currently have cooperative agreements on both climate change and energy conservation. The U.S.-USSR Agreement on Cooperation in the Field of Environmental Protection lists numerous projects on atmospheric science. The first U.S.-USSR Symposium on Energy Conservation was held in Moscow in June 1985 and several meetings followed, most recently in October 1988. A program of cooperative research has been developed, administered in the U.S. by the National Academy of Sciences (NAS, 1987; Memorandum of the Soviet-American Symposium, 1988). The Soviet Union has also independently expressed strong interest in global climate change problems. At the November 1988 meeting on global climate issues in Geneva sponsored by WMO and UNEP, the Soviet Union agreed to chair a panel on the effects of climate change.

U.S. LEADERSHIP TO PROMOTE INTERNATIONAL COOPERATION

The international and bilateral cooperation already in place has established a solid foundation for discussion of policy responses; however, there is much that needs to be done before this discussion and analysis leads to agreement on necessary actions. Several precedents suggest that U.S. leadership can help achieve agreement on response strategies. Two important sources of greenhouse gas emissions, the use of CFCs and tropical deforestation, have already been the subject of international agreements that should moderate global warming. In both cases, action was taken for reasons largely unrelated to climate change, but examination of the evolution of these agreements may suggest factors conducive to agreement on climate change.

Restricting CFCs to Protect the Ozone Layer

In September 1987, twenty-four nations signed an agreement in Montreal (the Montreal Protocol) to reduce emissions of CFCs 50 percent by 1999 because of growing concern about the effect of these chemicals on stratospheric ozone. The reductions are to be achieved through a phased process beginning with a freeze six months after the Protocol goes into effect (January 1, 1989, or shortly thereafter) and a 20 percent reduction by July, 1993. The U.S. ratified the Montreal Protocol on April 21, 1988. EPA issued regulations consistent with the requirements of the Protocol on August 1, 1988 (*Federal Register*, Aug. 12, 1988). For a description of the terms of the Protocol and

U.S. regulations, see **REGULATIONS AND STANDARDS in CHAPTER VIII**. These regulations also will go into effect at the time the Protocol takes effect.

Aside from its direct impact on reducing emissions of important greenhouse gases, the terms of the Protocol and the process that led up to the agreement offer some valuable insights for cooperation to address climate change. Developing nations, which currently use only about 15 percent of CFCs, were concerned that emissions reductions might hinder their economic growth. Therefore, to encourage their participation, Article 5 of the Protocol allows countries with consumption less than 0.3 kg per capita to delay compliance by 10 years so long as they do not exceed that amount. (U.S. CFC consumption was more than 1 kg/capita in 1985). The industrialized countries also agreed to provide technical assistance and financial aid in support of developing countries' efforts to adopt alternatives.

The Protocol was negotiated over a two-year period, but the foundations were laid over a period of a decade or more. Intergovernmental meetings were first held in 1977 and 1978 (Stoel et al., 1980). A consensus-building scientific process, the Coordinating Committee on the Ozone Layer, was created by the UNEP Governing Council in 1977. The Council decided to convene a working group of legal and technical experts in 1981, and the first meeting took place in January 1982. When agreement on proposals for action at first proved impossible, the participants came to an agreement in March 1985 that provided a framework for scientific cooperation: the Vienna Convention for Protection of the Ozone Layer. The Convention also committed the signatories to hold workshops and exchange information as the basis for further efforts to achieve a protocol.

Some of the factors that facilitated the Protocol may provide some insight into possible strategies for future efforts to achieve agreements to reduce greenhouse gases (Benedick, 1987). The existence of an international scientific consensus report prepared under WMO auspices, *Atmospheric*

Ozone 1985, helped avoid debate about the underlying seriousness of the problem. The workshop process and informal negotiations that they allowed proved to be very conducive to consensus building. UNEP served a valuable organizing role and provided a valuable objective international forum.

According to the chief U.S. negotiator, U.S. leadership had a major role. "The treaty as eventually signed was based upon the structure and concept advanced by the United States late last year." In addition to governmental efforts, such as a series of diplomatic initiatives, the U.S. role includes industry actions to take responsibility and express support for the proposed emission controls, efforts undertaken by public interest groups to inform the public, and threats of unilateral action and trade sanctions by members of Congress.

International Efforts to Halt Tropical Deforestation

There has also been substantial recent international cooperation on the problem of tropical deforestation. These efforts are of substantial interest because of some of the similarities between the obstacles to solving the tropical deforestation problem and the larger greenhouse problem. In both cases, there is a need for industrialized countries to help developing countries implement policy changes that may be costly and difficult. Mechanisms for achieving international cooperation must be found, substantial amounts of financial assistance must be provided to developing countries, and politically-difficult policy choices must be made.

The Tropical Forestry Action Plan is a promising response to this challenge (WRI, 1985; FAO, WRI, World Bank, UNDP, 1987). The Plan was developed by a consortium of institutions concerned about tropical deforestation, including the U.N. Food and Agriculture Organization (FAO), the U.N. Development Programme (UNDP), the World Bank, the World Resources Institute (WRI), and representatives of more than 60 tropical countries. The task force that drafted the Plan included Brazil's Secretary of the Environment and one of India's former Secretaries of the Environment. The broad and highly-visible sponsorship of the Plan has helped to highlight the important benefits provided by tropical forests and to draw attention to their accelerating loss. Equally important, the Plan offers the broad outlines of a solution, including regional and functional budget proposals. The total budget calls for $8 billion over a five-year period.

The Plan has become a focal point for cooperative efforts by bilateral aid agencies and has influenced World Bank policies (Wolf, 1988; Stowe, 1987).[2] Ultimately, the Plan's success is dependent on the cooperation and support of the affected developing countries, and many of them are preparing national action plans and increasing national support for reforestation (WRI, 1988).

The International Tropical Timber Organization (ITTO) is another important mechanism for north-south cooperation on tropical forest management. The ITTO was created by a March 1985 agreement reached under the auspices of the U.N. Conference on Trade and Development (UNCTAD), primarily as a commodity agreement to facilitate economic use of tropical timber; its purposes include "expansion and diversification of international trade," a "long-term increase in consumption," and greater access to international markets. However, unlike traditional commodity agreements, the ITTO explicitly recognizes the importance to conservation efforts and management policies of the premise of sustainable use. The connection has thus been made between conservation, economic development, and the export of resources to industrialized countries (Wolf, 1988; Forster, 1986).

---

[2] Periodic reports on progress implementing the Action Plan are available from the TFAP Coordinator, FAO Forestry Department, Via delle Terme di Caracalla, 00100 Rome, Italy.

The importance given to promoting cooperation between tropical nations and industrialized consumers of tropical hardwoods is evident in the ITTO organization; the headquarters is in Japan, a very large consumer, and the executive director is from Malaysia, an important producer. Another promising sign is that at the first meeting in April 1987, Japan announced a pledge of $2 million for research on reforestation and sustainable management.

The ITTO also affords a forum for addressing the linkage between industrial country policies and resource exploitation in developing countries. A recent WRI study concludes that "industrial-country trade barriers in the forest products sector have been partially responsible for inappropriate investments and patterns of exploitation in the Third World Forest industries. . . . [N]egotiations between exporting and importing countries should reduce tariff escalation and non-tariff barriers to processed wood imports from the tropical countries, and rationalize incentives to forest industries in the Third World" (Repetto, 1988).

Looking to the future, increased international cooperative efforts will be necessary if tropical deforestation is to be halted and reversed. A possible framework for a program to achieve this goal has recently been outlined by Professor Pedro Sanchez of North Carolina State University (Sanchez, 1988). Dr. Sanchez notes that 12 countries account for three-fourths of the net carbon emissions from clearing primary forests; 10 more account for much of the remainder (Table 9-12). Efforts could be made to engage the leaders of these countries in a dialogue for the purpose of bringing them to specific program agreements targeted to deforestation "hot spots," where technology transfer and government policies would be focused--an approach that is consistent with the Tropical Forest Action Plan's emphasis on national planning.

TABLE 9-12

Countries Responsible for Largest Share of Tropical Deforestation

Country	Net Carbon Emissions in 1980 From Primary Forests (million tons)
Brazil	207
Columbia	85
Indonesia	70
Malaysia	50
Côte d'Ivoire	47
Mexico	33
Thailand	33
Peru	31
Nigeria	29
Ecuador	28
Zaire	26
Philippines	21

Source: Sanchez, 1988.

Ongoing Efforts Toward International Cooperation

Some important first steps to promote international action have already been taken in the last few years and a foundation for international cooperation now exists. Several meetings without formal governmental status established a basis for international scientific cooperation. A conference on the status of the greenhouse problem, organized by the United Nations Environment Programme (UNEP), World Meteorological Organization (WMO), and International Council of Scientific Unions (ICSU), was held in October 1985 in Villach, Austria and was attended by experts from 29 countries, which included representatives from several U.S. agencies. Among their conclusions were recommendations that "scientists and policy-makers . . . begin an active collaboration to explore the effectiveness of alternative policies and adjustments." In response to the recommendations, a small task force was created to advise on needed domestic and international actions and to evaluate the need for efforts to initiate consideration of a global convention.

The government of Canada convened a non-governmental meeting on "The Changing Atmosphere" in June 1988, which was attended by more than 300 experts from 46 countries and several United Nations' organizations. The Conference Statement that resulted from the meeting urges development of an Action Plan for the Protection of the Atmosphere in addition to several specific proposals for government policy, including:

- directing energy R&D budgets to energy options that would greatly reduce CO_2 emissions;
- reducing CO_2 emissions by 20 percent of 1988 levels by the year 2005, half by reducing energy demand and half by changing the sources of energy;
- initiating development of a comprehensive global convention as a framework for protocols on the protection of the atmosphere; and
- establishing a World Atmosphere Fund, financed in part by a levy on fossil-fuel consumption in industrialized countries to help finance the Action Plan.

UNEP and WMO have continued their activities since the 1985 Villach meeting: An Intergovernmental Panel on Climate Change, organized by UNEP and WMO, met in November 1988 with representatives from __ countries. Participants agreed to establish three committees, the first to assess the state of scientific knowledge on the greenhouse issue, chaired by the United Kingdom, the second to assess social and economic effects from global warming, chaired by the Soviet Union, and the third to evaluate potential response strategies, chaired by the United States.

The U.S. also has some important bilateral cooperation with some of the other nations that emit large amounts of greenhouse gases, particularly the USSR and the People's Republic of China. (These three countries account for over 40 percent of the current commitment to global warming.) U.S./Soviet cooperative efforts include studies of future climates, climate studies in the Arctic, measurements of methane and ozone change in the polar regions, and analysis of possible response strategies. The joint communique released after the Reagan-Gorbachev Summit emphasized the high level of interest in cooperation on the greenhouse issue: "The two leaders approved a bilateral initiative to pursue joint studies in global climate and environmental changes through cooperation in areas of mutual concern . . . there will be a detailed study on the climate of the future. The two sides will continue to promote broad international and bilateral cooperation in the increasingly important area of global climate and environmental change."

The U.S. and Soviet Union also have some relevant cooperative activities designed to promote energy conservation. For example, the National Academy of Sciences and the Soviet Academy of Sciences have a cooperative program on energy conservation that includes some comparative evaluation of building codes and other energy conservation programs in addition to exchange of technical information.

The U.S. Government has also initiated cooperative research on climate change with the People's Republic of China, now the world's largest user of coal. Potential areas for cooperative research include exchange of information and development of data on future energy development paths, emissions from rice fields and other sources, and concentrations of trace gases in remote regions.

REFERENCES

Barnes, __., and __. Binswanger. 19__.

Benedick, R. 1987. Report for Montreal: A Landmark Global Treaty to Protect the Ozone Layer. In *Preparing for Climate Change*.

Bernardini, __. 198_.

Bhatia, __. 19__.

Chandler, W. 1986. *The Changing Role of the Market in National Economies.* Worldwatch Paper 72, Washington, D.C.

Chandler, W., H. Geller, and M. Ledbetter. 1988. *Energy Efficiency: A New Agenda.* ACEEE, Washington, D.C.

Dadi, Z. 1988. Energy Price Management and Economic System Reform. Paper presented at the China-U.S. symposium on energy markets and the future of energy demand, Nanjing, China, June 22, 1988.

Daffern, E. 1987. *World Bank Policy in the Promotion of Energy Producing Projects in Developing Countries*. Paper presented in Reading, England, April 7, 1987.

Federal Register. August 12, 1988.

FAO (Food and Agricultural Organization of the United Nations), WRI (World Resources Institute), World Bank, and UNDP (U.N. Development Programme). 1987. *The Tropical Forestry Action Plan.*

FAO . . . (as above). 1987. *The Tropical Forestry Action Plan: Investment Priorities and Policy Reforms for Sustainable Forest Land Use*. Keynote paper prepared for the Bellagio Strategy Meeting on Tropical Forests. July 1-2, 1987.

Flavin, C. 1988. Creating a Sustainable Energy Future. In Brown, et al., eds., *State of the World 1988.* New York, 22-40.

Forster, M. 1986. Two Cheers for the Agreement. *Environmental Policy and Law* 16(5): 142-143.

Geller, H., J. Goldemberg, J. Moreira, R. Hukai, C. Scarpinella, and M. Ysohizawa. 1988. Electricity Conservation in Brazil: Potential and Progress. *Annual Review of Energy* 13:469-483.

Goldemberg, J., T. Johansson, A. Reddy, and R. Williams. 1987. *Energy for Development.* World Resources Institute, Washington, (p. 79)

Goldemberg, J., T. Johansson, A. Reddy, and R. Williams, 1987. *Energy for a Sustainable World.* World Resources Institute, Washington, D.C.

Goldemberg, J., T. Johansson, A. Reddy, and R. Williams, 1988. *Energy for a Sustainable World.* Wiley Eastern, New Delhi.

Gray, J., W. Davis, and J. Harned. 1988. *Energy Supply and Use in Developing Countries.* University Press, New York.

Hill, ___. 1987.

Hindu Survey. 1988.

IDEA Inc. 1988. Forestry Project Database, Forestry Support Program.

IIED (International Institute for Environment and Development) and WRI (World Resources Institute). 1987. *World Resources 1987.* Basic Books, Inc., New York, 369pp.

IMF (International Monetary Fund). 1985. In Goldemberg et al., 1987.

Joyce, S., and B. Burwell. 1985. *Community Level Forestry Development: Options and Guidelines for Collaboration in PL 480 Programs.* Washington, D.C., Peace Corps and AID.

Leach, ___. 1987.

Leach, ___. 1988.

OECD/IEA (Organisation for Economic Co-operation and Development and International Energy Agency). 1987. *Energy Conservation in IEA Countries.* OECD/IEA, Paris.

Miller, A., I. Mintzer, and S. Hoagland. 1985. *Growing Power: Bioenergy for Development and Industry.* World Resources Institute, Washington, D.C.

Munasinghe, M. and R.J. Saunders. 1986. Electric Power Policy in Developing Countries. *Proceedings of the IAEE Conference.* Boston, Massachusetts, Williams 19??.

Repetto, R. 1988. *The Forest for the Trees? Government Policies and the Misuse of Forest Resources.* World Resources Institute, Washington, D.C.

Rich, B. 1985. Multilateral Development Banks, Environmental Policy, and the United States. *Ecology Law Quarterly* 12:681.

Rubinoff, I., 1985. *How to Improve the Conservation of Tropical Forest Ecosystems. Paper presented at the Tropical Forest Conservation Symposium sponsored by the Smithsonian Institution, Friends of the National Zoo, and the World Wildlife Fund-US, December 4-6, 1985, Washington, D.C.*

Sanchez, P. 1988. *Deforestation Reduction Initiative: An Imperative for World Sustainability in the Twenty-First Century.* Paper presented at the Bureau of Science and Technology, U.S. Aid, Washington, D.C., July 22, 1988.

Sathaye, J., A. Chirardi, and L. Schipper. 1987. Energy Demand in Developing Countries: A Sectoral Analysis of Recent Trends. in *Annual Review of Energy 1987* 12:259.

Sathaye, ___, et al.? 1988.

Schipper, ___. 198_.

Shea, C. 1988. "Shifting to Renewable Energy." In *State of the World 1988.* Worldwatch, New York: W.W. Norton.

Siddiqi, ___. 19__.

Stoel, T., A. Miller, and B. Milroy. 1980. *Fluorocarbon Regulation: An International Comparison*. Lexington, Massachusetts.

Stowe, R. 1987. United States Foreign Policy and the Conservation of Natural Resources: The Case of Tropical Deforestation. *Natural Resources Journal* 27 (Winter): 55-101.

U.S. AID (U.S. Agency for International Development). 1987. *Special Report: The Environment*. AID, Washington, D.C.

U.S. AID (U.S. Agency for International Development), 1988a. *Energy and the Environment: An Appraisal of Energy Assistance Strategies for Minimizing Environmental Impacts in Developing Countries*. AID, Washington, D.C.

U.S. AID (U.S. Agency for International Development). 1988b. *Power Shortages in Developing Countries: Magnitude, Impacts, Solution, and the Role of the Private Sector*. U.S. AID, Washington, D.C.

U.S. AID (U.S. Agency for International Development), 1988c. *Progress in Conserving Tropical Forests and Biological Diversity in Developing Countries*. U.S. AID, Washington, D.C.

Williams, R. 1988. *Are Runaway Energy Capital Costs a Constraint on Development?* Paper presented at the International Seminar on the New Era in the World Economy, Sao Paulo, Brazil, August 31 - September 2, 1988.

Williams, R., E. Larson, and M. Ross. 1987. Materials, Affluence, and Industrial Energy Use. *Annual Review of Energy* 12:99-144.

Wolf, E. 1988. Avoiding a Mass Extinction of Species. In Starke, L., ed., *State of the World 1988*. W.W. Norton, New York, 101-117.

World Bank. 1983. *The Energy Transition in Developing Countries*. World Bank, Washington, D.C.

World Bank. 1987. *Annual Report 1987*. Washington, D.C.: World Bank.

World Bank News. January 14, 1988.

WRI (World Resources Institute). 1985. *Tropical Forests: A Call for Action*. WRI, Washington, D.C.

WRI (World Resources Institute). 1988a. Summary of the WRI/EPA Workshop on Greenhouse Gas Emissions and Sustainable Development. Washington, D.C.

WRI (World Resources Institute). 1988b. *International Conservation Financing Project: Maintaining Biodiversity and Critical Ecosystems in Support of Sustainable Development, Preliminary Paper on Financing Mechanisms*. World Resources Institute, Washington, D.C.

WRI (World Resources Institute). 1987. *World Resources 1987*. Basic Books, Inc., New York, 369 pp.

WRI (World Resources Institute) and IIED (International Institute for Environment and Development), 1988. World Resources 1988-89. New York: Basic Books, Inc.

WRI (World Resources Institute), World Bank, and UNDP (United Nations Development Programme). 1985. *Tropical Forests: A Call for Action*. Washington, D.C.: World Resources Institute.

United Nations. 1986.

U.S. Treasury Department. 1988. *Progress Report on Implementation of Environmental Reform in Multilateral Development Banks*. Treasury Department, Washington, D.C.